BIOGEOGRAPHY AND ECOLOGY OF BULGARIA

MONOGRAPHIAE BIOLOGICAE

VOLUME 82

Series Editor

H. J. Dumont

Aims and Scope

The Monographiae Biologicae provide a forum for top-level, rounded-off monographs dealing with the biogeography of continents or major parts of continents, and the ecology of well individualized ecosystems such as islands, island groups, mountains or mountain chains. Aquatic ecosystems may include marine environments such as coastal ecosystems (mangroves, coral reefs) but also pelagic, abyssal and benthic ecosystems, and freshwater environments such as major river basins, lakes, and groups of lakes. In-depth, state-of-the-art taxonomic treatments of major groups of animals (including protists), plants and fungi are also elegible for publication, as well as studies on the comparative ecology of major biomes. Volumes in the series may include single-author monographs, but also multi-author, edited volumes.

The titles published in this series are listed at the end of this volume.

Biogeography and Ecology of Bulgaria

Edited by

VICTOR FET
Department of Biological Sciences
Marshall University
Huntington, West Virginia, USA

and

ALEXI POPOV
National Museum of Natural History
Bulgarian Academy of Sciences
Sofia, Bulgaria

Springer

A C.I.P. Catalogue record for this book is available from the Library of Congress.

ISBN 978-1-4020-4417-5 (HB)
ISBN 978-1-4020-5781-6 (e-book)

Published by Springer,
P.O. Box 17, 3300 AA Dordrecht, The Netherlands.

www.springer.com

Printed on acid-free paper

Captions to Cover Photos

Front Cover: Central Balkan National Park. Northern Djendem Reserve near Botev Peak. Photo: Georgi Stoyanov.

Rear Cover 1: *Onconotus servillei* (Orthoptera: Tettigoniidae). One of the most typical steppe animals from Northeastern Bulgaria. Included in the IUCN Red List of Threatened Species as "Vulnerable". Photo: Nikolay Simov and Dragan Chobanov.

Rear Cover 2: *Paralola buresi* (Opiliones: Phalangodidae). This 1.4 mm-long harvestman is one of the most remarkable animals in Bulgaria. The sole species of the local endemic monotypic genus, this troglobite Tertiary relict is found only in four caves near Lakatnik (Iskar Gorge, Western Stara Planina). Photo: Plamen Mitov and Ivan Yanchev.

Rear Cover 3: *Zamenis situla*, or Leopard Snake (Ophidia: Colubridae). Eastern Mediterranean species found in southern Bulgaria (Struma Valley, Black Sea Coast). Extinct in the Rhodopes. Photo: Boyan Petrov.

This book is dedicated to the memory of Dr. Ivan Buresch (1885–1980), Member of the Bulgarian Academy of Sciences, the founder of Bulgarian biodiversity studies.

Table of Contents

Contributors

Petar Beron
National Museum of Natural History,
Bulgarian Academy of Sciences
Tsar Osvoboditel Blvd. 1
1000 Sofia, Bulgaria
e-mail: beron@mail.bg

Venelin Beschovski
Institute of Zoology,
Bulgarian Academy of Sciences
Tsar Osvoboditel Blvd. 1
1000 Sofia, Bulgaria
e-mail: vbeschovski@zoology.bas.bg

Gergin Blagoev
Institute of Zoology,
Bulgarian Academy of Sciences
Tsar Osvoboditel Blvd. 1
1000 Sofia, Bulgaria
e-mail: gerginblagoev@yahoo.ca

Zlatozar Boev
National Museum of Natural History,
Bulgarian Academy of Sciences
Tsar Osvoboditel Blvd. 1
1000 Sofia, Bulgaria
e-mail: boev@nmnh.bas.bg

Malina Delcheva
Institute of Botany,
Bulgarian Academy of Sciences
Acad. G. Bonchev St. 23
1113 Sofia, Bulgaria
e-mail: malinad@bio.bas.bg

Christo Deltshev
Institute of Zoology,
Bulgarian Academy of Sciences
Tsar Osvoboditel Blvd. 1
1000 Sofia, Bulgaria
e-mail: cdeltshev@zoology.bas.bg

Victor Fet
Department of Biological Sciences,
Marshall University, Huntington,
West Virginia 25755-2510, USA
e-mail: fet@marshall.edu

Vassil Golemansky
Institute of Zoology,
Bulgarian Academy of Sciences
Tsar Osvoboditel Blvd. 1
1000 Sofia, Bulgaria
e-mail: golemansky@zoology.bas.bg

Borislav Guéorguiev
National Museum of Natural History,
Bulgarian Academy of Sciences
Tsar Osvoboditel Blvd. 1
1000 Sofia, Bulgaria
e-mail: bobivg@yahoo.com

Zdravko Hubenov
Institute of Zoology,
Bulgarian Academy of Sciences
Tsar Osvoboditel Blvd. 1
1000 Sofia, Bulgaria
e-mail: zhubenov@zoology.bas.bg

Mario Langourov
Institute of Zoology,
Bulgarian Academy of Sciences
Tsar Osvoboditel Blvd. 1
1000 Sofia, Bulgaria
e-mail: langourov@gmail.com

Stoyan Lazarov
Institute of Zoology,
Bulgarian Academy of Sciences
Tsar Osvoboditel Blvd. 1
1000 Sofia, Bulgaria
e-mail: slazarov@zoology.bas.bg

Milen Marinov
P.O. Box 134
1000 Sofia, Bulgaria
e-mail: mg_marinov@yahoo.com

Tenyo Meshinev
Institute of Botany,
Bulgarian Academy of Sciences
Akademik G. Bonchev St. 23
1113 Sofia, Bulgaria
e-mail: meshinev@bio.bas.bg

Boyan Milchev
Wildlife Management Department,
University of Forestry
Kliment Ohridski Blvd. 10
1756 Sofia, Bulgaria
e-mail: boyan.m@ltu.bg

Plamen Mitov
Department of Zoology and Anthropology,
Faculty of Biology,
St. Kliment Ohridski University of Sofia
Dragan Tsankov Blvd. 8
1164 Sofia, Bulgaria
e-mail: pl_mitov@yahoo.com

Ivan Pandourski
Institute of Zoology,
Bulgarian Academy of Sciences
Tsar Osvoboditel Blvd. 1
1000 Sofia, Bulgaria
e-mail: stygologia@gmail.com

Dimitar Peev
Institute of Botany,
Bulgarian Academy of Sciences
Acad. G. Bonchev St. 23
1113 Sofia, Bulgaria
e-mail: dpeev@bio.bas.bg

Boyan P. Petrov
National Museum of Natural History,
Bulgarian Academy of Sciences
Tsar Osvoboditel Blvd. 1
1000 Sofia, Bulgaria
e-mail: boyanpp@nmnh.bas.bg

Alexi Popov
National Museum of Natural History,
Bulgarian Academy of Sciences
Tsar Osvoboditel Blvd. 1
1000 Sofia, Bulgaria
e-mail: alpopov@bulinfo.net

Vasil Popov
Institute of Zoology,
Bulgarian Academy of Sciences
Tsar Osvoboditel Blvd. 1
1000 Sofia, Bulgaria
e-mail: popov@zoology.bas.bg

Vladimir Sakalian
Institute of Zoology,
Bulgarian Academy of Sciences
Tsar Osvoboditel Blvd. 1
1000 Sofia, Bulgaria
e-mail: sakalian@inagency.com

Michael E. Soleglad
P.O. Box 250
Borrego Springs,
California 92004, USA
e-mail: soleglad@la.znet.com

Tihomir Stefanov
National Museum of Natural History,
Bulgarian Academy of Sciences
Tsar Osvoboditel Blvd. 1
1000 Sofia, Bulgaria
e-mail: tisho@nmnh.bas.bg

Pavel Stoev
National Museum of Natural History,
Bulgarian Academy of Sciences
Tsar Osvoboditel Blvd. 1
1000 Sofia, Bulgaria
e-mail: stoev@nmnh.bas.bg

Milena Stoyanova
Episkop Konstantin Preslavski
University of Shumen
Universitetska St. 115
9712 Shumen, Bulgaria
e-mail: stoyanovamilena@abv.bg

Veliko Velikov
Episkop Konstantin Preslavski
University of Shumen
Universitetska St. 115
9712 Shumen, Bulgaria
e-mail: v.d.velikov@abv.bg

Notes from the Editors

Transliteration

When books or articles about Bulgaria are published in foreign languages, there is always a problem with spelling of geographical and proper names. This problem comes from the difference in alphabets. The Cyrillic alphabet was created in Bulgaria by St. Cyril and St. Methodius in 9th century A.D., and was adopted by a number of other Slavic languages. After Bulgaria's acceptance into European Union, the Cyrillic alphabet becomes the third (after Latin and Greek) official alphabet of EU. On this occasion, the Bulgarian Government plans to introduce official transliteration rules from Bulgarian to English, which will become mandatory.

Such attempts have been made in the past. During the Communist regime, the mandatory transliteration system of Bulgarian texts existed regardless of the target language. In this system, each Cyrillic character corresponded to a character of the Latin alphabet. Such principle, however, proved to be impractical since a number of diacritic characters had to be used. Thus, even when made mandatory within Bulgaria, this transliteration system was ignored by many Bulgarian zoologists when they published abroad. In the early 1990s, after the demise of the repressive regime and with broadened use of English, this transliteration system becomes defunct. At the same time, for uniformity of transliteration, certain rules have to be adopted.

In this book, the editors adopted the following rules of transliteration from Bulgarian to English.

Cyrillic	Latin	Cyrillic	Latin
а	a	п	p
б	b	р	r
в	v	с	s
г	g	т	t
д	d	у	u
е	e	ф	f
ж, дж	zh, dj	х	h
з	z	ц	ts
и	i	ч	ch
й	i	ш	sh
к	k	щ	sht
л	l	ъ	a
м	m	ь	y
н	n	ю	yu
о	o	я	ya

Of 30 characters of the Bulgarian Cyrillic aphabet, exactly one half (15) are transliterated identically in all existing systems, while others may have alternative possible transliterations. Among these alternatively transliterated characters, only in four cases we deviate from the system that is most commonly (but not always) used in Bulgaria:

дж = **dj** instead of **dzh**, e.g. Strandja Mts., not Strandzha Mts.;
й = **i** instead of **y**, e.g. Aitos, not Aytos;
у = **u** instead of **ou**, e.g. Uzundjovo, not Ouzoundzhovo;
ъ = **a** instead of **u**, e.g. Petar, not Petur.

The transliteration system adopted here has two benefits and two drawbacks. It benefits from simplified spelling and from being closer to actual pronunciation. Its first drawback is impossibility of reverse transliteration (from Latin to Cyrillic alphabet) in two cases: the Latin **i** is used for both Bulgarian **и** and **й** (these two sounds are almost identical); and the Latin **a** is used for both Bulgarian **a** and **ъ** (these sounds are close). A commonly used transliteration system also has one such drawback, when the Latin **y** is used for both Bulgarian **й** and **ь**. The second drawback is that the modern reference literature such as maps, atlases, and encyclopedias often transliterate Bulgarian **y** as **ou**, and Bulgarian **ъ** as **u**. At the same time, in many cases the reference literature uses the transliteration system adopted here.

This transliteration system was used for geographic names as well as for the titles of books and journals. In case of Russian names we used Russian transliteration rules, e.g. Russian **x** = **kh**; but we do not use **e** = **ye**.

Geographic Names

Some explanations are due on the spelling and translation of geographic names. The major latitudinal mountain range, which divides Bulgaria into its northern and southern parts, is called alternatively either Stara Planina (a Bulgarian name, meaning "the Old Mountain"), or Balkan (a Turkish name, meaning "a mountain area" in general). In this book, only Stara Planina name is used to avoid a confusion between the Balkan Mountains and English terms related to the Balkan Peninsula (such as "the Balkans", "West Balkans" etc.) and corresponding geographic ranges of taxa. For another major mountain range we used "the Rhodopes" instead of spellings Rodope, Rhodope, Rodopi, or Rhodopi, all of which can be found in literature. The adjective names of mountain ranges are translated as nouns with addition of "Mts.", e.g. Bulgarian name Maleshevska Planina is translated Maleshevska Mts. In cases when one mountain range in Bulgarian has alternative names (e.g. Osogovska Planina and Osogovo), we use the noun (Osogovo Mts.) although the adjective variant is more commonly used in Bulgaria. The traditional English spelling with "ia" is adopted for the names Bulgaria and Sofia, but in all other cases "ia" is transliterated as "iya", e.g. Kamchiya (not Kamchia) River.

Authors' Names

Regarding proper names of the authors we adopted the following principles of transliteration:

1. For those Bulgarian authors who have published their works in foreign languages using Latin alphabet (usually German, French, or English), we used transliteration traditionally adopted by these authors, e.g. I. Buresch; V. Guéorguiev, Christo Deltshev. This is why similar and even identical Bulgarian names sometimes have different transliteration, e.g. V. Beschovski and V. Golemansky; Vasil Popov and Vassil Golemansky.
2. For those Bulgarian authors whose names were transliterated *differently* in different publications, we used the most common transliteration; in the case of variable original transliteration, the most common variant was given in square brackets, e.g. [Peshev, G.] Peschev, G. This was necessary in order to bring together all works of the same author in the alphabetized reference list (e.g. spelled Cvetkov and Zvetkov) as well as in chronological order. In some curious cases, there are many alternative transliterations of the same author's name, e.g. Georgi Peshev was also transliterated as Pechev, Péchev, Pešev, and Peschev.
3. For those Bulgarian authors (usually the old ones) who published only in Bulgarian, we used the transliteration of their name given in the foreign-language summary; in case this was lacking, we used the closest possible transliteration. We did not use the spelling given in the foreign-language translation of journal's table of contents since it was usually arbitrary and often incorrect. Thus, it is possible that in different chapters of our book the same author's name is spelled differently (e.g. Jurinich, Jurinitch, or Iurinitsh) but the spelling is uniform within each chapter.

References

The information on books, book chapters, and journal articles written in Cyrillic alphabet is given in References according to the following rules. The obvious goal of this rather elaborate transliteration scheme is to facilitate bibliographic search for the titles of Bulgarian or Russian books and journals. The principles of authors' names spelling are described above.

Journal articles. The translated article title is taken from the foreign-language summary, if available, regardless of its language (usually German, French, or English). If such a summary is lacking, the title is translated to English without capitalized words. In both cases, translated title is given in square brackets to emphasize that it is not an original title. Journal titles are transliterated in Latin characters according to the transliteration rules described above. If translation of the journal title is present on the cover or title page, it is placed in parentheses after the transliterated title. If such a translation is lacking, the journal title is translated to English (with capitalized words) and placed in square brackets after the transliterated title. In rare cases when a translation of a journal title is given in two or three foreign languages using Latin characters, all translations are listed in parentheses consequently. The place of publication is placed in parentheses only when the translated title coincides with another existing journal, e.g. *Priroda* [*Nature*] (Sofia); *Ekologiya* (*Ecology*) (Sofia). All journal titles are given in full, non-abbreviated form.

Books (including dissertations). Book titles are transliterated in Latin characters, and followed by its translation. If translation of the book title is present on the cover or title page, it is placed in parentheses after the transliterated title. If such a translation is lacking, the book title is translated to English (with capitalized words) and placed in square brackets after the transliterated title. The title is followed by the publishers name and the place of publication. The publishers name is transliterated in Latin characters, and is followed by its translation in parentheses (if translated by the publishers) or square brackets (if translated by the editors of this book).

Book chapters. The title of the chapter is translated as for journal articles. The title and publishers of the book are translated as for books. Editor(s) are given except those cases when the book was edited by an editorial board rather than specific editor(s).

Zoogeographical Terminology

In addition to commonly accepted and frequently used zoogeographical categories, this book introduces a number of new terms, in some cases quite complex ones. This compelled us to formulate the following rules for construction and spelling of these terms in English.

1. A single word (no hyphenation) is used for those zoogeographical categories, which:
   (a) contain one geographic name, e.g. European, Pontian;
   (b) are derived from (a) via prefixes, e.g. Holomediterranean, Subboreal, Transpalearctic;
   (c) contain two unequal components, e.g. Pontomediterranean (which refers to the taxa from the Pontian part of the Mediterranean Subregion but not to the taxa from both Pontian and Mediterranean areas; those would be called Pontian–Mediterranean; see below);
   (d) are commonly used in the literature as a single word, but are derived from two geographic terms, e.g. Eurosiberian, Oreotundral, arctoalpine distribution.
2. Two separate words (no hyphenation) are used for those zoogeographical categories, which contain a single geographic term with a place adjective, e.g. Central European, Western Palearctic.
3. Two or more words or their combinations hyphenated (connected) by *en dash* (*not* hyphen) are used for those zoogeographical categories, which contain two or more geographic terms but are not commonly used in the literature. In this

case, the connected components can either be single words, e.g. Turanian–Mediterranean, Afrotropical–Palearctic, Macedonian–Rila–Rhodopean; or can also contain two separate words, e.g. European–Western Asian, Southeast European–Anatolian, North Mediterranean–Central Asian, Pontian–Caspian–Southeast European. *En dash* is also used in this book to designate value ranges (e.g. 100–200 m), year ranges, and page ranges.

4. Note that we do *not* use *hyphen* anywhere in geographic terminology, as opposed to the common practice of using hyphen to connect parts of adjectives, e.g. we use Southeastern but not South-Eastern.

The following example clarifies the necessity and significance of such elaborate conventions. In the present book, the term "Atlantic–Mediterranean" is used for marine species distributed in the Atlantic Ocean and Mediterranean Sea, while the term "Atlantomediterranean" is applied to terrestrial species distributed in Atlantic (Iberian) part of the Mediterranean Subregion.

The above mentioned rules were applied to any zoogeographical terminology independent of author's specific scheme of classification; it was not our goal to create standard or uniform zoogeographical classifications, but only to provide uniform spelling. In six chapters of this book, the authors use, or comment upon, the popular classification of zoogeographical chorotypes in the Western Palearctic introduced (in English) by Vigna Taglianti et al. (1999, *Biogeographia*, 20:31–59). Without discussing this scheme, we need to comment here that some of the proposed chorotype names were not formed correctly as English words, e.g. Sibero–European, Centralasiatic–Europeo–Mediterranean, Turano–Europeo–Mediterranean. Such names were changed in this book to make them appear more natural in English.

1 Introduction

Alexi Popov[1] and Victor Fet[2]

[1]*National Museum of Natural History, Bulgarian Academy of Sciences, Tsar Osvoboditel Blvd. 1, 1000 Sofia, Bulgaria, e-mail: alpopov@bulinfo.net*

[2]*Department of Biological Sciences, Marshall University, Huntington, West Virginia 25755-2510, USA, e-mail: fet@marshall.edu*

Europe grades to Asia through the Balkans. They act as a connecting corridor, a liberal barrier, and a mediating buffer. On top of this, humans with their several millennia of Balkan tenure have changed the biota here. Many creatures perished; others still fight back, meeting the anthropogenic challenge and coexistence with humans. A combination of geological factors, high mountains and river valleys, glacial processes, extensive karst regions with caves, forests and sea coast, as well as incessant human activity, made the Balkan Peninsula one of the two—along with the Iberian Peninsula—most interesting biogeographic regions in Europe. Biogeography, and specifically zoogeography, of the Balkans presents a fascinating reading for any specialist.

The interest in the Balkan biodiversity grew steadily during the last hundred years; its patterns and processes have been recently reviewed in an important volume (Griffiths et al., 2004). The authors of this volume addressed the Balkan Peninsula as "the great European hotspot" of biodiversity. As they commented, "The extremely high level of endemism in the Balkans compared to other parts of Europe is perhaps the most striking element of comparative pattern of biodiversity. ...The Balkans offer great potential at European scale for conserving the "last untouched" wilderness on the continent". These authors concluded, however, that "in many respects, we are far from achieving the goal of understanding pattern and process in Balkan biodiversity" (Griffiths et al., 2004).

Through a variety of geographical, historical, and cultural circumstances, Bulgaria is currently the best-studied of all countries in the Balkan Peninsula in regard of its biodiversity. Intensive research in this area started in Bulgaria immediately after this country's independence was restored in 1878. To study and preserve the rich nature and natural resources was an organic part of the young nation's patriotism, much as the study of its ethnographic, historical, and archeological heritage.

By the end of the 19th–beginning of the 20th century, already the best European expertise has been summoned to survey the still largely undisturbed Balkan landscapes of Bulgaria. The Royal (now National) Museum of Natural History, founded as early as in 1889, the oldest and the richest museum of this kind in the Balkans, served as a major focus of biodiversity research (Buresch, 1928; Atanasov, 1955; Popov, 1989), especially in the field of entomology

V. Fet and A. Popov (Eds.), Biogeography and Ecology of Bulgaria, 1–8.

(Buresch, 1924; Drenski, 1931). Another important center of such studies was the Physical-mathematical Faculty of the St. Kliment Ohridski University of Sofia. Studies of species diversity of the Black Sea commenced at the Marine Biological Station in Varna (Paspalew, 1933).

Numerous contributions and reviews on various taxonomic groups have been published in several professional Bulgarian journals: *Travaux de la Société bulgare des sciences naturelles* [*Trudove na Balgarskoto prirodoizpitatelno druzhestvo*], 18 volumes, 1900–1939; *Annuaire de l'Université de Sofia "St. Kliment Ohridski"* [*Godishnik na Sofiiskiya universitet "Sv. Kliment Ohridski"*], 95 volumes, since 1905; *Mitteilungen der Bulgarischen Entomologischen Gesellschaft in Sofia* [*Izvestiya na Balgarskoto entomologichno druzhestvo*], 12 volumes, 1924–1942; *Mitteilungen aus den Königlichen naturwissenschaftlichen Instituten in Sofia – Bulgarien* [*Izvestiya na Tsarskite prirodonauchni instituti v Sofia*], 16 volumes, 1928–1943 (Buresch, 1942).

It will be prudent to state that even under the repressive Communist regime of 1944–1989, which did not allow unrestricted exchange of ideas and specialists between Bulgaria and free Europe, a considerable effort and progress has been invested in biodiversity studies. Several generations of Bulgarian scientists with a great amount of expertise in taxonomy, systematics, biogeography, and ecology of various groups of organisms labored to amass the extensive knowledge (Popov, 1991; Golemansky, 1997). These studies have been conducted by the Institute of Zoology, Institute of Botany and National Museum of Natural History, all created in the Bulgarian Academy of Sciences on the basis of the former Royal Museum, as well as by biological faculties of the University of Sofia and Paisii Hilendarski University of Plovdiv, and by the Institute of Oceanology in Varna. The first Bulgarian natural history journals were followed by new periodicals: *Bulletin de l'Institut botanique* [*Izvestiya na Botanicheskiya institut*], 25 volumes, 1950–1974; *Bulletin de l'Institut de zoologie et musée* [*Izvestiya na Zoologicheskiya institut s muzei*], 41 volumes, 1951–1974; *Acta zoologica bulgarica*, 58 volumes, since 1975; *Ecology* [*Ecologiya*], 26 volumes, 1975–1994; *Hydrobiology* [*Hidrobiologiya*], 40 volumes, 1975–1996; *Fitologija*, 48 volumes, 1975–1996; *Phytologia Balcanica*, 12 volumes, since 1995; and *Historia naturalis bulgarica*, 17 volumes, since 1989 (Popov, 2000).

The Bulgarian science was also lucky to rely on the expertise of Soviet specialists, some of whom still preserved the brilliant academic traditions of the pre-1917 Russian school of zoology and botany. Similar expertise was drawn in 1950s–1980s from Eastern Europe; the specialists from Czechoslovakia, Poland, or East Germany took part in the surveys of this rich southeastern end of Europe. In addition, many Western European experts have always been keenly interested to bring together the pieces of the biogeographic and ecological puzzle which is the Balkan Peninsula.

Since 1989, many changes have occurred in the structure and practice of the Bulgarian academic research. With the collapse of the old socialist system, the state budgets designated for basic research, field expeditions, and purchase of modern research equipment decreased heavily. The number of personnel employed by basic research institutions was not steadily increasing anymore, which was the case during previous decades. The researcher's job ceased to be prestigious due to low salaries.

On the other hand, new opportunities have been emerging. Free research exchange became quite common, and possibilities to visit similar research institutions in Western Europe and North America increased significantly due to unrestricted access to foreign and international grant competition. The financial support by foreign governments and NGO allows maintaining activities in biodiversity conservation throughout Bulgaria, including those in protected territories.

As a result of these studies, several research initiatives have been recently designed and implemented: National Biological Diversity Conservation Strategy (Meine, 1998), with a special attention to endemic, relict, and rare species; CORINE Biotopes in Bulgaria; Biodiversity of Central Balkan and Rila National Parks (Sakalian, 2000a, 2000b; Popov & Meshinev, 2000); studies of biodiversity of some water bodies (Michev et al., 1998; Golemansky & Naidenov, 1998, 2000) and almost all nature parks. Currently, the research efforts are focused on creation of NATURA 2000 network in Bulgaria.

Still keep their importance, as milestones for the studies of Bulgarian biodiversity, the volumes of several ongoing periodical series: *Fauna bulgarica* [*Fauna na Balgariya*] (27 volumes, since 1950); *Flora Reipublicae Bulgaricae* [*Flora na Republika Balgariya*] (10 volumes, since 1963); and *Catalogus faunae bulgaricae* (five volumes, since 1992).

Another relevant development that has to be mentioned is the emergence of Pensoft, the new publishing house headed by Dr. Lyubomir Penev. In a very short time since the 1990s, Pensoft became the leading publisher of natural science books in Eastern Europe. While its production of research literature is very diverse, the favorite subject of Pensoft is biodiversity. It produced many splendidly published monographs on various systematic groups from all over the world, as well as periodical series, monographs, and reviews focused on Bulgarian biota. In addition to the abovementioned series, of which most are published by Pensoft, we need to mention two new series of this publishing house: *Zoocartographia Balcanica* (two volumes, since 2001) and *Biodiversity of Bulgaria* (together with National Museum of Natural History, two volumes, since 2001). The first volume of the latter (Beron, 2001) inaugurated an ambitious idea of regional surveys of the major natural and significant for conservation regions of Bulgaria, and is the first such project in the Balkan Peninsula. The second volume (Beron & Popov, 2004), is devoted to Eastern Rhodopes, and contains information on 4329 species of animals and 1962 species of vascular plants.

The most recent review of Bulgarian biodiversity (Petrova, 2005) addresses the current state of knowledge on animals, plants, fungi, and protists. Also, the second edition of the "Red Book of Bulgaria" is being prepared, which for the first time, along with vascular plants and vertebrates, will include bryophytes, fungi, and invertebrates, as well as habitats.

As Bulgaria is now heading for the admission to the European Union, new challenges present themselves. Not only academic biodiversity knowledge has to be conveniently assessed in modern databases, available at a click for anybody anywhere in the world, rendered in an inevitable *lingua franca* of English language. Further, the biodiversity data are of course inseparable from conservation policies, legislation, and its implementation in upcoming environmental surveys and assessments. Among others, the United Nations Development Program (UNDP), United States Agency of International Development (USAID), Bulgarian–Swiss Biodiversity Conservation Program (BSBCP), Regional Ecological Center for Southeastern Europe in Budapest (REC), and Bulgarian–French Foundation Le Balkan have been working in Bulgaria for years now, heavily relying on available local expertise, in order to help to determine and address priorities addressing biodiversity. An important role of mediators between these organizations and Bulgarian research community has been played by such NGOs as Bulgarian Society for Protection of Birds (BSPB), Wilderness Fund, and Green Balkans.

The total number of animal species currently recorded in Bulgaria is approximately 29,850. This figure results from combining the data of Hubenov (1996) on all animal groups, with the most recent data summarized in Petrova (2005); the latter, however, did not include several groups. Insects, at 20,574 species (Hubenov, 2005), comprise 69% of the fauna. Vertebrates include 792 species (2.7% of all animal species); however, this number includes all bird species recorded in Bulgaria even if they are occasional visitors. These data can be compared to the information from the *Fauna Europaea* project on animal species diversity in Europe (excluding all marine taxa and Protista). The number of species of terrestrial and freshwater animals (without protozoans) in Bulgaria is 26,655, which is 21% of ca. 130,000 animal species recorded in Europe. Since many animal groups are still to be investigated, the projected number of animal species for Bulgaria is ca. 56,000, with ca. 40,000 of insects (Hubenov, 1996).

The biodiversity census for Bulgaria can be further complemented by the data on non-animal taxa summarized in Petrova (2005). There are ca. 3840 species of seed plants, 60 species of pteridophytes, 719 species of bryophytes, 3063 species of algae (including Cyanoprokaryota), and ca. 4900 species of fungi. In total, Bulgarian biota includes ca. 42,400 identified species.

Due to the geographic location of Bulgaria and historical events that led to the formation of its fauna, one can observe two major, separate zoogeographical complexes: the northern complex of cold-tolerant species, and the southern complex of thermophilic species. Josifov (1988) named these two complexes, correspondingly, Eurosiberian and Mediterranean. These have been subdivided by Gruev & Kuzmanov (1994) into six complexes with 15 types of faunal elements. A varied combination of Siberian and Central European faunal elements on the one hand, and Mediterranean faunal elements on the other, expected for this transitional corner of Europe, correspond to two major zoogeographic complexes, and embrace majority of species in Bulgarian fauna and flora. They are often supplemented by a limited number of steppe elements in Northeastern Bulgaria; Anatolian and

Iranian (Iranian–Turanian) migrants in the southeastern part of this country; Pontian (Euxinian) species in Eastern Bulgaria, etc. A small but distinct group of species has a disjunctive Arctoalpine distribution (56 species according to Beron, 1969). Among the Mediterranean faunal elements, nearly all species are Holomediterranean and Pontomediterranean. In addition, Bulgarian fauna includes a large number of endemics. These are mainly arthropod species and subspecies, but a few endemic genera are also known. Two major groups of relicts are distinguished: preglacial (Tertiary) and glacial (Quaternary) relicts.

The ideas on zoogeographical subdivision of Bulgaria have been developed by several authors. Drenski (1946) correlated his division with that of the neighboring territories of the Balkan Peninsula. Buresch & Popov (1973) based their division on ecological-zoogeographical concepts. Guéorguiev (1982) reduced the number of regions and did not correlate with territories outside Bulgaria. Finally, Gruev (1988) based his subdivision scheme on general biogeographical principles. Current phytogeographical subdivision of Bulgaria includes 29 floristic regions and subregions (Assyov et al., 2002).

The current volume does not by any means embrace or even review all existing knowledge on Bulgarian biodiversity. However, animal groups selected for this volume's surveys are those presenting some of the best case studies in biogeography. Our goal was to address in detail the issues and interpretations concerning Bulgarian fauna and zoogeography. Some groups such as vertebrates are necessarily much better studied literally by the generations of academic researchers. This allows for in-depth analysis of distribution along ecological gradients of mammals (V. Popov) and birds (Boev, Milchev & V. Popov) as well as for the reviews of the Bulgarian herpetofauna (Petrov) and ichthyofauna (Stefanov).

Invertebrates, as always, present a mosaic of well-known and less-known groups, and a challenge of choice. Traditional, large insect groups of ground beetles (Guéorguiev) and Orthoptera (A. Popov) provide a broad-stroke approach to endemism and biogeographic trends, as well as selected, well-known groups of spiders (Deltshev; Blagoev; Lazarov). Additional information comes from more specialized insect groups such as Odonata (Beschovski & Marinov) and jewel beetles (Sakalian & Langourov). A review of all Bulgarian Myriapoda is presented by Stoev, and a chapter on harvestmen ecology, by Mitov. An account of scorpions, a small but interesting arachnid group, is given by Fet & Soleglad. An exhaustive analysis of mollusks by Hubenov provides both zoogeographical and ecological generalizations of this most important invertebrate group on land, in fresh waters, and in the Black Sea. A separate, detailed review of remarkable freshwater stygofauna is provided by Pandourski; intriguing, unique Protozoa of the Black Sea littoral are discussed by Golemansky. The book is supplemented by general reviews of the cave fauna (Beron) and the Bulgarian Black Sea fauna (Golemansky) as well as brief overviews of flora and vegetation (Peev & Delcheva; Meshinev) and landscapes of Bulgaria (Velikov & Stoyanova).

As its background, any study of biodiversity is impossible without regional surveys. Bulgaria has a great tradition of floristic and faunistic research.

Generalizations and theoretical treatments arrive as a product of those but only after the decades of groundwork. The case studies given in this volume provide a broad diversity of approaches to those generalizations. It was not our goal to unify all zoogegraphical schemes and patterns of evolution provided by the authors; on the contrary, one should appreciate the difference in those patterns among very different animal groups. At the same time, a few unifying themes sound clear through the mosaic of presented surveys. For instance, the difference and prevalence of mountain faunal elements is emphasized; among those, the endemism is prominent in the major ranges of Stara Planina and, especially, Rila–Rhodopes, but also expressed at much more local scale. Glacial processes of Pleistocene impacted heavily on many faunal elements, trapping boreal elements on the mountaintops and endemics in the caves. Similar zoogeographical trends are seen in Bulgaria in many other well-studied animal groups not included in our volume, e.g. Isopoda, Pseudoscorpiones, Ephemeroptera, Plecoptera, Heteroptera, Neuroptera, Trichoptera, Lepidoptera, and some families of Coleoptera (Elateridae, Cerambycidae, Chrysomelidae, Curculionidae).

It is the editors' wish to thank all the participants in this project, which incorporated 27 authors. We acknowledge the kind help of the Springer staff. We are especially grateful to Emilia Popova and Galina Fet for their help, patience, and understanding.

References

Assyov, B., D. Dimitrov, R. Vassilev & A. Petrova, 2002. Conspectus of the Bulgarian Vascular Flora. Distribution Maps and Floristic Elements. 2nd Revised and Enlarged Edition. Bulgarian-Swiss Biodiversity Conservation Programme, Sofia.

Atanassov, N., 1955. [The Natural History Museum at the Bulgarian Academy of Sciences]. Priroda [Nature] (Sofia) 4(1):98–105 (in Bulgarian).

Beron, P., 1969. Sur les éléments boréo-alpins de la faune bulgare. Izvestiya na Zoologicheskiya institut s muzei (Bulletin de l'Institut de zoologie et musée) 30:115–132.

Beron, P. (ed.), 2001. Bioraznoobrazie na Kresnenskiya prolom (Biodiversity of Kresna Gorge (SW Bulgaria)). Natsionalen prirodonauchen muzei i Institut po zoologiya pri BAN (National Museum of Natural History (Sofia) & Institute of Zoology, Bulgarian Academy of Sciences), Sofia (in Bulgarian).

Beron, P. & A. Popov (eds), 2004. Biodiversity of Bulgaria. 2. Biodiversity of Eastern Rhodopes (Bulgaria and Greece). Pensoft & National Museum of Natural History, Sofia.

Buresch, I., 1924. [History of entomological exploration of Bulgaria]. Trudove na Balgarskiya nauchen zemedelsko-stopanski institut [Proceedings of the Bulgarian Scientific Agricultural Institute] 8:1–136. (in Bulgarian).

Buresch, I., 1928. Die Naturwissenschaftlichen Institute Seiner Majestät des Königs der Bulgaren. Ziele und Aufgaben der "Mitteilungen der Königlichen Naturwissenschaftlichen Institute". Izvestiya na Tsarskite prirodonauchni instituti v Sofia (Mitteilungen aus den Königlichen naturwissenschaftlichen Instituten in Sofia – Bulgarien; Bulletin des Institutions royales d'histoire naturelle à Sofia – Bulgarie) 1:1–16.

Buresch, I., 1942 [Fünfzehn Jahre "Mitteilungen aus den Königlichen naturwissenschaftlichen Instituten in Sofia" (1928–1942). Systematischer Inhalt der in den Bänden I bis XV erschienenen Arbeiten und Verzeichnis der darin beschriebenen neuen Gattungen und Arten]. Izvestiya na Tsarskite prirodonauchni instituti v Sofia (Mitteilungen aus den Königlichen naturwissenschaftlichen Instituten in Sofia – Bulgarien; Bulletin des Institutions royales d'histoire naturelle à Sofia – Bulgarie) 15:257–304 (in Bulgarian).

Buresch, I. & A. Popov, 1973. [Zoogeographical regions]. In Galabov, Zh. (ed), Atlas Narodna republika Balgariya [Atlas of the People's Republic of Bulgaria]. Glavno upravlenie po geodeziya i kartografiya [General Direction of Geodesy and Cartography], Sofia: 91–92 (in Bulgarian).

Drenski, P., 1931. Geschichte und Tätigkeit der Königlichen Entomologischen Station in Sofia. Anlässlich ihres 25-jähriges Bestandes (1. September 1905–1. September 1930). Izvestiya na Tsarskite prirodonauchni instituti v Sofia (Mitteilungen aus den Königlichen naturwissenschaftlichen Instituten in Sofia – Bulgarien; Bulletin des Institutions royales d'histoire naturelle à Sofia – Bulgarie) 4:17–50.

Drenski, P., 1946. [Zoogeographical sketch of Bulgaria]. Godishnik na Sofiiskiya universitet. Fiziko-matematicheski fakultet (Annuaire de l'Université de Sofia. Faculté physico-mathématique) 42(3 – Natural History):109–161 (in Bulgarian).

Golemansky, V., 1997. [Half a century of Institute of Zoology]. In Golemansky, V. (ed), 50 godini Institut po zoologiya pri Balgarskata akademiya na naukite. 1947–1997 (50 years Institute of Zoology at the Bulgarian Academy of Sciences. 1947–1997). Akademichno izdatelstvo "Prof. Marin Drinov" ("Prof. Marin Drinov" Academic Publishing House), Sofia: 7–27 (in Bulgarian).

Golemansky, V. & W. Naidenov (eds), 1998. Biodiversity of Shabla Lake System. "Prof. Marin Drinov" Academic Publishing House, Sofia.

Golemansky, V. & W. Naidenov (eds), 2000. Biodiversity and Evolution of Glacial Water Ecosystems in the Rila Mountains. Institute of Zoology, Sofia.

Griffiths, H.I., B. Kryštufek & J.M. Reed (eds), 2004. Balkan Biodiversity. Pattern and Process in the European Hotspot. Kluwer Academic Publishers, Dordrecht.

Gruev, B., 1988. Obshta biogeografiya [General Biogeography]. Nauka i izkustvo [Science and Art], Sofia (in Bulgarian).

Gruev, B. & B. Kuzmanov, 1994. Obshta biogeografiya. Vtoro osnovno preraboteno izdanie [General Biogeography. Second Basic Revised Edition]. Universitetsko izdatelstvo "Sv. Kliment Ohridski" [St. Kliment Ohridski University Publishing House], Sofia (in Bulgarian).

Guéorguiev, V., 1982. [Zoogeographical regions on the basis of terrestrial fauna]. In Galabov, Zh. (ed), Geografiya na Balgariya. Fizicheska geografiya. Prirodni usloviya i resursi (Geography of Bulgaria. Physical Geography. Natural Conditions and Resources). Izdatelstvo na Balgarskata akademiya na naukite (Publishing House of the Bulgarian Academy of Sciences), Sofia: 472–477 (in Bulgarian).

Hubenov, Z., 1996. [Faunistic diversity of Bulgaria – invertebrates]. Historia naturalis bulgarica 6:11–16 (in Bulgarian).

Hubenov, Z., 2005. [Entomofaunistic diversity of Bulgaria]. In Petrova, A. (ed), Savremenno sastoyanie na bioraznoobrazieto v Balgariya – problemi i perspektivi (Current State of Bulgarian Biodiversity – Problems and Perspectives). Balgarska platforma za bioraznoobrazie, Ministerstvo na okolnata sreda i vodite (Bulgarian Biodiversity Platform, Ministry of Environment and Water), Sofia: 173–198 (in Bulgarian).

Josifov, M., 1988. Über den zoogeographischen Charakter der südeuropäischen Insektenfauna unter besonderer Berücksichtigung der Heteropteren. Berichte des naturwissenschaftlich-medizinischen Vereins, Innsbruck 75:177–184.

Meine, C. (ed), 1998. Bulgaria's Biological Diversity: Conservation Status and Needs Assessment. Volumes I and II. Pensoft, Sofia & Moscow.

Michev, T., B. Georgiev, A. Petrova & M. Stoyneva (eds), 1998. Biodiversity of the Srebarna Biosphere Reserve. Checklist and Bibliography. Context & Pensoft, Sofia.

Paspalew, G., 1933. Bulgarische Biologische Station und Aquarium in Varna am Schwarzen Meer. Izvestiya na Tsarskite prirodonauchni instituti v Sofia (Mitteilungen aus den Königlichen naturwissenschaftlichen Instituten in Sofia – Bulgarien; Bulletin des Institutions royales d'histoire naturelle à Sofia – Bulgarie) 6:1–33.

Petrova, A. (ed), 2005. Savremenno sastoyanie na bioraznoobrazieto v Balgariya – problemi i perspektivi (Current State of Bulgarian Biodiversity – Problems and Perspectives). Balgarska platforma za bioraznoobrazie, Ministerstvo na okolnata sreda i vodite (Bulgarian Biodiversity Platform, Ministry of Environment and Water), Sofia (in Bulgarian).

Popov, A., 1989. [The National Natural History Museum at the Bulgarian Academy of Sciences after 1974]. Historia naturalis bulgarica 1:5–13 (in Bulgarian).

Popov, A., 1991. [Faunistic entomological investigations in Bulgaria after the Second World War]. In Tsankov, G. (ed), Parva natsionalna konferentsiya po entomologiya, 28–30 oktomvri 1991 g., Sofia (First National Conference of Entomology, 28–30 October 1991, Sofia). Sayuz na uchenite v Balgariya, Balgarsko prirodoizpitatelno druzhestvo (Union of the Scientists in Bulgaria, Bulgarian Society of Natural Research), Sofia: 267–272 (in Bulgarian).

Popov, A., 2000 [Ten years of *Historia naturalis bulgarica*]. Historia naturalis bulgarica 11:147–159 (in Bulgarian).

Popov, A. & T. Meshinev (eds), 2000. High Mountain Treeless Zone of the Central Balkan National Park. Biological Diversity and Problems of its Conservation. Pensoft, Sofia.

Sakalian, M. (ed), 2000a. Biological Diversity of the Central Balkan National Park. Pensoft, Sofia.

Sakalian, M. (ed), 2000b. Biological Diversity of the Rila National Park. Pensoft, Sofia.

2 Terrestrial Mammals of Bulgaria: Zoogeographical and Ecological Patterns of Distribution

Vasil Popov
Institute of Zoology, Bulgarian Academy of Sciences, Tsar Osvoboditel Blvd. 1, 1000 Sofia, Bulgaria, e-mail: popov@zoology.bas.bg

Abstract: Our present knowledge of the terrestrial mammals of Bulgaria is reviewed on the basis of published records and original data collected in 1990–2002. The study is based on 93 species of terrestrial mammals. They are classified into four faunal complexes reflecting the influence of historical and environmental factors. The spatial differentiation of the mammalian fauna is considered within a network of landscape territorial units, or natural regions (NRs). Physiographic and climatic indices of each NR are used in the quantitative analyses in order to explore the environmental causes of the observed patterns. The variability of species richness across the NRs does not correlate with environmental variables and seems to depend on the completeness of regional inventory. However, the species richness of carnivores and artiodactyls shows patterns which could be due to the human impact in the lowlands and the greater habitat diversity in the mountains. Faunal differentiation of the Bulgarian mammals is primarily related to the altitudinal gradient and secondarily, to the humidity and continentality of the climate in the lowlands. Detailed analyses are performed on a set of 83 local assemblages of small epigean mammals (pitfall data) from 16 sampling areas. Each assemblage represents a specific habitat type within the sampling area hypothesized as perceived by the small mammals. Conditions within the habitats are described through the environmental variables. The ordination analyses reveal that, again, the composition and structure of the local assemblages are primarily affected by the altitudinal gradient, the secondary group of factors being local vegetation type and humidity. The classification analysis distinguishes five assemblage types. The structure of the local assemblages is affected primarily by the external environmental factors; the role of internal factors, such as competition, could be suggested only in one case. The environmental relationships of species change along the overall gradient. On the basis of the observed patterns we propose a zoogeographical subdivision of Bulgaria into two districts, with two subdistricts each. The changes of the zoogeographical structure of the mammalian fauna in Bulgaria are due to environmental differentiation, modified by human impact since the middle of Holocene

1 Introduction

To date, 95 species of wild terrestrial mammals have been recorded in Bulgaria. One species (*Mus musculus*) consists of two well defined forms, sometimes considered as semispecies: *Mus musculus musculus* and *M. musculus domesticus*. Three species of bats (*Pipistrellus pygmaeus* Leach, 1825, *Myotis aurascens* Kusjakin, 1935, and *M. alcathoe* Helversen et Heller, 2001) have been recently accepted as valid species in the European fauna. There are records of these species from Bulgaria as well

V. Fet and A. Popov (Eds.), Biogeography and Ecology of Bulgaria, 9–37.

(Benda et al., 2003; Peshev et al., 2004; Schunger et al., 2004). The first species is a sibling species of *Pipistrellus pipistrellus*, while the last two species belong to the group of *Myotis mystacinus*. Their distribution and systematic status need further clarification, so, they are not treated separately. Thus, the present study is based on 93 species of terrestrial mammals (the forms of *Mus musculus* are considered separately).

Main features of distribution of mammals in Bulgaria have been known for a long time (Boetticher, 1933), but because of the scarcity of detailed records, especially on the small mammals (insectivores, bats, and rodents), the knowledge remained sketchy. Later, using a greater amount of information, Zimina (1962) described general distributional patterns within Bulgaria, but defined it in rather general terms: "The characteristic features of the northern and mountainous areas of the country include typical European forest mammal species and the Eurasian (steppe and boreal) species, whereas the southern areas are characterized by West Asian and Mediterranean species".

The distributional maps in the recent *Atlas of European Mammals* (Mitchell-Jones et al., 1999) as well as those in *The National Biological Diversity Conservation Strategy* (Meine, 1994) show that even today there is no detailed information on the species occurrence and distribution for many areas of Bulgaria. Since effective implementation of conservation measures needs such detailed information, active regional surveys have been conducted in the various areas of Bulgaria. Total information has been recently summarized by Peshev et al. (2004).

Explaining the factors and mechanisms affecting patterns of distribution and diversity is one of the most important goals in the ecological and biogeographical studies. The climate-species relationship has been well documented at the macrogeographical scale. However, its role at the regional and local scales is still poorly known despite of its theoretical importance for the detailed zoogeographical subdivision, which in turn has practical applications for species management and conservation.

The goals of the present study are: (1) to analyze the entire available faunistic data in order to reveal the distributional patterns of (non-marine) mammals in Bulgaria, and to suggest the possible causes for these patterns; (2) to present a more detailed analysis based on the landscape (ecological) determinants for the quantitative structure of the small mammal assemblages representative for various terrestrial habitats across Bulgaria; (3) to propose a zoogeographical subdivision of Bulgaria; and (4) to characterize the mammalian associations in each zoogeographical unit, providing a base for further studies, as well as for conservation and land development.

2 Material and Methods

Faunistic data. It may be expected that, besides geographical regularities, the major spatial determinants of distribution on a limited territory such as Bulgaria are ecological. In this respect, in order to bring together the distribution of mammals and environmental spatial heterogeneity, the system of 24 landscape subprovinces (Fig. 1), proposed by Petrov (1997), was used as a reference system to record the presence or absence of species. These subprovinces are relatively homogeneous territories with a particular combination of relief, climate, and vegetation; they are referred to in further analysis as natural regions (NRs). This approach offers a possibility to accumulate the often disparate faunistic records in an environmentally meaningful context and to reveal environmental aspects of mammalian distribution. Additionally, this approach allows recognizing ecologically significant spatial boundaries.

The environmental data used at this scale of analysis are based on indices, which characterize each of the 24 NRs (Petrov, 1997). They comprised the range and mean values of 24 quantitative environmental variables. The following climatic variables are used in the analyses: v1. Annual temperature amplitude, expressed as a difference between the mean annual temperatures of July and January (°C); v2. Sum of average monthly temperatures above 0 °C; v3. Sum of precipitation (mm) for months with average temperature above 0 °C; v4. Coefficient of air humidity; v5. Annual precipitation (mm); v6. Annual evaporation (mm).

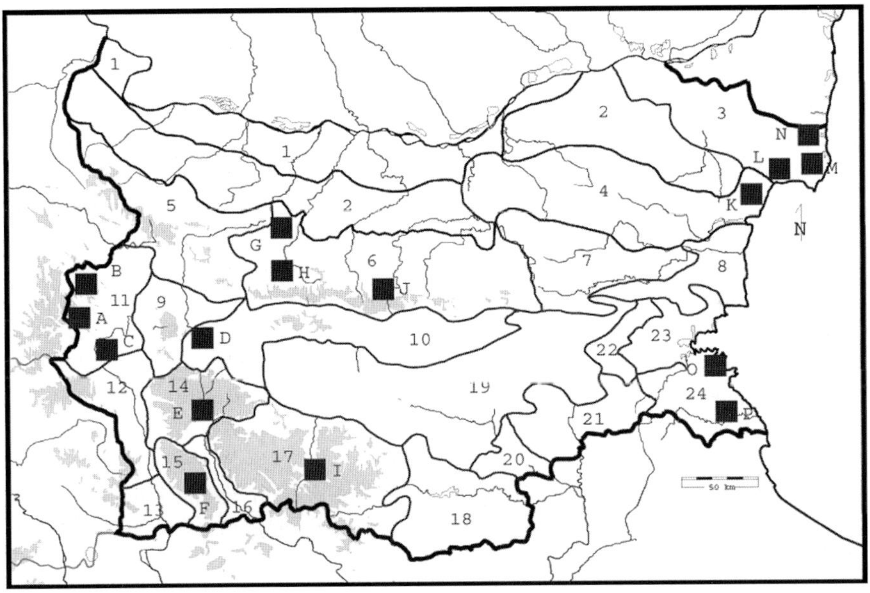

Fig. 1 Map of Bulgaria showing the natural regions (1–24, from Petrov, 1997, modified) and sampling areas (A–P). Natural regions: 1, Northern Danubian Plain; 2, Southern Danubian Plain; 3, Southern Dobrudja; 4, Popovo–Shumen–Franga; 5, Western Stara Planina; 6, Central Stara Planina; 7, Eastern Stara Planina; 8, Maritime Stara Planina; 9, Vitosha–Ihtiman; 10, Sredna Gora–Podbalkan; 11, Kraishte; 12, Osogovo–Middle Struma; 13, Southern Struma; 14, Rila; 15, Pirin; 16, Middle Mesta; 17, Western Rhodopes; 18, Eastern Rhodopes; 19, Upper Thrace; 20, Lower Thrace; 21, Sakar–Dervent; 22, Bakadjik–Hisar; 23, Burgas–Aitos; 24, Strandja.

Literature, museum materials, and unpublished data were the sources of the distribution records. The majority of this information was summarized in Peshev et al. (2004). In some instances, species that were not recorded in a particular subprovince but occur in portions of the provinces outside of the Bulgarian borders were also included in the analyses. In particular, these include bat records from Turkish Thrace, near the Bulgarian–Turkish boundary (Benda & Horáček, 1998), and a dubious observation of *Myotis dasycneme* (Limpens, 2001). Some sibling species (*Myotis mystacinus/M. aurascens, Pipistrellus pipistrellus/P. pygmaeus*) are considered together. In total, this analysis includes 93 species of mammals.

Assemblage data. The more detailed quantitative analysis of distributional patterns of mammals in Bulgaria was based on studies of the specific small epigean mammal assemblages. Small mammals are especially suitable for this kind of analysis since the climate-species relationship is particularly important in shaping the distribution of small endotherms. Moreover, this group of animals provides a possibility to sample quantitatively and simultaneously a large number of species. During the period of 1990–2002, within the framework of various projects concerning the inventory of the biodiversity of important areas defined on the basis of the National Biological Diversity Conservation Strategy, we had an opportunity of sampling 16 areas (Fig. 1, A–P), thus collecting more than 10,000 small mammal specimens. To sample small epigean mammals, we used a protocol based on removal pitfall trapping (for more details see Popov & Krusteva, 1999; Popov et al., 2000; and Popov, 2000a). This method permitted a cheap, large-scale quantitative assessment of local assemblages. In each area, a variable number of trap transects was established in order to cover as completely as possible the regional habitat diversity as perceived by the investigator. Further, in order to delimit uniform spatial assemblages of particular habitats hypothesized as perceived by the small mammals themselves, individual pitfall samples within each area were subjected to ordination analysis. The samples with similar scores on the first two axes and located in close spatial proximity in the field were pooled and considered a sample from a local habitat assemblage (LHA). Examples of this approach are presented in Popov (2000a) and Minkova & Popov (2002). As a result, 83 LHAs were distinguished and analyzed.

We used the term "assemblage" to denote a group of populations belonging to different species that occur in the same area regardless of their interactions. For the descriptive purpose of the present study, this term is more suitable than the term "community", which presumes an association of interacting populations belonging to different species (Whittaker, 1975).

The environmental data associated with each assemblage consisted of two groups of variables. The first group comprised the same climatic variables as used in the analyses of faunistic data (v1–v6, see above). The estimates for each assemblage were based on interpolations between isopleths on appropriate climatic maps. The second group comprised the characteristics of local vegetation (v7) and humidity (v8). The local vegetation component of a habitat was described in terms of three distinct types: (1) open, grassland, cultivated fields; (2) shrubs, forest edges;

(3) forests. The local humidity was coded according to the following categories: (1) dry, far above local erosion basis; (2) moderately dry (depending on a season); (3) moist, near water basins. In our figures, the individual LHAs are denoted by the letter of the sampling area (Fig. 1) followed by the codes of local vegetation (v7) and local humidity (v8).

The comparative data on distribution and habitat selection, presented in this paper, may suggest possible competitive interactions between coexisting, closely related species. In order to emphasize this aspect of the data and to overcome difficulties in the comparison of pitfall catches between species with different morphological, trophic, and ethological features, which cause differences in their trappability, we divided recorded small mammal species for further comparison into three "guilds" (Root, 1967; Schröpfer, 1990), comprising the most abundant terrestrial small mammals in pitfall catch, (1) shrews, (2) voles, and (3) mice.

Analytical procedures. The faunistic and assemblage data were treated using two complementary types of multivariate analyses, classification and ordination. The main classification method used to reveal the structure of the data sets was Two-Way-Indicator Species Analysis, or TWINSPAN (Hill, 1979). This analysis resulted in a double classification of samples (NRs or LHAs) and species, in which samples with similar species composition and/or structure (and therefore presumably similar environment) are grouped together. The Jaccard index was used to quantify the faunal similarity between NRs. The resulting similarity matrix was simplified by cluster analysis using the unweighted pair group method (UPGMA).

Ordinations were made in order to visualize the structure of the datasets and to help interpreting the observed patterns in terms of environmental gradients, along which small mammal assemblages vary in structure and composition. Ordinations were done by canonical correspondence analysis (CCA) and redundancy analysis (RDA), constrained versions of correspondence and principal component analysis, respectively (Braak, 1987). Prior to analysis the quantitative species data were log $(x+1)$-transformed to reduce the skewness. The ordinations were performed using CANOCO Version 3.12 (Braak, 1985, 1990).

For independent examination of species richness, rarefaction values (Simberloff, 1978; James & Rathbun, 1981) were calculated using BIODIV (Baev & Penev, 1995). In addition, Simpson's Inverse Index (N2) was applied to evaluate the combination between species richness and evenness (= diversity, = heterogeneity). The value of N2 can roughly be interpreted as the equivalent number of species if they were evenly abundant.

A two-way ANOVA and post-hoc Scheffe's pairwise comparisons were used to test for the influence of habitat (categorial variable) and altitude (independent variable) on species diversity.

When the variables showed non-normal and asymmetric distribution, relationships were examined by means of Spearman rank correlation coefficient (Rs).

In statistical programs used in the analyses (BIODIV, CANOCO, TWINSPAN) species names are shown as abbreviations, consisting of three letters of the generic name and three of the species name.

3 Results

3·1 Faunistic analysis

Zoogeographical composition of the mammalian fauna of Bulgaria. The present-day species ranges are the result of species evolution within the context of the paleogeographical events (historical factors) and the recent environmental differentiation (ecological factors). In order to address the role of both groups of factors in forming the recent mammalian fauna of Bulgaria, the species were classified in several groups, based on the similarity of their distributional history and likeness of the present-day ranges. Such groups, where species have common geographical origin and common ecological preferences, are usually called *faunal complexes* (Udvardy, 1969). The data on the evolution and distributional history of species is based predominantly on the fossil record from Europe and Bulgaria, summarized in Peshev et al. (2004). Details of their history reflected in present-day ranges are expressed in terms of a lower category, a *faunal element*.

(1) *Boreal faunal complex* (14 species). Cold resistant species with a broad temperature tolerance; they or their close ancestors invaded Europe probably from Siberia during the Pleistocene.
  - (a) Eurosiberian forest element: *Sorex minutus, S. araneus, Neomys fodiens, Myotis dasycneme, Plecotus auritus, Eptesicus nilssonii, Vespertilio murinus, Clethrionomys glareolus, Arvicola terrestris, Martes martes, Lutra lutra, Capreolus capreolus*;
  - (b) Eurosiberian forest-steppe element: *Lepus capensis, Microtus arvalis* (sensu lato).

(2) *Nemoral complex* (38 species). Thermotolerant species associated with mesophilous broad-leaved Palearctic forests. The majority of these species have persisted in the western (Europe) and eastern (East Asia) portions of Palearctic since the Tertiary. In particular, species combined in this complex or their close ancestors lived or appeared in Europe during the Pliocene and the first half of the Pleistocene, but became rare during the extreme climatic intervals of the Pleistocene (except of some mountain species, which during the glacials were forced to inhabit lower altitudes and thus were widely distributed in southern Europe during the second half of the Pleistocene).
  - (a) Transpalearctic or Holarctic element (species associated with forested areas): *Myotis nattereri, M. mystacinus, M. brandtii, M. daubentonii, Nyctalus noctula, Pipistrellus pipistrellus, P. nathusii, Sciurus vulgaris, Micromys minutus, Apodemus agrarius, Canis lupus, Vulpes vulpes, Nyctereutes procyonoides, Ursus arctos, Felis silvestris, Meles meles, Mustela nivalis, Sus scrofa, Cervus elaphus*;
  - (b) South (East) European and (West) Asian element: *Erinaceus concolor, Crocidura leucodon, Nyctalus lasiopterus, N. leisleri, Dryomys nitedula, Sylvaemus uralensis, Microtus epiroticus*;

(c) European element: *Talpa europaea, Neomys anomalus, Myotis bechsteinii, Barbastella barbastellus, Glis glis, Muscardinus avellanarius, Sylvaemus flavicollis, S. sylvaticus, Microtus subterraneus, Mustela putorius*;
(d) South European and West Asian mountain element: *Chionomys nivalis, Rupicapra rupicapra*.

(3) *Mediterranean faunal complex* (26 species). Thermophilous species associated with southern parts of Europe, Asia Minor, and Central Asia since the Pliocene; they disappeared from southern Europe during the cold intervals of the second half of Pleistocene, but reappeared during the interglacials, and especially during Holocene. This complex also includes some species with a wide distribution in South Asia and Africa, which invaded Europe during Holocene.
(a) Widely distributed in southern areas: *Suncus etruscus, Rhinolophus blasii, Miniopterus schreibersii, Canis aureus*;
(b) Submediterranean element (species widely distributed outside the Mediterranean region *sensu stricto*, but tending to occur in thermoxerophilous landscapes): *Crocidura suaveolens, Rhinolophus ferrumequinum, Rh. hipposideros, Myotis blythii, M. emarginatus, Plecotus austriacus, Hypsugo savii, Eptesicus serotinus, Tadarida teniotis, Martes foina, Vormela peregusna*;
(c) Mediterranean element: *Rhinolophus euryale, Rh. mehelyi, Myotis capaccinii, Pipistrellus kuhlii*;
(d) West Mediterranean element: *Myotis myotis, Oryctolagus cuniculus*;
(e) East Mediterranean element (*sensu lato*, species with ranges that can include Iran, Asia Minor, and the Balkan Peninsula): *Myomimus roachi, Sylvaemus mystacinus, Mus macedonicus, Microtus guentheri*;
(f) Euxinian element (species associated with the relict broad-leaved forest of the Caucasian region, invading the Balkan Peninsula via the Pontic Mts.): *Talpa levantis*.

(4) *Eurasian Steppe complex* (eight species). Species associated with the periphery of the arid core of Eurasia, which invaded Europe from the East during Pleistocene; the majority of them were widely distributed in the East and Southeast Europe during the continental and arid climatic intervals of this period.
(a) East European–West Asian element: *Cricetulus migratorius*;
(b) East European–Kazakhstanian element: *Sicista subtilis, Cricetus cricetus, Mustela eversmanni*;
(c) East European element: *Mus spicilegus*;
(d) Balkan (and Southeast European) element: *Spermophilus citellus, Mesocricetus newtoni, Nannospalax leucodon*.

Introduced (*Ondatra zibethicus, Myocastor coypus*, and *Dama dama*) and synanthropic (*Mus musculus musculus, M. musculus domesticus, Rattus rattus*, and *R. norvegicus*) species were not included in this classification.

Although the placement of some species into any of the above groups is not always unequivocal, the presented classification provides a summary of the main aspects of the formation and origin of the recent mammalian fauna of Bulgaria.

Additionally, it presents a characteristic of each species, which incorporates its evolutionary and biogeographical features and will be useful in the interpretation of the results of the further analyses presented below.

3·2 Spatial structure of the mammalian fauna of Bulgaria

Patterns of faunal diversity. The diversity of mammals (species richness, the number of species in a given NR) is relatively non-homogeneous across the 24 NRs. The number of species per area varies from 37 to 64. The following descriptive and correlation statistics (significant Rs, at $P < 0.05$) may help to interpret these initial data:

All mammal species: mean species richness, $X = 52.4$, coefficient of variation, $CV = 16.6$; correlation with insectivores, $Rs = 0.61$; with bats, $Rs = 0.93$; with rodents, $Rs = 0.70$; with carnivores, $Rs = 0.70$; with artiodactyls, not significant.

Insectivores: $X = 7.0$, $CV = 15.1$; correlation with bats, $Rs = 0.47$; with rodents, $Rs = 0.39$; with carnivores, $Rs = 0.44$.

Bats: $X = 16.7$, $CV = 32.7$; correlation with rodents, $Rs = 0.50$; with carnivores, $Rs = 0.59$; with artiodactyls, not significant.

Rodents: $X = 15.6$, $CV = 15.9$; correlation with carnivores, $Rs = 0.49$; with artiodactyls, not significant.

Carnivores: $X = 10.0$, $CV = 18.0$.

Artiodactyls: $X = 3$, $CV = 18.0$.

The correlations of species richness by order with environmental variables are as follows (significant Rs at $P < 0.05$): insectivores, no significant coefficients; bats, with size of area $Rs = 0.41$; rodents, no significant coefficients; carnivores, with v3, $Rs = -0.43$; artiodactyls, with v5, $Rs = 0.48$.

These data reveal the existence of positive correlations between species richness of insectivores, bats, and rodents and the lack of correlation with any of the analyzed environmental variables. For large mammals (carnivores and artiodactyls), however, there is a well pronounced and environmentally related pattern in species richness; the obtained correlations show that their species richness tends to be higher in the NRs with lower temperatures and higher precipitation, i. e. in the mountainous NRs.

Spatial differentiation of the mammalian fauna. In order to provide an insight into the environmental determinants of the territorial differentiation of the mammalian fauna in Bulgaria, the matrix consisting of distributional data (Appendix 1) and environmental indices associated with each NR, was analyzed by CCA. The following environmental variables were selected as statistically significant on the basis of the Monte Carlo permutation tests (in parentheses, the proportion of variance in species data explained by these variables): v2, sum of temperatures above 0 °C (0.10); v6, annual amplitude (0.05); v3, sum of precipitation for months with temperatures above 0 °C (0.03); v5, sum of annual precipitation (0.03). The first two and the fourth variables were highly correlated with the first ordination axis ($R = -0.75$, -0.93, and 0.82, respectively). The second and third variables were correlated with the second axis ($R = -0.42$ and 0.72, respectively). No variables were correlated with the third and fourth axes. The fraction of the total variance in the environmental data extracted by species axis 1 was 16.9%, while axis 2 extracted 11.0%. This result indicated that only part of the species was affected by the environmental gradients. Nevertheless, the overall Monte Carlo permutation test showed that the species are significantly related to these environmental variables (99 random permutations, $P < 0.01$).

The correlations described above and the ordination diagrams (Figs. 2–3) indicate that the first axis represents the effect of the altitudinal gradient on mammalian distribution. It is connected with changes in temperatures, humidity, and continentality of the climate. The second axis mirrors the effect of the humidity (precipitation) during the vegetation season and the continentality of the climate on the mammalian distribution in the lowlands. In particular, it contrasts the northern NRs with dry

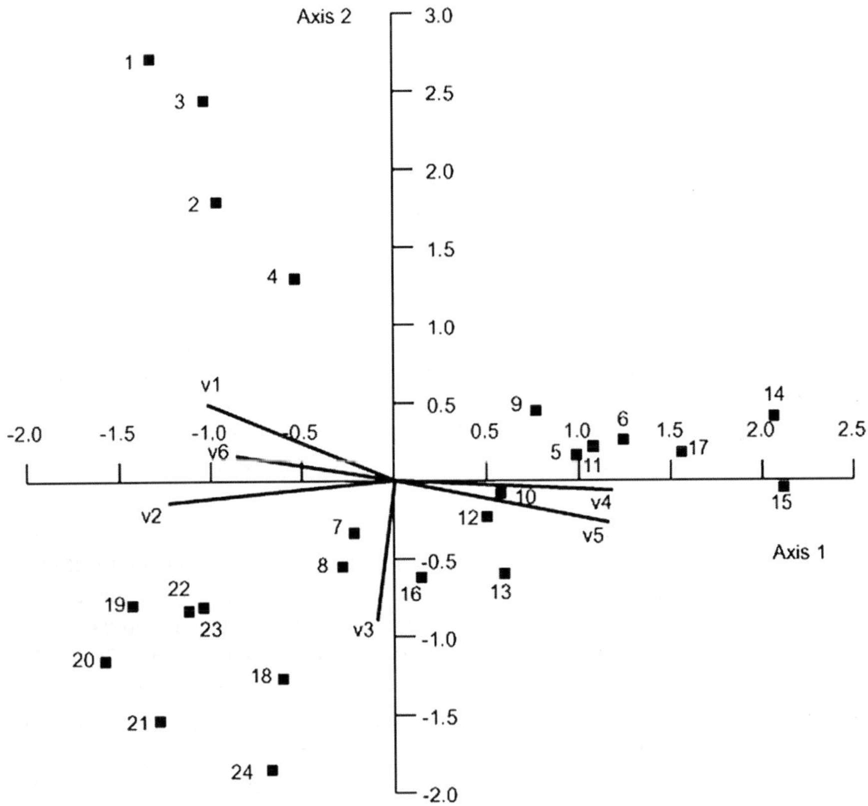

Fig. 2 CCA ordination diagrams (Axis 1 vs. Axis 2). Biplot of natural areas (S1–S24) and environmental variables (v1–v6).

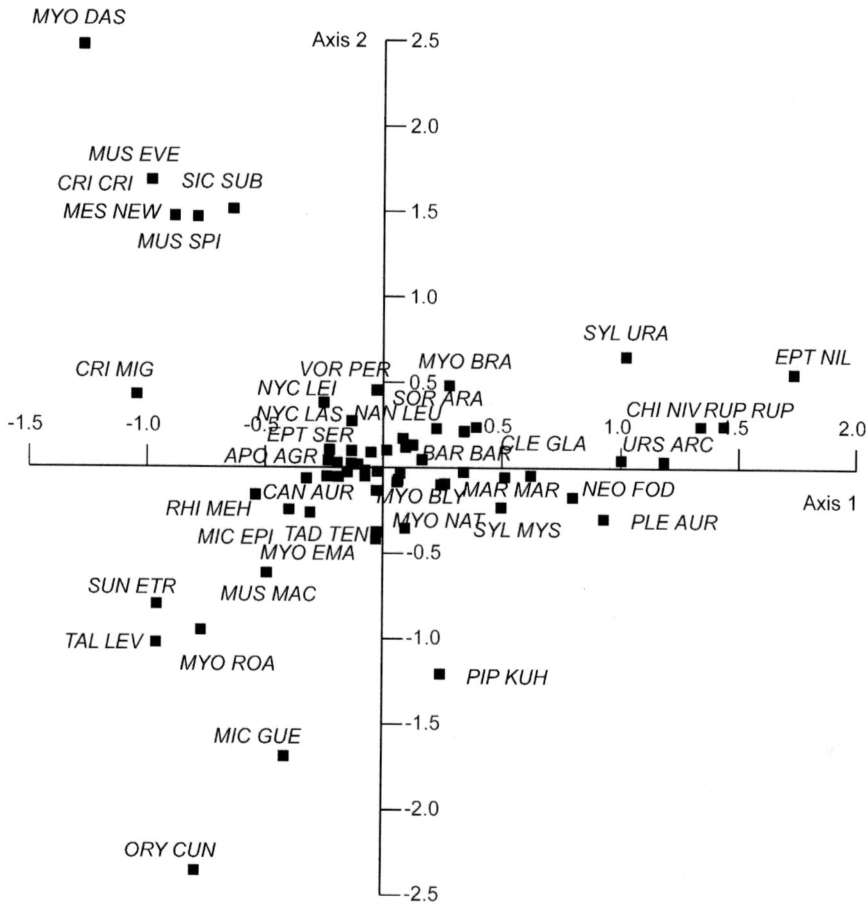

Fig. 3 CCA ordination diagrams (Axis 1 vs. Axis 2). Scatter diagram of species.

and continental climate to the Strandja NR, which is characterized by a relatively even and humid climate.

In order to provide a basis for a territorial subdivision of Bulgaria according to the patterns revealed above, the initial data matrix (Appendix 1) was subjected to classification analyses.

The first analysis was based on the Jaccard coefficient of faunistic similarity. The resulting dendrogram (Fig. 4) presents a picture of the species similarity between the NRs. Five main clusters can be recognized at 0.7 similarity level. The first cluster consists of NRs 20 and 21; it is probably an artifact resulting from the small number of species in these NRs (37 species each). The second cluster is composed by the NRs 10 and 11. These NRs have a rather heterogeneous species

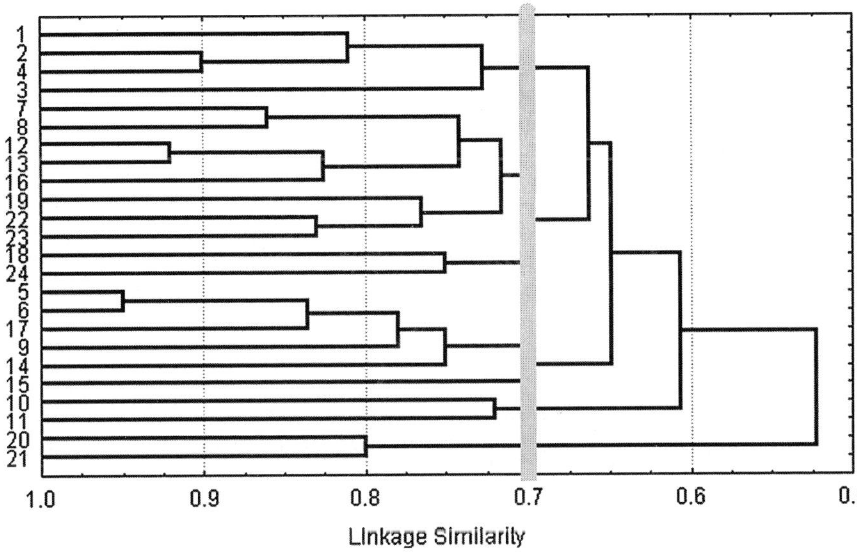

Fig. 4 Cluster analysis dendrogram of natural areas (NRs 1–24), based on Jaccard coefficient of similarity among species lists.

composition with regard to the separation recognized above, since they comprise some low mountain and some semimountainous areas and basins. The third cluster, consisting of NRs 15, 14, 9, 17, 6, and 5, represents the mountainous areas. The fourth cluster comprises the lowland and semimountainous NRs from southern and central-eastern Bulgaria. The fifth cluster embraces the NRs from North Bulgaria (NRs 1–4). At the lower level (0.65) the dendrogram indicates that the mammalian fauna of Bulgaria forms two well defined territorial units: mountains (cluster 3) and lowlands (clusters 4 and 5).

In order to avoid the discrepancies associated with agglomerative clustering techniques, the above data were clustered by TWINSPAN, a divisive classification technique. This analysis revealed similar relationships (Fig. 5). The NRs were classified into six end-groups according to their mammalian fauna. The obtained classification corresponds to NRs grouping of mountainous areas (NRs 5, 6, 14, 15, 17), semimountainous areas or areas located close to the large mountain ranges (NRs 10, 11 and NRs 9, 12, 13, 16), the lowlands of South Bulgaria (NRs 7, 8, 18, 19, 22, 23, 24 and NRs 20, 21), and the lowlands of North Bulgaria (NRs 1, 2, 3, 4).

The left branch of the primary division was characterized by *Neomys fodiens* and *Clethrionomys glareolus*, species belonging to the Boreal complex and the right branch by *Rhinolophus mehelyi*, attributed to the Mediterranean complex. The indicator species for the second level of division were *Microtus epiroticus* and *Mus spicilegus*, while indicator species for the third level were *Myotis nattereri* and *M. myotis* (Fig. 5).

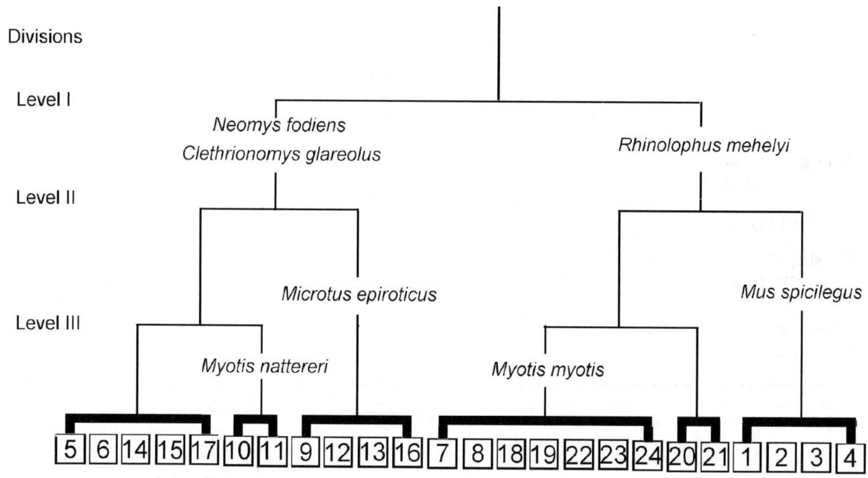

Fig. 5 TWINSPAN dendrogram of natural areas (NRs 1–24), based on presence/absence data, with indicator species in each division.

Clearly, the first TWINSPAN division reflected the effect of the altitudinal gradient on the distribution of mammals, while the second level discerned the residual effects of environmental variables operating within semimountainous and lowland areas. The third level matched some specific features related to the altitudinal heterogeneity of certain areas (NRs 10, 11, 9, 12, 13, 16) or to the incomplete faunal inventory (NRs 20, 21), already identified above.

Ecological analysis: spatial patterns of local assemblages of small epigean mammals. In order to provide more detailed data on the ecological aspects of the patterns revealed above, we conducted an analysis of concrete terrestrial small mammal assemblages. As can be seen from Fig. 1, the sampling areas are not evenly distributed within Bulgaria; nevertheless, we consider this dataset to be suitable to conduct a biogeographical analysis on the basis of local samples taken from a range of habitats since the sampling areas characterize each of the clusters of NRs as defined above.

The data on the structure of 83 LHAs were related to the variation in environment using RDA (Figs. 6–7). The correlation coefficients of sums of temperatures above 0 °C, air humidity, total annual precipitation, and, to a lesser extent, evaporation obtained the highest absolute values on the first axis (Table 1). The second axis was moderately negatively correlated with local vegetation and local humidity. The fraction of the total variance in the environmental data extracted by the species axis 1 was 30.6%, while axis 2 extracted 6.4%. The first axis separates the LHAs from mountainous areas from these from lowlands. High mountain LHAs score on the positive half of this axis, while the LHAs from open, dry habitats occupy the opposite end of this axis. Well-pronounced positive correlations with this axis show such mesophilous species from the Nemoral and Boreal faunal complexes as

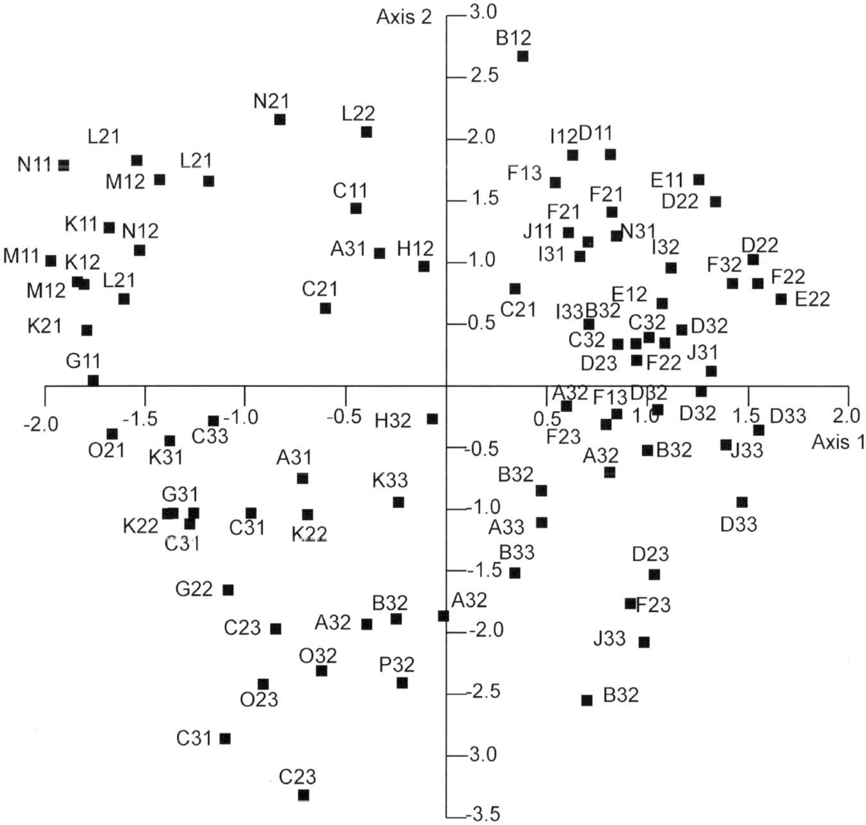

Fig. 6 RDA ordination diagrams (Axis 1 vs. Axis 2). Scatter diagram of LHAs; each LHA is denoted by the letter of the sampling area, followed by the scores of local vegetation and local humidity (for details see text).

Microtus subterraneus, Sorex minutus, S. araneus, Talpa europaea, Clethrionomys glareolus, etc. (Fig. 7). The second axis contrasts LHAs from the forested sites (lower part of the diagram) against these from the more or less open sites (upper part of the diagram). The LHA from Strandja occupies an extreme position on this axis, showing similarity with the situation revealed by the analysis of the faunistic data.

These results show that the first species axis was mainly related to the changes in temperature, precipitation, humidity, and continentality along the altitudinal gradient, while the second axis was related mainly to local conditions in terms of vegetation type (forest-shrubs-open) and local humidity. The Monte Carlo permutation test showed that the shares of species in local assemblages were significantly related to the environmental variables under study (99 random permutations,

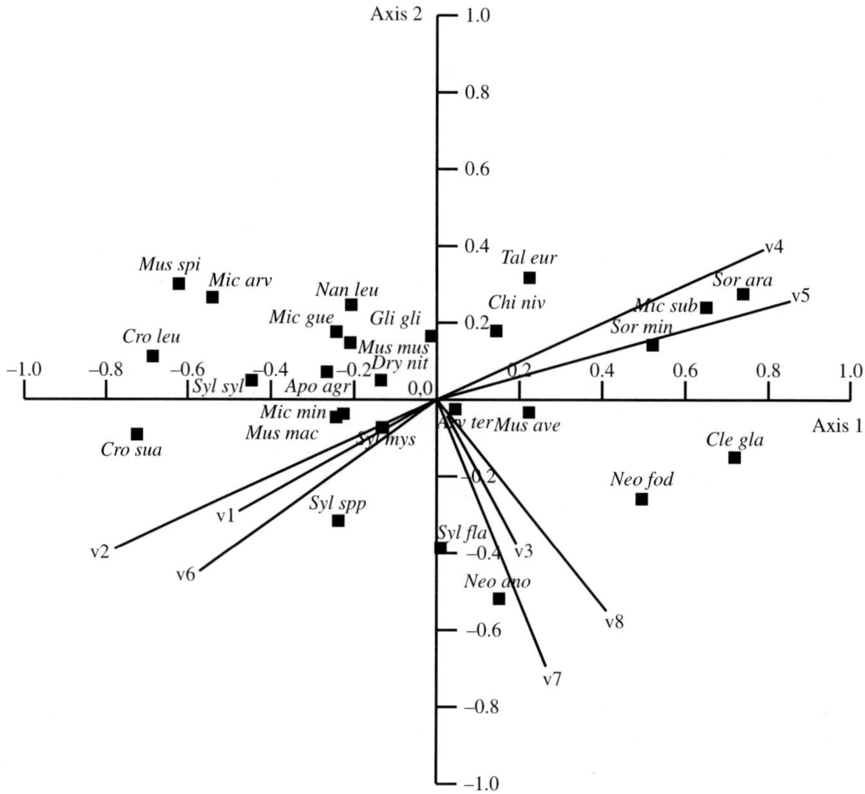

Fig. 7 RDA ordination diagrams (Axis 1 vs. Axis 2). Biplot of species and environmental variables (v1–v8).

Table 1 Correlations of significant environmental variables with the first two ordination axes. Definitions of environmental variables, see "Material and Methods"

Variable	Axis 1	Axis 2
v1	–0.45	–0.22
v2	–0.72	–0.28
v3	0.14	–0.22
v4	0.68	–0.23
v5	0.74	0.18
v6	–0.53	–0.32
v7	0.27	–0.54
v8	0.37	–0.44

$P < 0.01$). This analysis confirms the pattern already observed on the basis of regional faunas (presence/absence data).

Based on TWINSPAN analysis, the 83 LHAs were classified into five end-groups (Table 2), which were considered the assemblage types. The first group comprises high mountain LHAs from variable habitats (sampling areas F, D, J, E, B, I, C, A). The second group comprises LHAs from meadows and shrubs in middle mountain belt from areas A, C, I, H, B. The third group consists of local associations from forested habitats in lowlands (areas A, B, C, K, O). The fourth group is composed by associations from shrubs and forest edges in lowlands (areas G, O, C, K, P). The fifth group is composed by LHAs from dry steppe habitats in the northeastern Bulgaria (areas K, L, M, N).

The mean values for relative abundance of each species, for assemblage heterogeneity, and for each environmental variable are shown on Table 2. Generally, the revealed pattern is similar to those obtained via RDA. Some of species were found only in the mountain sites (*Neomys fodiens, Clethrionomys glareolus*), while others also occur rarely in humid lowland sites (*Sorex araneus, Sorex minutus, Microtus subterraneus*). These species belong to the Boreal and Nemoral faunal complexes. Typical species for the lowland assemblages are *Crocidura suaveolens, Mus* spp., etc. belonging to the Mediterranean and Steppe complexes. Elevation appears particularly important in limiting distribution of some predominantly lowland species of the Nemoral complex, such as *Crocidura leucodon, Micromys minutus*, and *Sylvaemus sylvaticus*, which occur in the Vitosha Mts. and in the Rhodopes only in the middle mountain belt in open or shrub habitats. According to the data presented in Table 2, *Dryomys nitedula* seems to prefer lowlands. Only two species, *Sylvaemus flavicollis* and *Microtus arvalis* (sensu lato), occur in suitable habitats of both groups of sites, showing in this way a rather broad adaptability and a relative independence of general climatic factors, involved in the composite environmental variable “altitude”.

The data in Table 2 indicate that the species heterogeneity (N2) is greater in the mountainous assemblages (types 1 and 2). A comparison of N2 index among assemblage types shows significant differences (ANOVA, $F0.01[4, 78] = 5.64$, $P = 0.0004$). Post hoc Scheffe’s comparisons reveal significant differences between groups 1/5, and 2/5, indicating that the assemblages in the extreme ends of the overall gradient differ in species heterogeneity.

The significant ($P < 0.001$) Spearman correlation coefficients (Rs) for N2 and environmental characteristics within the data set of 83 LHAs are as follows: with altitude above sea level, $Rs = 0.34$; with v2, $Rs = -0.33$; with v3, $Rs = 0.28$; with v4, $Rs = 0.38$; with v5, $Rs = 0.38$. Although these correlations are relatively low they follow the general trend identified above: N2 is positively affected by the altitudinal gradient, i.e. it is negatively associated with temperature and positively with precipitation.

In order to evaluate the effect of the habitat complexity, we compared the frequency distributions of the Czekanovsky–Sørensen similarity indices between individual LHAs within the two main groups defined at the first level of TWINSPAN

Table 2 (A) Mean relative abundance of species; (B) mean values of some LHA characteristics; (C) mean values of some environmental variables across five types of LHA's as classified by TWINSPAN. Definitions of environmental variables, see "Material and Methods"

			TWINSPAN groups of LHAs				
		Level I					
		Level II					
		Level III					
		TWINSPAN species groups	1	2	3	4	5
A	1	*Talpa europaea*	0.17	0.07	–	–	–
		Sorex araneus	22.71	11.29	–	1.23	3.69
		Neomys fodiens	4.50	4.64	–	–	–
		Clethrionomys glareolus	14.53	7.21	4.37	0.16	–
		Glis glis	0.19	–	–	0.05	–
		Microtus subterraneus	11.27	5.55	0.38	0.89	0.05
	2	*Sorex minutus*	18.89	13.21	2.46	2.08	9.04
		Muscardinus avellanarius	0.78	0.90	–	0.67	–
		Arvicola terrestris	1.00	0.93	–	0.21	–
		Chionomys nivalis	0.65	–	–	–	–
	3	*Neomys anomalus*	4.58	7.20	1.78	14.53	0.44
	4	*Sylvaemus flavicollis*	15.88	13.52	62.45	29.73	8.91
		Microtus arvalis	3.10	21.55	1.90	14.53	12.59
		Nannospalax leucodon	0.01	–	–	0.03	–
	5	*Crocidura leucodon*	1.15	5.21	16.29	4.50	21.79
		Mus musculus	–	0.30	–	–	1.91
	6	*Dryomys nitedula*	0.29	0.10	0.18	1.08	0.04
		Micromys minutus	–	2.62	0.18	1.15	0.32
		Apodemus agrarius	–	0.80	–	1.68	–
	7	*Crocidura suaveolens*	0.21	4.24	6.94	21.43	17.70
		Sylvaemus sylvaticus	0.02	0.57	–	1.67	1.21
	8	*Suncus etruscus*	–	–	–	0.41	–
		Mus macedonicus	–	–	–	2.04	–
		Sylvaemus mystacinus	–	–	3.03	–	–
		Microtus guentheri	–	–	–	0.03	–
	9	*Mus spicilegus*	–	–	–	1.81	22.26
B		Mean species heterogeneity (N2)	7.33	7.75	5.51	6.43	5.09
		Number of local associations	39	9	9	15	11.00
		Mean species richness (ES100)	9.8	10.2	–	9.8	7.7
		Number of local associations presented by more than 100 specimens	19	7	–	6	3
C		Mean altitude above sea level [m]	1371.79	927.77	568.88	241.66	31.81
		v1	19.05	21.64	21.10	20.29	21.59
		v2	78.99	116.37	122.82	127.46	130.81
		v3	583.64	604.88	563.88	631.60	452.45
		v4	7.09	3.00	1.13	1.37	0.07
		v5	906.76	759.00	622.66	688.93	531.27
		v6	373.91	519.95	567.43	544.94	521.17
		v7	2.38	2.11	3.00	2.13	1.45
		v8	1.97	2.00	1.55	1.93	1.27

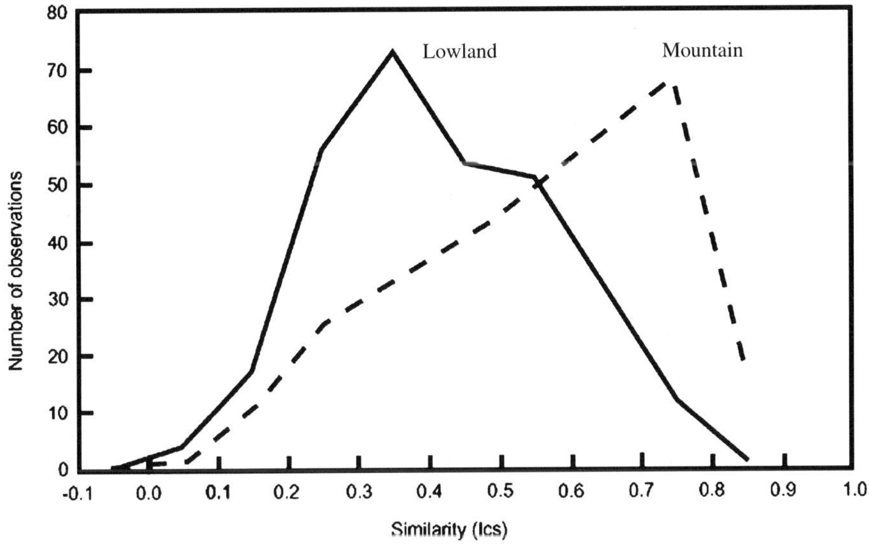

Fig. 8 Frequency distributions of the similarity indices (Ics) between LHAs within lowland and mountain typological units, identified at the first level of TWINSPAN division.

division (Fig. 8). The results (Fig. 8) show that in the mountains the assemblages tend to be more similar, suggesting a lower degree of spatial differentiation (heterogeneity). The differences between these distributions were statistically highly significant (Kruskal–Wallis test: $H\ (1,\ N = 600) = 77.75,\ P < 0.0001$).

The species richness, on the other hand, is little affected by the altitudinal gradient. This parameter, evaluated by rarefaction (ES100) for 35 LHAs presented by more than 100 specimens (Table 2), shows only two significant correlation coefficients: with v7, $Rs = 0.45$; and with v8, $Rs = 0.37$.

These data indicate that in more forested and humid habitats the LHAs tend to have higher species number. This statement is confirmed by the fact that the assemblage type 5, composed by the LHAs from the driest steppe-like habitats in the lowland of northeastern Bulgaria has relatively low species richness (Table 2).

4 Discussion

4·1 Patterns in faunal species richness and diversity

The analyses described above revealed the existence of positive correlations between species richness of insectivores, bats, and rodents, and the lack of correlation with any of the analyzed environmental variables. These results most probably reflect the effect of factors other than the measured environmental characteristics. These taxonomic groups show low number of species in NRs 10, 20, 21, 7, and high species number in NRs 9, 12, 13, 19, 23, 24, 1, 2, 4. For two groups (bats and artiodactyls), this pattern is related to the size of the NRs. For others, it is most probably caused by some subjective factors, e.g. biases in the interests of the researchers, proximity to large cities with research centers, etc. The bat fauna tends to be poorly known in areas lacking caves. Thus, it can be supposed that the main part of variability of species richness results from the completeness of regional species inventory. To a lesser extent it is positively related to the size of the area.

The obtained correlations for large mammals (carnivores and artiodactyls) show that their species richness was higher in the NRs with lower temperatures and higher precipitation, i. e. in the mountainous NRs. Most probably this circumstance reflects the human impact on these animals in lowlands and the greater habitat diversity in mountainous areas. The negative correlation of carnivores with the sum of precipitation for months with temperatures above 0 °C mirrors the co-occurrence of *Mustela eversmanni* and *Vormela peregusna* in the NRs of North Bulgaria.

4·2 Territorial subdivision of Bulgaria on the basis of mammals

The ordination and classification analyses described above provide a basis for zoogeographical subdivision of Bulgaria according to the mammalian fauna. Besides showing the territorial differentiation, the dendrogram of the faunistic similarity (Fig. 4) provides a basis for employing some objective criteria for a hierarchical subdivision. The following threshold levels have been proposed in this respect (Hagmeier & Stults, 1964): 0–5%, for defining a region; 20–25%, for a subregion; 40–45%, for a superprovince; and 60–65%, for a province. According to these criteria, it is evident (Fig. 4) that the territory of Bulgaria is rather homogeneous from zoogeographical point of view and should be assigned at most to a single zoogeographical province. Thus, according to the criteria presented above, the obtained subdivision should be interpreted in terms of spatial categories lower in hierarchy than a zoogeographical province.

On the basis of the TWINSPAN classification two main types of territorial units can be distinguished, which according to our analysis can be defined as *districts* (Fig. 9):

I. Mountainous district of western, southwestern, and central Bulgaria;
II. Lowland district of northern, northeastern, and southern Bulgaria.

Subsequently, according to the second level of the TWINSPAN classification they can be subdivided into second-order territorial units, which can be defined as *subdistricts* (Fig. 9):

Ia. Western and Central mountainous subdistrict: Western and Central Stara Planina (including the Predbalkan region), Sredna Gora, Kraishte, and the Rila–Rhodopes mountain massif;

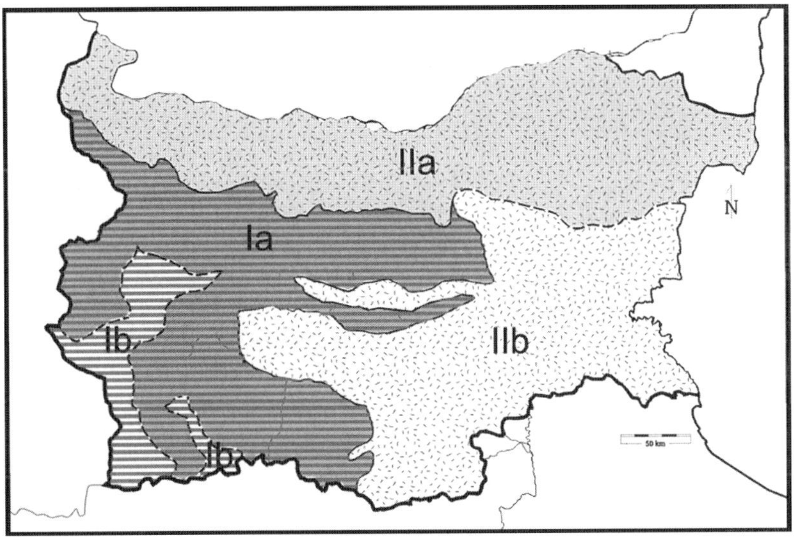

Fig. 9 Zoogeographical subdivision of Bulgaria based on mammals. I, Mountainous district; Ia, Western and Central mountainous subdistrict; Ib, Southwestern semimountainous subdistrict; II, Lowland district; IIa, Northern lowland subdistrict; IIb, Southern lowland subdistrict.

Ib. Southwestern semimountainous subdistrict: western and southwestern semimountainous areas and Struma and Mesta valleys;

IIa. Northern lowland subdistrict: lowlands and hilly areas of North Bulgaria (Danubian Plain, Ludogorie, and Dobrudja);

IIb. Southern lowland subdistrict: lowlands of South Bulgaria (including Podbalkan basins).

The obtained division reflects the main pattern of spatial differentiation of the mammalian fauna of Bulgaria. The analyses and results presented above clearly show that the faunistic differentiation of the country is primarily determined by the landscape variability across the territory and secondarily, by the geographical location of the country on the crossroads of different faunistic stocks.

The zoogeographical subdivision of Bulgaria on the basis of distribution of mammals generally corresponds to the biogeographical subdivision proposed by Gruev (1988). This author separated the country into five regions: (1) South Bulgarian, (2) Central Bulgarian, (3) North Bulgarian, (4) Black Sea coast, and (5) Mountain Region. This subdivision was based predominantly on the differentiation of the Bulgarian biota according to climate and topography. These two subdivisions differ in the lower number of territorial units in our subdivision based on mammals. This is due to some specific features of mammals. Being endotherms, they are relatively independent of climatic conditions and can occupy various habitats. Moreover, many species (large mammals, bats) have good dispersal and colonizing abilities, and therefore are not restricted to particular territories.

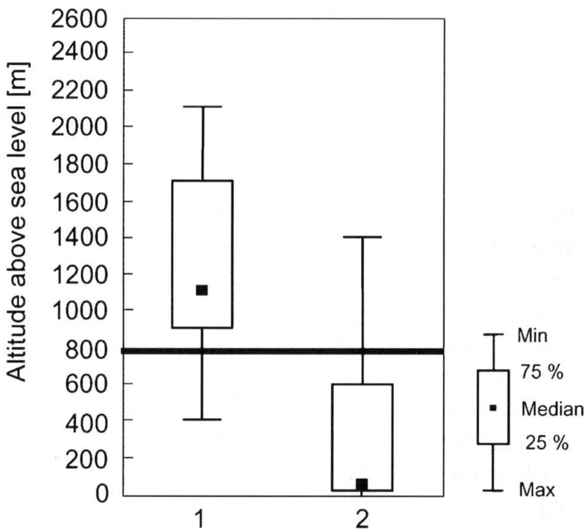

Fig. 10 Altitudinal distribution of LHAs within the TWINSPAN groups, identified at the first level of division. The black line represents the boundary between the mountainous and lowland districts (see Fig. 9).

Some of the delineated territorial units differ in their boundaries from those defined on the basis of the entire biota. The main reason for this is due to the fact that the mammalian species with a restricted distribution form a small portion of the respective regional faunas. As a result, the statistical analyses underestimate the distinctiveness of these faunas. In particular, this shortcoming concerns the occurrence of some Mediterranean species (*Sylvaemus mystacinus, Myomimus roachi, Microtus guentheri, Talpa levantis*) in the southernmost portions of subdistricts Ib and IIb. It seems, however, that the obtained subdivision is objective and meaningful because many other Mediterranean species, especially bats, have a wider distribution in Bulgaria.

On the other hand, there is a good agreement in the altitudinal boundary between mountainous and lowland territorial units. Gruev (1988) places the boundary between the Mountain Region and the remaining lowland regions at about 700 m a.s.l. The analysis of local habitat associations of small epigean mammals corresponds fairly well to this boundary. A summary of the altitudinal ranges of LHAs in the two main types resulting from the TWINSPAN classification (mountain and lowland) is shown in Fig. 10. It is clear that the majority of the mountain type LHAs occur above 900 m, while those from the lowlands occur below 700 m.

4·3 Origin and spatial zoogeographical structure of the mammalian fauna

The proposed classification of recent mammals into faunal complexes provides a summary of their geographical patterns of distribution both from historical and ecological perspectives, and allows outlining some aspects of their history and evolution. It is evident that this faunal composition is a result of a long process, which started at least in the Pliocene. The presented classification allows recognizing the following patterns in origin and present composition of the mammalian fauna of Bulgaria.

According to the classification proposed by the Russian phytogeographer A. Krasnov (Penev et al., 2003), the regional floras (and, respectively, faunas), can be divided into three groups: relict, orthoselective, and migratory. Relict faunas have persisted *in situ* since the Tertiary; orthoselective faunas consist of species, which evolved *in situ* under the pressure of large-scale, directed environmental changes; and migratory faunas consist of different chronological layers and comprise species of different origin and age. As could be expected on the basis of the geographical position of Bulgaria within the context of the paleogeographical and paleoclimatic evolution in this part of Europe, the proposed biogeographical classification shows that, as a whole, the mammalian fauna of Bulgaria can be attributed to the third category. It should be mentioned, however, that the majority of Bulgarian species belongs to the Nemoral complex, which can be considered a "relict" component of the fauna. In contrast, the Boreal and Steppe complexes comprise the "migratory" components of the fauna, associated with the Pleistocene climatic changes. The Mediterranean complex occupies an intermediate position in this respect since it involves both component types. The migratory component of the last complex is related to the Holocene paleoenvironmental changes. We should note, however, that, according to the existing paleontological data, the majority of the species of all complexes show a well-pronounced evolution, related to the gradual cooling of the climate since the Miocene, i.e. they comprise an orthoselective component as well.

As a whole, the mammalian fauna of Bulgaria is composed predominantly of species belonging to the Nemoral complex (38 species). Species of the Mediterranean complex (26 species) occupy a secondary position. The Boreal complex is relatively well represented (14 species), while the representatives of the Steppe complex form only a very small portion of the fauna (eight species). This structure corresponds to the Holocene shift in climate, sea level, and vegetation in that part of Europe. Since the beginning of the Holocene, the climate has become warmer, less continental, and more humid. As a result, the forests become widespread across the modern territory of Bulgaria, leading to the reorganization of the Late Pleistocene mammalian fauna, which in Bulgaria was dominated by the species of Steppe and Boreal (forest-steppe element) complexes (Popov, 2000b). Many of these species disappeared; others became rare, some of them now showing a scattered, restricted, and relict distribution in the lowland northern and eastern portions of Bulgaria, where the continentality of the climate remains more pronounced (namely *Cricetus cricetus, Mesocricetus newtoni, Sicista subtilis, Cricetulus migratorius*). In contrast, expansion of the forests favored spreading of the species belonging to the Nemoral faunal complex, and, to a lesser extent, some species forming the forest element of the Boreal complex. The second half of the Holocene is characterized by a somewhat drier climate. This coincides with the great human-caused destruction of lowland forests and establishment of agriculture over large areas. As a result, some xerothermophilous Mediterranean species have extended their ranges northward.

The territorial differentiation of the zoogeographical structure of the mammalian fauna in Bulgaria (Table 3) could reflect the details of the Holocene faunal development. Species of the Nemoral faunal complex are better represented in the mountainous district than in the lowland one. This pattern corresponds to a relatively drier climate during the second half of the Holocene and the great anthropogenic impact on lowlands, leading to the destruction of forests and increasing areas of cultivated land. In addition, human activity in the lowlands leads to the direct extermination of some large mammals (carnivores and game herbivores). As a result, some species of the Nemoral complex disappeared from the lowlands and became restricted to the mountains. New conditions in the lowlands were suitable for some ubiquitous species, especially small mammals of the Mediterranean, Boreal (forest-steppe element) and to a lesser extent the Steppe complexes, which have became widespread and abundant, especially in North Bulgaria (*Crocidura suaveolens, Microtus arvalis, Mus spicilegus*, etc.).

Table 3 Faunal elements of the Bulgarian zoogeographical units (terrestrial mammals)

Faunal elements	Districts			
	I		II	
	Subdistricts			
	a	b	a	b
s1a Eurosiberian forest elements	15.71	13.43	12.12	13.23
s1b Eurosiberian forest-steppe elements	2.85	2.98	3.03	2.94
Total Boreal complex	18.56	16.41	15.15	16.17
s2a Transpalearctic or Holarctic	25.71	25.37	24.24	23.52
s2b South (East) European and (West) Asian	8.57	8.95	7.57	7.35
s2c European	14.28	14.92	15.15	14.70
s2d South European and West Asian mountain	2.85	1.49	–	–
Total Nemoral complex	51.41	50.73	46.96	45.57
s3a Widely distributed in southern areas	4.28	4.47	4.54	5.88
s3b Submediterranean	15.71	16.41	15.15	14.70
s3c Mediterranean	4.28	4.47	4.54	4.41
s3d West Mediterranean	1.42	1.49	1.51	2.94
s3e East Mediterranean	1.42	2.98	–	4.41
s3f Euxinian	–	–	–	1.47
Total Mediterranean complex	27.11	29.82	25.74	33.81
s4a East European–West Asian	–	–	1.51	1.47
s4b East European–Kazakhstanian	–	–	4.54	–
s4c East European	–	–	1.51	–
s4d Balkan (and Southeast European)	2.85	2.98	4.54	2.94
Total Steppe complex	2.85	2.98	12.10	4.41

Shares of the Mediterranean complex species are relatively uniformly distributed across the territorial units. This unusual circumstance is related to the fact that the main part of this complex consists of bats, which regularly occur in the karst areas of both mountains (Western and Central Stara Planina, Pirin, the Rhodopes) and lowlands. Nevertheless, the greatest percentages of this complex are associated with the southern lowland and semimountainous subdistricts (IIb, Ib), while the share of this species group is the lowest in the northern lowland subdistrict.

These patterns correspond fairly well with the geographical position of Bulgaria, which is located in the southeastern corner of Europe, adjacent to Asia Minor. It also conforms to the ecological affinities of species forming the faunal complexes, and confirms the result, obtained at species level analyses, that the spatial differentiation of the Bulgarian mammalian fauna is environmentally determined. In addition, these patterns reflect the greater human impact on the mammalian fauna in the lowlands.

4·4 Local assemblages

The overall assemblage order generated by TWINSPAN (Table 2) represents a composite gradient which demonstrates in a greater detail the effect of environmental factors (altitude, moisture, vegetation cover) on the distribution of small mammals. Composition of the local habitat assemblages (LHAs) largely correlates with temperature and humidity, and to a lesser degree with the local humidity and the structure of vegetation. The TWINSPAN classification of the LHAs represents the effect of these factors and can be considered as an overall gradient. The obtained LHA groups can be considered as representative for relatively homogeneous sectors along this composite gradient. Thus, analyses within each type will allow demonstrating some low-scale relationships, which were obscured by spatial changes revealed by the broad-scale analyses. The among-type comparisons of the within-type relationships will allow tracing the changes of species-environment relationships along the general gradient. This analysis was conducted via Spearman rank correlation coefficients concerning two aspects: (1) species-habitat relationships, and (2) relationships among species. In the first case, the correlation of individual species with two environmental variables was analyzed: local vegetation cover (V) and local humidity (H). In the second case, analyses were conducted within guilds, which facilitated the interpretation of the observed relationships (see above). Having in mind that the assemblages forming a particular type represent relatively homogeneous conditions, it may be expected that the correlations between dominant species of each guild within the assemblage types may represent the small-scale factors affecting the structure of local assemblages. Positive correlations most probably will mirror the similarity in the ecological requirements of the compared species, while negative correlations should reflect either opposite ecological requirements, or the effect of competition, or both. These analyses will allow inferring in a greater detail the factors responsible for shaping the concrete assemblages of small epigean mammals. In particular, it may help to evaluate the role of external versus internal factors in assemblage organization. The term "external" means factors other than interspecific relationships within the assemblage, e.g. role of climatic factors and/or habitat characteristics in determining the composition and structure of an assemblage. In contrast, the term "internal" comprises the effect of interspecific relationships in shaping an assemblage.

Role of external factors. The data in Table 4 indicate that the share of *Sorex minutus* is positively correlated with humidity only in the lowland local assemblages, existing under relatively dry continental climate. In the mountainous assemblages, the relative proportions of *Sorex minutus* and *S. araneus* were not correlated with any of the two habitat characteristics under study. Both species were negatively affected by forests in assemblages of type 2, associated with relatively humid semimountainous habitats. In contrast, within the assemblages of type 5, existing in more or less dry conditions, *Sorex araneus* was positively correlated with forests. As may be expected, the shares of *Neomys fodiens* and *N. anomalus* were positively associated with local humidity. In assemblages of type 2, however, both species were also positively correlated with forested habitats. As a rule, in the lowland assemblages, where *Crocidura leucodon* and *C. suaveolens* were dominant species among shrew guilds, they were negatively affected by local humidity (*Crocidura leucodon*) and forests (both species). *Muscardinus avellanarius* in the assemblages of type 2 was positively affected by local humidity and forested habitats. Similar relationships were exhibited by *Sylvaemus* spp., *S. flavicollis*, *Apodemus agrarius*, and *Clethrionomys glareolus*. In contrast, the correlations show that *Microtus arvalis* avoids humid habitats (types 1, 4) and forests (type 2). In the mountains, *Microtus subterraneus* avoids forests, while in the lowlands it prefers more forested habitats.

Table 4 Significant Spearman rank correlation coefficients ($P < 0.05$) of species quantitative representation in LHAs and local conditions as expressed by humidity (H) and vegetation (V) within a particular assemblage type (within type 3 there are no significant correlations)

Species	type 1		type 2		type 4		type 5	
	H	V	H	V	H	V	H	V
Sorex minutus				−0.61			0.66	
Sorex araneus				−0.69				0.66
Neomys anomalus	0.49		0.81	0.69	0.83			
Neomys fodiens	0.64		0.77	0.62				
Crocidura leucodon					−0.69	−0.57		
Crocidura suaveolens								−0.75
Muscardinus avellanarius			0.61	0.62				
Sylvaemus spp.	0.34	0.52	0.58			0.56		
Sylvaemus flavicollis				0.78				
Apodemus agrarius						0.50		
Clethrionomys glareolus		0.56	0.58	0.69				
Microtus arvalis	−0.34			−0.61	−0.63			
Microtus subterraneus		−0.34				0.48		

Generally, these data agree with the existing knowledge on the ecological requirements of these mammal species. On the other hand, they show that the habitat demands can change according to the general conditions, being more restricted in less optimal environment.

Role of internal factors. Only two statistically significant ($P < 0.05$) Spearman rank correlations were obtained, both within assemblage of type 1: *Sorex araneus / S. minutus* ($Rs = -0.36$), and *Clethrionomys glareolus / Microtus subterraneus* ($Rs = -0.51$). Taking into account the similarity in ecological requirements of both shrew species, it can be suggested that their low negative correlation reflects the effect of competition. Most probably, the negative correlation between the two vole species mirrors the differences in the preferred habitats.

These analyses show that the structure of local assemblages is affected primarily by environmental factors external to the assemblage, while the role of internal factors, such as competition, could be suggested only in one case.

The analyses at the level of local associations reveal some details in the patterns of diversity and species richness. They show that these two aspects of the diversity of local small mammal assemblages are affected, although at a low level, by the environmental factors related to altitudinal gradient. It was shown that the species richness was positively correlated with local vegetation and precipitation. Within the study data set, both variables are positively related to the altitude. This trend is also pronounced in heterogeneity, but related to more variables: it is negatively associated with temperatures and positively with precipitation. In general, such changes in diversity have been variously ascribed to the patterns of temperature, habitat complexity, and food availability (Brown, 2001). Usually, precipitation is considered a measure of productivity (O'Brien, 1998; Whittaker & Field, 2000). Thus, the data are consistent with the third cause (food availability). However, the second cause might also contribute to the revealed pattern. In particular, precipitation and evapotranspiration determine the structural heterogeneity of vegetation, which in turn provides elements of habitat structure used for foraging, nesting, or avoiding predation (Brown, 2001). In the presented analyses, habitat complexity can also be interpreted in terms of spatial, coarse- or fine-grained habitat heterogeneity. Within the Bulgarian territory, this dichotomy is associated with altitude. It can be suggested that under relatively humid conditions and pronounced relief variability, causing a mosaic of different but closely situated habitat patches, the local small mammal assemblages are spatially less segregated forming in this way more heterogeneous assemblages. In contrast, in the lowlands, under relatively dry and more continental climate, the territorial habitat differentiation is greater and conditions within each habitat type differ greatly. As a result, the LHAs are spatially segregated and each LHA is ecologically uniform, including species with respective environmental requirements. In addition, deforestation and wide distribution of the cultivated fields accentuate the spatial segregation of habitat associations and contribute to the overall lower diversity of local small mammal assemblages in the lowland areas.

These patterns may serve as a basis for comparison in studying specific ecological situations, especially those related to the evaluation of the human impact on a particular landscape.

5 Appendix 1

5·1 Distribution of terrestrial mammals of the Bulgarian fauna according to the natural regions

Erinaceus concolor Martin, 1838: 1–24; *Talpa europaea* Linnaeus, 1758: 1–24; *Talpa levantis* Thomas, 1906: 22–24; *Sorex araneus* Linnaeus, 1758: 1–17, 22–24; *Sorex minutus* Linnaeus, 1766: 1, 3–6, 8–9, 11–15, 17, 23–24; *Neomys fodiens* (Pennant, 1771): 5–6, 9, 11–17; *Neomys anomalus* Cabrera, 1907: 1–24; *Suncus etruscus* (Savi, 1822): 19–24; *Crocidura suaveolens* (Pallas, 1811): 1–13, 16–24; *Crocidura leucodon* (Hermann, 1780): 1–24; *Rhinolophus ferrumequinum* (Schreber, 1774): 1–24; *Rhinolophus hipposideros* (Bechstein, 1800): 1–24; *Rhinolophus blasii* Peters, 1867: 1–2, 4–6, 9, 11–13, 16–19, 21–24; *Rhinolophus euryale* Blasius, 1853: 1–2, 4–19, 21–24; *Rhinolophus mehelyi* (Matschie, 1901): 1–2, 4–8, 18–24; *Myotis myotis* (Borkhausen, 1797): 1–2, 4–19, 22–24; *Myotis blythii* (Tomes, 1857): 1–2, 4–9, 11–19, 22–24; *Myotis bechsteinii* (Kuhl, 1817): 2–5, 8, 10, 12–19, 21–24; *Myotis nattereri* (Kuhl, 1817): 4–6, 12–17, 23–24; *Myotis emarginatus* (Geoffroy, 1806): 4–8, 12–13, 15–17, 19, 21–24; *Myotis mystacinus* (Kuhl, 1817): 4–6, 9–12, 14, 17, 19, 22–24; *Myotis brandtii* (Eversmann, 1845): 1, 4–6, 10, 17; *Myotis daubentonii* (Kuhl, 1817): 1–11, 14, 17, 23–24; *Myotis capaccinii* (Bonaparte, 1837): 1–2, 4–6, 9, 17–19, 23–24; *Myotis dasycneme* (Boie, 1825): 1; *Plecotus austriacus* (Fischer, 1829): 1–6, 8–9, 12–13, 15–17, 19, 22–24; *Plecotus auritus* (Linnaeus, 1758): 5–6, 10, 14–15, 17, 24; *Barbastella barbastellus* (Schreber, 1774): 2–6, 8, 12–15, 17, 24; *Nyctalus noctula* (Schreber, 1774): 1–9, 12–13, 15–16, 18–20, 23–24; *Nyctalus lasiopterus* (Schreber, 1780): 1–2, 4, 9, 14, 17, 19, 23–24; *Nyctalus leisleri* (Kuhl, 1817): 1–4, 6, 8–9, 12–13, 24; *Pipistrellus pipistrellus* (Schreber, 1774): 1–6, 8–9, 12–13, 15–16, 18–21, 23–24; *Pipistrellus nathusii* (Keyserling et Blasius, 1839): 9, 12–14, 16, 19, 23–24; *Pipistrellus kuhlii* (Kuhl, 1817): 13, 20; *Hypsugo savii* (Bonaparte, 1837): 1–2, 4–6, 9–17, 19; *Eptesicus serotinus* (Schreber, 1774): 1–6, 8–9, 12–14, 16–20, 23–24; *Eptesicus nilssonii* (Keyserling et Blasius, 1839): 14; *Vespertilio murinus* Linnaeus, 1758): 1–2, 4–6, 9, 12–15, 17, 19, 24; *Miniopterus schreibersii* (Kuhl, 1817): 1–9, 11–14, 16–19, 21–24; *Tadarida teniotis* (Rafinesque, 1814): 13, 17, 19; *Oryctolagus cuniculus* (Linnaeus, 1758): 24; *Lepus capensis* Linnaeus, 1758: 1–24; *Sciurus vulgaris* Linnaeus, 1758: 1–17, 19, 22–24; *Spermophilus citellus* (Linnaeus, 1766): 1–14, 16, 18–24; *Glis glis* (Linnaeus, 1766): 1–14, 16–24; *Muscardinus avellanarius* (Linnaeus, 1758): 1–9, 11, 13–17, 19, 22–24; *Dryomys nitedula* (Pallas, 1778): 1–24; *Myomimus roachi* (Bate, 1937): 18–21, 23–24; *Sicista subtilis* (Pallas, 1773): 3; *Nannospalax leucodon* (Nordmann, 1840): 1–14, 16–17, 19, 22–24;

Micromys minutus (Pallas, 1771): 1, 6–10, 13, 16, 19–24; *Apodemus agrarius* (Pallas, 1771): 1–8, 10, 12–13, 16, 19, 22–24; *Sylvaemus uralensis* (Pallas, 1811): 9, 11; *Sylvaemus sylvaticus* (Linnaeus, 1758): 1–24; *Sylvaemus flavicollis* (Melchior, 1834): 1–24; *Sylvaemus mystacinus* (Danford et Alston, 1877): 11–13, 15–16; *Mus spicilegus* Petenyi, 1882: 1–4; *Mus macedonicus* (Petrov et Ruzic, 1983): 7–8, 12–13, 16, 18–24; *Mus musculus musculus* (Linnaeus, 1758): 1–6, 9; *Mus musculus domesticus* (Schwarz et Schwarz, 1943): 9, 11, 13, 17–24; *Rattus norvegicus* (Berkenhout, 1769): 1–13, 16–24; *Rattus rattus* (Linnaeus, 1758): 1–13, 16–24; *Cricetulus migratorius* (Pallas, 1773): 1–2, 4, 19–20, 22–23; *Cricetus cricetus* (Linnaeus, 1758): 1–4; *Mesocricetus newtoni* (Nehring, 1898): 1–4; *Ondatra zibethicus* (Linnaeus, 1758): 1–4; *Clethrionomys glareolus* (Schreber, 1780): 5–17, 19; *Arvicola terrestris* (Linnaeus, 1758): 1–17, 19–24; *Microtus arvalis* (Pallas, 1778): 1–24; *Microtus epiroticus* Ondrias, 1966: 2, 4, 7–9, 12–13, 16, 18–19, 22–24; *Microtus guentheri* (Danford et Alston, 1880): 18, 24; *Microtus subterraneus* (de Selys-Longchamps, 1836): 1–15, 17, 19, 22–24; *Chionomys nivalis* (Martins, 1842): 5–6, 9, 14–15, 17; *Myocastor coypus* (Molina, 1782): 4, 23; *Canis lupus* Linnaeus, 1758: 2–18, 20–21, 24; *Canis aureus* Linnaeus, 1758: 1–9, 12–13, 16, 18–24; *Vulpes vulpes* (Linnaeus, 1758): 1–24; *Nyctereutes procyonoides* (Gray, 1834): 1–6, 9, 23–24; *Ursus arctos* Linnaeus, 1758: 5–6, 9, 13–15, 17; *Martes martes* (Linnaeus, 1758): 2, 4–15, 19, 24; *Martes foina* (Erxleben, 1777): 1–24; *Mustela nivalis* Linnaeus, 1766: 1–24; *Mustela putorius* Linnaeus, 1758: 1–24; *Mustela eversmanni* Lesson, 1827: 1–3; *Vormela peregusna* (Güldenstaedt, 1770): 1–6, 9, 12–14, 19, 22–23; *Meles meles* (Linnaeus, 1758): 1–24; *Lutra lutra* (Linnaeus, 1758): 1–9, 11–14, 16–17, 19–24; *Felis silvestris* Schreber, 1777: 1–24; *Sus scrofa* Linnaeus, 1758: 1–24; *Cervus elaphus* Linnaeus, 1758: 2–15, 17–19, 22–24; *Dama dama* (Linnaeus, 1758): 2, 4–6, 9, 11–12, 17, 19, 22–24; *Capreolus capreolus* (Linnaeus, 1758): 1–24; *Rupicapra rupicapra* (Linnaeus, 1758): 6, 14–15, 17.

References

Baev, P. & L.D. Penev, 1995. BIODIV. Program for Calculating Biological Diversity Parameters, Similarity, Niche Overlap, and Cluster Analysis. Version 5.1. Pensoft, Sofia & Moscow.

Benda, P. & I. Horáček, 1998. Bats (Mammalia: Chiroptera) of the Eastern Mediterranean. Part 1. Review of distribution and taxonomy of bats in Turkey. Acta Societatis Zoologicae Bohemicae 62:255–313.

Benda, P., T. Ivanova, I. Horáček, V. Hanák, J. Červený, J. Gaisler, A. Gueorguieva, B. Petrov & V. Vohralik, 2003. Bats (Mammalia: Chiroptera) of the Eastern Mediterranean. Part 3. Review of bat distribution in Bulgaria. Acta Societatis Zoologicae Bohemicae 67:245–357.

Boetticher, H. von, 1933. Die Elemente der bulgarischen Säugetierfauna und ihre geographischen und ökologischen Grundlagen. Izvestiya na Tsarskite prirodonauchni instituti v Sofia (Mitteilungen aus den Königlichen naturwissenschaftlichen Instituten in Sofia – Bulgarien; Bulletin des Institutions royales d'histoire naturelle à Sofia – Bulgarie) 6:43–45.

Braak, C.J.F. ter, 1985. CANOCO: A FORTRAN Program for Canonical Correspondence Analysis and Detrended Correspondence Analysis. IWIS-TNO, Wageningen, The Netherlands.

Braak, C.J.F. ter, 1987. Ordination. In Jongman, R.H., C.J.F. ter Braak & O.F.R. van Tongeren (eds), Data Analysis in Community and Landscape Ecology. Pudoc, Wageningen, The Netherlands: 91–169.

Braak, C.J.F. ter, 1990. Update Notes: CANOCO Version 3.10. Agricultural Mathematics Group, Wageningen, The Netherlands.

Brown, J.H., 2001. Mammals on mountainsides: elevational patterns of diversity. Global Ecology and Biogeography 10:101–109.

Gruev, B., 1988. Obshta biogeografiya [General Biogeography]. Nauka i izkustvo [Science and Art], Sofia (in Bulgarian).

Hagmeier, E.M. & C.D. Stults, 1964. A numerical analysis of the distributional patterns of North American mammals. Systematic Zoology 13:125–155.

Hill, M.O., 1979. TWINSPAN: a FORTRAN Program for Arranging Multivariate Data in an Ordered Two-way Table by Classification of the Individuals and Attributes. Cornell University Press, Ithaca, New York.

James, F.C. & S. Rathbun, 1981. Rarefaction, relative abundance, and diversity of avian communities. The Auk 98:785–800.

Limpens, H.J.G.A., 2001. Assessing the European distribution of the pond bat (*Myotis dasycneme*) using bat detectors and other survey methods. Nietoperze 2:169–178.

Meine, C., 1994. Conserving Biological Diversity in Bulgaria: The National Biological Diversity Conservation Strategy. Biodiversity Support Program c/o World Wildlife Fund, Washington, DC.

Minkova, T.V. & V.V. Popov, 2002. Spatial patterns of terrestrial small mammal communities in Central Western Bulgaria (Mammalia: Insectivora, Rodentia). Acta zoologica bulgarica 54:55–74.

Mitchell-Jones, A., G. Amori, W. Bogdanowicz, B. Krystufek, P. Reijnders, F. Spitzenberger, M. Stubbe, J. Thissen, V. Vohralik & J. Zima, 1999. Atlas of European Mammals. Academic Press, London.

O'Brien, E.M., 1998. Water-energy dynamics, climate, and prediction of plant species richness: an interim general model. Journal of Biogeography 25:379–398.

Penev, L., A. Casale & H. Turin, 2003. Biogeography. In Turin, H., L. Penev & A. Casale (eds), The genus *Carabus* in Europe. A synthesis. Pensoft, Sofia & Moscow: 327–426.

Peshev, Ts., D. Peshev & V. Popov, V., 2004. [Mammalia]. In Fauna na Balgariya. 27 (Fauna bulgarica. 27). Akademichno izdatelstvo "Marin Drinov" (Editio academica "Marin Drinov"), Sofia (in Bulgarian).

Petrov, P, 1997. [Landscape structure]. In Yordanova, M. & D. Donchev (eds), Geografiya na Balgariya. Fizicheska geografiya. Sotsialno-ikonomicheska geografiya (Geography of Bulgaria. Physical Geography. Socio-Economic Geography). Akademichno izdatelstvo "Prof. Marin Drinov" (Prof. Marin Drinov Academic Publishing House), Sofia: 340–356 (in Bulgarian).

Popov, V., 2000a. Epigeobiont animal assemblages from two landscapes of the Bulgarian Black Sea coast: relationship to environmental gradients, assemblage structure, and biodiversity. III. Small mammals (Insectivora, Rodentia). Acta zoologica bulgarica 52(3):75–87.

Popov, V., 2000b. The small mammals (Mammalia: Insectivora, Chiroptera, Lagomorpha, Rodentia) from Cave 16 (North Bulgaria) and the paleoenvironmental changes during the Late Pleistocene. In Ginter, B., J.K. Kozlowski & K. Laville, (eds), Temnata Cave. Excavations in Karlukovo Karst Area, Bulgaria. Volume 2, part 1. Jagellonian University Press, Krakow: 13–65.

Popov, V., Ch. Deltshev, G. Blagoev, I. Krusteva & D. Deltshev, 2000. Epigeobiont animal assemblages from two landscapes of the Bulgarian Black Sea coast: relationship to environmental gradients, assemblage structure, and biodiversity. II. Spiders (Araneae). Acta zoologica bulgarica 52(1):51–88.

Popov, V. & I. Krusteva, I., 1999. Epigeobiont animal assemblages from two landscapes of the Bulgarian Black Sea coast: relationship to environmental gradients, assemblage structure, and biodiversity. I. Ground beetles (Coleoptera: Carabidae). Acta zoologica bulgarica 51(1):81–114.

Root, R.B., 1967. The niche exploitation pattern of the blue-gray gnatcatcher. Ecological Monographs 37:317–349.

Schröpfer, R., 1990. The structure of European small mammal communities. Zoologische Jahrbücher, Abteilung für Systematik 117:355–367.

Schunger, C., D. Dietz, D. Merdschanova, K. Merdschanov, I. Christov, S. Borissov, S. Staneva, B. Petrov, 2004. Swarming of bats (Chiroptera, Mammalia) in the Vodnite Dupki Cave (Central Balkan National Park, Bulgaria). Acta zoologica bulgarica 56(3):323–330.

Simberloff, D.S., 1978. Use of rarefaction and related methods in ecology. In Dickson, K.L., J. Cairns, Jr. & R.J. Livingston (eds), Biological Data in Water Pollution Assessment: Quantitative and Statistical Analyses. American Society for Testing and Materials, Philadelphia: 150–165.

Udvardy, M.D.F., 1969. Dynamic Zoogeography, with Special Reference to Land Animals. Van Nostrand Reinhold, New York.

Whittaker, R.H., 1975. Communities and Ecosystems, 2nd edition. Macmillan, London.

Whittaker, R.J. & R. Field, 2000. Tree species richness modelling: an approach of global applicability? Oikos 89:399–402.

Zimina, R.P., 1962. [The mammals of Bulgaria, their faunal composition and peculiarities of their distribution.] Zoologicheskiy zhurnal [Zoological Journal] 41:1226–1237 (in Russian).

3 Fauna, Zoogeography, and Ecology of Birds in Bulgaria

Zlatozar Boev[1], Boyan Milchev[2] and Vasil Popov[3]

[1]*National Museum of Natural History, Bulgarian Academy of Sciences, Tsar Osvoboditel Blvd. 1, 1000 Sofia, Bulgaria, e-mail: boev@nmnh.bas.bg*

[2]*Wildlife Management Department, University of Forestry, Kliment Ohridski Blvd. 10, 1756 Sofia, Bulgaria, e-mail: boyan.m@ltu.bg*

[3]*Institute of Zoology, Bulgarian Academy of Sciences, Tsar Osvoboditel Blvd. 1, 1000 Sofia, Bulgaria, e-mail: popov@zoology.bas.bg*

Abstract: This paper is the first overview and evaluation of the exploration (1744 to present), composition (records of 96 Neogene taxa, 160 Pleistocene taxa, and 399 recent species), formation (since the Middle Miocene), and recent development of the Bulgarian bird fauna. We discuss composition and changes of the recent avifauna; changes in the composition of the breeding species in the last 100 years; species, which have disappeared or are declining; species with extended breeding range; and new species recently recorded for Bulgaria. We also outline the importance of the Bulgarian territory for the migration and wintering of the birds in Europe, the migration routes, wintering grounds, and the network of the Important Bird Areas. A classification of the avian assemblages in Bulgaria is given using the two-way indicator species analysis, while the application of the ordination-correspondence analysis allowed to study the spatial differentiation of the recent breeding avifauna, and to characterize the bird communities of the main landscapes in Bulgaria

1 Introduction

The Balkan Peninsula is considered one of the richest regions in Europe in the biodiversity of terrestrial vertebrates, mainly birds (Blondel, 1982, 1987a, 1987b, 1997). Bulgaria's avifaunistic diversity so far has not been evaluated in regard to the specifics of local physical geography. Bulgarian rich and highly varied recent avifauna is a result of both complex history and diversified landscapes, as well as the position of the country's territory on the crossroads between Europe, Asia, and Africa.

Some ornithogeographical aspects of the modern bird fauna in Bulgaria have been outlined by Simeonov et al. (1990), Georgiev & Simeonov (1992), Boev (1997), and Nankinov (1997). The present paper's goal is to present an overview and to evaluate the exploration history, composition, formation, recent development, and the zoogeographical and ecological affinities of the Bulgarian bird fauna.

V. Fet and A. Popov (Eds.), Biogeography and Ecology of Bulgaria, 39–84.

2 Material and Methods

We treat Triassic to Pleistocene avian records as the fossil fauna; Early Holocene to Late Holocene ones, as the subfossil; and the records of ca. last 200 years, as the recent fauna. Although we traced the composition of the breeding species over the last 100 years, we discuss only the distribution of the breeding birds since 1990 to present (approximately, the last decade) following the literature sources (Table 1). Data on the breeding status of birds (Table 1) have been partly compiled after the following sources: Petrov (1980), Simeonov (1986), Simeonov & Baeva (1988), Simeonov & Delov (1989), Georgiev (1989), Donchev (1994), Simeonov & Dimitrov (1994), Simeonov & Marinov (1994), Georgiew (1998), Ivanov (1998), Ivanov et al. (1998), Milchev (1998), Milchev & Kovachev (1998), Georgiev & Milchev (2000), Karaivanov (2000), Michev & Petrov (2000), Baumgart (2001a, 2001b), Menzel & Miltschev (2001), Nankinov & Dalaktchieva (2001), Stojanov (2001a), Hristov (2002), Miltschev & Tschobanov (2002), Nikolov (2002), and our unpublished data for southeastern Bulgaria.

We do not comment on the reasons of the cited authors for listing a species as breeding, "probably breeding", or a non-breeding one. Three species without firm proofs of their breeding and later unconfirmed as nesting are not included in the country's list of the breeding birds: *Glaucidium passerinum* (Spiridonov 2000a, 2000b), *Calandrella rufescens* (Nankinov, 2001b), and *Acrocephalus melanopogon* (Nankinov, 2001c).

New species, not covered by Voous (1960) but included in the recent lists (Hagemeijer & Blair, 1997; Snow & Perrins, 1998), are: *Alectoris chukar, Oenanthe pleschanka*, and *O. isabellina* (Turkestanian), *Alectoris graeca* (Mediterranean), and *Larus cachinnans* (Sarmatian).

The system of 24 landscape subprovinces (Fig. 1), proposed by Petrov (1997), was used as a reference system to record the presence or absence of each species. These subprovinces are relatively homogeneous territories with a particular combination of relief, climate, and vegetation; consequently they are considered in further analysis as *natural regions* (NRs). This approach offers a possibility to accumulate the often disparate faunistic records in an environmentally meaningful context and to reveal the environmental aspects of bird distribution. Each NR was scored for presence (nesting) or absence of each species and, as a result, regional species lists were generated. As the most appropriate techniques to reveal the bird distributional patterns across these NRs we have used two complementary approaches, based on ordination-correspondence analysis (CA) and two-way indicator species analysis (TWINSPAN). In general, the ordination analyses reveal the patterns of coincidence of many taxa along spatial environmental gradients and may be helpful in revealing the species/environmental relationships by recognizing some generalized distributional patterns, which in turn are suggestive and predictable on the basis of known ecological requirements and general distributions of the included species.

Table 1 TWINSPAN classification of the NRs according to species composition of the breeding avifauna (breeding species are marked with "+")

TWIN-SPAN levels	Division 1		1																2								TWIN-SPAN end-group
	Division 2		1.1								1.2										2.1				2.2		
	Division 3		1.1.1	1.1.2							1.2.1						1.2.2			2.1.1				2.2.1			
	Groups		1	2							3						4			5				6			
No.	Species	Abbreviation	9	5	6	7	12	14	15	17	8	10	11	13	16	18	4	21	24	2	19	20	22	1	3	23	
1.	*Falco naumanni*	Fal nau							+	+						+											
2.	*Falco peregrinus*	Fal per	+	+	+	+		+	+	+		+				+											A
3.	*Bonasa bonasia*	Bon bon	+	+	+			+	+	+		+															
4.	*Tetrao urogallus*	Tet uro	+	+	+			+	+	+		+															
5.	*Alectoris graeca*	Ale gra	+	+	+	+	+	+	+	+		+	+	+	+		+										
6.	*Scolopax rusticola*	Sco rus		+	+					+		+															
7.	*Actitis hypoleucos*	Act hyp		+	+	+		+	+	+										+							
8.	*Strix uralensis*	Str ura			+																						
9.	*Aegolius funereus*	Aeg fun	+		+			+		+																	

continued

Table 1 continued

TWIN-SPAN levels	Division 1		1																2							TWIN-SPAN end-group
	Division 2		1.1								1.2								2.1				2.2			
	Division 3		1.1.1	1.1.2							1.2.1						1.2.2			2.1.1			2.2.1			
	Groups		1	2							3						4			5			6			
No.	Species	Abbreviation	9	5	6	7	12	14	15	17	8	10	11	13	16	18	4	21	24	2	19	20	22	1	3	23	
10.	*Apus pallidus*	Apu pal	+	+			+	+		+				+		+											
11.	*Dryocopus martius*	Dry mar	+	+	+	+	+	+	+	+	+	+		+		+	+		+								
12.	*Dendrocopos leucotus*	Den leu	+	+	+	+		+	+	+		+							+								
13.	*Picoides tridactylus*	Pic tri						+	+	+																	
14.	*Eremophila alpestris*	Ere alp	+	+	+			+	+			+															
15.	*Anthus trivialis*	Ant tri	+	+	+	+	+	+	+	+		+	+	+		+	+		+								
16.	*Anthus spinoletta*	Ant spi	+	+	+			+	+	+		+															
17.	*Motacilla cinerea*	Mot cin	+	+	+	+	+	+	+	+	+	+	+	+		+		+	+								
18.	*Cinclus cinclus*	Cin cin	+	+	+	+	+	+	+	+		+	+	+		+			+								

19.	*Prunella modularis*	Pru mod	+	+	+	+	+	−	+	+									+
20.	*Prunella collaris*	Pru col	+	+	+		+	+	+	+									
21.	*Saxicola rubetra*	Sax rub	+	+	+	+		+	+	+		+		+			+		
22.	*Monticola saxatilis*	Mon sax	+	+	+	+	+	+	+	+		+		+	+	+	+	+	+
23.	*Turdus torquatus*	Tur tor	+	+	+	+	+	+	+	+									
24.	*Phylloscopus sibilatrix*	Phy sib	+	+	+	+	+	+	+	+	+	+	+			+	+		+
25.	*Regulus regulus*	Reg reg	+	+	+			+	+	+		+							
26.	*Regulus ignicapillus*	Reg ign	+	+	+			+	+	+									
27.	*Ficedula parva*	Fic par	+		+														
28.	*Ficedula albicollis*	Fic alb		+															
29.	*Parus montanus*	Par mon	+	+	+	+		+	+	+		+					+		+
30.	*Parus cristatus*	Par cri						+	+	+									
31.	*Parus ater*	Par ate	+	+	+	+	+	+	+	+		+	+	+					
32.	*Tichodroma muraria*	Tic mur		+	+			+	+	+				+			+		
33.	*Certhia famil-iaris*	Cer fam	+	+	+	+	+	+	+	+		+	+	+			+		+

continued

Table 1 continued

TWIN-SPAN levels	Division 1		1																2							TWIN-SPAN end-group
	Division 2		1.1								1.2									2.1				2.2			
	Division 3		1.1.1	1.1.2							1.2.1						1.2.2			2.1.1				2.2.1			
	Groups		1	2							3						4			5				6			
No.	Species	Abbreviation	9	5	6	7	12	14	15	17	8	10	11	13	16	18	4	21	24	2	19	20	22	1	3	23	
34.	*Nucifraga caryocatactes*	Nuc car	+	+	+		+	+	+	+																	
35.	*Pyrrhocorax graculus*	Pyr gra		+	+			+	+	+																	
36.	*Serinus serinus*	Ser ser	+	+	+	+	+	+	+	+		+	+	+		+	+										
37.	*Carduelis spinus*	Car spi	+	+	+	+		+	+	+																	
38.	*Loxia curvirostra*	Lox cur	+	+	+		+	+	+	+																	
39.	*Pyrrhula pyrrhula*	Pyr pyr	+	+	+	+	+	+	+	+		+		+													
40.	*Pernis apivorus*	Per api	+	+	+	+	+	+		+	+	+	+			+	+	+	+	+							
41.	*Gyps fulvus*	Gyp ful														+											B
42.	*Aegypius monachus*	Aeg mon														+											

43.	*Accipiter nisus*	Acc nis	+	+	+	+	+	+	+	+	+	+	+	+	+	+	+	+	+	+			+		+
44.	*Aquila chrysaetos*	Aqu chr		+	+	+		+	+	+	+	+	+	+		+	+		+	+					
45.	*Hieraaetus pennatus*	Hie pen	+	+	+			+	+	+		+				+	+	+	+	+				+	
46.	*Falco biarmicus*	Fal bia														+									
47.	*Falco cherrug*	Fal che		+	+				+	+	+	+				+							+		
48.	*Columba oenas*	Col oen	+	+	+	+	+	+	+	+		+	+	+		+			+				+		
49.	*Clamator glandarius*	Cla gla														+			+						
50.	*Strix aluco*	Str alu	+	+	+	+	+	+	+	+	+	+	+	+	+	+	+	+	+	+			+	+	
51.	*Troglodytes troglodytes*	Tro tro	+	+	+	+	+	+	+	+	+	+	+	+	+	+	+	+	+	+					
52.	*Phoenicurus ochruros*	Pho och	+	+	+	+	+	+	+	+	+	+	+	+	+	+	+			+	+			+	
53.	*Phoenicurus phoeni-curus*	Pho pho	+	+	+	+	+		+	+		+	+	+	+	+	+		+						
54.	*Monticola solitarius*	Mon sol							+					+		+									
55.	*Turdus viscivorus*	Tur vis	+	+	+	+	+	+	+	+	+	+		+		+	+	+	+			+			
56.	*Sylvia cantillans*	Syl can					+		+					+	+	+									
57.	*Sylvia melanocephala*	Syl mel							+					+	+	+									

continued

Table 1 continued

TWIN-SPAN levels	Division 1		1																2							TWIN-SPAN end-group
	Division 2		1.1								1.2								2.1				2.2			
	Division 3		1.1.1	1.1.2							1.2.1						1.2.2			2.1.1				2.2.1		
	Groups		1	2							3						4			5				6		
No.	Species	Abbreviation	9	5	6	7	12	14	15	17	8	10	11	13	16	18	4	21	24	2	19	20	22	1	3	23	
58.	*Phylloscopus orientalis*	Phy ori				+			+	+	+	+		+	+	+		+	+								
59.	*Ficedula semitorquata*	Fic sem	+	+	+	+		+			+		+				+		+						+		
60.	*Sitta neumayer*	Sit neu							+					+		+											
61.	*Certhia brachy-dactyla*	Cer bra	+	+	+	+		+	+	+	+	+	+		+	+	+	+	+								
62.	*Petronia petronia*	Pet pet												+		+											
63.	*Emberiza cia*	Emb cia	+	+	+	+	+	+	+	+		+	+	+	+	+	+	+	+	+					+		
64.	*Ciconia ciconia*	Cic cic	+	+	+	+	+	+	+	+	+	+	+	+	+	+	+	+	+	+	+	+	+	+	+	+	
65.	*Neophron percnopterus*	Neo per		+	+	+			+	+	+					+	+	+	+	+		+			+		C
66.	*Circaetus gallicus*	Cir gal	+	+	+	+	+	+		+	+	+	+			+	+	+	+	+		+	+		+	+	

67.	*Accipiter gentilis*	Acc gen	+	+	+	+	+	+	+	+	+	+	+	+	+	−	+	+	+	+	+	+	+	+	+	+
68.	*Buteo buteo*	But but	+	+	+	+	+	+	+	+	+	+	+	+	+	−	+	+	+	+	+	+	+	+	+	+
69.	*Buteo rufinus*	But ruf	+	+	+	+		+	+	+	+	+	+	+		−	+	+	+	+	+	+			+	+
70.	*Aquila pomarina*	Aqu pom	+	+	+	+	+	+	+	+	+	+	+	+	+	−	+	+	+	+	+	+	+	+	+	+
71.	*Falco tinnun-culus*	Fal tin	+	+	−	+	+	+	+	+	+	+	+	+	+	−	+	+	+	+	+	+	+	+	+	+
72.	*Perdix perdix*	Per per	+	+	−	+	+	+	+	+	+	+	+	+	+	−	+	+	+	+	+	+	+	+	+	+
73.	*Coturnix coturnix*	Cot cot	+	+	−	+	+	+	+	+	+	+	+	+	+	−	+	+	+	+	+	+	+	+	+	+
74.	*Phasianus colchicus*	Pha col	+	+	−	+		+	+	+	+	+	+	+		−	+	+	+	+	+	+	+	+	+	+
75.	*Crex crex*	Cre cre	+	+	−	+		+	+	+		+	+	+	+	−	+	+	+	+	+			+	+	+
76.	*Charadrius dubius*	Cha dub	+	+	−	+	+	+	+	+	+	+	+	+	+	−	+	+	+	+	+	+	+	+	+	+
77.	*Columba livia*	Col liv	+	+	−	+	+	+	+	+	+	+	+	+	+	−	+	+	+	+	+	+	+	+	+	+
78.	*Columba palumbus*	Col pal	+	+	−	+	+	+	+	+	+	+	+	+	+	−	+	+	+	+	+	+	+	+	+	+
79.	*Streptopelia decaocto*	Str dec	+	+	−	+	+	+	+	+	+	+	+	+	+	−	+	+	+	+	+	+	+	+	+	+
80.	*Streptopelia turtur*	Str tur	+	+	−	+	+	+	+	+	+	+	+	+	+	−	+	+	+	+	+	+	+	+	+	+
81.	*Cuculus canorus*	Cuc can	+	+	−	+	+	+	+	+	+	+	+	+	+	−	+	+	+	+	+	+	+	+	+	+

continued

Table 1 contunied

TWIN-SPAN levels	Division 1		1																2								TWIN-SPAN end-group
	Division 2		1.1								1.2										2.1				2.2		
	Division 3		1.1.1	1.1.2							1.2.1						1.2.2				2.1.1				2.2.1		
	Groups		1	2							3						4				5				6		
No.	Species	Abbreviation	9	5	6	7	12	14	15	17	8	10	11	13	16	18	4	21	24	2	19	20	22	1	3	23	
82.	*Otus scops*	Otu sco	+	+	+	+	+	+	+	+	+	+	+	+	+	+	+	+	+	+	+	+	+	+	+	+	
83.	*Bubo bubo*	Bub bub	+	+	+	+	+	+	+	+	+	+	+	+		+	+	+	+	+	+	+	+	+	+	+	
84.	*Athene noctua*	Ath noc	+	+	+	+	+	+	+	+	+	+	+	+	+	+	+	+	+	+	+	+	+	+	+	+	
85.	*Asio otus*	Asi otu	+	+	+	+	+	+	+	+	+	+	+	+	+	+	+	+	+	+	+	+	+	+	+	+	
86.	*Caprimulgus europaeus*	Cap eur	+	+	+	+	+	+	+	+	+	+	+	+	+	+	+	+	+	+	+	+	+	+	+	+	
87.	*Apus apus*	Apu apu	+	+	+	+	+	+	+	+	+	+	+	+	+	+	+	+	+	+	+	+		+	+	+	
88.	*Tachymarptis melba*	Tac mel	+	+	+	+			+	+	+		+	+		+	+			+				+	+		
89.	*Alcedo atthis*	Alc att	+	+	+	+	+	+	+	+	+	+	+	+	+	+	+	+	+	+	+	+		+	+	+	
90.	*Merops apiaster*	Mer api	+	+	+	+	+	+	+	+	+	+	+	+	+	+	+	+	+	+	+	+	+	+	+	+	

91.	*Upupa epops*	Upu epo	+	+	+	+	+	+	+	+	+	+	+	+	+	+	+	+	+	+	+	+	+	+	+	+
92.	*Jynx torquilla*	Jyn tor	+	+	+	+	+	+	+	+	+	+	+	+	+	+	+	+	+	+	+	+	+	+	+	+
93.	*Picus canus*	Pic can	+	+	+	+	+	+	+	+	+	+	+	+	+	+	+	+	+	+	+			+	+	+
94.	*Picus viridis*	Pic vir	+	+	+	+	+	+	+	+	+	+	+	+	+	+	+	+	+	+	+	+	+	+	+	+
95.	*Dendrocopos major*	Den maj	+	+	+	+	+	+	+	+	+	+	+	+	+	+	+	+	+	+	+	+	+	+	+	+
96.	*Dendrocopos syriacus*	Den syr	+	+	+	+	+	+	+	+	+	+	+	+	+	+	+	+	+	+	+	+	+	+	+	+
97.	*Dendrocopos medius*	Den med	+	+	+	+	+	+	+	+	+	+	+	+	+	+	+	+	+	+	+			+	+	+
98.	*Dendrocopos minor*	Den min	+	+	+	+	+	+	+	+	+	+	+	+	+	+	+	+	+	+	+	+	+	+	+	+
99.	*Galerida cristata*	Gal cri	+	+	+	+	+	+	+	+	+	+	+	+	+	+	+	+	+	+	+	+	+	+	+	+
100.	*Lullula arborea*	Lul arb	+	+	+	+	+	+	+	+	+	+	+	+	+	+	+	+	+	+	+	+	+	+	+	+
101.	*Alauda arvensis*	Ala arv	+	+	+	+	+	+	+	+	+	+	+	+	+	+	+	+	+	+	+	+	+	+	+	+
102.	*Ptyonoprogne rupestris*	Pty rup	+	+	+	+	+	+	+	+	+	+	+	+		+	+			+				+	+	
103.	*Hirundo rustica*	Hir rus	+	+	+	+	+	+	+	+	+	+	+	+	+	+	+	+	+	+	+	+	+	+	+	+
104.	*Hirundo daurica*	Hir dau	+	+	+	+	+	+	+	+	+	+	+	+	+	+	+	+	+	+	+	+	+	+	+	+
105.	*Delichon urbica*	Del urb	+	+	+	+	+	+	+	+	+	+	+	+	+	+	+	+	+	+	+	+	+	+	+	+

continued

Table 1 continued

TWIN-SPAN levels	Division 1		1																2							TWIN-SPAN end-group
	Division 2		1.1							1.2									2.1				2.2			
	Division 3		1.1.1	1.1.2						1.2.1						1.2.2			2.1.1				2.2.1			
	Groups		1	2						3						4			5				6			
No.	Species	Abbreviation	9	5	6	7	12	14	15	17	8	10	11	13	16	18	4	21	24	2	19	20	22	1	3	23
106.	*Motacilla flava*	Mot fla	+	+	+	+	+	+	+	+	+	+	+	+	+	+	+	+	+	+	+	+	+	+	+	+	
107.	*Motacilla alba*	Mot alb	+	+	+	+	+	+	+	+	+	+	+	+	+	+	+	+	+	+	+	+	+	+	+	+	
108.	*Erithacus rubecula*	Eri rub	+	+	+	+	+	+	+	+	+	+	+	+	+	+	+	+	+	+	+	+	+	+	+	+	
109.	*Luscinia megarhynchos*	Lus meg	+	+	+	+	+	+	+	+	+	+	+	+	+	+	+	+	+	+	+	+	+	+	+	+	
110.	*Saxicola torquata*	Sax tor	+	+	+	+	+	+	+	+	+	+	+	+	+	+	+	+	+	+	+	+	+	+	+	+	
111.	*Oenanthe oenanthe*	Oen oen	+	+	+	+	+	+	+	+	+	+	+	+	+	+	+	+	+	+	+	+	+	+	+	+	
112.	*Turdus merula*	Tur mer	+	+	+	+	+	+	+	+	+	+	+	+	+	+	+	+	+	+	+	+	+	+	+	+	
113.	*Turdus philomelos*	Tur phi	+	+	+	+	+	+	+	+	+	+	+	+	+	+	+	+	+	+	+	+	+	+	+	+	
114.	*Acrocephalus arundinaceus*	Acr aru	+	+	+	+	+	+		+	+	+	+	+	+	+	+	+	+	+	+	+	+	+	+	+	

115.	*Sylvia nisoria*	Syl nis	+	+	+	+	+	+	+	+	+	+	+	+	+	–	+	+	+	+	+	+	+	+	+	+
116.	*Sylvia curruca*	Syl cur	+	+	+	+	+	+	+	+	+	+	+	+	+	–	+	+	+	+	+	+	+	+	+	+
117.	*Sylvia communis*	Syl com	+	+	+	+	+	+	+	+	+	+	+	+	+	–	+	+	+	+	+	+	+	+	+	+
118.	*Sylvia borin*	Syl bor	+	+	+	+	+	+	+	+		+	+	+	+		+			+	+		+	+		
119.	*Sylvia atricapilla*	Syl atr	+	+	+	+	+	+	+	+	+	+	+	+	+	–	+	+	+	+	+	+	+	+	+	+
120.	*Phylloscopus collybita*	Phy col	+	+	+	+	+	+	+	+	+	+	+	+	+	–	+	+	+	+	+	+	+	+	+	+
121.	*Muscicapa striata*	Mus str	+	+	+	+	+	+	+	+	+		+	+	+	–	+	+	+	+	+		+	+	+	
122.	*Aegithalos caudatus*	Aeg cau	+	+	+	+	+	+	+	+	+	+	+	+	+	–	+	+	+	+	+	+	+	+	+	+
123.	*Parus palustris*	Par pal	+	+	+	+	+	+	+	+	+	+	+	+	+	–	+	+	+	+	+		+	+	+	
124.	*Parus lugubris*	Par lug	+	+	+	+	+	+	+	+	+	+	+	+	+	–	+	+	+	+	+	+	+	+	+	+
125.	*Parus caeruleus*	Par cae	+	+	+	+	+	+	+	+	+	+	+	+	+	–	+	+	+	+	+			+	+	+
126.	*Parus major*	Par maj	+	+	+	+	+	+	+	+	+	+	+	+	+	+	+	+	+	+	+	+	+	+	+	+
127.	*Sitta europaea*	Sit eur	+	+	+	+	+	+	+	+	+	+	+	+	+	+	+	+	+	+	+	+	+	+	+	+
128.	*Oriolus oriolus*	Ori ori	+	+	+	+	+	+	+	+	+	+	+	+	+	+	+	+	+	+	+	+	+	+	+	+
129.	*Lanius collurio*	Lan col	+	+	+	+	+	+	+	+	+	+	+	+	+	+	+	+	+	+	+	+	+	+	+	+

continued

Table 1 continued

TWIN-SPAN levels	Division 1		1																2							TWIN-SPAN end-group
	Division 2		1.1								1.2								2.1				2.2			
	Division 3		1.1.1	1.1.2							1.2.1						1.2.2		2.1.1				2.2.1			
	Groups		1	2							3						4		5				6			
No.	Species	Abbre-viation	9	5	6	7	12	14	15	17	8	10	11	13	16	18	4	21	24	2	19	20	22	1	3	23
130.	*Lanius senator*	Lan sen	+	+			+	+	+	+	+	+	+	+	+	+	+	+	+		+	+	+			+
131.	*Garrulus glandarius*	Gar gla	+	+	+	+	+	+	+	+	+	+	+	+	+	+	+	+	+	+	+	+	+	+	+	+
132.	*Pica pica*	Pic pic	+	+	+	+	+	+	+	+	+	+	+	+	+	+	+	+	+	+	+	+	+	+	+	+
133.	*Corvus monedula*	Cor mon	+	+	+	+	+	+	+	+	+	+	+	+	+	+	+	+	+	+	+	+	+	+	+	+
134.	*Corvus corone*	Cor cor	+	+	+	+	+	+	+	+	+	+	+	+	+	+	+	+	+	+	+	+	+	+	+	+
135.	*Corvus corax*	Cor cox	+	+	+	+	+	+	+	+	+	+	+	+	+	+	+	+	+	+	+	+	+	+	+	+
136.	*Sturnus vulgaris*	Stu vul	+	+	+	+	+	+	+	+	+	+	+	+	+	+	+	+	+	+	+	+	+	+	+	+
137.	*Passer domesticus*	Pas dom	+	+	+	+	+	+	+	+	+	+	+	+	+	+	+	+	+	+	+	+	+	+	+	+
138.	*Passer hispan-iolensis*	Pas his	+	+	+	+	+	+	+	+	+	+	+	+	+	+	+	+	+	+	+	+	+	+	+	+

139.	*Passer montanus*	Pas mon	+	+	−	+	+	+	+	+	+	+	+	+	+	+	+	+	+	+	+	+	+	+	+	+	
140.	*Fringilla coelebs*	Fri coe	+	+	−	+	+	+	+	+	+	+	+	+	+	+	+	+	+	+	+	+	+	+	+	+	
141.	*Carduelis chloris*	Car chl	+	+	−	+	+	+	+	+	+	+	+	+	+	+	+	+	+	+	+	+	+	+	+	+	
142.	*Carduelis carduelis*	Car car	+	+	+	+	+	+	+	+	+	+	+	+	+	+	+	+	+	+	+	+	+	+	+	+	
143.	*Carduelis cannabina*	Car can	+	+	+	+	+	+	+	+	+	+	+	+	+	+	+	+	+	+	+	+	+	+	+	+	
144.	*Carpodacus erythrinus*	Car ery			+			+							+											+	
145.	*Coccothraustes coccothraustes*	Coc coc	+	+	+	+	+	+	+	+	+	+	+	+	+	+	+	+	+	+	+	+	+	+	+	+	
146.	*Emberiza citrinella*	Emb cit	+	+	+	+	+	+	+	+	+	+	+	+	+	−	+	+	+	+	+	+	+	+	+	+	
147.	*Emberiza cirlus*	Emb cir	+	+	+	+	+	+	+	+	+	+	+	+	+	−	+	+	+	+	+	+	+	+	+	+	
148.	*Emberiza hortulana*	Emb hor	+	+	+	+	+	+	+	+	+	+	+	+	+	−	+	+	+	+	+	+	+	+	+	+	
149.	*Miliaria calandra*	Mil cal	+	+	+	+	+	+	+	+	+	+	+	+	+	−	+	+	+	+	+	+	+	+	+	+	
150.	*Lanius nubicus*	Lan nub							+			+		+		−		+	+		+	+				+	
151.	*Aquila heliaca*	Aqu hel		+		+						+				−		+				+	+				D
152.	*Coracias garrulus*	Cor gar	+	+			+		+	+	+	+	+	+	+	+	+	+	+	+	+	+	+	+	+	+	
153.	*Calandrella brachy-dactyla*	Cal bra	+	+						+		+		+	+	+	+				+				+	+	

continued

Table 1 continued

TWIN-SPAN levels	Division 1		1																2						TWIN-SPAN end-group
	Division 2		1.1								1.2								2.1				2.2		
	Division 3		1.1.1	1.1.2							1.2.1						1.2.2		2.1.1				2.2.1		
	Groups		1	2							3						4		5				6		
No.	Species	Abbreviation	9	5	6	7	12	14	15	17	8	10	11	13	16	18	4	21	24	2	19	20	22	1	3	23	
154.	*Cettia cetti*	Cet cet												+	+		+	+	+					+		+	
155.	*Hippolais pallida*	Hip pal	+	+			+	+	+	+	+	+		+	+	+	+	+	+	+	+	+	+	+	+	+	
156.	*Hippolais icterina*	Hip ict							+	+		+	+			+	+	+	+	+				+	+	+	
157.	*Lanius minor*	Lan min	+	+			+		+	+	+	+	+	+	+	+	+	+	+	+	+	+	+	+	+	+	
158.	*Tachybaptus ruficollis*	Tac ruf	+	+						+	+		+			+	+	+	+	+	+	+	+	+	+	+	
159.	*Podiceps cristatus*	Pod cri	+									+	+				+		+	+				+	+	+	E
160.	*Ciconia nigra*	Cic nig	+	+	+	+				+	+	+				+	+	+	+	+	+	+	+	+	+	+	
161.	*Anas platyrhynchos*	Ana pla	+	+		+	+			+	+	+	+	+		+	+	+	+	+	+	+	+	+	+	+	
162.	*Accipiter brevipes*	Acc bre								+						+	+	+	+	+	+	+			+		

163.	*Falco subbuteo*	Fal sub	+	+	+				+	+						+	+	+	+		+	+	+	+	+	+
164.	*Gallinula chloropus*	Gal chl	+	+		+	+			+	+	+	+	+		+	+	+	+	+	+	+	+	+	+	+
165.	*Fulica atra*	Ful atr	+				+				+	+	+	+			+	+	+	+	+		+	+	+	+
166.	*Burhinus oedic-nemus*	Bur oed	+											+	+	+	+	+	+		+	+	+		+	+
167.	*Vanellus vanellus*	Van van	+							+		+	+	+		+	+	+	+	+	+	+	+	+	+	+
168.	*Tringa ochropus*	Tri och	+									+							+						+	+
169.	*Anthus campestris*	Ant cam	+	+				+	+	+	+	+		+	+	+	+	+	+	+	+	+	+	+	+	+
170.	*Oenanthe hispanica*	Oen his	+	+					+	+	+	+		+	+	+	+	+	+		+	+	+	+	+	+
171.	*Acrocephalus schoenobaenus*	Acr sch	+							+		+	+	+	+	+			+		+			+	+	+
172.	*Hippolais olive-torum*	Hip oli						+	+			+		+	+	+		+	+		+	+	+		+	+
173.	*Sylvia hortensis*	Syl hor							+			+		+	+	+		+	+		+	+	+			+
174.	*Remiz pendulinus*	Rem pen	+				+		+			+	+	+	+		+	+	+		+	+		+	+	+
175.	*Corvus frugilegus*	Cor fru	+				+						+	+	+		+			+	+			+	+	
176.	*Emberiza schoeniclus*	Emb sch	+											+			+		+					+	+	+

continued

Table 1 continued

TWIN-SPAN levels	Division 1		1																2							TWIN-SPAN end-group
	Division 2		1.1								1.2									2.1				2.2			
	Division 3		1.1.1	1.1.2							1.2.1						1.2.2			2.1.1				2.2.1			
	Groups		1	2							3						4			5				6			
No.	Species	Abbreviation	9	5	6	7	12	14	15	17	8	10	11	13	16	18	4	21	24	2	19	20	22	1	3	23	
177.	*Emberiza melanocephala*	Emb mel	+					+	+	+	+	+		+	+	+	+	+	+		+	+	+		+	+	
178.	*Podiceps griseigena*	Pod gri	+																+						+	+	
179.	*Botaurus stellaris*	Bot ste	+																+					+	+	+	F
180.	*Ixobrychus minutus*	Ixo min	+									+	+	+			+	+	+			+	+	+	+	+	
181.	*Nycticorax nycticorax*	Nyc nyc	+									+				+			+		+			+	+	+	
182.	*Ardea cinerea*	Ard cin					+					+				+				+	+	+		+	+	+	
183.	*Cygnus olor*	Cyg olo															+		+					+	+	+	
184.	*Tadorna ferruginea*	Tad fer															+		+					+	+	+	
185.	*Tadorna tadorna*	Tad tad															+		+						+	+	

186.	*Anas querquedula*	Ana que	+					+			+		+			+	+	+
187.	*Anas clypeata*	Ana cly	+			+										+	+	+
188.	*Aythya nyroca*	Ayt nyr	+	+			+		+			+	+			+	+	+
189.	*Milvus migrans*	Mil mig				+		+		+	+		+	+		+	+	+
190.	*Circus aerugi-nosus*	Cir aer	+	+		+		+	+		+		+			+	+	+
191.	*Circus pygargus*	Cir pyg								+	+		+	+	+			+
192.	*Falco vesper-tinus*	Fal ves						+									+	
193.	*Alectoris chukar*	Ale chu			+			+		+	+		+	+	+			+
194.	*Rallus aquaticus*	Ral aqu	+	+	+	+	+	+	+		+	+	+	+	+	+	+	+
195.	*Porzana porzana*	Por por	+			+			+		+	+	+		+	+	+	+
196.	*Porzana parva*	Por par	+			+			+				+			+	+	+
197.	*Porzana pusilla*	Por pus	+						+				+			+	+	+
198.	*Himantopus himan-topus*	Him him				+			+								+	+
199.	*Recurvirostra avosetta*	Rec avo							+								+	+

continued

Table 1 continued

TWIN-SPAN levels	Division 1		1																2							TWIN-SPAN end-group
	Division 2		1.1								1.2								2.1				2.2			
	Division 3		1.1.1	1.1.2							1.2.1						1.2.2		2.1.1				2.2.1			
	Groups		1	2							3						4		5				6			
No.	Species	Abbreviation	9	5	6	7	12	14	15	17	8	10	11	13	16	18	4	21	24	2	19	20	22	1	3	23	
200.	*Glareola pratincola*	Gla pra	+											+											+	+	
201.	*Tringa totanus*	Tri tot	+									+									+			+	+	+	
202.	*Larus cachinnans*	Lar cac	+								+						+		+	+	+			+	+	+	
203.	*Sterna hirundo*	Ste hir														+			+		+			+	+	+	
204.	*Tyto alba*	Tyt alb	+													+		+	+		+	+	+	+	+	+	
205.	*Melanocorypha calandra*	Mel cal											+			+	+	+	+		+	+	+		+	+	
206.	*Riparia riparia*	Rip rip	+	+			+		+			+		+		+				+	+	+		+	+	+	
207.	*Oenanthe isabellina*	Oen isa	+							+		+				+		+	+		+	+	+		+	+	
208.	*Oenanthe pleschanka*	Oen ple															+								+		

209.	*Acrocephalus palustris*	Acr pal	+	+				+	+	+	+	+	+	+	+	
210.	*Acrocephalus scirpaceus*	Acr sci	+		+	+		+	+	+		+	+	+	+	
211.	*Panurus biarmicus*	Pan bia						+		+			+	+	+	
212.	*Podiceps nigricollis*	Pod nig												+	+	
213.	*Phalacrocorax carbo*	Pha car											+	+	+	G
214.	*Phalacrocorax aristotelis*	Pha ari												+		
215.	*Phalacrocorax pygmeus*	Pha pyg										+	+	+	+	
216.	*Pelecanus onocro-talus*	Pel ono												+		
217.	*Pelecanus crispus*	Pel cri												+		
218.	*Ardeola ralloides*	Ard ral										+	+	+	+	
219.	*Egretta garzetta*	Egr gar					+					+	+	+	+	
220.	*Egretta alba*	Egr alb											+	+		
221.	*Ardea purpurea*	Ard pur								+			+	+	+	
222.	*Plegadis falcinellus*	Ple fal											+	+	+	
223.	*Platalea leucorodia*	Pla leu											+	+	+	

continued

Table 1 continued

TWIN-SPAN levels	Division 1		1																	2							TWIN-SPAN end-group
	Division 2		1.1								1.2									2.1				2.2			
	Division 3		1.1.1	1.1.2							1.2.1						1.2.2			2.1.1				2.2.1			
	Groups		1	2							3						4			5				6			
No.	Species	Abbreviation	9	5	6	7	12	14	15	17	8	10	11	13	16	18	4	21	24	2	19	20	22	1	3	23	
224.	*Anser anser*	Ans ans																						+	+		
225.	*Anas strepera*	Ana str																						+	+	+	
226.	*Anas crecca*	Ana cre																							+	+	
227.	*Anas acuta*	Ana acu																								+	
228.	*Netta rufina*	Net ruf																						+	+	+	
229.	*Aythya ferina*	Ayt fer															+							+	+	+	
230.	*Haliaeetus albicilla*	Hal alb																	+					+	+	+	
231.	*Otis tarda*	Oti tar																							+		
232.	*Haematopus ostralegus*	Hae ost														+					+			+		+	

233.	*Charadrius alexandrinus*	Cha ale							+	+
234.	*Larus melanocephalus*	Lar mel								+
235.	*Larus ridibundus*	Lar rid						+	+	+
236.	*Larus genei*	Lar gen								+
237.	*Gelochelidon nilotica*	Gel nil								+
238.	*Thalasseus sandvicensis*	Tha san								+
239.	*Sterna albifrons*	Ste alb						+	+	+
240.	*Chlidonias hybridus*	Chl hyb						+	+	
241.	*Chlidonias niger*	Chl nig		+		+		+	+	+
242.	*Turdus pilaris*	Tur pil							+	
243.	*Locustella fluviatilis*	Loc flu				+		+	+	+
244.	*Locustella luscinioides*	Loc lus	+		+	+		+	+	+
245.	*Acrocephalus agricola*	Acr agr								+
246.	*Sturnus roseus*	Stu ros				+	+		+	+

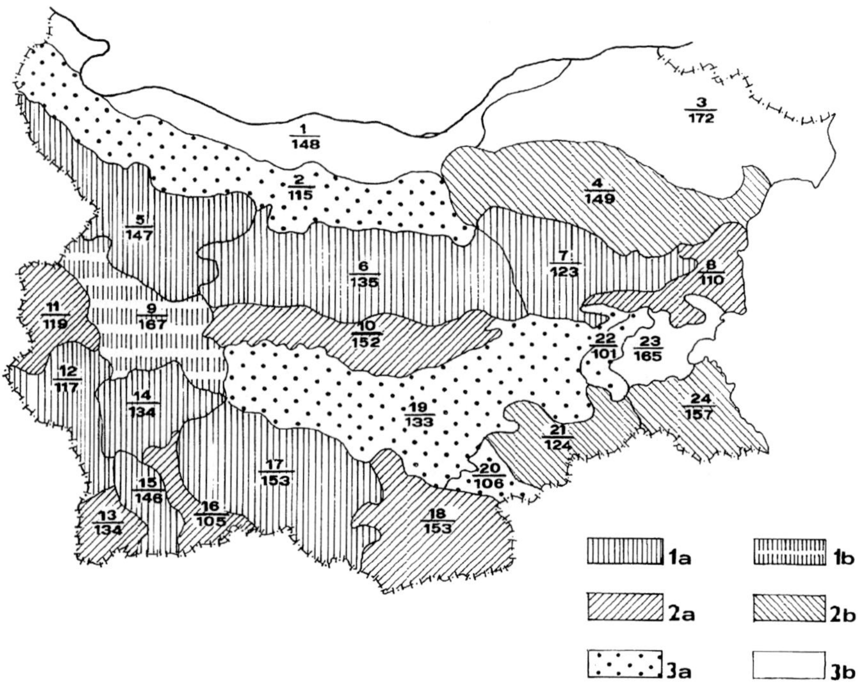

Fig. 1 Natural regions (NR) in Bulgaria (after Petrov, 1997). (1) High mountains: (1a) The characterizing species belong to the coniferous and mixed (coniferous–broadleaved) forests and alpine zone; (1b) Vitosha–Ihtiman NR (distinguished from 1a by the inclusion of the plain and aquatic species of the Sofia Basin); (2) Lower and moderately high mountains, hill farmlands, and river valleys: (2a) Western and Southern groups (excluding the Emine Mts.); (2b) Eastern group; (3) Hilly lands in the lowland plains dominated by the farmlands and pastures: (3a) Large inland water bodies and the Black Sea coast excluded; (3b) Large inland water bodies and the Black Sea coast included. We do not divide the country's territory into Northern and Southern Bulgaria, or distinguish the territory of a stronger Mediterranean impact. Numerator: number of the NR (1, Northern Danubian Plain; 2, Southern Danubian Plain; 3, Southern Dobrudja; 4, Popovo–Shumen–Franga; 5, Western Stara Planina; 6, Central Stara Planina; 7, Eastern Stara Planina; 8, Maritime Stara Planina; 9, Vitosha–Ihtiman; 10, Sredna Gora–Podbalkan; 11, Kraishte; 12, Osogovo–Middle Struma; 13, Southern Struma; 14, Rila; 15, Pirin; 16, Middle Mesta; 17, Western Rhodopes; 18, Eastern Rhodopes; 19, Upper Thrace; 20, Lower Thrace; 21, Sakar–Dervent; 22, Bakadjik–Hisar; 23, Burgas–Aitos; 24, Strandja). Denominator: number of the breeding species in the NR.

The correspondence analysis (CA) was conducted in two variants. The first variant was based on all 246 bird species. In the second variant, the bird species associated with water basins (Podicipediformes, Pelecaniformes, Ciconiiformes, Anseriformes, and Charadriiformes) were excluded in order to eliminate the effect of their concentration along Danube and Black Sea shores on the ordination results. The second analysis was based on 176 species. Since the eigenvalues of the third and fourth axes in both variants of analysis were very low, they were ignored. The obtained ordination diagrams show a well-pronounced arch effect (see below). Although various methods of detrending have been developed in order to correct this "fault" of the CA (Hill & Gauch, 1980, but see also Pielou, 1984, p. 197), the detrended correspondence analysis (DCA) was not employed here because the trial analyses revealed that the interpretation of the ordination diagrams remained the same, but the percentages of the explained variance of species data were lower. The analyses were implemented using the computer program CANOCO 3.10 (Braak, 1990). In both variants of analysis the option of "downweighting of rare species" was implied in order to minimize the effect of rare species occurring only in one or two regions.

The goal of TWINSPAN analysis was to provide a hierarchical classification of the NRs and the species (Hill, 1979). This method classifies the NRs according to their species composition and then uses this classification to classify species according to their occurrence in each NR (Tongeren, 1987). Following this an ordered two-way summary table was produced (Table 1) allowing to trace the distributional patterns of individual species across NRs arranged according to their similarity in species lists. Thus, the arrangement of the NRs represents the effect of the main environmental gradients on the avifauna across the country. The *a priori* defined maximum number of levels of division was three.

3 Exploration of the Bulgarian Avifauna

First publications containing faunistic data on the recent Bulgarian bird fauna appeared back in the 18th century (Marsigli, 1744). Their number slowly increased during the 19th century, when a series of papers outlined the faunistic diversity and distribution of the Bulgarian birds (Farman, 1868, 1869; Homeyer, 1877; Radakoff, 1879; Alléon, 1886; Reiser, 1894; etc.). The first Bulgarian publication on the country's avifauna appeared in the last decade of the 19th century (Hristovich, 1890). A series of comprehensive studies have been published in the 20th century (Klein, 1909; Patev, 1950; Balát, 1962; Peshev & Boev, 1962, Botev & Peshev, 1985; Simeonov & Michev, 1991; Iankov, 1996; Kostadinova, 1997). The most recent and complete data have been included in the first two volumes (Simeonov et al., 1990; Nankinov et al., 1997) of the planned three-volume "Fauna of Bulgaria. Birds" monograph. A more exhaustive review of the development of Bulgarian ornithology is given by Boev (1990).

The following new bird taxa have been described from Bulgaria: one recent species (*Streptopelia decaocto* Frivaldszky, 1838), two fossil genera (*Balcanas* Boev, 1998 and *Chauvireria* Boev, 1997), and 16 fossil species (*Phalacrocorax serdicensis* Burchak-Abramovich et Nikolov, 1984; *Balcanas pliocaenica* Boev, 1998; *Anser thraceiensis* Burchak-Abramovich et Nikolov, 1984; *Cygnus verae* Boev, 2000; *Geronticus balcanicus* Boev, 1998; *Buteo spassovi* Boev, 1998; *Falco bakalovi* Boev, 1998; *Chauvireria balcanica* Boev, 1997; *Lagopus balcanicus* Boev, 1995; *Tetrao rhodopensis* Boev, 1998; *Gallinula balcanica* Boev, 1999; *Actitis balcanica* Boev, 1998; *Regulus bulgaricus* Boev, 1999; *Coccothraustes simeonovi* Boev, 1998; *Coccothraustes balcanicus* Boev, 1998; and *Loxia patevi* Boev, 1999).

Large regions of the country have been covered by a series of elaborate avifaunistic researches in the second half of the 20th century (Michev & Jankov, 1998): (1) Mountains: Stara Planina, Vasilyovska, Lyulin, Vitosha, Lozenska, Sredna Gora, Konyavska, Osogovska, Ograzhden, Pirin, Rhodopes, Chernatitsa, Strandja, etc.; (2) Lakes and swamps: Dragomansko, Choklyovo, Garvan, Srebarna, Durankulak, Shabla, Pomoriisko, Atanasovsko, etc.; (3) Plains and lowlands: Botevgrad, Thrace, Burgas; (4) River valleys: Danube, Tsibritsa, Rozova, Ropotamo; (5) Gorges: Chernelka, Emen, Iskar, Zemen, Kresna; (6) Cities: Sofia, Pleven, Veliko Tarnovo, Gabrovo, Burgas.

Studies on the Bulgarian fauna of fossil and subfossil birds appeared much later in the 1980s (Bochenski, 1982; Burchak-Abramovich & Nikolov, 1984). More than 100 localities, 12 of them from Neogene (Boev, 2002) and 18 from Pleistocene (Boev, 2001a), revealed a large number of fossil bird taxa (see below).

Avian collections in Bulgaria have a long history starting back in the middle of the 19th century, when the oldest mounted specimens of birds have been collected. They are still preserved in the richest and the oldest natural history museum in the Balkans, the National Museum of Natural History of the Bulgarian Academy of Sciences in Sofia. Its collections contain 30,192 specimens of about 1980 recent and fossil taxa (number of taxa in parentheses): fossil and subfossil bones, 12,154 (349); stuffed skins, 11,237 (1298); mounted skins, 4271 (1116); partial and complete skeletons, 1995 (358); eggs, 259 (70); feather sets, 129 (97); nests, 122 (56), wet preserved specimens, 25 (14); and fossil egg shells, 1 (1) (Boev, 2003a). More than 340 specimens (stuffed and mounted skins) of 56 species belong to the extinct, endangered, and vulnerable species and subspecies (Boev, 2003b).

4 Historical Formation of the Bulgarian Avifauna

The history of the Bulgarian avifauna could be traced back to the Middle Miocene (MN zone 5–8) (Boev, 1999, 2002). The fossil/subfossil record of the country's birds recently consists of at least 349 taxa, 105 of those established in the fossil record. They are referred to 46 recent families and 19 orders (Boev, unpublished data).

4·1 Neogene avifaunas

The Neogene fossil remains throughout the country have recorded a total of 96 taxa belonging to 24 families and 12 orders, among them several new species and genera (see above). Among the Miocene fossils are the oldest remains of *Melanocorypha* and one of the oldest in the world fossil findings of *Buteo* (Boev, 2002), as well as the first European record of Bucerotidae (Boev & Kovachev, in progress). The Early Pliocene faunas revealed a so far unknown avian association, "*Lagopus–Pavo*" (*Lagopus* cf. *atavus–Pavo bravardi*; Boev, 2001b). The Late Pliocene avifaunas are well-documented in two Villanyian localities, Varshets and Slivnitsa. The former has provided numerous remains of over 62 taxa, thus being the riches European site of that age. It is the type locality for nine avian species (see above). The most numerous of them is *Chauvireria balcanica* Boev, 1997, a partridge-like species of Perdicinae (Boev, 2002). The Late Pliocene avian remains also provided the oldest world record of 15 recent genera (*Gyps, Circaetus, Porzana, Actitis, Galerida, Lullula, Erithacus, Regulus, Sturnus, Pyrrhula, Carduelis, Fringilla, Coccothraustes,* and *Loxia*), as well as the first European Tertiary record of following eleven genera: *Gyps, Circaetus, Accipiter, Actitis, Galerida, Lullula, Erithacus, Regulus, Sturnus, Parus, Carduelis, Coccothraustes, Loxia*, and *Pyrrhula* (Boev, 1999).

4·2 Quaternary avifaunas

The avian Pleistocene record of Bulgaria includes 160 species, or 40.0% of the country's recent bird fauna. Two-thirds of the material originate from the Wurmian deposits. It contains seven disappeared species: *Lagopus mutus, Lagopus lagopus, Tetrao tetrix, Phasianus colchicus* (native subspecies *colchicus* form), *Glaucidium passerinum, Nyctea scandiaca*, and *Pyrrhocorax pyrrhocorax*. The Late Pleistocene avifaunas of Bulgaria included two fossil taxa: *Perdix palaeoperdix* and *Pyrrhocorax graculus vetus*. The Early Pleistocene bird faunas are known by 26 taxa, while the Middle Pleistocene ones include only seven recognized taxa (Boev, 2001 a).

The Late Pleistocene avian faunas are the best-studied, and include 134 taxa. Most of the 50 recent genera recorded in the Late Pleistocene deposits could be unquestionably referred to the group of the dendrophilous, or arboreal birds. A smaller part (*Alectoris, Apus, Riparia*, etc.) includes obligatory thermophilous, chiefly petrophilous elements. Thermophilous fauna in the Late Pleistocene was represented by the following genera: *Pernis, Apus, Caprimulgus, Merops, Riparia, Oenanthe* (entomophagous), *Coturnix, Crex, Porzana* (phyto- and zoophagous), *Streptopelia* (phytophagous), *Circaetus* (herpetophagous), etc. Some Mediterranean species (*Tachymarptis melba, Hirundo daurica*) extended their ranges northward in the interglacials of the Late Pleistocene, reaching northern parts of Bulgaria.

The occurrence of some aquatic species of a more "northern" distribution type in the Early to Late Holocene has been also established. Some of the records from the southern Black Sea coast and North Bulgaria include *Oxyura leucocephala, Aythya nyroca, A. ferina, A. fuligula*, and *Phalacrocorax aristotelis* (Boev, 1995; unpubl. data). Most of these species are now either only winter visitors to Bulgaria or have a different (more limited) range. Some migrants that now visit Bulgaria only in winter are known from numerous fossil (and subfossil) remains. These species are: *Gavia arctica, G. immer, G. stellata, Cygnus olor, Anser fabalis, A. albifrons, A. erythropus, Anas penelope, A. crecca, Bucephala clangula, Melanitta nigra, Buteo lagopus, Circus cyaneus, Tringa nebularia, T. glareola*, etc. During the Middle and Late Pleistocene these species have been visiting the territory of Bulgaria but we do not know their residency status. Their breeding range now lies far north from Bulgaria.

The Würmian avifauna of Bulgaria shows the whole Boreomountain complex, consisting of at least 16 woodland, steppe, and hydrophilous species. These species are: *Lagopus lagopus, L. mutus, Nyctea scandiaca, Asio flammeus, Glaucidium passerinum, Aegolius funereus, Strix* aff. *nebulosa, Buteo lagopus, Numenius phaeopus/tenuirostris, N. torquatus, Branta ruficollis, Melanitta nigra, Bucephala clangula, Eremophila alpestris, Pyrrhocorax graculus, P. pyrrhocorax*. All these records also confirm for the first time for Bulgaria the southward extension of the ranges of boreal elements during the cold phases of the Pleistocene.

The so-called "mixed" avifaunas have been widespread in Bulgaria during the Pleistocene. The mountain chains of the Apennines and the Balkans played a key role for the meridional faunal interchange in the Quaternary. That is why, because of the high diversity of the relief and the mosaic character of the landscape, in the neighboring parts of a territory such different bird types as the thermophilic "Mediterranean" *Sylvia atricapilla*, and the cold-loving "Boreal" *Aegolius funereus* occurred together. *S. atricapilla* coexisted in the Early Pleistocene with *Dendrocopos minor* and *Pyrrhula pyrrhula*, while *Sylvia communis* in the Late Pleistocene coexisted with *Tetrao tetrix, Lagopus lagopus, Bonasa bonasia, Glaucidium passerinum*, and *Aegolius funereus*. Similar are the associations of *"Nyctea scandiaca–Pyrrhocorax graculus"*, "*Nyctea scandiaca–Alectoris* sp.", "*Lagopus lagopus–Hirundo daurica"*, "*Melanocorypha* sp.–*Pyrrhocorax pyrrhocorax/P. graculus"*, *"Lagopus mutus–Alectoris graeca"*, *"Aegypius monachus–Alectoris graeca"*, "*Aegypius monachus–Nucifraga caryocatactes*", and "*Pyrrhocorax pyrrhocorax–Lagopus lagopus*". Those examples of "mixed" faunas, composed of species that nowhere in the recent faunas occur together, could only be explained by the uniqueness of physical-geographic characteristics of the Balkans being a kaleidoscope of landscapes in the Pleistocene.

Besides the controversial statements for the origin of *Phasianus colchicus colchicus* in Bulgaria, the Pleistocene to Holocene record proves its autochthonous origin. At present, the native subspecies has disappeared from the bird fauna of Bulgaria, and all over the Europe. Another disappeared galliform species, *Tetrao tetrix* (no records in the recent Bulgarian bird fauna), was widespread throughout Bulgaria in the Pleistocene. Its remains document its former distribution during the last 80,000 years (Boev, 1999, 2001a).

5 Composition and Changes of the Recent Avifauna

5·1 Bulgarian avifauna as a part of the world, Palearctic, and European avifaunas

Out of 159 families of the class Aves, 50 are represented in the recent bird fauna of Bulgaria. This fauna includes 391 to 399 species (Kostadinova, 1997; Nankinov, 1997), which represent 20 zoogeographic types (Voous, 1960) according to their present distribution (Boev, 1997). Most of the species have Palearctic (102 species, or 25.6%), Holarctic (38 species, or 9.5%), or Arctic (33 species, or 8.3%) types of distribution (Nankinov, 1997). The Bulgarian avifauna comprises 50.8% of the Western Palearctic and 73% of the European bird fauna.

It has been believed for a long time that *Picoides tridactylus* and *Turdus torquatus* are the only species with the Arctoalpine type of distribution within Bulgaria (Beron, 1969) although such a statement does not fit in the categorization of Voous (1960). *Troglodytes troglodytes* is the only species of Nearctic origin (Austin, 1983). On the contrary, *Certhia familiaris* is considered an autochthonous species with a relict distribution on the Balkans (Stepanyan & Butyev, 1967). Paleornithological data show that at least several dozen species have an autochthonous origin in Bulgaria and in the Balkans. Their remains, or remains of their closely related ancestors, have been also found in the fossil deposits from the end of the Late Pliocene (Boev, 2002; see below).

5·2 Changes in the composition of the breeding species in the last 100 years

5·2·1 Disappeared and declining species

Nankinov (1997) listed 13 disappeared breeding species (last breeding recorded in 19th or 20th century) but without a firm proof that all of them were formerly breeding: *Pelecanus onocrotalus, Aythya fuligula, Bucephala clangula, Haliaetus albicilla, Gypaetus barbatus, Aquila rapax, Tetrao tetrix, Grus leucogeranus, Grus grus, Anthropoides virgo, Tetrax tetrax, Otis tarda, Pyrrhocorax pyrrhocorax*. In addition, *Gallinago gallinago* is also considered a disappeared breeding species in Bulgaria (Botev & Peshev, 1985).

However, *Pelecanus onocrotalus* had a successful breeding event at Srebarna Lake in 2000 after a period of 37 years (Simeonov et al., 2001). *Haliaetus albicilla* also nests in several sites in Bulgaria (Kostadinova, 1997). A nesting locality of *Otis tarda* has been confirmed in 1997 in northeastern Bulgaria (Anonymous, 1998), while a breeding pair of *Falco biarmicus* on the nest has been recorded in 2002 in the Rhodopes (Hristov, 2002). Three species are presently unknown as breeding in Bulgaria: *Aquila rapax* has not been observed as breeding since 1941; and no data exists on *Grus leucogeranus* and *Tetrao tetrix* (Simeonov et al., 1990). Our data confirms the total disappearance of the nominate (southeastern European) subspecies of the Colchic (common) pheasant (*Phasianus colchicus colchicus*) from Bulgaria (Boev, 1999).

5·2·2 Species with the extended breeding range

A number of species have considerably extended their breeding range throughout the country in the second half of 20th century: *Buteo rufinus, Larus cachinnans, Thalasseus sandvicensis, Streptopelia decaocto, Apus pallidus, Clamator glandarius,*

Aegolius funereus, Dryocopus martius, Hirundo daurica, Lanius nubicus, Oenanthe isabellina, Locustella fluviatilis (Nankinov, 2000), *Acrocephalus agricola, Passer hispaniolensis* (Micev & Simeonov, 1985), *Haematopus ostralegus* (Boev, 1991), *Sylvia hortensis* (Milchev & Kovachev, 2000), *Emberiza melanocephala* (Shurulinkov et al., 2001), *Tyto alba* (Miltschev et al., 2002; we accumulated more detailed information on this secretive species, until now insufficiently known throughout Bulgaria), and *Carpodacus erythrinus* (Shurulinkov et al., 2003).

5·2·3 Recently recorded new species of the Bulgarian avifauna

Micev & Simeonov (1985) listed 48 species recorded for the first time in Bulgaria between 1950 and 1984. Five more species have been later recorded for the first time in the country: *Phylloscopus borealis* (Nankinov et al., 1989), *Apus pallidus* (Fuhrmann, 1987; Iankov, 1990), *Lagopus mutus* (Miltschew & Georgiewa, 1998), *Ficedula semitorquata* and *Larus cachinnans* (last two separated from *Ficedula albicollis* and *Larus argentatus*, respectively). In addition, at least seven new species have been recorded and published since 1999: *Vanellus leucurus* (Hristov et al., 1999), *Phylloscopus proregulus* (Nankinov & Dalaktčijeva, 1999), *Egretta gularis* (Georgiewa, 1999), *Oenanthe deserti* (Shurulinkov & Tsonev, 2001), *Sylvia rueppelli* (Stoyanov, 2001b), *Larus relictus* (Nankinov, 2001a), and *Mycteria ibis* (Kovachev, 2002).

6 Importance of the Bulgarian Territory for the Migration and Wintering of the European Birds

Bulgaria is one of the six European countries preserving the largest number of species of the European conservation significance, 210 of the total of 287 species (Kostadinova, 1997). Moreover, the recent Bulgarian avifauna (399 breeding, wintering, passage, and vagrant species per the territory of ca. $111,000\,km^2$) is the richest in Europe.

6·1 Migration routes

Two main migration intercontinental routes of meridional orientation pass through the Bulgaria's territory, Via Pontica (Western Black Sea flyway) along the Bulgarian Black Sea coast, and Via Aristotelis along the Struma River in the western part of the country. In addition, a third intensively used route exists along the Maritsa River valley. The Atanasovsko Lake nature reserve near the city of Burgas is situated on Via Pontica; it is the site of the highest number of the migrating flocks in Europe for *Pelecanus onocrotalus* (maximum 31,665 individuals in 1993), *P. crispus* (maximum 432 individuals in 1980), *Circus aeruginosus* (maximum 1530 individuals in 1981) and *Falco vespertinus* (maximum 3110 individuals in 1993) (Michev et al., 1999).

6·2 Wintering grounds

The southern geographic position of Bulgaria allows for the migrating populations of many birds to use the advantages of the winter shelters. Some of these sites have been also used as refugia during the glacials in the Pleistocene (see below).

At least 95 species of aquatic birds have been found to winter throughout Bulgaria in the five-year period from 1997 to 2001 (Kostadinova & Dereliev, 2001). Their numbers varied from 703,782 individuals in 1997 to 186,350 individuals in 2001. Three of the twelve wetland areas, sheltering important number of aquatic birds, are Ramsar Convention sites: Durankulak, Shabla, and Atanasovsko lakes. Eight world threatened species winter in Bulgaria; two of them, *Pelecanus crispus* (up to 606 individuals) and *Branta ruficollis* (up to 50,209 individuals), form wintering populations at the internationally important levels.

The most numerous populations of the wintering species in Bulgaria in 1997–2001 were those of *Phalacrocorax carbo* (up to 23,500 individuals), *Anser albifrons* (up to 398,115 individuals), *Anas platyrhynchos* (up to 148,601 individuals), *Aythya ferina* (up to 32,301 individuals), and *Fulica atra* (up to 82,019 individuals) (Kostadinova & Dereliev, 2001).

7 Important Bird Areas

A total of 50 Important Bird Areas (IBA) have been recognized in Bulgaria. They cover 7000 km^2 (6.3% of the country's territory). The IBA cover breeding localities of 19 bird species belonging to the world threatened category (of total of 35 in Europe). Ten species, because of their higher population concentration, are covered by the specially designated IBA: *Phalacrocorax pygmeus, Pelecanus crispus, Anser erythropus, Branta ruficollis, Aythya nyroca, Oxyura leucocephala, Aegypius monachus, Aquila heliaca, Crex crex*, and *Numenius tenuirostris*. Among the Bulgarian IBA, 48 are of continental importance, while 34 of them have world importance; 27 IBA (54%) are protected areas according to the Bulgarian Law on the Protection of Nature (Kostadinova,1997).

8 Spatial Differentiation of the Recent Avifauna: Characteristic Bird Communities of the Main Landscapes (Ordination–Correspondence Analysis)

The goal of this analysis is to reveal the main environmental gradients responsible for the patterns of bird distribution and to provide a basis for the biogeographical subdivision of Bulgaria. The results of this analysis are of a particular importance for landscape planning and nature conservation management.

The results of the correspondence analysis (CA) based on all bird species are shown in Figs. 2–3. Together, the first two ordination axes account for 45.5% of the total variation of the distribution of the species among NRs. The first ordination axis, which accounts for 32.5% of the variance, identifies an altitudinal gradient. The plain and lowland NRs (1, 3, 23) with large wetlands and rich breeding avifauna of Podicipediformes, Pelecaniformes, Ciconiiformes, Anseriformes, and Charadriiformes form a separate group on the right side of the diagram (Fig. 2).

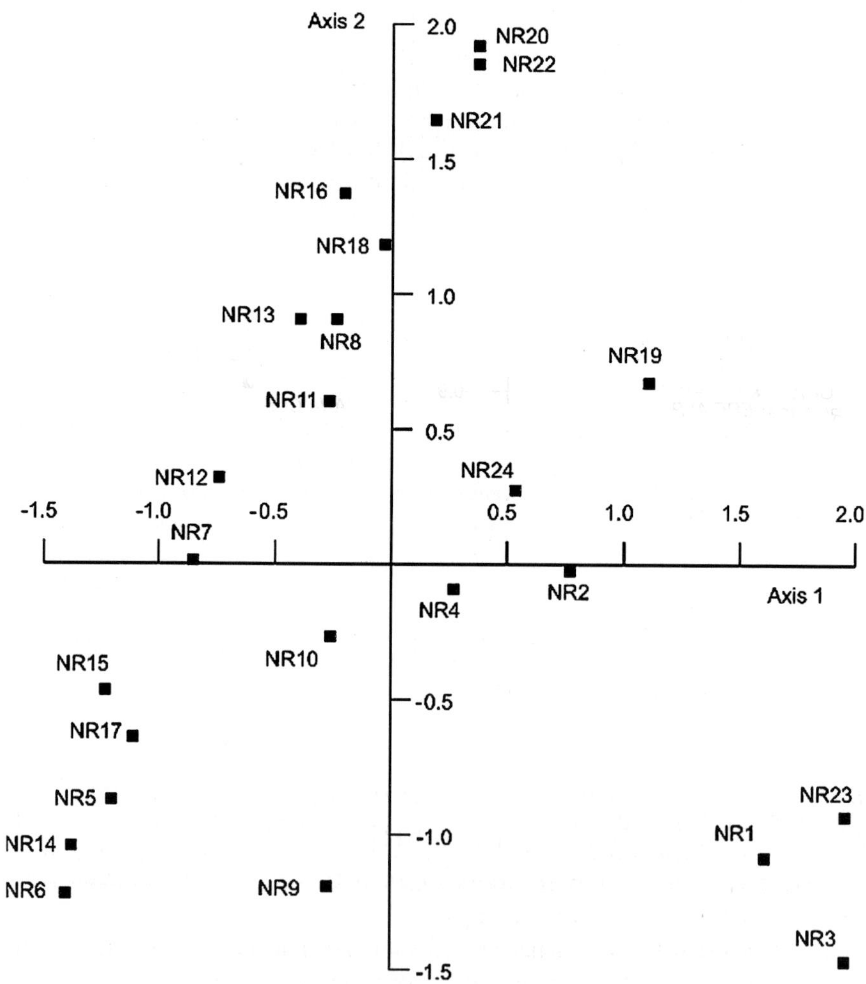

Fig. 2 Ordination diagrams (Axis 1 and Axis 2) based on correspondence analysis of the distribution of all bird species (246 species) across 24 NRs. Natural regions (numeration, see Fig. 1).

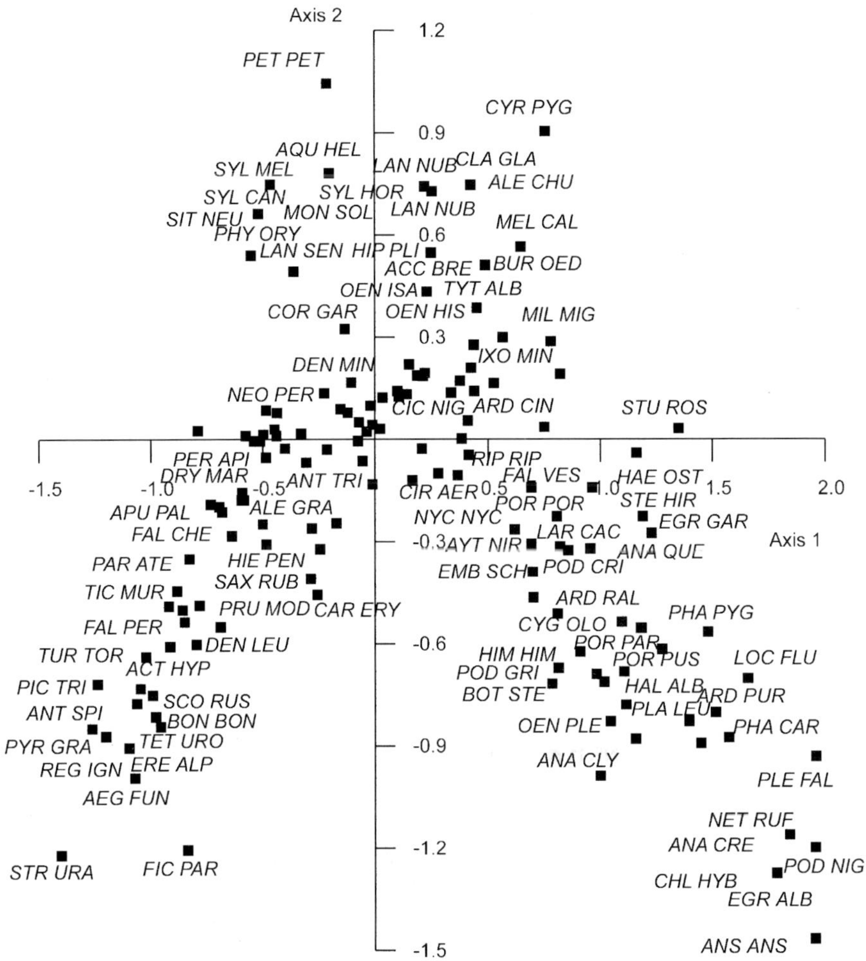

Fig. 3 Ordination diagrams (Axis 1 and Axis 2) based on correspondence analysis of the distribution of all bird species (246 species) across 24 NRs. Species (the abbreviations consist of the first three letters of the generic name and the first three letters of the species name).

In contrast, the highland NRs (5, 6, 14, 15, and 17) (Fig. 2) with their typical Boreal avifauna (*Aegolius funereus, Regulus ignicapillus, Ficedula parva, Tetrao urogallus, Bonasa bonasia, Scolopax rusticola, Picoides tridactylus*, etc.) (Fig. 3) score on the left side of the ordination plot.

The second ordination axis separates some southern NRs (16, 18, 20, 21, and 22) (Fig. 2) with the Mediterranean climatic influence from the remaining NRs. This could indicate a weak effect of the temperature gradient on the spatial differentiation of the avifauna.

The results of CA excluding the wetland avian complex (freshwater aquatic habitats), are shown in Figs. 4–5. Together, the first two ordination axes account for 44.7% of the total variation of the distribution of bird species among NRs. The first axis contrasts mountainous and semimountainous NRs (5, 6, 9, 14, 17) with lowland NRs (3, 19, 20, 21, 22, 23, etc.) (Fig. 4). This axis represents the altitudinal gradient. This gradient explains the majority of variance of distributional data (34.3%), showing that temperature, humidity, and the vegetation type are the main differentiating factors. The second axis separates the lowland NRs in the southwestern part of Bulgaria (11, 12, 13, 16) from those in the southern and southeastern parts

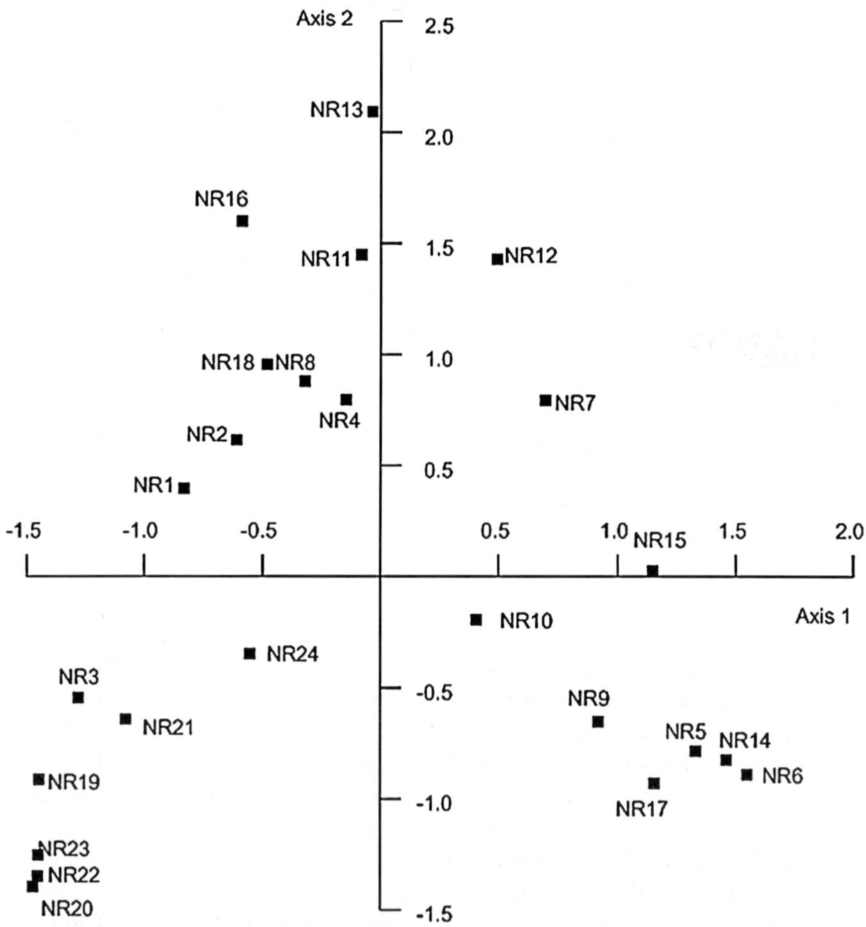

Fig. 4 Ordination diagrams (Axis 1 and Axis 2) based on correspondence analysis of the distribution of "non-aquatic" bird species (176 species) across 24 NRs. Natural regions (numeration, see Fig. 1).

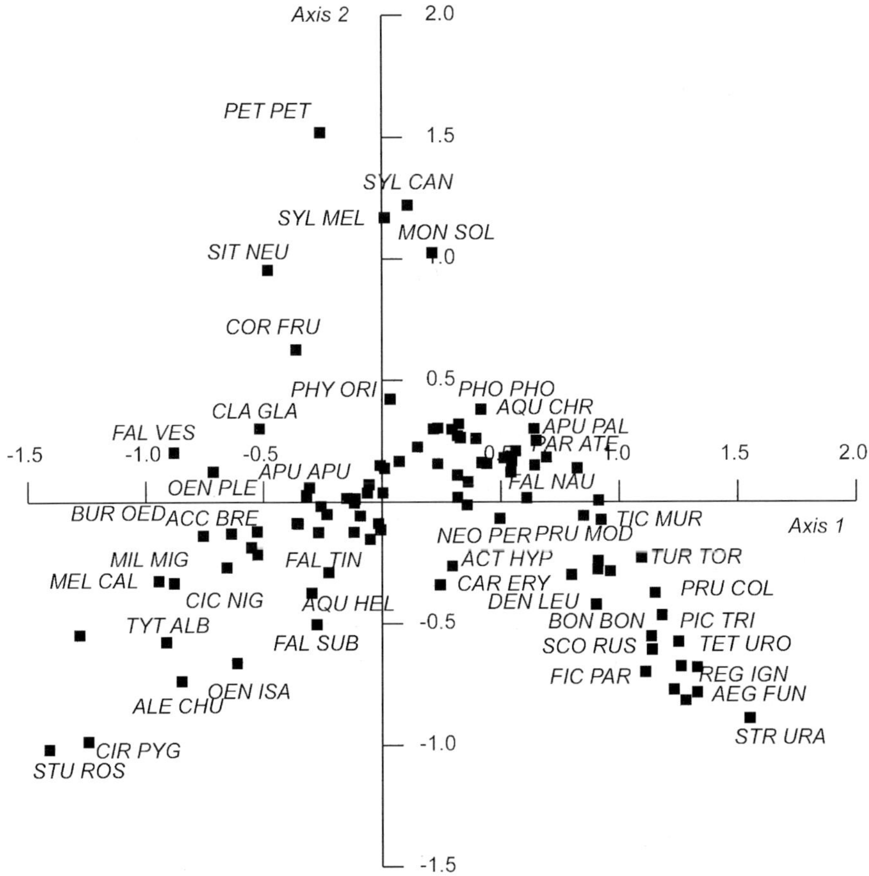

Fig. 5 Ordination diagrams (Axis 1 and Axis 2) based on correspondence analysis of the distribution of "non-aquatic" bird species (176 species) across 24 NRs. Species (the abbreviations consist of the first three letters of the generic name and the first three letters of the species name).

(19, 20, 21, 22, 23) (Fig. 4). The first group of NRs is characterized by a Submediterranean climate and a sporadic distribution of large rocky habitats. These NRs have a higher number of Mediterranean species (*Petronia petronia, Sylvia melanocephala, S. cantillans, Monticola solitarius*, and *Sitta neumayer*) (Fig. 5) inhabiting shrubs and bushy rocky habitats in the dry, warmest sunny slopes. The other group of NRs consists of plain and hilly areas with large farmlands and pastures seving as breeding habitats for some characteristic species (*Circus pygargus, Alectoris chukar, Oenanthe isabellina, Sturnus roseus, Tyto alba*) (Fig. 5).

In the two variants of analysis the eigenvalues of the first two axes were: all species, 0.196 and 0.079; "non-aquatic" species, 0.136 and 0.041. The eigenvalue is a measure of importance of an ordination axis; values over 0.40 denote a good separation of the taxa along the axis (Braak, 1987). The obtained low eigenvalues show that, as a whole, the separation within data sets was low, indicating a high homogeneity of the avifauna across Bulgaria. These specific features of data sets are pronounced in the ordination diagrams as well. In both variants of analysis there is a distinct arch effect (Figs. 2–5). According to Hill & Gauch (1980), the arch effect is a mathematical artifact, which does not correspond to any real structure in the data.

Regardless of the overall spatial homogeneity of the avifauna, the ordination results represent some lines of biogeographical differentiation of the Bulgarian avifauna, suggesting that it is influenced primarily by the altitude. Most likely, the altitude affects the birds through geomorphology, temperature, humidity, and vegetation. The role of geomorphology is related to the concentration of large bodies of water in the lowlands, especially near the Danube and the Black Sea shore. In addition, the lowland areas have relatively high temperatures, low precipitation, and highly fragmented forest cover. The mountain areas, in contrast, have lower temperatures, higher precipitation, and well-developed forests.

The ordination results also show that the secondary factors affect predominantly the avifauna of lowland areas. When all birds were considered, it was evident that the areas of the extreme south of the country are rather specific because of the occurrence of xerophilous and petrophilous species with the Mediterranean affinities. It can be suggested that this feature reflects the effect of a latitudinal gradient, which, however, is of low importance. In the case of the analysis in which the "aquatic" species are excluded, the second axis reveals a weakly pronounced effect of the prevailing habitat type in lowland areas, especially in the south.

9 Classification of Avian Assemblages: Two-way Indicator Species Analysis

By application of TWINSPAN to the presence/absence data matrix (246 bird species across 24 NRs) the NRs have been classified into six end-groups according to their nesting avifauna, numbered from left to right (Table 1, Fig. 1).

(1) The first level of division separates the bird assemblages of the mountainous and semimountainous (low mountains, hilly areas) NRs from those of the lowland ones. Indicator species for the first group of NRs within this division are *Alectoris graeca, Serinus serinus*, and *Motacilla cinerea*. They breed in the mountain habitats: *A. graeca*, in mountain rocky terrains; *S. serinus*, in the outskirts of mountain forests; and *M. cinerea*, in the habitats along the fast-running rivers, streams, and mountain lakes. No indicator species are defined for the second group of NRs. This dichotomy clearly corresponds to the first ordination axis in CA.

(2) At the second level of division, the NRs form four groups. The first two groups belong to the "mountainous" group of NRs defined at the first level of division. The respective bird assemblages are separated on the basis of the

landscape. The first group within this level of division consists of the high mountain NRs. Indicator species for this group is *Turdus torquatus*, which breeds only in the forests of the highest mountains. The second group consists of the bird assemblages of low mountains and hilly NRs. Indicator species are not identified for this group. Within the second group defined at the first level of division, the NRs form two second-level groups. The first second-level group includes bird assemblages of lowland NRs without large water bodies, while the second second-level group embraces the three lowland NRs with large water basins (NRs 1, 3, 23). Due to the effect of the Black Sea shore and the large inland continental water bodies, the nesting bird assemblages of the Northern Danubian and the Southern Dobrudja NRs are positioned close to the Burgas–Aitos NR (end-group 6) and distinguished from other lowland NRs (end-group 5). The sixth end-group is indicated by *Botaurus stellaris*, which breeds mainly in the marshlands along the Danube River and the Black Sea coast.

(3) The third level of division reveals some specifics of the avifauna within the mountainous–semimountainous group of NRs defined at the first level. Four groups are distinguished here. The first one (end-group 1) consists of one NR only (NR 9; Table 1). The avian complex of this region (Vitosha–Ihtiman) shows stronger differences compared with complexes of the remaining high-mountain NRs because of its habitat heterogeneity. It includes breeding birds of the mountain forests and the alpine zone, as well as birds of the open plain habitats, wetlands, and various anthropogenic landscapes in the Sofia Basin. The remaining eight NRs of higher mountains, containing specific high-mountain and alpine avifauna, form end-group 2.

The low-mountain and hilly NRs have been separated according to their nesting avian assemblages into two groups (end-groups 3 and 4). Six of these NRs, chiefly from the southern and the western part of South Bulgaria (end-group 3) include low mountains, foothills, and valleys. The other group consists of three NRs from the eastern Bulgaria (end-group 4), characterized by the predominance of hills. It groups closer to the large plain NRs and farther away from the high-mountain NRs. The indicator species for this group is *Acrocephalus palustris*, which predominantly breeds in the valleys in the neighborhood of the plain NRs.

The TWINSPAN classification of species (Table 1) produced seven end-groups (A–G). Group A consists of 39 species associated with mountains. Group B is formed by 24 species, most of which have wider distribution but tend to occur in mountainous and semimountainous areas. The most numerous group C (87 species) is comprised of widely distributed bird species occurring in all natural regions as well as some rare species whose distributions across the country do not show any particular regularity. Group D involves eight species with similar distribution. The next two groups (E and F) include 20 and 34 species, respectively, tending to avoid mountains and mountainous areas. The last group (G) embraces 35 species associated with large water basins in lowlands. These results confirm the high faunal homogeneity across the country. About 38% of the species (groups C and D)

Table 2 Zoogeographical structure of the bird complexes across Natural Regions: number of bird species (% in parentheses). Zoogeographical types (after Voous, 1960): HOL, Holarctic; SIB–CAN, Siberian–Canadian; SIB, Siberian; EUR, European; MON–TIB, Montane–Tibetian; PALX–MON, Paleoxeric–Montane; PALMON, Paleomontane; PAL, Palearctic; NAT, North Atlantic; PALXE, Paleoxeric; COS, Cosmopolitan; OW, Old World; SAR, Sarmatian; ETH, Ethiopian; MED, Mediterranean; TUR–MED, Turkestanian–Mediterranean; IND–AF, Indian–African; EUR–TUR, European–Turkestanian; TUR, Turkestanian

NATURAL REGION	HOL	SIB–CAN	SIB	EUR	MON–TIB	PALX–MON	PALMON	PAL	NAT	PALXE	COS	OW	SAR	ETH	MED	TUR–MED	IND–AF	EUR–TUR	TUR
9	15 (8.9)	–	1 (0.6)	24 (14.3)	–	3 (1.8)	3 (1.8)	60 (35.7)	–	2 (1.2	4 (2.4)	9 (5.4)	1 (0.6)	–	10 (6.0)	10 (6.0)	3 (1.8)	22 (13.1)	1 (0.6)
5	15 (10.2)	–	2 (0.7)	24 (16.3)	1 (0.7)	3 (2.0)	5 (3.4)	49 (33.3)	–	2 (1.4)	2 (1.4)	6 (4.1)	–	–	11 (7.5)	7 (4.8)	2 (1.4)	19 (12.9)	–
6	13 (9.6)	–	1 (1.5)	21 (15.6)	1 (0.7)	3 (2.2)	5 (3.7)	50 (37.0)	–	2 (1.5)	1 (0.7)	5 (3.7)	–	–	7 (5.2)	6 (4.4)	2 (1.5)	17 (12.6)	–
7	10 (8.1)	–	–	22 (17.9)	–	3 (2.4)	2 (1.6)	45 (36.6)	–	2 (1.6)	2 (1.6)	5 (4.1)	–	–	7 (5.7)	6 (4.9)	2 (1.6)	17 (13.8)	–
12	10 (8.5)	–	–	17 (14.5)	–	3 (2.6)	3 (2.6)	43 (36.8)	–	–	1 (0.9)	5 (4.3)	–	–	10 (8.5)	4 (3.4)	2 (1.7)	19 (16.2)	–
14	13 (9.7)	1 (0.7)	1 (0.7)	22 (16.4)	–	3 (2.2)	5 (3.7)	46 (34.3)	–	2 (1.5)	1 (0.7)	5 (3.7)	–	–	10 (7.5)	6 (4.5)	2 (1.5)	17 (12.7)	–
15	13 (8.9)	1 (0.7)	1 (0.7)	23 (15.8)	1 (0.7)	4 (2.7)	5 (3.4)	47 (32.2)	–	2 (1.4)	1 (0.7)	5 (3.4)	–	–	15 (10.3)	9 (6.2)	1 (0.7)	18 (12.3)	–
17	13 (8.5)	1 (0.7)	1 (0.7)	24 (15.7)	1 (0.7)	3 (2.0)	5 (3.3)	50 (32.7)	–	2 (1.3)	2 (1.3)	6 (3.9)	–	–	11 (7.2)	9 (5.9)	3 (2.0)	20 (13.1)	2 (1.3)
8	8 (7.3)	–	–	18 (16.4)	1 (0.9)	2 (1.8)	–	37 (33.6)	–	2 (1.8)	1 (0.9)	6 (5.5)	1 (0.9)	–	8 (7.3)	6 (5.5)	2 (1.8)	18 (16.4)	–
10	13 (8.5)	–	1 (0.7)	23 (15.0)	1 (0.7)	3 (2.0)	1 (0.7)	55 (35.9)	–	2 (1.3)	4 (2.6)	8 (5.2)	–	–	12 (7.8)	7 (4.6)	2 (1.3)	20 (13.1)	1 (0.7)

continued

11	9 (7.6)	–	–	20 (16.8)	–	2 (1.7)	1 (0.8)	41 (34.5)	–	2 (1.7)	1 (0.8)	8 (6.7)	–	–	8 (6.7)	6 (5.0)	2 (1.7)	19 (16.0)	–
13	10 (7.5)	–	–	18 (13.4)	–	4 (3.0)	2 (1.5)	43 (32.1)	–	3 (2.2)	1 (0.7)	6 (4.5)	–	–	15 (11.2	10 (7.5)	2 (1.5)	20 (14.9)	–
16	6 (5.7)	–	–	18 (17.1)	–	2 (1.9)	–	36 (34.3)	–	–	–	5 (4.8)	–	–	11 (10.5)	9 (8.6)	1 (1.0)	17 (16.2)	–
18	10 (6.6)	–	–	20 (13.2)	2 (1.3)	4 (2.6)	1 (0.7)	46 (30.3)	–	3 (2.0)	5 (3.3)	8 (5.3)	–	1 (0.7)	16 (10.5)	11 (7.2)	3 (2.0)	20 (13.2)	2 (1.3)
4	9 (6.0)	–	–	24 (16.1)	–	3 (2.0)	2 (1.3)	49 (32.9)	–	3 (2.0)	2 (1.3)	9 (6.0)	2 (1.3)	–	11 (7.4)	12 (8.1)	3 (2.0)	19 (12.8)	1 (0.7)
21	7 (5.6)	–	–	19 (15.3)	–	1 (0.8)	–	41 (33.1)	–	2 (1.6)	2 (1.6)	8 (6.5)	–	–	12 (9.7)	8 (6.5)	3 (2.4)	19 (15.3)	2 (1.6)
24	11 (7.1)	–	–	24 (15.4)	–	1 (0.6)	1 (0.6)	52 (33.3)	–	3 (1.9)	3 (1.9)	9 (5.8)	2 (1.3)	1 (0.6)	12 (7.7)	8 (5.1)	4 (2.6)	23 (14.7)	2 (1.3)
2	9 (8.2)	–	–	16 (14.2)	–	2 (1.5)	–	39 (33.6)	–	3 (2.2)	2 (1.5)	10 (9.0)	2 (1.5)	1 (0.7)	6 (5.2)	7 (6.0)	3 (2.2)	16 (14.2)	–
19	9 (6.8)	–	–	17 (12.8)	–	1 (0.8)	–	44 (33.1)	–	2 (1.5)	4 (3.0)	9 (6.8)	2 (1.5)	1 (0.8)	11 (8.3)	8 (6.0)	2 (1.5)	20 (15.0)	3 (2.3)
20	7 (6.6)	–	–	11 (10.4)	–	–	–	36 (34.0)	–	2 (1.9)	2 (1.9)	8 (7.5)	–	–	12 (11.3)	6 (5.7)	3 (2.8)	17 (16.0)	2 (1.9)
22	6 (5.9)	–	–	13 (12.9)	–	–	–	35 (34.7)	–	1 (1.0)	2 (2.0)	5 (5.9)	–	–	10 (9.9)	6 (5.9)	2 (2.0)	17 (16.8)	3 (3.0)
1	11 (7.4)	–	–	18 (12.2)	1 (0.7)	1 (0.7)	–	54 (36.5)	–	2 (1.4)	6 (4.1)	14 (9.5)	4 (2.7)	1 (0.7)	6 (4.1)	8 (5.4)	2 (1.4)	20 (13.5)	–
3	13 (7.8)	–	1 (0.6)	19 (10.8)	–	1 (0.6)	–	54 (31.3)	1 (0.6)	3 (1.8)	7 (4.2)	17 (9.6)	6 (3.6)	1 (0.6)	9 (5.4)	11 (6.6)	5 (3.0)	20 (11.4)	3 (1.8)
23	13 (7.9)	–	–	17 (10.3)	–	–	–	53 (32.1)	–	3 (1.8)	9 (5.5)	14 (8.5)	7 (4.2)	1 (0.6)	11 (6.7)	10 (6.1)	4 (2.4)	19 (11.5)	4 (2.4)

are widespread, and their distribution is not related to the environmental gradients identified at this scale of analysis. The best-defined groups (A and G) comprise about 30% of the nesting avifauna of Bulgaria (15.9% and 14.2%, respectively).

Although this analysis reveals regularities similar to those demonstrated by the CA, it is useful because it provides a base for a general avifaunal division of Bulgaria. The inspection of the TWINSPAN table indicates that only one more or less spatially coherent unit can be identified: the mountainous areas of central and western Bulgaria (NRs 5, 6, 7, 9, 12, 14, 15, 17). This outcome corresponds to the results presented by Georgiev & Simeonov (1992) who found that Stara Planina, Rila, Pirin, and the Rhodopes form a well-defined region with a specific faunal composition. More detailed subdivision of the country could be possible on the basis of quantitative data within the context of landscape heterogeneity.

10 Zoogeographical Types

The recent breeding avifauna of Bulgaria includes representatives of 19 zoogeographical types (after Voous, 1960) (Table 2). Several groups of these zoogeographical types have been distinguished for the avifauna of the natural regions arranged according to the TWINSPAN classification presented above.

The mountain NRs show larger presence of the first seven types (Holarctic, Siberian–Canadian, Siberian, European, Montane–Tibetian, Paleoxeric–Montane, Paleomontane), that include mainly woodland species of broad distribution in Europe, Siberia, and even the entire Holarctic. Next three types (Palearctic, North Atlantic, and Paleoxeric) are represented evenly.

The widespread birds of the Cosmopolitan, Old World, Sarmatian, and Ethiopian types have higher representation in the plain, lowland, and hilly NRs, which are rich in wetland habitats (NRs 1, 3, 23).

The species that breed mainly to the south (the Mediterranean) and to the east (Asia Minor and Southwest Asia) from the Bulgarian territory are better represented in the South Bulgaria and along the Black Sea coast, respectively. The Mediterranean species are represented mainly in the NRs of the southwestern Bulgaria and the Eastern Rhodopes, while the Turkestanian species breed chiefly in the southeastern NRs and this corresponds to the results of the analysis given above. Georgiev & Simeonov (1992) and Boev (1997) also concluded that the Mediterranean type is the most represented after the European type.

References

Alléon, A., 1886. Memoire sur les oiseaux dans la Dobrudja et la Bulgarie. Ornithologisches Jahrbuch 2:397–428.

Anon., 1998. Information récentes sur la reproduction de la grande outarde (*Otis tarda*) en Bulgarie. Branta 3:26.

Austin, O., 1983. Birds of the World. Optimum Books, Montreal, Quebec.

Balát, F., 1962. Contribution to the knowledge of the avifauna of Bulgaria. Acta Academiae scientiarum Čechoslovenicae, Basis Brunensis (Práce Brněnské základny ČSAV) 34(10), 445–492.

Baumgart, W., 2001a. Europas Geier: Flugriesen im Aufwind. AULA-Verlag (Sammlung Vogelkunde), Wiebelsheim, 144 pp.

Baumgart, W., 2001b. Die Bestandssituation des Schleiadlers *Aquila pomarina* und Probleme ihrer Ermittlung in Bulgarien. In Meyburg, B.-U. & R. Chancellor (eds), Eagle Studies. WWGBP, Berlin, London & Paris: 327–336.

Beron, P., 1969. Sur les éléments boréo-alpins de la faune bulgare. Izvestiya na Zoologicheskiya institut s muzei (Bulletin de l'Institut de zoologie et musée) 30:115–133.

Blondel, J., 1982. Caractérisation et mise en place des avifaunes dans le bassin méditérranéen. Ecologia Mediterranea 8(1/2):253–272.

Blondel, J., 1987a. History and development of bird faunas in the Mediterranean region. Documents de Laboratoire de Géologie de Lyon 99:231–238.

Blondel, J., 1987b. Avifaune forestière méditérranéenne: histoire des peuplements. Le Bulletin Aves, no. special: 24.

Blondel, J., 1997. Evolution and history of the European bird fauna. In Hagemeijer, W.J.M. & M.L. Blair (eds), The EBCC Atlas of European Breeding Birds. Their Distribution and Abundance. Poyser, London: CXXIII–CXXVI.

Bochenski, Z., 1982. Aves. In Kozlowski, J. (ed), Excavation in the Bacho Kiro Cave (Bulgaria). Final Report. Panstwowe Wydawnictwo Naukowe, Warszawa:31–38.

Boev, Z., 1990. [One hundred years of the ornithological researches in Bulgaria]. Historia naturalis bulgarica 2:25–35 (in Bulgarian).

Boev, Z., 1991. [Distribution and status of the oystercatcher (*Haematopus ostralegus* L., 1758) (Haematopodidae – Aves) in Bulgaria]. Historia naturalis bulgarica 3:75–91 (in Bulgarian).

Boev, Z., 1995. Eneolithic and Early Bronze Age birds from the sunken settlement at the Sozopol Bay (Bulgarian Black Sea coast). Historia naturalis bulgarica 5:51–60.

Boev, Z., 1997. [On some ornithofaunistical and ornithogeographical peculiarities of Bulgaria]. Historia naturalis bulgarica 8:147–154 (in Bulgarian).

Boev, Z., 1999. Neogenski i kvaternerni ptitsi (Aves) ot Balgariya [Neogene and Quaternary Birds (Aves) from Bulgaria]. Disertatsiya za doktor na naukite [DrSc Thesis]. Natsionalen prirodonauchen muzei [National Museum of Natural History], Sofia (in Bulgarian).

Boev, Z., 2001a. Birds over the mammoth's head in Bulgaria. In: Cavaretta, G., P. Gioia, M. Mussi, M. & M.R. Palombo (eds), The World of Elephants. Proceedings of the 1st International Congress. Roma, 16–20 Ottobre 2001: 180–186.

Boev, Z., 2001b. Early Pliocene avifauna of Muselievo (Central Northern Bulgaria). Acta Zoologica Cracoviensia 44(1):37–52.

Boev, Z., 2002. Neogene avifauna of Bulgaria. In Zhou, Z. & F. Zhang (eds), Proceedings of the 5th Symposium of the Society of Avian Palaeontology and Evolution, Beijing, 01–04.06.2000. Science Press, Beijing: 29–40.

Boev, Z., 2003a. The ornithological collections of the National Museum of Natural History of the Bulgarian Academy of Sciences: their history and scientific value. In Rheinwald, G. (ed), Proceedings of the Second European Conference "Bird Collections in Europe: Promotion of Science in Ornithology. The Challenge of Mutual Cooperation", 9–12 September 2001, Bonn. Bonner Zoologische Beiträge 51(2/3):157–163.

Boev, Z., 2003b. Specimens of extinct and threatened birds in the collections of the National Museum of Natural History in Sofia, Bulgaria. In Collar, N., C. Fisher & Ch. Feare (eds), Why Museums Matter: Avian Archives in Age of Extinction. British Ornithologists' Club, 123A, Suppl. 2003:234–245.

Botev, B. & Tz. Peshev (eds), 1985. Chervena kniga na NR Balgariya. Tom 2. Zhivotni (Red Data Book of the People's Republic of Bulgaria. Volume 2. Animals). Izdatelstvo na Balgarskata akademiya na naukite (Publishing House of the Bulgarian Academy of Sciences), Sofia (in Bulgarian).

Braak, C.J.F. ter, 1987. Ordination. In Jongman, R.H., C.J.F. ter Braak & O.F.R. van Tongeren (eds), Data Analysis in Community and Landscape Ecology. Pudoc, Wageningen, The Netherlands: 91–169.

Braak, C.J.F. ter, 1990. Update Notes: CANOCO Version 3.10. Agricultural Mathematics Group, Wageningen, The Netherlands.

Burchak-Abramovich, N.I. & I. Nikolov, 1984. [Fossil birds *Phalacrocorax serdicensis* sp. n. and *Anser thraceiensis* sp. n. from Bulgaria]. Paleontologiya, stratigrafiya i litologiya [Paleontology, Stratigraphy and Lithology] 19:23–27 (in Russian).

Donchev, S., 1994. [Birds]. In: Chervena kniga na Shumenskiya region (Red Book of the Shumen Region). Slavcho Nikolov & Co., Shumen: 93–173 (in Bulgarian).

Farman, C., 1868. On some of the birds of prey of Central Bulgaria. Ibis 2(6):406–414.

Farman, C., 1869. On some of the birds of prey of Central Bulgaria. Ibis 3(5):199–204.

Fuhrmann, P., 1987. [Occurrence of the pallid swift (*Apus pallidus* (Shelley)) in the town of Sandanski in 1986]. Ornitologicheski informatsionen byuletin [Ornithological Information Bulletin] 21/22:40–41 (in Bulgarian).

Georgiev, V., 1989. Ptitsite na Lyulin planina [The Birds of the Lyulin Mts.]. Diplomna rabota [Diploma Thesis]. Sofiiski universitet "Sv. Kliment Ohridski", Biologicheski fakultet [St. Kliment Ohridski University of Sofia, Biological Division], Sofia (in Bulgarian).

[Georgiev, V.] Georgiew, V., 1998. Erste Angaben über hybride Schleiereulen *Tyto alba* in Bulgarien. Ornithologische Mitteilungen 50(8/9):219–220.

Georgiev, V. & B. Milchev, 2000. Birds of the Vratza mountains. II. Breeding bird atlas. Godishnik na Sofiiskiya universitet "Sv. Kliment Ohridski". Biologicheski fakultet (Annuaire de l'Université de Sofia "St. Kliment Ohridski". Faculté de Biologie) 91(1 – Zoology):83–109.

Georgiev, V. & S. Simeonov, 1992. [Zoogeographical characteristic of the breeding birds in Bulgaria]. Godishnik na Sofiiskiya universitet "Sv. Kliment Ohridski". Biologicheski fakultet (Annuaire de l'Université de Sofia "St. Kliment Ohridski". Faculté de Biologie) 80(1 – Zoology):200–210 (in Bulgarian).

Georgiewa, U., 1999. Erstbeobachtung eines Küstenreiher *Egretta gularis* in Bulgarien. Ornithologische Mitteilungen 51:351–352.

Hagemeijer, W.J.M. & M.L. Blair (eds), 1997. The EBCC Atlas of European Breeding Birds: Their Distribution and Abundance. Poyser, London.

Hill, M.O., 1979. TWINSPAN: a FORTRAN Program for Arranging Multivariate Data in an Ordered Two-way Table by Classification of the Individuals and Attributes. Cornell University Press, Ithaca, New York.

Hill, M.O. & H.G. Gauch, 1980. Detrended correspondence analysis, an improved ordination technique. Vegetatio 42:47–58.

Homeyer, E., 1877. Bemerkungen zur Ornis Bulgariens mit Rücksicht auf den Bericht der Gebruder Sintensis und der Reissergebnisse von Dr. Finsch im J. Orn., 1859, 1–378. Journal für Ornitologie 25:69–74.

Hristov, H., 2002. [Nest of the lanner falcon found]. Za ptitsite [On birds] 2:6 (in Bulgarian).

Hristov, I., B. Nikolov & M. Sarov, 1999. The white-tailed plover (*Vanellus leucurus* Lichtenstein 1823) – a new species for Bulgaria. Acta zoologica bulgarica 51(2/3):131–132.

Hristovich, G., 1890. [Materials for the exploration of the Bulgarian fauna. A list of the birds of some localities in Bulgaria]. Sbornik za narodni umotvoreniya, nauka i knizhnina [A Collection of Folklore, Science and Literary Works] 2:185–225 (in Bulgarian).

Iankov, P., 1990. [The pallid swift (*Apus pallidus* (Shelley)) – a nesting species in Bulgaria]. Ekologiya (Ecology) (Sofia) 23:44–46 (in Bulgarian).

Iankov, P., 1996. Where to Watch Birds in Bulgaria. Pensoft, Sofia.

Ivanov, B., 1998. The birds of the Shabla and Tuzlata lakes. In Golemansky, V. & W. Naidenov (eds), Biodiversity of Shabla Lake System. Prof. Marin Drinov Academic Publishing House, Sofia: 129–141.

Ivanov, B., N. Karaivanov & S. Nonev, 1998. Breeding bird communities in the steppe habitats of Dobrudja, Bulgaria. Acta zoologica bulgarica 50(2/3):67–77.

Karaivanov, N., 2000. New data on nesting birds in the region of the quarter Kremikovtzi. Godishnik na Sofiiskiya universitet "Sv. Kliment Ohridski". Biologicheski fakultet (Annuaire de l'Université de Sofia "St. Kliment Ohridski". Faculté de Biologie) 91(1 – Zoology): 117–122.

Klein, E., 1909. Nashi ptitsi, opisani za zemledeltsi, lesnichei, lovtsi i lyubiteli na prirodata [Our birds, described for the farmers, forestmen, hunters, and naturalists]. Pchela [Bee], Plovdiv (in Bulgarian).

Kostadinova, I. (ed), 1997. [Important bird areas in Bulgaria] Conservation Series. Bulgarian Society for the Protection of Birds (Sofia) 1 (in Bulgarian).

Kostadinova, I. & S. Dereliev, 2001. [Results from the mid-winter counts of waterbirds in Bulgaria for the period 1997–2001]. Conservation Series. Bulgarian Society for the Protection of Birds (Sofia) 3 (in Bulgarian).

Kovachev, A., 2002. [New species for the Bulgarian ornithofauna – yellow-billed stork]. Informatsionen byuletin na zashtitenata mestnost "Poda" [Information Bulletin of Poda Preserve] (Burgas) 11:1 (in Bulgarian)

Marsigli, L.F. de, 1744. Description du Danube depuis la montagne de Kalenberg en Autriche jusqu'au confluent de la rivière Jantra dans la Bulgarie. Vol. V: Des oiseaux qui fréquentent les bords du Danube. Traduit du latin. La Haye.

Menzel, J. & B. Miltschev, 2001. Über eine späte Brut des Triels *Burhinus oedicnemus* in Bulgarien. Ornithologishe Mitteilungen 53(11):384–385.

Michev, T. & P. Jankov, 1998. The Bulgarian ornithofauna. In Meine, C. (ed), Bulgaria's Biological Diversity: Conservation Status and Needs Assessment. Volumes I and II. Pensoft, Sofia & Moscow: 411–438.

Michev, T. & Ts. Petrov, 2000. Ptitsite na Rodopite [The Birds of the Rhodope Mountains]. Balgarski sayuz za zashtita na Rodopite [Bulgarian Union for the Protection of Rhodopes] (in Bulgarian).

Michev, T., L. Profirov, M. Dimitrov & K. Nyagolov, 1999. [The birds of the Atanasovsko Lake. Status and checklist]. Burgas Wetlands Publications Series. Bulgarian-Swiss Biodiversity Conservation Program (Burgas) 1 (in Bulgarian).

[Michev, T.] Micev, T. & S. Simeonov, 1985. Changes in bird fauna of Bulgaria over the last thirty-five years (1950–1984). In International Symposium "Protection of natural areas and the genetic fund they contain". Project No. 8 on the Program "Man and Biosphere" (MAB) of UNESCO, 23–28.09.1985, Blagoevgrad. Publishing House of the Bulgarian Academy of Sciences, Sofia: 203–216.

Milchev, B., 1998. The birds in the Bulgarian part of the Strandja Mountains. I. Species list and status. Godishnik na Sofiiskiya universitet "Sv. Kliment Ohridski". Biologicheski fakultet (Annuaire de l'Université de Sofia "St. Kliment Ohridski". Faculté de Biologie) 88/90(1 – Zoology): 25–43.

[Milchev, B.] Miltschev, B., V. Georgiev & A. Kovatschev, 2002. Brutbestand und Brutplatzwahl der Schleiereule (*Tyto alba*) in Südost-Bulgarien. Egretta 45:114–120.

[Milchev, B.] Miltschew, B. & U. Georgiewa, 1998. Erstbeobachtung des Alpenschneehuhns *Lagopus mutus* in Bulgarien. Ornithologische Mitteilungen 50(2):43–44.

Milchev, B. & A. Kovachev, 1998. A contribution to the bird fauna of the Sakar mountains. Godishnik na Sofiiskiya universitet "Sv. Kliment Ohridski". Biologicheski fakultet (Annuaire de l'Université de Sofia "St. Kliment Ohridski". Faculté de Biologie) 88/90(1 – Zoology): 45–53.

Milchev, B. & A. Kovachev, 2000. Habitat, distribution and breeding biology of the orphean warbler (*Sylvia hortensis*) in the Southeastern Bulgaria. Sandgrouse 22:87–92.

[Milchev, B.] Miltschev, B. & D. Tschobanov, 2002. Brutverluste und Nahrung des Rosenstars *Sturnus roseus* in Südost-Bulgarien im Jahr 2000. Vogelwelt 123:99–103.

Nankinov, D., 1997. [Birds]. In Yordanova, M. & D. Donchev (eds), Geografiya na Balgariya. Fizicheska geografiya. Sotsialno-ikonomicheska geografiya (Geography of Bulgaria. Physical Geography. Socio-Economic Geography). Akademichno izdatelstvo "Prof. Marin Drinov" (Prof. Marin Drinov Academic Publishing House), Sofia: 326–328 (in Bulgarian).

Nankinov, D., 2000. Range expansion and current breeding distribution of river warbler *Locustella fluviatilis* in Bulgaria. Sandgrouse 22(1):50–54.

Nankinov, D., 2001a. Die Lönnbergmöwe *Larus relictus*: eine neue Art für die Fauna der Westlichen Paläarktis. Ornithologische Mitteillungen 53(4):120–121.

Nankinov, D., 2001b. Die Lebensräume der Stummellerche *Calandrella rufescens* auf der Balkan-Halbinsel. Ornithologische Mitteillungen 53(8):260–263.

Nankinov, D., 2001c. Der Status des Mariskensängers in Bulgarien. Ornithologische Mitteillungen 53(9):306–313.

Nankinov, D. & S. Dalaktčijeva, 1999. First records of the Pallas's Warbler *Phylloscopus proregulus* in Bulgaria and reasoning about its influx into Europe. Acrocephalus 20(93):53–55.

Nankinov, D. & S. Dalaktčijeva, 2001. State of the stone curlew *Burhinus oedicnemus* in Bulgaria. Russian Journal of Ornithology 152:603–613.

Nankinov, D., S. Kirilov & K. Popov, 1989. Encountering the arctic warbler (*Phylloscopus borealis* Blasius, 1858) in Bulgaria for the first time. Larus (Zagreb) 40:163–166.

Nankinov, D., S. Simeonov, T. Mičev & B. Ivanov, 1997. [Aves. Part II]. In: Fauna na Balgariya. 26 (Fauna bulgarica. 26). Akademichno izdatelstvo "Prof. Marin Drinov" & Pensoft (Editio academica "Prof. Marin Drinov" & Pensoft), Sofia (in Bulgarian).

Nikolov, S., 2002. Izsledvaniya na ornitofaunata na Ponor planina [Study of the Ornithofauna of Ponor Mts.]. Diplomna rabota [Diploma Thesis]. Sofiiski universitet "Sv. Kliment Ohridski", Biologicheski fakultet [St. Kliment Ohridski University of Sofia, Biological Division], Sofia (in Bulgarian).

Patev, P., 1950. Ptitsite v Balgariya [The Birds of Bulgaria]. In Buresch, I. (ed), Fauna na Balgariya. 1 [Fauna bulgarica. 1). Balgarska akademiya na naukite [Bulgarian Academy of Sciences], Sofia (in Bulgarian).

Peshev, Ts. & N. Boev, 1962. Fauna na Balgariya – grabnachni. Kratak opredelitel [Fauna of Bulgaria – Vertebrates (A Short Field Guide)]. Narodna prosveta [National Education], Sofia (in Bulgarian).

Petrov, P., 1997. [Landscape structure]. In Yordanova, M. & D. Donchev (eds), Geografiya na Balgariya. Fizicheska geografiya. Sotsialno-ikonomicheska geografiya (Geography of Bulgaria. Physical Geography. Socio-Economic Geography). Akademichno izdatelstvo "Prof. Marin Drinov" (Prof. Marin Drinov Academic Publishing House), Sofia: 340–356 (in Bulgarian).

Petrov, Ts., 1980. [The birds of the Sredna Gora Mts.] Izvestiya na muzeite ot Yuzhna Balgariya [Proceedings of the Museums of South Bulgaria] 7:9–49 (in Bulgarian).

Pielou, E.C., 1984. The Interpretation of Ecological Data. A Primer on Classification and Ordination. Wiley, New York.

Radakoff, W.N., 1879. Ornithologische Bemerkungen über Bessarabien, Moldau, Walachei, Bulgarien und Ostrumelien. Bulletin de la Société imperiale des naturalistes de Moscou 13(53):150–178.

Reiser, O., 1894. Materialen zu einer Ornis balcanica. II. Bulgarien. Carl Gerold's Sohn, Wien.

Shurulinkov, P., B. Nikolov, G. Stoyanov & I. Nikolov, 2003. Erstes sicheres Brüten des Karmingimpels in Bulgarien. Ornithologishe Mitteilungen, 55:122–127.

Shurulinkov, P., B. Nikolov & R. Tsonev, 2001. On the distribution of the black-headed bunting (*Emberiza melanocephala*) in Bulgaria. In Tryjanowski, P., T. Osiejuk, & M. Kupczyk (eds), Proceedings 3rd International Ortolan Bunting Symposium (Bunting Studies in Europe). Bogucki Wydawnictwo Naukowe, Poznan: 81–87.

Shurulinkov, P. & R. Tsonev, 2001. First observation of the desert wheatear *Oenanthe deserti* in Bulgaria. Acrocephalus 22(104/105):53.

Simeonov, S., 1986. [The birds of the Pirin Mts.]. In Fauna na Yugozapadna Balgariya. Chast 1 (Fauna of Southwestern Bulgaria. Part 1). Izdatelstvo na Balgarskata akademiya na naukite (Publishing House of the Bulgarian Academy of Sciences), Sofia: 61–81 (in Bulgarian).

Simeonov, S. & V. Baeva, 1988. [Birds in the Ograzhden Mts.]. In Fauna na Yugozapadna Balgariya. Chast 2 (Fauna of Southwestern Bulgaria. Part 2). Izdatelstvo na Balgarskata akademiya na naukite (Publishing House of the Bulgarian Academy of Sciences), Sofia: 7–22 (in Bulgarian).

Simeonov, S. & V. Delov, 1989. [Birds of the Konyavska Mts.]. Acta zoologica bulgarica 38:65–81 (in Bulgarian).

Simeonov, S. & D. Dimitrov, 1994. [Contemporary state of the Lozen Mts. ornithofauna]. Godishnik na Sofiiskiya universitet "Sv. Kliment Ohridski". Biologicheski fakultet (Annuaire de l'Université de Sofia "St. Kliment Ohridski". Faculté de Biologie) 85(1 – Zoology): 253–262 (in Bulgarian).

Simeonov, S. & Y. Marinov, 1994. [The birds of the Osogova Mts.]. Godishnik na Sofiiskiya universitet "Sv. Kliment Ohridski". Biologicheski fakultet (Annuaire de l'Université de Sofia "St. Kliment Ohridski". Faculté de Biologie) 85(1 – Zoology): 237–252 (in Bulgarian).

Simeonov, S. & T. Michev, 1991. Ptitsite na Balkanskiya poluostrov. Polevi opredelitel [The Birds of the Balkan Peninsula. A Field Guide]. Petar Beron, Sofia (in Bulgarian).

Simeonov, P., T. Mitchev, G. Dobler & A. Crivelli, 2001. White Pelican breeding in Bulgaria. British Birds 94:390–392.

Simeonov, S., T. Mičev & D. Nankinov, 1990. [Aves. Part I]. In: Fauna na Balgariya. 20 (Fauna bulgarica. 20). Izdatelstvo na Balgarskata akademiya na naukite (Aedibus Academiae scientiarum bulgaricae), Sofia (in Bulgarian).

Snow, D. & C. Perrins, 1998. The Birds of the Western Palearctic. Concise Edition. Volume 1: Non-Passerines; Volume 2: Passerines. Oxford University Press, Oxford.

Spiridonov, G., 2000a. The ornithofauna of the Central Balkan National Park and its conservation value. In Sakalian, M. (ed), Biological Diversity of the Central Balkan National Park. Pensoft, Sofia: 393–423.

Spiridonov, G., 2000b. The ornithofauna of the Rila National Park and its conservation value. In Sakalian, M. (ed), Biological Diversity of the Rila National Park. Pensoft, Sofia: 361–388.

Stepanyan, L.S. & V.T. Butyev, 1967. [New data on the distribution of the short-toed treecreeper (*Certhia brachydactyla*) in the Caucasus]. Zoologicheskiy zhurnal [Zoological Journal] 46(6):960–961 (in Russian).

[Stoyanov, G.] Stojanov, G., 2001a. [Contribution to the ornithofauna of the Shumensko and Provadiisko Plateaus]. In Mircheva, M. & Z. Iliev (eds), 40 godini peshteren klub grad Shumen, 1961–2001 [Forty Years of the Cave Club of the Town of Shumen, 1961–2001]. Iskar, Sofia: 40–47 (in Bulgarian).

Stoyanov, G., 2001b. Die Maskengrasmücke – eine neue Art für bulgarische Ornithofauna. Ornithologische Mitteilungen 53(10):336–338.

Tongeren, O.F.R. van, 1987. Cluster analysis. In Jongman, R.H., C.J.F. ter Braak & O.F.R. van Tongeren (eds), Data Analysis in Community and Landscape Ecology. Pudoc, Wageningen, The Netherlands: 174–206.

Voous, K.H., 1960. Atlas van de Europese Vogels. Elsevier, Amsterdam & Brussel.

4 Amphibians and Reptiles of Bulgaria: Fauna, Vertical Distribution, Zoogeography, and Conservation

Boyan P. Petrov

National Museum of Natural History, Bulgarian Academy of Sciences, Tsar Osvoboditel Blvd. 1, 1000 Sofia, Bulgaria, e-mail: boyanpp@nmnh.bas.bg

Abstract: The geographical position, mild climate, mountainous landscape, and habitat heterogeneity determine Bulgaria's rich diversity of herpetofauna. Currently, 17 species of amphibians (20 subspecies) and 36 species of reptiles (45 subspecies) are known from Bulgaria. Due to its transitional position, Bulgaria lies at the southern or northern distribution limits of many species of amphibians and reptiles. Vertical distribution is presented for all species found in Bulgaria. The richest in terms of diversity and abundance are the areas below 400 m, where nearly 95% of taxa occur. The geographical ranges of almost all amphibians and reptiles in Bulgaria are well-known, although on a local scale the information is often limited. Zoogeographical analysis demonstrates that amphibian species belong to nine chorotypes, of which the Eastern Mediterranean and the European chorotypes are represented with three species each. Out of the ten chorotypes recognized for the reptiles, 51.4% of the species belong to the Eastern Mediterranean (25.7%) and Turanian–Mediterranean (25.7%) chorotypes. Only two subspecies are Bulgarian endemics (*Salamandra salamandra beschkovi* and *Mediodactylus kotschyi rumelicus*). Due to the high diversity and mosaic distribution of habitats in Bulgaria, the herpetofauna has a rather continuous occurrence and high relative abundance. Although 80% of the Bulgarian herpetofauna is legally protected by the Biodiversity Protection Act (2002) and other conventions, some species are threatened through collection for trade or habitat changes due to construction. Several sites in the southern part of Bulgaria are very rich in herpetofauna (nine to ten species of amphibians and 19 to 21 species of reptiles)

1 Introduction

Bulgaria is situated in the center of the eastern part of the Balkan Peninsula. Compared to other Balkan countries, Bulgaria has a relatively large territory, 110,993 km^2. The relief is very diverse starting from the sea level (Black Sea coast), through the hilly landscape to low and high mountains up to 2925 m high (Musala Peak in Rila Mts.). The territories below 200 m occupy 31.4% of Bulgaria; between 200 m and 600 m, 41%; between 600 and 1000 m, 15.5%; between 1000 and 1600 m, 9.8%; the mountains above 1600 m occupy only 2.5%. Stara Planina Mts. (the Balkan Range) divide Bulgaria into southern and northern parts. This partition directly influences the climate rather than the species composition and distribution of the herpetofauna (Beshkov, 1984b). Three main types of climate are characteristic for Bulgaria: continental (in the northernmost part of the country),

V. Fet and A. Popov (Eds.), Biogeography and Ecology of Bulgaria, 85–107.

transitional-continental (in the central part), and Submediterranean (in the south and along the Black Sea coast). All these environmental factors determine Bulgaria's rich biodiversity, including a diverse and abundant herpetofauna.

2 Brief History of the Herpetological Research in Bulgaria

Herpetological research in Bulgaria started relatively late, at the end of the 19th century when Hristovich (1892) published brief records for the amphibians and reptiles encountered during his natural surveys. New species and localities were published from time to time in the beginning of the 20th century. The first attempt to summarize the herpetological diversity of Bulgaria was by Kovachev (1912) who reported 13 species of amphibians and 26 species of reptiles. Two years later, Chichkoff (1914) found and published *Triturus alpestris* and *Typhlops vermicularis* as new species for the Bulgarian fauna. The most comprehensive data for the distribution of amphibians and reptiles in Bulgaria, and in the entire Balkan Peninsula, were published by Buresch and Zonkow (1933, 1934, 1941, 1942). These works became the foundation of the current herpetological knowledge for this region of Europe. In the next 50 years, the records of *Telescopus fallax, Rana esculenta* (Beshkov, 1959, 1965), and *Platyceps collaris* (= *Coluber rubriceps*) (Bartošik and Beshkov, 1979) were published as new for Bulgaria. Meanwhile, V. Beshkov, the most active Bulgarian herpetologist, published many papers on distribution, taxonomy, ecology, and conservation of the Bulgarian amphibians and reptiles. The most important contributions dealing with the Bulgarian herpetofauna are those by Beškov and Beron (1964) and Beshkov and Nanev (2002). Distribution and ecology of some species were studied in detail: *Salamandra salamandra* (Beshkov and Tsonchev, 1963), *Rana graeca* (Beshkov, 1970a, 1970b, 1972b), *R. temporaria* (Beshkov and Angelova, 1981), *Bombina variegata* (Beshkov and Jameson, 1980), land tortoises *Testudo* spp. (Beshkov, 1984a, 1993, 1997), *Anguis fragilis* (Beshkov, 1966), *Zamenis longissimus* (Beshkov, 1975, 1976), *Vipera ursinii* (Beshkov, 1973), *V. aspis* (Buresch and Beškov, 1965), and *V. ammodytes* (Beshkov, 1977; Christov and Beshkov, 1999).

In the recent years, herpetologists studied the composition and distribution of the herpetofauna in some regions along the southern border of Bulgaria (Stoev, 2000; Petrov et al., 2001; Tzankov, 2003). Other current contributions reported some rare species from previously unknown localities (e.g. *Telescopus fallax* from the Eastern Rhodopes; Petrov et al., 2002). The recent report by Hristov et al. (2004) of *Vipera ursinii*, considered extinct in Bulgaria, proved to be a hoax.

The majority of the papers on the Bulgarian herpetofauna were published by local herpetologists. However, many researchers from other countries also participated in the herpetological research on certain species. They studied taxonomic aspects, e.g. for *Triturus vulgaris graecus* (Obst and Geissler, 1982), *Ophisaurus apodus thracius* (Obst, 1978), *Mediodactylus kotschyi* (Rösler, 1999), *Platyceps collaris* (= *Coluber rubriceps*) *thracius* (Rehák, 1985); distribution (e.g. Cyrén, 1933, 1941; Geissler and Bruhl, 1980; Chlebicki, 1985; Noellert and Ritter, 1986a, 1986b), or ecology, e.g. of *Lacerta trilineata* (Peters, 1962, 1963) and *L. vivipara* (Guillaume et al., 1997).

3 Species Composition and Distribution

Bulgaria is inhabited by 17 species of amphibians (five species of tailed amphibians and 12 species of anurans) and 36 species of reptiles (four species of turtles, two species of tortoises, 13 species of lizards, and 17 species of snakes). At the subspecific level, amphibians are represented by 20 recognized subspecies (Table 1), and reptiles, by 45 subspecies (Table 2). While no amphibians became recently extinct in Bulgaria, some species of reptiles have not been found in this country for a long time. Two species of sea turtles (*Caretta caretta* and *Chelonia mydas*) are very rarely found at the Black Sea coast and thus are considered vagrant, non-native species. There is no contemporary field record of *Vipera aspis balcanica*, which was found in 1933 close to Harmanli, in the Bulgarian Eastern Rhodopes (Buresch and Zonkow, 1934). Its occurrence and systematic position were discussed by Buresch and Beškov (1965) who concluded that this snake could be considered extinct in Bulgaria.

3·1 Amphibians

Among the tailed amphibians, the most common within the aquatic habitats of Bulgaria are *Triturus vulgaris* and *T. karelinii*. The sister species of the latter, *T. dobrogicus* is known only along Danube. *Triturus vulgaris graecus* was found only in the southern parts of the Struma Valley (southwestern Bulgaria). *Salamandra salamandra beschkovi* is known only from its type locality. The nominate subspecies of the fire salamander is common in the hilly and mountainous regions but has not been found in the vast parts of the Danubian Plain and Thracian Lowland, along the Black Sea coast, and in Strandja Mts. *Triturus alpestris* and *Rana temporaria* occur only in the mountains. Species of *Bombina* very rarely occur sympatrically: *B. variegata* is very common in the hilly and mountainous areas, while *B. bombina* occurs only along rivers and wetlands in the plains and along the sea coast. The most widespread and abundant anurans in Bulgaria are *Rana ridibunda, R. dalmatina, Bufo viridis, B. bufo*, and *Hyla arborea*. Distribution of *Rana graeca* is restricted to the western and southern border mountains reaching the Eastern Rhodopes to the east. *R. esculenta* is found only at several sites along Danube but further research is needed to clarify its distribution. *Pelobates fuscus* is rarely found in the Danubian Plain and Sofia Basin. *P. syriacus* occurs in southwestern Bulgaria (south of Kresna), Thracian Lowland along Maritsa River, Strandja Mts., Black Sea coast, and in scattered localities along Danube. In conclusion, distribution of most amphibian species in Bulgaria is relatively well known. Detailed data are lacking for the distribution of *Bufo bufo spinosus, Rana esculenta, Triturus karelinii, T. vulgaris vulgaris, T. vulgaris graecus*, and *Salamandra salamandra beschkovi*.

3·2 Reptiles

Among turtles, *Emys orbicularis* is a common species in the lowlands of Bulgaria, and rarely is found above 1000 m. *Mauremys rivulata* occurs only in the southernmost parts of the country (southwestern Bulgaria, Eastern Rhodopes, Sakar Mts., Strandja Mts., and along the Black Sea coast south of Tsarevo). Both species of tortoises (*Testudo* spp.) occur mainly in the lowlands. A recent survey in the Eastern Rhodopes revealed that tortoises are still common in about one-third of the mountain territory due to the high variety and mosaic distribution of suitable habitats (Petrov, 2004). This region is therefore known to shelter the richest and most compact population of both tortoise species in Bulgaria (Petrov et al., 2004). It is worth mentioning that the record-sized *Testudo graeca* (carapace length, 389 mm; weight, 5.86–7 kg) and *T. hermanni* (carapace length, ca. 357 mm) known to date were found in Bulgaria (Beshkov, 1997).

Several species of lizards have scattered distribution and relatively low population densities, e.g. *Ablepharus kitaibelii* and *Darevskia praticola*, or inhabit disjunct ranges, e.g. *Ophisaurus apodus, Ophisops elegans* (Fig. 1), and *Lacerta vivipara*. The most common and widespread snakes in Bulgaria are *Natrix natrix* and *Dolichophis caspius*. Restricted ranges are known for *Platyceps collaris* (southern Bulgarian Black Sea coast, Fig. 2), *Elaphe*

Fig. 1 *Ophisops elegans* from Dolno Lukovo, Eastern Rhodopes; the snake-eyed lizard reaches here the northernmost limits of its European distribution.

Fig. 2 *Platyceps collaris* from Ropotamo; the red-headed whip-snake is found in Europe only in a very narrow range along the southern Bulgarian and Turkish Black Sea coast.

quatuorlineata quatuorlineata (southwestern Bulgaria, south of Kresna Gorge), and *E. quatuorlineata sauromates* (scattered in the eastern part of the country). *Zamenis longissimus* is known to occur in many localities but it is considered threatened by habitat changes. Due to its aggressive behavior and adaptability, *Malpolon monspessulanus* has notably increased its population density and probably its range, though it is restricted only to the south of the country (Beshkov, 1998).

Distribution of the majority of species is not limited by specific ecological constraints. Only few species require special types of habitats. The gecko *Mediodactylus kotschyi* occurs in Bulgaria almost entirely in human settlements. Microclimatic features are considered to confine its distribution. This species was found in the wild only in two places, in the rocks along the southern Black Sea coast. Occurrence of *Typhlops vermicularis* and *Eryx jaculus* within their small ranges is limited by the presence of dry and sandy soils. *Telescopus fallax* was mainly found only in the areas with screes and numerous rocks, and rarely in the sandstone terrain. The geographical ranges of most reptiles in Bulgaria are well known. However, almost no data are available for the local distribution of the rare and endangered species (e.g. *Ophisaurus apodus, Eryx jaculus, Elaphe quatuorlineata quatuorlineata, Zamenis situla, Vipera ursinii*). Regional abundance and population ecology of the latter species are poorly studied. National and local quantitative assessments were conducted in the past only for some species (e.g. *Mauremys rivulata, Testudo* spp.). Large areas in the Danubian Plain, Dobrudja, Eastern Stara Planina, Thracian Lowland, and the border mountain ranges Slavyanka and Belasitsa are inadequately studied.

Diversity of the habitats in Bulgaria is extremely high, and the distribution of herpetofauna within the landscapes is highly mosaic. Such a mosaic habitat pattern is among the basic factors determining a rather continuous distribution of the herpetofauna in Bulgaria. Some of the geographical districts have low herpetological diversity and population densities due to intensive agriculture, habitat fragmentation, or development of large open mining areas (e.g. in parts of the Upper Thracian Lowland and Dobrudja).

There are no introduced species of amphibians and reptiles in Bulgaria, which would have become widespread or locally adapted. At present, this could be explained by the weakly developed market of terrarium-held herpetospecies and a generally negative attitude towards these groups as pet animals. However, introduction of native species in the areas or altitudes where they do not naturally occur is known (e.g. release of homebred *Testudo* spp. in the mountains above 1500 m). With few exceptions of species that still could be found (e.g. *Rana balcanica* in southwestern Bulgaria, *Eremias arguta* in Bulgarian Dobrudja, *Vipera xanthina* in Bulgarian Thrace), no other species of amphibians and reptiles are expected to be discovered in Bulgaria.

4 Altitudinal Distribution

Vertical distribution of the amphibians and reptiles in Bulgaria has never been summarized in detail. Beškov and Beron (1964) outlined the basic data known by that time. Few regional attempts revealed some patterns at a local scale (e.g. Beshkov, 1974; Beshkov and Stoyanov, 2000). Beshkov (1989) presented brief data on the occurrence of some species in the mountains. The most recent descriptive review was done by Beshkov and Nanev (2002).

Tables 1 and 2 show distribution and relative abundance of all species and subspecies found in Bulgaria. Distribution is presented along 14 (Amphibia) and 15 (Reptilia) 200 m altitudinal belts. The species are sorted according to their abundance within these belts. Abundance is relative for each belt and is based upon the field records published in the literature. The systematic arrangement follows Arnold and Ovenden (2002). Question marks indicate presumable distribution. Zeros indicate absence of the taxon from a given altitudinal belt. High altitude occurrence of some taxa (e.g. *Rana dalmatina, Lacerta agilis, Darevskia praticola, Ablepharus kitaibelii*) is based on the unpublished records of Vladimir Beshkov and Nikolay Tzankov.

4.1 Amphibians (Table 1)

The regions below 200 m are the richest in species and their abundance (18 taxa, 15 species). Five taxa, *Triturus vulgaris graecus, T. dobrogicus, Rana* kl. *esculenta, Pelobates syriacus*, and *Bombina bombina*, are known to occur only below 400 m. Eight taxa (40%) do not climb higher than 600 m. Up to 1200 m, ten species are still abundant and common. Five species are common between 1200 m and

Table 1 Bulgarian amphibians: relative abundance and altitudinal distribution. 0, absent; 1, rare; 2, common; 3, occasional; ?, no data. Altitude in meters

Taxa	0–200	200–400	400–600	600–800	800–1000	1000–1200	1200–1400	1400–1600	1600–1800	1800–2000	2000–2200	2200–2400	2400–2600	2600–2800
Triturus vulgaris graecus	1	?	?	?	0	0	0	0	0	0	0	0	0	0
Triturus dobrogicus	2	?	0	0	0	0	0	0	0	0	0	0	0	0
Rana kl. *esculenta*	2	0	0	0	0	0	0	0	0	0	0	0	0	0
Pelobates syriacus balcanicus	2	1	0	0	0	0	0	0	0	0	0	0	0	0
Bombina bombina bombina	2	1	?	0	0	0	0	0	0	0	0	0	0	0
Bufo bufo spinosus	2	2	1	?	0	0	0	0	0	0	0	0	0	0
Pelobates fuscus fuscus	2	2	1	0	0	0	0	0	0	0	0	0	0	0
Salamandra salamandra beschkovi	1	2	2	?	?	?	0	0	0	0	0	0	0	0
Rana graeca	2	2	2	2	2	1	0	0	0	0	0	0	0	0
Triturus vulgaris vulgaris	2	2	2	2	2	2	1	?	0	0	0	0	0	0

continued

Table 1 continued

Taxa	0–200	200–400	400–600	600–800	800–1000	1000–1200	1200–1400	1400–1600	1600–1800	1800–2000	2000–2200	2200–2400	2400–2600	2600–2800
Triturus karelinii	2	2	2	2	2	2	1	?	0	0	0	0	0	0
Hyla arborea arborea	2	2	2	2	2	1	1	1	0	0	0	3	0	0
Rana dalmatina dalmatina	2	2	2	2	2	2	1	1	1	0	0	0	0	0
Rana ridibunda	2	2	2	2	2	2	2	1	1	1	0	0	0	0
Bufo viridis viridis	2	2	2	2	2	2	2	1	1	1	0	3	0	0
Salamandra salamandra salamandra	1	1	2	2	2	2	2	2	1	1	3	3	0	0
Bufo bufo bufo	2	2	2	2	2	2	1	1	1	1	3	0	0	0
Bombina variegata scabra	2	2	2	2	2	2	1	1	1	1	1	0	0	0
Rana temporaria temporaria	0	1	0	1	1	1	2	2	2	2	2	2	1	0
Triturus alpestris	0	0	0	0	0	?	2	2	2	2	2	2	1	0

1400 m, and only two, above 1600 m. Species reaching the highest altitudes are *Triturus alpestris* (2500 m) and *Rana temporaria* (ca. 2500 m), which are considered glacial relicts in the fauna of Bulgaria. Unusual vertical breeding migration from 1400 m down to 360 m and back was recorded for *Rana temporaria* (Beshkov and Angelova, 1981). Findings of *Salamandra salamandra salamandra* and *Bufo bufo* above 2000 m are occasional, and accidental for *Hyla arborea* and *Bufo viridis* at 2300 m. *Salamandra salamandra salamandra, Bufo bufo bufo, Bombina variegata*, and *Rana temporaria* have the greatest vertical tolerance and occur in 11 belts, i.e. within a range of 2200 m. *Bombina variegata* and *Rana ridibunda* are the only two species which can be found in sympatry with all other anurans recorded in Bulgaria (Beshkov, 1972a).

4·2 Reptiles (Table 2)

Beshkov (1974) assessed diversity of snakes in a region located in south-western Bulgaria between the altitudes of 180 m and 600 m where he has found 11 species. High diversity of six lacertid lizard species was observed in a marginal area of the Thracian Lowland (village Mezek) between 180 m and 500 m (Tzankov, 2003).

In general, abundance of the species (e.g. *Mediodactylus kotschyi, Ophisops elegans, Typhlops vermicularis*) for which Bulgaria lies in the northern border of their European distribution is higher below 400 m of altitude. The species belonging to the Turanian–Mediterranean chorotype complex live in the lowest, warmest parts of the country, and only some occasionally reach 1000–1400 m. While the species of the Mediterranean complex (*Eryx jaculus* and *Malpolon monspessulanus*) occur strictly below 500 m, most of the species of the South European chorotype complex (*Testudo hermanni, Lacerta viridis, Podarcis muralis*, and *Zamenis longissimus*) have considerably greater altitudinal range (0 to 1400–2000 m).

The species diversity of the reptiles in Bulgaria is the greatest below 400 m where nearly 95% of the taxa occur (sea turtles are not considered). *Platyceps collaris, Mauremys rivulata, Mediodactylus kotschyi rumelicus, M. kotschyi danilevskii, Typhlops vermicularis*, and *Eryx jaculus* have never been found above 400 m. Species richness and abundance between 400 m and 600 m is still very high (22 taxa). Slight decrease in the abundance of reptiles is visible between 800 m and 1000 m where only eight taxa have rich populations. The reptiles have low abundance above 1400 m (only four taxa are common at these altitudes). The poorest region in terms of the species richness and abundance are the mountains above 2000 m, where only *Vipera berus, Lacerta agilis bosnica, Podarcis muralis muralis* (occasional), and *Lacerta vivipara* occur. The latter is the species which reaches the highest altitude in Bulgaria (2900 m). The most widely occurring taxa are *P. muralis muralis* and *L. vivipara* which are found in 11 altitudinal belts (i.e. in the range of 2200 m).

Table 2 Bulgarian reptiles: relative abundance and altitudinal distribution. 0, absent; 1, rare; 2, common; 3, occasional; ?, no data. Altitude in meters

Taxa	0–200	200–400	400–600	600–800	800–1000	1000–1200	1200–1400	1400–1600	1600–1800	1800–2000	2000–2200	2200–2400	2400–2600	2600–2800	2900
Caretta caretta	?	0	0	0	0	0	0	0	0	0	0	0	0	0	0
Chelonia mydas	?	0	0	0	0	0	0	0	0	0	0	0	0	0	0
Platyceps collaris thracius	2	0	0	0	0	0	0	0	0	0	0	0	0	0	0
Mauremys rivulata	2	?	0	0	0	0	0	0	0	0	0	0	0	0	0
Mediodactylus kotschyi rumelicus	2	1	0	0	0	0	0	0	0	0	0	0	0	0	0
Typhlops vermicu-laris	2	2	0	0	0	0	0	0	0	0	0	0	0	0	0
Eryx jaculus turcicus	1	2	0	0	0	0	0	0	0	0	0	0	0	0	0
Mediodactylus kotschyi danilevskii	2	1	0	0	0	0	0	0	0	0	0	0	0	0	0
Mediodactylus kotschyi bibroni	2	1	1	0	0	0	0	0	0	0	0	0	0	0	0
Anguis fragilis colchicus	2	2	1	0	0	0	0	0	0	0	0	0	0	0	0

Ophisaurus apodus thracius	2	2	1	0	0	0	0	0	0	0	0	0	0	0	0
Elaphe quatuorlineata quatuorlineata	2	2	1	0	0	0	0	0	0	0	0	0	0	0	0
Elaphe quatuorlineata sauromates	2	2	1	0	0	0	0	0	0	0	0	0	0	0	0
Zamenis situla	2	2	1	0	0	0	0	0	0	0	0	0	0	0	0
Malpolon monspessulanus insignitus	2	2	1	0	0	0	0	0	0	0	0	0	0	0	0
Lacerta trilineata dobrogica	2	2	1	0	0	0	0	0	0	0	0	0	0	0	0
Telescopus fallax fallax	2	2	1	1	0	0	0	0	0	0	0	0	0	0	0
Ophisops elegans macrodactylus	2	2	1	1	0	0	0	0	0	0	0	0	0	0	0

continued

Table 2 continued

Taxa	0–200	200–400	400–600	600–800	800–1000	1000–1200	1200–1400	1400–1600	1600–1800	1800–2000	2000–2200	2200–2400	2400–2600	2600–2800	2900
Lacerta trilineata trilineata	2	2	1	1	0	0	0	0	0	0	0	0	0	0	0
Podarcis taurica taurica	2	2	2	?	0	0	0	0	0	0	0	0	0	0	0
Platyceps najadum dahlii	2	2	2	?	0	0	0	0	0	0	0	0	0	0	0
Natrix natrix persa	2	2	2	3	0	0	0	0	0	0	0	0	0	0	0
Lacerta viridis merid-ionalis	2	2	2	1	?	0	0	0	0	0	0	0	0	0	0
Lacerta agilis cherson-ensis	1	1	2	1	0	0	0	0	0	0	0	0	0	0	0
Darevskia praticola pontica	1	2	2	1	1	0	0	0	0	0	0	0	0	0	0
Vipera ursinii moldavica	0	1	2	2	0	0	0	0	0	0	0	0	0	0	0
Testudo graeca ibera	2	2	2	1	1	1	1	0	0	0	0	0	0	0	0

Testudo hermanni boettgeri	2	2	2	1	1	1	1	1	0	0	0	0	0	0	0
Emys orbicularis	2	2	2	1	1	1	?/3	0	0	0	0	0	0	0	0
Natrix tessellata tessellata	2	2	2	2	1	1	0	0	0	0	0	0	0	0	0
Vipera ammodytes merid-ionalis	2	2	2	2	1	1	1	0	0	0	0	0	0	0	0
Podarcis erhardii riveti	2	2	2	2	1	1	1	1	0	0	0	0	0	0	0
Podarcis muralis albanica	2	2	2	1	3	3	3	3	3	0	0	0	0	0	0
Lacerta viridis viridis	2	2	2	2	2	1	1	1	1	0	0	0	0	0	0
Ablepharus kitaibelii stepaneki	2	2	2	2	2	1	1	1	0	0	0	0	0	0	0
Vipera ammodytes ammodytes	1	2	2	2	1	1	1	1	0	0	0	0	0	0	0

continued

Table 2 continued

Taxa	0–200	200–400	400–600	600–800	800–1000	1000–1200	1200–1400	1400–1600	1600–1800	1800–2000	2000–2200	2200–2400	2400–2600	2600–2800	2900
Dolichophis caspius	2	2	2	2	1	1	1	1	0	0	0	0	0	0	0
Natrix natrix natrix	1	2	2	2	2	1	1	1	1	1	0	0	0	0	0
Coronella austriaca austriaca	1	2	2	2	2	2	1	1	1	1	0	0	0	0	0
Zamenis longis-simus longis-simus	2	2	2	2	2	2	1	1	1	1	0	0	0	0	0
Podarcis muralis muralis	1	2	2	2	2	1	1	1	1	1	1	0	0	0	0
Anguis fragilis fragilis	0	1	1	2	2	2	2	2	1	1	0	0	0	0	0
Lacerta agilis bosnica	0	?	1	1	2	2	2	2	1	1	1	1	1	0	0
Vipera berus	0	0	0	?	1	1	1	2	2	2	2	2	1	1	0
Lacerta vivipara	0	0	0	0	1/?	1	2	2	2	2	2	2	2	1	1

5 Zoogeographical Analysis of the Herpetofauna of Bulgaria

There are only a few conflicting attempts to present zoogeographical classification of the Bulgarian herpetofauna. Buresch and Zonkow (1933, 1934, 1941, 1942) provided brief zoogeographical characteristics for each species reported at that time in Bulgaria. Beškov and Beron (1964) have classified all herpetospecies but did not discuss their relations and affinities. Popov (1997) has grouped Bulgarian amphibians and reptiles according to their ecological preferences in two main complexes:

Mesophilous forest fauna. Boreal elements: *Vipera berus, Lacerta vivipara, Rana temporaria, Triturus vulgaris*; Nemoral elements: *Bufo bufo, B. viridis, Hyla arborea, Anguis fragilis, Natrix natrix, Coronella austriaca*; Steppe elements: *Vipera ursinii, Lacerta agilis*; Broad-leaved deciduous forest elements: *Salamandra salamandra, Triturus cristatus, T. alpestris, Pelobates fuscus, Bombina bombina, B. variegata*; South European elements: *Rana dalmatina, Lacerta viridis, Zamenis longissimus, Elaphe quatuorlineata, Testudo hermanni*; Balkan, Pontian and Southeast European elements: *Darevskia praticola, Podarcis taurica, Dolichophis caspius*.

Thermoxerophilous forest fauna: *Rana ridibunda, R. graeca, Pelobates syriacus, Mauremys rivulata, Emys orbicularis, Testudo graeca, Mediodactylus kotschyi, Podarcis muralis, P. erhardii, Ablepharus kitaibelii, Lacerta trilineata, Ophisaurus apodus, Zamenis situla, Telescopus fallax, Malpolon monspessulanus, Eryx jaculus, Platyceps najadum, P. collaris, Typhlops vermicularis, Natrix tessellata, Vipera ammodytes*.

Our present analysis follows the chorotype classification of Vigna Taglianti et al. (1999), which among the other considered groups was based upon the distribution of many Western Palearctic species of amphibians and reptiles. We also considered the approach of Sindaco et al. (2000) who have further clarified and detailed the chorotypes of the reptiles of Anatolia. General distribution of the species is based on the data of Gasc et al. (1997) and Engelmann et al. (1993). Zoogeographical classification of the Bulgarian herpetofauna is presented in Table 3.

5·1 Amphibians

The Bulgarian amphibians are classified into nine main chorotypes (Table 3). Compared to the species richness of the Bulgarian reptiles, the amphibians belong to greater number of chorotypes considering the total number of species (17 species, i.e. 1.89 species per chorotype). Six species are classified into Eastern Mediterranean and European chorotypes (three species, or 17.6%, each). Eight species are grouped in four chorotypes (two species, or 11.8%, each): Turanian–European–Mediterranean, European–Mediterranean, South European, and Central European.

Table 3 Chorotype classification of Bulgarian amphibians and reptiles. Chorotype names after Vigna Taglianti et al. (1999) with modifications

Chorotype	Number of species amphibians	reptiles	Species
Eurosiberian	1	2	*Rana temporaria, Lacerta vivipara, Vipera berus*
Turanian–European–Mediterranean	2	1	*Rana ridibunda, Bufo viridis, Emys orbicularis*
Central Asian–European–Mediterranean	–	1	*Natrix natrix*
Central Asian–European	1	3	*Pelobates fuscus, Lacerta agilis, Natrix tessellata, Vipera ursinii*
Turanian–Mediterranean	1	9	*Pelobates syriacus, Testudo graeca, Mauremys rivulata, Ophisaurus apodus, Ophisops elegans, Typhlops vermicularis, Elaphe quatuorlineata, Dolichophis caspius, Platyceps najadum, Telescopus fallax*
European–Mediterranean	2	–	*Salamandra salamandra, Hyla arborea*
European	3	2	*Triturus alpestris, T. vulgaris, Bufo bufo, Anguis fragilis, Coronella austriaca*
South European	2	4	*Bombina variegata, Rana dalmatina, Testudo hermanni, Lacerta viridis, Podarcis muralis, Zamenis longissimus*
Central European	2	–	*Bombina bombina, Rana* kl. *esculenta*
Mediterranean	–	2	*Eryx jaculus, Malpolon monspessulanus*
Eastern Mediterranean	3	9	*Triturus dobrogicus, T. karelinii, Rana graeca, Mediodactylus kotschyi, Lacerta trilineata, Darevskia praticola, Podarcis erhardii, P. taurica, Ablepharus kitaibelii, Platyceps collaris, Zamenis situla, Vipera ammodytes*
Cosmopolitan	–	2	*Caretta caretta, Chelonia mydas*
Total species	17	35	

Three species belong to Eurosiberian (*Rana temporaria*), Central Asian–European (*Pelobates fuscus*), and Turanian–Mediterranean (*Pelobates syriacus*) chorotypes. Bulgaria and its adjacent border areas with Greece and Turkey are the southernmost limits of the ranges of *Bombina bombina, Pelobates fuscus*, and *Rana temporaria*. The frog species *Rana graeca* is endemic to the southern Balkan Peninsula.

5·2 Reptiles

The Bulgarian reptiles are classified in ten main chorotypes (Table 3). With regard to the total number of species, the chorotype diversity (3.5 species per chorotype) is lower compared to the amphibians. Since only the main chorotypes were taken into account, some species were assigned to a much broader chorotype than they actually inhabit. For instance, *Darevskia praticola* has a Pontian–Caucasian distribution (areas around the Black Sea) and thus the species was placed into the Eastern Mediterranean chorotype. *Podarcis erhardii* was also classified in the latter although it is endemic for the southern part of the Balkan Peninsula (southern Albania, Macedonia, southern Bulgaria, Greece, and Crete). *Lacerta viridis bilineata* (i.e. western populations of *L. viridis*) is considered a subspecies (not as in Sindaco et al., 2000). The analysis does not account for *Vipera aspis* because it is considered extinct in Bulgaria. Both species of sea turtles are included as cosmopolitan species.

Due to the geographic position of Bulgaria, it is not surprising that 51.4% of the species belong to the Eastern Mediterranean (25.7%) and Turanian–Mediterranean chorotypes (25.7%) (Table 3). Most of the species belonging to these chorotypes occur in the Southern zoogeographical region of Bulgaria (including Struma–Mesta, Thracian, Strandja, and Black Sea coast subregions) as defined in the zoogeographical division of Guéorguiev (1997). Being the warmest part of the country, these subregions are the richest in terms of the species richness and abundance of reptiles in Bulgaria. The remaining eight chorotypes are represented by one to four species each (2.9% to 11.4%). The only two Eurosiberian species, *Lacerta vivipara* and *Vipera berus*, are common only in the high mountains above 1200 m. The Bulgarian Western Rhodopes are the southernmost point within the range of *Lacerta vivipara*.

The endemic Bulgarian taxa are *Salamandra salamandra beschkovi* Obst, 1981 and *Mediodactylus kotschyi rumelicus* (Müller et al., 1939). The first was described from the vicinity of Sandanski (southwestern Bulgaria), the second from the town of Plovdiv. Although both taxa are regarded as valid, *S. salamandra beschkovi* is considered a subspecies whose status needs further clarification. *Mediodactylus kotschyi rumelicus* is one of the three known subspecies of the Kotschy's gecko, which are entirely allopatric in Bulgaria. *Ophisaurus apodus thracius* Obst, 1978 and *Platyceps collaris thracius* Rehák, 1985 are Bulgarian subendemics. Both taxa occur also in European Turkey; the first one is found also in Greece, Macedonia, Albania, Montenegro, Croatia, and Bosnia and Herzegovina.

6 Conservation of the Bulgarian Herpetofauna

Although 80% of the Bulgarian herpetofauna is legally protected by the 2002 Biodiversity Protection Act (Table 4), some species are severely threatened. Both tortoise species are endangered at the national scale because of the rapid population decline due to collection for consumption and trade (Petrov et al., 2004). The public

Table 4 Protected status of the Bulgarian herpetofauna according to the national and international legislation (marine turtles excluded)

Species	BG	Number of protected species					
		BPA	RDB	92/43	BERN	CITES	IUCN
Salamanders and newts	5	5	1	2	5	–	2
Frogs and toads	12	12	1	11	12	–	2
Turtles	2	2	1	2	2	–	1
Tortoises	2	2	–	2	2	2	2
Lizards	13	6	2	10	13	–	–
Snakes	16	13	6	12	16	1	2
Total	50	40	11	39	50	3	9

BG, number of species found in Bulgaria; BPA, 2002 Biodiversity Protection Act [Bulgaria]; RDB, Red Data Book of Bulgaria, 1985; 92/43, Habitats Directive 92/43/EEC; BERN, Bern Convention; CITES, Convention on the International Trade in Endangered Species of Wild Fauna and Flora; IUCN, 2003 IUCN Red List of Threatened Animals (www.redlist.org)

concern and awareness toward tortoises were raised again in the recent years and this led to confiscation of 200–400 adult individuals annually from poachers and merchants. For many years tortoises were known to be very common all over the country's lowlands and hilly areas (Beshkov, 1984a, 1993). At present most of the populations became isolated and the number of mature individuals tends to be critically low.

Some species of snakes (e.g. *Zamenis situla, Eryx jaculus, Telescopus fallax*) with restricted distribution in Bulgaria are regularly collected for trade, and some of their populations were hunted to the brink of extinction (e.g. *Zamenis situla* in the vicinity of Sozopol). On the other hand, although listed in the 2003 IUCN Red List of Threatened Animals, *Bombina bombina, Hyla arborea*, and *Emys orbicularis* still have high population densities nearly in all their Bulgarian localities. The Red Data Book of Bulgaria is rather old (published in 1985) and its outdated species list is based on surveys made before 1984. A new, revised edition is currently under preparation.

Amphibians are mostly threatened by loss and fragmentation of proper aquatic habitats. The herpetofauna in general is vulnerable to nearly all kinds of construction activities that alter natural habitats and destroy their microhabitats and breeding sites in particular. The latter is a serious threat along the Black Sea coast where at many places the rapid development of recreation infrastructure led to habitat fragmentation and loss of connectivity between the sites with herpetological importance. Construction of new roads and highways (e.g. in Kresna Gorge, southwestern Bulgaria) without detailed study of their environmental impact could be a serious threat for the herpetofauna, especially in narrow gorges and river valleys with steep slopes.

In Bulgaria, so far only one natural wetland site (Muhalnitsa Swamp, located in the Western Stara Planina) was declared protected because of their importance for the breeding of *Rana temporaria* and *Triturus alpestris*. Several relatively small sites (ca. 3–5 km^2) in the southern half of Bulgaria are known to hold great herpetological richness. The latter is mostly due to the very high habitat heterogeneity and the mosaic distribution of proper microhabitats. In the Kresna Gorge (southwestern Bulgaria), ten species of amphibians and 21 species of reptiles are known (Petrov and Beshkov, 2001). The valley of Byala Reka River in the Eastern Rhodopes is another site where, respectively, nine and 19 species were found (Petrov et al., 2001). The third most important site is Ropotamo–Maslen Nos (Southern Black Sea coast) which is inhabited by nine species of amphibians and 19 species of reptiles (Petrov, in press). Species diversity of these sites is very high not only at the national and Balkan level but on the European scale as well. In order to protect the rich biodiversity, parts of these sites were declared as reserves (Tisata Reserve in Kresna Gorge; Ropotamo Reserve at the Black Sea coast) or protected areas (Meanders of Byala Reka in the Eastern Rhodopes). However, none of these sites was specially designated for protection of vulnerable or endangered amphibian or reptile species. In order to set priorities for further conservation efforts, a List of Important Herpetofaunal Areas in Bulgaria was recently prepared (Petrov, in press).

Acknowledgments

I thank Vladimir Beshkov (Sofia) for his support in my herpetological studies and for reviewing the vertical distribution of the herpetofauna. The help of Pavel Stoev and Nikolay Tzankov (both NMNH, Sofia) in data analyses and compilation is greatly acknowledged.

References

Arnold, E. & D. Ovenden, 2002. A Field Guide to the Reptiles and Amphibians of Britain and Europe. Harper Collins, London.

Bartošik, M. & V. Beshkov, 1979. [A new species for the fauna of Bulgaria]. Priroda i znanie [Nature and Knowledge] 30(6):39 (in Bulgarian).

Beshkov, V., 1959. [A new snake for the fauna of Bulgaria]. Priroda [Nature] (Sofia) 8(1):89 (in Bulgarian).

Beshkov, V., 1965. [Discovery of the edible frog (*Rana esculenta* L.) in Bulgaria]. Izvestiya na Zoologicheskiya institut s muzei (Bulletin de l'Institut de zoologie et musée) 19:45–54 (in Bulgarian).

Beshkov, V., 1966. [Research on the taxonomy and distribution of the slow worm (*Anguis fragilis* L.) in Bulgaria]. Izvestiya na Zoologicheskiya institut s muzei (Bulletin de l'Institut de zoologie et musée) 21:185–201 (in Bulgarian).

Beshkov, V., 1970a. [Biology and distribution of the stream frog (*Rana graeca* Blgr.) in Bulgaria. I. Study of the food and feeding]. Izvestiya na Zoologicheskiya institut s muzei (Bulletin de l'Institut de zoologie et musée) 31:5–17 (in Bulgarian).

Beshkov, V., 1970b. [Biology and distribution of the stream frog (*Rana graeca* Blgr.) in Bulgaria. II. Study of the reproduction and larvae]. Izvestiya na Zoologicheskiya institut s muzei (Bulletin de l'Institut de zoologie et musée) 32:159–180 (in Bulgarian).

Beshkov, V., 1972a. [Interspecific interactions between the frogs and toads in Bulgaria]. Izvestiya na Zoologicheskiya institut s muzei (Bulletin de l'Institut de zoologie et musée) 34:85–95 (in Bulgarian).

Beshkov, V., 1972b. [Biology and distribution of the stream frog (*Rana graeca* Blgr.) in Bulgaria. III. Study of ecology and distribution]. Izvestiya na Zoologicheskiya institut s muzei (Bulletin de l'Institut de zoologie et musée) 36:126–136 (in Bulgarian).

Beshkov, V., 1973. [On the subspecific status of Bulgarian *Vipera ursinii* (Bonaparte, 1835) (Serpentes)]. Izvestiya na Zoologicheskiya institut s muzei (Bulletin de l'Institut de zoologie et musée) 37:103–112 (in Bulgarian).

Beshkov, V., 1974. [Vertical distribution of the snakes in a region with specific species composition in Southwestern Bulgaria]. Izvestiya na Zoologicheskiya institut s muzei (Bulletin de l'Institut de zoologie et musée) 40:167–173 (in Bulgarian).

Beshkov, V., 1975. [A study of biology and ecology of the snakes in Maleshevska Planina Mts. (Southwestern Bulgaria). I. Reproduction of the Aesculapian snake (*Elaphe longissima* (Laur.)]. Ekologiya (Ecology) (Sofia) 1:75–83 (in Bulgarian).

Beshkov, V., 1976. [A study of biology and ecology of the snakes in Maleshevska Planina Mts. (Southwestern Bulgaria). II. Food and feeding of the Aesculapian snake (*Elaphe longissima* (Laur.)]. Ekologiya (Ecology) (Sofia) 2:34–42 (in Bulgarian).

Beshkov, V., 1977. [A study of biology and ecology of the snakes in Maleshevska Planina Mts. (Southwestern Bulgaria). III. Food and reproduction of the Nose-horned viper (*Vipera ammodytes meridionalis* Boulenger)]. Ekologiya (Ecology) (Sofia) 4:3–12 (in Bulgarian).

Beshkov, V., 1984a. [On the distribution, relative abundance and protection of tortoises in Bulgaria]. Ekologiya (Ecology) (Sofia) 14:14–34 (in Bulgarian).

Beshkov, V., 1984b. [The effect of the Balkan range on the distribution of the herpetofauna of Bulgaria]. Acta zoologica bulgarica 25:9–15 (in Bulgarian).

Beshkov, V., 1989. [Amphibians and reptiles]. In Mishev, K. (ed), Prirodniyat i ikonomicheskiyat potentsial na planinite v Balgariya. Tom I. Priroda i resursi (Natural and Economic Potential of the Mountains in Bulgaria. Volume I. Nature and Resources). Izdatelstvo na Balgarskata akademiya na naukite (Publishing House of the Bulgarian Academy of Sciences), Sofia: 397–398 (in Bulgarian).

Beshkov, V., 1993. On the distribution, relative abundance and protection of tortoises in Bulgaria. Chelonian Conservation & Biology 1(1):53–62.

Beshkov, V., 1997. Record-sized tortoises, *Testudo graeca ibera* and *Testudo hermanni boettgeri*, from Bulgaria. Chelonian Conservation & Biology 2(4), 593–596.

Beshkov, V., 1998. Bulgaria's amphibians and reptiles. In Meine, C. (ed), Bulgaria's Biological Diversity: Conservation Status and Needs Assessment. Volumes I and II. Pensoft, Sofia & Moscow: 395–409.

Beshkov, V. & B. Angelova, 1981. [An unusual reproductive migration of the common frog (*Rana temporaria* L.)]. Ekologiya (Ecology) (Sofia) 8:34–42 (in Bulgarian).

[Beshkov, V.] Beškov, V & P. Beron, 1964. *Catalogue et bibliographie des amphibiens et des reptiles en Bulgarie*. Éditions de l'Académie bulgare des sciences, Sofia.

Beshkov, V. & D. Jameson, 1980. Movement and abundance of the yellow-bellied toad *Bombina variegata*. Herpetologica 36(4):365–370.

Beshkov, V. & K. Nanev, 2002. Zemnovodni i vlechugi v Balgariya [Amphibians and Reptiles in Bulgaria]. Pensoft, Sofia & Moscow (in Bulgarian).

Beshkov, V. & A. Stoyanov, 2000. The amphibians of the glacial lakes in the Rila Mountains. In Golemansky, V. & W. Naidenow (eds), Biodiversity and Evolution of Glacial Water Ecosystems in the Rila Mountains. Institute of Zoology, Sofia: 79–82.

Beshkov, V. & T. Tsonchev, 1963. [The fire salamander (*Salamandra salamandra* L.) in Vitosha Mts.]. Izvestiya na Zoologicheskiya institut s muzei (Bulletin de l'Institut de zoologie et musée) 13:79–91 (in Bulgarian).

Buresch, I. & V. Beškov, 1965. [Wird die Giftschlange *Vipera aspis* L. in Bulgarien angetroffen?] Izvestiya na Zoologicheskiya institut s muzei (Bulletin de l'Institut de zoologie et musée) 18:5–30 (in Bulgarian).

Buresch, I. & J. Zonkow, 1933. [Untersuchungen über die Verbreitung der Reptilien und Amphibien in Bulgarien und auf der Balkanhalbinsel. I Teil: Schildkrötten (Testudinata) und Eidechsen (Sauria)]. Izvestiya na Tsarskite prirodonauchni instituti v Sofia (Mitteilungen aus den Königlichen naturwissenschaftlichen Instituten in Sofia – Bulgarien; Bulletin des Institutions royales d'histoire naturelle à Sofia – Bulgarie) 6:150–207 (in Bulgarian).

Buresch, I. & J. Zonkow, 1934. [Untersuchungen über die Verbreitung der Reptilien und Amphibien in Bulgarien und auf der Balkanhalbinsel. II Teil: Schlangen (Serpentes)]. Izvestiya na Tsarskite prirodonauchni instituti v Sofia (Mitteilungen aus den Königlichen naturwissenschaftlichen Instituten in Sofia – Bulgarien; Bulletin des Institutions royales d'histoire naturelle à Sofia – Bulgarie) 7: 106–188 (in Bulgarian).

Buresch, I. & J. Zonkow, 1941. [Untersuchungen über die Verbreitung der Reptilien und Amphibien in Bulgarien und auf der Balkanhalbinsel. III Teil: Schwanzlurche (Amphibia, Caudata)]. Izvestiya na Tsarskite prirodonauchni instituti v Sofia (Mitteilungen aus den Königlichen naturwissenschaftlichen Instituten in Sofia – Bulgarien; Bulletin des Institutions royales d'histoire naturelle à Sofia – Bulgarie) 14:171–237 (in Bulgarian).

Buresch, I. & J. Zonkow, 1942. [Untersuchungen über die Verbreitung der Reptilien und Amphibien in Bulgarien und auf der Balkanhalbinsel. IV Teil: Froschlurche (Amphibia, Salientia)]. Izvestiya na Tsarskite prirodonauchni instituti v Sofia (Mitteilungen aus den Königlichen naturwissenschaftlichen Instituten in Sofia – Bulgarien; Bulletin des Institutions royales d'histoire naturelle à Sofia – Bulgarie) 15:68–145 (in Bulgarian).

Chichkoff, G., 1914. [Two new species for the Bulgarian herpetology]. Godishnik na Sofiiskiya universitet. Fiziko-matematicheski fakultet (Annuaire de l'Université de Sofia. Faculté physico-mathématique) 8–9(2):1–9 (in Bulgarian).

Chlebicki, A., 1985. Notaki herpetologiczne z gor Sakar (Thracija). Przeglad Zoologiczny 29(2):193–198.

Christov, K. and Beshkov, V., 1999. On the subspecies morphological characteristics of the sandviper (*Vipera ammodytes*) from different localities in Bulgaria. Acta zoologica bulgarica 51(2–3):61–67.

[Christov, K.] Hristov, K., G. Russev, A. Westerström, A. & V. Beshkov, 2004. New record of *Vipera ursinii* (Bonaparte, 1835) in Bulgaria. Historia naturalis bulgarica 16:119–121.

Cyrén, O., 1933. Lacertiden der südöstlichen Balkanhalbinsel. Izvestiya na Tsarskite prirodonauchni instituti v Sofia (Mitteilungen aus den Königlichen naturwissenschaftlichen Instituten in Sofia – Bulgarien; Bulletin des Institutions royales d'histoire naturelle à Sofia – Bulgarie) 6:219–240.

Cyrén, O, 1941. Beiträge zur Herpetologie der Balkanhalbinsel. Izvestiya na Tsarskite prirodonauchni instituti v Sofia (Mitteilungen aus den Königlichen naturwissenschaftlichen Instituten in Sofia – Bulgarien; Bulletin des Institutions royales d'histoire naturelle à Sofia – Bulgarie) 14:36–152.

Engelmann, W.-E., J. Fritzsche, R. Günther, R. & F.-J. Obst, 1993. Lurche und Kriechtiere Europas. Neumann Verlag, Radebeul.

Gasc, J.P., A. Cabela, J. Crnobrnja-Isailovic, D. Dolmen, K. Grossenbacher, P. Haffner, J. Lescure, H. Martens, J.P. Martínez Rica, H. Maurin, M.E. Oliveira, T.S. Sofianidou, M. Veith & A. Zuiderwijk (eds), 1997. Atlas of Amphibians and Reptiles in Europe. Societas Europaea Herpetologica, Muséum National d'Histoire Naturelle, Paris.

Geissler, L. & A. Bruhl, 1980. Zur Herpetofauna Südwestbulgariens. Aquarium-Terrarium 27:246–247.

Guéorguiev, V., 1997. [Zoogeographical division]. In Yordanova, M. & D. Donchev (eds), Geografiya na Balgariya. Fizicheska geografiya. Sotsialno-ikonomicheska geografiya (Geography of Bulgaria. Physical Geography. Socio-Economic Geography). Akademichno izdatelstvo "Prof. Marin Drinov" (Prof. Marin Drinov Academic Publishing House), Sofia: 330–333 (in Bulgarian).

Guillaume, C.-P., B. Heulin & V. Beshkov, 1997. Biogeography of *Lacerta* (*Zootoca*) *vivipara*, reproductive mode and enzyme phenotypes in Bulgaria. Ecography 20:240–246.

Hristovich, G., 1892. [Materials towards the study of the fauna of Bulgaria]. Sbornik za narodni umotvoreniya, nauka i knizhnina [A Collection of Folklore, Science and Literary Works] 8:413–428 (in Bulgarian).

Kovachev, V., 1912. Herpetologichnata fauna na Balgariya (vlechugi i zemnovodni) [The Herpetological Fauna of Bulgaria (Reptiles and Amphibians)]. Hr. G. Danov, Plovdiv (in Bulgarian).

Noellert, A. & A. Ritter, 1986a. Einige Beobachtungen zur Herpetofauna der bulgarischen Schwarzenmeerküste und Südwestbulgariens. I. Amphibia. Herpetofauna 8(42):23–27.

Noellert, A. & A. Ritter, 1986b. Einige Beobachtungen zur Herpetofauna der bulgarischen Schwarzenmeerküste und Südwestbulgariens. II. Reptilia. Herpetofauna 8(44):30–34.

Obst, F.-J., 1978. Zur geographischen Variabilität des Scheltopusik, *Ophisaurus apodus* (Pallas). Zoologische Abhandlungen aus dem Staatlichen Museum für Tierkunde in Dresden 35(8):129–140.

Obst, F.-J. & L. Geissler, 1982. *Triturus vulgaris graecus* (Wolterstorff, 1905) neu für Bulgarien. Faunistische Abhandlungen aus dem Staatlichen Museum für Tierkunde in Dresden 9:117–119.

Peters, G., 1962. Studien zur Taxionomie, Verbreitung und Ökologie der Smaragdeidechsen. I. *Lacerta trilineata, viridis* und *strigata* als selbständige Arten. Mitteilungen aus dem Zoologischen Museum in Berlin 38:127–152.

Peters, G., 1963. Studien zur Taxionomie, Verbreitung und Ökologie der Smaragdeidechsen. II. Ökologische Notizen über einige ostbulgarische Populationen von *Lacerta trilineata*. Mitteilungen aus dem Zoologischen Museum in Berlin 39:203–222.

Petrov, B., 2004. The herpetofauna (Amphibia and Reptilia) of the Eastern Rhodopes (Bulgaria and Greece). In Beron, P. & A. Popov (eds), Biodiversity of Bulgaria. 2. Biodiversity of Eastern Rhodopes (Bulgaria and Greece). Pensoft & National Museum of Natural History, Sofia: 863–879.

Petrov, B., in press. Important herpetofaunal areas in Bulgaria. In Corbett, K. & A. Stumpel (eds), Important Herpetofaunal Areas in Europe. Societas Europaea Herpetologica & Netherlands' Ministry of Agriculture, Nature Managament and Fisheries (EC-LNV).

Petrov, B. & V. Beshkov, 2001. [Amphibians (Amphibia) and reptiles (Reptilia) in Kresna Gorge (SW Bulgaria)]. In Beron, P. (ed), Bioraznoobrazie na Kresnenskiya prolom (Biodiversity of Kresna Gorge (SW Bulgaria)). Natsionalen prirodonauchen muzei i Institut po zoologiya pri BAN (National

Museum of Natural History (Sofia) & Institute of Zoology, Bulgarian Academy of Sciences), Sofia: 297–303 (in Bulgarian).

Petrov, B., V. Beshkov, G. Popgeorgiev & D. Plachiiski, 2004. Plan za deistvie za zashtita na suhozemnite kostenurki v Balgariya [Action Plan for Conservation of Tortoises in Bulgaria]. Natsionalen prirodonauchen muzei, Balgarsko druzhestvo za zashtita na ptitsite, Balgarska fondatsiya Bioraznoobrazie [National Museum of Natural History, Bulgarian Society for Protection of Birds, Bulgarian Biodiversity Foundation], Sofia (in Bulgarian).

Petrov, B., S. Hristova & H. Hristov, 2002. First record of the cat snake *Telescopus fallax* Fleischmann, 1831 (Reptilia: Serpentes) in the Eastern Rhodopes Mt., Bulgaria. Historia naturalis bulgarica 15:143–146.

Petrov, B., P. Stoev & V. Beshkov, 2001. [Review of the species composition and distribution of amphibians (Amphibia) and reptiles (Reptilia) in the Eastern Rhodope Mts.]. Historia naturalis bulgarica 13:127–153 (in Bulgarian).

Popov, V., 1997. [Amphibians and reptiles]. In Yordanova, M. & D. Donchev (eds), Geografiya na Balgariya. Fizicheska geografiya. Sotsialno-ikonomicheska geografiya (Geography of Bulgaria. Physical Geography. Socio-Economic Geography). Akademichno izdatelstvo "Prof. Marin Drinov" (Prof. Marin Drinov Academic Publishing House), Sofia: 325–326 (in Bulgarian).

Rehák, I., 1985. *Coluber rubriceps thracius* ssp. n. from Bulgaria (Reptilia, Squamata, Colubridae). Věstnik Československé společnosti zoologické 49:276–280.

Rösler, H., 1999. Zur Taxonomie und Verbreitung von *Cyrtopodion kotschyi* (Steindachner, 1870) in Bulgarien (Sauria, Gekkonidae). Gekkota 2:3–19.

Sindaco, R., A. Venchi, G.M. Carpaneto & M. Bologna, 2000. The reptiles of Anatolia, a checklist and zoogeographical analysis. Biogeographia 21:441–554.

Stoev, P., 2000. On the distribution, biology and ecology of the amphibians and reptiles in the Derventski Heights and Sakar Mountain, South-East Bulgaria. Historia naturalis bulgarica 1:59–69.

Tzankov, N., 2003. Sympatric distribution of six lacertid lizards. In Abstracts of 10th Scientific Session, 20–21.11.2003, University of Sofia, Faculty of Biology. Bulvest, Sofia: 41.

Vigna Taglianti, A., P. Audisio, M. Biondi, M. Bologna, G. Carpaneto, A. De Biase, S. Fattorini, E. Piattella, R. Sindaco, A. Venchi & M. Zapparoli, 1999. A proposal for a chorotype classification of the Near East fauna, in the framework of the Western Palearctic region. Biogeographia 20:31–59.

5 Fauna and Distribution of Fishes in Bulgaria

Tihomir Stefanov
National Museum of Natural History, Bulgarian Academy of Sciences, Tsar Osvoboditel Blvd. 1, 1000 Sofia, Bulgaria, e-mail: tisho@nmnh.bas.bg

Abstract: The ichthyofauna of Bulgaria is reviewed based on the published and original data. A total of 217 fish species belonging to 58 families have been reported. Of these, 194 species are native for Bulgaria and 23 are introduced. More than 140 species were reported for the Bulgarian fresh and brackish waters, and a total of 134 species for the Bulgarian part of the Black Sea. A full faunistic list with notes on origin, distribution, and ecology of the fish species is given

1 Introduction

Until 1920, the Bulgarian ichthyofauna has been studied mainly by foreign specialists (Heckel, Kner, and others). The first publications by the Bulgarian authors were those by Kovatcheff (1909, 1921a, 1921b, 1923) and Moroff (1931); they, however, discussed only the freshwater ichthyofauna. Drensky (1923b) published the first paper on the ichthyofauna of the Bulgarian part of the Black Sea. Later Drensky (1948, 1951) summarized all the data about the Bulgarian fish species diversity and provided information about 182 species (Drensky, 1951). According to Drensky, 101 fish species inhabited the fresh and brackish waters of Bulgaria, and other 81 species, the Bulgarian part of the Black Sea. Later, Stoyanov et al. (1963) reported 135 species for the Bulgarian Black Sea coast. Karapetkova et al. (1998) reviewed all available data on the Bulgarian freshwater ichthyofauna and reported 122 species. Prodanov et al. (1998) reported 126 species for the Bulgarian Black Sea coastal waters. The most recent monographic work concerning Bulgarian fish fauna listed 207 species (Karapetkova and Zivkov, 1995).

Many authors conducted faunistic, biological, ecological, ecotoxicological, and parasitological studies on fishes in different types of water bodies in Bulgaria (the references are cited below).

The goal of the present work is to provide a review of the published literature concerning the ichthyological studies in Bulgaria and based on these results, as well as on the original data, to characterize the composition, origin, distribution, and ecology of the Bulgarian ichthyofauna.

V. Fet and A. Popov (Eds.), Biogeography and Ecology of Bulgaria, 109–139.

2 Material and Methods

A total of 223 published sources (more than the cited in the references) were used in the present study, as well as the original data, which included field surveys in 1997–2003 as well as a revision of the ichthyological collections of the National Museum of Natural History in Sofia.

The classification of families followed Nelson (1994). The taxonomy of freshwater fishes was based on the review of Kottelat (1997). The Latin names of the marine species (except those belonging to the family Syngnathidae) were given according to Fischer et al. (1987), and the taxonomy of Syngnathidae, according to Whitehead et al. (1986).

3 Discussion

The current Bulgarian ichthyofauna includes 217 species (Table 1), 142 of which inhabit fresh and brackish waters. We discuss these species according to the basin and type of water body which they inhabit. The ichthyofauna of the Bulgarian part of Black Sea includes 134 species, which are discussed separately.

3·1 The reservoirs and fish ponds

The total surface area of the reservoirs in Bulgaria is about 39,500 ha. Large reservoirs with a surface over 200 ha are most common (Karapetkova et al., 1998). There are also about 3000 small reservoirs used for irrigation and fish farming.

A few reservoirs are situated in the higher parts of the mountains (with an altitude over 1800 m a.s.l.). Data on the fish populations in the reservoirs Beli Iskar and Belmeken (Rila Mts.) were published by Karapetkova et al. (1995), Raikova-Petrova and Zivkov (1996), Raikova-Petrova et al. (1996, 1997, 1998), Raikova-Petrova (2000) and Beron et al. (2000). These reservoirs are inhabited mainly by *Phoxinus phoxinus, Salmo trutta*, and *Salvelinus fontinalis*.

Most numerous are the mountain reservoirs (from 600 to 1800 m a.s.l.), including Batak, Dospat, Iskar, Krichim, etc. The ichthyofauna of these reservoirs is well studied. Beltscheva (1959), Naidenow (1964), Zhivkov (1974, 1975, 1976, 1980), Zhivkov and Stoyanova (1976), Michajlova and Marinov (1979), Zivkov and Grupcheva (1979), Zivkov and Raikova-Petrova (1983), Zivkov and Petrova (1984), Dikov and Zhivkov (1985), Zivkov (1987), Naumova and Zivkov (1988), Petrova and Zivkov (1988), Raikova-Petrova and Zivkov (1988, 2000), Raikova-Petrova et al. (1993a) provided data on the fish species composition, as well as age composition, growth rate, feeding, fecundity, and maturity of dominant fish species of the reservoirs Dospat, Batak, Golyam Beglik, etc. Data on the ichthyofauna of the reservoirs in the Rhodope Mts. were summarized by Zivkov and Dobrev (2001) and Stefanov et al. (2003). The fish species composition in the mountain

Table 1 Fish species distribution in Bulgaria. Columns: 1, Reservoirs and small dams. 2, Fish ponds. Aegean Sea basin (3–5): 3, Upper portions of the rivers; 4, Middle portions of the rivers; 5, Lower portions of the rivers. Danube River basin (6–10): 6, Adjacent marshlands; 7, Bulgarian sector of the Danube River; 8, Upper portions of the tributaries; 9, Middle portions of the tributaries; 10, Lower portions of the tributaries. Black Sea basin (11–14): 11, Coastal lakes; 12, Upper portions of the rivers; 13, Middle portions of the rivers; 14, Lower portions of the rivers. Black Sea (15–17): 15, Pontian–Caspian relict; 16, Boreal–Atlantic relict; 17, Mediterranean immigrant. N, native; I, introduced

Taxa	Distribution																
			Aegean Sea basin			Danube River basin					Black Sea basin				Black Sea		
	1	2	3	4	5	6	7	8	9	10	11	12	13	14	15	16	17
CEPHALASPIDOMORPHI																	
PETROMYZONTIDAE																	
Eudontomyzon danfordi Regan, 1911							N		N	N							
Eudontomyzon mariae (Berg, 1931)							N		?	N							
CHONDRICHTHYES																	
SQUALIDAE																	
Squalus acanthias Linnaeus, 1758																N	
Squalus blainvillei (Risso, 1826)																N	
RAJIDAE																	
Raja clavata Linnaeus, 1758																N	
DASYATIDAE																	
Dasyatis pastinaca (Linnaeus, 1758)																N	
OSTEICHTHYES																	
ACIPENSERIDAE																	
Acipenser baerii Brandt, 1869		I															
Acipenser gueldenstaedti Brandt et Ratzeburg, 1833		I					N								N		
Acipenser nudiventris Lovetzky, 1828							N								N		
Acipenser ruthenus Linnaeus, 1758		I					N			N					N		

continued

Table 1 continued

Taxa	Distribution																
	Aegean Sea basin					Danube River basin					Black Sea basin				Black Sea		
	1	2	3	4	5	6	7	8	9	10	11	12	13	14	15	16	17
Acipenser stellatus Pallas, 1771		I			N		N			?					N		
Acipenser sturio Linnaeus, 1758					N		N									N	
Huso huso (Linnaeus, 1758)		I					N								N		
POLYODONTIDAE																	
Polyodon spathula (Walbaum, 1792)	I	I															
ANGUILLIDAE																	
Anguilla anguilla (Linnaeus, 1758)				N	N	N	N		N	N	N		?	N		N	
CONGRIDAE																	
Conger conger (Linnaeus, 1758)																	N
CLUPEIDAE																	
Alosa fallax (La Cepéde, 1803)																	N
Alosa immaculata Bennett, 1835							N			N	N			N	N		
Alosa maeotica (Grim, 1901)															N		
Alosa tanaica (Grim, 1901)[1]							N				N			N	N		
Clupeonella cultriventris (Nordmann, 1840)											N				N		
Sardina pilchardus (Walbaum, 1792)																	N
Sardinella aurita Valenciennes, 1847																	N
Sprattus sprattus (Linnaeus, 1758)																N	
ENGRAULIDAE																	
Engraulis encrasicholus (Linnaeus, 1758)														N			N
CYPRINIDAE																	
Abramis ballerus (Linnaeus, 1758)							N			N							
Abramis bjoerkna (Linnaeus, 1758)						N	N			N	N			N			
Abramis brama (Linnaeus, 1758)	N	I			N	N	N		N	N	N			N			

Abramis sapa (Pallas, 1814)						N	N			N				
Alburnoides bipunctatus (Bloch, 1782)			N	N	N			N	N			N	N	
Alburnus alburnus (Linnaeus, 1758)	N			N	N	N	N		N	N			N	N
Aspius aspius (Linnaeus, 1758)	N			N	N	N	N			N				N
Barbus barbus (Linnaeus, 1758)	N						N		N	N				
Barbus cyclolepis Heckel, 1837	N		N	N	N									
Barbus peloponnesius Valenciennes, 1842								N	N			N	N	
Barbus tauricus Kessler, 1877												N	N	
Carassius gibelio (Bloch, 1782)	I	I		I	I	I	I		I	I	I		I	I
Carassius carassius (Linnaeus, 1758)	N				?	N	N			N	?			?
Chalcalburnus chalcoides (Güldenstädt, 1772)					?	N	N			N	N			N
Chondrostoma nasus (Linnaeus, 1758)	N						N		N	N				
Chondrostoma vardarense Karaman, 1928	N			N	N									
Ctenopharyngodon idella Valenciennes, 1844	I	I				I	I			I	I			
Cyprinus carpio Linnaeus, 1758	N	I			I	I	N			N	N			I
Gobio albipinnatus Lukasch, 1933							N			N				
Gobio gobio (Linnaeus, 1758)	N			N	N		N		N	N	N		N	N
Gobio kesslerii Dybowski, 1862									N	N				
Gobio uranoscopus (Agassiz, 1828)								N	N					
Hypophthalmichthys molitrix (Valenciennes, 1844)	I	I			I	I	I			I	I			
Hypophthalmichthys nobilis (Richardson, 1845)	I	I				I	I			I	I			
Leucaspius delineatus (Heckel, 1843)		?			?	N					N			
Leuciscus borysthenicus (Kessler, 1859)				N	N								N	N
Leuciscus cephalus (Linnaeus, 1758)	N			N	N		N		N	N	N		N	N
Leuciscus idus (Linnaeus, 1758)						N	N			N				
Leuciscus souffia Risso, 1826[2]														

continued

Table 1 continued

Taxa	Distribution																
	Aegean Sea basin					Danube River basin					Black Sea basin				Black Sea		
	1	2	3	4	5	6	7	8	9	10	11	12	13	14	15	16	17
Mylopharyngodon piceus (Richardson, 1846)	I	I															
Pelecus cultratus (Linnaeus, 1758)							N			N							
Phoxinus phoxinus (Linnaeus, 1758)	N		N	N				N				N					
Pseudorasbora parva (Temminck et Schlegel, 1842)	I	I		I	I	I			I	I	I			I			
Rhodeus amarus (Bloch, 1782)	N			N	N	N	N		N	N	N		N	N			
Rutilus frisii (Nordmann, 1840)														N			
Rutilus rutilus (Linnaeus, 1758)	N			N	N	N	N		N	N	N		N	N			
Scardinius erythrophthalmus (Linnaeus, 1758)	N			N	N	N	N	N		N	N			N			
Tinca tinca (Linnaeus, 1758)	N	I		N	N	N	N		N	N	N			N			
Vimba melanops (Heckel, 1837)	N			N	N												
Vimba vimba (Linnaeus, 1758)							N			N			N	N			
COBITIDAE																	
Cobitis elongata (Heckel et Kner, 1858)							N		N	N							
Cobitis elongatoides Bacescu et Maier, 1969	N					N	N		N	N				N			
Cobitis strumicae Karaman, 1955				N	N		?			?	N		N	N			
Misgurnus fossilis (Linnaeus, 1758)					N	N	N			N	N						
Sabanejewia balcanica (Karaman, 1922)				N				N	N								
Sabanejewia bulgarica (Drensky, 1928)							N			N							
BALITORIDAE																	
Barbatula barbatula (Linnaeus, 1758)								N	N								
Barbatula bureschi (Drensky, 1928)			N	N													

CATOSTOMIDAE																
Ictiobus bubalis Rafinesque, 1810	I	I														
Ictiobus cyprinellus Valenciennes, 1844	I	I														
Ictiobus niger Rafinesque, 1810	I	I														
ICTALURIDAE																
Ameiurus nebulosus (Lesueur, 1814)	I	I														
Ictalurus punctatus (Rafinesque, 1818)	I	I														
SILURIDAE																
Silurus glanis Linnaeus, 1758	N	I		N	N	N	N		N	N	N		N	N		
ESOCIDAE																
Esox lucius Linnaeus, 1758	N	I		N	N	N	N		N	N	N		N	N		
UMBRIDAE																
Umbra krameri Walbaum, 1792						N										
COREGONIDAE																
Coregonus[3] *albula* (Linnaeus, 1758)	I															
Coregonus lavaretus (Linnaeus, 1758)	I															
Coregonus peled (Gmelin, 1788)	I															
SALMONIDAE																
Hucho hucho (Linnaeus, 1758)	I															
Oncorhynchus mykiss (Walbaum, 1792)	I	I	I	I				I				I				
Salmo labrax Pallas, 1814							N						N	N	N	
Salmo macedonicus (Karaman, 1924)			N													
Salmo salar Linnaeus, 1758	I	I														
Salmo trutta Linnaeus, 1758[4]	I	I	I	I				I				I				
Salvelinus fontinalis (Mitchill, 1814)	I	I	I													
Thymallus thymallus (Linnaeus, 1758)	I		I													
GADIDAE																
Gaidropsarus mediterraneus (Linnaeus, 1758)											N					N
Lota lota (Linnaeus, 1758)							N			N						
Merluccius merluccius (Linnaeus, 1758)																N

continued

Table 1 continued

Taxa	Distribution																
	Aegean Sea basin					Danube River basin					Black Sea basin				Black Sea		
	1	2	3	4	5	6	7	8	9	10	11	12	13	14	15	16	17
Merlangius merlangus (Linnaeus, 1758)											N					N	
OPHIDIIDAE																	
Ophidion rochei Müller, 1845											N						N
LOPHIIDAE																	
Lophius piscatorius Linnaeus, 1758																	N
GOBIESOCIDAE																	
Apletodon dentatus (Facciola, 1887)																	N
Diplecogaster bimaculata (Bonnaterre, 1788)																	N
Lepadogaster lepadogaster (Bonnaterre, 1788)																	N
Lepadogaster candollei Risso, 1810																	N
ATHERINIDAE																	
Atherina boyeri Risso, 1810											N			N			N
Atherina hepsetus Linnaeus, 1758											N						N
POECILIDAE																	
Gambusia holbrooki Girard, 1859	I	I			I	I					I			I			
BELONIDAE																	
Belone belone (Linnaeus, 1761)																	N
ZEIDAE																	
Zeus faber Linnaeus, 1758																	N
GASTEROSTEIDAE																	
Gasterosteus aculeatus Linnaeus, 1758					?						N		N	N		N	
Pungitius platygaster (Kessler, 1859)						N					N			N	N		

SYNGNATHIDAE													
Nerophis ophidion (Linnaeus, 1758)									N		N	N
Syngnathus abaster Risso, 1826					N	N				N		N	N
Syngnathus schmidti Popov, 1928												N	N
Syngnathus tenuirostris Rathke, 1837													N
Syngnathus typhle Linnaeus, 1758										N		N	N
Syngnathus variegatus Pallas, 1811													N
Hippocampus ramulosus Leach, 1814										N			N
SCORPAENIDAE													
Scorpaena porcus Linnaeus, 1758										N			N
TRIGLIDAE													
Eutrigla gurnardus (Linnaeus, 1758)													N
Trigla lucerna Linnaeus, 1758													N
COTTIDAE													
Cottus gobio Linnaeus, 1758 [5]							N						
SERRANIDAE													
Dicentrarchus labrax (Linnaeus, 1758)												N	N
Serranus cabrilla (Linnaeus, 1758)													N
Serranus scriba (Linnaeus, 1758)													N
CENTRARCHIDAE													
Lepomis gibbosus (Linnaeus, 1758)	I	I	I	I	I	I		I	I	I	I	I	
PERCIDAE													
Gymnocephalus baloni Holcik et Hensel, 1974						N			N				
Gymnocephalus cernuus (Linnaeus, 1758)	I					N			N			N	
Gymnocephalus schraetser (Linnaeus, 1758)						N			N			N	
Perca fluviatilis Linnaeus, 1758	N		N	N	N	N		N	N	N	N	N	
Sander lucioperca (Linnaeus, 1758)	N				N	N			N	N		N	
Sander volgensis (Gmelin, 1788)						N			N				
Zingel streber (Siebold, 1863)						N			N				

continued

Table 1 continued

Taxa	Distribution																
	Aegean Sea basin					Danube River basin					Black Sea basin				Black Sea		
	1	2	3	4	5	6	7	8	9	10	11	12	13	14	15	16	17
Zingel zingel (Linnaeus, 1766)							N			N							
POMATOMIDAE																	
Pomatomus saltator (Linnaeus, 1766)											N			N			N
ECHENEIDAE																	
Echeneis naucrates Linnaeus, 1758																	N
CARANGIDAE																	
Lichia amia (Linnaeus, 1758)																	N
Naucrates ductor (Linnaeus, 1758)																	N
Trachurus mediterraneus (Steindachner, 1868)											N			N			N
Trachurus trachurus (Linnaeus, 1758)																	N
SPARIDAE																	
Boops boops (Linnaeus, 1758)																	N
Dentex dentex (Linnaeus, 1758)																	N
Diplodus annularis (Linnaeus, 1758)											N			N			N
Diplodus puntazzo (Cetti, 1777)																	N
Diplodus sargus (Linnaeus, 1758)																	N
Diplodus vulgaris (Geoffroy Saint-Hilaire, 1817)																	N
Lithognathus mormyrus (Linnaeus, 1758)																	N
Oblada melanura (Linnaeus, 1758)																	N
Pagellus erythrinus (Linnaeus, 1758)																	N
Sarpa salpa (Linnaeus, 1758)																	N
Sparus aurata Linnaeus, 1758																	N
Spondyliosoma cantharus (Linnaeus, 1758)																	N

CENTRACANTHIDAE			
Spicara flexuosa Rafinesque, 1810			N
Spicara maena (Linnaeus, 1758)			N
Spicara smaris (Linnaeus, 1758)	N	N	N
SCIAENIDAE			
Sciaena umbra Linnaeus, 1758		N	N
Umbrina cirrosa (Linnaeus, 1758)			N
MULLIDAE			
Mullus barbatus Linnaeus, 1758	N	N	N
MUGILIDAE			
Chelon labrosus Risso, 1826			N
Liza aurata (Risso, 1810)	N	N	N
Liza ramada (Risso, 1810)			N
Liza saliens (Risso, 1810)	N	N	N
Mugil cephalus Linnaeus, 1758	N	N	N
Mugil soiuy Basilewsky, 1855 [6]	I		
POMACENTRIDAE			
Chromis chromis (Linnaeus, 1758)			N
LABRIDAE			
Coris julis (Linnaeus, 1758)			N
Ctenolabrus rupestris (Linnaeus, 1758)			N
Labrus viridis Linnaeus, 1758			N
Symphodus cinereus (Bonnaterre, 1788)	N	N	N
Symphodus ocellatus (Forsskal, 1775)	N	N	N
Symphodus roissali (Risso, 1810)		N	N
Symphodus rostratus (Bloch, 1797)			N
Symphodus tinca (Linnaeus, 1758)		N	N
AMMODYTIDAE			
Gymnammodytes cicerelus (Rafinesque, 1810)	N		N
TRACHINIDAE			
Trachinus draco Linnaeus, 1758		N	N

continued

Table 1 continued

Taxa	Distribution																
	Aegean Sea basin					Danube River basin					Black Sea basin				Black Sea		
	1	2	3	4	5	6	7	8	9	10	11	12	13	14	15	16	17
URANOSCOPIDAE																	
Uranoscopus scaber Linnaeus, 1758														N			N
BLENNIIDAE																	
Aidablennius sphynx (Valenciennes, 1836)											N						N
Blennius ocellaris (Linnaeus, 1758)																	N
Coryphoblennius galerita (Linnaeus, 1758)																	N
Lipophrys pavo (Risso, 1810)											N			N			N
Parablennius sanguinolentus (Pallas, 1811)											N			N			N
Parablennius tentacularis (Brunnich, 1768)											N						N
Parablennius zvonimiri (Kolombatovic, 1892)											N						N
CALLIONYMIDAE																	
Callionymus pusillus Delaroche, 1809																	N
Callionymus risso Lesueur, 1814																	N
ODONTOBUTIDAE																	
Percottus glenii Dybowski, 1877							I										
GOBIIDAE																	
Aphia minuta (Risso, 1810)											N						N
Benthophiloides brauneri Beling et Iljin, 1927											N				N		
Benthophilus stellatus (Sauvage, 1874)							N										
Chromogobius quadrivittatus (Steindachner, 1863)																	N
Gobius bucchichi Steindachner, 1870																	N
Gobius cobitis Pallas, 1814																	N

Gobius niger Linnaeus, 1758						N		N		N
Gobius paganellus Linnaeus, 1758										N
Knipowitschia caucasica (Berg, 1916)						N		N	N	
Knipowitschia longecaudata (Kessler, 1877)						N			N	
Mesogobius batrachocephalus (Pallas, 1814)						N		N	N	
Neogobius cephalargoides Pinchuk, 1976				N		N		N	N	
Neogobius fluviatilis (Pallas, 1814)			N	N	N	N	N	N	N	
Neogobius gymnotrachelus (Kessler, 1857)				N	N	N	N	N	N	
Neogobius kessleri (Günther, 1861)				N	N	N				
Neogobius melanostomus (Pallas, 1814)			N	N	N	N		N	N	
Neogobius platyrostris (Pallas, 1814)						N			N	
Neogobius ratan (Nordmann, 1840)									N	
Neogobius syrman (Nordmann, 1840)						N		N	N	
Pomatoschistus marmoratus (Risso, 1810)						N				N
Pomatoschistus minutus (Pallas, 1770)						N				N
Proterorhinus marmoratus (Pallas, 1814)	N	N		N		N		N	N	
Zosterisessor ophiocephalus (Pallas, 1814)						N		N		N
Sphyraenidae										
Sphyraena sphyraena (Linnaeus, 1758)										N
Xiphiidae										
Xiphias gladius Linnaeus, 1758										N
Scombridae										
Euthynnus alletteratus (Rafinesque, 1810)										N
Sarda sarda (Bloch, 1793)						N				N
Scomber japonicus Houttuyn, 1782										N
Scomber scombrus Linnaeus, 1758						N				N
Thunnus thynnus (Linnaeus, 1758)										N

continued

Table 1 continued

Taxa	Distribution																
	Aegean Sea basin					Danube River basin					Black Sea basin				Black Sea		
	1	2	3	4	5	6	7	8	9	10	11	12	13	14	15	16	17
SCOPHTHALMIDAE																	
Psetta maxima (Linnaeus, 1758)											N			N			N
Scophthalmus rhombus (Linnaeus, 1758)																	N
BOTHIDAE																	
Arnoglossus kessleri Schmidt, 1915																	N
PLEURONECTIDAE																	
Platichthys flesus Linnaeus, 1758											N			N		N	
SOLEIDAE																	
Solea nasuta (Pallas, 1814)											N			N			N

Notes

[1] According to Kottelat (1997), *Alosa bulgarica* described by Drensky (1934) is a synonym of *A. tanaica* but because of its separated distribution and morphological features it is probably a different species; more data are needed.

[2] *Leuciscus souffia* was found only in the Lower Kremensko Lake (a glacial lake in Pirin Mts.); only two specimens are known. The morphological features of these specimens show a little difference from the typical *L. souffia* (Karapetkova & Marinov, 1991).

[3] The three *Coregonus* species in Bulgaria were introduced from the former USSR; there are no data about their natural distribution as well as their systematics. In the Bulgarian ichthyological literature they are known as *Coregonus albula*, *C. lavaretus*, and *C. peled*.

[4] There are no data about the origin of the introduced *Salmo* species in the Bulgarian fish ponds and they were always listed as *Salmo trutta*.

[5] Populations of this species have been described in Bulgaria as endemic *Cottus gobio haemusi* by Marinov & Dikov (1986). However, their morphological features show a difference between them and the typical *C. gobio* and thus it could be a different species; more data are needed.

[6] *Mugil soiuy* was introduced in the Black Sea and now it is widely distributed along the Bulgarian coast.

reservoirs is comparatively rich. Initially, the ichthyofauna of the reservoirs has been formed by the existing native rheophilic species (*Barbus cyclolepis, Gobio gobio, Chondrostoma vardarense*, and *Leuciscus cephalus*). Later, the reservoirs were widely stocked with *Cyprinus carpio, Salmo trutta, Oncorhynchus mykiss, Coregonus albula, Coregonus peled*, and *Sander lucioperca*. Attempts for the introduction of *Salmo salar, Salvelinus fontinalis*, and *Thymallus thymallus* to some reservoirs were made. In addition, some fish species have been unintentionally introduced (e.g. *Carassius gibelio, Rutilus rutilus, Tinca tinca*, and *Perca fluviatilis*). Most numerous species in this type of reservoirs are *Alburnus alburnus, Carassius gibelio, Cyprinus carpio, Leuciscus cephalus, Rutilus rutilus*, and *Perca fluviatilis*.

The lowland reservoirs (below 600 m a.s.l.) have high water temperature and high biological productivity. Their ichthyofauna is comparatively well studied. Data on species diversity, age composition, growth rate, feeding, sex ratio, fertility, and mortality of dominant fish species in some of these reservoirs are published by Dimitrov (1957, 1960), Dimitrov and Lyudskanova (1967), Marinov and Bojadschiev (1967), Zivkov and Grupcheva (1987), Pavlov et al. (1988), Raikova-Petrova and Zivkov (1993), Raikova-Petrova et al. (1993b, 2001), Zivkov and Raikova-Petrova (1996a, 1996b), Zivkov et al. (1996), Pehlivanov (2000b), Raikova-Petrova (2001), Pehlivanov and Vassilev (2003), and Stefanov and Trichkova (2004). Comparative biological studies concerning the fish populations in different types of reservoirs were conducted by Zivkov (1993), Zivkov and Petrova (1995), Raikova-Petrova and Zivkov (1998), Zivkov and Raikova-Petrova (2001), and Zivkov et al. (2003). The lowland reservoirs are characterized by high species richness. Their ichthyofauna was also formed initially by the native fish species and then by the introduced commercial and incidental species. The most frequently occurring species in the lowland reservoirs are *Abramis brama, Alburnus alburnus, Carassius gibelio, Cyprinus carpio, Hypophthalmichthys molitrix, H. nobilis, Leuciscus cephalus, Pseudorasbora parva, Rutilus rutilus, Lepomis gibbosus, Perca fluviatilis*, and *Sander lucioperca*.

There are 141 fish ponds in Bulgaria (official records of the Ministry of Agriculture and Forestry). The main fish species cultured in these ponds are *Acipenser gueldenstaedti, Huso huso, Ctenopharyngodon idella, Cyprinus carpio, Hypophthalmichthys molitrix, H. nobilis, Mylopharyngodon piceus, Ictalurus punctatus, Silurus glanis, Esox lucius, Salmo salar, S. trutta, Oncorhynchus mykiss*, and *Salvelinus fontinalis*. Studies on fish culture were conducted by Kamburova and Kechaijov (1967), Dimitrov and Lyudskanova (1970), Grupcheva (1975), and Uzunova et al. (2002). The main goals of fish farming in Bulgaria are production of stock material and fish for consumption. Some species were introduced and bred in fish farms with different purposes: algae control (*Hypophthalmichthys molitrix*), control of submerged aquatic vegetation (*Ctenopharyngodon idella*), mosquito control (*Gambusia holbrooki*), recreational fishery (*Oncorhynchus mykiss,*

Salmo salar, Salvelinus fontinalis, Thymallus thymallus, and *Coregonus* spp.), and experimental aquaculture (*Acipenser baerii*).

3·2 The Aegean Sea basin

The basin of the Aegean Sea in Bulgaria consists of the valleys of Maritsa, Mesta, and Struma and their tributaries; however, the lower courses of these rivers are situated outside of the Bulgarian borders.

The first data about the ichthyofauna of the Bulgarian rivers flowing to the Aegean Sea comes from the works of Kovatcheff (1921a, 1921b). Later, Drensky (1923b, 1926, 1928, 1948, 1951), Chichkoff (1937a, 1939b, 1940a, 1940b, 1943), Marinov (1964, 1973a, 1973b, 1975), Michajlova (1965a, 1965b, 1970), and others published data on morphological characteristics, taxonomic status, and distribution of fishes in the Bulgarian part of the Aegean Sea basin.

The ichthyofauna of the Bulgarian rivers flowing to the Aegean Sea includes 39 species (27.5% of all freshwater species) (Table 1). Seven of these species are not typical for Bulgarian waters but are introduced: *Carassius gibelio, Hypophthalmichthys molitrix, Pseudorasbora parva, Oncorhynchus mykiss, Salmo trutta, Gambusia holbrooki*, and *Lepomis gibbosus*. Species with abundant and widespread populations are *Alburnus alburnus, Barbus cyclolepis, Chondrostoma vardarense, Gobio gobio, Leuciscus cephalus, Phoxinus phoxinus, Rhodeus amarus, Cobitis strumicae, Perca fluviatilis* as well as the introduced *Carassius gibelio, Pseudorasbora parva*, and *Lepomis gibbosus*. Other numerous species are *Alburnoides bipunctatus, Rutilus rutilus, Vimba melanops, Sabanejewia balcanica, Esox lucius*, and *Silurus glanis*.

A typical complex of species for the upper portions of these rivers includes *Alburnoides bipunctatus, Barbus cyclolepis, Phoxinus phoxinus, Barbatula bureschi* as well as the introduced *Salmo trutta, Oncorhynchus mykiss, Salvelinus fontinalis, Thymallus thymallus*, and *Coregonus* spp. *Leuciscus souffia* was caught only once (Karapetkova and Marinov, 1991) in the Dolno Kremensko Lake (the sources of the Mesta River), a glacial lake in the Pirin Mts.

The rivers of the Aegean Sea basin in Bulgaria house a relatively large number of endemic forms. Two species, *Chondrostoma vardarense* and *Vimba melanops*, are endemic for the Aegean Sea basin. *Cobitis strumicae, Sabanejewia balcanica*, and *Salmo macedonicus* are endemics of the Balkan Peninsula, and *Barbatula bureschi* is distributed only in the Struma and Mesta rivers. It is interesting that Maritsa River is *terra typica* for *Barbus cyclolepis* and *Vimba melanops*, and Struma River is *terra typica* for *Barbatula bureschi*.

Some species, like *Eudontomyzon hellenicus* Vladykov, Renaud, Kott et Economidis, 1982, *Acipenser stellatus, Chalcalburnus chalcoides, Leucaspius delineatus, Cobitis punctilineata* Economidis et Nalbant, 1997, and others, are typical for the lower (Greek) portions of these rivers but have not been found in their Bulgarian parts.

3·3 The Danube River basin

There are 13 tributaries of the Danube River in Bulgaria. Most of them run from the northern slope of Stara Planina Mts. with the exception of the Iskar River; this longest river in Bulgaria originates from Rila Mts. The first data about the ichthyofauna of these rivers come from the works of Kovatcheff (1909, 1923), Drensky (1921, 1923a, 1926, 1928, 1932, 1943, 1948, 1951), and Chichkoff (1929, 1937a, 1937b, 1939a, 1939c, 1940a, 1940b, 1941). Later, intensive ichthyological studies have been conducted on Danube River and its tributaries (Manolov-Gheorghiev 1966; Karapetkova, 1972, 1979, 1985, 1998; Sivkov, 1979, 1985, 1987a, 1987b, 1988a, 1988b, 1989a, 1989b, 1991a, 1991b; Karapetkova and Dikov, 1986; Karapetkova and Undžian, 1988; etc.). Many faunistic and taxonomic data about the ichthyofauna of the Bulgarian sector of Danube and its tributaries are provided by Oundjian (1964), Marinov (1966a, 1966b, 1967, 1978, 1979, 1987), Naidenov and Marinov (1974), Marinov and Dikov (1986), and others. The quantity, productivity, and biological characteristics of different fish species have been studied by Dikov and Zhivkov (1985), Dikov et al. (1988), Jankov and Zivkov (1988), etc. The ichthyofauna of the Srebarna Lake, the main Danube riverside lake in Bulgaria, have been studied by Bulgurkov (1958) and Pehlivanov (2000a).

The ichthyofauna of the Bulgarian tributaries of Danube River consist of 73 species (Table 1), or ca. 51.4% of all freshwater species in Bulgaria. Nine of them are introduced: *Carassius gibelio, Ctenopharyngodon idella, Hypophthalmichthys molitrix, H. nobilis, Pseudorasbora parva, Oncorhynchus mykiss, Salmo trutta, Gambusia holbrooki*, and *Lepomis gibbosus*. Typical for the upper courses of the rivers are *Alburnoides bipunctatus, Barbus peloponnesius, Gobio uranoscopus, Phoxinus phoxinus, Barbatula barbatula, Cottus gobio* as well as the introduced *Salmo trutta* and *Oncorhynchus mykiss*. Species with widespread and numerous populations in the middle and lower portions of the rivers are *Abramis* spp., *Alburnus alburnus, Barbus barbus, Chondrostoma nasus, Gobio gobio, G. kesslerii, Leuciscus cephalus, Rhodeus amarus, Rutilus rutilus, Tinca tinca, Esox lucius, Gymnocephalus* spp., *Perca fluviatilis, Sander lucioperca*, and others. Populations of *Alosa immaculata, Chalcalburnus chalcoides, Aspius aspius, Pelecus cultratus, Scardinius erythrophthalmus, Vimba vimba*, some Cobitidae, *Silurus glanis, Neogobius fluviatilis, Neogobius melanostomus*, etc. are also stable.

The Bulgarian sector of the Danube River lies between the 375th and 845th kilometers of the river. Of 83 fish species inhabiting the entire Danube (Karapetkova et al., 1998), 67 are found in its Bulgarian section and the adjacent marshlands. This is ca. 47.2% of all freshwater ichthyofauna of Bulgaria. The Danube basin, including the Bulgarian sector, is a center of speciation (Karapetkova et al., 1998) with such endemic forms as *Eudontomyzon danfordi, Sabanejewia bulgarica, Umbra krameri, Hucho hucho, Gymnocephalus baloni*, etc. Marinov and Dikov (1986) described *Cottus gobio haemusi* as an endemic subspecies from the Bulgarian Danube tributaries. According to its morphology, it could represent a distinct species but this issue has to be further studied. The Bulgarian part of the Danube River contains some rare fish species with limited populations. These are species of Acipenseridae (*Acipenser gueldenstaedti, A. ruthenus, A. stellatus, Huso huso), Alosa tanaica, Carassius carassius, Salmo labrax, Lota lota, Pungitius platygaster, Sander volgensis*, etc. Several species can be considered extremely rare and on the edge of extinction: *Eudontomyzon danfordi, E. mariae, Acipenser nudiventris, Umbra krameri*, and *Hucho hucho*. The species *Eudontomyzon vladykovi* Oliva et Zanandrea, 1959; *Rutilus meidingeri* (Heckel, 1851); *R. pigus* (La Cepede, 1803); and others inhabit Danube and its tributaries but they were never found in the Bulgarian section of this river.

3·4 The Black Sea basin

The Black Sea basin in Bulgaria includes two types of water bodies: Black Sea tributaries and coastal lakes. Their ichthyofauna includes 87 species (Table 1), which fall into two categories: 48 are the permanent residents of the fresh waters, while 39 species are immigrants from the sea, which reside temporarily in the lakes and rivers.

Nineteen rivers belong to this basin in Bulgaria. Twelve of them flow directly to the Black Sea and other seven flow into the coastal lakes. The Black Sea tributaries in Bulgaria are characterized with the relatively low altitude of their basins, the small declination of the river beds, and a well-developed branching network in their lower portions. The mouths of these rivers are strongly affected by the sea due to the eastern winds gathering alluvial sands in front of some of these rivers and blocking their outflow.

The ichthyofauna of the Black Sea tributaries in Bulgaria is comparatively well studied. Data about the species composition, distribution and origin were published by Pechev (1965, 1966, 1969, 1970), Apostolov (1972), Karapetkova and Pechev (1973), Sivkov (1974), and Karapetkova (1974, 1975, 1976).

A total of 75 species are found in the rivers of the Black Sea basin. Of these, 47 are permanent residents, and other 28 reside temporarily in the rivers after entering from the sea. The most numerous permanent inhabitants are *Barbus tauricus, Carassius gibelio, Chalcalburnus chalcoides, Gobio gobio, Leuciscus borysthenicus, L. cephalus, Rhodeus amarus, Rutilus rutilus, Vimba vimba, Neogobius fluviatilis, N. gymnotrachelus*, and *N. melanostomus*. Some other fish species also have abundant populations but are restricted to short sections of some of the rivers: *Alburnoides bipunctatus, Rutilus frisii, Scardinius erythrophthalmus*, and *Pungitius platygaster*. The temporary river residents primarily have the sea origin; most of them are Mediterranean immigrants. Fish species found in the majority of rivers include *Atherina boyeri*, three species of Mugilidae, *Platichthys flesus*, and others (Karapetkova et al., 1998).

A few endemic forms can be found also in the Bulgarian Black Sea tributaries. *Cobitis strumicae* is an endemic of the Balkan Peninsula, and *Salmo labrax* is an endemic of the Black Sea basin. Drensky (1934) described *Alosa bulgarica* as an endemic species of the Bulgarian Black Sea tributaries south from Burgas. According to Kottelat (1997), this species is a synonym of *A. tanaica*. In the past, this name was also listed as *A. caspia bulgarica* by different authors. The specific morphology and distinct distribution of this form indicate that it could be an endemic species, but more data are needed to determine its status.

There are about 18 lakes along the Bulgarian Black Sea coast (Kopralev et al., 2002). Their ichthyofauna was first studied by Valkanov (1935a, 1935b, 1936). Later, Manolov-Gueorguiev (1967) published a full list of the fish species found in Bulgarian Black Sea coastal lakes. Many other authors studied the ichtyofauna in different lakes: Stoyanov (1946), Alexandrova (1965, 1967), Beshovski (1967), Sivkov (1979), Vassilev (1998), and Mikhov (2000a, 2000b). Studies on the biology and ecology of the fish populations in the larger lakes were conducted by Nechaev (1932), Paspaleff (1941), Valkanov (1941), Zvetkov (1955), Saschev and Angelov (1959), Mihailova-Neikova (1961), Ivanov et al. (1980), and Pehlivanov (1998).

A total of 79 fish species are recorded in the Bulgarian Black Sea coastal lakes (Table 1). Of these, 50 species inhabit the lakes permanently while others migrate from the sea occasionally or regularly for feeding, reproduction, and wintering. Four group of species can be distinguished according to their origin: freshwater species (24 species), Pontian–Caspian relicts (17 species), Boreal–Atlantic relicts (five species), and Mediterranean immigrants (33 species) (see Table 1). The most abundant species in the coastal lakes are *Carassius gibelio, Cyprinus carpio, Pseudorasbora parva, Rutilus rutilus, Scardinius erythrophthalmus, Silurus glanis, Atherina boyeri, Gambusia holbrooki, Gasterosteus aculeatus, Lepomis gibbosus, Perca fluviatilis, Sander lucioperca, Knipowitschia caucasica, Neogobius melanostomus*, and *Proterorhinus marmoratus*.

3·5 The Black Sea

Drensky (1923b) published the first paper on the ichthyofauna of the Bulgarian part of the Black Sea, listing 92 species found near the coast. Later, Drensky (1948, 1951) summarized all data about the Bulgarian fish species diversity and published information about 182 species (Drensky, 1951). According to Drensky, 101 species inhabited the fresh and brackish waters of the country, and other 81 species, the Bulgarian part of the Black Sea. After Drensky, Georgieff (1954) reported for the first time a finding of *Sarpa salpa* in Bulgarian coastal waters. Martino (1954) found *Spondyliosoma cantharus* for the first time in the Bulgarian part of the Black Sea. Later, Georgiev and Kolarov (1958) published the finding of *Alosa fallax* in Bulgaria, and Gheorghiev (1960) reported *Euthynnus alletteratus* as new for the Bulgarian coastal waters. Karapetkova (1960) added *Arnoglossus kessleri* to the Bulgarian ichthyofauna. Stoyanov et al. (1963) reported 135 species for Bulgarian Black Sea coastal waters. Manolov-Gheorghiev (1966) revised the species of Gobiidae found in the Bulgarian waters, and listed 23 species. Manolov (1970) listed four new species of Sparidae for the Bulgarian Black Sea waters: *Dentex*

dentex, Diplodus vulgaris, Oblada melanura, and *Pagellus erythrinus*. Recently, Prodanov et al. (1998) published data about 126 species from 46 families found in the coastal waters of the Bulgarian Black Sea. According to Prodanov et al. (1998), 66 species have pelagic spawn; other 49 species spawn in bottom waters. Six species spawn in the autumn-winter season, and all the rest, during spring and summer. Many ichthyologists studied the major biological features (reproduction, growth in length and mass with age, population dynamics) of the commercial species in the Black Sea coastal waters as well as faunistic and taxonomic research.

A total of 134 species belonging to 44 families are found in the Bulgarian Black Sea coastal waters (Table 1). This constitutes 61.8% of all fish species reported for Bulgaria. The majority of these (100 species) are Mediterranean immigrants, 23 species are Pontian–Caspian relicts, and ten species are Boreal–Atlantic relicts. One fish species (*Mugil soiuy*) was introduced into Black Sea and now is widely distributed along the Bulgarian coast. A number of species reported from the Black Sea were never found in the Bulgarian coastal waters. These are *Balistes carolinensis* Gmelin, 1788; *Arnoglossus laterna* (Walbaum, 1792); *Callionymus lyra* Linnaeus, 1758; *C. fasciatus* Valenciennes, 1837; *Scorpaena notata* Rafinesque, 1810; *Buglossidium luteum* (Risso, 1810); *Solea vulgaris* Quensel, 1806; *Aspitrigla cuculus* (Linnaeus, 1758); and others.

The main commercial species for the Bulgarian fishing industry in the Black Sea are *Sprattus sprattus* with annual catches about 2650 tons (Prodanov et al., 1998), *Trachurus mediterraneus* (164 tons), *Sarda sarda, Alosa immaculata, Squalus acanthias*, and species of Gobiidae. In the past, very important commercial fish species for Bulgaria were also *Scomber scombrus, Pomatomus saltator, Psetta maxima*, and *Engraulis encrasicholus*, but their populations dramatically decreased in the late 1970s – early 1980s.

4 Conclusions

The review of the available literature and original data shows that 217 fish species belonging to 58 families were reported from Bulgaria. The faunistic survey shows that 194 species are native to Bulgaria. Seven of these are endemic to the Balkan Peninsula: *Chondrostoma vardarense, Vimba melanops, Cobitis strumicae, Sabanejewia balcanica, S. bulgarica, Barbatula bureschi*, and *Salmo macedonicus*. Twenty-three of the reported species (10.6%) are not typical for the Bulgarian ichthyofauna. Most of these are reared in fish ponds and cages and then introduced to the natural water bodies. Others were introduced accidentally together with the stocking material (*Pseudorasbora parva*) or as aquarium fish in Europe (*Lepomis gibbosus, Carassius gibelio*). Most of the introduced species are from North America (nine species), seven species are from Asia, and seven from other parts of Europe.

A total of 134 species were reported from the Bulgarian Black Sea coastal waters. Most of those (100 species) are the Mediterranean immigrants, other 23 species are Pontian–Caspian relicts and 10 species are Boreal–Atlantic relicts. One species (*Mugil soiuy*) was introduced in Black Sea and it is now widely distributed along the Bulgarian coast.

All this information demonstrates the comparatively high species diversity of fishes in Bulgaria and importance of its protection.

References

Alexandrova, K., 1965. [The ichthyoplankton of Varna Lake during the summer months of 1956–1964]. Izvestiya na Nauchnoizsledovatelskiya institut za ribno stopanstvo i okeanografiya – Varna (Proceedings of the Research Institute of Fisheries and Oceanography – Varna) 6:111–119 (in Bulgarian).

Alexandrova, K., 1967. [Mullets in the lakes along the Bulgarian Black Sea coast]. Izvestiya na Nauchnoizsledovatelskiya institut za ribno stopanstvo i okeanografiya – Varna (Proceedings of the Research Institute of Fisheries and Oceanography – Varna) 8:263–293 (in Bulgarian).

Apostolov, A., 1972. [The ichthyofauna of the Karaach River]. Ribno stopanstvo [Fisheries] 19(3):8–9 (in Bulgarian).

Beltscheva, R., 1959. [Untersuchungen im Stausee "W. Kolarow" zwecks seiner Ausnutzung für der Fischwirtschaft]. Godishnik na Sofiiskiya universitet. Biologo-geologo-geografski fakultet (Annuaire de l'Université de Sofia, Faculté de Biologie, Géologie et Géographie) 51(1 – Biology):301–338 (in Bulgarian).

Beron, P., V. Beshkov, V. Popov, M. Vassilev, R. Pandurska & T. Ivanova, 2000. Biodiversity of small vertebrates (Pisces, Amphibia, Reptilia, Mammalia – Insectivora, Chiroptera, Lagomorpha and Rodentia) in the Rila National Park. In Sakalian, M. (ed), Biological Diversity of the Rila National Park. Pensoft, Sofia: 33–360.

Beshovski, V., 1967. [Carp and perch-pike feed in the Bourgasko, Mandrensko and Blatnishko lakes along the Bulgarian Black Sea coast]. Izvestiya na Nauchnoizsledovatelskiya institut za ribno stopanstvo i okeanografiya – Varna (Proceedings of the Research Institute of Fisheries and Oceanography – Varna) 8:195–209 (in Bulgarian).

Bulgurkov, K., 1958. [Hydrological peculiarities of the reserve of Lake Srebarna and composition of its fish fauna]. Izvestiya na Zoologicheskiya institut (Bulletin de l'Institut zoologique de l'Académie des sciences de Bulgarie) 7:251–268 (in Bulgarian).

Chichkoff, G., 1929. [Sur un nouveau poisson du gènre *Gobio* Cuvier, *Gobio similis* n. sp.] Godishnik na Sofiiskiya universitet. Fiziko-matematicheski fakultet (Annuaire de l'Université de Sofia. Faculté physico-mathématique) 25(3 – Natural History):158–174 (in Bulgarian).

Chichkoff, G., 1937a. [Est-t'il present dans les eaux douces de la Bulgarie *Leuciscus leuciscus*?] Ribarski pregled [Fishing Review] 7(2):18–20 (in Bulgarian).

Chichkoff, G., 1937b. [Les poissons de la rivière Vit]. Ribarski pregled [Fishing Review] 7(5):66–68 (in Bulgarian).

Chichkoff, G., 1939a. [Poissons nouveaux et peu connus des eaux douces de la Bulgarie]. Godishnik na Sofiiskiya universitet. Fiziko-matematicheski fakultet (Annuaire de l'Université de Sofia. Faculté physico-mathématique) 35(3 – Natural History):91–199 (in Bulgarian).

Chichkoff, G., 1939b. [Les poissons des rivières du bassin de la Mer Egée]. Ribarski pregled [Fishing Review] 9(4):1–3 (in Bulgarian).

Chichkoff, G., 1939c. [Sur la pêche de la rivière Iskar]. Ribarski pregled [Fishing Review] 9(8):4–7 (in Bulgarian).

Chichkoff, G., 1940a. [Etudes morphologiques sur le vairon Phoxinus phoxinus L. habitant les eaux douces de la Bulgarie]. Godishnik na Sofiiskiya universitet. Fiziko-matematicheski fakultet (Annuaire de l'Université de Sofia. Faculté physico-mathématique) 36(3 – Natural History):21–71 (in Bulgarian).

Chichkoff, G., 1940b. [Etude morphologique sur les spirlins (*Alburnoides*) des eaux douces de la Bulgarie]. Godishnik na Sofiiskiya universitet. Fiziko-matematicheski fakultet (Annuaire de l'Université de Sofia. Faculté physico-mathématique) 36(3 – Natural History):73–146 (in Bulgarian).

Chichkoff, G., 1941. [Etude morphologique sur l'ablette commune (*Alburnus alburnus*) habitant les eaux douces de la Bulgarie]. Godishnik na Sofiiskiya universitet. Fiziko-matematicheski fakultet (Annuaire de l'Université de Sofia. Faculté physico-mathématique) 37(3 – Natural History):31–91 (in Bulgarian).

Chichkoff, G., 1943. [Etudes morphologiques sur *Leuciscus cephalus* et *Rutilus rutilus* répandus dans nos eaux douces]. Godishnik na Sofiiskiya universitet. Fiziko-matematicheski fakultet (Annuaire de l'Université de Sofia. Faculté physico-mathématique) 39(3 – Natural History):1–67 (in Bulgarian).

Dikov, C., J. Jankov & S. Jocev, 1988. [Ichthyofauna composition, number and biomass of the different species in the river Palakarija, a tributary of the river Iskar]. Hidrobiologiya (Hydrobiology) (Sofia) 33:59–68 (in Bulgarian).

Dikov, Ts. & M. Zhivkov, 1985. [A comparative analysis of the growth of the barbel (*Barbus tauricus cyclolepis* Heckel) in the Dzerman River and the Dospat Dam]. *Hidrobiologiya (Hydrobiology)* (Sofia) 26:81–90 (in Bulgarian).

Dimitrov, M., 1957. [Hydrological and hydrobiological characteristics of Al. Stamboliiski Dam Lake]. Nauchni trudove. Nauchno-izsledovatelski institut po ribarstvo i ribna promishlenost – Varna (Travaux scientifiques. Institut de recherches scientifiques sur la pêche et les industries s'y rattachant – Varna) 1:157–197 (in Bulgarian).

Dimitrov, M., 1960. [Hydrologische und hydrobiologische Charakteristik der Talsperre "G. Dimitrov"]. Trudove na Tsentralniya nauchnoizsledovatelski institut po ribovadstvo i ribolov. Sektsiya Sladkovodno ribarstvo – Plovdiv (Arbeiten des Zentralen Forschungsinstitutes für Fischzucht und Fischerei. Abteilung Süsswasserfischerei – Plowdiw (Bulgarien)) 1:87–122 (in Bulgarian).

Dimitrov, M. & Zh. Lyudskanova, 1967. [Feeds and feeding correlation of commercial fish in the "G. Dimitrov" Dam]. Izvestiya na Nauchnoizsledovatelskiya institut za ribno stopanstvo i okeanografiya – Varna (Proceedings of the Research Institute of Fisheries and Oceanography – Varna) 8:339–358 (in Bulgarian).

Dimitrov, M. & Zh. Lyudskanova, 1970. [The role of grass carp in the productivity of carp-breeding ponds]. Izvestiya na Stantsiyata po sladkovodno ribarstvo, Plovdiv [Mitteilungen der Versuchsstation für Süsswasserfischzucht, Plovdiv; Proceedings of the Freshwater Fishery Research Station, Plovdiv), 7:35–43 (in Bulgarian).

Drensky, P., 1921. [The fishing in the Iskar River]. Svedeniya po zemledelieto [Information on Agriculture] 9:5–16 (in Bulgarian).

Drensky, P., 1923a. [*Eupomotis gibbosus* L. one new fish in the Lake Svishtovsko]. Svedeniya po zemledelieto [Information on Agriculture] 11:6–7 (in Bulgarian).

Drensky, P., 1923b. [Contribution à l'étude des poissons de la Mer Noire, récoltes sur le cotes bulgares]. Spisanie na Balgarskata akademiya na naukite. Klon prirodo-matematichen [Journal of the Bulgarian Academy of Sciences. Natural-Mathematical Series] 25:59–112 (in Bulgarian).

Drensky, P., 1926. [Neue und seltene Fische aus Bulgarien]. Trudove na Balgarskoto prirodoizpitatelno druzhestvo (Travaux de la Société bulgare des sciences naturelles) 12:121–150 (in Bulgarian).

Drensky, P., 1928. [Die Fische der Familie Cobitidae in Bulgarien]. Izvestiya na Tsarskite prirodonauchni instituti v Sofia (Mitteilungen aus den Königlichen naturwissenschaftlichen Instituten in Sofia – Bulgarien; Bulletin des Institutions royales d'histoire naturelle à Sofia – Bulgarie) 1:156–181 (in Bulgarian).

Drensky, P., 1932. [Fishes and fishery of Osam River]. Lovech i Lovchansko [Lovech and surroundings] 4:171–200 (in Bulgarian).

Drensky, P., 1934. [*Alosa bulgarica* nov. sp., ein bis jetzt unbekannter Fisch aus den Flüssen der bulgarischen Küste des Schwarzen Meeres, südlich von Burgas]. Izvestiya na Tsarskite prirodonauchni instituti v Sofia (Mitteilungen aus den Königlichen naturwissenschaftlichen Instituten in Sofia – Bulgarien; Bulletin des Institutions royales d'histoire naturelle à Sofia – Bulgarie) 7:79–87 (in Bulgarian).

Drensky, P., 1943. [*Chalcalburnus chalcoides* Guld. in Bulgarien]. Godishnik na Sofiiskiya universitet. Fiziko-matematicheski fakultet (Annuaire de l'Université de Sofia. Faculté physico-mathématique) 39(3 – Natural History):343–360 (in Bulgarian).

Drensky, P., 1948. [Synopsis and distribution of fishes in Bulgaria]. Godishnik na Sofiiskiya universitet. Fiziko-matematicheski fakultet (Annuaire de l'Université de Sofia. Faculté physico-mathématique) 44(3 – Natural History):11–71 (in Bulgarian).

Drensky, P., 1951. [The Fishes of Bulgaria]. In: Buresch, I. (ed.). Fauna na Balgariya. 2 [Fauna bulgarica. 2]. Balgarska akademiya na naukite [Bulgarian Academy of Sciences], Sofia (in Bulgarian).

Fischer, W., M. Schneider, M. & M.-L. Bauchot (eds), 1987. Mediterranée et Mer Noire. Zone de pêche 37. Volume II. Vertebres. FAO, CEE, Organisation des Nations Unies pour l'Alimentation et l'Agriculture, Rome: 761–1530.

Georgieff, G.; Georgiev, J.; Gheorghiev, J. see Manolov-Gueorguiev, J.

Grupcheva, G., 1975. [Protozoan parasites of the rainbow trout (*Salmo gairdneri* Richardson) in some fish-breeding farms in Bulgaria]. Acta zoologica bulgarica 2:70–82 (in Bulgarian).

Ivanov, E., A. Rozhdestvenski, T. Marinov, V. Petrova-Karadjova, A. Konsulov & S. Mitev, 1980. [On the hydrochemical conditions, nutritive base and fish stocks in Mandra dam]. Izvestiya. Institut po ribni resursi – Varna (Proceedings. Institute of Fisheries – Varna) 18:7–70 (in Bulgarian).

Jankov, J. & M. Zivkov, 1988. [Maturing and size-sex ratio of the brown trout populations (*Salmo trutta fario* L.) in the main trout rivers in Bulgaria]. Hidrobiologiya (Hydrobiology) (Sofia) 32:68–84 (in Bulgarian).

Kamburova, S. & E. Kechaijov, 1967. [Versuche zur Einführung ergänzender Fischarten in den Karpfenzuchtwirtschaften]. Izvestiya na Nauchnoizsledovatelskiya Institut za ribno stopanstvo i okeanografiya [Proceedings of the Research Institute of Fisheries and Oceanography], Varna, 8:399–406 (in Bulgarian).

Karapetkova, M., 1960. [A new species for our Black Sea ichthyofauna, *Arnoglossus kessleri* Schmidt]. Priroda [Nature] (Sofia) 9(1):82–84 (in Bulgarian).

Karapetkova, M., 1972. [L'ichthyofaune de la rivière Yantra]. Izvestiya na Zoologicheskiya institut s muzei (Bulletin de l'Institut de zoologie et musée) 36:149–182 (in Bulgarian).

Karapetkova, M., 1974. [Ichthyofaune de la rivière Kamtchia]. Izvestiya na Zoologicheskiya institut s muzei (Bulletin de l'Institut de zoologie et musée) 39:85–98 (in Bulgarian).

Karapetkova, M., 1975. [Ichthyological characteristics of the Veleka River]. Hidrobiologiya (Hydrobiology) (Sofia) 1:54–64 (in Bulgarian).

Karapetkova, M., 1976. [Composition et caractère de l'ichthyofaune des rivières bulgares se jetant dans la Mer Noire]. Hidrobiologiya [Hydrobiology] (Sofia) 4:52–59 (in Bulgarian).

Karapetkova, M., 1979. Übersicht der Ichthyofauna der bulgarischen Donauzuflüsse. In: Limnologische Berichte der 21. Tagung der IAD. Novi Sad, Yugoslavia, pp. 378–384.

Karapetkova, M., 1985. [Some characteristic peculiarities of the Bulgarian freshwater fish fauna and the problems of its protection]. In Natsionalen simpozium po biologiya, Mihailovgrad, septemvri 1984 [National Biological Symposium, Mihailovgrad, September 1984]: 54–58 (in Bulgarian).

Karapetkova, M., 1998. Class Pisces (Fishes). In Michev, T., B. Georgiev, A. Petrova & M. Stoyneva (eds). Biodiversity of the Srebarna Biosphere Reserve. Checklist and bibliography. Context & Pensoft, Sofia: 84–85.

Karapetkova, M. & C. Dikov, 1986. [On the composition, distribution number and biomass of the ichthyofauna in the Vit River]. Hidrobiologiya (Hydrobiology) (Sofia) 28:3–14 (in Bulgarian).

Karapetkova, M. & B. Marinov, 1991. [A new species for ichthyofauna in Bulgaria, *Leuciscus souffia* Risso, 1826]. Hidrobiologiya (Hydrobiology) (Sofia) 37:75–78.

Karapetkova, M. & I. Pechev., 1973. [L'ichthyofaune de la rivière Ropotamo]. Izvestiya na Zoologicheskiya institut s muzei (Bulletin de l'Institut de zoologie et musée) 38:183–194 (in Bulgarian).

Karapetkova, M. & E. Undžian, 1988. [The ichthyofauna of the river valley of the Roussenski Lom River]. Hidrobiologiya (Hydrobiology) (Sofia) 32:44–50 (in Bulgarian).

Karapetkova, M. & M. Zivkov, 1995. Ribite v Balgariya [Fishes in Bulgaria]. Geya-Libris [Gea-Libris], Sofia (in Bulgarian).

Karapetkova, M., M. Zivkov & K. Alexandrova-Kolemanova, 1998. Freshwater fish of Bulgaria. In Meine, C. (ed), Bulgaria's Biological Diversity: Conservation Status and Needs Assessment. Volumes I and II. Pensoft, Sofia & Moscow: 347–374.

Karapetkova, M., M. Zivkov, G. Raikova-Petrova & E. Nikolova, 1995. Content of toxic elements in bioindicatory species of fish and amphibians from Rila Mountain. Observatoire de montagne de Moussala, OM2, 3:212–217.

Kopralev, I., M. Yordanova & Ch. Mladenov (eds), 2002. Geografiya na Balgariya. Fizicheska geografiya. Sotsialno-ikonomicheska geografiya [Geography of Bulgaria. Physical Geography. Socio-Economic Geography]. ForCom, Sofia (in Bulgarian).

Kottelat, M., 1997. European freshwater fishes. An heuristic checklist of the freshwater fishes of Europe (exclusive of former USSR), with an introduction for non-systematists and comments on nomenclature and conservation. Biologia 52(Suppl. 5):1–271.

Kovatcheff, V., 1909. [Contribution à la ichthyofaune bulgare]. Godishnik na Rusenskata mazhka gimnaziya (Annuaire du gymnase de garçons de Rousse):3–7 (in Bulgarian).

Kovatcheff, V., 1921a. [Les populations des poissons de la rivière Strouma]. Trudove na Balgarskoto prirodoizpitatelno druzhestvo (Travaux de la Société bulgare des sciences naturelles) 9:87–89 (in Bulgarian).

Kovatcheff, V., 1921b. [Une étude sur la faune ichthyologique de la rivière Mariza et ses affluents]. Trudove na Balgarskoto prirodoizpitatelno druzhestvo (Travaux de la Société bulgare des sciences naturelles) 9:90–94 (in Bulgarian).

Kovatcheff, V., 1923. [Faune des poissons d'eau douce en Bulgarie]. Arhiv na Ministerstvo na zemledelieto (Archives du Ministère de l'agriculture) 3:1–163 (in Bulgarian).

[Manolov-Gueorguiev, J.] Georgieff, G., 1954. [Nouvelle espèce pour notre ichthyofaune de la Mer Noire]. Izvestiya na Zoologicheskiya institut (Bulletin de l'Institut zoologique de l'Académie bulgare des sciences) 3:245–250 (in Bulgarian).

[Manolov-Gueorguiev, J.] Gheorghiev, J., 1960. [Une nouvelle espèce pour l'ichthyofaune bulgare *Euthynnus alletteratus* (Raf.)]. Trudove na Nauchnoizsledovatelskiya institut po ribarstvo i ribna promishlenost – Varna (Travaux de l'Institut de recherches scientifiques sur la pêche et les industries s'y rattachant – Varna (Bulgarie)) 2:35–40 (in Bulgarian).

[Manolov-Gueorguiev, J.] Manolov-Gheorghiev, J., 1966. [Composition d'espèce et caractéristique des Gobiides (Pisces) en Bulgarie]. Izvestiya na Nauchnoizsledovatelskiya institut za ribno stopanstvo i okeanografiya – Varna (Proceedings of the Research Institute of Fisheries and Oceanography – Varna) 7:159–228 (in Bulgarian).

Manolov-Gueorguiev, J., 1967. [Composition d'espèce de l'ichthyofaune des lacs bulgares de la Mer Noire]. Izvestiya na Nauchnoizsledovatelskiya institut za ribno stopanstvo i okeanografiya – Varna (Proceedings of the Research Institute of Fisheries and Oceanography – Varna) 8:211–227 (in Bulgarian).

[Manolov-Gueorguiev, J.] Manolov, J., 1970. [Aperçu sur la composition d'espèce de la famille Sparidae (Pisces) dans les eaux du littoral bulgare de la Mer Noire]. Izvestiya na Instituta po okeanografiya i ribno stopanstvo – Varna (Proceedings of the Institute of Oceanography and Fisheries – Varna) 10:165–189 (in Bulgarian).

[Manolov-Gueorguiev, J.] Georgiev, J. & P. Kolarov, 1958. [Eine neue Fischart in der bulgarischen Fauna – *Alosa fallax nilotica* Geoffroy]. Izvestiya na Zoologicheskiya institut (Bulletin de l'Institut zoologique de l'Académie des sciences de Bulgarie) 7:351–356 (in Bulgarian).

Marinov, B., 1964. [Ein neuer Fundort von *Leuciscus borysthenicus* (Kessler) im System des Ägäischen Meeres]. Godishnik na Sofiiskiya universitet "Kliment Ohridski". Biologicheski fakultet (Annuaire de l'Université de Sofia "Kliment Ohridski", Faculté de Biologie) 56(1 – Zoology):227–231 (in Bulgarian).

Marinov, B., 1966a. [Über die Fischfauna des bulgarischen Donausektors]. Izvestiya na Zoologicheskiya institut s muzei (Bulletin de l'Institut de zoologie et musée) 20:139–155 (in Bulgarian).

Marinov, B., 1966b. [*Cobitis elongata* Heckel et Kner, 1858 – eine für die Ichthyofauna Bulgariens neue Art]. Izvestiya na Zoologicheskiya institut s muzei (Bulletin de l'Institut de zoologie et musée) 21:153–156 (in Bulgarian).

Marinov, B., 1967. [*Gobio albipinnatus* Lukasch, 1933, a fish unknown to Bulgaria]. Godishnik na Sofiiskiya universitet "Kliment Ohridski". Biologicheski fakultet (Annuaire de l'Université de Sofia "Kliment Ohridski", Faculté de Biologie) 59(1 – Zoology):39–41 (in Bulgarian).

Marinov, B., 1973a. [Über die systematische Lage der Unterarten von *Gobio gobio* (L., 1758) aus Bulgarien]. Godishnik na Sofiiskiya universitet "Kliment Ohridski". Biologicheski fakultet (Annuaire de l'Université de Sofia "Kliment Ohridski", Faculté de Biologie) 65(1 – Zoology):101–115 (in Bulgarian).

Marinov, B., 1973b. [Über die systematische Lage der Unterarten von *Alburnoides bipunctatus* (Bloch, 1782) aus Bulgarien]. Godishnik na Sofiiskiya universitet "Kliment Ohridski". Biologicheski fakultet (Annuaire de l'Université de Sofia "Kliment Ohridski", Faculté de Biologie) 65(1 – Zoology):117–125 (in Bulgarian).

Marinov, B., 1975. [Über die systematische Stellung der Unterarten der Elritze – *Phoxinus phoxinus* (L.), 1758 (Pisces, Cyprinidae) von Bulgarien]. Godishnik na Sofiiskiya universitet "Kliment Ohridski". Biologicheski fakultet (Annuaire de l'Université de Sofia "Kliment Ohridski", Faculté de Biologie) 67(1 – Zoology), 1–14 (in Bulgarian).

Marinov, B., 1978. Die Fischfauna im bulgarischen Donauabschnitt und ihre Bedeutung für die Wirtschaft. In Russev, B. & W. Naidenow (eds), Limnologie des bulgarischen Donauabschnitts. Bulgarische Akademie der Wissenschaften, Sofia: 201–228.

Marinov, B., 1979. [*Pseudorasbora parva* (Schlegel, 1942) (Pisces, Cyprinidae) – a new representative of the ichthyofauna of Bulgaria]. Hidrobiologiya (Hydrobiology) (Sofia) 8:75–78 (in Bulgarian).

Marinov, B., 1987. [Growth of the asp (*Aspius aspius* L.) from the Bulgarian sector of the Danube]. In Botev, B. (ed), Savremenni postizheniya na balgarskata zoologiya. Materiali ot chetvartata yubileina nauchna konferentsiya po zoologiya [Modern Achievements in the Bulgarian Zoology. Materials of the Fourth Jubilee Scientific Conference on Zoology]. Izdatelstvo na Balgarskata akademiya na naukite [Publishing House of the Bulgarian Academy of Sciences], Sofia: 111–116 (in Bulgarian).

Marinov, B. & A. Bojadschiev, 1967. [Über das Anwachsungstempo der wirtschaftlich wertvollen Fischarten im Stausee "G. Dimitrov"]. Izvestiya na Nauchnoizsledovatelskiya institut za ribno stopanstvo i okeanografiya – Varna (Proceedings of the Research Institute of Fisheries and Oceanography – Varna) 8:359–369 (in Bulgarian).

Marinov, B. and Dikov, C., 1986. *Cottus gobio haemusi* subsp. n. (Pisces, Cottidae) from Bulgaria. Acta zoologica bulgarica 31:18–23.

Martino, K., 1954. [A new species for the Black Sea ichthyofauna, *Spondyliosoma cantharus* (L.), fam. Sparidae, Pisces]. Izvestiya na Zoologicheskiya institut (Bulletin de l'Institut zoologique de l'Académie bulgare des sciences) 3:251–255 (in Bulgarian).

Michajlova, L., 1965a. [Untersuchungen über die Fischfauna des Struma-Flusses]. Izvestiya na Zoologicheskiya institut s muzei (Bulletin de l'Institut de zoologie et musée) 19:55–71 (in Bulgarian).

Michajlova, L., 1965b. [Über die Ichthyofauna Thrakiens]. In Fauna na Trakiya. Sbornik ot statii. Chast II (Die Fauna Thrakiens. Sammelwerk. Band II). Izdatelstvo na Balgarskata akademiya na naukite (Verlag der Bulgarischen Akademie der Wissenschaften), Sofia: 265–289 (in Bulgarian).

Michajlova, L., 1970. [Ichthyofauna in the rivers of the Aegean watershed]. Priroda [Nature] (Sofia) 19(4):62–65 (in Bulgarian).

Michajlova, L. & B. Marinov, 1979. [Additional data on the species composition of the ichthyofauna of the Batak Dam]. Hidrobiologiya (Hydrobiology) (Sofia) 8:70–74 (in Bulgarian).

Mihailova-Neikova, M., 1961. [Hydrobiological research of the Mandra Lake with regard to its importance as a fishing ground]. Godishnik na Sofiiskiya universitet. Biologo-geologo-geografski fakultet (Annuaire de l'Université de Sofia, Faculté de Biologie, Géologie et Géographie) 53(1 – Biology), 57–122 (in Bulgarian).

Mikhov, S. 2000a. [New data about fishes in the Lake Mandra]. Burgas Wetlands Publications Series. Bulgarian-Swiss Biodiversity Conservation Program (Burgas) 4:1–6 (in Bulgarian).

Mikhov, S., 2000b. [Lake Atanassovsko – high conservation status of the ichthyofauna]. Burgas Wetlands Publications Series. Bulgarian-Swiss Biodiversity Conservation Program (Burgas) 5:1–5 (in Bulgarian).

Moroff, Th., 1931. Sladkovodnite ribi v Balgariya [The Freshwater Fishes of Bulgaria]. Balgarski ribarski sayuz [Union of Fishermen of Bulgaria], Sofia (in Bulgarian).

Naidenow, W., 1964. [The influence of zooplankton on the quantitative and qualitative development of the ichthyofauna in Batak Reservoir]. Zhivotnovadni Nauki [Animal Husbandry] 1(1):53–62 (in Bulgarian).

[Naidenow, W.] Naidenov, V. & B. Marinov, 1974 [Condition and perspective of the fishery from the Bulgarian section of the Danube]. Ribno stopanstvo [Fisheries] 5/6:21–22 (in Bulgarian).

Naumova, S. & M. Zivkov, 1988. [Mutual influence of the intensive trout rearing in cage and the hydrochemical and hydromicrobiological properties of the water in the "Dospat" reservoir]. Hidrobiologiya (Hydrobiology) (Sofia) 33:45–58 (in Bulgarian).

Nechaev, A., 1932. [The fish mortality in the Vaya-Kyoi Swamp in the autumn 1931]. Ribarski pregled [Fishing Review] 2(6):88–89 (in Bulgarian).

Nelson, J., 1994. Fishes of the World. Wiley-Interscience, New Work.

Oundjian, E., 1964. [Über einen Fund von Neunaugen aus der Donau]. Izvestiya na Narodniya muzei – Ruse (Bulletin de Musée National à Rousse) 1:227–232 (in Bulgarian).

Paspaleff, G., 1941. [The mortality of fishes in the Varna Lake]. Ribarski pregled [Fishing Review] 11(2):4–5 (in Bulgarian).

Pavlov, D.S, V.N. Mikheev, M.V. Vassilev & L.Z. Pehlivanov, 1988. Pitanie, raspredelenie i migratsiya molodi ryb iz vodokhranilishcha "Alexandar Stamboliiski" [Feeding, Distribution and Migration of Young Fish in Alexandar Stamboliiski Reservoir]. Nauka [Science], Moscow (in Russian).

Pechev, I., 1965. [Contribution à l'ichthyofaune de la rivière Batova]. Izvestiya na Narodniya muzei – Varna (Bulletin de Musée national de Varna) 1(16):127–134 (in Bulgarian).

Pechev, I., 1966. [Revue de l'ichthyofaune de la rivière d'Elechnitza]. Izvestiya na Narodniya muzei – Varna (Bulletin de Musée national de Varna) 2(17):179–191 (in Bulgarian).

Pechev, I., 1969. [Au sujet de l'ichthyofaune de quelques rivières bulgares se jetant dans la Mer Noire]. Izvestiya na Narodniya muzei – Varna (Bulletin de Musée national de Varna) 5(20):213–220 (in Bulgarian).

Pechev, I., 1970. [La faune ichthyologique de certaines rivières de Bulgarie de l'Est]. Izvestiya na Narodniya muzei – Varna (Bulletin de Musée national de Varna) 6(21):143–155 (in Bulgarian).

Pehlivanov, L., 1998. On the role of zooplankton in food of some numerous fish species in lake system Shabla–Ezerets (Northeastern Bulgaria). In Golemansky, V. & W. Naidenow (eds), Biodiversity of Shabla Lake System. Prof. Marin Drinov Academic Publishing House, Sofia: 107–113.

Pehlivanov, L., 2000a. Ichthyofauna in the Srebarna Lake, the Danube Basin, state and significance of the management and conservation strategies of this wetland. International Association for Danube Research 33:317–322.

Pehlivanov, L., 2000b. Ichthyofauna of the East Rhodopes (South Bulgaria): composition and distribution. Acta zoologica bulgarica 52(3):45–53.

Pehlivanov, L. & M. Vassilev. 2003. Feeding of age-1 perch (*Perca fluviatilis*) with a view to the diurnal dynamics of its distribution in the Alexander Stamboliiski Reservoir, Bulgaria. Acta zoologica bulgarica 55(2):81–90.

Prodanov, K., K. Dencheva & L. Ivanov, 1998. Fish of the Bulgarian coastal waters. In Meine, C. (ed), Bulgaria's Biological Diversity: Conservation Status and Needs Assessment. Volumes I and II. Pensoft, Sofia & Moscow: 375–393.

Raikova-Petrova, G., 2000. The ichthyofauna in glacial waters of the Rila Mountains. In Golemansky, V. & W. Naidenow (eds), Biodiversity and Evolution of Glacial Water Ecosystems in the Rila Mountains. Institute of Zoology, Sofia: 75–78.

Raikova-Petrova, G., 2001. Age and growth rate of perch (*Perca fluviatilis* L.) in Ovcharitsa Reservoir (Bulgaria). Arhiv bioloških nauka (Belgrade) 53(3–4):81–85.

Raikova-Petrova, G., Ts. Belomatcheva & M. Zivkov, 2001. [First maturity, sex ratio and fecundity of ruffe (Gymnocephalus cernuus L.) in Ovcharitsa Cooling-Reservoir]. In Dokladi. Deseti yubileen mezhdunaroden simpozium "Ekologiya 2001", Burgas (Reports. 10th Jubilee International Symposium "Ecology 2001", Bourgas). Burgas: 413–420 (in Bulgarian).

Raikova-Petrova, G., L. Pehlivanov & M. Zivkov, 1993a. [The food of pike-perch (*Stizostedion lucioperca* (L.)) in Batak Reservoir]. Hidrobiologiya (Hydrobiology) (Sofia) 38:45–56 (in Bulgarian).

Raikova-Petrova, G., L. Pehlivanov & M. Zivkov, 1993b. [The food of pike-perch (*Stizostedion lucioperca* (L.)) in Ovcaritsa Cooling-Reservoir]. Hidrobiologiya (Hydrobiology) (Sofia) 38:57–66 (in Bulgarian).

[Raikova-Petrova, G.] Petrova, G. & M. Zivkov, 1988. Regularities in the growth rate of pikeperch *Stizostedion lucioperca* (L.) in the Batak Dam, Bulgaria. Věstník Československé společnosti zoologické 52:31–43.

Raikova-Petrova, G. & M. Zivkov, 1988. [Introduction and biological significance of the pikeperch (*Stizostedion lucioperca* L.) in the "Batak" reservoir]. Hidrobiologiya (Hydrobiology) (Sofia) 32:58–67 (in Bulgarian).

Raikova-Petrova, G. & M. Zivkov, 1993. [Age and growth rate of pikeperch (*Stizostedion lucioperca* (L.)) in Ovcharitsa Cooling-Reservoir]. Hidrobiologiya (Hydrobiology) (Sofia) 38:67–81 (in Bulgarian).

Raikova-Petrova, G. & M. Zivkov, 1996. Age and growth of minnow, *Phoxinus phoxinus* (L.), in the Beli Iskur Reservoir and Grunchar Lake of Rila National Park (Bulgaria). In National Parks and their Role in Biodiversity Protection on the Balkan Peninsula, Ohrid: 127–129.

Raikova-Petrova, G. & M. Zivkov, 1998. Maturity, spawning, and sex ratio of pikeperch, *Stizostedion lucioperca* (L.) in two Bulgarian reservoirs as compared to other European habitats. Journal of Applied Ichthyology 14:31–35.

Raikova-Petrova, G. & M. Zivkov, 2000. [Age and growth rate of perch (*Perca fluviatilis* L.) in Batak Reservoir (Bulgaria)]. In Dokladi. Deveti mezhdunaroden simpozium "Ekologiya 2000", 8–10 yuni, Burgas, Balgariya (Reports. 9th International Symposium "Ecology 2000", 8–10 June, Bourgas, Bulgaria). Burgas: 295–300 (in Bulgarian).

Raikova-Petrova, G., M. Zivkov, M. Karapetkova & E. Nikolova, 1996. Toxic elements in zoo-monitory species of fish and amphibians from Rila mountain. Observatoire de montagne de Moussala, OM2, 4:178–183.

Raikova-Petrova, G., M. Zivkov, M. Karapetkova, T. Trichkova, E. Nikolova & A. Artinjan, 1998. Content of toxic elements in fish and amphibians from the Rila mountain in the period 1994-1996. Observatoire de montagne de Moussala, OM2, 7:211–214.

Raikova-Petrova, G., M. Zivkov, E. Nikolova & A. Artinjan, 1997. Toxic elements in zoomonitory species of fish and amphibians from Rila Mountain in 1996. Observatoire de montagne de Moussala, OM2, 6:133–136.

Saschev, G. & A. Angelov, 1959. [Untersuchungen über den Burgas-See (Waja) in Beziehung zur Verbesserung seiner fischwirtschaftlichen Ausbeutung]. Godishnik na Sofiiskiya universitet. Biologo-geologo-geografski fakultet (Annuaire de l'Université de Sofia, Faculté de Biologie, Géologie et Géographie) 51(1 – Biology):161–210 (in Bulgarian).

Sivkov, Y., 1974. [Ichthyofauna des Flusses Devnja]. Izvestiya na Narodniya muzei – Varna (Bulletin de Musée National de Varna) 10(25):374–377 (in Bulgarian).

Sivkov, Y., 1979. [Morphological characteristics of the common perch *Perca fluviatilis* L., from the Danube and the Shabla Lake]. Izvestiya na Narodniya muzei – Varna (Bulletin de Musée National de Varna) 15(30):149–157 (in Bulgarian).

Sivkov, Y., 1985. [Morphological characteristics of *Gymnocephalus baloni* Holcik et Hensel, 1974 (Pisces, Percidae), a new species to the Bulgarian ichthyofauna]. Acta zoologica bulgarica 27:10–20 (in Russian).

Sivkov, Y., 1987a. [Morphometric characteristic of *Zingel streber* Siebald, 1863 (Pisces, Percidae) from the Bulgarian section of the Danube]. Izvestiya na Narodniya muzei – Varna (Bulletin de Musée National de Varna) 23(38):198–204 (in Bulgarian).

Sivkov, Y., 1987b. [Morphometric characteristic of *Stizostedion volgense* (Gmelin, 1788) (Pisces, Percidae) from the Bulgarian section of the Danube]. Izvestiya na Narodniya muzei – Varna (Bulletin de Musée National de Varna) 23(38):204–211 (in Bulgarian).

Sivkov, Y., 1988a. [Morphological characteristic of the pikeperch *Stizostedion lucioperca* (L.) from the Danube River and the Shabla Lake]. Izvestiya na Narodniya muzei – Varna (Bulletin de Musée National de Varna) 24(39):174–179 (in Bulgarian).

Sivkov, Y., 1988b. [Morphological characteristics of the ruff *Gymnocephalus cernuus* (Linnaeus, 1758) (Pisces, Percidae) from the Bulgarian section of the Danube River]. Izvestiya na Narodniya muzei – Varna (Bulletin de Musée National de Varna) 24(39):180–185 (in Bulgarian).

Sivkov, Y., 1989a. [Morphological characterization of the gudgeon (*Gobio albipinnatus* Lukasch, 1933) (Pisces, Cyprinidae) from the Bulgarian section of the Danube River]. Acta zoologica bulgarica 38:11–15 (in Bulgarian).

Sivkov, Y., 1989b. [Morphological characterization of the *Cobitis elongata* Heckel et Kner, 1858 (Cobitidae) from Bulgaria]. Acta zoologica bulgarica 38:16–23 (in Bulgarian).

Sivkov, Y., 1991a. [Morphological characteristic of the Danubian loach *Sabanejewia bulgarica* (Drensky, 1928) (Pisces, Cobitidae). Acta zoologica bulgarica 42:34–43 (in Bulgarian).

Sivkov, J., 1991b. [Morphological characteristics of the Balkan barbel *Barbus peloponnesius* Valenciennes, 1842 from Bulgaria]. Acta zoologica bulgarica 42:44–52 (in Bulgarian).

Stefanov, T. & T. Trichkova, 2004. Fish species diversity in the Eastern Rhodopes (Bulgaria). In Beron, P. & A. Popov (eds), Biodiversity of Bulgaria. 2. Biodiversity of Eastern Rhodopes (Bulgaria and Greece). Pensoft & National Museum of Natural History, Sofia: 849–861.

Stefanov, T., T. Trichkova & M. Zivkov, 2003. Changes in the ichthyofauna of the Rhodope Mountains. In Kolarov, D. (ed), Sbornik. Nauchni dokladi. Mezhdunarodna nauchna konferentsiya "50 godini

Lesotehnicheski universitet" (Proceedings. Scientific Papers. International Scientific Conference "50 Years University of Forestry"). Lesotehnicheski universitet (University of Forestry), Sofia: 227–231.

Stoyanov, S., 1946. [On the *Carassius carassius* from the lake of Gebedje]. Ribarski pregled [Fishing Review] 15(4/5):15–16 (in Bulgarian).

Stoyanov, S., J. Georgiev, L. Ivanov, D. Hristov, P. Kolarov, K. Alexandrova & M. Karapetkova, 1963. Ribite v Cherno more (balgarskoto kraibrezhie) [Fishes of the Black Sea (Bulgarian Coast)]. Darzhavno izdatelstvo [State Publishing House], Varna (in Bulgarian).

Uzunova, E., K. Yorov, K. & B. Velkov, 2002. Growth and first maturation of *Salmo salar* Sebago introduced in fish farm "Nishava". Bulgarian Journal of Agricultural Science 8:629–633.

Valkanov, A., 1935a. [The Varna Lake]. Ribarski pregled [Fishing Review] 5(3):39–41 (in Bulgarian).

Valkanov, A., 1935b. [The Pomorie Lake]. Ribarski pregled [Fishing Review] 5(6):84–86 (in Bulgarian).

Valkanov, A., 1936. [The Burgas Lake]. Ribarski pregled [Fishing Review] 6(5):69–71 (in Bulgarian).

Valkanov, A., 1941. [Where do the fishes of the Varna Lake spend the winter?] Ribarski pregled [Fishing Review] 11(4):5–7 (in Bulgarian).

Vassilev, M., 1998. Alteration of the ichthyofauna in the Shabla and Ezerets lakes. In Golemansky, V. & W. Naidenow (eds), Biodiversity of Shabla Lake System. Prof. Marin Drinov Academic Publishing House, Sofia: 101–106.

Whitehead, P.J.P., M.-L. Bauchot, J.-C. Hureau, J. Nielsen & E. Tortonese, E., 1986. Fishes of the North-eastern Atlantic and the Mediterranean. Volume 3. UNESCO, Paris.

[Zivkov, M.] Zhivkov, M., 1974. [Dynamics of the numerical strength of fish populations in the Batak Dam. I. Age structure and increase in the population of *Leuciscus cephalus* L.]. Izvestiya na Zoologicheskiya institut s muzei (Bulletin de l'Institut de zoologie et musée) 40:203–216 (in Bulgarian).

[Zivkov, M.] Zhivkov, M., 1975. [Dynamics of the numerical strength of fish populations in the Batak Dam Lake. II. Age composition and growth of the carp population (*Cyprinus carpio* L.)]. Hidrobiologiya (Hydrobiology) (Sofia) 2:32–45 (in Bulgarian).

[Zivkov, M.] Zhivkov, M., 1976. [Dynamics of the numerical strength of fish populations in the Batak Dam. IV. Age composition and growth of the incidental species]. Hidrobiologiya (Hydrobiology) (Sofia) 4:36–51 (in Bulgarian).

[Zivkov, M.] Zhivkov, M., 1980. [Numerical dynamics of fish population in the Batak Dam. V. Rate of sexual maturation and sex ratio]. Hidrobiologiya (Hydrobiology) (Sofia) 12:49–62 (in Bulgarian).

Zivkov, M., 1987. [Ichthyofauna and fish-farming in the "Dospat" Reservoir]. Hidrobiologiya (Hydrobiology) (Sofia) 30:15–22 (in Bulgarian).

Zivkov, M., 1993. Comparative analysis of the age composition, growth rate and condition of carp (*Cyprinus carpio* L.) in three Bulgarian reservoirs. Ichthyologia (Belgrade) 25(1):7–18.

Zivkov, M., Ts. Belomacheva & T. Trichkova, 2003. Growth rate and condition of rudd, *Scardinius erythrophthalmus* (L.), in two Bulgarian reservoirs as compared to other European water bodies. Acta zoologica bulgarica 55(1):93–105.

Zivkov, M. & D. Dobrev, 2001. Ribite, zemnovodnite i vlechugite na Rodopite [Fishes, Amphibians and Reptiles of the Rhodope Mountains]. Balgarski sayuz za zashtita na Rodopite [Bulgarian Union for the Protection of Rhodopes], Sofia, 128 pp. (in Bulgarian).

Zivkov, M. & G. Grupcheva, 1979. [On the presence and relative numerical strength of the redeye *Scardinius erythrophthalmus* L. in the Batak Dam]. Hidrobiologiya (Hydrobiology) (Sofia) 10:39–42 (in Bulgarian).

Zivkov, M. & G. Grupcheva, 1987. [Specificities of the hydrochemical state, formation of the ichthyofauna and the fish-farming in the reservoir-cooler "Ovcharitsa"]. Hidrobiologiya (Hydrobiology) (Sofia) 30:23–36 (in Bulgarian).

Zivkov, M. & G. Petrova, 1984. An attempt to study the effect of the water volume on the fish growth in the Batak Reservoir, Bulgaria. Věstník Československé společnosti zoologické 48:141–159.

Zivkov, M. & G. Petrova, 1995. Fecundity of pikeperch (*Stizostedion lucioperca*) in two Bulgarian reservoirs as compared to other European habitats. Acta Societatis Zoologicae Bohemicae 59:127–136.

Zivkov, M. & G. Raikova-Petrova, 1983. [Numerical dynamics of fish populations in the Batak Dam. VI. Fecundity]. Hidrobiologiya (Hydrobiology) (Sofia) 18:10–22 (in Bulgarian).

Zivkov, M. & G. Raikova-Petrova, 1996a. Age composition, growth rate and mortality of the bream (*Abramis brama*) in the cooling-reservoir Ovcharitsa (Bulgaria). Acta Universitatis Carolinae, Biologica 40:203–212.

Zivkov, M. & G. Raikova-Petrova, 1996b. Maturity, sex ratio and spawning of bream, *Abramis brama* (L.) in the cooling-reservoir Ovcharitsa (Bulgaria). Acta Universitatis Carolinae, Environmentalica 10:53–61.

Zivkov, M. & G. Raikova-Petrova, 2001. Comparative analysis of age composition, growth rate and condition of roach, *Rutilus rutilus* (L.), in three Bulgarian reservoirs. Acta zoologica bulgarica 53(1):47–60.

Zivkov, M., G. Raikova-Petrova & T. Trichkova, 1996. [Effect of thermally stressed waters on the reproductive biology of bream *Abramis brama* (L.) in the cooling-reservoir Ovcharitsa]. In Dokladi. Peti mezhdunaroden simpozium "Ekologiya '96", 4–6 septemvri 1996, Burgas (Reports. 5th International Symposium "Ecology '96", 4–6 September, 1996, Bourgas). Burgas, pp. 107–110 (in Bulgarian).

[Zivkov, M.] Zhivkov, M. & L. Stoyanova, 1976. [Dynamics of the numerical strength of fish populations in the Batak Dam. III. Age composition and growth of goldfish (*Carassius auratus gibellio* (Bloch.)) population]. Hidrobiologiya (Hydrobiology) (Sofia) 3:20–30 (in Bulgarian).

Zvetkov, L., 1955. [Untersuchungen über das Futter der Fische im Beloslavsko-See]. Izvestiya na Zoologicheskiya institut (Bulletin de l'Institut zoologique de l'Académie des sciences de Bulgarie) 4/5:330–352 (in Bulgarian).

6 Fauna and Zoogeography of Marine, Freshwater, and Terrestrial Mollusks (Mollusca) in Bulgaria

Zdravko Hubenov

Institute of Zoology, Bulgarian Academy of Sciences, Tsar Osvoboditel Blvd. 1, 1000 Sofia, Bulgaria, e-mail: zhubenov@zoology.bas.bg

Abstract: A total of 445 species of mollusks has been recorded from Bulgaria, of which 111 are marine; 96, freshwater; and 238, terrestrial species. They belong to three classes, 18 orders, and 94 families. The Black Sea mollusks are divided into 14 zoogeographical categories; most of them have Atlantic–Mediterranean origin (95 species, or 85.5%). Typical for the Black Sea is the zone composed of *Mytilus galloprovincialis* (down to 70 m) and *Modiolula phaseolina* (70–140 m). The freshwater mollusks are represented by two classes: Gastropoda (71 species) and Bivalvia (24 species). The highest diversity of freshwater mollusks (77 species, or 80.1%) is found in the lowlands (below 200 m above sea level). The freshwater mollusks are divided into 22 zoogeographical categories; they include 20 endemic species (21.1%); 11 freshwater species are considered relicts (preglacial, glacial, or Caspian). The terrestrial mollusks are represented by two subclasses of Gastropoda. The highest diversity of terrestrial mollusks (181 species, or 76.0%) is found in the hills and mountains. The terrestrial mollusks are divided into 36 zoogeographical categories; they include 68 endemic species (28.5%), 21 preglacial relicts (8.8%), and one glacial relict. In Bulgaria, 206 mollusk species (42.3%) are considered important for conservation

1 Introduction

The Bulgarian mollusks have been studied for 170 years (Rossmässler, 1835). In the 19th century, malacological studies were done by Rossmässler (1835–1859), Mousson (1859), Jickeli (1874), and Westerlund (1884–1890). In the beginning of the 20th century, A. Wagner (1906, 1915), Wohlberedt (1911), and Hesse (1911, 1913, 1914, 1916) worked on the Bulgarian mollusk fauna. Between the two world wars, the studies of A. Wagner (1919–1920, 1927), Petrbok (1925, 1940, 1941), Büttner (1928), J. Wagner (1933, 1934a, 1934b, 1940), and Knipper (1939, 1941) have been published. Most explorers mentioned above studied aquatic as well as terrestrial mollusks. The first Bulgarian author who published works on mollusks was Iurinitsh (1906, 1908). In the decades after the World War II, along with the well-known foreign experts, such as Petrbok (1947, 1948, 1950), Jaeckel (1954a, 1954b), Jaeckel et al. (1957), Riedel (1957, 1963, 1967, 1972, 1975, 1978, 1985, 1988, 1996), Urbański (1960, 1964, 1969, 1977, 1978, 1979), Hudec (1964, 1975), Pinter (1968a, 1968b), Nordsieck (1973, 1974, 1977, 1978, 1979), Hudec and Vašatko (1971, 1973), Körnig (1983), Wiktor (1983, 1987, 1997),

V. Fet and A. Popov (Eds.), Biogeography and Ecology of Bulgaria, 141–198.

and Kroupa (1988, 1994), worked the Bulgarian malacologists: Drensky (1947) and Angelov (1956, 1959, 1972, 1976, 1983, 1998a, 1998b, 2000a, 2000b), on the freshwater mollusks; Kaneva-Abadjieva (1957, 1958, 1960, 1962, 1968, 1974, 1975) and Kaneva-Abadjieva and Marinov (1960, 1966, 1967, 1984), on the marine mollusks; Ošanova (1964, 1968, 1970, 1972) and Damjanov (see Damjanov and Pintér, 1969; Damjanov and Likharev, 1975), on the terrestrial mollusks. Many zoologists who published catalogues of Bulgarian fauna made a contribution to the study of Bulgarian mollusks, e.g. Valkanov (1957), on the Black Sea and brackish water species; Guéorguiev and Beron (1962), Beron and Guéorguiev (1967), and Beron (1972, 1994), on subterranean mollusks. Hydrobiological investigations of the Black Sea (Caspers, 1951, 1952; Kaneva-Abadjieva and Marinov, 1960, 1966, 1967, 1984; Marinov et al., 1983; Marinov, 1990; Marinov and Stoykov, 1995) and the Bulgarian rivers (Russev, 1957, 1964, 1966a, 1966b, 1967, 1978; Russev et al., 1981, 1994, 1998) have a great importance for the studies of aquatic mollusks. Currently, Bulgarian malacologists Zaprjanov, Dedov, and Irikov work on the terrestrial mollusks (Zaprjanov, 1987, 1989, 1990, 1991; Dedov, 1998; Dedov and Mitov, 1998; Irikov, 1999, 2001, 2002; Dedov and Penev, 2000; Irikov and Georgiev, 2002).

Different groups of mollusks are studied to a different degree. Faunistic composition of the Black Sea is known exhaustively. Long-term hydrobiological research obtained data on mollusks' distribution in marine ecosystems and coastal basins. In spite of the complete coenotic characteristic of the zoobenthos of Bulgarian sector of the Black Sea, an essential omission is the lack of a generalized faunistic work on the marine mollusks. An annotated list of the Bulgarian Black Sea gastropods was published recently (Wilke, 1996). The freshwater mollusks have been systematically studied all over the territory of Bulgaria, and their fauna is known exhaustively. More specific is the situation with crenobiotic and stygobiotic gastropods, where homomorphy and difficult collecting of living forms cause problems for the faunistic analysis. A current catalogue of the Bulgarian freshwater mollusks has been published (Angelov, 2000a). A generalized monograph on the terrestrial gastropods has been also published (Damjanov and Likharev, 1975) as well as a more recent annotated list (Dedov, 1998); however, the status of many gastropod subspecies is unclear.

A total of 445 species of mollusks (111 marine, 96 freshwater, and 238 terrestrial) has been recorded in Bulgaria. They belong to three classes (Polyplacophora, two species; Gastropoda, 371; and Bivalvia, 72), 18 orders and 94 families.

The goal of this work is to present the malacofauna of Bulgaria and some zoogeographical and ecological features of the Bulgarian mollusks.

2 Material and Methods

The literature data on Bulgaria's mollusks for the last 200 years are generalized in this work. The original label information on species distribution from the collections of the National Museum of Natural History in Sofia has been included.

A system of differentiated natural territories is used to represent the taxa distribution in Bulgaria (Hubenov, 1997). This is a hierarchical system (Fig. 1), which allows using larger or smaller number of territorial units. The first letter in the abbreviations corresponds to the region; the second, to the subregion; and the third, to a smaller territorial unit. To present distribution of a specific species in all subterritories of a certain region, it is enough to use only the first letter of the abbreviation, which applies to the region. To address certain subregions, it is necessary to use the first two letters.

For the marine mollusks, the maximal depth at which they have been established in the Bulgarian Black Sea is given (Table 1). If there are no data from the Bulgarian coast for some species, data from other areas of the Black Sea are presented. In the case of considerable differences in data in Bulgarian literature and data, reported for other sea parts, the corresponding foreign data for the Black Sea are also given. For many mollusks, the maximal depth at which they have been found in the Bulgarian sector of Danube River is presented as well. Vertical distribution of both freshwater and terrestrial mollusks in Bulgaria is also reported.

Zoogeographical categorization of the species (Table 1) is given separately for the marine, freshwater, and terrestrial mollusks on the basis of their distribution. Ecological data (Table 1) are from Bulgarian malacological literature. Only if there are no data from Bulgaria, foreign data are included for some species. The communities of Black Sea in which the mollusks dominate are addressed. Attention

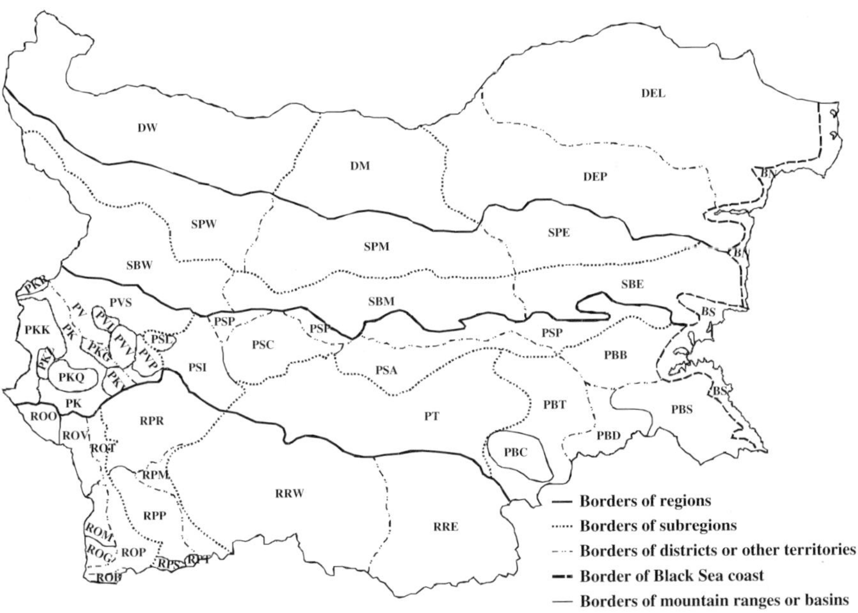

Fig. 1 Natural geographical territorial units in Bulgaria. Abbreviations, see text.

Table 1 Faunistic and ecological diversity of Mollusca in Bulgaria. Abbreviations, see text

Taxa	Distribution: Horizontal	Distribution: Vertical	Zoogeographical category	Ecological data and conservation importance
POLYPLACOPHORA				
ISCHNOCHITONIDA				
Ischnochitonidae				
Lepidochitona cinerea (Linnaeus, 1767)	B	−10(−30)	mls	M, se
Lepidochitona corrugata (Reeve, 1848)	B		ml	M, se
GASTROPODA				
PROSOBRANCHIA				
ARCHAEOGASTROPODA				
Patellidae				
Patella ulyssiponensis Gmelin, 1791	B	−5	ml	M, se
Neritidae				
Theodoxus fluviatilis (Linnaeus, 1758)	D^r, PBS, B	0–300	e	L, b-5‰, po, sw, t^0
Theodoxus danubialis (C. Pfeiffer, 1828)	D^r, D, SP, BN	0–200, −9	Rc, pcsee	L, rh, po, sw, li, N
Theodoxus euxinus (Clessin, 1885)	BN	0, (−10)	Sf, Rc, p	L, b-5‰, N
Theodoxus pallasi Lindholm, 1924	BN, DEL	0, −10	Rc, pc	L, b, (M), se, N
Theodoxus pilidei (Tournouer, 1879)	B		Sf, Rc, p	(M), EX
Theodoxus transversalis (C. Pfeiffer, 1828)	D^r, D	0–200, −11	csee	L, po, li, DD, W
Trochidae				
Gibbula adriatica (Philippi, 1844)	B	(−50)	m	M, se
Gibbula albida (Gmelin, 1791)	B	(−40)	m	M, se
Gibbula divaricata (Linnaeus, 1758)	B	−20	ml	M, se
Tricoliidae				
Tricolia pullus (Linnaeus, 1758)	B	−20(−50)	ml	M, se
MESOGASTROPODA				
Viviparidae				
Viviparus acerosus (Bourguignat, 1862)	D^r	0–30	see	L, po, LR/nt, W
Viviparus contectus (Millet, 1813)	D^r, DM, DEL	0–10	wes	L, po, sw, ph
Viviparus danubialis (Bourguignat, 1862)	D^r	0–30	Eb, pb	L, eu, pe, E
Viviparus mamillatus Küster, 1852	?		Ebs	L, po, sw, EX, W
Viviparus viviparus (Linnaeus, 1758)	D^r, D, PVS, PT	0–600, −8	e	L, po, sw, pe
Valvatidae				
Valvata cristata (O. F. Müller, 1774)	DEL, S, PT, BN	0–200	wes	L, po, sw, ph, r, N
Valvata naticina Menke, 1845	D^r, DM, DEL	0–50	cee	L, po, sw, li, r, N
Valvata piscinalis (O. F. Müller, 1774)	D, S, P, R, B	0–1100, −9	wcp	L, rh, po, sw, ph
Valvata pulchella Studer, 1820	D^r, DM	0–50	wes	L, sw, ph, r, N
Cerithiidae				
Cerithium vulgatum Bruguière, 1792	B	−15(−25)	ml	M, se

continued

Table 1 continued

Taxa	Distribution Horizontal	Distribution Vertical	Zoogeographical category	Ecological data and conservation importance
Bittium reticulatum (da Costa, 1778)	B	−110	mls	M, eb
Bittium submamillatum (De Rayneval et Ponzi, 1854)	B	(−20, −140)	ml	M, mb–eb
Triphoridae				
Marshallora adversa (Montagu, 1803)	BS	(−15, −80)	mls	M, mb
Cerithiopsidae				
Cerithiopsis minima (Brusina, 1865)	BS	(−30)	mm	M, se
Cerithiopsis tuberculatus (Montagu, 1803)	B	−110(−85)	mls	M, eb, pe
Aclididae				
Graphis albida (Kanmacher, 1798) (syn.?)	BS	(0,−50)	ml	M, se, r, N
Epitoniidae				
Epitonium turtonis (Turton, 1819)	B	−40(−60)	mls	M, mb
Eulimidae				
Vitreolina incurva (Bucquoy, Dautzenberg et Dollfus, 1883)	B	−100(−35)	mls	M, eb, pe
Littorinidae				
Melarhaphe neritoides (Linnaeus, 1758)	B	0–2	mls	M, se
Rissoidae				
Rissoa euxinica Milaschewitch, 1909 (syn.?)	B	−10(−20)	ml	M, se, ph
Rissoa membranacea (J. Adams, 1800)	B	−10(−10)	mls	M, b-6‰, ph, se
Rissoa splendida Eichwald, 1830	B	−10(−12)	ml	M, se, ph
Pusillina lineolata (Michaud, 1832) ?	B	(−40)	m	M, mb
Setia turriculata Monterosato, 1884 (syn.?)	BS		mm	M, se
Alvania lactea (Michaud, 1832)	B		ml	M, se
Assimineidae				
Assiminea rufostrigata Hesse, 1916 (syn.?)	BN		m	M, b, se
Caecidae				
Caecum armoricum De Folin, 1869	BN	(−16)	m	M, se
Caecum trachea (Montagu, 1803)	B	−15(−50)	ml	M, se
Hydrobiidae				
Hydrobia acuta (Draparnaud, 1805)	B	(−20)	m	M, b-1–60‰, se, sw
Hydrobia ventrosa (Montagu, 1803)	B	−111, (−20)	ml	(M), b, mb
Bythinella austriaca (Frauenfeld, 1857)	SBW, PK, PV, R	500–1720	Rg, cee	L, cr, po, x, N
Belgrandiella angelovi Pinter, 1968	SBM	450–500	El	L, cr, st, W
Belgrandiella bulgarica Angelov, 1972	SPM	400–450	El	L, cr, st, r, W
Belgrandiella bureschi Angelov, 1976	PKK	850–900	El	(L), cr, st, t^0, W
Belgrandiella hessei A. Wagner, 1927	SBW	400–450	El	(L), ps, st, W
Belgrandiella pusilla Angelov, 1959	SBW	450–500	El	L, cr, st, r, W
Pontobelgrandiella nitida (Angelov, 1972)	SPM	380–450	El	L, cr, st, r, W
Cavernista zaschevi (Angelov, 1959)	SBW	400–500	El	L, cr, st, r, W
Insignia macrostoma Angelov, 1972	SPM	400–450	El	(L), cr, st, r, W
Iglica acicularis Angelov, 1959	SBW	400–500	El	(L), cr, ps, st, r, W
Paladilhiopsis bureschi A. Wagner, 1927	SBW	400–450	El	(L), ps, st, W
Plagigeyeria procerula Angelov, 1965	PVS	550–600	El	(L), cr, st, r, W
Pseudamnicola codreanui (Grossu, 1946)	BN	10–50	Er	L, cr, W

continued

Table 1 continued

Taxa	Distribution		Zoogeographical category	Ecological data and conservation importance
	Horizontal	Vertical		
Pseudamnicola consociella (Frauenfeld, 1863)	DE, SB, RR, BN	0–900	Eb	L, cr, E
Horatia lucidulus Angelov, 1967	DEL, BN	0–20	El	L, cr, st, r, W
Sadleriana virescens (Küster, 1852)	PT, PBS, RRW	150–350	Eb	L, cr, rh, E
Lithoglyphidae				
Lithoglyphus naticoides (C. Pfeiffer, 1828)	D^r, PT, BN	0–160, −12	? Rc, psee	L, b, (M), po, li, N
Lithoglyphus pyramidatus O. F. Müller, 1774	D^r, DW	0–30	Ebs	L, po, rh, li, r, E
Bithyniidae				
Bithynia leachi (Sheppard, 1823)	D^r, PKK, PVS, BN	0–600	wp	L, po, rh, sw
Bithynia tentaculata (Linnaeus, 1758)	D^r, BN	0–3	wp	L, b, po, sw, ph
Micromelaniidae				
Micromelania lincta Milaschewitch, 1908	D^r, B	0, −20, −148	Sf, Rc, p	L, b, (M), sw, N
Clessiniola variabilis (Eichwald, 1838)	D^r, DEL	0, −7	?Sf, Rc pc	(L), b, N
Melaniidae				
Amphimelania holandri (C. Pfeiffer, 1828)	D^r, SBW	0–650	Eb	L, rh, po, t^0, li, E
Fagotia acicularis (Férussac, 1823)	D^r, D, SP	0–400, −6.5	Rc, psee	L, cr, rh, po, li, N
Fagotia esperi (Férussac, 1823)	D^r, DM	0–200, −7	Rc, psee	L, cr, po, li, DD, W
Melanopsis parreyssi Philippi, 1848	PT	160	Rp, csee	L, cr, t^0, N
Tornidae				
Tornus subcarinatus (Montagu, 1803)	BS		ml	M, se, r, N
Calyptraeidae				
Calyptraea chinensis (Linnaeus, 1758)	B	−40(−70)	ml	M, mb
Pomatiasidae				
Pomatias elegans (O. F. Müller, 1774)	SB, P, R, B	0–800	Rp, am	T, me, ca, N
Pomatias rivulare (Eichwald, 1829)	D, S, P, R, B	0–1200	Rp, pm	T, me, ca, N
Aciculidae				
Platyla polita (Hartmann, 1840)	PBS	200	e	T, me–hg, r, N
Platyla orthostoma (Jackiewicz, 1979)	BS	0–20	Er	T, me–hg, r, W
Acicula similis (Reinhardt, 1880)	D, S, PT, R	0–800	see	T, hg, ca, r, N
NEOGASTROPODA				
Muricidae				
Trophon muricatus (Montagu, 1803)	B	−36, −138	ml	M, sh, pe
Rapana venosa (Valenciennes, 1846)	B	−70	j, i	M, eb
Nassariidae				
Nassarius reticulatus (Linnaeus, 1758)	B	−63(−60)	mls	M, mb, ps–pe
Cyclope neritea (Linnaeus, 1758)	B	−54(−50)	ml	M, mb, ps–pe
Cyclope westerlundi (Brusina, 1900) (syn.?)	B	−15(−100)	ap	M, sh, r, N
Conidae				
Bela nebula (Montagu, 1803)	BN	−15(−20)	mls	M, se, r, N
Mangelia costata (Donovan, 1804)	B	−15(50)	mls	M, se

continued

Table 1 continued

Taxa	Distribution: Horizontal	Distribution: Vertical	Zoogeographical category	Ecological data and conservation importance
Omalogyridae				
Omalogyra atomus (Philippi, 1841)	BS		mls	M, se
Ammonicera fischeriana (Monterosato, 1869)	BS		wimm	M, ?eb
OPISTHOBRANCHIA				
PYRAMIDELLOMORPHA				
Pyramidellidae				
Chrysallida emaciata (Brusina, 1866)	BS		m	M, se
Chrysallida incerta (Milaschewitch, 1916) (syn.?)	BS	(−18)	m	M, se
Chrysallida obtusa (Brown, 1827)	B	(−80)	mls	M, mb
Chrysallida terebellum (Philippi, 1844)	B	−10, −42	m	M, se, pe
Eulimella acicula (Philippi, 1836)	BS	(−18)	ml	M, se, r, N
Odostomia erjaveciana Brusina, 1869	BS		m	M, se
Odostomia eulimoides Hanley, 1844	BS	(−70)	ml	M, mb
Odostomia plicata (Montagu, 1803)	BS	(−20)	ml	M, se
Odostomia scalaris MacGyllivray, 1843	B	−15, −92	mls	M, mb
Noemiamea dolioliformis (Jeffreys, 1848)	BS		ml	M, se, r, N
Turbonilla delicata Monterosato, 1874	BN	(−35)	ml	M, se
Turbonilla pusilla (Philippi, 1844)	BS		ml	M, se, r, N
Ebalidae				
Ebala pointeli (De Folin, 1868)	BN	−15(−20)	ml	M, se
CEPHALASPIDEA				
Retusidae				
Retusa truncatula (Bruguière, 1792)	B	−111(−140)	mls	M, eb, pe
Cylichnina umbilicata (Montagu, 1803)	B	−50(−50)	ml	M, mb
SACOGLOSSA				
Stiligeridae				
Calliopaea bellula d'Orbigny, 1837	BN		ml	M, se
Limapontidae				
Limapontia capitata (O. F. Müller, 1774)	BN		mls	M, ?se
ACOCHLIDIOIDEA				
Microhedylidae				
Parahedyle tyrtowii (Kowalevsky, 1901)	B		m	M, se
NUDIBRANCHIA				
Corambidae				
Corambe obscura (Verrill, 1870)	B	−2(−5)	vwi, i	M, se
Pseudovermidae				
Pseudovermis paradoxus Perejaslawtzeva, 1891	B		ap	M, se
Tergipedidae				
Embletonia pulchra (Alder et Hancock, 1844)	B	(−40)	ml	M, mb
PULMONATA				
BASOMMATOPHORA				
Ellobiidae				
Carychium minimum O. F. Müller, 1774	D, S, P, R, B	0–1000	wes	T, hg
Carychium tridentatum (Risso, 1826)	D, S, P, R, B	0–1600	e	T, me–hg

continued

Table 1 continued

Taxa	Distribution		Zoogeographical category	Ecological data and conservation importance
	Horizontal	Vertical		
Myosotella myosotis (Draparnaud, 1801)	B	0–2	Rp, am	T, ?M, se, hd, r, N
Acroloxidae				
Acroloxus lacustris (Linnaeus, 1758)	DEL, PVS, PT	0–1000	hes	L, ph, sw
Lymnaeidae				
Myxas glutinosa (O. F. Müller, 1774) ?	DW	30	cne	L, rh, sw, o, β, DD, ESC, BC–1, CORINE, W
Limnaea stagnalis (Linnaeus, 1756)	D^r, PV, PT	0–900, −4	h	L, eu
Stagnicola corvus (Gmelin, 1791)	D^r, PT, BN	0–200	hp	L, sw
Stagnicola palustris (O. F. Müller, 1774)	D^r, PVS, PT, RR	0–550	h	L, eu, ph, pe
Stagnicola turricula (Held, 1836)	D^r, DW	0–200	hp	L, sw, pe, r, N
Radix auricularia (Linnaeus, 1758)	D^r, SPW, P, B	0–550	hp	L, b-3‰, rh, po, sw, ph, pe
Radix ovata (Draparnaud, 1801)	D^r, PVS, RP, RR	(0–2500)	hp	L, b-2.5‰, sw
Radix peregra (O. F. Müller, 1774)	D, S, P, R, B	0–2300, −11	hp	L, eu, ph, pe
Galba truncatula (O. F. Müller, 1774)	D, S, P, R, B	0–2000, −6	h	L, eu, pe, ph
Ancylidae				
Ancylus fluviatilis O. F. Müller, 1774	D, S, P, R, B	0–2300	wp	L, cr, po, rh
Ancylus recurvus (Küster, 1852) (?= *A. fluviatilis*)	SBM	500	wp	L, cr, r, N
Ferrissia clessiniana (Jickeli, 1882)	PVS, B	(0–600)	a, i	L, sw, ph, r, N
Planorbidae				
Planorbis carinatus O. F. Müller, 1774	D, S, P, R, B	(0–1200)	wes	L, ph, r, N
Planorbis planorbis (Linnaeus, 1758)	D, S, P, R, B	(0–1000)	wcp	L, b-1.5‰, eu, sw, ph, pe
Anisus leucostomus (Millet, 1813)	DEL, PK, PT	(0–1700)	wces	L, sw, ph
Anisus septemgyratus (Rossmässler, 1835)	DEL, SBE, BN	0–230 (500)	wes	L, b-8‰, sw, ph, r, α – β, N
Anisus spirorbis (Linnaeus, 1758)	D^r, PVV, BN	0–1200	wcp	L, sw, ph, α–β, r, N
Anisus vortex (Linnaeus, 1758)	D^r, PVS, PT, B	(0–1000)	wces	L, b-5–8‰, po, sw, ph, α–β, r, N
Anisus vorticulus (Troschel, 1834)	DEL, PVS, B	0–550	wes	L, sw, ph, r, BC–1, E
Bathyomphalus contortus (Linnaeus, 1758)	PVS, PT	(0–1800)	hes	L, b-8‰, sw, ph, r, α–β, N
Gyraulus albus (O. F. Müller, 1774)	D^r, PT, RRW	(0–1100)	wces	L, b-8‰, po, rh, sw, ph, r, N

continued

Table 1 continued

Taxa	Distribution		Zoogeographical category	Ecological data and conservation importance
	Horizontal	Vertical		
Gyraulus crista (Linnaeus, 1758)	D, PV, PT, BN	0–1200	h	L, b-1.5‰, eu, sw, po, ph, α–β
Gyraulus laevis (Alder, 1838)	D[r], PVS, PT	0–550	e	L, ph, r, N
Gyraulus piscinarum Bourguignat, 1852	PT	160	h, ?wp	L, po, sw, ph
Hippeutis complanatus (Linnaeus, 1758)	DE, P, ROP, B	0–550	wcp	L, ph, sw, α, r, N
Segmentina nitida (O. F. Müller, 1774)	DE, P, RR, BS	0–1100	wcp	L, sw, po, ph, α–β, W VU, ESC, CORINE
Planorbarius corneus (Linnaeus, 1758)	D[r], PVS, PT, B	0–500, −9	wces	L, b-8‰, sw, po, α–β
Physidae				
Physa fontinalis (Linnaeus, 1758)	D[r], PV, PT, B, RR	0–500 (−26)	tp	L, ph, α–β, r, N
Physella acuta (Draparnaud, 1801)	D, S, P, R, B	0–1300	nmt, ?set	L, eu, pe, tx, α–β
Aplexa hypnorum (Linnaeus, 1758)	SBE, PVS, PT	0–550	des	L, sw, α, r, N
STYLOMMATOPHORA				
Cochlicopidae				
Cochlicopa lubrica (O. F. Müller, 1774)	D, S, P, R, B	0–1300	h	T, me–eu
Cochlicopa lubricella (Rossmässler 1835)	SB, P, RO, RPP, B	0–1750	h	T, me–eu, ca
Cochlicopa nitens (Gallenstein, 1852)	SBM, B	0–900	tp	T, me, r, N
Pyramidulidae				
Pyramidula rupestris (Draparnaud, 1801)	S, PK, R, B	0–2700	mca	T, me, ca
Vertiginidae				
Vertigo pusilla O. F. Müller, 1774	SB, PT, R, B	0–1600	eca	T, me
Vertigo substriata (Jeffreys, 1830)	PT, BN	0–160	wes, ?bm	T, me–hg, r, N
Vertigo antivertigo (Draparnaud, 1801)	SB, P, R, BN	0–1000	wcp, ?pat	T, me–hg
Vertigo moulinsiana (Dupuy, 1849)	D, PT, B	0–600	e	T, eu, ca, ESC, $HD\!-_{II}$ LR/cd, CORINE, W
Vertigo pygmaea (Draparnaud, 1801)	D, S, P, R, B	0–1500	h	T, me, eu
Vertigo alpestris Alder, 1838	SB, PV, R	1000–2300	Rg, des	T, me–hg, r, N
Vertigo angustior Jeffreys, 1830	PT, B	0–160	et	T, me-hg, ESC, $HD\!-_{II}$ LR/cd, CORINE, W
Truncatellina claustralis (Gredler, 1856)	SB, PT, R, B	0–1000	nm, mom	T, me–xph, ca
Truncatellina cylindrica (Férussac, 1821)	D, S, P, R, B	0–1850	wp, ?e	T, xph–me, ca, li

continued

Table 1 continued

Taxa	Distribution		Zoogeographical category	Ecological data and conservation importance
	Horizontal	Vertical		
Truncatellina callicratis (Scacchi, 1833)	RR	300–800	mca, mo	T, xph–me, ca, r, N
Truncatellina costulata (Nilsson, 1822)	PBS	0–350	et	T, me, ca, r, N
Columella edentula (Draparnaud, 1805)	RPR, BN	0–1500	h	T, me–hg, r, N
Pupillidae				
Pupilla muscorum (Linnaeus, 1758)	D, S, P, R, B	0–1200	h	T, me–eu
Pupilla triplicata (Studer, 1820)	SBE	0–800	nm(?e)ca	T, me, ca, r, N
Pupilla sterri (Voith, 1838)	SBW, PK, BN	0–1200	eca, ?mo	T, xph–me, ca
Pupilla bogdanovi Urbanski, 1960 (syn.?)	BS	0–3	El	T, r, W
Pupilla valkanovi Urbanski, 1960 (syn.?)	BN	0–3	El	T, r, W
Lauria cylindracea (da Costa, 1778)	PBS, BS	0–350	Rp, am	T, me, r, N
Leiostyla anglica (Wood, 1828)	BS	0–50	Rp, ase	T, me–hg, r, N
Argna macrodonta (Hesse, 1916)	D, S, P, R, B	0–1200	Eb	T, me, ca, li, tr, E
Argna truncatella (L. Pfeiffer, 1841)	SP	300–1000	see, mo	T, me, ca, r, N
Argna parreyssi (L. Pfeiffer, 1848)	?		see	T, me, ca, r, N
Orculidae				
Orculella bulgarica (Hesse, 1915)	DEL, PVV, BN	(0–2000)	Rp, nm	T, me–hg, r, N
Orculella zilchi Urbanski, 1960	PBS, BS	0–300	Er	T, me–hg, r, W
Sphyradium doliolum (Bruguière, 1792)	D, S, P, R, B	0–1900	nmca, mo	T, me, eu, ca
Pagodulina subdola (Gredler, 1856)	PT, RRW	0–1200	• cseean	T, me, r, me–hg, E
Pagodulina pagodula (Desmoulins, 1830) ?	RR	0–1200	seean	T, me
Chondrinidae				
Granaria frumentum (Draparnaud, 1801)	SB, PT	500–1500	csee	T, me, ca
Chondrina avenacea (Bruguière, 1792)	SB, PK, PV, RR	500–1800	se, mom	T, xph, ca, li
Chondrina clienta (Westerlund, 1883)	S, PK, PV, RR	0–1900	• nm, mo	T, xph, ca, li, r, E
Chondrina spelta (Beck, 1837) ?	?	?	Eb	T, E
Valloniidae				
Vallonia costata (O. F. Müller, 1774)	D, S, P, R, B	0–1300	h	T, me, eu
Vallonia pulchella (O. F. Müller, 1774)	D, S, P, R, B	0–1200	h	T, me, eu
Vallonia enniensis (Gredler, 1856)	P, RR, B	0–500	wp	T, me, DD, W
Vallonia excentrica Sterki, 1892	SB, PK, PT, B	0–1200	h	T, me, r, N
Acanthinula aculeata (O. F. Müller, 1774)	D, S, P, R, B	0–1500	wp	T, me
Spelaeodiscus triaria (Rossmässler, 1839)	SBW	600–1000	see	T, me, ca, r, N
Enidae				
Merdigera obscura (O. F. Müller, 1774)	D, S, P, R, B	0–2000	wp	T, me

continued

Table 1 continued

Taxa	Distribution		Zoogeographical category	Ecological data and conservation importance
	Horizontal	Vertical		
Merdigera montana (Draparnaud, 1801)	S, PV, RP, RR	500–2000	e	T, me–hg
Zebrina detrita (O. F. Müller, 1774)	D, S, P, R, B	0–2500	• cseean	T, xph, ca, E
Zebrina kindermanni (L. Pfeiffer, 1850)	PBB, BS	0–100	• ban	T, me–xph, E
Zebrina varnensis (L. Pfeiffer, 1847)	BN	0–100	Er	T, xph, ca, W
Chondrus tournefortianus (Férussac, 1821)	BN	0–20	ban, ?p	(T), xph, r, N
Chondrus zebra (Olivier, 1801)	RRW	300–2000	pm	T, xph, r, N
Multidentula ovularis (Olivier, 1801)	DEL, B	0–300	em	T, me
Multidentula squalina (Rossmässler, 1848)	B	0–50	ban, ?p	T, me, ca, r, N
Pseudochondrula seductilis (Rossmässler, 1848)	SP, PT, B	0–900	ban, ?pm	T, me, ca
Chondrula tridens (O. F. Müller, 1774)	D, S, P, R, B	0–1900	et, ?nmt	T, xph–me
Chondrula microtragus (Rossmässler, 1848)	D, S, P, R, B	0–1000	pm, ?nem	T, xph–me, ca
Chondrula macedonica A. Wagner, 1915	ROT, ROP	500–1000	Eb	T, xph–me, ca, r, E
Chondrula bicallosa (L. Pfeiffer, 1847)	PT, PBS, RRW, B	0–700	ban, ?Ebs	T, me–xph, r, N
Mastus carneolus (Mousson, 1863)	PT, BS	0–150	ban, ?pm	T, me–xph, ca
Mastus rossmaessleri (L. Pfeiffer, 1846)	SBW, PT, RR, B	0–600	pm	T, me–xph
Mastus etuberculatus (Frauenfeld, 1867) (? = *M. pupa* L.)	PT, RO, RP, RRW, B	0–1000	em, ?pm	T, me–xph, ca
Clausiliidae				
Serrulina serrulata (L. Pfeiffer, 1847)	PT, B	0–50	Rp, p	T, me–hg, r, N
Dobatia goettingi (Brandt, 1961)	BS	20	ban, p	T, me, r, N
Carinigera schuetti Brandt, 1962	RPS	200–600	Eb	T, me–xph, ca, r, E
Carinigera buresi (A. Wagner, 1927)	RPM	600	Eb	T, me, r, E
Cochlodina laminata (Montagu, 1803)	D, S, P, RP, RRW, B	200–2000	e	T, me, ca
Macedonica marginata (Rossmässler, 1835)	D, S, P, R	400–1900	• Ebs	T, me–xph, ca, E
Macedonica macedonica (Rossmässler, 1839) ?	ROP, RP	150–2000	Eb	T, me–xph, ca, E
Macedonica frauenfeldi (Rossmässler, 1856)	SBW, PK, PV	400–800	Eb	T, me, ca, E
Macedonica marthae Sajo, 1968	RPP	2400	El	T, me, ca, W
Macedonica pinteri Sajo, 1968	? SPW, PK	375–800	Eb	T, me, ca, E
Macedonica hartmuti Irikov, 2003	RRW	1300	El	T, me–xph, ca, W
Macedonica zilchi Urbanski, 1971	RRW	1200	Er	T, me, ca, W

continued

Table 1 continued

Taxa	Distribution		Zoogeographical category	Ecological data and conservation importance
	Horizontal	Vertical		
Macedonica brabeneci Nordsieck, 1977	RRW	1600–1700	Er	T, me, ca, W
Idyla castalia (Roth, 1856)	RPP	1000–2600	Eb	T, me–xph, ca, E
Euxina circumdata (L. Pfeiffer, 1848)	PBS, BS	0–350	Rp, ban, p	T, me, N
Euxina persica (Boettger, 1879)	PT, PBS, BS	0–350	• p	T, me–hg, E
Euxina pontica (Retowski, 1887)	? PT, RR, B	0–200	ban, p	T, me–hg
Galeata schwerzenbachi (L. Pfeiffer, 1848)	PBS, BS	0–350	Rp, ban, p	T, me, N
Ruthenica filograna (Rossmässler, 1836)	SPM	250	cee	T, me
Micridyla pinteri (Nordsieck, 1973)	RRW	700–1200	Er	T, me, W
Clausilia pumila C. Pfeiffer, 1828	RPP	500–1900	see	T, me
Mentissela rebeli (Sturany, 1897)	DEP, SBE, B	0–500	Ebg	T, me–xph, ca, E
Laciniaria plicata (Draparnaud, 1801)	D, S, P, R, B	100–1500	• cee	T, me, ca, E
Laciniaria macilenta (Rossmässler, 1842)	RRW	250–700	Er	T, me, ca, li, W
Laciniaria bajula (A. Schmidt, 1868)	? SB, RRW	350–1500	Er	T, me–hg, ca, r, W
Balea biplicata (Montagu, 1803)	SB, PVV, R	0–2300	csee	T, me
Balea vratzatica (Likharev, 1972)	SBW	370–1000	El	T, tph, xph, ca, W
Balea wagneri (Wohlberedt, 1911)	SBM, ? RRW	350–700	Er	T, me, W
Balea kaeufeli (Brandt, 1962)	SBM	1200–1350	El	T, me, W
Balea perversa (Linnaeus, 1758)	SBM	600 (2400)	e	T, me–xph, ca, V, ESC, CORINE, W
Balea serbica (Moellendorff, 1873)	ROO	1350–2000	Er	T, me, W
Balea mursalicae (Urbański, 1969) (? = *Laciniaria bajula*)	RRW	1100–1800	Er	T, me, W
Balea petrohanica Urbański, 1969 (syn.?)	SBW	1400	El	T, me, W
Vestia ranojevici (Pavlović, 1912)	SB, PV, RP, RR	350–1800	• see	T, me
Vestia roschitzi (Brancsik, 1889)	SPW	550	• Ebs	T, me, E
Vestia petrovici (Pavlović, 1912)	SPW	550	Eb	T, me, E
Bulgarica vetusta (Rossmässler, 1836)	SPW, PVV	550–1000	see	T, me–hg, r, N
Bulgarica varnensis (L. Pfeiffer, 1848)	D, S, PB, R, B	0–1350	Eb	T, me, ca, E
Bulgarica fraudigera (Rossmässler, 1839)	? SBV, RRW	350–1000	Eb	T, me, ca, li, E
Bulgarica urbanski Nordsieck, 1973	SBE	300–1100	Er	T, me–xph, W
Bulgarica hiltrudae Nordsieck, 1974	SBW	300	El	T, me, ca, W
Bulgarica fritillaria (Rossmässler, 1839)	SB, PT, R, BN	0–1000	Ebg	T, me, E
Bulgarica bulgariensis (L. Pfeiffer, 1848)	SPM, SBM, RRW	200–500	Ebg	T, me, ca, E

continued

Table 1 continued

Taxa	Distribution		Zoogeographical category	Ecological data and conservation importance
	Horizontal	Vertical		
Bulgarica rugicollis (Rossmässler, 1836) ?	SPM	250	see, ?Ebs	T, me, ca
Bulgarica denticulata (Olivier, 1801)	P, R, B	0–1600	Eb	T, me, E
Succineidae				
Succinea putris (Linnaeus, 1758)	BN	0–20 (1800)	hes	T, hg–hd, r, N
Succinea oblonga (Draparnaud, 1801)	D, S, P, R, B	0–1300	wces	T, hg
Oxyloma elegans (Risso, 1826)	S, PT, RP, B	0–700	h	T, hg–hd
Oxyloma sarsii (Esmark, 1886) ?	?		cne	T, hd, CORINE, ESC, E
Ferussaciidae				
Cecilioides acicula (O. F. Müller, 1774)	D, S, P, R, B	0–1000	ean, ?et	T, me, ca
Cecilioides jani (De Betta et Martinati, 1855)	S, PT, R, B	0–1000	Eb, ?nem	T, me, ca, E
Cecilioides spelaeus (A. Wagner, 1914)	PT, RR	100–700	Eb	T, me, ca, E
Subulinidae				
Rumina decollata (Linnaeus, 1758)	BN	(0–950)	Rp, hm, ?i	T, me, r, N
Oleacinidae				
Poiretia algira (Bruguière, 1792)	PT, ROP, RP	70–600	Rp, hm	T, hg, r, N
Arionidae				
Arion lusitanicus Mabille, 1868	SBW, SBM	350–450	e, ?se, ?i	T, me–hg
Arion subfuscus (Draparnaud, 1801)	S, P, R	150–2900	e	T, me
Arion silvaticus Lohmander, 1937	S, P, R	200–1600	e	T, me–hg
Arion hortensis Férussac, 1819	PVS	500–600	c, ?h	T, me, eu, r, N
Arion fasciatus (Nilsson, 1822) ?	PVV, PVL, PVS	750–900	cne	T, me, eu
Euconulidae				
Euconulus fulvus (O. F. Müller, 1774)	D, S, P, R, B	0–1500	h	T, me
Vitrinidae				
Vitrina pellucida (O. F. Müller, 1774)	SBW, P, R, B	0–1800	h	T, me–hg
Eucobresia diaphana (Draparnaud, 1805)	PK, PVV, RP	1500–2600	e, mo	T, me–hg
Phenacolimax annularis (Studer, 1820)	SB, RP, RR, B	0–2800	et, mom	T, me
Semilimacella reitteri (O. Boettger, 1880)	RRW	1450	see	T, me
Zonitidae				
Vitrea diaphana (Studer, 1820)	SB, PV, PB, R	300–1300	csee, ?mo	T, me, tx
Vitrea transsylvanica (Clessin, 1877)	SPM	200–500	see, ?kb	T, me, r, N
Vitrea subrimata (Reinhardt, 1871)	SBW, SBM	500–2000	cse, ?hm	T, me
Vitrea bulgarica Damjanov et Pinter, 1969 (syn.?)	SPM, R	250–1900	Eb	T, me, E

continued

Table 1 continued

Taxa	Distribution		Zoogeographical category	Ecological data and conservation importance
	Horizontal	Vertical		
Vitrea neglecta Damjanov et Pinter, 1969	SB, RO, RRW	250–1200	Ebg	T, me, E
Vitrea contracta (Westerlund, 1871)	SB, P, R, B	0–1400	wp	T, me, ca
Vitrea riedeli Damjanov et Pinter, 1969	RRE, BS	0–500	ban	T, me, r, N
Vitrea pygmaea (O. Boettger, 1880)	SB, PB, RR, B	0–1400	Rp, hm	T, me–xph, N
Vitrea vereae Irikov, Georgiev et Riedel, 2004	PSA	300–650	Er	T, xph–me, r, W
Spinophallus uminskii (Riedel, 1960)	SBM	500–1500	El	T, tr, r, W
Aegopinella pura (Alder, 1830)	D, S, P, R, B	0–2400	wes	T, me
Aegopinella minor (Stabile, 1864)	SB, PV, R, B	0–1200	csee, ?e	T, me–xph
Aegopinella nitens (Michaud, 1831)	PVS	550 (2500)	csee	T, me–hg
Balcanodiscus frivaldskyanus (Rossmässler, 1848)	SPM, SBM, RR	250–1100	Eb	T, me, ca, tr, E
Oxychilus translucidus (Mortillet, 1854)	PVS, PT	150–600	pt, i	T, me
Oxychilus deilus (Bourguignat, 1857)	DE, SB, RR, B	0–350	Rp, p	T, me–hg, N
Oxychilus hydatinus (Rossmässler, 1838)	SPM, P, R, B	0–1100	Rp, hm	T, me, tr, N
Oxychilus draparnaudi (Beck, 1837)	PKK, PVS	550	we, ?i	T, me–xph, tr
Oxychilus investigatus Riedel, 1993	SBE, PBS	200–500	Rp, ban, p	T, me, tx, N
Oxychilus glaber (Rossmässler, 1835)	D, S, P, R, B	0–1850	cse, mo	T, me–hg, tr
Oxychilus urbanskii Riedel, 1963	SBE, PB, BS	0–800	Rp, ban, p	T, me, tr, N
Oxychilus inopinatus (Uličný, 1887)	SB, P, R, BN	0–2000	csee, mo	T, me–hg, ca
Oxychilus depressus (Sterki, 1880)	SB, PVV, R	200–2600	csee, mo	T, me–hg, tr
Oxychilus camelinus (Bourguignat, 1852) ?	PT	150	em, i	T, me
Nesovitrea hammonis (Ström, 1765)	SBW, PV, RP	400–1500	wces	T, me–xph
Daudebardia rufa (Draparnaud, 1805)	D, S, P, RP, B	0–1500	csean, mo	T, me–hg, tx
Daudebardia brevipes (Draparnaud, 1805)	S, PVV, R, BS	0–1500	csean, mo	T, me–hg, tx
Daudebardia wiktori Riedel, 1967	SBW, RR	350–1400	ban, p	T, me–hg, tx
Carpathica bielawskii Riedel, 1963	PBS, BS	0–350	Ebg	T, me, E
Carpathica stussineri (A. Wagner, 1895)	SBW, PBS, PV, R	0–1400	see	T, me–hg
Zonitoides nitidus (O. F. Müller, 1774)	SB, PK, RR, B	0–1200	h	T, hg–hd
Milacidae				
Milax parvulus Wiktor, 1968	SB, P, RR, BS	0–1300	Ebg	T, me–hg, ca, E

continued

Table 1 continued

Taxa	Distribution		Zoogeographical category	Ecological data and conservation importance
	Horizontal	Vertical		
Milax verrucosus Wiktor, 1969	SBW, SBM	300–1200	Er	T, me, W
Tandonia kusceri (H. Wagner, 1931)	D, S, P, R, B	0–2000	Eb	T, me, eu, tx, E
Tandonia serbica (H. Wagner, 1931)	SBW, P, R	100–1800	Eb	T, me, ca, E
Tandonia budapestensis (Hazay, 1881)	SBW, PVV, R	250–2200	see	T, me, eu
Tandonia piriniana Wiktor, 1983	RPP	300–700	Er	T, me, r, W
Tandonia cristata (Kaleniczenko, 1851)	S, P, RR, B	0–1500	pm	T, me
Tandonia pinteri (Wiktor, 1975)	RRW	700–1500	Er	T, me, r, W
Tandonia rustica (Millet, 1843) ?	PT, RPR, RR	200–1600	cse	T, me
Tandonia totevi (Wiktor, 1975)	RRW	750	Eb	T, me, r, E
Punctidae				
Punctum pygmaeum (Draparnaud, 1801)	D, S, P, R, B	0–1500	h	T, me
Discus ruderatus (Férussac, 1821)	PVV, RP, RR	300–2000	tp, bm	T, me, r, N
Discus perspectivus (Megerle v. Mühlfeld, 1816)	SPM, SBM	400–1500	csee, mo	T, me, ca, r, N
Limacidae				
Limax maximus Linnaeus, 1758	D, S, P, R, B	0–1500	e, ?wp	T, me
Limax subalpinus Lessona, 1880	RPR	1400–1600	se	T, me, r, N
Limax cinereoniger Wolf, 1803	DE, S, P, R, B	0–2500	e	T, me
Limax punctulatus Sordelli, 1870	SBM, RRW	400–1500	cse	T, me
Limax macedonicus Hesse, 1928	SB, P, R, B	0–2000	Eb	T, me, E
Limax conemenosi O. Boettger, 1882	SB, PV, RO, RPP	150–1300	Eb	T, me–hg, E
Limax flavus Linnaeus, 1758	SB, P, ROP, B	0–800	Rp, p, nm	T, me, N
Limax maculatus (Kaleniczenko, 1851)	PBS, B	0–350	nm	T, me
Lehmannia marginata (O. F. Müller, 1774) ?	S, P, R	300–1500	e	T, me
Lehmannia nyctelia (Bourguignat, 1861)	S, P, R	250–2700	?csee, ?kb	T, me
Lehmannia brunneri (H. Wagner, 1931)	SBM, PK, PV, R	700–2650	Eb	T, me–hg, E
Lehmannia horezia Grossu et Lupu, 1962	SBM	500–2200	kb	T, me, r, N
Lehmannia sarmizegetusae Grossu, 1970	PVV	(800–2000)	kb, mo	T, me, r, N
Malacolimax tenellus (O. F. Müller, 1774)	SBW, RR	300–700	e	T, me, r, N
Agriolimacidae				
Deroceras laeve (O. F. Müller, 1774)	S, PV, RPR, RR	250–2500	h	T, me–hg
Deroceras sturanyi (Simroth, 1894)	S, P, R, B	0–800	e	T, me–hg
Deroceras panormitanum (Lessona et Pollonera, 1882)	RPP	250–1500	e	T, me

continued

Table 1 continued

Taxa	Distribution		Zoogeographical category	Ecological data and conservation importance
	Horizontal	Vertical		
Deroceras turcicum (Simroth, 1894)	S, P, R, B	0–2000	see	T, me–hg
Deroceras reticulatum (O. F. Müller, 1774)	SBW, PVS, BS	0–1300	e	T, me–hg
Deroceras thersites (Simroth, 1886)	SPE, PB, RRE, B	0–400	Eb	T, me, r, E
Deroceras agreste (Linnaeus, 1758)	SBW, PB, RP	0–2600	• wcp	T, me, E
Deroceras pageti Grossu, 1972 ?	DEP	50	Eb	T, me, r, E
Deroceras zilchi Grossu, 1969	SPW, SBW	350–450	Ebs	T, me, tx, r, E
Deroceras bulgaricum Grossu, 1969	PBT	50	El	T, me, r, W
Deroceras bureschi (H. Wagner, 1934) (syn.?)	SB, PV, PSC, R	400–2000	Eb	T, me–hg, tx, E
Krynickillus urbanskii (Wiktor, 1971)	DEP, SB, PBS, B	0–1300	Rp, ban, p	T, tx, N
Helicidae				
Arianta arbustorum (Linnaeus, 1758)	DEL, SPW, SPM	100–1500	e	T, me–hg, r, N
Arianta pelia (Hesse, 1912)	SB, PVV, RPR	1000–2700	Ebg	T, me–hg, ?ca, E
Helicigona trizona (Rossmässler, 1835)	DEL, S, PV, PT, R	150–1700	• see	T, me–hg, ca, E
Helicigona haberhaueri (Sturany, 1897)	SB, RP, RR	800–2000	Ebg	T, me–hg, r, li, E
Helicigona kattingeri (Knipper, 1941)	ROP	300	Eb	T, me, r, E
Wladislawia polinskii (A. Wagner, 1927)	RPP	1800–2914	El	T, me–hg, ca, W
Wladislawia sztolcmani (A. Wagner, 1927)	RPP	2000–2700	El	T, me, ca, W
Eobania vermiculata (O. F. Müller, 1774)	PT, B	0–150	hm	T, xph, r, N
Cepaea vindobonensis (Férussac, 1821)	D, S, P, R, B	0–1500	pm	T, xph
Cryptomphalus aspersus (O. F. Müller, 1774)	PT, BN	0–150	we, i	T, me, r, N
Helix albescens Rossmässler, 1839	DE, PVS, T, RO, RP	150–650	Rp, p	T, xph–me, N
Helix lucorum Linnaeus, 1758	D, S, P, R, B	0–1200	• nmt	T, xph–me, eu, E
Helix pomatia Linnaeus, 1758	D, S, P, R, B	0–1900	• e	T, me–hg, CORINE, ESC, BC-3, HD_{V}, E
Helix pomacella Mousson, 1854	SBE, PBS, BS	0–1000	Rp, ban, p	T, me, N
Helix figulina Rossmässler,1839	SBE, PT, ROP, RR	50–400	Rp, pm, p	T, me, N
Bradybaenidae				
Bradybaena fruticum (O. F. Müller, 1774)	D, S, PV, PT, R	0–1770	hes	T, me–hg
Helicodontidae				
Soosia diodonta (Férussac, 1821)	PVV	1300	kb	T, me–hg, r, N

continued

Table 1 continued

Taxa	Distribution		Zoogeographical category	Ecological data and conservation importance
	Horizontal	Vertical		
Lindholmiola girva (Frivaldszky, 1835)	D, S, P, R, B	0–2000	• pm	T, me–xph, E
Hygromiidae				
Cochlicella acuta (O. F. Müller, 1774)	BS	0–50	am, ?i	T, me, r, N
Trichia hispida (Linnaeus, 1758)	PKK, RR	700–1000	e	T, me, r, N
Trichia erjaveci (Brusina, 1870)	SB, PK, PB, RR, B	0–1230	Ebs	T, me–hg, E
Xerolenta obvia (Menke, 1828)	D, S, P, R, B	0–1900	• seean	T, xph, E
Helicella macedonica Hesse, 1928	RPP, RPS	1200–2000	Eb	T, xph–me, E
Helicella spiruloides (A. Wagner, 1916)	B	0–150	Ebg	T, xph, r, E
Helicopsis striata (O. F. Müller, 1774)	DM, PSP, PT, B	0–300	cseean	T, xph–me
Helicopsis rhabdotoides (A. Wagner 1927)	S, PKK, PVS, BN	0–800	Eb	T, xph, ca, E
Helicopsis dejecta (Jan, 1832)	DEL, ROS	100–300	pm	T, xph, r, N
Helicopsis instabilis (Rossmässler, 1835)	DEL, SPM, BN	0–300	see	T, xph, r, N
Xeropicta krynickii (Krynicki, 1833)	PT, B	0–150	pt, ?nemt	T, xph, r, N
Xeropicta derbentina (Krynicki, 1833)	DEL	50–200	em	T, xph, r, N
Pseudotrichia rubiginosa (Rossmässler, 1838)	D, S, P, R, B	0–500	hes, ?tp	T, me–hg
Monachoides incarnatus (O. F. Müller, 1774)	S, PV, PS, RP, RR, B	0–1600	csee, ?mo	T, me–hg
Cernuella virgata (da Costa, 1778)	PT, B	0–150	am	T, xph, tph
Cernuella jonica (Mousson, 1854)	SB, PT, B	0–800	nem	T, xph
Monacha carascaloides (Bourguignat, 1855)	S, PT, B	0–1000	ban	T, me–xph, ca
Monacha pilosa Pinter, 1969 ?	PBS, BS	0–250	Er	T, me–xph, W
Monacha cartusiana (O. F. Müller, 1774)	D, S, P, R, B	0–1000	nm	T, me–xph
Euomphalia strigella (Draparnaud, 1801)	D, S, P, R, B	0–1600	e	T, me–hg, ?ca
BIVALVIA				
PTERIOMORPHA				
ARCOIDA				
Arcidae				
Anadara inaequivalvis (Bruguière, 1789)	B	−25	im, i	M, se
Noetiidae				
Striarca lactea (Linnaeus, 1758)	B	−15	mlg	M, se
MYTILOIDA				
Mytilidae				
Mytilus galloprovincialis Lamarck, 1819	B	−70	ml	M, eb
Mytilaster lineatus (Gmelin, 1790)	B	−30(−50)	ml	M, b-5‰, se
Modiolus adriaticus (Lamarck, 1819)	B	−20, −40	ml	M, mb, pe
Modiolula phaseolina (Philippi, 1844)	B	−30, −140	mls	M, sh, pl
OSTREOIDA				
Pectinidae				
Chlamys glabra (Linnaeus, 1758)	B	−40	m	M, se

continued

Table 1 continued

Taxa	Distribution		Zoogeographical category	Ecological data and conservation importance
	Horizontal	Vertical		
Anomiidae				
Anomia ephippium (Linnaeus, 1758)	BN	(−30)	mls	M, se
Ostreidae				
Ostrea edulis Linnaeus, 1758	B	−20(−30)	mls	M, se
Ostrea lamellosa Brocchi, 1814 (syn.?)	B	−10	ml	M, se
PALAEOHETERODONTA				
UNIONIDA				
Unionidae				
Unio crassus Retzius, 1783	D^r, D, SPW, P	0–400	e	L, rh, po, ps–pe, BC–1, LR/nt, $HD_{II,IV}$, W
Unio pictorum (Linnaeus, 1758)	D^r, D, SP, P, B	0–500, −11	e	L, b-2‰,rh, pe–ps, po
Unio tumidus Philipsson, 1788	D^r, DE, PBT, BN	0–100, −9	e	L, rh, po, ar, pe–ps
Anodonta cygnaea (Linnaeus, 1758)	D^r, PVS, PT, BS	0–800, −17	e	L, b-2‰,rh, po,sw, pe
Anodonta anatina (Linnaeus, 1758)	D^r, DEL, PVS	0–500, −3	wes	L, po, sw, pe
Pseudanodonta complanata (Rossmässler, 1835)	D^r, DEL, BS	0–50, −9	e	L, po, sw, pe–ps, LR/nt, r, W
Colletopterum letourneuxi Bourguignat, 1881	D^r	0–30	p	L, ps–pe, r, N
HETERODONTA				
VENEROIDA				
Lucinidae				
Loripes lacteus (Linnaeus, 1758)	B	−25	ml	M, se, ps
Lucinella divaricata (Linnaeus, 1758)	B	−30	ml	M, se
Leptonidae				
Hemilepton nitidum (Turton, 1822)	B	(−25)	mls	M, se, ps
Montacutidae				
Mysella bidentata (Montagu, 1803)	BS	(−50)	mls	M, se
Cardiidae				
Acanthocardia paucicostata (Sowerby, 1841)	B	−30, −80	ml	M, mb, pe
Parvicardium exiguum (Gmelin, 1791)	B	−10, −120	mls	M, b-3‰, eb, pe
Plagiocardium papillosum (Poli, 1795)	B	−30(−90)	ml	M, eb, pe
Cerastoderma edule (Linnaeus, 1758)	B	−80	mls	M, eb, pe
Cerastoderma glaucum (Poiret, 1789)	B	−10, −50	ml	M, b-3‰, eb, ps–pe
Hypanis colorata (Eichwald, 1829)	B	−13, −148	Sf, Rc, p	(M), b, se, N
Hypanis plicta (Eichwald, 1829)	B	−35, −92	• Sf, Rc,pc	(M), b, se, E
Mactridae				
Spisula subtruncata (da Costa, 1778)	B	−10, −100	mls	M, eb, pe
Mesodesmatidae				
Donacilla cornea (Poli, 1791)	B	−0.2, −2	ml	M, se, ps
Solenidae				
Solen marginatus Pulteney, 1799	B	−10	mls	M, se, ps

continued

Table 1 continued

Taxa	Distribution		Zoogeographical category	Ecological data and conservation importance
	Horizontal	Vertical		
Tellinidae				
Gastrana fragilis (Linnaeus, 1758)	B	−36	mls	M, se
Tellina donacina (Linnaeus, 1758)	B	−20(−30)	ml	M, se, ps–pe
Tellina tenuis da Costa, 1778	B	−24	mls	M, se, ps
Tellina fabula Gmelin, 1791	B	−10(−40)	mls	M, se, ps, r, N
Donacidae				
Donax semistriatus Poli, 1795	B	−15	m	M, se, ps
Donax trunculus (Linnaeus, 1758)	B	−10	• mls	M, se, ps, E
Semelidae				
Abra alba (Wood, 1802)	B	−135(−170)	mls	M, eb, pe
Abra nitida (O. F. Müller, 1776)	B	−90	• ml	M, eb, pe, E
Abra ovata (Philippi, 1836) (syn.?)	B	(−15)	ml	M, b-4‰, se
Dreissenidae				
Dreissena polymorpha (Pallas, 1771)	D^r, PVS, PT, B	0–500, −15	Rc, e	L, b-3‰, sw, po, li
Dreissena rostriformis (Deshayes, 1838)	B	−50, −200	• Sf, Rc, pc	(M), b, ? mb
Corbiculidae				
Corbicula fluminea (O. F. Müller, 1774)	D^r	0–50	spo, i	L, b, po
Sphaeriidae				
Pisidium amnicum (O. F. Müller, 1774)	D^r, PV, PT, RP	0–2000, −7	hp	L, rh, po, sw
Pisidium henslowanum (Sheppard, 1825)	D^r, D	0–660, −5	hp	L, eu, pe
Pisidium supinum A. Schmidt, 1851	BS	20	wes	L, sw, ph, ?EX, r, E
Pisidium milium Held, 1836?	PVS, PVV	450–700	h	L, b-0.5‰, sw, r, N
Pisidium subtruncatum Malm, 1855	PVS, PT, RPR	150–1000	wcp	L, sw, r, N
Pisidium nitidum Jenyns, 1832	PVS	550 (2650)	hp	L, sw, ps, r, N
Pisidium pseudosphaerium Schlesch, 1947	DEL	20	wes	L, sw, ph
Pisidium obtusale (Lamarck, 1818)	PVS, RPR	500–1600	hes	L, po, sw, o, r, N
Pisidium personatum Malm, 1855	DE, SB, PV, RP, BN	0–2000	wces	L, po, sw, cr, tx
Pisidium casertanum (Poli, 1791)	D^r, S, PVV, RP	0–2370, −8	hp	L, po, sw, ps, tx
Pisidium moitessierianum Paladilhe, 1866	D^r	0–230	e	L, po, sw, r, N
Pisidium bulgaricum Odhner, 1929 (syn.?)	PVV	2000	? El	L, sw, r, W
Sphaerium corneum (Linnaeus, 1758)	D^r, SP, PV, PT, R	0–1100	hp	L, b-5‰, sw, eu, pe, ph, r, N
Sphaerium rivicola (Lamarck, 1818)	D^r, D, PT	0–500, −4	cee, ?, e	L, po, t^0, ps, pe
Musculium lacustre (O. F. Müller, 1774)	D^r, PV, PT, RR, B	0–1100, −4	tp	L, b-3‰, sw
Veneridae				
Chamelea gallina (Linnaeus, 1758)	B	−30(−55)	• mls	M, se–mb, E
Gouldia minima (Montagu, 1803)	B	−25(−70)	mls	M, se, ps–pe
Pitar rudis (Poli, 1795)	B	−11, −40	• mlg	M, se–mb, E
Irus irus (Linnaeus, 1758)	B	−10	ml	M, se, li
Paphia aurea (Gmelin, 1791) (complex)	B	−60	• mls	M, mb, pe, E
Paphia petalina (Lamarck, 1818) (syn.?)	B	−14, −60	m	M, mb, pe, r, N

continued

Table 1 continued

Taxa	Distribution		Zoogeographical category	Ecological data and conservation importance
	Horizontal	Vertical		
Petricolidae				
Petricola lithophaga (Retzius, 1786)	B	−10(−26)	ml	M, se, li
MYOIDA				
Myidae				
Mya arenaria Linnaeus, 1758	B		cb, i	M, b-5‰, se, ps
Corbulidae				
Lentidium mediterraneum (Costa, 1829)	B	−10(−20)	• m	M, b-7‰, se, E
Pholadidae				
Pholas dactylus Linnaeus, 1758	B	−10	mls	M, se, li, B, BC, E
Barnea candida (Linnaeus, 1758)	B	(−15)	mls	M, se, li, r, N
Teredinidae				
Teredo navalis Linnaeus, 1758	B	−70	k	M, se–mb
Teredo utriculus Gmelin, 1791	B	−70	mls	M, se–mb
ANOMALODESMATA				
PHOLADOMYOIDA				
Thraciidae				
Thracia papyracea (Poli, 1791)	B	−15	mls	M, se, r, N

is paid to the immigrants from remote seas that significantly changed the Black Sea communities in the second half of 20th century.

The conservation value of taxa (Table 1) is determined regarding their populations inhabiting Bulgaria. For local endemics, 100% of their populations are localized in Bulgaria, therefore they are given the highest conservation category (world importance). This category also includes regional endemics because of its restricted distribution and species from the IUCN Red List. Taxa of European importance include Balkan endemics and subendemics as well as the species from ESC Red List, Bern Convention, Habitats Directive, and CORINE (Coordination of Information on the Environment) lists. Relicts and rare taxa (if not listed under other category) form the group of national importance.

2·1 Abbreviations used

Horizontal distribution: Territorial units: D, DANUBIAN PLAIN; DW, Western Danubian Plain; DM, Middle Danubian Plain; DE, Eastern Danubian Plain; DEL, Ludogorie–Dobrudja District; DEP, Popovo–Provadiya District. S, STARA PLANINA RANGE; SP, *Predbalkan*; SPW, Western Predbalkan; SPM, Middle Predbalkan; SPE, Eastern Predbalkan; SB, *Stara Planina (Balkan) Mts.*; SBW, Western Stara Planina Mts.; SBM, Middle Stara Planina Mts.; SBE, Eastern Stara Planina Mts. P, TRANSITIONAL REGION; PK, *Kraishte–Konyavo District*; PKR, Rui Mts.; PKG, Golo Bardo Mts.; PKV, Verila Mts.; PKK, Kraishte; PKZ,

Zemenska Planina Mts.; PKQ, Konyavska Planina Mts.; PV, *Vitosha District*; PVS, Sofia Basin; PVL, Lyulin Mts.; PVV, Vitosha Mts.; PVP, Plana Mts.; PS, *Srednogorie–Podbalkan Subregion*; PSP, Podbalkan Basins; PSS, Sredna Gora Mts.; PSI, Ihtimanska Sredna Gora; PSL, Lozenska Planina Mts.; PSC, Sashtinska Sredna Gora; PSA, Sarnena Gora; PT, Thracian Lowland; PB, *Tundja–Strandja Subregion*; PBT, Sakar–Tundja District; PBC, Sakar Mts.; PBB, Bakadjik–Burgas District; PBD, Strandja–Dervent District; PBS, Strandja Mts. R, RILA–RHODOPE MASSIF; RO, *Osogovo–Belasitsa Group*; ROO, Osogovska Planina Mts.; ROV, Vlahina Planina Mts.; ROM, Maleshevska Planina Mts.; ROG, Ograzhden Mts.; ROB, Belasitsa Mts.; ROS, Srednostrumska Valley; ROT, Boboshevo–Simitli Valley; ROP, Krupnik–Sandanski–Petrich Valley; RP, *Rila–Pirin Group*; RPR, Rila Mts.; RPP, Pirin Mts.; RPS, Slavyanka Mts.; RPT, Stargach Mts.; RPM, Mesta Valley; RR, *Rhodope Mts.*; RRW, Western Rhodope Mts.; RRE, Eastern Rhodope Mts. B, BLACK SEA COAST; BN, northern Black Sea coast; BS, southern Black Sea coast. D^r, the Danube River and its marshes.

Vertical distribution: " - ", the sign before the figure is indication of depth; (), figures in parentheses present foreign data; ?, questionable or missing data.

Zoogeographical categories: a, Australian; am, Atlantomediterranean; ant, Anatolian–Turanian; ap, Aegean–Pontian; ase, Atlantic–South European; ban, Balkan–Anatolian; bm, Boreomontane; cb, Circumboreal; cee, Central and East European; cne, Central and North European; cse, Central and South European; csean, Central and South European–Anatolian; csee, Central and Southeast European; cseean, Central and Southeast European–Anatolian; des, Disjunct Eurosiberian; e, European; ean, European–Anatolian; Eb, Balkan endemic; Ebg, Bulgarian endemic; Ebs, Balkan subendemic; eca, European–Central Asian; ee, East European; El, local endemic; Em, Black Sea endemic; em, East Mediterranean; Er, regional endemic; et, European–Turanian; h, Holarctic; hes, Holoeurosiberian; hm, Holomediterranean; hp, Holopalearctic; i, introduced species (immigrants); im, Indo-Malayan; j, Japanese; k, Cosmopolitan; kb, Carpathian–Balkan; m, Mediterranean; mca, Mediterranean–Central Asian; ml, Mediterranean–Lusitanian; mlg, Mediterranean–Lusitanian–Guinean; mls, Mediterranean–Lusitanian–Scandinavian; mm, Mediterranean–Mauritanian; mo, montane; mom, Montane Mediterranean; nem, Northeast Mediterranean; nemt, Northeast Mediterranean–Turanian; nm, North Mediterranean; nmca, North Mediterranean–Central Asian; nmt, North Mediterranean–Turanian; p, Pontian; pat, Palearctic–Afrotropical; pb, Pontian–Balkan; pc, Pontian–Caspian; pcsee, Pontian–Caspian–Southeast European; pm, Pontomediterranean; psee, Pontian–Southeast European; pt, Pontian–Turanian; Rc, Caspian relict; Rg, glacial relict; Rp, preglacial relict; se, South European; see, Southeast European; seean, Southeast European–Anatolian; set, South European–Turanian; Sf, subfossil; spo, South Palearctic–Oriental; tp, Transpalearctic; vwi, Virginian–West Indian; wces, West and Central Eurosiberian; wcp, West and Central Palearctic; we, West European; wes, West Eurosiberian; wimm, West Indian–Mediterranean–Mauritanian; wp, West Palearctic; •, occurrence of endemic subspecies.

Ecological data and conservation importance: α, α-mesosaprobic; α-β, α-β-mesosaprobic; ar, argillophilous; b, brackish; b-‰, limiting freshwater level for marine and salinity level for the freshwater forms; BC, Bern Convention; ca, calciphilous; cr, crenobiont; E, European importance; eb, eurybathic; ESC, the Red List of threatened animals and plants in Europe; eu, eurybiont; HD, Habitats Directive; hd, hydrophilous; hg, hygrophilous; IUCN categories: DD, Data Defficiency, EX, Extinct, LR, Lower Risk, nt, Near Threatened, VU, Vulnerable; L, freshwater; li, lithophilous; M, marine; mb, mesobathic; me, mesophile; N, national importance; o, oligosaprobic; pe, pelophilous; ph, phytophilous; po, potamophilous; ps, psammophilous; r, rare; rh, rhithrophilous; sb, stenobathic; se, stenoepibathic; sh, stenohypobathic; st, stygobiont; sw, stagnant water; T, terrestrial; tb, troglobiont; tph, thermophile; tr, troglophile; tx, trogloxene; t°, thermal water; W, world importance; x, xenosaprobic; xph, xerophile; (), species known only from shells.

3 Black Sea Mollusks

A total of 111 mollusk species (24.9% of the Bulgarian malacofauna) have been recorded for the Bulgarian portion of the Black Sea. They belong to three classes: Polyplacophora (two species), Gastropoda (61 species), and Bivalvia (48 species) (Table 1). The malacofauna of Bulgaria includes 51.2% of the total known 217 species for the Black Sea and Sea of Azov. Many authors accept *Myosotella myosotis* as a marine form because of its occurrence in supralittoral zone; we list is as a terrestrial snail.

Most marine mollusks are recorded along the entire Bulgarian Black Sea coast. There are only 24 species (21.6%), which are distributed either in the northern or in the southern part of the coast; of these, 22 species belong to Gastropoda (Table 1), mainly small-sized and poorly known Opisthobranchia (14 species). The recently reported new species of mollusks for the Bulgarian fauna (Wilke, 1995, 1996, 1997) could be added to this group.

The Black Sea below the depths of 180–200 m is enriched with released H_2S, which makes the real deep-sea life impossible. The groups of steno-, meso-, and eurybathic mollusk species are presented in Table 2.

Table 2 Vertical distribution of the Black Sea mollusks

Categories	Polyplacophora		Gastropoda		Bivalvia		All Mollusca	
	number	%	number	%	number	%	number	%
Stenoepibathic	2	100	39	63.9	34	70.8	75	67.6
Stenohypobathic			2	3.3	1	2.1	3	2.7
Mesobathic			14	23.0	5	10.4	19	17.1
Eurybathic			6	9.8	8	16.7	14	12.6

The most numerous are the *stenoepibathic* species (67.6% of all marine mollusks). The representatives of Polyplacophora and *Melarhaphe neritoides*, the inhabitant of the supralittoral zone, belong to this group as well. The gastropods *Patella ulyssiponensis, Tricolia pullus, Gibbula divaricata, Rissoa euxinica*, and *R. splendida* constantly occur permanently in the shallow coastal zone (to 5–10 m); all of them, except *P. ulyssiponensis*, are also found at higher depths. From class Bivalvia, similar stenoepibathic species are *Ostrea lamellosa, Tellina tenuis, Donacilla cornea* (for the surf zone), *Solen marginatus, Donax trunculus, Irus irus, Lentidium mediterraneum*, and *Pholas dactylus*. Some stenoepibathic forms such as *Caecum trachea, Cerithium vulgatum, Mangelia costata, Eulimella acicula*, and *Ebala pointeli* (Gastropoda) and *Anadara inaequivalvis, Ostrea edulis, Lucinella divaricata, Loripes lacteus, Tellina donacina, Donax semistriatus*, and *Gouldia minima* (Bivalvia) occur at the depths to 15–30 m and are approaching mesobathic species. An intermediate niche, closer to mesobathic forms, is occupied by *Chlamys glabra* and some Veneridae (*Pitar rudis* and *Chamelea gallina*), which reach higher depths.

Typical *stenohypobathic* species are a gastropod *Trophon muricatus* (found between 36 and 138 m of depth, most numerous between 65–80 m) and a bivalve *Modiolula phaseolina* (62 to 140 m, rarely at smaller depths up to 25 m; highest density at 110–115 m of depth).

The *mesobathic* species, mostly gastropods, form the second largest group; they reach the depth over 40 m. Typical mesobathic species are gastropods *Calyptraea chinensis, Epitonium turtoni, Cyclope neritea, Nassarius reticulatus*, and *Cylichnina umbilicata*, and bivalves *Modiolus adriaticus, Acanthocardia paucicostata, Paphia aurea*, and *P. petalina*. Some mussels as *Teredo navalis* and *T. utriculus* could be considered mesostenobathic forms.

The group of *eurybathic* species includes many numerous Black Sea species such as gastropods *Bittium reticulatum, Cerithiopsis tuberculatus*, and the Japanese snail *Rapana venosa* (which appeared in 1956) and bivalves *Mytilus galloprovincialis, Cerastoderma edule, Parvicardium exiguum*, and *Spisula subtruncata*. Other species of this group such as *Vitreolina incurva, Retusa truncatula, Abra alba, A. nitida*, etc. are not so numerous but are often found in both little and great depths. Most eurybathic forms reach the depth of 80–110 m, while *Parvicardium exiguum* and *Abra alba* reach the greatest depths, 120 and 135 m, respectively.

Formation of the Black Sea malacofauna is connected with the origin of the Black Sea basin itself. The Upper Miocene Sarmatian Sea (a descendant of Tethys) gave rise to the Pontian Sea-Lake, from which two separate basins were formed later, the Black Sea and the Caspian Sea. Initially, the Black Sea basin had been inhabited by fauna similar to the Caspian one (time of the Ancient Euxinian Sea). Then, it had been connected with the Mediterranean Sea and became saline, so the Mediterranean fauna penetrated into it, whereas the Caspian fauna retreated to the brackish coastal parts (time of the Uzunlar Sea). Later, the connection with the Mediterranean Sea had been severed, and the brackish basin (the New Euxinian Sea) originated, where the Mediterranean fauna disappeared. Recently (7000 to

10,000 years ago), this basin had been again connected with the Mediterranean Sea, the marine fauna invaded it, and the current Black Sea has been formed. Then, for the last time, a connection with the Caspian basin arose (via Manych channel), and Caspian postglacial immigrants invaded the Black Sea (Mordukhay-Boltovskoy, 1960). Most authors accept these species as Caspian relicts (known also as Sarmatian, Pontian, Pontian–Caspian, or autochthonous faunal elements). They are concentrated mainly in the coastal lakes-firths and the mouths of the Black Sea rivers. Nine Caspian relicts have been recorded in Bulgaria (Table 1), of which six are subfossils for the sea itself but still inhabit brackish basins along the coast and the Caspian Sea (Golikov and Starobogatov, 1966; Valkanov et al., 1978). *Theodoxus euxinus, Clessiniola variabilis*, and *Dreissena polymorpha* inhabit the freshwater and brackish basins. Shells of the latter species have been found in the sea as well. The Caspian relicts usually have Pontian or Pontian–Caspian ranges.

There is no unanimity among the experts about the zoogeographical status of the Black Sea, which is either considered an independent subregion or is unified with the Mediterranean Sea (and Lusitanian Subregion). The zoogeographical scheme used here (Table 3) is based on the works of Guryanova (1964), de Lattin (1967), Golikov and Starobogatov (1968), Golikov (1982), and Nesis (1982). The zoogeographical categorization of species (Table 1) is done on the basis of data for their distribution, taken from Grossu (1962, 1986), Nordsieck (1968, 1969), Golikov and Starobogatov (1972), Skarlato and Starobogatov (1972), Salvini-Plawen (1983), Abbott and Dance (1991), Anistratenko and Stadnichenko (1994), and Hayward et al. (1996). The Black Sea mollusks are divided into 15 zoogeographical categories (Table 3).

The main portion of the Black Sea mollusk fauna (95 species, or 85.5%) has an Atlantic–Mediterranean origin and represents the impoverished Mediterranean–Lusitanian fauna. As this fauna was becoming impoverished, the stenobiotic species were eliminated, so this category is defined by the eurybiotic forms, often distributed along the European coast up to Scotland. Thus an impression is created of the atlantization of this fauna. This feature is stronger expressed for Bivalvia, where 22 species (45.8%) have Mediterranean–Lusitanian–Scandinavian distribution, whereas for Gastropoda 15 species (24.5%) belong to this category. This group also includes some high-latitude boreal forms as *Limapontia capitata, Anomia ephippium*, etc., which are not placed in a separate category. There are more Mediterranean species in Gastropoda (11 species, or 18.0%) than in Bivalvia (five species, or 10.4%). There are also single tropical, subtropical, Pacific, Atlantic–Pacific, and Cosmopolitan forms (a total of 8 species, or 7.2%), of which four are introduced during the last 50 years: *Rapana venosa* (Japanese), *Corambe obscura* (Virginian–West Indian), *Anadara inaequivalvis* (Indo-Malayan), and *Mya arenaria* (Circumboreal). Some authors (Hayward et al., 1996) accept *Teredo navalis* as a Mediterranean–Lusitanian species with recent cosmopolitan distribution. The West Indian–Mediterranean–Mauritanian species *Ammonicera fischeriana* is recorded for the Black Sea recently (Wilke, 1996, 1997). The data on the distribution of the two Mediterranean–Lusitanian–Guinean species, *Striarca lactea* and *Pitar rudis*, are inconsistent (Nordsieck, 1969; Skarlato and Starobogatov, 1972; Hayward et al.,

Table 3 Zoogeographical characteristics of the marine mollusks of the Bulgarian Black Sea. Abbreviations, see text

	Arealographical categories													
Zoogeographical categories and main taxa	k	cb	j	im	wimm	vwi	mls	ml	mm	mlg	m	ap	p	pc
ATLANTIC														
Atlantic–Boreal														
Virginian						+								
Celtic							+							
Atlantic–Mediterranean														
Lusitanian					+		+	+		+				
Mediterranean														
Holomediterranean					+		+	+	+	+	+			
East Mediterranean												+		
Black Sea													+	+
Mauritanian					+				+	+				
Atlantic–Subtropical														
Carolinian						+								
Angolan–Namibian										+				
Atlantic–Tropical														
West Indian					+	+								
Guinean										+				

continued

Table 3 continued

Zoogeographical categories and main taxa	Arealographical categories													
	k	cb	j	im	wimm	vwi	mls	ml	mm	mlg	m	ap	p	pc
PACIFIC														
West Pacific														
Japanese			+											
Indo-Malayan				+										
ATLANTIC–PACIFIC														
Amphiboreal		+												
COSMOPOLITAN	+													
Polyplacophora (2 species)							1	1						
Gastropoda (61 species)			1		1	1	15	25	2		11	2	1	2
Bivalvia (48 species)	1	1		1			22	13		2	5		1	2
Total MOLLUSCA (111 species)	*1*	*1*	*1*	*1*	*1*	*1*	*38*	*39*	*2*	*2*	*16*	*2*	*2*	*4*

1996). The complex of subendemics and endemics (eight species, or 7.2%) is formed by Caspian relicts and two species with Aegean–Pontian distribution (*Cyclope westerlundi* and *Pseudovermis paradoxus*). The small number of the Black Sea endemic species, of which there were more in the old literature (Kaneva-Abadjieva recorded 24 species in 1960) is notable. Today, many of the former endemics they are brought to synonyms or downgraded to subspecies. Twelve species of Bivalvia have Black Sea endemic subspecies (Table 1).

Typical for the Black Sea is the almost continuous zone between 40 and 140 m, composed of *Mytilus galloprovincialis* (to 70 m) and *Modiolula phaseolina* (70–140 m). Nowhere *Mytilus galloprovincialis* reaches such quantities as in Black Sea. *Spisula subtruncata* is a characteristic species for the soft slimy bottom where it reaches the depth of 100 m. It has two maximums of density and biomass (at 40 m and 60–65 m) and is the third most common mussel along the Bulgarian coast. *Abra alba* has the widest range of distribution, from 5 to 135 m of depth, and a lot of maximums of the density and biomass, of which those at 95 and 105 m are the best exhibited (Kaneva-Abadjieva, 1962). *Chamelea gallina* is the most abundant between 15 and 30 m depth. The distribution of the mollusk biomass along the Bulgarian coast shows that in the south it usually does not exceed 50 g/m^2, but in many regions north of the Kamchiya River the biomass can be higher. In the south half of the Bulgarian shelf higher biomass is confined to 50 m isobath, whereas in the northern half it reaches the 70–80 m isobaths. In all cases when the biomass is higher than 500 g/m^2, it is mainly due to *Mytilus galloprovincialis*. Biomass as high as 3354 g/m^2 was recorded near Tsarevo; 2694 g/m^2 near Pomorie; and 4865 g/m^2 near Burgas (Marinov, 1990; Stoykov and Uzunova, 1999). The higher biomass (200–500 g/m^2) between 50 and 100 m isobaths is due mostly to *Modiolula phaseolina*. In southern direction between these isobaths the species is poorly represented; its average biomass is often below 50 g/m^2. The data presented are referred to the sublittoral bottom, located below 10 m isobath. Between the shore and this isobath the mollusks are also of great importance in the formation of zoobenthos biomass, mostly by phyto- and psammophilous species. In the localities overgrown with algae, these are *Tricolia pullus, Rissoa splendida*, and *Mytilaster lineatus*, the biomass of which reaches to a few hundred g/m^2. In the sandy bottom, the biomass is formed mostly of *Donacilla cornea, Donax trunculus, Tellina tenuis, Chamelea gallina, Lentidium mediterraneum*, etc. (Marinov, 1990).

Three main zones are established in the Bulgarian Black Sea shelf: supralittoral, mediolittoral, and sublittoral. In these zones, from eight to 12 biocoenoses and a great number of series are differentiated, most of which are characterized by definite mollusk species (Kaneva-Abadjieva, 1960, 1962; Valkanov et al., 1978; Marinov, 1990; Konsulov, 1998).

Rocky supralittoral. A characreristic marine species is *Melarhaphe neritoides*, found up to 2–3 m above the water in the crannies of the rocks. The highest settled specimens are active only during rough sea. The terrestrial snail *Myosotella myosotis* is found sometimes under the stones in that biocoenosis.

Rocky mediolittoral. A typical inhabitant is *Patella ulyssiponensis*, which is comparatively rare; in the older literature, it was listed as occurring in great quantity along the southern Bulgarian coast (Kaneva-Abadjieva, 1960). The presence of *Mytilus galloprovincialis* is also constantly observed.

Sandy mediolittoral. A very typical and abundant species is *Donacilla cornea*, the maximum density of which at Alepu reaches 9800 specimens/m^2. A comparatively high density is established at Stomoplo (6000 specimens/m^2) and Ahtopol (2000 specimens/m^2).

Cystoseira *overgrowing coenosis.* An exceptionally rich biocoenosis, connected to the algae *Cystoseira barbata* and *C. crinita* that develop on the rocky bottom from 0.5 to 20 m of depth. Eight mollusk species develop on the thallus of these algae, the majority of which are *Tricolia pullus, Bittium reticulatum, Rissoa splendida, Mytilaster lineatus* (which densely overgrows the base of the thallus), and *Mytilus galloprovincialis.* The maximum density reaches 1,040,000 specimens/m^2, and the biomass, 400 g/m^2. The density and biomass are at their highest in the spring and summer, as the biomass is formed mainly from *Mytilaster* and *Mytilus.*

Rocky sublittoral. This biocoenosis spreads in depth from 0.5 to 30 m. Here, 36 mollusk species have been found, of which the mussels *Mytilus galloprovincialis* and *Mytilaster lineatus* occur in large quantities. The most abundant snails are*Tricolia pullus, Gibbula divaricata*, and *Rissoa splendida*. The karst and mergel rocks are pierced by the holes of *Petricola lithophaga, Pholas dactylus*, and *Barnea candida*. The species *Lepidochitona cinerea, Rapana venosa, Ostrea edulis*, and *O. lamellosa* are permanent inhabitants of the rocky bottom.

Sandy sublittoral. Extends to 10–17 m of depth. Thirty mollusk species have been recorded here, of which *Chamelea gallina, Lucinella divaricata, Abra alba*, and *Nassarius reticulatus* dominate. The average density of mollusks is 398 specimens/m^2; the highest is the denisty of *Lentidium mediterraneum* (168 specimens/m^2), *Chamelea gallina* (65 specimens/m^2) and *Lucinella divaricata* (60 specimens/m^2). The mollusks comprise 92.7% of the average biomass in this zoocoenosis (126.5 g/m^2); the biomass is the highest for *Ch. gallina* (89.9 g/m^2) and *Paphia aurea* (9.6 g/m^2). The maximum biomass is established for *Chamelea gallina* (1542 to 1787 g/m^2; density to 2700 specimens/m^2 according to Kaneva-Abadjieva and Marinov, 1966; Marinov, 1990), and maximum density, for *Lentidium mediterraneum* (5380 to 21,000 specimens/m^2 according to Valkanov et al., 1978; Marinov, 1990). *Loripes lacteus, Solen marginatus, Tellina donacina, T. tenuis, T. fabula, Donax trunculus*, and *Gouldia minima* are typical psammophiles. The introduced species *Rapana venosa* (biomass up to 400 g/m^2), *Anadara inaequivalvis* (biomass up to 4282 g/m^2) and *Mya arenaria* (biomass up to 4596 g/m^2) are also represented here.

Coastal silt. Extends from 15–20 to 30–40 m depth. Mollusk fauna includes 25 species, mainly *Caecum trachea, Calyptraea chinensis, Cyclope neritea, Nassarius reticulatus, Lucinella divaricata, Spisula subtruncata, Abra alba, A. nitida, Gouldia minima, Chamelea gallina*, and *Paphia aurea*. At higher depths, *Acanthocardia paucicostata, Spisula subtruncata, Abra nitida*, and *Pitar rudis* dominate.

S. subtruncata and *Ch. gallina* have the highest density. *Ch. gallina* and *P. aurea* have the highest biomass, to 56.1 g/m^2 and 84.7 g/m^2 respectively. Mollusks form from 72.1 to 77.9% of the biomass of this zoocoenosis.

Zone of Mytilus silt. This zone begins to the north of Kaliakra Cape from the depth of 45 m and reaches the depth ca. 70 m; at the Emine Cape, 80 m. Width of this zone varies from 2–3 miles to 10–15 miles. Ahead of Burgas Bay the zone is the widest and begins from the depth of 15–20 m. There are 23 mollusk species (six of Gastropoda and 17 of Bivalvia). The main dominant is *Mytilus galloprovincialis*; its biomass reaches from 3354 to 5900 g/m^2 in some places (Kaneva-Abadjieva, 1962; Marinov and Stoykov, 1995). The biomass of *M. galloprovincialis* shows three maximums, of which this one at 45 m depth is the biggest. *Spisula subtruncata, Abra alba* and *Paphia aurea* have a very high density. Biomass of the zoocoenosis is formed mainly by the mollusks (94.3%). *M. galloprovincialis* alone gives 63.6% from the biomass; however, as a result of its patchy distribution, the coefficient of permanency of this species is only 25.6%, followed by *S. subtruncata* with a coefficient of permanency 79.2% but with twice less biomass.

Zone of Phaseolina silt. This zone extends from 60–65 m to 184 m of depth. In the southern region, due to the larger slope of the bottom, this zone is narrower. It is inhabited by 10 mollusk species, of which only *Trophon muricatus* is a gastropod. This conspicuous predator feeds on the mussels with thinner shells such as *Modiolula phaseolina, Parvicardium exiguum, Plagiocardium papillosum, Spisula subtruncata, Abra alba*, and *A. nitida*. *M. phaseolina* is a dominant with a coefficient of permanency from 22.2% (south of Kamchiya River) to 96.5% (in the north). Its density and biomass show many peaks, of which the density maximum (4963 specimens/m^2) is at 110 m of depth and the biomass maximum (389 g/m^2), at 65 m (Kaneva-Abadjieva and Marinov, 1960). Recently, this species was recorded at the depth of 55 m at the level of Kaliakra Cape with a density up to 13,040 specimens/m^2 and biomass up to 995.6 g/m^2 (Marinov and Stoykov, 1995). The mollusks account for 86.5% of the whole biomass, and *M. phaseolina* alone, for 79.2% (Marinov, 1990).

Considerable changes in the Black Sea biocoenoses are caused by three mollusk species introduced in the last 50 years: *Rapana venosa, Anadara inaequivalvis*, and *Mya arenaria* (Cvetkov and Marinov, 1986). The first specimen of *R. venosa* in the Bulgarian part of the Black Sea was found in 1956 in Varna Bay near Galata Cape (Kaneva-Abadjieva, 1958). Development of this snail in the rocky sublittoral had a substantial impact on *Mytilus* and *Ostrea*, and in the sand sublittoral, on *Chamelea gallina*. The great eurybiontness, high fecundity, and lack of competitors allowed this predator to reach mass development in the Black Sea and aroused discussion for eventual measures for a struggle with it. In a single trawling, up to 1500 specimens have been caught, and in some regions between Balchik and Kavarna the entire bottom was covered with Rapa whelks. During the last years, gathering of the snail for food with all possible means causes a rapid decrease of its quantity.

Mya arenaria has been first reported for the Bulgarian coast in the Bay of Burgas in front of Marinka River in 1973 (Kaneva-Abadjieva, 1974). The species inhabits the sandy sublittoral and reaches the wash zone. It has a high ecological plasticity and easy endures variations of the salinity and temperature, and oxygen deficiency. It reaches a high density (over 300–400 specimens/m^2) in the bays in front of the river mouths. *M. arenaria* is found along the beaches all over the Bulgarian coast but the greatest number of it occurs in front of Durankulak and Albena, in the Varna Bay, Varnensko Lake, and Burgas Bay (Stoykov, 1983; Cvetkov and Marinov, 1986; Marinov, 1990).

The first specimens of *Anadara inaequivalvis* for the Bulgarian coast were found in 1982 in Varna Bay (Marinov et al., 1983; Kaneva-Abadjieva and Marinov, 1984). Much later, a high density of the species has been found in Burgas Bay (up to 400 specimens/m^2). In a short time, this mussel became a significant element of psammo- and pelophilous zoocoenoses, and started to displace some local species. As a result, the "*Chamelea gallina*" group in front of Balchik, Varna, and Burgas transforms into "*Anadara inaequivalvis*" group. The distribution of this species in the Bulgarian part of the Black Sea is restricted from Balchik to the south part of Burgas Bay (Cvetkov and Marinov, 1986; Marinov, 1990; Konsulov, 1998).

4 Freshwater Mollusks

The freshwater mollusks (21.6% of the Bulgarian malacofauna) belong to two classes, Gastropoda (72 species) and Bivalvia (24 species). They are divided in two groups, supraterranean and subterranean (stygobiotic) forms. The supraterranean (above-ground) mollusks (59 species of Gastropoda and 24, of Bivalvia) are well studied; an insignificant number of new taxa for Bulgaria probably could be still found in karstic springs. This group includes predominantly mollusks with wide geographic ranges (Holarctic, Palearctic, and Eurosiberian). Smaller number of species with South European, Southeast European, Balkan, Pontian–Balkan, and Pontian distribution belong to the same group as well. The group of stygobionts, represented by the family Hydrobiidae of class Gastropoda, includes 12 local endemics. This family is likely insufficiently studied in Bulgaria (half of the stygobiotic species are known only from their shells), and current research should considerably enrich its species list. The listed data on the number of species include all studies historically conducted in Bulgaria. Currently, the species composition is poorer because of the increasing domestic and industrial waste and use of underground water.

Three watersheds are differentiated in Bulgaria: Danube, Black Sea, and Aegean Sea. The water bodies in these watersheds are divided into running (rivers and channels) and stagnant (lakes, marshes, and dams). The malacofauna is discussed according to the watersheds and a type of the water body. Eleven species are stagnophilous (*Valvata pulchella, Acroloxus lacustris, Stagnicola corvus, S. turricula, Radix ovata, Hippeutis complanatus, Segmentina nitida, Aplexa hypnorum, Pisidium subtruncatum, P. nitidum*, and *P. obtusale*) but it is possible for them to be found in running waters as well (Table 1). Most Bulgarian mollusks do not exhibit a clear stagnophily and inhabit various water basins.

Danube River. From the Bulgarian sector of the Danube River, 52 mollusk species are known, and 23 species have been found in the channels and marshes of the Danube islands (Russev, 1978; Russev et al., 1998). Five zoocoenoses are distinguished for the zoobenthos, in three of which the mollusks are always found (Russev, 1966a, 1967, 1978). *Theodoxus transversalis* (96 specimens/m^2 and 26 g/m^2), *Lithoglyphus naticoides* (up to 45.4 g/m^2) and *Dreissena polymorpha* (up to 867 specimens/m^2 and 83.9 g/m^2) dominate in the *lithorheophilic* zoocoenosis. It is the most densely inhabited zoocoenosis with average density of mollusks 30 specimens/m^2 and biomass 65.9 g/m^2. The *pelorheophilic* zoocoenosis is characterized by *Unio pictorum, U. tumidus* and *Pseudanodonta complanata.* These species are often found in great numbers along the shore. In the river, their density reaches 46 specimens/m^2, and the biomass, 682 g/m^2. The average biomass of all mollusks is 45.1 g/m^2. *U. tumidus* and *L. naticoides* occur in the *argillorheophilic* zoocoenosis where the mollusk biomass is 29.8 g/m^2. The order Gastropoda forms 79.9%, and Bivalvia, 17.5% of the total Danube benthos biomass. Sometimes the mollusk biomass exceeds 1 kg/m^2. *Th. transversalis, Th. danubialis, L. naticoides, U. tumidus, U. pictorum* and *D. polymorpha* are the most abundant here. These species (except *Unio*) prefer rubble, rocks, cinder, and stones around the pontoons. *L. naticoides* inhabits the Corophium silt as well, where it comprises 25.7% of all mollusks by number, and 36.6% of the mollusk biomass. *Th. danubialis* does not occur on silt and clay bottom, whereas the mussels *U. pictorum* and *U. tumidus* are not found on sand bottom. *U. pictorum* often inhabits clay substrate, where it comprises 45.0% of all mollusks by number, and 36.5% of the biomass. The slow current is more favourable for the development of mollusks. *Th. danubialis* and *L. naticoides* have the highest average biomass between 1 and 4 m depth; *D. polymorpha*, between 4 and 7 m; and *Th. transversalis* and *U. tumidus*, below 7 m. The many-year average biomass of the mollusks in the Bulgarian portion of Danube is 27 g/m^2 (Russev, 1978). Recently, appearance of the clam *Corbicula fluminea* (Hubenov, 2001), the density of which exceeds 1000 specimens/m^2 in some ares, causes big changes in the benthic ecosystems of Danube.

Bulgarian tributaries of the Danube River. Sixteen rivers enter the Danube within the Bulgarian territory, of which 12 have source in the Stara Planina Mts. or Predbalkan; three, in the Danubian Plain; and one (Iskar River), in the Rila Mts. These rivers form the Danube catchment area; from here, 47 mollusk species (49.5%) have been recorded, differently presented in separate rivers (Russev et al., 1994). The malacofauna of the rivers Iskar (31 species) and Yantra (30 species) is the richest. This is connected with their large watersheds, appreciable water quantity, and high level of study. Iskar River is the best investigated both as a river and as catchment area. The number of species could be considerably greater if mollusks from the stagnant basins and underground waters, captured from the river, were included. In other rivers, the species number varies from three to 24, depending on the level of study and river size. The northwestern rivers Topolovets, Voinishka, Vidbol, Archar, Skomlya, Lom, Tsibritsa, and Ogosta have been poorly investigated, or their sizes are smaller, and therefore their species diversity is poorer, from three to ten species. In the rivers Iskar, Vit, Osam, Yantra, and Rusenski Lom the number of recorded mollusk species is from 15 to 31. In the large part of the middle and lower portions of the rivers Vit, Yantra, and Osam, fauna became impoverished due to strong anthropogenic impact. Iskar River, which receives the waste waters of Sofia, is in the most unfavorable condition.

Bulgarian rivers of the Black Sea watershed. Nineteen rivers enter the Bulgarian part of the Black Sea, of which 12 flow directly to the sea and eight, to the coastal lakes (Beloslavsko, Burgasko, and Mandra). The portions of the Black Sea rivers located closed to their mouths, are influenced by the sea, and their estuaries often have firth character, with many oligo- or euryhaline forms represented in their malacofauna. Therefore, it is difficult to discuss the fauna of these rivers separately from that of the coastal stagnant basins. The Provadiiska River is the most mollusk-rich (35 species). In other rivers the number of the species varies from one to 15 (Uzunov et al., 1998). More mollusk species are known in the rivers Kamchiya (11 species) and Ropotamo (15 species). It has to be noted that the Ropotamo River is well studied; moreover, the firth type of this river allows many marine and brackish forms to occur there. The rivers Veleka and Rezovska have considerable catchment areas as well as firths, and undoubtedly have a rich malacofauna, which is not yet sufficiently investigated. A total of 43 mollusk species have been recorded in the Bulgarian Black Sea rivers.

Bulgarian rivers of the Aegean watershed. This area is drained by the greatest Bulgarian rivers that collect their waters from South Bulgaria. The greatest number of mollusks, 48 species and subspecies, have been established in Maritsa River, which is followed by the rivers Tundja (28 taxa), Arda (14), Struma (12), and Mesta (six) (Uzunov et al., 1998). It has to be mentioned that Maritsa River is the best investigated and has the largest watershed. Moreover, with the accumulation of data, some subspecies taxa and species from the stagnant basins along Maritsa River have been included here as well. The actual number of species is smaller, and it especially decreased in the recent years due to the strong anthropogenic impact (Russev, 1966b; Russev et al., 1981; Uzunov and Kovachev, 1985). Widespread gastropods of the families Valvatidae, Lymnaeidae, Ancylidae, Planorbidae, and Physidae are present but do not reach high numbers. Similar is the condition of Arda River that is comparatively well studied (Russev, 1966b; Russev and Yaneva, 1975; Yaneva, 1989; Yaneva et al., 1998, 2001). Common species in this river are *Valvata piscinalis, Galba truncatula, Radix peregra, Ancylus fluviatilis*, and *Physella acuta*. Single representatives of the bivalve genera *Anodonta* and *Pisidium* also were found in Arda. All main rivers in this region are under anthropogenic pressure, and are saprogenic, inert, toxic or mixed polluted, which led to the impoverishment of their fauna (Uzunov et al., 1998). A total of 43 mollusk species have been recorded in the Aegean watershed.

Coastal basins. There are 31 lakes, marshes, and areas flooded by rivers along the Black Sea coast (Varbanov, 2002). Most common are firth lakes (lower part of an estuary, a strait) and lagoon lakes (areas separated from the open sea). The presence of brackish elements is a special feature of their fauna. A "saline wedge" is formed at the lower parts of the rivers, which is situated below the fresh waters. In this "wedge" the bottom inhabitants are marine or brackish, whereas those in upper water layers are freshwater species. A specific malacofauna inhabits the lakes, firths, and marshes along the Black Sea coast. The marine brackish species *Hydrobia acuta* and *H. ventrosa* endure water down to 1‰ salinity; at the same time, freshwater forms *Theodoxus fluviatilis, Th. pallasi, Lithoglyphus naticoides, Bithynia tentaculata, Micromelania lincta, Radix auricularia, R. ovata, Planorbis planorbis, Anisus vortex, Bathyomphalus contortus, Gyraulus albus, Planorbarius corneus, Unio pictorum, Anodonta cygnaea, Dreissena polymorpha, Sphaerium corneum*, and *Musculium lacustre* withstand water salinization from 1.5 to 8‰. Many euryhaline sea species, such as *Bittium reticulatum, Rissoa membranacea, Nassarius reticulatus, Mytilus galloprovincialis, Mytilaster lineatus, Cerastoderma edule, Parvicardium exiguum, Abra ovata*, and *Lentidium mediterraneum* together with the taxa mentioned above, form malacofauna of the coastal lakes and firths. This fauna, depending on the water salinity in different basins, could vary from marine to freshwater. In the coastal basins, 26 freshwater species (27.4%) and 17 marine species (15.3%) of mollusks have been found (Valkanov, 1935, 1936; Drensky, 1947; Petrbok, 1947; Kaneva-Abadjieva, 1957, 1975; Zaschev and Angelov, 1959; Mihailova-Neikova, 1961; Kaneva-Abadjieva and Marinov, 1967; Marinov, 1990). After the connection of Varnensko Lake with the sea, the freshwater taxa have disappeared there, and the recent malacofauna is formed by euryhaline marine and brackish species (Kaneva-Abadjieva, 1957). The following species dominate: *Hydrobia ventrosa* (up to 24,520 specimens/m^2), *Mytilaster lineatus* (up to 37,400 specimens/m^2 and 620 g/m^2), *Mytilus galloprovincialis* (to 1837 specimens/m^2 and 6023 g/m^2), *Cerastoderma edule* (to 968 specimens/m^2), and *Abra ovata* (to 1177 specimens/m^2). The reserves of *M. galloprovincialis* in Varnensko Lake are estimated as ca. 5000 tons, and are of great economical importance. In the last decades, this lake underwent profound anthropogenic transformations, and its malacofauna changed. In Burgasko Lake, *A. ovata* in 1954 reached average density 5544 specimens/m^2 and biomass 790 g/m^2; its total biomass was estimated as 16,957 tons (Zaschev and Angelov, 1959). Later, because the salinity decreased from 11 to 0.7–1.9‰, this mussel completely disappeared. The average density of *Hydrobia ventrosa* during the same period was 3256 specimens/m^2, biomass 10.4 g/m^2, and the total biomass, 216 tons. For *Cerastoderma edule* these data are, respectively: 1840 specimens/m^2, 201.8 g/m^2 and 3081 t. No living mollusks have been found currently in Burgasko Lake (Pandourski, 2001). In Mandra Lake, before the dam construction, many marine forms such as *C. edule, A. ovata* and *H. ventrosa* have been recorded, which can be found now in the lakes Uzungeren and Poda, remaining outside of the dam (Valkanov, 1936; Mihailova-Neikova, 1961; Stoykov, 1979). The mass development of the mussel *Dreissena polymorpha* that reaches to 3–4 m depth and forms a ring around the shore is characteristic for the lakes Durankulak, Ezerets, and Shabla (Kaneva-Abadjieva and Marinov, 1967; Valkanov et al., 1978; Kovachev and Uzunov, 1981; Kovachev et al., 1999, 2002; Petrova and Stoykov, 2002). In the firth of Veleka River, *D. polymorpha* covers the shells of *Pseudanodonta, Anodonta*, and *Unio*. In most of the coastal lakes and firths a considerable number of *Theodoxus fluviatilis* is observed (and often also *Planorbis planorbis*). *Nassarius reticulatus, Hydrobia ventrosa, Abra ovata*, and *Cerastoderma edule* are representatives of the marine forms in these lakes. These species have been found also in the supersaline lagoon lakes Pomoriisko and Atanasovsko, where salinity sometimes reaches 100–250‰ (Ivanov et al., 1964; Stoykov, 1986). Some of the coastal basins were explored comparatively long ago and the investigations do not reflect the recent condition of their malacofauna (Shabla Tuzla, Batovsko Swamp, Beloslavsko Lake, Nesebarsko Swamp, Mandra Lake, the swamps Alepu, Arkutino, Stomoplo, and Dyavolsko). Furthermore, the coast is exposed to strong anthropogenic presence and changes that require a continuous update of the information.

Glacial lakes. These lakes were formed during the Quaternary glacials in Rila Mts. and Pirin Mts., and are located between 1858 and 2710 m a.s.l.. In total, their number is 330, of which 259 are persistent (Varbanov, 2002). Typically, these lakes are oligotrophic, with poorly represented living organisms. Three gastropod species (*Radix ovata, R. peregra*, and *Galba truncatula*) and three bivalves (*Pisidium amnicum, P. personatum*, and *P. casertanum*) have been found there (Valkanov, 1932, 1934; Angelov, 1976, 2000a). These are eurybiotic forms, widespread in Bulgaria, with Holarctic, Palearctic, and Eurosiberian ranges (Table 1).

Tectonic lakes. These lakes are located in depressions formed due to the earth's crust movement. They are not deep and lake pans are rapidly filled with alluvia. There are several similar lakes in Bulgaria, of which the Rabisha Lake (transformed into a reservoir now) was studied. During the research, the lake has been in critical ecological condition, and no living mollusks were found. Shells of 17 species (14 gastropods and three bivalves), which inhabited the lake prior to the study, have been found (Angelov, 1956). No malacological research has been done after the transformation of this lake into a reservoir.

Karst lakes. These lakes have been developed in karst regions, and are either surface or underground lakes. The surface karst lakes are highly eutrophicated. Usually their malacofauna is rich, but there is a lack of current studies. Near Sofia are situated the drying swamps Aldomirovsko and Dragomansko. A large number of karst lakes (about 40) are located in Devetashko Plateau (Ivanov et al., 1964). In the underground lakes and springs, 15 endemics (stygo- and crenobionts) of Hydrobiidae (Willmann and Pieper, 1978; Radoman, 1983; Bole and Velkovrh, 1986) and two trogloxenes of Sphaeriidae have been recorded (Table 1).

Landslide lakes. They are spread along the northern Black Sea coast (Nanevska Tuzla, Shabla Tuzla, lakes around Aladja Monastery), West Rhodopes (Smolyan Lakes), and Rila Mts. (Suhoto Lake). Smolyan Lakes are comparatively well studied, where seven gastropod species (*Stagnicola palustris, Radix peregra, Galba truncatula, Planorbis planorbis, Gyraulus albus, Physa fontinalis*, and *Physella acuta*) and one species of genus *Pisidium* of Bivalvia (Russev and Yaneva, 1975) have been recorded. These mollusks (except *Ph. acuta*) have ranges of Holarctic, Palearctic, and Eurosiberian types (Table 1).

Lakes and marshes along the rivers. These lakes are located in old meanders abandoned by a river, and have a rich fauna. Most of them are drained. Srebarna Lake is the best studied (Angelov, 1998a, 1998b; Uzunov et al., 2001); its malacofauna includes 29 gastropod species from seven families (Viviparidae, Valvatidae, Lithoglyphidae, Bithyniidae, Acroloxidae, Lymnaeidae, Planorbidae, and Physidae) and 12 bivalve species from three families (Unionidae, Dreissenidae, and Sphaeriidae). Two species are accepted as Caspian relicts and 13 are of conservation importance.

Reservoirs. In most cases, successional processes are disturbed by the regime of water usage, thus the malacofauna (especially along the shores) is poor. After the autogenous succession, mainly eurybiotic gastropods of family Lymnaeidae and pelophiles such as *Anodonta cygnaea* are established. In some cases, *A. cygnaea* develops at considerable depth (below 17 m in Iskar Reservoir) and with high density (Pancharevo Reservoir). Some species reach great numbers, e.g. *Viviparus viviparus* in the microreservoir near Botunets at the vicinity of Sofia, and *Dreissena polymorpha* in Ovcharitsa Reservoir reach 300,000 specimens/m^2 under favorable conditions.

Technogenous lakes. These lakes are formed under excavation works for quarries. They have a slow water exchange and fast eutrophication, and their malacofauna usually is rather poor. It includes widespread species of Lymnaeidae, Planorbidae, Physidae, and Unionidae (genus *Anodonta*), which do not reach high numbers. Lately, *Dreissena polymorpha* was introduced in the lakes near Chepintsi Village, where it quickly increases in number.

The largest number of freshwater mollusk species (77 species, or 80.1%) has been found in the lowlands (below 200 m a.s.l.) and in the hill zone (200–600 m) (62 species, or 65.3%). In the lower mountains (600–1000 m), 30 species (31.6%) have been recorded; in the middle mountain zone (1000–1600 m), 19 species (12.0%); and in the high mountain zone (above 1600 m), 11 species (11.6%). All mollusks found above 1000 m a.s.l. are eurybiotic forms, widespread in Bulgaria, with Holarctic, Palearctic, and Eurosiberian ranges (Table 1). The exceptions are *Bythinella austriaca*, a glacial relict inhabiting cold waters, that not found below 500 m a.s.l., and *Pisidium bulgaricum*, a local endemic described from Vitosha Mts. peatbogs (about 2000 m) (many authors accept this taxon as questionable). According to their vertical distribution, freshwater mollusks can be divided into three groups: those distributed in a single altitudinal zone (41 species, or 43.2%); those distributed in more than one altitudinal zone (44 species, or 46.3%); and those distributed in all zones (10 species, or 10.5%). These groups are tentative to some extent and depend on both the specific features of certain mollusk species and the level of study.

The zoogeographical categorization of the species (Table 1) is done on the basis of data for their distribution, according to Zhadin (1952), Grossu (1962, 1987, 1993), Starobogatov (1970), Willmann and Pieper (1978), Radoman (1983), Stadnichenko (1984, 1990), Bole and Velkovrh (1986), Frank et al. (1990), Bănărescu (1990), Anistratenko and Stadnichenko (1994), Kornyushin (1996), and Angelov (2000a). The freshwater mollusks are divided in 22 zoogeographical categories, combined into three large groups (Table 4).

Species distributed both in Palearctic and beyond it. This group (eight species, or 8.4%) includes three zoogeographical categories, in two of which the introduced species *Ferrissia clossiniana* and *Corbicula fluminea* are included. The rest six Holarctic species belong to the families Lymnaeidae, Planorbidae, and Sphaeriidae. *Galba truncatula* and *Gyraulus crista* are widely distributed in Bulgaria. This group is poorly presented in the Black Sea rivers. It is not important for zoogeographical characteristic of the freshwater mollusks because of its heterogeneity and small number of species.

Species distributed only in Palearctic but in more than one subregion. A total of 22 species (23.2%) from this group have been recorded in Bulgaria. Its character is determined by the Holopalearctic species that are the most numerous, followed by the West and Central Palearctic ones, in all three watersheds (Table 4). Three paleolimnetic forms (*Valvata piscinalis, Bithynia leachi*, and *B. tentaculata*) of gastropod subclass Prosobranchia belong to this group. Most common are the representatives of gastropod subclass Pulmonata and bivalve family Sphaeriidae. This group is poorly represented in the Black Sea rivers. *Radix peregra* and *Ancylus fluviatilis*, which occur all over the country, have the widest distribution (Table 1). The group includes five species (50%) of the freshwater species found in all altitudinal zones, and three species (*Bithynia tentaculata, Stagnicola corvus*, and *S. turricula*) of those found only in one zone. The most represented families are Lymnaeidae (five species) and Sphaeriidae (seven species).

Species distributed within the boundaries of one Palearctic subregion. This group in Bulgaria includes 46 species (48.4%) with Eurosiberian and Mediterranean distributions. There are 35 (36.8%) Eurosiberian species, of which the most numerous are European and West Eurosiberian ones. Seventeen taxa have exclusively European distribution, of which ten are widespread in Europe, and seven, in its separate areas (Table 1). The glacial relict *Bythinella austriaca* (Hydrobiidae) has European distribution. The representatives of the family Unionidae also have a similar distribution. Of a great interest is the finding of a rare Central and North European species *Myxas glutinosa* in Skomlya River in Northwest Bulgaria (Russev et al., 1994). According to Stadnichenko (1990), *Aplexa hypnorum* has a longitudinal disjunction of its range (absent in Siberia). Zhadin (1952) and Frank et al. (1990) accept this species as Holarctic. Nine species have been reported only for the Danube region (Table 1). *Theodoxus transversalis* and *Viviparus acerosus* are considered Danube–Don endemics (Zhadin, 1952; Starobogatov, 1970). The mollusks in the Danube watershed are twice more numerous than those in other watersheds. This group includes three species (*Anisus leucostomus, Bathyomphalus contortus*, and *Pisidium*

Table 4 Zoogeographical characteristics of the Bulgarian freshwater mollusks

Zoogeographical categories	Total	Danube Basin	Black Sea Basin	Aegean Basin
Species distributed in Palearctic and beyond it	*8*	*7*	*3*	*5*
Australian (introduced)	1	1	1	
South Palearctic–Oriental (introduced)	1	1		
Holarctic	6	5	2	5
Species distributed only in Palearctic	*68*	*66*	*36*	*35*
PALEARCTIC TYPE	22	22	14	16
Holopalearctic	10	10	4	7
Transpalearctic	2	2	2	2
West and Central Palearctic	6	6	5	5
West Palearctic	4	4	3	2
EUROSIBERIAN TYPE	35	35	16	17
Holoeurosiberian	3	3		2
West and Central Eurosiberian	5	5	3	3
West Eurosiberian	9	9	6	2
Disjunct Eurosiberian	1	1	1	1
European	10	10	6	7
Central and North European	1	1		
Central and East European	3	3		2
Central and Southeast European	2	2		
Southeast European	1	1		
MEDITERRANEAN TYPE	11	9	6	2
Pontian–Caspian–Southeast European	1	1	1	
Pontian–Southeast European	3	3	1	1
Pontian–Caspian	2	1	1	
Pontian	3	2	2	
Pontian–Balkan	1	1		
North Mediterranean–Turanian	1	1	1	1
Endemics and subendemics, total	*20*	*17*	*4*	*3*
local	13	11	1	1
regional	1	1	1	
Balkan	4	3	2	2
subendemics	2	2		
GASTROPODA	72	67	34	32
BIVALVIA	24	23	9	11
Total MOLLUSCA	*96*	*90*	*43*	*43*
Relicts, total	*11*	*8*	*6*	*4*
Caspian	9	7	6	2
preglacial	1			1
glacial	1	1		1

personatum) from the taxa established in all altitudinal zones and ten species from those found only in one zone. The family Planorbidae is the richest with Eurosiberian species (8 species, or 53.3%). *Dreissena polymorpha* is a Caspian relict with secondary anthropogenic range; it is already introduced to North America as well.

Mediterranean species include 11 taxa (11.6%); all of these (except the North Mediterranean–Turanian *Physella acuta*) have Pontomediterranean ranges. Nine species of the families Neritidae, Melaniidae, Lithoglyphidae, and Micromelaniidae, distributed in the lowest portions of the rivers belonging to Danube and Black Sea watersheds, are accepted as Caspian relicts (Table 1). The lack of these species in the rivers of the Aegean basin is explained by their origin and connection with the Pontian–Caspian basin (Zhadin, 1952; Starobogatov, 1970; Bănărescu, 1990).

Endemics. This category of freshwater mollusks includes taxa, which are not distributed outside the Balkan Peninsula. The endemics are divided into: Balkan (found in more than one Balkan country), Bulgarian (found in Bulgaria only), regional (known from more than one locality of certain region), and local (known only from one locality). Endemics are of high conservation value for the evaluation of any territory, and reflect the unique character of a fauna. The endemics belong to paleo- and mesolimnic families. Twenty endemic taxa have been established (21.1% of the freshwater species) (Table 1). Most numerous are the local endemics (13 species, or 65.0%), followed by the Balkan endemics and subendemics (six species, or 30.0%). One regional endemic exists, *Pseudamnicola codreanui*, typical for the coastal Dobrudja. Most of the local endemics (nine species) have been found in the karst massifs of Stara Planina and Predbalkan (Table 1), and one local endemic species is found in each of the following regions: Dobrudja, Kraishte, and Sofia Basin. Among Balkan endemics, *Viviparus mamillatus* is accepted as extinct for the Bulgarian fauna; *V. danubialis* is found only along the Danube River; *Amphimelania holandri* is known along Danube and Nishava rivers, where its numbers reach 3000 specimens/m^2; *Pseudamnicola consociella* is widespread in Bulgaria; *Sadleriana virescens* is established in West Rhodopes, Thracian Lowland, and Strandja Mts.; *Lithoglyphus pyramidatus* is found along the Danube River at town of Lom. The family Hydrobiidae is the richest in endemics (15 species, or 83.3% of the family taxa), of which 12 local stygobiotic endemics were described as new (Angelov, 2000a). They belong to the genera *Belgrandiella, Pontobelgrandiella, Cavernista, Insignia, Iglica, Paladilhiopsis, Plagigeyeria*, and *Horatia* (Table 1). *Pisidium bulgaricum* (Sphaeriidae) was described from the wetlands in the high parts of Vitosha Mts. (Odhner, 1929) but Kuiper (1963) and Kornyushin (1996) did not accept presence of such an endemic in Bulgaria. *P. bulgaricum* is known from three specimens and has not been found again (Angelov, 1976). Most new taxa (nine species) are recorded in the western and central parts of Stara Planina and Predbalkan (Table 1). *Belgrandiella bureschi* is described from Kraishte; *Plagigeyeria procerula* is found west of Sofia; and *Horatia lucidulus*, along the Dobrudja coast at Taukliman.

Relicts. The relict mollusk fauna is a result of complex paleoclimatic and paleogeographical changes from Tertiary to present. According to their origin, Bulgarian relict freshwater mollusks are preglacial (*Melanopsis parreyssi*, found in a thermal spring at Pyasachnik Reservoir), glacial (*Bythinella austriaca*, established in the mountains of West Bulgaria), and Caspian (nine species, connected with Pontian–Caspian brackish region (Starobogatov, 1970; Bănărescu, 1990). Their ranges are: Pontian–Caspian–Southeast European (*Theodoxus danubialis*), Pontian–Southeast European (*Fagotia acicularis, F. esperi*, and *Lithoglyphus naticoides*), Pontian–Caspian (*Theodoxus pallasi* and *Clessiniola variabilis*), and Pontian (*Theodoxus euxinus* and *Micromelania lincta*). Only *Dreissena polymorpha* has European distribution (due to anthropogenic introduction). Relicts are of high conservation importance and contribute to the specificity and uniqueness of the fauna. They belong to mesolimnic taxa such as those from the families Neritidae, Melaniidae, Hydrobiidae, Lithoglyphidae, Micromelaniidae and Dreissenidae.

Rare (stenotopic) species. This category includes taxa with scarce populations or known from single localities only. Rare species, found all over Bulgaria, usually occupy specific habitats. Negative changes of microclimate, environmental pollution, and habitat disturbance cause extinction of species or decrease of their range. A total of 35 mollusk species (36.8%) are considered rare in Bulgaria, of which ten are endemics at the same time. *Anisus vorticulus* is listed in the Bern Convention; *Pseudanodonta complanata* is included in IUCN Red List; and *Pisidium supinum* probably is extinct (Table 1).

Threatened species. European Red Lists are made mainly for Central and West Europe but species distributed in Bulgaria are included there as well. Nine freshwater mollusk species are included in these lists (Table 1): (1) IUCN Red List: *Theodoxus transversalis, Viviparus acerosus, V. mamillatus, Fagotia esperi, Myxas glutinosa, Segmentina nitida, Unio crassus*, and *Pseudanodonta complanata*; (2) ESC Red List: *M. glutinosa* and *S. nitida*; (3) Bern Convention: *M. glutinosa, Anisus vorticulus*, and *U. crassus*; (4) Habitats Directive: *U. crassus*; (5) CORINE: *M. glutinosa* and *S. nitida*. In Bulgaria the distribution of these species is always connected with Danube River, Danubian Plain, and less commonly with Predbalkan and other territories. The stygobiotic forms from underground lakes and springs, among which are the local endemics of family Hydrobiidae, are protected under the Bulgarian legislation.

Taxa with high conservation value. Among the freshwater mollusks, 60 taxa (63.2%) are considered important for conservation (Table 1). Of these, 22 species (23.2%) are of world importance; seven species (7.7%) are of European importance; and 31 species (32.6%) are of national importance. The conservation value of species differs. It is especially high when more than one criterion are combined for one taxon (rare and endemic, rare and relicts, etc.). Seventeen such taxa (17.9%) are known among the Bulgarian freshwater mollusks. The Caspian relict *Dreissena polymorpha*, which quickly expands its range and creates serious economical problems, is excluded from this category.

Indicator species. Fourteen species are considered indicators (Table 1). Most of them (11 species) are α- and β-mesosaprobic, comparatively widespread in Bulgaria Holarctic, Palearctic, and Eurosiberian taxa of the families Lymnaeidae, Planorbidae, and Physidae. Quite rare is oligo- to β-mesosaprobic *Myxas glutinosa*. The species *Bythinella austriaca* is considered a xenosaprobic glacial relict with Southeast European distribution, found in Western Stara Planina, mountains of Kraishte, and Rhodopes. *Pisidium obtusale* is considered an oligosaprobic Eurosiberian species, known from Sofia Basin and Rila Mts.

5 Terrestrial Mollusks

Terrestrial mollusks (53.5% of the Bulgarian malacofauna) include two subclasses of class Gastropoda (snails): Prosobranchia with order Mesogastropoda (families Pomatiasidae and Aciculidae), and Pulmonata with orders Basommatophora (family Ellobiidae) and Stylommatophora (the remaining 25 families). The largest families are Clausiliidae (44 species), Zonitidae (31 species), and Hygromiidae (20 species); together they comprise 39.9% of terrestrial species. Considerably rich are Vertiginidae, Enidae, Limacidae, Agriolimacidae, and Helicidae that comprise 29.4% of the terrestrial malacofauna and contain from 12 to 17 species each.

The territory of Bulgaria is not equally studied, although terrestrial mollusks have been investigated here for a long time. One can immediately notice the small number of species known from the Danubian Plain (56, or 23.5%) and Eastern Rhodopes (66, or 27.7%). Most mollusks (145 to 154 species, or 60.9 to 64.7%) have been recorded in Stara Planina, Transitional region, and Rila–Rhodopes Massif. The malacofauna of the Black Sea coast is poorer because of the specific intrazonal habitats, and includes 119 species (50% of all terrestrial snails). Only 32 species have been found in all regions throughout the country. Most of those (23 species) have Holarctic, Palearctic, and Eurosiberian distribution. Seven species have Mediterranean type of distribution, and two species (*Argna macrodonta* and *Tandonia kusceri*) are Balkan endemics (Table 1).

The largest number of terrestrial mollusk species (181 species, or 76.0%), have been found in the hill zone. In the low mountains, 154 species (64.7%) have been recorded; in the middle mountains, 124 species (52.1%); and in the high mountains, 62 species (26.0%). In the lowlands, 146 species (61.3%) have been found (20% less than number of freshwater forms). There are 37 species of terrestrial mollusks, which reaching altitude of 2000 m a.s.l.; this is a heterogeneous group, that combine 18 zoogeographical categories. Most of them (21 species) are Palearctic and Eurosiberian species, 12 species are endemics, and four are Mediterranean species (Table 1). *Vertigo alpestris, Oxychilus inopinatus, Discus ruderatus*, and *Lehmannia sarmizegetusae* have Montane or Boreomontane distribution and rarely are found below 600–1000 m a.s.l. If the mountain malacofauna found above 1600 m is considered, the number of zoogeographical categories in this group reaches 26, it includes 12 Boreomontane or Montane species. Some species such as *Sphyradium doliolum, Phenacolimax annularis, Oxychilus glaber*, and *Monachoides incarnatus* are predominantly living in the mountains but are found in the lowlands as well (Damjanov and Likharev, 1975; Kerney et al., 1983; Schileyko, 1984; Schütt, 1996). According to their vertical distribution, the terrestrial mollusks can be divided into three groups: those distributed in a single altitudinal zone (46 species, or 19.3%); those distributed in more than one altitudinal zone (164 species, or 68.9%); and those distributed in all zones (28 species, or 11.8%). These groups are tentative to some extent and depend on both the specific features of certain mollusk species and the level of study.

The zoogeographical categorization of the species (Tables 1 and 5) is done on the basis of their distributional data, according to Damjanov and Likharev (1975), Riedel (1975, 1996), Nordsieck (1977, 1979, 1987), Kerney et al. (1983), Grossu (1983, 1986, 1987), Wiktor (1983, 1987, 1997), Schileyko (1984), Boeters et al. (1989), and Schütt (1996). The terrestrial mollusks are divided into 36 zoogeographical categories, combined in three large groups (Table 5).

Species distributed in the Palearctic and beyond it. This group (14 species, or 5.9%) includes species with Holarctic distribution, of which seven are represented in all regions of the country. This group includes mesophilous, hydrophilous, and eurybiotic forms, often with broad vertical distribution (*Cochlicopa lubricella, Vitrina pellucida*, and *Deroceras laeve*). Only *Oxyloma elegans* is a hydrophilous species. Schütt (1996) considered *Vertigo antivertigo* as Holarctic–Afrotropical species, but most authors accept it as a Palearctic one.

Species distributed within the Palearctic but in more than one subregion. A total of nine species (3.8%) from this group has been recorded in Bulgaria. This group includes three categories, ratio of which remains the same in various regions and altitudinal zones. *Truncatellina cylindrica, Merdigera obscura* and *Deroceras agreste* have the widest vertical distribution (Table 1). They are presented in all altitudinal zones, and the first two taxa are found in all regions of Bulgaria. *Discus ruderatus* has a Boreomontane distribution. This group consists of mesophilous forms; however, *T. cylindrica* and *Vitrea contracta* are calciphilous at the same time.

Species distributed within one subregion of the Palearctic. A total of 148 species (62.2%) with Eurosiberian and Mediterranean type of distribution belong to this group. The Mediterranean–Central Asian species are also included here according to many authors who combine Mediterranean, Iranian–Turanian and Central Asian subregions. Thus the species with Mediterranean type of distribution are accepted in a general way and include elements (Submediterranean, Atlantic, Subiranian, Pontian, and Euxinian), that can be considered separately as well (Gruev and Kuzmanov, 1994; Irikov, 2002).

Eurosiberian. There are 86 (36.1%) Eurosiberian species of terrestrial mollusks, of which the best represented are European (19 species, or 22.1%), Central and Southeast European (12 species, or 13.9%), and Southeast European (16 species, or 18.6%). There are 63 exclusively European taxa, of which 19 are found all over Europe and 44, in its separate areas (Table 5). *Vertigo alpestris* has both a longitudinal disjunction of its range in regard to Siberia and latitudinal one as a Boreomontane species. Most Eurosiberian species have been recorded in Stara Planina (61 species, or 70.9%), Transitional region (65 species, or 76.6%), the hill zone (69 species, or 80.2%) and lower mountains (64 species, or 74.4%). The Eurosiberian species are mainly mesohygrophilous and, rarely, xeromesophilous and calciphilous (Table 1). Only *Succinea putris* and *S. oblonga* are hygrohydrophilous. Seventeen species from 11 families are accepted as calciphilous. *Zebrina detrita* and *Chondrula tridens* are xerophilous but belong to genera of West Asian origin (Damjanov and Likharev, 1975). The family Zonitidae contains the greatest number of Eurosiberian species (14 species, or 45.2%). Probably introduced in Bulgaria

Table 5 Distribution and zoogeographical characteristics of Bulgarian terrestrial gastropods

Zoogeographical categories	Total number	Danubian Plain (D)	Stara Planina (S)	Transitional region (P)	Rila–Rhodopes Massif (R)	Black Sea coast (B)	Lowlands (below 200 m)	Hills (200–600 m)	Low mountains (600–1000 m)	Middle mountains (1000–1600 m)	High mountains (above 1600 m)
Species distributed in Palearctic and beyond it											
Holarctic	14	7	14	14	13	12	13	13	14	12	3
Species with Palearctic distribution	*157*	*43*	*96*	*116*	*92*	*86*	*107*	*124*	*103*	*80*	*40*
PALEARCTIC TYPE	9	3	7	8	8	7	8	9	8	6	4
Transpalearctic	2		1	1	1	1	1	2	2	1	1
West and Central Palearctic	3	1	3	3	3	2	3	3	3	2	2
West Palearctic	4	2	3	4	4	4	4	4	3	3	1
EUROSIBERIAN TYPE	86	25	61	65	53	35	45	69	64	58	29
Holoeurosiberian	3	2	2	2	2	3	3	2	1	1	1
West and Central Eurosiberian	2	1	2	2	2	1	1	2	2	2	
West Eurosiberian	3	2	2	3	2	3	3	2	2	1	1
Disjunct Eurosiberian	1		1	1	1					1	1
European–Central Asian	2		2	2	1	2	2	2	2	1	
European–Turanian	4	1	2	3	2	3	4	3	2	2	2
European–Anatolian	1	1	1	1	1	1	1	1	1		
Central and South European–Anatolian	2	1	2	2	2	2	2	2	2	2	
Central and Southeast European–Anatolian	3	2	1	3	2	2	3	3	2	2	1
Southeast European–Anatolian	2	1	1	1	2	1	2	2	2	2	1
European	19	9	16	17	15	9	11	18	17	15	6
West European	2			2		1	1	1			
Central and North European	2			1					1	1	
Central and East European	2	1	2	1	1	1	2	2	1	1	
Central and South European	4	1	2	2	3	1	1	4	3	3	1
Central and Southeast European	12		10	10	8	3	3	9	10	10	5
Southeast European	16	3	13	7	9	2	5	13	11	10	7
Carpathian–Balkan	6		2	5			1	3	5	4	3
MEDITERRANEAN TYPE	62	15	28	43	31	44	54	46	31	16	7
Mediterranean–Central Asian	2		1	1	2	1	1	2	2	1	1

North Mediterranean–Central Asian	2	1	2	1	1	1	2	2	2	1	1
North Mediterranean–Turanian	1	1	1	1	1	1	1	1	1	1	
Atlantomediterranean	5		1	3	2	4	5	3	2		
Atlantic–South European	1					1	1				
Holomediterranean	5		2	4	3	4	5	4	3	2	
East Mediterranean	4	2		2	1	2	4	2	1		
North Mediterranean	5	2	3	5	3	3	5	5	4	2	2
South European	2		1	1	2			1	1	2	1
Northeast Mediterranean	1		1	1		1	1	1	1		
Pontomediterranean	9	5	7	7	9	6	8	9	6	5	2
Pontian–Turanian	2			2		1	2	1			
Pontian (+Euxinian)	5	2	2	3	3	4	5	4	2		
Balkan–Anatolian	18	2	7	12	4	15	14	11	6	2	
Endemics and subendemics, total	*67*	*6*	*35*	*24*	*45*	*21*	*26*	*44*	*37*	*32*	*19*
local	11		5	1	3	2	3	4	3	3	3
regional	16		3	2	10	4	3	7	8	8	3
Bulgarian	8	1	6	5	6	5	5	6	3	4	2
Balkan	28	4	17	14	24	9	14	23	21	14	9
subendemics	4	1	4	2	2	1	1	4	2	3	2
Total Mollusca	*238*	*56*	*145*	*154*	*150*	*119*	*146*	*181*	*154*	*124*	*62*
Relicts											
preglacial	21	5	10	15	8	16	19	18	10	4	1
glacial	1		1	1	1					1	1
Rare species	67	7	19	25	26	27	38	37	29	19	7
Troglophiles and trogloxenes	16	4	14	12	11	7	7	15	13	12	4

are *Oxyloma sarsii* and *Arion fasciatus* (Central and North European species) and *Oxychilus draparnaudi* and *Cryptomphallus aspersus* (West European species). Most Eurosiberian species originated in the West Palearctic centers of speciation (Schileyko, 1978, 1984).

Mediterranean species number 62 (26.0%); their number rapidly decreases with the altitude. This group is complex; it is divided into many subgroups with different origin, distribution, and ecology. This complexity contributes to varied zoogeographical grouping. This group includes preglacial relicts (21 species, or 33.9%). The Balkan–Anatolian (18 species, or 29.0%) and Pontomediterranean (9 species, or 14.5%) species are most numerous. An exclusively Mediterranean distribution is recorded for 51 species, of which five are widespread throughout this subregion and 46, in its separate regions (Table 5). The great number of Mediterranean species has been recorded from the Transitional region (43 species, or 69.4%), Black Sea coast (44 species, or 70.9%), and lowlands (54 species, or 87.1%). In this group, mesophilous, xerophilous, and mesoxerophilous species predominate (Table 1). *Myosotella myosotis* is a hydrophilous Atlantomediterranean species that inhabits wet stones of the seashore. The Holomediterranean species *Poiretia algira* is accepted as hygrophilous. Thirteen species from eight families are clearly calciphilous. The family Enidae is the richest in Mediterranean species (10 species, or 58.8%). Twelve (70.6%) of the species from this family are xerophilous. Most of xerophilous taxa in Enidae have Anatolian origin, represented in Bulgaria by Balkan–Anatolian species. Many Euxinian species have a similar distribution. They are mesophilous forms, originated during Neogene in the forests near the Black Sea and usually are preglacial relicts. Some of them such as *Pomatias rivulare* and *Vitrea pygmaea* have a wider distribution. The Pontian species are similar to the Euxinian ones in their distribution, and they sometimes are considered in the same group. They are mainly xerophilous with Pontomediterranean origin and centers of distribution north of the Black Sea. *Chondrina avenacea* has a South European montane range and is accepted as a Montane Mediterranean species (South European mountain sub-element according to Gruev and Kuzmanov, 1994). Damjanov and Likharev (1975) list ten species with montane North Mediterranean ranges, mainly calciphilous and xerothermic, which belong to this category. However, not all of those can be currently accepted as Montane Mediterranean (Table 1).

Endemics. The Bulgarian terrestrial mollusk fauna includes 67 endemic species (28.2%) (Tables 1 and 5). In addition, there is a considerable number of endemic subspecies, which are not discussed in the present work since many of them have an unclear status. The Balkan endemics predominate (28 species, or 41.8%), followed by the regional endemics (16 species, or 23.9%), and the local endemics (11 species, or 16.4%). The high number of endemics (45 species, or 67.2%) has been established in the Rila–Rhodopes Massif and the hill zone (44 species, or 65.7%). A number of local endemics (five species, or 7.6%) has been found in the karst massifs of Stara Planina. Local and regional endemics are absent only in the Danubian Plain. The Balkan endemics are not uniform in distribution, and the Bulgarian fauna includes only a few endemic forms, which are found all over the Balkan Peninsula. This fact is historically connected with the isolated evolution of the malacofauna in the eastern and western portions of the Balkan Peninsula. The family Clausiliidae (28 species, or 63.6%) is the richest in endemics, of which 14 local and regional endemics are described as new (Table 1).

Relicts. The Bulgarian terrestrial mollusk fauna includes 21 preglacial relict species (8.8%). From 80 to 95% of those relict species are found localized along the Black Sea coast, lowlands and hill zone (Table 5). All relicts have Mediterranean distribution and are divided into eight zoogeographical categories; these are mainly Euxinian species. The Balkan–Anatolian, Pontian, and Holomediterranean forms dominate and include 14 species (Table 1). Eleven families of terrestrial mollusks include relicts but the richest is Zonitidae (four species). The mesophilous (13 species) and xeromesophilous (five species) taxa predominate. Only one hydrophilous (*Myosotella myosotis*), one hygrophilous (*Poiretia algira*) and one mesohygrophilous (*Oxychilus deilus*) species have been recorded. *Vertigo alpestris* can be accepted as a glacial relict, disjunct Eurosiberian, Boreomontane species, distributed in the mountains above 1000 m a.s.l.

Rare species. In Bulgaria, 67 species (28.2%), of which 19 are endemics and eight are relicts, are considered rare. They are mainly stenotopic forms, connected with specific habitats and ecology. The mesophilous species predominate (32), but considerable number of mesohygrophilous (13), xerophilous (10), and calciphilous (13) taxa occur as well. Rare species are distributed throughout the country but more taxa (27 to 38) have been found along the Black Sea coast and lowlands. The least number of rare species is found in the Danubian Plain and high mountains (Tables 1 and 5).

Threatened species. Five terrestrial mollusk species are included in the different Red Lists (Table 1): IUCN Red List: *Vertigo moulinsiana* and *V. angustior*; ESC Red List: *V. moulinsiana, V. angustior, Balea perversa, Oxyloma sarsii*, and *Helix pomatia*; Bern Convention: *H. pomatia*; Habitats Directive: *V. moulinsiana* and *V. angustior*; and CORINE: *V. moulinsiana, V. angustior, B. perversa, O. sarsii*, and *H. pomatia*. Distribution of these species includes the Danubian Plain, Thracian Lowland, Black Sea coast, and Central Stara Planina. *H. pomatia* is found all over the country (sometimes up to 1900 m). There are no typical troglobionts among the Bulgarian terrestrial mollusks, and they often inhabit edaphic habitats. At the same time they are elements of the subterranean ecosystems and are protected in the caves at any rate. Eight troglophilic and eight trogloxenic taxa have been recorded (Table 1). The greatest number of subterranean gastropods has been established in Stara Planina and the hill zone.

Taxa with high conservation value. Among the Bulgarian terrestrial mollusks, 145 taxa (60.9%) can be accepted as important for conservation (Table 1). Of these, 31 species (13.0%) are of world importance; 54 species (22.7%), of European importance; and 60 species (25.2%), of a national importance. The conservation value of the species is especially high when more than one criterion are combined for one taxa (rare and endemic, rare and relicts, etc.). Of those, 26 taxa (10.9%) are known among the terrestrial mollusks of Bulgaria.

Table 6 Freshwater and terrestrial mollusks of some Bulgarian mountains

	Total in Bulgaria	Western Stara Planina	Central Stara Planina	Vitosha	Strandja	Rila	Pirin	Western Rhodopes	Eastern Rhodopes
Families	28	34	25	21	36	18	27	27	27
Species, total	*333*	*117*	*90*	*67*	*85*	*61*	*89*	*110*	*66*
freshwater	95	27	3	18	10	4	6	12	4
terrestrial	238	90	87	49	75	57	83	98	62
Endemics and subendemics, total	*87*	*23*	*21*	*7*	*14*	*10*	*23*	*36*	*11*
local endemics	24	6	3		1	2	6	11	1
Bulgarian endemics	25	4	8	2	6	5	2	6	2
Balkan endemics	32	12	10	5	7	3	13	19	8
Balkan subendemics	6	1					2		
Relicts, total	*33*	*2*	*1*	*2*	*13*	*1*	*5*	*8*	*9*
preglacial relicts	22	1		2	12	1	5	7	9
glacial relicts	2	1	1					1	
Caspian relicts	9				1				
Subterranean species	33	6	1		2		6	5	4
Rare species	102	12	6	9	10	5	3	12	3
Conservation status									
Protected in Bulgaria	16	6							
IUCN	10	1	2	1	1	1		1	1
ESC	7	1			1		1	1	1
BC	4	2					1	1	1
HD	3	2					1	1	1
CORINE	7	1	2	1	1	1		1	1

6 Mollusks in the Bulgarian Mountains

The summarized data on the malacofauna of Bulgarian mountains (territories of national and natural parks) are presented in Table 6. The small number of species known from Rila Mts. in spite of its large area and good level of study is notable; it is connected with the lack of calciphilous forms because of the silicate character of this mountain massif. Eastern Rhodopes largely remain unexplored due to the late onset of the investigations; lack of karst massifs attractive for visitors; they are generally poorly visited compared to other mountains. In Western Stara Planina, considerable number of aquatic mollusks occur, a part of which are local stygobiotic endemics. This mountain range has the richest malacofauna. The faunistic richness of the Western Rhodopes is similar to the Western Stara Planina, but number of relicts and endemics is higher in Western Rhodopes. Vitosha Mts. is the best studied Bulgarian mountain massif; however, its small area and insignificant karst (only in the southern part) make it poor in mollusk species diversity in spite of the occurrence of many aquatic forms. Strandja Mts. is the richest area with preglacial relicts. Faunas of Central Stara Planina and Pirin Mts. are comparatively rich, well-studied and have similar species composition and number of endemics. There are reviews on the mollusk fauna of Central Stara Planina, Rila Mts., Pirin Mts., and partially for Western Rhodopes (Dedov and Mitov, 1998; Deltshev et al., 2000a, 2000b; Irikov, 2002).

References

Abbott, R. & R. Dance, 1991. Compendium of Seashells. Charles Letts & Co., London.

Angelov, A., 1956. [Hydrologische und hydrobiologische Untersuchungen über den Rabischa-See]. Godishnik na Sofiiskiya universitet. Biologo-geologo-geografski fakultet (Annuaire de l'Université de Sofia, Faculté de Biologie, Géologie et Géographie) 49(1 – Biology): 1–30 (in Bulgarian).

Angelov, A., 1959. Neue Gastropoden aus den unterirdischen Gewässern Bulgariens. Archiv für Molluskenkunde 88(1/3):51–54.

Angelov, A., 1972. Neue Hydrobiidae aus Höhlengewässern Bulgariens. Archiv für Molluskenkunde 102(1/3):107–112.

Angelov, A., 1976. [Revision der Fam. Pisididae Gray, 1857 (Bivalvia – Mollusca) in Bulgarien]. Godishnik na Sofiiskiya universitet "Kliment Ohridski". Biologicheski fakultet (Annuaire de l'Université de Sofia "Kliment Ohridski", Faculté de Biologie) 69(1 – Zoology): 109–119 (in Bulgarian).

Angelov, A., 1983. [*Ferrissia wautieri* (Mirolli) (Gastropoda, Ancylidae) – a new representative of the Bulgarian freshwater fauna]. Acta zoologica bulgarica 21:95–97 (in Bulgarian).

Angelov, A., 1998a. Aquatic gastropods. In Michev, T., B. Georgiev, A. Petrova & M. Stoyneva (eds), Biodiversity of the Srebarna Biosphere Reserve. Checklist and bibliography. Context & Pensoft, Sofia: 63.

Angelov, A., 1998b. Class Bivalvia. In Michev, T., B. Georgiev, A. Petrova & M. Stoyneva (eds), Biodiversity of the Srebarna Biosphere Reserve. Checklist and bibliography. Context & Pensoft, Sofia: 64.

Angelov, A., 2000a. Mollusca (Gastropoda et Bivalvia) aquae dulcis. In Catalogus faunae bulgaricae. 4. Pensoft & Backhuys Publishers BV, Sofia & Leiden.

Angelov, A., 2000b. Freshwater molluscs (Mollusca, Gastropoda and Bivalvia) from the Srebarna Lake, Northeastern Bulgaria. Historia naturalis bulgarica 11:133–138.

Anistratenko, V.V. & A.P. Stadnichenko, 1994. [Littoriniformes, Rissoiformes]. In Fauna Ukrainy. Tom 29. Mollyuski. Vypusk 1, kniga 2. [Fauna of Ukraine. Vol. 29. Mollusks. No. 4, 2]. Naukova dumka [Scientific Thought], Kiev (in Russian).

Bănărescu, P., 1990. Zoogeography of Fresh Waters. 1. General Distribution and Dispersal of Freshwater Animals. Aula-Verlag, Wiesbaden.

Beron, P., 1972. Essai sur la faune cavernicole de Bulgarie. III. Résultats des recherches biospéologiques de 1966 à 1970. International Journal of Speleology 4:285–349.

Beron, P., 1994. Résultats des recherches biospéléologiques en Bulgarie de 1971 à 1994 et liste des animaux cavernicoles bulgares. Tranteeva (Sofia) 1:1–137.

Beron, P. & V. Guéorguiev, 1967. Essai sur la faune cavernicole de Bulgarie.II. Résultats des recherches biospéologiques de 1961 à 1965. Izvestiya na Zoologicheskiya institut s muzei (Bulletin de l'Institut de zoologie et musée) 24:151–212.

Boeters, H., E. Gittenberger & P. Subai, 1989. Die Aciculidae (Mollusca, Gastropoda Prosobranchia). Zoologische Verhandelingen 252:1–234.

Bole, J. & F. Velkovrh, 1986. Mollusca from continental subterranean aquatic habitats. In: Botosaneanu, L. (ed.). Stygofauna Mundi. E. J. Brill & Dr. W. Backhuys, Leiden: 177–208.

Büttner, K., 1928. Beitrag zur Molluskenfauna Bulgariens. Jahresbericht des Vereins für Naturkunde zu Zwickau: 12–20.

Caspers, H., 1951. Quantitative Untersuchungen über die Bodentierwelt des Schwarzen Meeres im bulgarischen Küstenbereich. Archiv für Hydrobiologie 45:1–191.

Caspers, H., 1952. Untersuchungen über die Tierwelt von Meeressalinen an der bulgarischen Küste des Schwarzen Meeres. Zoologischer Anzeiger 148(5/8):243–259.

Cvetkov, L. & T. Marinov, 1986. [Faunistic enrichment of the Black Sea and changes in its benthic ecosystems]. Hidrobiologiya (Hydrobiology) (Sofia) 27:3–20 (in Russian).

Damjanov, S. & I.M. Likharev, 1975. [Gastropoda terrestria]. In Fauna na Balgariya. 4 (Fauna bulgarica. 4). Izdatelstvo na Balgarskata akademiya na naukite (Aedibus Academiae scientiarum bulgaricae), Sofia (in Bulgarian).

Damjanov, S. & L. Pintér, 1969. Neue Vitreini aus Bulgarien (Gastropoda, Euthyneura). Archiv für Molluskenkunde 99(1/2):35–40.

Dedov, I., 1998. Annotated check-list of the Bulgarian terrestrial snails (Mollusca, Gastropoda). Linzer biologische Beiträge 30(2):745–765.

Dedov, I. & P. Mitov, 1998. Species composition of the terrestrial snails (Mollusca, Gastropoda) from coniferous and alpine areas of the Northern Pirin Mountains, Bulgaria. Historia naturalis bulgarica 9:19–26.

Dedov, I. & L. Penev, 2000. Species composition and origins of the terrestrial gastropod fauna of Sofia City, Bulgaria. Ruthenica 10(2):121–131.

Deltshev, Ch., P. Beron, G. Blagoev, V. Golemansky, V. Peneva, P. Stoev, P., M. Todorov & Z. Hubenov, 2000a. Faunistic diversity of invertebrates (non Insecta) of the Rila National Park. In Sakalian, M. (ed.). Biological Diversity of the Rila National Park. Pensoft, Sofia: 249–284, 429–525, 619–622.

Deltshev, Ch., P. Beron, G. Blagoev, V. Golemansky, V. Peneva, P. Stoev, P., M. Todorov & Z. Hubenov, 2000b. Faunistic diversity of invertebrates (non Insecta) in Central Balkan National Park. In Sakalian, M. (ed), Biological Diversity of the Central Balkan National Park. Pensoft, Sofia: 289–317, 491–538, 586–590.

Drensky, P., 1947. [Synopsis and distribution of freshwater Mollusca in Bulgaria]. Godishnik na Sofiiskiya universitet. Fiziko-matematicheski fakultet (Annuaire de l'Université de Sofia. Faculté physico-mathématique) 43(3):33–54 (in Bulgarian).

Frank, C., J. Jungbluth & A. Richnovszky, 1990. Die Mollusken der Donau vom Schwarzwald bis zum Schwarzen Meer. Akaprint, Budapest.

Golikov, A.N., 1982. [On principles of regionalization and unification of terminology in marine biogeography]. In Kusakin, O.G. (ed), Morskaya biogeografiya, predmet, metody, printsipy rayonirovaniya [Marine Biogeography, Its Subject, Methods, and Principles of Regionalization]. Nauka [Science], Moscow: 94–98 (in Russian).

Golikov, A.N. & Ya.I. Starobogatov, 1966. [Ponto–Caspian gastropod molluscs in the Azov–Black Sea basin]. Zoologicheskii zhurnal 45(3):352–362 (in Russian).

Golikov, A.N. & Ya.I. Starobogatov, 1968. [Zoogeographical characteristics of the gastropod mollusks of Black and Azov Seas]. In Biologicheskie issledovaniya Chernogo morya i ego promyslovykh resursov [Biological Studies of the Black Sea and Its Fishery Resources]. Nauka [Science], Moscow: 70–83 (in Russian).

Golikov, A.N. & Ya.I. Starobogatov, 1972. [Class Gastropoda]. In Opredelitel' fauny Chernogo i Azovskogo morey [The Keys to the Fauna of Black and Azov Seas]. 3. Naukova dumka [Scientific Thought], Kiev: 65–166 (in Russian).

Grossu, A., 1962. [Bivalvia (Scoici)]. In Fauna Republicii Populare Romîne. Mollusca. Volumul III. Fascicula 3. Editura Academiei Republicii Populare Romîne, Bucharest.

Grossu, A., 1983. Gastropoda Romanie. 4. Ordo Stylommatophora. Litera, Bucharest.

Grossu, A., 1986. Gastropoda Romanie. 1. Prosobranchia et Opistobranchia. Litera, Bucharest.

Grossu, A., 1987. Gastropoda Romanie. 2. Pulmonata. Litera, Bucharest.

Grossu, A., 1993. The Catalogue of the Molluscs from Romania. Travaux du Museum National d'histoire Naturelle "G. Antipa" 33:291–366.

Gruev, B. & B. Kuzmanov, 1994. Obshta biogeografiya. Vtoro osnovno preraboteno izdanie [General Biogeography. Second Basic Revised Edition]. Universitetsko izdatelstvo "Sv. Kliment Ohridski" [St. Kliment Ohridski University Publishing House], Sofia (in Bulgarian).

Guéorguiev, V. & P. Beron, 1962. Essai sur la faune cavernicole de Bulgarie. Annales de Spéléologie 17(2/3):285–441.

Guryanova, Ye.F., 1964. [Zoogeographic regionalization of the benthic fauna of the World Ocean]. In Fiziko-geograficheskiy atlas mira [Physico-geographical Atlas of the World], Moscow, Map 68B: 291–292 (in Russian).

Hayward, P., T. Nelson-Smith & C. Shields, 1996. Sea Shore of Britain and Europe. Collins, London.

Hesse, P., 1911. Zur Kenntnis der Molluskenfauna von Ostrumelien. Nachrichtsblatt der Deutschen malakozoologischen Gesellschaft 43:142–155.

Hesse, P., 1913. Zur Kenntnis der Molluskenfauna von Ostrumelien. II. Nachrichtsblatt der Deutschen malakozoologischen Gesellschaft 45(1):1–16; 45(2):69–74.

Hesse, P., 1914. Zur Kenntnis der Molluskenfauna von Ostrumelien. III Nachrichtsblatt der Deutschen malakozoologischen Gesellschaft 46(2):49–58.

Hesse, P., 1916. Zur Kenntnis der Molluskenfauna von Ostrumelien. IV. Nachrichtsblatt der Deutschen malakozoologischen Gesellschaft 48:113–122.

Hubenov, Z., 1997. Possibilities for using of a system from the really defined natural territories for the faunistic researches in Bulgaria. Acta zoologica bulgarica 49:5–9.

Hubenov, Z., 2001. Corbiculidae – a new family for the Bulgarian recent malacofauna (Mollusca, Bivalvia). Acta zoologica bulgarica 53(3):61–66.

Hudec, V., 1964. *Milax* (*Milax*) *bojanensis* n. sp. aus Bulgarien. Archiv für Molluskenkunde 93(5/6):187–191.

Hudec, V., 1975. Zum Vorkommen der Schnecke *Jaminia* (*Euchondrus*) *ovularis* (Olivier, 1801) in Bulgarien. Časopis Národního Musea, oddíl přírodovědný (Prague) 144(1/4):73–76.

Hudec, V. & J. Vašatko, 1971. Beitrag zur Molluskenfauna Bulgariens. Acta scientiarum naturalium Academiae Scientiarum Bohemoslovacae Brno 5(2):1–38.

Hudec, V. & J. Vašatko, 1973. Zur Kenntnis der Molluskenfauna Bulgariens. Acta scientiarum naturalium Academiae Scientiarum Bohemoslovacae Brno 7(9):1–33.

Irikov, A., 1999. [New localities of terrestrial snail (Gastropoda, Pulmonata) in Bulgaria]. Nauchni trudove na Plovdivskiya universitet "Paisii Hilendarski" (Travaux scientifiques de l'Université de Plovdiv "Paisii Hilendarski"), Animalia, 35(6):29–32 (in Bulgarian).

Irikov, A., 2001. [To the knowledge of the anatomy and taxonomy of two species from genus *Vitrea* Fitzinger, 1833 (Gastropoda, Zonitidae)]. Nauchni trudove na Plovdivskiya universitet "Paisii Hilendarski" (Travaux scientifiques de l'Université de Plovdiv "Paisii Hilendarski"), Animalia, 37(6):35–38 (in Bulgarian).

Irikov, A., 2002. [Species composition and zoogeographical characterization of the terrestrial shell malacofauna (Mollusca, Gastropoda) in the Dobrostanski ridge of the Western Rhodopes with ecological notes]. Historia naturalis bulgarica 14:89–102 (in Bulgarian).

Irikov, A. & D. Georgiev, 2002. [Species composition, habitat distribution, zoogeographical structure and origin of the malacofauna in town Stara Zagora (Bulgaria)]. Nauchni trudove na Plovdivskiya universitet "Paisii Hilendarski" (Travaux scientifiques de l'Université de Plovdiv "Paisii Hilendarski"), Animalia, 38(6):5–16 (in Bulgarian).

Iurinitsh, S., 1906. Notes sur les Bivalves d'eau douce de Bulgarie. Godishnik na Sofiiskiya universitet. Fiziko-matematicheski fakultet (Annuaire de l'Université de Sofia. Faculté physico-mathématique) 2:61–67.

Iurinitsh, S., 1908. [Sur les Helicides de la Bulgarie]. Godishnik na Sofiiskiya universitet. Fiziko-matematicheski fakultet (Annuaire de l'Université de Sofia. Faculté physico-mathématique) 3/4:1–21 (in Bulgarian).

Ivanov, K., A. Sotirov, A. Rozhdestvensky & D. Vodenicharov, 1964. [The lakes of Bulgaria]. Izvestiya na Instituta po hidrologiya i meteorologiya [Proceedings of the Institute of Hydrology and Meteorology] 16:1–243 (in Bulgarian).

Jaeckel, S., 1954a. Zur Systematik und Faunistik der Mollusken der nördlichen Balkanhalbinsel. Mitteilungen aus dem Zoologischen Museum in Berlin 30:54–95.

Jaeckel, S., 1954b. Zur Kenntnis der Meeres- und Brackwasser-Mollusken von Varna (Bulgarien). Hidrobiologia. Acta hydrobiologica, hydrographica et limnologica 6:70–82.

Jaeckel, S., W. Klemm & W. Meise, 1957. Die Land- und Süsswasser-Mollusken der nördlichen Balkanhalbinsel. Abhandlungen und Berichte aus dem Staatlichen Museum für Tierkunde in Dresden 23(2):141–205.

Jickeli, C., 1874. Verzeichnis der auf meiner Reise nach Rothen Meere in Europa aufgegriffenen Mollusken. Nachrichtsblatt der Deutschen malakozoologischen Gesellschaft 6:8–11.

Kaneva-Abadjieva, V., 1957. [Mollusca and Malacostraca im Varnasee]. Trudove na Morskata biologichna stantsiya vav Varna (Arbeiten aus der Biologischen Meeresstation in Varna, Bulgarien) 19:127–154 (in Bulgarian).

Kaneva-Abadjieva, V., 1958. [A new harmful snail on our coast of Black Sea]. Priroda [Nature] (Sofia) 7(3):89–91 (in Bulgarian).

Kaneva-Abadjieva, V., 1960. [Materials to the study of the mollusc fauna in the Back Sea at the Bulgarian shores]. Trudove na Nauchnoizsledovatelskiya institut po ribarstvo i ribna promishlenost – Varna (Travaux de l'Institut de recherches scientifiques sur la pêche et les industries s'y rattachant – Varna (Bulgarie)) 2:149–172 (in Bulgarian).

Kaneva-Abadjieva, V., 1962. [Répartition des mollusques de la Mer Noire dans les eaux littorales bulgares]. Izvestiya na Tsentralniya nauchnoizsledovatelski institut po ribovadstvo i ribolov – Varna (Bulletin de l'Institut central de recherche scientifique de pisciculture et de pêcherie – Varna) 2:67–79 (in Bulgarian).

Kaneva-Abadjieva, V., 1968. Distribution des mollusques subfossiles devant le littoral bulgare de la Mer Noire. Rapport. Comité internationale Mer Méditerranée 19(2):213–215.

Kaneva-Abadjieva, V., 1974. [A new mussel species for our Black Sea Coast] Priroda [Nature] (Sofia) 23(2):64–65 (in Bulgarian).

Kaneva-Abadjieva, V., 1975. [On the alimentary zoobenthos dynamics in Bourgas Lake and Mandra Dam during the 1967–1974 period]. Izvestiya. Institut po ribni resursi – Varna (Proceedings. Institute of Fisheries – Varna) 14:43–56 (in Bulgarian).

Kaneva-Abadjieva, V. & T. Marinov, 1960. [Zoobenthos distribution in the Bulgarian Black Sea area]. Trudove na Tsentralniya nauchnoizsledovatelski institut po ribovadstvo i ribolov – Varna (Arbeiten des Zentralen Forschungsinstitutes für Fischzucht und Fischerei Varna (Bulgarien)) 3:117–166 (in Bulgarian).

Kaneva-Abadjieva, V. & T. Marinov, 1966. [Distribution of the zoobenthos in the sand biocoenosis of the Bulgarian Black Sea coast]. Izvestiya na Nauchnoizsledovatelskiya institut za ribno stopanstvo i okeanografiya – Varna (Proceedings of the Research Institute of Fisheries and Oceanography – Varna) 7:69–95 (in Bulgarian).

Kaneva-Abadjieva, V. & T. Marinov, 1967. [Dynamics of the zoobenthos in the lakes along the Bulgarian Black Sea coast in the 1964–1966 period]. Izvestiya na Nauchnoizsledovatelskiya institut za ribno stopanstvo i okeanografiya – Varna (Proceedings of the Research Institute of Fisheries and Oceanography – Varna) 8:177–194 (in Bulgarian).

Kaneva-Abadjieva, V. & T. Marinov, 1984. [A new mollusk species for the Black Sea, *Cunearca cornea* (Reeve)]. *Priroda* [*Nature*] (Sofia) 33(5):63–64 (in Bulgarian).

Kerney, M., R. Cameron & J. Jungbluth, 1983. Die Landschnecken Nord- und Mitteleuropas. Parey, Hamburg & Berlin.

Knipper, H., 1939. Systematische, anatomische, ökologische und tiergeographische Studien an südosteuropäischen Heliciden (Moll. Pulm.). Archiv für Naturgeschichte N.F. 8(3/4):327–517.

Knipper, H., 1941. Nachträge zur südosteuropäischen Helicidenfauna (Moll. Pulm.). I. Archiv für Molluskenkunde 73:177–186.

Konsulov, A., 1998. Black Sea Environmental Series. 5. Black Sea Biological Diversity (Bulgaria). UN Publications, Istanbul.

Körnig, G., 1983. Beitrag zur Ökologie und Zoogeographie bulgarischer Landgastropoden. Malakologische Abhandlungen. Staatliches Museum für Tierkunde Dresden 9(5):31–52.

Kornyushin, A.V., 1996. Dvustvorchatye mollyuski nadsemeistva Pisidioidea v Palearktike (Fauna, sistematika, filogeniya) [Bivalve Mollusks of the Superfamily Pisidioidea in the Palearctic Region (Fauna, systematics, phylogeny)]. Natsional'naya akademiya nauk Urkainy [National Academy of Sciences of Ukraine], Kiev (in Russian).

Kovachev, S., V. Hainadjieva & S. Stoichev, 2002. Quantitative characteristics of the zoobenthos of several coastal lakes in North-East Bulgaria. Godishnik na Sofiiskiya universitet "Sv. Kliment Ohridski". Biologicheski fakultet (Annuaire de l'Université de Sofia "St. Kliment Ohridski". Faculté de Biologie) 93/94(1 – Zoology):19–22.

Kovachev, S., S. Stoichev & V. Hainadjieva, 1999. The zoobenthos of several lakes along the Northern Bulgarian Black Sea coast. Lauterbornia 35:33–38.

Kovachev, S. & Y. Uzunov, 1981. [The macrozoobenthos of the Durankulak Lake. Structure and dynamics of numbers and biomass]. Hidrobiologiya [Hydrobiology] (Sofia) 15:74–80 (in Bulgarian).

Kroupa, O., 1988. Kritische Revision der Weichtiertypenmaterialien in den Sammlungen des Nationalmuseums in Prague. I. Sbornik Národniho Musea v Praze 44:105–126.

Kroupa, O., 1994. Zur Molluskenfauna Bulgariens. 3. Über *Arianta pelia* (Hesse, 1912) (Stylommatophora, Helicidae). Nachrichtenblatt der Ersten vorarlberger malakologischen Gesellschaft 2:22–28.

Kuiper, I., 1963. Hauptzüge der Verbreitung des Genus *Pisidium* in Europa. Archiv für Molluskenkunde 92(5/6):247–252.

Lattin, G. de, 1967. Grundriss der Zoogeographie. VEB Gustav Fischer Verlag, Jena.

Marinov, T., 1990. Zoobentosat ot balgarskiya sektor na Cherno more (The Zoobenthos from the Bulgarian Sector of the Black Sea). Izdatelstvo na Balgarskata akademiya na naukite (Publishing House of the Bulgarian Academy of Sciences), Sofia (in Bulgarian).

Marinov, T. & S. Stoykov, 1995. [Zoobenthos distribution in Bulgarian continental shelf of the Black Sea]. Izvestiya na Instituta po ribni resursi – Varna (Proceedings of the Institute of Fisheries – Varna) 23:119–137 (in Bulgarian).

Marinov, T., S. Stoykov & M. M'Barek, 1983. [The zoobenthos of the sublittoral and muddy bottom of bay of Varna]. Izvestiya na Instituta po ribni resursi (Proceedings of the Institute of Fisheries) (Varna) 20:109–133 (in Bulgarian).

Mihailova-Neikova, M., 1961. [Hydrobiological research of the Mandra Lake with regard to its importance as a fishing ground]. Godishnik na Sofiiskiya universitet. Biologo-geologo-geografski fakultet (Annuaire de l'Université de Sofia, Faculté de Biologie, Géologie et Géographie) 53(1 – Zoology): 57–123 (in Bulgarian).

Mordukhay-Boltovskoy, F.D., 1960. Kaspiyskaya fauna v Azovsko-Chernomorskom basseyne [The Caspian Fauna in the Azov–Black Sea Basin]. Izdatel'stvo Akademii Nauk SSSR [Publishing House of the Academy of Sciences of the USSR], Moscow & Leningrad (in Russian).

Mousson, A., 1859. Coquilles terrestres et fluviatiles requeillis dans l'Orient par M. le Dr. Alexandre Schläfli. V. La Bulgarie. Vierteljahresschrift der Naturforschenden Gesellschaft in Zürich 4:253–297.

Nesis, K.N., 1982. [Zoogeographical position of the Mediterranean Sea]. In Kusakin, O.G. (ed), Morskaya biogeografiya. Predmet, metody, printsipy rayonirovaniya [Marine Biogeography, Its Subject, Methods, and Principles of Regionalization]. Nauka [Science], Moscow: 270–299 (in Russian).

Nordsieck, F., 1968. Die europäischen Meeres-Gehäuseschnecken (Prosobranchia). G. Fischer, Stuttgart.

Nordsieck, F., 1969. Die europäischen Meeresmuscheln (Bivalvia). G. Fischer, Stuttgart.

Nordsieck, H., 1973. Zur Anatomie und Systematik der Clausilien. XIII. Neue Balkan-Formen der Mentissoideinae und Baleinae (mit taxonomischer Revision der zugehörigen Gruppen). Archiv für Molluskenkunde 103(4/6):179–208.

Nordsieck, H., 1974. Zur Anatomie und Systematik der Clausilien. XV. Neue Clausilien der Balkan-Halbinsel (mit taxonomischer Revision einiger Gruppen der Alopinae und Baleinae). Archiv für Molluskenkunde 104(4/6):123–170.

Nordsieck, H., 1977. Zur Anatomie und Systematik der Clausilien. XVIII. Neue Taxa rezenter Clausilien. Archiv für Molluskenkunde 108(1/3):73–107.

Nordsieck, H., 1978. Zur Anatomie und Systematik der Clausilien. XX. Die rezenten Arten der Serrulininae und der Gattung Caspiophaedusa. Archiv für Molluskenkunde 109(1/3):91–101.

Nordsieck, H., 1979. Zur Anatomie und Systematik der Clausilien. XXI. Das System der Clausilien, II: Die rezenten europäischen Clausilien. Archiv für Molluskenkunde 109(4/6):249–275.

Nordsieck, H., 1987. Revision des Systems der Helicoidea (Gastropoda: Stylommatophora). Archiv für Molluskenkunde 118(1/3):9–50.

Odhner, N., 1929. Sphaerien und Pisidien aus Bulgarien. Archiv für Molluskenkunde 61:177–185.

Ošanova, N., 1964. [Über die Biologie und Ökologie von zwei für die Fauna Bulgariens neuen Arten der Familie Arionidae]. Izvestiya na Zoologicheskiya institut s muzei (Bulletin de l'Institut de zoologie et musée) 15:203–214 (in Bulgarian).

Ošanova, N., 1968. Zur Verbreitung der Arioniden in Bulgarien. Malakologische Abhandlungen. Staatliches Museum für Tierkunde Dresden 2(17):235–241.

Ošanova, N., 1970. Die Nacktschnecken im westlichen Teil des Balkan-Gebirges (Bulgarien). Malakologische Abhandlungen. Staatliches Museum für Tierkunde Dresden 3(7):71–79.

Ošanova, N., 1972. [Nacktschnecken (Familien Arionidae, Limacidae und Milacidae) aus dem Vitoscha-Gebirge]. Izvestiya na Zoologicheskiya institut s muzei (Bulletin de l'Institut de zoologie et musée) 35:139–154 (in Bulgarian).

Pandourski, I., 2001. Recherches hydrobiologiques des zones humides de la côte bulgare de la Mer Noire. I. Le lac de Vaja. Rivista di Idrobiologia 40(2/3):321–334.

Petrbok, J., 1925. K problému marinniho plistocenu u Varny. Věstník Státního geologického ústavu Československé republiky 1:1–6.

Petrbok, J., 1940. Zweiter Beitrag zur Kenntnis der Mollusken-Fauna von Bulgarien. Archiv für Molluskenkunde 72: 85–88.

Petrbok, J., 1941. Posttertiaria nonmarina mollusca bulgarica. Věstník Královské České Společnosti Nauk, Třída matematicko-přírodovìdecká (Prague): 1–39.

Petrbok, J., 1947. [The freshwater molluscs of the lakes of Varna and of Gebedže]. Trudove na Morskata biologichna stantsiya vav Varna (Arbeiten aus der Biologischen Meeresstation in Varna, Bulgarien) 13: 71–75 (in Bulgarian).

Petrbok, J., 1948. A contribution to the knowledge of the post-Tertiary molluscs of Bulgaria. Sbornik Národniho Musea v Praze 4B(3), 1–28.

Petrbok, J., 1950. Mollusca bulgarica nova. Sbornik Národniho Musea v Praze 6B(1), zool. 1, 1–16.

Petrova, E. & S. Stoykov, 2002. The zoobenthos from the coastal lakes on the northern Bulgarian Black Sea coast during 2000–2001 years. In 34th Conference of the International Association for Danube Research, Tulcea, Romania, August 2002: 337–343.

Pinter, L., 1968a. Über bulgarische Mollusken. Malakologische Abhandlungen. Staatliches Museum für Tierkunde *Dresden* 2(2):209–230.

Pinter, L., 1968b. Eine neue Wasserschnecke aus Bulgarien. Archiv für Molluskenkunde 98(1/2):61–63.

Radoman, P., 1983. Hydrobioidea, a Superfamily of Prosobranchia (Gastropoda). I. Systematics. Serbian Academy of Sciences, Belgrade.

Riedel, A., 1957. Materialien zur Kenntnis der paläarktischen Zonitidae (Gastropoda). I. Annales zoologici 16(19):325–332.

Riedel, A., 1963. Zwei neue Zonitidae (Gastropoda) aus Sudostbulgarien. Annales zoologici 20(22):474–485.

Riedel, A., 1967. Daudebardiidae (Gastropoda, Zonitidae) Bulgariens. Annales zoologici 24(8):463–483.

Riedel, A., 1972. Die Untergattung *Schistophallus* A. J. Wagner in Europa und Kleinasien (Gastropoda, Zonitidae). Annales zoologici 24(7):181–207.

Riedel, A., 1975. Die Zonitiden-Fauna Bulgariens (Gastropoda), ihre Herkunft und Verbreitung. Fragmenta faunistica 20(11):157–177.

Riedel, A., 1978. Kritische Bemerkungen und Ergänzungen zur Kenntnis der Subfamilie Daudebardiinae (Gastropoda, Zonitidae) mit Verzeichnis aller akzeptierten Arten. Annales zoologici 34(8):139–204.

Riedel, A., 1985. Zur Kenntnis der Gattung *Balcanodiscus* (Gastropoda, Stylommatophora, Zonitidae). Malakologische Abhandlungen. Staatliches Museum für Tierkunde Dresden 10(14):139–146.

Riedel, A., 1988. Zur Kenntnis der Gattung *Balcanodiscus*, II (Gastropoda, Stylommatophora, Zonitidae). Malakologische Abhandlungen. Staatliches Museum für Tierkunde Dresden 13(10):93–104.

Riedel, A., 1996. Die in West-Paläarktis unterirdisch lebenden Zonitidae sensu lato (Gastropoda, Stylommatophora). Fragmenta faunistica 39(24):363–390.

Rossmässler, E., 1835–1859. Iconographie der Land- und Süßwasser-Mollusken, mit vorzüglicher Berücksichtigung der europäischen, noch nicht abgebildeten Arten. 1-3. Dresden, Leipzig & Wiesbaden (Fortges. von W. Kobelt u. a.).

Russev, B., 1957. [Sur la vie benthique du Danube pres de la côtes bulgares]. Priroda [Nature] (Sofia) 6(2):44–49 (in Bulgarian).

Russev, B., 1964. [Hydrobiologischen Untersuchungen der Arda und einiger ihrer Nebenflüsse]. Izvestiya na Zoologicheskiya institut s muzei (Bulletin de l'Institut de zoologie et musée) 17:5–49 (in Bulgarian).

Russev, B., 1966a. [Das Zoobenthos der Donau zwischen dem 845. und 375. Flusskilometer. I. Zusammensetzung, Verteilung und Ökologie]. Izvestiya na Zoologicheskiya institut s muzei (Bulletin de l'Institut de zoologie et musée) 20:55–131 (in Bulgarian).

Russev, B., 1966b. [Hydrobiologische Untersuchungen der Marica. I]. In: Fauna na Trakiya. Sbornik ot statii. Chast III (Die Fauna Thrakiens. Sammelwerk. Band III). Izdatelstvo na Balgarskata akademiya na naukite (Verlag der Bulgarischen Akademie der Wissenschaften), Sofia: 231–291 (in Bulgarian).

Russev, B., 1967. [Das Zoobenthos der Donau zwischen dem 845. und 375. Flusskilometer. II. Biozönologie und Dinamik]. Izvestiya na Zoologicheskiya institut s muzei (Bulletin de l'Institut de zoologie et musée) 23:33–78 (in Bulgarian).

Russev, B., 1978. [Besonderheiten und Bedeutung des Zoobenthos der Donau zwischen Strom-km 845 und 357]. In Limnologiya na balgaskite dunavski pritotsi [Limnologie der bulgarischen Donauzuflüsse]. Izdatelstvo na Balgarskata akademiya na naukite (Verlag der Bulgarischen Akademie der Wissenschaften), Sofia: 145–200 (in Bulgarian).

Russev, B., A. Petrova, I. Yaneva & S. Andreev, 1998. Diversity of zooplankton and zoobenthos in the Danube River, its tributaries, and adjacent water bodies. In Meine, C. (ed), Bulgaria's Biological Diversity: Conservation Status and Needs Assessment. Volumes I and II. Pensoft, Sofia & Moscow: 263–292.

Russev, B., J. Uzunov, S. Kovachev, I. Yaneva & L. Ivanova, 1981. [Tendencies of the changes in the saprobic conditions of the Maritsa River]. Hidrobiologiya [Hydrobiology] (Sofia) 14:51–64 (in Bulgarian).

Russev, B. & I. Yaneva., 1975. [Hydrofaunistische Erforschungen einiger rhodopischer Gewässer]. In: Fauna na Rodopite. Materiali (La faune des Rhodopes. Materiaux). Izdatelstvo na Balgarskata akademiya na naukite (Éditions de l'Académie bulgare des sciences), Sofia: 11–39 (in Bulgarian).

Russev, B., I. Yaneva, R. Detcheva & M. Karapetkova, 1994. [Zusammensetzung der Hydrofauna]. In Russev, B.K. (ed), Limnologiya na balgarskite dunavski pritotsi [Limnologie der bulgarischen Donauzuflüsse]. Knizhen Tigar [Paper Tiger], Sofia: 130–174 (in Bulgarian).

Salvini-Plawen, L., 1983. Stamm Mollusca (Weichtiere). In Riedl, R. (ed), Fauna und Flora des Mittelmeeres. Parey, Hamburg & Berlin: 248–390.

Schileyko, A.A., 1978. [Terrestrial mollusks of the superfamily Helicoidea]. In Fauna SSSR. Mollyuski. Tom 3, vypusk 6. [Fauna of the USSR. Mollusks. Vol. 3, No. 6]. Nauka [Science], Leningrad (in Russian).

Schileyko, A.A., 1984. [Terrestrial mollusks of the suborder Pupillina in the USSR fauna (Gastropoda, Pulmonata, Geophila)]. In Fauna SSSR. Mollyuski. Tom 3, vyp. 3. [Fauna of the USSR. Mollusks. Vol. 3, No. 3]. Nauka [Science], Leningrad (in Russian).

Schütt, H., 1996. Landschnecken der Türkei. Acta Biologica Benrodis, Suppl., 4:1–482.

Skarlato, O.A. & Ya.I. Starobogatov, 1972. [Class of Bivalve Mollusks – Bivalvia]. In Opredelitel' fauny Chernogo i Azovskogo morei. 3 [The Key for the Fauna of the Black and Azov Seas. 3]. Naukova dumka [Scientific Thought], Kiev: 178–249 (in Russian).

Stadnichenko, A.P., 1984. [Unionidae, Cycladidae]. In Fauna Ukraini. Tom 29. Molyuski. Vipusk 9. [Fauna of Ukraine. Vol. 29. Mollusks. No. 9]. Naukova dumka [Scientific Thought], Kiiv (in Ukrainian).

Stadnichenko, A.P., 1990. [Lymnaeiformes: Physidae, Bulinidae, Planorbidae]. In Fauna Ukrainy. Tom 29. Mollyuski. Vypusk 4. [Fauna of Ukraine. Vol. 29. Mollusks. No. 4)]. Naukova dumka [Scientific Thought], Kiev (in Russian).

Starobogatov, Ya.I., 1970. Fauna mollyuskov i zoogeograficheskoe rayonirovanie kontinentalnykh vodoyomov zemnogo shara [Fauna of Mollusks and Zoogeographic Regionalization of the Continental Water Bodies of the World]. Nauka [Science], Leningrad (in Russian).

Stoykov, S., 1979. [On the zoobenthos dynamics in Bourgas Lake and Mandra dam in the period 1975–1977]. Izvestiya. Institut po ribni resursi – Varna (Proceedings. Institute of Fisheries – Varna) 17:47–53 (in Bulgarian).

Stoykov, S., 1983. [Preliminary data on age and growth of the Soft Clam, *Mya arenaria* L., of the Bulgarian coast]. Izvestiya na Instituta po ribni resursi – Varna (Proceedings of the Institute of Fisheries – Varna) 20:161–167 (in Bulgarian).

Stoykov, S. & S. Uzunova, 1999. Dynamics of macrozoobenthos from the Bourgas Bay (Bulgarian Black Sea coast) during the period 1993–1995. Izvestiya na Instituta po ribni resursi – Varna (Proceedings of the Institute of Fisheries – Varna) 25:153–154.

Urbański, J., 1960. Beiträge zur Molluskenfauna Bulgariens (excl. Clausiliidae). (Systematische, zoogeographische und ökologische Studien über die Mollusken der Balkan-Halbinsel. V.). Bulletin de la Societé des Amis des Sciences et des Lettres Poznan, Ser. D, 1:69–110.

Urbański, J., 1964. Beiträge zur Kenntnis balkanischer Stylommatophoren. Bulletin de la Societé des Amis des Sciences et des Lettres Poznan, Ser. D, 8:19–56.

Urbański, J., 1969. Bemerkenswerte balkanische Stylommatophoren. Bulletin de la Societé des Amis des Sciences et des Lettres Poznan, Ser. D, 9:225–261.

Urbański, J., 1977. Bemerkenswerte Clausiliiden (Mol., Pulm.) der nördlichen Balkan-Halbinsel. Bulletin de la Societé des Amis des Sciences et des Lettres Poznan, Ser. D, 17:235–251.

Urbański, J., 1978. Bemerkungen über balkanischen Helicigonen (Gastrop., Pulm.). Bulletin de la Societé des Amis des Sciences et des Lettres Poznan, Ser. D, 18:139–149.

Urbański, J., 1979. Zwei neue Stylommatophoren (Gastropoda, Pulmonata) von der Balkan-Halbinsel. (Systematische, zoogeographische und ökologische Studien über die Mollusken der Balkan-Halbinsel, XVII). Bulletin de la Societé des Amis des Sciences et des Lettres Poznan, Ser. D, 19:109–115.

Uzunov, Y. & S. Kovachev, 1985. [Macroinvertebrate communities structures in the Maritsa River under human activity impact]. Hidrobiologiya [Hydrobiology] (Sofia) 24:33–47 (in Bulgarian).

Uzunov, Y., S. Kovachev, K. Kumanski & J. Ludskanova-Nikolova, 1998. Aquatic ecosystems of the Aegean and Black Sea basins. In Meine, C. (ed), Bulgaria's Biological Diversity: Conservation Status and Needs Assessment. Volumes I and II. Pensoft, Sofia & Moscow: 292–346.

Uzunov, Y., V. Tzavkova, I. Todorov, I. & E. Varadinova, 2001. The macrozoobenthic fauna of the Biosphere reserve Srebarna Lake in North-Eastern Bulgaria. Lauterbornia 40:43–51.

Valkanov, A., 1932. [Notizen über das Leben in den Hochgebirgsseen Bulgariens]. Trudove na Balgarskoto prirodoizpitatelno druzhestvo (Travaux de la Société bulgare des sciences naturelles) 15/16:207–225 (in Bulgarian).

Valkanov, A., 1934. [Beitrag zur Kenntnis der Hydrofauna Bulgariens]. Private publication, Sofia (in Bulgarian).

Valkanov, A., 1935. [Notizen über die Brackwässer Bulgariens. I]. Godishnik na Sofiiskiya universitet. Fiziko-matematicheski fakultet (Annuaire de l'Université de Sofia. Faculté physico-mathématique) 31(3):249–303 (in Bulgarian).

Valkanov, A., 1936. [Notizen über die Brackwässer Bulgariens. II. Versuch einer hydrographischen und biologischen Erforschung derselben]. Godishnik na Sofiiskiya universitet. Fiziko-matematicheski fakultet (Annuaire de l'Université de Sofia. Faculté physico-mathématique) 32(3):209–341 (in Bulgarian).

Valkanov, A., 1957. [Katalog unserer Schwarzmeerfauna]. Trudove na Morskata biologichna stantsiya vav Varna (Arbeiten aus der Biologischen Meeresstation in Varna, Bulgarien) 19:1–62 (in Bulgarian).

Valkanov, A., V. Petrova, A. Roshdestvenski, T. Marinov & V. Naidenov, 1978. [Black Sea lakes]. In Cherno more [The Black Sea]. G. Bakalov, Varna: 262–283 (in Bulgarian).

Varbanov, M., 2002. [Lakes and swamps]. In: Kopralev, I., M. Yordanova & Ch. Mladenov (eds), Geografiya na Balgariya. Fizicheska geografiya. Sotsialno-ikonomicheska geografiya [Geography of Bulgaria. Physical Geography. Socio-Economic Geography]. ForCom, Sofia: 237–242 (in Bulgarian).

Wagner, A., 1906. Bemerkungen zum Genus *Daudebardia* Hartmann. Nachrichtsblatt der Deutschen malakozoologischen Gesellschaft 38:177–186.

Wagner, A., 1915. Beiträge zur Anatomie und Systematik der Stylommatophoren aus dem Gebiete der Monarchie und der angrenzenden Balkanländer. Denkschriften der kaiserlichen Akademie der Wissenschaften in Wien, Mathematisch-naturwissenschaftliche Klasse, 91:429–498.

Wagner, A., 1919–1920. Zur Anatomie und Systematik der Clausiliiden. Nachrichtsblatt der Deutschen malakozoologischen Gesellschaft 51/52(1–13):97–108, 129–158.

Wagner, A., 1927. Studien zur Molluskenfauna der Balkanhalbinsel mit besonderer Berücksichtigung Bulgariens und Thraziens, nebst monographischer Bearbeitung einzelner Gruppen. Annales zoologici Musei Polonici Historiae Naturalis 6(4):263–399.

Wagner, J., 1933. Die in die Unterfamilie Parmacellinae gehörenden Formen des Naturhistorischen Museums in Wien. Annalen des Naturhistorischen Museums in Wien 46:57–71.

Wagner, J., 1934a. Die Nacktschnecken des Königlichen Naturhistorischen Museums in Sofia. Izvestiya na Tsarskite prirodonauchni instituti v Sofia (Mitteilungen aus den Königlichen naturwissenschaftlichen Instituten in Sofia – Bulgarien; Bulletin des Institutions royales d'histoire naturelle à Sofia – Bulgarie) 7:51–60.

Wagner, J., 1934b. Über einige von Herrn Dr. B. Rensch in der bulgarischen Gebirgen gesammelte Nacktschnecken. Izvestiya na Tsarskite prirodonauchni instituti v Sofia (Mitteilungen aus den Königlichen naturwissenschaftlichen Instituten in Sofia – Bulgarien; Bulletin des Institutions royales d'histoire naturelle à Sofia – Bulgarie) 7:88–90.

Wagner, J., 1940. Neue Beiträge zur Kenntnis der Nacktschneckenfauna der Balkanhalbinsel, mit besonderer Berücksichtigung der griechischen Arten. Annales Musei Nationalis Hungarici, Zool. 33:137–152.

Westerlund, C., 1884–1890. Fauna der in der Paläarktischen Region lebenden Binnenconchylien. I–VII. Lund & Berlin.

Wiktor, A., 1983. The slugs of Bulgaria (Arionidae, Milacidae, Limacidae, Agriolimacidae – Gastropoda, Stylommatophora). Annales zoologici 37(3): 71–206.

Wiktor, A, 1987. Milacidae (Gastropoda, Pulmonata) – systematic monograph. Annales zoologici 41:153–319.

Wiktor, A., 1997. Endemism of slugs within the Balkan Peninsula and adjacent islands (Gastropoda, Pulmonata, Arionidae, Milacidae, Limacidae, Agriolimacidae). Genus 8:205–221.

Wilke, T., 1995. *Hemilepton nitidum* (Turton, 1822) and *Mysella bidentata* (Montagu, 1803), two bivalve species new for the Bulgarian Black Sea coast. Izvestiya na Instituta po ribni resursi – Varna (Proceedings of the Institute of Fisheries – Varna) 23:138–147.

Wilke, T., 1996. Annotated check-list of the marine gastropods of the Bulgarian Black Sea coast. Izvestiya na Instituta po ribni resursi – Varna (Proceedings of the Institute of Fisheries – Varna) 24:144–166.

Wilke, T., 1997. Gastropods of the Black Sea. Identification of species new for the Bulgarian Coast. Mitteilungen aus dem Zoologischen Museum in Berlin 73(1): 3–15.

Willmann, R. & H. Pieper, 1978. Gastropoda et Lamellibranchia. In: Illies, J. (ed.) Limnofauna Europaea (2. Auflage). G. Fischer, Stuttgart: 118–137.

Wohlberedt, O., 1911. Zur Molluskenfauna von Bulgarien. Abhandlungen und Berichte der Naturforschenden Gesellschaft zu Görlitz 27:167–234.

Yaneva, I., 1989. [Changes in the saprobiological state of the Èerna River and their effect on the composition and structure of the benthic organisms]. Hidrobiologiya (Hydrobiology) (Sofia) 34:20–29 (in Bulgarian).

Yaneva, I., L. Pehlivanov, Y. Vidinova, S. Stoichev, S. & V.Tyufekchieva, 1998. A comparative ecological characterisation of lotic bental zoocoenoses from two streams under different anthropogenic influence. In: Ecomonitoring in Rozen and Srednogorie. Bulgaria. Ministry of Environment, Sofia: 101–112.

Yaneva, I., Vidinova, Y. and Tyufekchieva, V., 2001. Contemporary saprobiological characteristics of Arda River in the section of future "Gorna Arda" cascade building. Acta zoologica bulgarica 53(3):37–46.

Zaprjanov, L., 1987. [Biometrical study of the land mollusc *Bradybaena fruticum* (Müller, 1774) (Gastropoda, Bradybaenidae) from North Bulgaria]. Izvestiya na Narodniya muzei – Varna (Bulletin de Musée National de Varna) 23(38):211–215 (in Bulgarian).

Zaprjanov, L., 1989. [Morphologische Charakteristik der Landschnecke *Helix pomacella* Mouson, 1854 (Gastropoda, Helicidae) der bulgarischen Schwarzmeerküste]. Izvestiya na Narodniya muzei – Varna (Bulletin de Musée National de Varna) 25(40):174–176 (in Bulgarian).

Zaprjanov, L., 1990. Morphological characteristic of the land snail *Helix figulina* Rossmässler, 1839 (Gastropoda, Helicidae) from Bulgaria. Izvestiya na Narodniya muzei – Varna (Bulletin de Musée National de Varna) 26(41):254–258.

Zaprjanov, L., 1991. [Morphological characteristic of the land snail *Cepea vindobonensis* Ferussac, 1821 (Gastropoda, Helicidae) in Bulgaria]. Izvestiya na Narodniya muzei – Varna (Bulletin de Musée National de Varna) 27(42):232–240 (in Bulgarian).

Zaschev, G. & A. Angelov, 1959. [Untersuchungen über den Burgas-See (Waja) in Beziehung zur Verbesserung seiner fischwirtschaftlichen Ausbeutung]. Godishnik na Sofiiskiya universitet. Biologo-geologo-geografski fakultet (Annuaire de l'Université de Sofia, Faculté de Biologie, Géologie et Géographie) 51 (1 – Biologie):161–210 (in Bulgarian).

Zhadin, V.I., 1952. Mollyuski presnykh i solonovatykh vod SSSR [Mollusks of the Fresh and Brakish Waters of the USSR]. Akademiya nauk SSSR [Academy of Sciences of the USSR], Moscow & Leningrad (in Russian).

7 Fauna, Ecology, and Zoogeography of Dragonflies (Insecta: Odonata) of Bulgaria

Venelin Beschovski[1] and Milen Marinov[2]

[1]*Institute of Zoology, Bulgarian Academy of Sciences, Tsar Osvoboditel Blvd. 1, 1000 Sofia, Bulgaria, e-mail: vbeschovski@zoology.bas.bg*

[2]*P.O.Box 134, 1000 Sofia, Bulgaria, e-mail: mg_marinov@yahoo.com*

Note. Sections 7.2 and 7.3 are written by V. Beschovski; sections 7.4 to 7.6, by M. Marinov.

Abstract: Sixty-eight dragonfly species, belonging to 28 genera and ten families, have been reported from Bulgaria. According to their larval habitat, they are divided in two ecological complexes: rheophilous and limnophilous. Adaptations of both complexes are discussed. Zoogeographical characteristics are given at the genus and species level. Their vertical and horizontal distribution is outlined. Seven phenological groups are established. Their habitats are divided in 12 groups; for each group, key species and co-occurring species are listed

1 Fauna

The dragonflies of Europe have been well studied (Corbet, 1962, 1999; Aguesse, 1968; Dreyer, 1986; Askew, 1988) and this also holds true for Bulgaria (Petkov, 1921; Beschovski, 1994a, 1994b; Marinov, 1995, 1999, 2001a, 2001b, 2001c, 2001d; Marinov and Simov, 2004). The Odonata of Bulgaria and neighboring Balkan countries are therefore suitable for an ecological and zoogeographical review. The total of 68 dragonfly species currently known belong to 28 genera and ten families (Table 1).

2 Ecology

Dragonflies are amphibious insects, with their immature stage aquatic, and adults airborne. The larvae are closely tied to a certain environment, and cannot freely migrate in search of suitable conditions as does the adult. Although it is possible for dragonflies to fly over large territories, adults have to be close to a certain type of water body suitable for their larvae. According to larval habitat, dragonflies can be divided in two ecological complexes: rheophilous and limnophilous (Beschovski, 1967, 1968), i.e. inhabiting running or stagnant waters.

V. Fet and A. Popov (Eds.), Biogeography and Ecology of Bulgaria, 199–231.

Table 1 Systematic list of the Bulgarian Odonata

ZYGOPTERA
Calopterygidae
Calopteryx splendens (Harris, 1782)
Calopteryx virgo (Linnaeus, 1758)
Euphaeidae
Epallage fatime (Charpentier, 1840)
Lestidae
Chalcolestes parvidens (Artobolevski, 1929)
Chalcolestes viridis (Vander Linden, 1825)
Lestes barbarus (Fabricius, 1798)
Lestes dryas Kirby, 1890
Lestes macrostigma (Eversmann, 1836)
Lestes sponsa (Hansemann, 1823)
Lestes virens (Charpentier, 1825)
Sympecma fusca (Vander Linden, 1820)
Platycnemidae
Platycnemis pennipes (Pallas, 1771)
Coenagrionidae
Pyrrhosoma nymphula (Sulzer, 1776)
Erythromma lindeni (Sélys, 1840)
Erythromma najas (Hansemann, 1823)
Erythromma viridulum (Charpentier, 1840)
Coenagrion hastulatum (Charpentier, 1825)
Coenagrion ornatum (Sélys, 1850)
Coenagrion puella (Linnaeus, 1758)
Coenagrion pulchellum (Vander Linden, 1825)
Coenagrion scitulum (Rambur, 1842)
Enallagma cyathigerum (Charpentier, 1840)
Ischnura elegans (Vander Linden, 1820)
Ischnura pumilio (Charpentier, 1825)
ANISOPTERA
Aeshnidae
Caliaeschna microstigma (Schneider, 1845)
Aeshna affinis Vander Linden, 1820
Aeshna cyanea (Müller, 1764)
Aeshna isosceles (Müller, 1767)
Aeshna juncea (Linnaeus, 1758)
Aeshna mixta Latreille, 1805
Aeshna subarctica Walker, 1908
Anax imperator Leach, 1815
Anax parthenope (Sélys, 1839)
Hemianax ephippiger (Burmeister, 1839)
Brachytron pratense (Müller, 1764)
Gomphidae
Gomphus flavipes (Charpentier, 1825)
Gomphus vulgatissimus (Linnaeus, 1758)
Onychogomphus forcipatus (Linnaeus, 1758)
Ophiogomphus cecilia (Fourcroy, 1785)
Cordulegastridae
Cordulegaster bidentata Sélys, 1843
Cordulegaster heros Theischinger, 1979
Cordulegaster insignis Schneider, 1845
Cordulegaster picta Sélys, 1854
Corduliidae
Cordulia aenea (Linnaeus, 1758)
Somatochlora arctica (Zetterstedt, 1840)
Somatochlora borisi Marinov, 2001
Somatochlora flavomaculata (Vander Linden, 1825)
Somatochlora meridionalis Nielsen, 1935
Somatochlora metallica (Vander Linden, 1825)
Libellulidae
Libellula depressa Linnaeus, 1758
Libellula fulva Müller, 1764
Libellula quadrimaculata Linnaeus, 1758
Orthetrum albistylum (Sélys, 1848)
Orthetrum brunneum (Fonscolombe, 1837)
Orthetrum cancellatum (Linnaeus, 1758)
Orthetrum coerulescens (Fabricius, 1798)
Crocothemis erythraea (Brullé, 1832)
Sympetrum depressiusculum (Sélys, 1841)
Sympetrum flaveolum (Linnaeus, 1758)
Sympetrum fonscolombei (Sélys, 1840)
Sympetrum meridionale (Sélys, 1841)
Sympetrum pedemontanum (Allioni, 1766)
Sympetrum sanguineum (Müller, 1764)
Sympetrum striolatum (Charpentier, 1840)
Sympetrum vulgatum (Linnaeus, 1758)
Leucorrhinia dubia (Vander Linden, 1825)
Leucorrhinia pectoralis (Charpentier, 1825)
Selysiothemis nigra (Vander Linden, 1825)

2·1 Limnophilous species

This group includes the Coenagrionidae, Lestidae, Aeshnidae (except *Caliaeschna*), Corduliidae (except some *Somatochlora*), and Libellulidae. By larval habitat, limnophilous dragonflies are either *limnophilous epiphytic* or *limnophilous benthic*

(Beschovski, 1968). The term "epiphytic" is used here when larvae live only on submerged vegetation, not *sensu* Corbet (1962, 1999) for oviposition.

Limnophilous epiphytic species include Lestidae, Coenagrionidae, and Aeshnidae. Their larvae develop on vegetation in stagnant waters, usually permanent but sometimes temporary.

The limnophilous epiphytic species inhabiting permanent water include the Coenagrionidae *Pyrrhosoma nymphula, Erythromma najas, E. viridulum, E. lindeni, Coenagrion hastulatum, C. ornatum, C. puella, C. pulchellum, C. scitulum, Enallagma cyathigerum, Ischnura elegans, I. pumilio*; and the Aeshnidae *Aeshna affinis, A. cyanea, A. juncea, A. mixta, A. isosceles, A. subarctica, Anax imperator, A. parthenope, Hemianax ephippiger*, and *Brachytron pratense*. Their larvae are found mainly along the margins of waters with amphibious and submerged vegetation.

The limnophilous epiphytic species of stagnant, temporary waters include eight Lestidae: *Sympecma fusca, Chalcolestes parvidens, C. viridis, Lestes barbarus, L. dryas, L. macrostigma, L. sponsa*, and *L. virens*. Their larvae are found in temporary water or where water level fluctuates considerably. These temporary basins can be divided into two groups. The habitats with dense hygrophilous vegetation look like meadows inundated with spring water. Here, the larvae of *Lestes* and *Sympetrum* are found even above water level in the grass. Another group of temporary habitats are shallow pools and flooded areas next to permanent water bodies, often overgrown with *Ranunculus repens, Eleocharis palustris*, and *Marsilea quadrifolia*. Here, larvae of *Lestes* are predominant (especially *L. barbarus*) as well as *Sympetrum striolatum, S. sanguineum*, and *S. meridionale*.

Limnophilous benthic species include Corduliidae and Libellulidae whose larvae inhabit the bottom communities. Limnophilous benthic species inhabiting permanent water bodies include the Corduliidae *Cordulia aenea, Somatochlora alpestris, S. borisi* (?), *S. flavomaculata, S. metallica, S. meridionalis* (?), and the Libellulidae *Libellula fulva, L. depressa, L. quadrimaculata, Orthetrum cancellatum, O. albistylum, O. brunneum, O. coerulescens, Crocothemis erythraea, Leucorrhinia dubia*, and *L. pectoralis*. Their larvae are found in shallow, warm pelophilous habitats rich in detritus. In large swamps and lakes, they are found only in shallow coastal sectors; they are absent from thick reeds in such lakes and swamps as Srebarna, Nova Cherna, Malak Preslavets, Orsoya, Shabla, Durankulak, Atanasovsko Lake, as well as in deep artificial reservoirs. In smaller, shallow water basins with sparse vegetation, these larvae are found everywhere on the bottom. For example, in the Second Smolyan Lake, the larvae of *Libellula quadrimaculata* are almost absent, since water temperature even at midday (May) rarely reaches 6–8 °C. However, they are found in large numbers next to this lake, in the shallow swamps and small pools over sphagnum moss, where water is 10–15 cm deep (midday temperature in May 16–18 °C).

The limnophilous benthic species of temporary waters include all eight species of *Sympetrum* (Libellulidae). Their larvae prefer benthic habitats with dense hydrophilous vegetation; if the bottom does not get warm enough, they move higher and coexist with the true epiphytic larvae of *Lestes*. In some areas, many *Sympetrum striolatum, S. sanguineum*, and *S. meridionale* larvae were collected in hydrophilous vegetation together with *Lestes barbarus*. In permanent waters these larvae are found either in the marginal sections with seasonal water fluctuations, or inside the water bodies if water level decreases in summer and early fall. In flood or temporary water bodies with sparse vegetation, *Sympetrum* larvae are found at the bottom. Their adherence to temporary water bodies determines the habitat preference of the adults. In the end of summer and early fall, adults can be seen flying over these waters or over their dry bottom to lay eggs, which will become submerged later in fall and winter.

Shortened larval stage in limnophilous dragonflies of temporary waters. The larval stage of most dragonflies lasts from one to three years. However, in those inhabiting temporary water, it lasts from six to seven months. Such a shortened larval stage is an adaptation to the short existence of temporary waters; it is combined with a prolonged egg stage (Fischer, 1964). Adult *Sympetrum* lay eggs in late summer and early fall on an almost dry bottom (Beschovski, 1960). Their larvae hatch in the next spring. By June, larvae complete their development, and adults fly in large numbers around stagnant water. These dragonflies are at risk as they try to emerge from the pools and avoid predation by birds. We observed prematurely dried water bodies or their marginal portions with many larvae on the dry surface; some were dead, others were still alive in the moist cracks of the bottom. In the Choklyovo swamp near Kyustendil, many larvae were trapped in small pools on a dried peat bottom; tracks of white storks (*Ciconia ciconia*) on the mud showed that the birds walked from pool to pool, picking insects. Pools without stork tracks had *Sympetrum* and *Lestes* larvae.

2·2 Rheophilous species

This group includes 13 species of Euphaeidae, Calopterygidae, Platycnemididae, Gomphidae, Cordulegastridae, and Aeshnidae (only one genus, *Caliaeschna*); it makes up ca. 20% of all species of running waters. Some of them, e.g. *Platycnemis pennipes*, can be found in large, open water bodies such as Durankulak and Shabla lakes near the Black Sea coast.

Rheophilous epiphytic species. Phytophilous communities are a specific type of benthic communities (Zhadin, 1950), together with litho-, pelo-, psammo-, and argillorheophilous types. Four species belong to the subgroup rheophilous epiphytic (phytorheophilous) dragonflies: *Calopteryx virgo, C. splendens, Platycnemis pennipes*, and *Caliaeschna microstigma*. Their larvae are found among vegetation. The most common are *P. pennipes* and *Calopteryx splendens*. *C. virgo festiva* is found in the upper parts of rivers with clean oxygenated water. This group includes two interesting taxa: (a) *Caliaeschna microstigma*, a unique representative of Anisoptera and Aeshnidae in this kind of community in Bulgaria. It is found in small rivers and even in little streams, which dry up in summer or early fall, leaving only isolated bottom pools (Beschovski, 1964). Along the Black Sea coast and in South Bulgaria, such small rivers are a refugial habitat. (b) *Calopteryx virgo meridionalis*, a subspecies isolated from *C. v. festiva* geographically and ecologically, is found in southeastern Bulgaria (Strandja Mts.), in a deep gulch with a small stream near the Black Sea coast. The trees and bushes make a tunnel above the stream. A visitor has the exotic impression of moving in a gallery, surrounded with blue-winged dragonflies flying "trustfully" and closely, such that one can touch them. Their larval stage is vulnerable: the streams in which they live are small and open to human influence. In the Ahtopol district, in two or three little streams this species is missing due to waste disposal from pigsty or chicken farms.

Rheophilous benthic species. The larvae of this group inhabit the remaining rheophilous benthic communities: lithorheophilous, psammorheophilous, pelorheophilous, and more rarely the argillorheophilous (Beschovski, 1967). They belong to Euphaeidae, Gomphidae, and Cordulegastridae.

Larvae living in lithorheophilous communities are found on the underside of stones (*Epallage fatime*), or under stones in sand and mire (*Onychogomphus forcipatus* and *Gomphus vulgatissimus*). *E. fatime*, a typical lithorheobiont found rather rarely in Bulgaria, prefers streams with stony bottom in the Mediterranean and transitional Mediterranean climate. The most frequent and widespread is *O. forcipatus*, which prefers smaller rivers with average annual flow below $10\,m^3/sec$ and with gravel bottom. *G. vulgatissimus* is also found in this community under stones in sand or mire bottoms.

The pelorheophilous community is best developed in the lentic zone of the large Danubian tributaries and the Kamchiya, Maritsa, and Tundja rivers. It includes *Gomphus flavipes, Cordulegaster bidentata, C. heros, C. insignis*, and *C. picta. G. flavipes* prefers pelorheophilous community of larger rivers with average annual flow above $10\,m^3/sec$. Its presence indicates a well-developed pelorheophilous community; it is the most common dragonfly in the Danube (Russev, 1962) along its Bulgarian side. Larvae of *Gomphus vulgatissimus* are also sometimes present in this community. *Cordulegaster* larvae are found in small rivers and streams, mainly in their meanders or mouths.

In the psammorheophilous community, larvae are found sporadically. The larvae of *Gomphus vulgatissimus* and *Onychogomphus forcipatus* are usually connected with stones and small gravel on sandy bottoms. Larvae of the rare *Ophiogomphus cecilia* are found only in two rivers: Iskar (sandy-mire bottom) and Maritsa (sandy bottom). *Gomphus vulgatissimus* is possibly more adapted to different habitats: it is found in pelo-, psammo-, and lithorheophilous communities. *Onychogomphus forcipatus* is found in the argillorheophilous (clay) community only in Cherni Lom River near Cherven Village (Ruse district); its larvae prefer spots of mire or sand on clay. In general, rheophilous benthic larvae clearly prefer litho-, psammo-, and pelorheophilous river communities.

Transitional species. These include all *Cordulegaster* and some *Somatochlora*. *Cordulegaster* larvae live mainly in slowly running water, meanders, or swamps having influx of water from forest streams or rivers. These larvae lack specific adaptations. Such kind of habitat is also preferred by some Corduliinae (*Somatochlora meridionalis, S. borisi*). These habitats, as well as those of *Calopteryx virgo meridionalis* and *Caliaeschna microstigma*, have a refugial character: small running water bodies with variable flow in the Mediterranean and Submediterranean regions. At the present time, such habitats are vulnerable.

Larval adaptations of the rheophilous species. Living conditions in running waters are more specific compared to those of stagnant basins. Specific adaptations of rheophilous species include modified body shape as well as modified morphology of body parts related to moving and feeding. In this aspect, larvae of dragonflies are similar to other hydrobionts. Already Zernov (1949) and Zhadin (1950) listed specific adaptive features of rheophilous species: a dorsoventrally flattened body, an enlarged surface adhering to the substrate, and a decreased to fully lost swimming ability. In addition, Beschovski (1967, 1968) found that rheophilous dragonfly larvae possess important adaptations to support themselves against the current through modifications of their body and antennae. Below, we discuss these adaptations in rheophilous larvae.

(a) *Dorsoventral flattening of the body.* The dorsoventrally flattened body of aquatic invertebrates is not always an adaptation to life in running water, but in rheophilous dragonfly larvae it clearly is adaptive. We compared the thorax shape in rheophilous and limnophilous larvae as well as in corresponding adults. The thoracic index (TI; the ratio of maximal thorax height to its maximal width; Beschovski, 1967) in rheophilous larvae varied from 0.579 (*Onychogomphus forcipatus*) to 1.280 (*Lestes barbarus*) while in adults it was ca. 1.0. This reflects a clear thorax flattening, approaching that of typical rheobiotic larvae such as *Perla* sp. (TI = 0.509). The TI of limnophilous larvae is higher than that of rheophilous ones (0.774 to 1.128). For limnophilous species, the TI is lower in benthic larvae (0.774 to 0.927) than in the epiphytic ones (0.896 to 1.28); in some epiphytic species it is approximately the same as in adults. The TI in the larvae of a widely distributed *Platycnemis pennipes* is close to that of limnophilous benthic larvae; larvae of *P. pennipes* also have other characters uncommon in rheophilous species. The TI in the adults is always higher than 1, and does not depend on the degree of dorsoventral flattening of the larva. Since Odonata have hemimetabolous metamorphosis, the larval features are expected to be close to those of the adults. There is a significant difference in the thorax shape between the benthic larvae of limnophilous (*Orthetrum brunneum*) and rheophilous (*Onychogomphus forcipatus*) species (Fig. 1), a difference almost absent in adults.

(b) *Increased surface adhering to the substratum.* This adaptation increases the body width/height ratio addressed in the previous paragraph (Fig. 1b). Such adherence is also provided by the abdominal gills in *Epallage fatime*, and by the almost flat ventral surface of caudal lamellae (Fig. 3.4). These larval features do not affect the morphology of adults.

(c) *Decreased motor activity.* In the laboratory, most rheophilous epiphytic larvae do not exhibit a special "desire" to part with substratum and float. This is

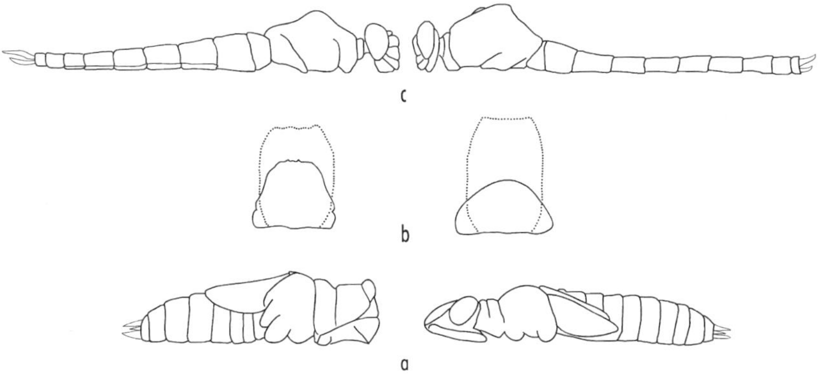

Fig. 1 Body shape of dragonflies from two ecological groups: limnophilous benthic (*Orthetrum brunneum*, left) and rheophilous benthic (*Onychogomphus forcipatus*, right); (a) larva in profile; (b) thorax cross-section, adult (stippled line) and larva (solid line); (c) adult in profile.

well seen if they are kept together with limnophilous species. The larvae of *Calopteryx* and *Platycnemis* always cling to the vegetation; with their long legs, they move slowly and carefully. Without the substratum, they make clumsy and uncertain floating movements. Their body and caudal lamellae are comparatively hard, and do not provide good floating. Caudal lamellae in these species are more chitinized and have a different shape. In *Calopteryx*, lateral caudal lamellae are triangular in cross-section (Fig. 3.3). In *Platycnemis*, only the middle rib of the lateral lamellae is more thickened than in limnophilous Zygoptera. The so-called cross suture dividing the caudal lamellae into a harder basal part and a more flexible distal part is missing in both *Calopteryx* and *Platycnemis*. Limnophilous larvae float better than rheophilous ones. Both their body and caudal lamellae are considerably softer, and can bend when floating. Their caudal lamellae are wider than in rheophilous larvae, only three times longer than wide. In rheophilous larvae, the caudal lamellae are narrower and longer, six to ten times longer than wide. Larvae of Gomphidae are also regarded as "nonswimmers" (Corbet, 1999). Larvae of *Epallage fatime* stay in one place and do not leave the underside of a stone, even when it is taken out of water.

The specific motor behavior of *Caliaeschna microstigma* larvae has to be mentioned here. In the laboratory, given a chance to escape the experimental basin (a tub with standing water) via a stick, they crawl out. Such a behavior could help in searching for suitable pools along a stream bottom, when these basins dry up in summer and fall (Beschovski, 1964), as is common in the Mediterranean region.

(d) *Body and antennal stabilization (support)*. As rheophilous organisms position themselves against the water current, their hardened body, especially its caudal part, serves as a rudder. In dragonfly larvae, caudal lamellae serve as both rudders and stabilizers supporting the larva against the current and minimizing its impact. Antennae of rheophilous larvae have several adaptations absent in limnophilous larvae, which allow supporting the antennae in a stable position, directed against the water current. This is achieved by a considerable elongation of one of the antennal segments while some other segments are reduced (Beschovski, 1967). The rheophilous epiphytic *Calopteryx* has an elongated first (basal) antennal segment (Fig. 2.1), followed by six small distal segments, which bend in the current, but always in front of the head. The basal segment is

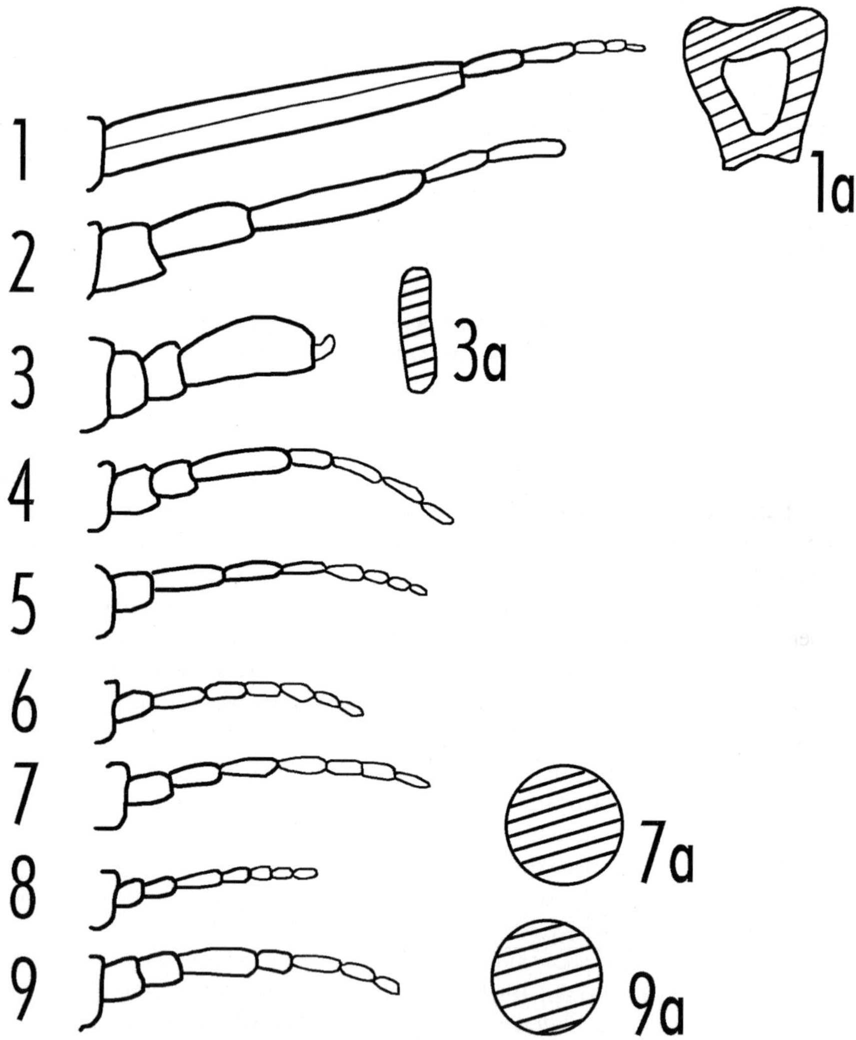

Fig. 2 Antennal segments in dragonfly larvae from different ecological groups: (1) *Calopteryx splendens*, (2) *Caliaeschna microstigma*, (3) *Gomphus vulgatissimus*, (4) *Cordulegaster bidentata*, (5) *Epallage fatime*, (6) *Platycnemis pennipes*, (7) *Lestes barbarus*, (8) *Aeshna affinis*, (9) *Libellula depressa*. 1, 2, 6: rheophilous epiphytic; 3, 4, 5: rheophilous benthic; 7, 8, 9: limnophilous; 1a, 3a, 7a, 9a: cross-section of the basal (1a) and third (3a, 7a, 9a) antennal segments.

trapezoid in cross-section, with thickened edges (Fig. 2.1a). Normally, dragonfly larvae have six to seven antennal segments, similar in size (Figs. 2.7–2.9). In rheophilous benthic species, the number of antennal segments decreases: in Gomphidae (*Gomphus, Onychogomphus, Ophiogomphus*) it is reduced to four; the third segment is elongated and dorsoventrally flattened; the fourth segment is reduced (Fig. 2.3). The larval adaptive character of these features is obvious since the antennal segments in adults are not modified.

Caliaeschna microstigma is the only Aeshnid that has rheophilous epiphytic larvae. In adults, the antenna has seven segments, comparable in size with that of other Aeshnidae. However, in larvae the number of segments is reduced to five, with the three basal ones elongated (Fig. 2.2); this degree of modification is intermediate between such rheophilous forms as benthic Gomphidae and epiphytic *Calopteryx*.

In *Cordulegaster*, the third antennal segment is weakly elongated (Fig. 2.4), and the number of segments is not reduced. The same is true for *Somatochlora meridionalis* and *S. borisi*. Their body is also only weakly dorsoventrally flattened. These species inhabit slowly running waters with a detritus bottom, as in stagnant basins.

No stabilizing antennal modifications are found in the larvae of *Platycnemis pennipes* and *Epallage fatime*; there are no elongated segments, and their number is high: seven in *P. pennipes* (Fig. 2.6), and eight in *E. fatime* (Fig. 2.5). In the latter, larvae cling to the underside of stones with their soft, elastic antennae more or less steady, hanging in running water. Larvae of *P. pennipes* lack many other rheophilous adaptations: their caudal lamellae are not sclerotized, and body is not flattened dorsoventrally. This species can be considered a newcomer in running waters; its larvae also inhabit large open lakes such as Shabla and Durankulak.

(e) *Leg features* (Fig. 4). Dragonfly larvae crawl well, especially in limnophilous benthic species, both under water and on dry bottoms. In adults, all tarsi have three segments (tarsomeres), and the apex of the tibia is not modified. However, rheophilous benthic larvae inhabiting sandy-stone substrata show adaptations such as the development of an apicolateral spur on the tibia and a reduction of the tarsomere number. The digging apicolateral spur on tibia I is present in

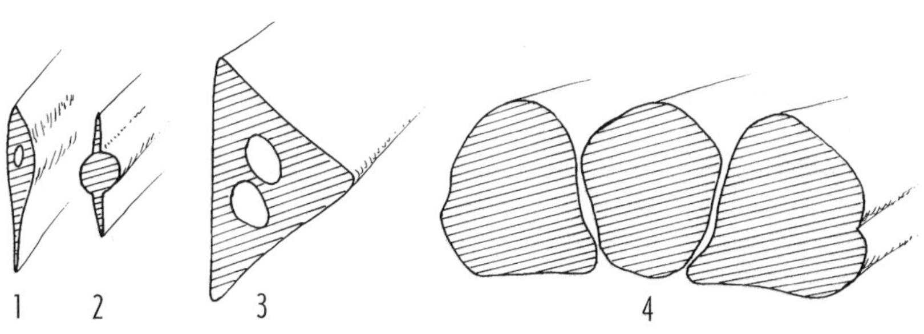

Fig. 3 The cross-section of the external caudal lamella in dragonfly larvae from different ecological groups: limnophilous epiphytic: (1) *Ischnura elegans*; rheophilous epiphytic: (2) *Platycnemis pennipes*, (3) *Calopteryx splendens*; lithorheophilous: (4) *Epallage fatime*.

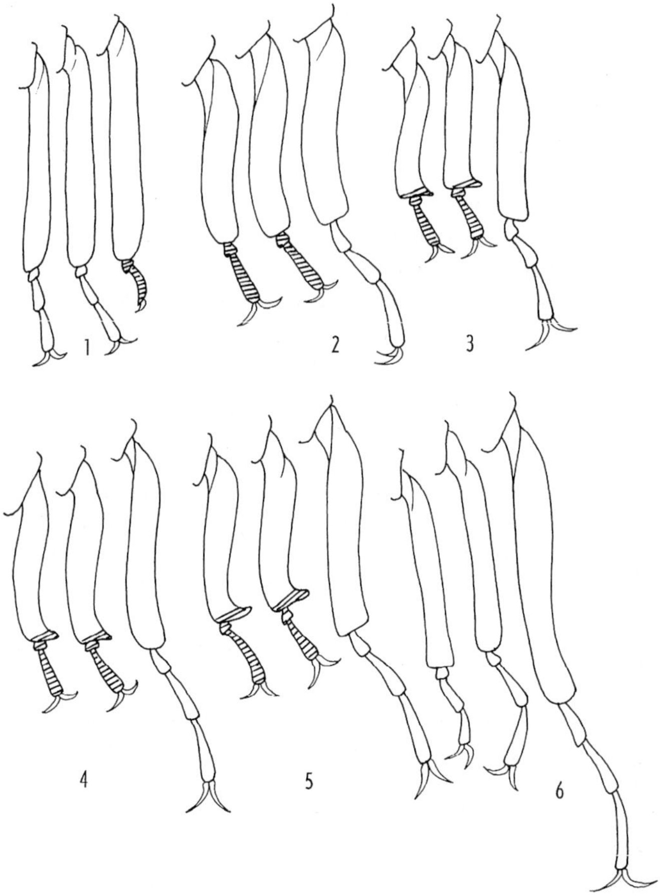

Fig. 4 Leg morphological adaptations of dragonfly larvae from different ecological groups: (1) lithorheobiont: *Epallage fatime*; (2) pelorheobiont: *Gomphus flavipes*; burrowing lithorheobionts: (3) *Onychogomphus forcipatus* and (4) *Gomphus vulgatissimus*; (5) burrowing psammorheobiont: *Ophiogomphus cecilia*; (6) pelolimnobiont: *Libellula quadrimaculata.*

larvae burrowing into more or less hard bottoms: *Onychogomphus forcipatus, Gomphus vulgatissimus*, and *Ophiogomphus cecilia* (Figs. 4.3–4.5). The number of tarsomeres of legs I and II, used for burrowing, is reduced. Larvae of *Gomphus flavipes* (Fig. 4.2) inhabit soft river bottom in slow current and river mouth areas; they do not possess the digging spur, but have reduced number of tarsomeres I and II. Larvae of *Epallage fatime*, which inhabit the lithorheophilous community, have no digging spur and normal number of tarsomeres on legs I and II; however, the number of tarsomeres on leg III is reduced (Fig. 4.1). These could support the larva against the current. The adults of all abovementioned species have three tarsomeres as well as the larvae of all other species (Fig. 4.6).

3 Zoogeography

A zoogeographical review of dragonfly fauna including some theoretical aspects was published by Belyshev and Kharitonov (1981, 1983). On the basis of this work, Beschovski (1994a, 1994b) provided a zoogeographical review of Odonata fauna of Bulgaria.

3·1 Zoogeographical characteristics of the genera

Following Belyshev and Kharitonov (1981), all 28 Bulgarian genera can be divided into six groups and two complexes, a Boreal and a Meridional one.

Boreal complex of genera. This complex includes 21 genera (75.0%) that fall into three groups: Sonoran, Eurosiberian, and Mediterranean.

(1) Sonoran group: nine genera (32.1%), *Enallagma, Aeshna, Gomphus, Ophiogomphus, Cordulegaster, Cordulia, Somatochlora, Libellula*, and *Sympetrum*. The richest genus is *Sympetrum* (eight species) followed by *Aeshna* (six species).

(2) Eurosiberian group: six genera (21.4%), *Calopteryx, Erythromma, Coenagrion, Ischnura, Brachytron*, and *Leucorrhinia*. The genus *Coenagrion* is the best represented (six species).

(3) Mediterranean group: six genera (21.4%), *Epallage, Chalcolestes, Sympecma, Pyrrhosoma, Caliaeschna*, and *Selysiothemis*. Each of these genera is represented by a single species (Beschovski (1994a). The East Mediterranean *Epallage* and *Caliaeschna* are limited to the southern Balkan Peninsula. Their northern boundaries pass through South and Southeast Bulgaria and North Greece, and continue to the west through Macedonia and Dalmatia.

Meridional complex of genera. This complex includes seven genera (25%) and three groups.

(1) Paleotropical group: three genera (10.7%), *Anax, Onychogomphus*, and *Orthetrum*. These are widespread genera that originated from both Paleotropical regions and penetrated to the Palearctic. Their origin (Oriental or Afrotropical) cannot be established.

(2) Afrotropical group: three genera (10.7%), *Platycnemis, Hemianax*, and *Crocothemis*, represented by one species each.

(3) Oriental group: only *Lestes* (3.6%). Ranges of its species lie mainly to the northwest of the Balkan Peninsula; they are found in all neighboring countries.

The boreal genera are dominant in the fauna of Bulgaria (75.0%). As a rule, the genera for which the range boundaries pass through the territory of the Balkan Peninsula, are monotypic or represented by not more than two species.

3·2 Zoogeographical characteristics of the species

Currently, 68 species of Odonata are recorded for Bulgaria (Table 1). Their subdivision into complexes and groups according to their origin and recent distribution is given below.

Boreal complex. This includes 67 species (98.5%), divided in two faunal types.

1. *Eurosiberian type*: well represented with 43 species (63.2%) and five groups.
   1.1. *Holarctic group*: five species (7.4%), *Lestes dryas, Enallagma cyathigerum, Aeshna juncea, Ae. subarctica*, and *Libellula quadrimaculata. Lestes dryas* to the south reaches European Turkey (Yazicioglu, 1982). *Aeshna subarctica* has been found only once in the Balkan Peninsula, in the Rila Mts. of Bulgaria, at 2000 m a.s.l. (Beschovski, 1960). This record has to be confirmed.
   1.2. *Holopalaearctic group*: nine species (13.2%), viz. *Calopteryx virgo, Lestes sponsa, Ischnura elegans, Aeshna mixta, Anax parthenope, Sympetrum striolatum, S. vulgatum, S. flaveolum*, and *S. pedemontanum*.
   1.3. *Eurosiberian group*: 14 species (20.6%), the most important component of the Bulgarian dragonfly fauna: *Erythromma najas, Coenagrion hastulatum, C. pulchellum, Gomphus flavipes, Ophiogomphus cecilia, Cordulia aenea, Somatochlora arctica, S. metallica, S. flavomaculata, Orthetrum cancellatum, O. albistylum, Sympetrum depressiusculum, Leucorrhinia dubia*, and *L. pectoralis*.
   1.4. *European group*: four species (5.9%), *Libellula fulva, L. depressa, Aeshna cyanea*, and *Gomphus vulgatissimus*. The southern boundary of *A. cyanea* passes through Bulgaria.
   1.5. *European–Mediterranean group*: 11 species (16.2%), *Calopteryx splendens, Platycnemis pennipes, Pyrrhosoma nymphula, Coenagrion ornatum, C. puella, Erythromma lindeni, Aeshna affinis, Ae. isosceles, Brachytron pratense, Cordulegaster bidentata*, and *Sympetrum sanguineum*. The southern limits of their range pass through the south Balkan Peninsula.

The European–Mediterranean and Eurosiberian groups are the largest components of the Eurosiberian fauna in Bulgaria.

2. *Mediterranean faunal type*: 24 species (35.3%) in five groups.
   2.1. *Holomediterranean group*: 12 species (17.6%), *Chalcolestes viridis, Ch. parvidens, Lestes barbarus, L. virens, L. macrostigma, Sympecma fusca, Erythromma viridulum, Coenagrion scitulum, Ischnura pumilio, Onychogomphus forcipatus, Orthetrum brunneum, Sympetrum meridionale. Lestes macrostigma* is rare and associated with brackish water along the Black Sea coast.
   2.2. *East Mediterranean group*: four species (5.9%), *Epallage fatime, Caliaeschna microstigma, Cordulegaster insignis*, and *Selysiothemis nigra. C. microstigma*, through southwestern Bulgaria and Macedonia, reaches Montenegro (Beschovski, 1966; Peters and Hackenthal, 1986;

Peters, 1987). *E. fatime* in Bulgaria is limited to the southeast. *S. nigra* occurs in North Greece, southwestern Bulgaria, and Macedonia (Campion, 1918, 1921; Galletti and Pavesi, 1983, 1987; Beschovski and Gastarov, 1997).

2.3. West Mediterranean group: two species (2.9%), *Somatochlora meridionalis* and *Orthetrum coerulescens*.

2.4. Balkan–Mediterranean group: three species (4.4%), *Cordulegaster picta, C. heros*, and *Somatochlora borisi* (Theischinger, 1979; Beutler, 1987a, 1987b; Verschuren, 1989; Marinov, 2001c). *C. heros* and *S. borisi* have a refugial distribution in the Balkan.

2.5. Mediterranean–Afrotropical group: two species (2.9%), *Anax imperator* and *Crocothemis erythraea*, both widespread in the Palaearctic and Afrotropical regions.

2.6. Mediterranean–Paleotropical group: one species (1.5%), *Sympetrum fonscolombei*, distributed in the European and Mediterranean regions of the Palearctic, and the Afrotropical and Oriental regions (Belyshev and Kharitonov, 1981).

Most authors treat the species included in the last two groups as Meridional or Paleotropical elements penetrating northward to Europe, since the larger part of their range lies in the Afrotropical region. Their larvae, however, are well-adapted to a temperate climate; thus, it would be possible to assume that these species have originated in the Mediterranean. Later, being eurythermal, they could have dispersed southward. This hypothesis is emphasized here in the names of faunal groups: the adjective "Mediterranean" (belonging to the Palaearctic) comes first, followed by "Paleotropical" or "Afrotropical". *Hemianax ephippiger* can be mentioned for comparison: its adults occur almost throughout Europe, but larvae have been found only in the eastern and northern Mediterranean (Marmels, 1975; Askew, 1988; Dumont and Desmet, 1990). The thermophilous East Mediterranean *Epallage fatime* and *Caliaeschna microstigma* are found only in the southern part of the Balkan Peninsula; along the Bulgarian Black Sea coast, *C. microstigma* does not cross northward the January isotherm 0–1 °C (Beschovski, 1964), regardless of its Boreal origin.

Belyshev and Kharitonov (1983), in their notes of the presence of *Anax imperator, Crocothemis erythraea*, and *Sympetrum fonscolombei* in the Cape region of Africa, point out that these are Holarctic elements, which have dispersed southward.

The Meridional complex includes a single species, *Hemianax ephippiger* (1.5%), separated as *Paleotropical faunal type* and *Afrotropical–Mediterranean group*. Its adults are widespread in Europe, but its range is Afrotropical. (In our interpretation of the Paleotropical faunal type, the adjective "Afrotropical" comes first, pointing at the origin of the species in the Meridional realm. Thus the Afrotropical–Mediterranean group is not the same as Mediterranean–Afrotropical mentioned above; the latter includes species of Boreal origin.

4 Distribution Patterns in Bulgaria

4·1 Ecological features

Due to their aquatic larval stage, adult dragonflies depend on water. Their distribution is limited by the hypersaline lagoons along the Black Sea, fast rivers in their upstream parts, high mountain lakes lacking vegetation, basins with regulated water regime (reservoirs and fish farms), and water pollution. Still, certain species survive in (or even are adapted to) these extreme conditions. E.g., *Lestes macrostigma* is often found in coastal and inland lakes with increased alkalinity (Dévai et al., 1976). In Bulgaria, its larvae live in temporary pools with salinity ca. 5‰. They survive in warmer periods even when salinity rises to 10–13‰ due to evaporation. Other species living in such conditions are *L. barbarus, Ischnura elegans, Anax parthenope, Orthetrum cancellatum, Sympetrum striolatum. Cordulegaster bidentata* is one of the dragonfly species adapted to water current in narrow (sometimes ca. 1 m wide) upstream rivers. In Bulgaria, *C. bidentata* is found mostly next to river sources up to 1200 m a.s.l., but is reported from lower zones as well. It can be seen flying in shady places over spring banks with almost no vegetation.

Another limit to dragonfly distribution are stony bottoms and shores of high mountain lakes: no species reproduces in such lakes or in the rivers originating from them. A typical habitat for high-mountain dragonflies are vegetated littoral zones, covered by *Sparganium angustifolium*; and closer to the banks *Juncus filiformis, Eleocharis palustris*, and *Carex acuta*. Larvae of the most common species in these lakes, *Aeshna juncea*, are epiphytic, and probably survive predation by fish (e.g. minnows, *Phoxinus phoxinus*) among dense plants. Such habitats were observed as high as 2512 m a.s.l. (Tevnoto Ezero Lake in Pirin Mts.), but no evidence of breeding was recorded.

Other habitats of special importance at these altitudes are peatbogs, where *Leucorrhinia* develop. *Sphagnum* mosses in shallow water are favored by *Somatochlora arctica*; there, *Aeshna subarctica* could be rediscovered for Bulgaria. As for reservoirs and fish farms, their regulated regime reduces the survival of dragonfly larvae. Only species with higher ecological valence exist there, e.g. *Ischnura elegans* and *Lestes barbarus* (which develop in temporarily flooded areas). *I. elegans* is a pioneer in newly constructed water bodies (Chovanec and Raab, 1997). It also resists pollution up to a certain level.

4·2 Vertical distribution

Three species are excluded from the classification given below. *Chalcolestes viridis* and *Somatochlora flavomaculata* are known from few records only; for *S. flavomaculata* there is no reliable data for the last 30 years. *C. viridis* is found mainly in the lowlands; *S. flavomaculata* is known from two points at 200–250 m a.s.l. The third species excluded is the Paleotropical *Hemianax ephippiger*. Its larvae develop in the arid and semiarid zones of North Africa; the adults enter Europe, crossing the Mediterranean (Dumont and Desmet, 1990). Their invasion through the East Mediterranean is discussed by Marinov (2001e). In Bulgaria, *H. ephippiger* uses corridors along the Black Sea coast and Struma Valley. It is found mainly in the lowlands but during migration it flies over peaks as high as 862 m a.s.l. (Petkov leg.) and 1800 m a.s.l. (Beron and Petrov leg.).

According to their vertical distribution, species inhabiting Bulgaria can be divided into two main types and seven groups. (The terms *Northern* and *Southern distribution type* are used in a broad sense; both include species from Eurosiberian and Mediterranean zoogeographic types. Eurosiberian species of the *Southern distribution type* have ranges continuing southward beyond Bulgaria).

1. Northern distribution type. Usually, their southern boundaries pass through the Balkan Peninsula. In Bulgaria, they are found in the northern lowlands and southern mountains, and, rarely, in the lowlands along the Black Sea coast. Peaks in the vertical distribution of this species are upward from 1000 m a.s.l. This type is divided into four groups:

1.1. Species absent in the lowlands: *Somatochlora metallica* (Fig. 5), *S. arctica, Coenagrion hastulatum, Aeshna subarctica, Leucorrhinia pectoralis, L. dubia*. (*A. subarctica* and *L. pectoralis* have not been confirmed for Bulgarian fauna for over 40 years). All species belong to the Eurosiberian zoogeographic type.

1.2. Species with a secondary peak in lowland areas: Aeshna juncea (Fig. 6), *A. cyanea, Pyrrhosoma nymphula, Cordulia aenea, Libellula quadrimaculata, Sympetrum vulgatum, S. flaveolum*. All belong to Eurosiberian zoogeographic type. They are found also along the Black Sea coast, or recorded in the lowlands without clear evidence of reproducing there.

1.3. Species equally distributed below and above 1000 m a.s.l.: Enallagma cyathigerum (Fig. 7), *Lestes virens, L. sponsa, L. dryas, Coenagrion puella*. Four species (80.0%) belong to Eurosiberian zoogeographic type and one (20.0%), to the Mediterranean. These species are believed to be evenly

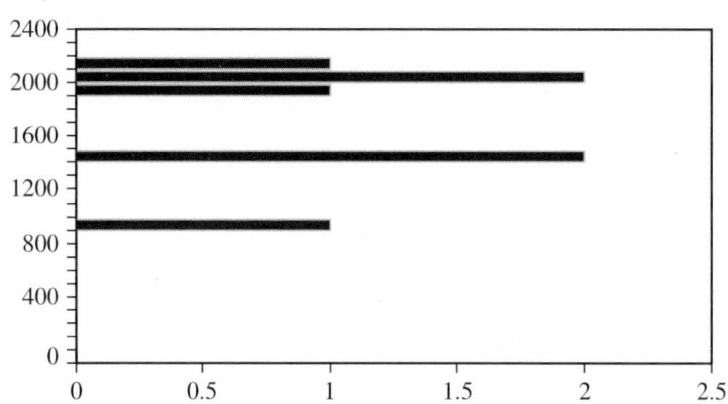

Fig. 5 Vertical distribution of *Somatochlora metallica*.

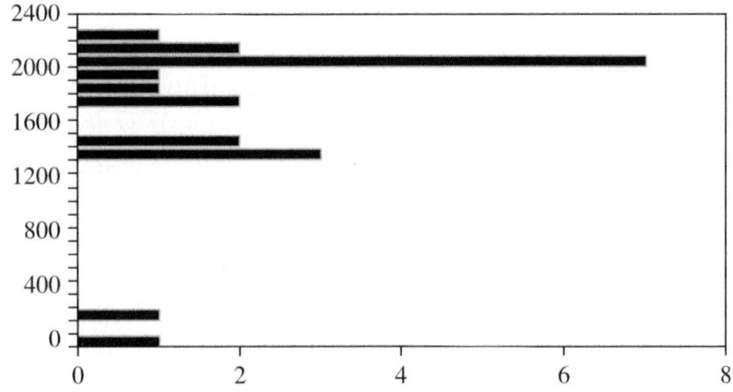

Fig. 6 Vertical distribution of *Aeshna juncea*.

distributed vertically; larger peaks in the lowlands are due to more intensive studies and more wetlands available there.

1.4. Species from the mountain regions with insufficient information. These are species with a few records for Bulgaria both from the lowlands and mountains: *Erythromma najas, Brachytron pratense, Cordulegaster heros, C. bidentata*. Three species (75.0%) belong to Eurosiberian zoogeographic type and one (25.0%), to the Mediterranean.

2. *Southern distribution type.* Found mostly in the lowlands of South Bulgaria (below 100 m a.s.l.), but occupying mountains as well; some have a secondary peak of distribution above 1000 m a.s.l. but no evidence of reproduction. They also penetrate to the north, mostly found along the Black Sea coast or the Danube. This type is divided in three groups:

2.1. Species found only in the lowlands. Lestes macrostigma (Fig. 8), *Cordulegaster insignis, Somatochlora borisi*, and *Selysiothemis nigra*, all of the Mediterranean zoogeographic type.

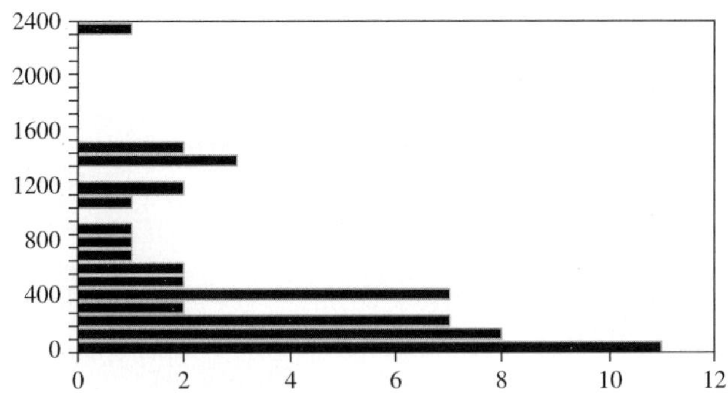

Fig. 7 Vertical distribution of *Enallagma cyathigerum*.

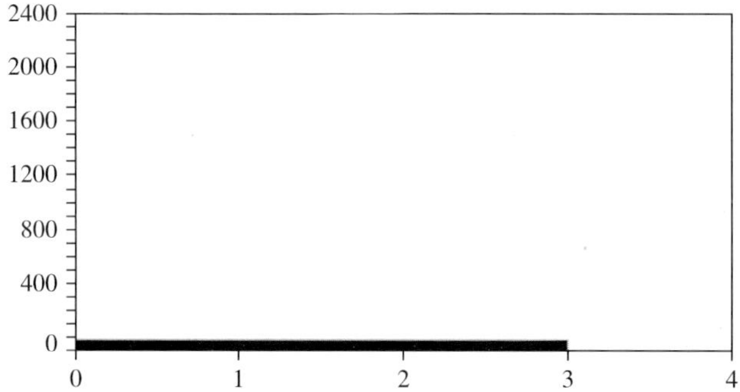

Fig. 8 Vertical distribution of *Lestes macrostigma.*

2.2. *Species most common in the lowlands, records gradually decreasing vertically.* Some have peaks above 1000 m a.s.l., usually in mountains with Mediterranean climatic influence. Includes *Anax parthenope* (Fig. 9), *Calopteryx splendens, Chalcolestes parvidens, Lestes barbarus, Sympecma fusca, Platycnemis pennipes, Erythromma viridulum, Coenagrion pulchellum, Ischnura pumilio, I. elegans, Aeshna mixta, Ae. affinis, Ae. isosceles, Anax imperator, Gomphus flavipes, G. vulgatissimus, Ophiogomphus cecilia, Onychogomphus forcipatus, Somatochlora meridionalis, Orthetrum cancellatum, O. albistylum, O. brunneum, O. coerulescens, Crocothemis erythraea, Sympetrum striolatum, S. meridionale, S. sanguineum, S. depressiusculum, S. pedemontanum.* Seventeen species (58.6%) belong to Eurosiberian zoogeographic type, and twelve (41.4%), to the Mediterranean.

2.3. *Species evenly distributed up to 500–600 m a.s.l.* Some have peaks at higher altitudes, but without evidence of reproduction. Includes *Coenagrion scitulum* (Fig. 10), *Calopteryx virgo, Epallage fatime, Coenagrion*

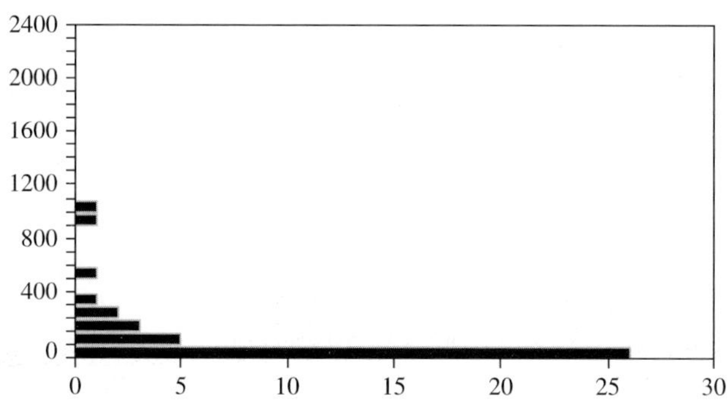

Fig. 9 Vertical distribution of *Anax parthenope.*

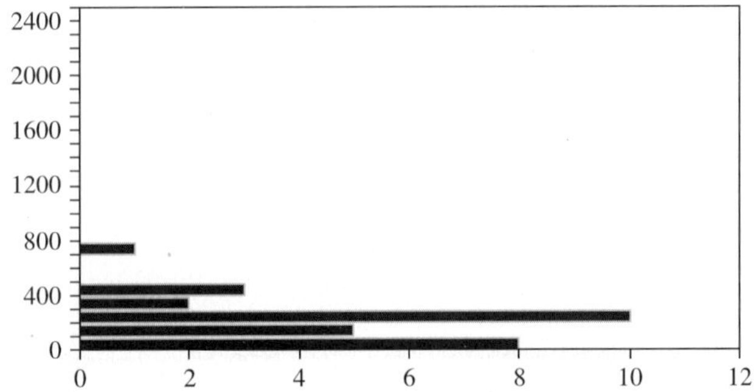

Fig. 10 Vertical distribution of *Coenagrion scitulum*.

ornatum, Erythromma lindeni, Caliaeschna microstigma, Cordulegaster picta, Libellula fulva, L. depressa, Sympetrum fonscolombei. Five species (50.0%) belong to Eurosiberian zoogeographic type, and five (50.0%), to the Mediterranean.

Some species, e.g. *Lestes dryas*, develop well from sea level to ca. 2400 m a.s.l. At a first look, their distribution graphs do not show this, and usually a peak in the lowlands is present. This is, however, due to more intensive studies in the lowlands, and to more suitable wetlands available there.

4·3 Horizontal distribution

Distribution maps (Figs. 11–16) are given for selected species whose vertical distribution is illustrated in Figs. 5–10. Localities are plotted on an UTM-map where the altitude of 1200 m is marked (proposed by Abadjiev, 2001).

5 Phenology

Dragonfly flying periods in the Balkans differ from those in Central and West Europe. Seven phenological groups of Odonata are found in Bulgaria as listed below (question marks indicate undefined affiliation; such species can be in more than one group). In Figs. 17–23, typical examples are given for each group. Individual species graphs show their flying periods (bars representing the existing record by decades for each species) compared with records of all species for Bulgaria (a line).

1. *Overwintering species*. Adults survive through winter and appear next spring after the first warm days. The activity peak around the end of June is due to the appearance of a new generation. Only *Sympecma fusca* (Fig. 17).
2. *Spring species*. Activity peaks in May–June. *Aeshna isosceles* (Fig. 18), *Epallage fatime, Lestes macrostigma, Pyrrhosoma nymphula, Coenagrion scitulum, C. ornatum, C. puella, C. pulchellum, Ischnura pumilio* (?), *Brachytron pratense,*

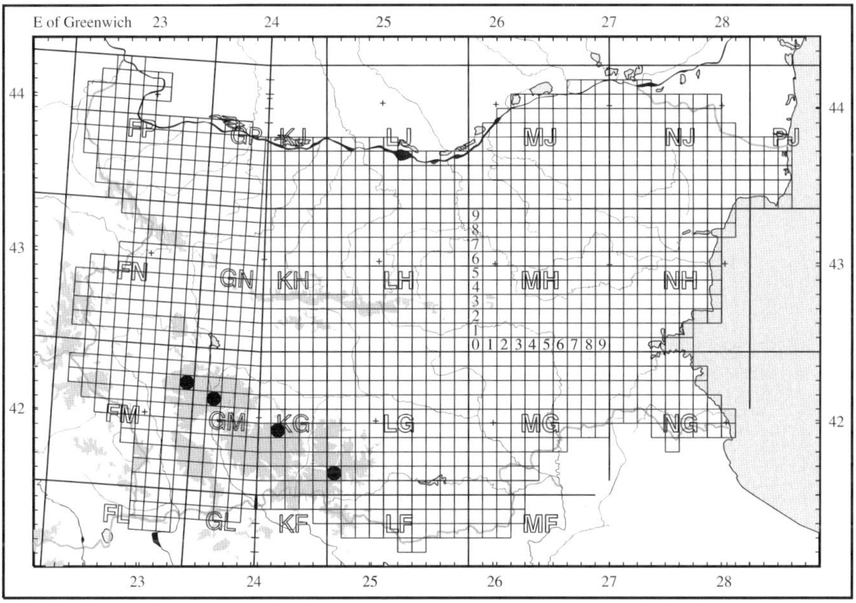

Fig. 11 Horizontal distribution of *Somatochlora metallica.*

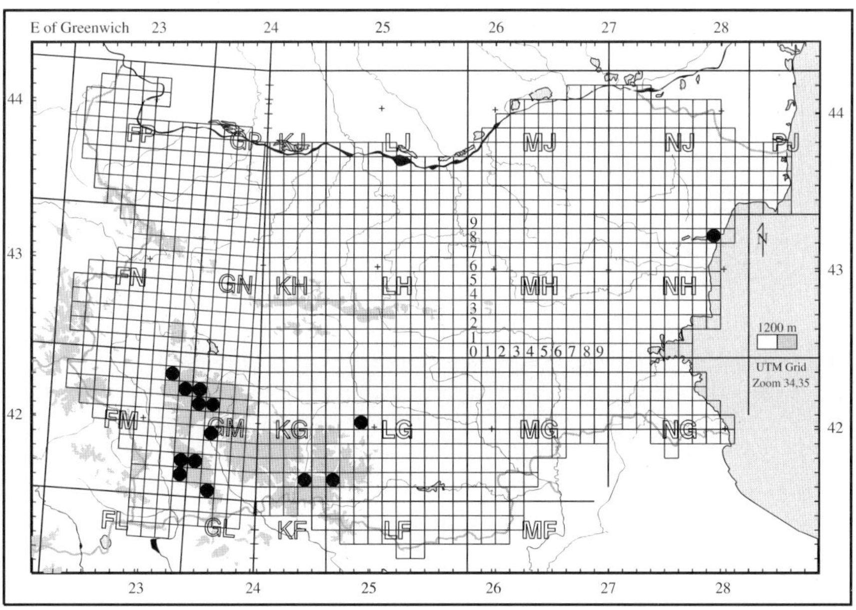

Fig. 12 Horizontal distribution of *Aeshna juncea.*

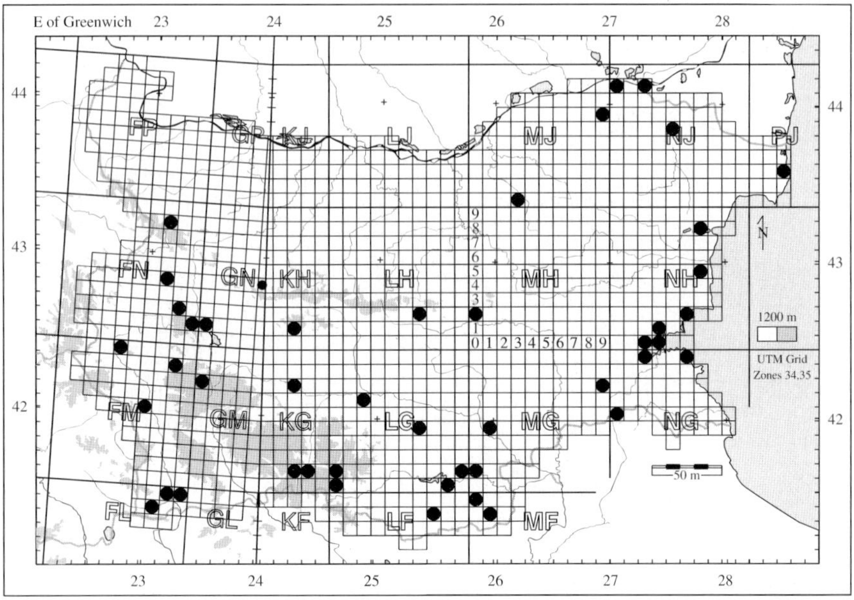

Fig. 13 Horizontal distribution of *Enallagma cyathigerum.*

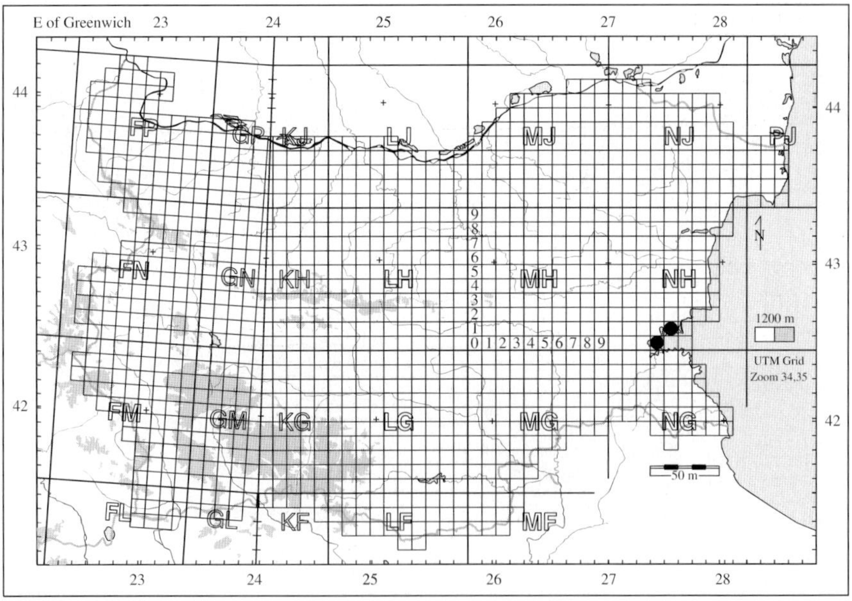

Fig. 14 Horizontal distribution of *Lestes macrostigma.*

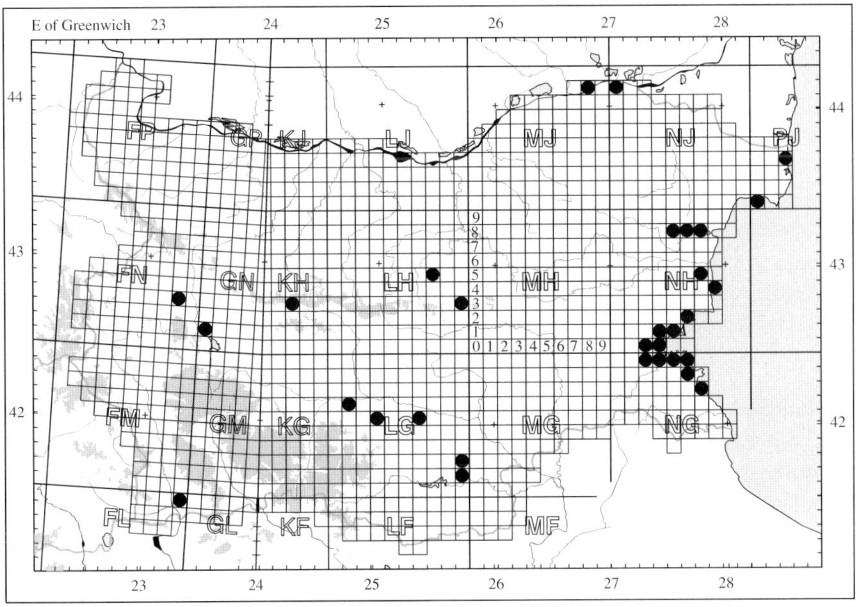

Fig. 15 Horizontal distribution of *Anax parthenope.*

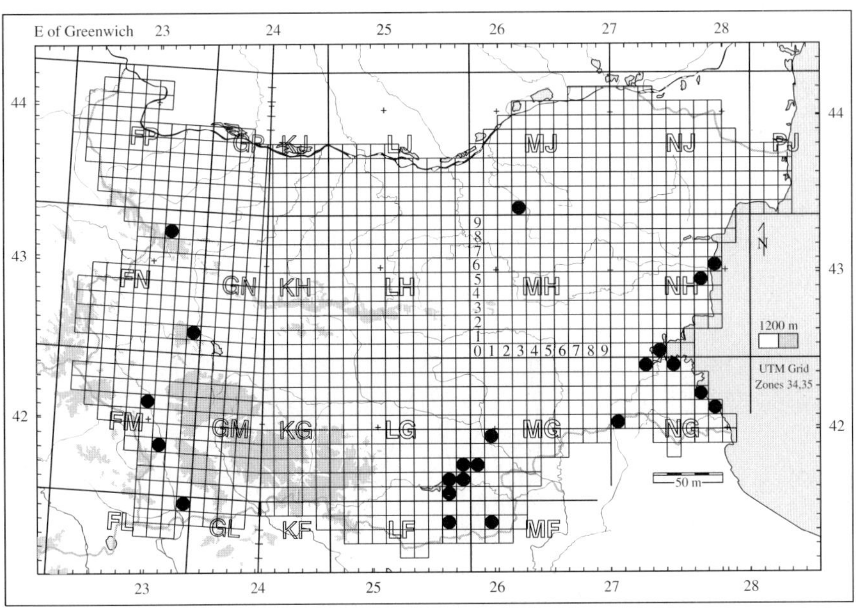

Fig. 16 Horizontal distribution of *Coenagrion scitulum.*

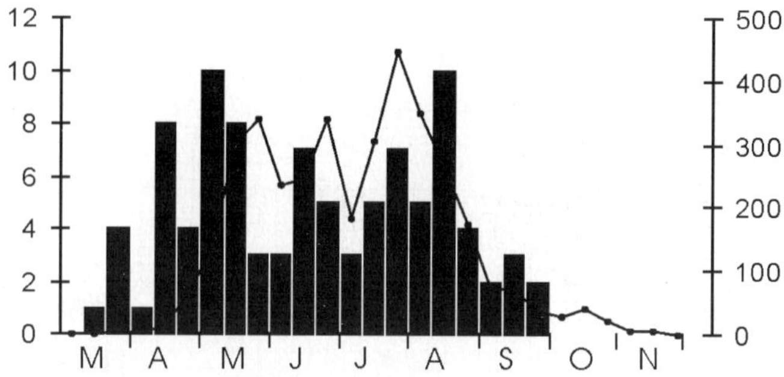

Fig. 17 Flying period of *Sympecma fusca.*

Gomphus vulgatissimus, Cordulia aenea, Somatochlora borisi, Libellula fulva, L. depressa.

3. *Spring–summer species.* Two activity peaks: late May–early June and middle to late July–early August. *Ischnura elegans* (Fig. 19), *Calopteryx virgo* (?), *C. splendens* (?), *Epallage fatime* (?), *Lestes dryas, Platycnemis pennipes, Enallagma cyathigerum, Ischnura pumilio, Anax imperator, A. parthenope, Caliaeschna microstigma, Cordulegaster insignis, C. bidentata* (?), *Libellula quadrimaculata, Orthetrum cancellatum, O. albistylum, Crocothemis erythraea, Sympetrum fonscolombei* (?).
4. *Summer species.* Activity peak in July–August. *Sympetrum sanguineum* (Fig. 20), *Calopteryx virgo, C. splendens, Lestes barbarus, L. sponsa, Coenagrion hastulatum, Erythromma lindeni, E. viridulum, Aeshna juncea, A. affinis, A. cyanea, Gomphus flavipes, Onychogomphus forcipatus, Cordulegaster picta, C. bidentata, Somatochlora metallica, S. meridionalis, Orthetrum brunneum, O. coerulescens, Sympetrum meridionale, S. fonscolombei, S. flaveolum, S. depressiusculum, S. pedemontanum.*
5. *Summer–fall species.* Activity peak in August–late September and mid-October. The weaker peaks in September and October are most probably due to

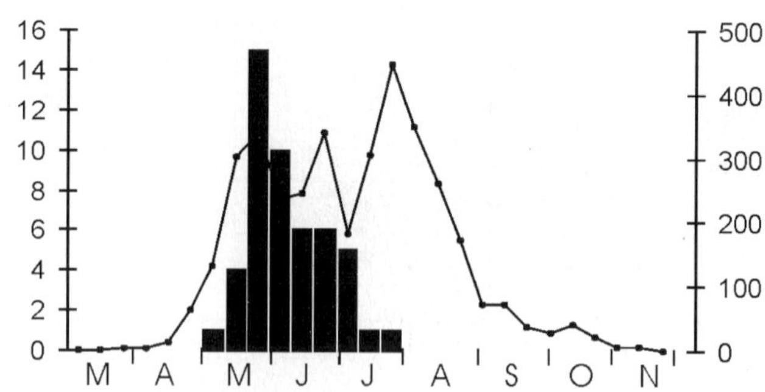

Fig. 18 Flying period of *Aeshna isosceles.*

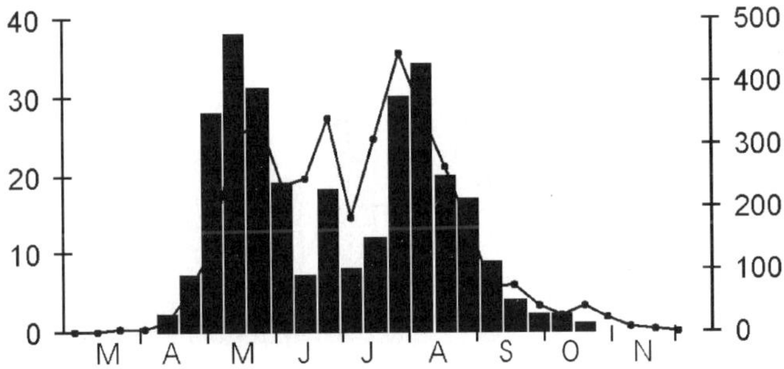

Fig. 19 Flying period of *Ischnura elegans*.

low numbers of observations during these months. *Aeshna mixta* (Fig. 21), *Chalcolestes parvidens*, and *Lestes virens*.

6. *Fall species*. Activity peak in October. An additional peak in August mainly in the mountains. Only *Sympetrum vulgatum* (Fig. 22).
7. *Spring–summer–fall species*. Activity peaks in all three seasons. Only *Sympetrum striolatum* (Fig. 23).

6 Habitat Classification

We establish 12 habitat types in Bulgaria following Jacob (1969), with some modifications. Fig. 24 illustrates habitat selection by dragonflies. As there is no clear definition of "fast" or "slow" water current, these terms are used in a broad sense. *Fast* current is associated with rivers running down a considerable slope, with current well-visible. *Slow* waters flow through nearly flat areas, at a first look appearing almost still. Fast and slow currents alternate along a river depending on the terrain.

(H0) Vegetation on the banks of glacial water bodies. This habitat type does not have a key species: all listed dragonfly species are also found in other

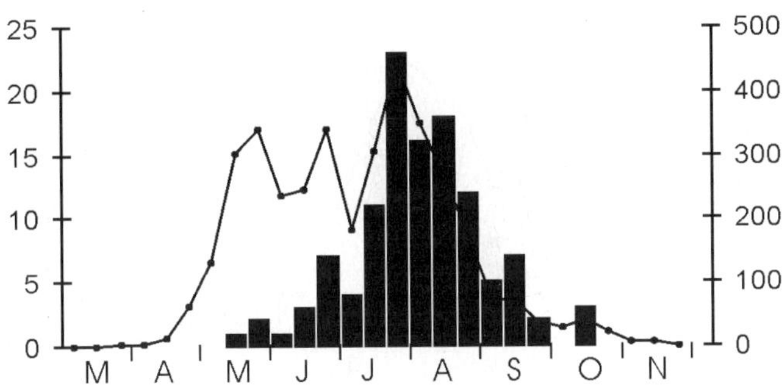

Fig. 20 Flying period of *Sympetrum sanguineum*.

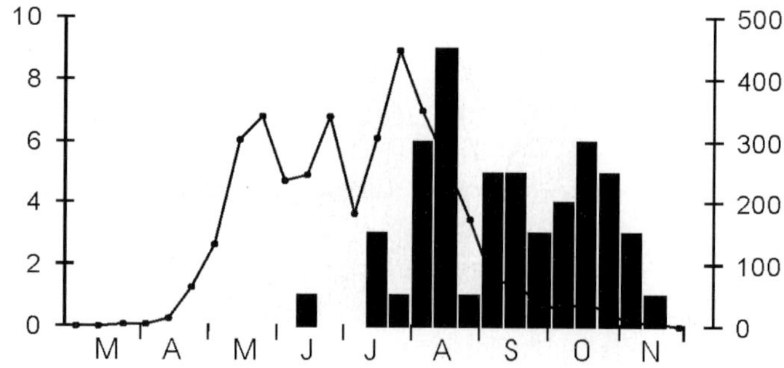

Fig. 21 Flying period of *Aeshna mixta.*

habitats (mostly in peatbogs). The water bodies of glacial origin are located in high mountains. In their shallow parts (to 1 m), the surface is covered by floating leaves of *Sparganium angustifolium*; their banks, down to a depth of ca. 25 cm are overgrown by Juncaceae and Cyperaceae (*Juncus filiformis, Eleocharis palustris, Carex acuta*). Co-occurring species: *Aeshna juncea, Somatochlora metallica, Aeshna cyanea, Lestes dryas, Coenagrion puella, Pyrrhosoma nymphula, Enallagma cyathigerum, Ischnura pumilio.* Only *S. metallica* larvae belong to the limnophilous benthic subgroup; all others are limnophilous epiphytic (Beschovski, 1968).

(H1) Peatbog vegetation. Peatbogs of Bulgaria and their dragonfly fauna are poorly studied. These water bodies are 0.5 to 5 m deep. *Sphagnum* mosses overgrow the surface and create thick layers over it or on the bottom. These are still water basins, but in some peatbogs water balance is supported by seepage water. In summer, their shallow portions can dry up. Key species: *Leucorrhinia dubia, Somatochlora arctica, Coenagrion hastulatum*; for these, the shallow portion of the bogs is of great importance. Wildermuth (1986) gives 1.7 m as the maximal depth of *S. arctica* habitats in Swiss high mountains. Co-occurring species: *Aeshna juncea,*

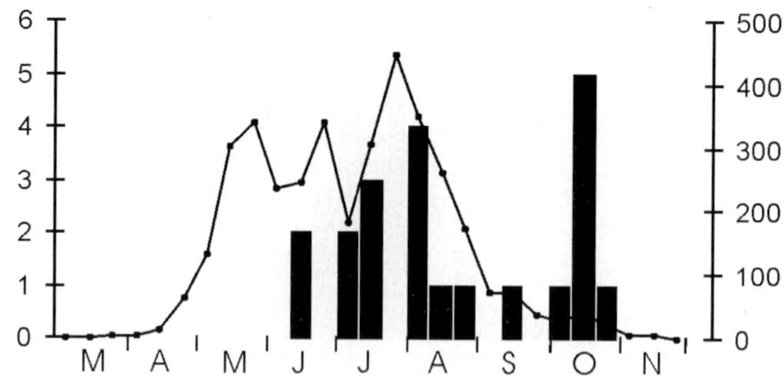

Fig. 22 Flying period of *Sympetrum vulgatum.*

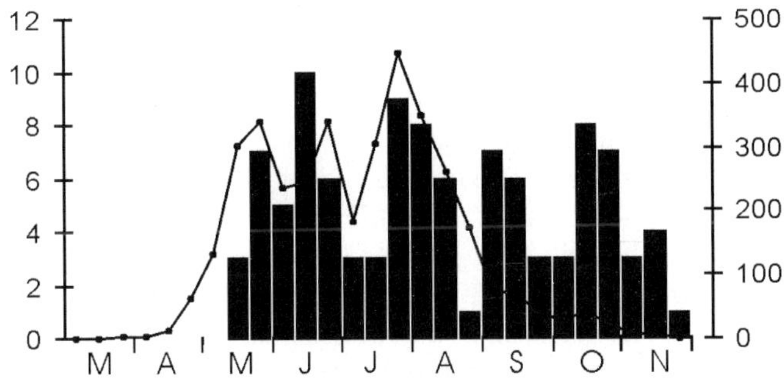

Fig. 23 Flying period of *Sympetrum striolatum.*

A. cyanea, Libellula quadrimaculata, Somatochlora metallica, Sympetrum flaveolum, S. vulgatum, Cordulia aenea, Erythromma najas, Pyrrhosoma nymphula, Lestes sponsa. They also occur elsewhere in Bulgaria; some (*Lestes sponsa, Libellula quadrimaculata, S. flaveolum, P. nymphula, Ae. juncea*) are reported along the Black Sea coast, probably due to the specific temperature regime.

(H2) Mostly shady areas along water bodies with fast to slow current; banks or surface can be overgrown. Usually 0.5 to 3 m wide and 10 to 50 cm

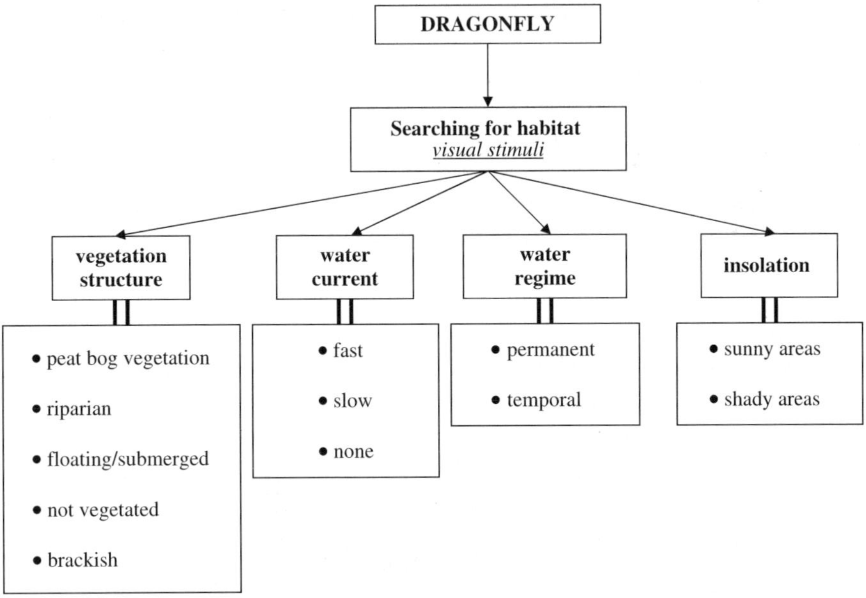

Fig. 24 Visual cues (physical parameters of the habitat) important for habitat selection by dragonfly species.

deep; water level can fluctuate more than 1 m. The shady areas are formed by trees (*Alnus* sp., *Ulmus* sp., *Salix* sp., *Ostrya carpinifolia, Carpinus* sp.) or because the water body itself is on the bottom of a gully with high slopes (up to 10 m). Aquatic vegetation: mostly *Berula erecta*. Key species: *Caliaeschna microstigma, Somatochlora borisi, S. meridionalis, Cordulegaster picta*, and *C. bidentata*. Their larvae are either rheophilous epiphytic and benthic (*Caliaeschna microstigma*) or only benthic (other species) (Beschovski, 1967; Marinov, 2000; Wildermuth, 2001). There is not enough information on *S. borisi*, listed as a key species due to its very specific habitat. Shore vegetation is not a characteristic for H2 habitat. As water level fluctuates during the active flying period (May–August), most of plants remains away from water. The H2 species lay eggs on the wet bank devoid of vegetation, close to the water (*Caliaeschna microstigma*), in the substrate of some shallow areas (10 to 50 cm) (*Cordulegaster picta, C. bidentata*) or directly on the water surface (*Somatochlora borisi, S. meridionalis*). Exuviae were found on various substrates: leaves of surrounding plants (*Cordulegaster picta, S. meridionalis*), roots or tree bark up to 2.15 m high (*Caliaeschna microstigma*), stones or bridges (*Cordulegaster picta, Caliaeschna microstigma*), and macrophytes in water (*C. microstigma*). In Bulgaria, *Cordulegaster bidentata* is found mainly in the upper portions of rivers. *Caliaeschna microstigma* and *Cordulegaster picta* are also found there, but only in the southern part of the country and at the Black Sea coast. Co-occurring species: *Calopteryx splendens, Chalcolestes viridis, Cordulegaster heros, C. insignis*. They also can be found in vegetation within sunny areas of H2; such areas are used for egg-laying (*C. insignis*), copulation (*Chalcolestes viridis*), or rest (*Calopteryx splendens*).

(H3) Vegetated banks of water bodies with fast current passing through sunny areas. Size of H3 water bodies varies more than in H2. They can be more than 10 m wide and more than 1 m deep. Sunny areas exist because trees are located farther away from the banks than in H2. Vegetation along the banks: *Carex, Lythrum*, etc. Water depth and fast current usually do not allow plant growth on the water surface. In the shallower parts (up to 50 cm depth) *Ranunculus* sp. could develop. Key species: *Epallage fatime* and *Calopteryx virgo* (?); their larvae represent epiphytic (*C. virgo*) and benthic (*E. fatime*) rheophilous types (Beschovski, 1967). Vegetation along the banks is a main characteristic since adults of both species hide or roost there. Although *E. fatime* larvae are benthic, its adults use plants in the shallow areas for egg-laying (Buchholtz, 1955). In Bulgaria, *C. virgo* is found usually in upper parts of streams, but also along the Black Sea coast and in South Bulgaria. There, adults are found in more or less shady areas, but prefer to perch on plants mostly exposed to the sun. Co-occurring species: *Platycnemis pennipes* and *Calopteryx splendens*, both widely distributed in Bulgaria. Usually associated with fast flowing rivers, but also found along water bodies with slow current, especially *P. pennipes*.

(H4) *Water bodies with slow current and overgrown banks; shade from the surrounding bushes/trees does not completely cover the habitat*. Usually not wider than 2 m and not deeper than 1 m, these habitats are often located around the inflow and outflow parts of still water bodies. They are also formed due to the water movement in areas without an expressed slope or artificial channels. Vegetation: flooded hygrophytes (*Sparganium erectum, Iris pseudacorus, Phragmites australis, Typha* sp.), emergent hygrophytes (*Epilobium hirsutum, Eupatorium cannabinum, Lythrum salicaria*), other plants (*Cirsium* sp., *Urtica dioica*). Key species: *Coenagrion ornatum, Orthetrum coerulescens, O. brunneum*. Although *C. ornatum* larvae were first described by Haymer and Plattner (1969), their clear diagnostic features were published only recently (Bellmann, 1993; Heidemann and Seidenbusch, 1993; Gerken and Sternberg, 1999), and therefore their biology and ecology are not well studied. *C. ornatum* is listed in H4 tentatively; the water current (even a slow one) is probably important for its larval development. Other species have limnophilous benthic larvae (Beschovski, 1968). The H4 species are common in Bulgaria, reaching 1250 m a.s.l. (*O. coerulescens*). Co-occurring species: *Erythromma lindeni, Calopteryx splendens, Platycnemis pennipes, Coenagrion puella, Ischnura elegans, Pyrrhosoma nymphula, Sympetrum sanguineum, S. depressiusculum, S. pedemontanum*. Of these, only *Erythromma lindeni* prefers slow water current, but it could be found in still water as well. *Calopteryx splendens* larvae are typical rheophiles but larvae develop also in waters with almost no current. All other larvae (excluding *P. pennipes*) are typical limnophiles but they could also develop in water with very slow current. *S. depressiusculum* and *S. pedemontanum* typically occupy temporary waters next to the main water body.

(H5) *Temporary vegetated water bodies, either separate or formed around larger ones, mostly in sunny areas*. Varying greatly in size and shape, these water bodies most typically are shallow (to 0.5 m), and partially or totally drying in summer. The H5 habitats are often formed around larger water bodies in the littoral flooded in spring. These could also represent separate, shallow waters adjacent to, or removed from the main water bodies. The surface is covered by macrophytes such as *Phragmites australis, Typha* sp., *Schoenoplectus* sp., *Eleocharis* sp., *Juncus* sp., *Carex* sp. Key species: *Lestes sponsa, Sympetrum vulgatum, S. flaveolum*. Classification schemes usually list almost all *Lestes* and *Sympetrum* species for this type of habitat. Here, most of them are listed as co-occurring species for other habitats (cf. H6 and H11). Larvae of key species are typically limnophilous, with *L. sponsa* having epiphytic and two other, benthic larvae (Beschovski, 1968). H5 habitat is found throughout Bulgaria even above 1000 m a.s.l. *S. flaveolum* is found up to 2400 m a.s.l. where it probably inhabits the H0 habitats. We, however, list it as a key species for H5 habitat since its ecology in Bulgarian high mountains is poorly studied. Co-occurring species: *Lestes barbarus, L. virens, L. dryas,*

Sympetrum striolatum, S. meridionale, S. sanguineum, S. depressiusculum, S. pedemontanum, distributed throughout Bulgaria along temporary water bodies.

(H6) Shady areas with trees/bushes close to, or above water of temporary/ permanent water bodies. Differs from H5 by reduced insolation; this habitat does not always dry up due to tree shadow (*Salix* sp., *Populus* sp., etc.) and slow currents. The vegetation is similar to H5. Key species: *Chalcolestes parvidens*. This type of habitat is introduced due to the high preference of tree canopies by *C. parvidens*. Females lay eggs in the branches of the surrounding trees and bushes, where adults hide or roost as well. *C. parvidens* is an East Mediterranean species, distributed throughout Bulgaria but not higher than 300 m a.s.l. Co-occurring species: *Lestes virens, Coenagrion scitulum, Ischnura elegans*. For these species, H6 type serves mainly as a maturation or roosting place. *C. scitulum* couples were found here but an affinity with this habitat is not confirmed.

(H7) Open sunny areas (sandy banks, single stones) along running water with fast current. These waterways usually are 0.5 to 5.0 m wide and 10 to 50 cm deep; depth is important in this case. In wider water basins, this habitat is typical for areas with full insolation and scarce vegetation. Key species: *Onychogomphus forcipatus* and *Gomphus vulgatissimus*, both with rheophilous benthic larvae (Beschovski, 1967). Open space along the banks is important during emergence. *O. forcipatus* larvae select stones or wet ground exactly at the water borderline so that during the whole emergence process the exuviae stay moist. In some cases, *G. vulgatissimus* larvae move out of water up to 1.9 m and climb up to 2.4 m onto tree barks or bridge walls. Surrounding vegetation serves as secondary substrate during emergence. These places are also used by adults during maturation but *G. vulgatissimus* is still listed as a key species since open areas probably play a key role in its habitat selection. Species of this group are distributed throughout Bulgaria along medium and lower stream sections. *G. vulgatissimus* is recorded up to 800 m a.s.l., and *O. forcipatus*, up to 1440 m a.s.l. Co-occurring species: *Erythromma lindeni* (see also H4).

(H8) Open areas (sandy banks, single stones) among surrounding vegetation (bushes, trees) along slow waters flowing through sunny areas. Includes mainly the middle and lower sections of the largest Bulgarian rivers. Due to level fluctuations up to 5 m, surrounding trees and bushes are far from the water during the flying period of key species (June–August); the vegetation varies. Key species: *Gomphus flavipes* and *Ophiogomphus cecilia*, both with rheophilous benthic larvae (Beschovski, 1967). *G. flavipes* exuviae are always found exposed to sun on the surface (ground or stones), but adults choose surrounding trees and bushes for maturation. *O. cecilia* is known from Bulgaria mainly as larvae; adults are rarely collected, in upper sections of rivers (Petkov, 1921; Beutler, 1987b). Co-occurring species: *Platycnemis pennipes, Calopteryx splendens, Libellula depressa, Anax imperator,*

Ischnura elegans, found mainly along running water. *L. depressa* and *A. imperator* larvae are limnophilous (Beschovski, 1968). They survive in slow currents, and are also found at river mouths and in the upper stream sections (up to 600 m a.s.l.).

(H9) *Open sunny areas among macrophytes inside or along the banks of permanent water bodies*. Varies in size, strongly depending on density of macrophytes, usually *Phragmites australis* and *Typha* sp. around or inside water bodies with maximal depth 1.0–1.5 m. Macrophytes can form small islands, which in case of high water can float and move with the wind. Thus the key species could be found both around the shore and inside the water body. Key species: *Orthetrum albistylum* and *Aeshna isosceles*, with benthic or epiphytic limnophilous larvae, respectively (Beschovski, 1968). Adults were observed mostly in the small open areas between macrophytes, sheltered from wind but exposed to the sun. *A. isosceles* is known to fly over long distances (Marinov, 1996); in Bulgaria, some individuals were found several km away from the closest place appropriate for egg laying. *A. isosceles* probably breeds also in H11. Both species are widely distributed in Bulgaria, mainly below 100–200 m. Co-occurring species: *Libellula depressa, Orthetrum cancellatum, Crocothemis erythraea, Anax imperator, A. parthenope, Cordulia aenea, Aeshna mixta, Enallagma cyathigerum, Ischnura pumilio*. The H9 habitat corresponds to the "*Orthetrum–Libellula depressa*" community of Jacob (1969). However, key species proposed by Jacob (*O. cancellatum* and *L. depressa*) are listed here as co-occurring since in Bulgaria both could be found in other habitats as well: *O. cancellatum* in H11, and *L. depressa* in H8. Other species are typical for still waters.

(H10) *Floating or submerged vegetation in still or slow waters exposed to sun*. Size of this habitat is much less important than the presence of typical floating vegetation: *Nymphaea alba, Nuphar lutea, Potamogeton* sp., *Nymphoides peltata, Callitriche* sp., *Ceratophyllum* sp., *Utricularia* sp., etc. These plants usually develop in still water but could be also found in meandering rivers. Key species: *Erythromma viridulum* and *E. najas*, with limnophilous larvae (Beschovski, 1968). Both lay eggs on the submerged plant parts. *E. viridulum* usually perches on leaves over water, and *E. najas* chooses higher parts of the plants several centimeters above the surface. Adults use areas with bushes or trees several meters (in some cases more than 100 m) away from water. *E. viridulum* is reported across Bulgaria up to 700 m a.s.l. For *E. najas*, there are only five records in Bulgaria (the southern boundary of its range). Co-occurring species: *Anax imperator, Sympecma fusca, Coenagrion pulchellum*. Jacob (1969) included *A. imperator* as key species for this habitat type. We list it as co-occurring since it is also found in H8 and H9; there, *A. imperator* usually depends on plants typical for H10.

(H11) *Brackish water vegetation in temporary or permanent water bodies with still or slow waters in sunny areas*. This type of habitat is common along the Bulgarian Black Sea coast. The combined influence (direct or through infiltration) of both Black Sea and fresh water entering the sea is of great importance. Typical plants are *Bolboschoenus maritimus, Juncus maritimus*, and *Phragmites australis*. Vegetation density varies with water composition and hydrological regime. In summer, some habitats dry up completely, and salinity increases to over 10‰. Key species: *Lestes macrostigma*. Its larvae, first discovered in Bulgaria in 1997, are epiphytic limnophiles as other Lestidae. *L. macrostigma* is closely associated with *Bolboschoenus maritimus*; egg laying is only on this plant, but exuviae have been also found on reed stems. For maturation, adults prefer dense reeds (*Phragmites*), which may be only 5 to 10 m from the water. Reed stems are stronger and higher than those of *B. maritimus* and probably offer hiding places for adults or refuges to resist strong breezes, typical in these areas around noon. In Bulgaria, *L. macrostigma* is reported only from two places, but it probably occurs along the whole Black Sea coast. Co-occurring species: *Anax parthenope, Sympetrum striolatum, Crocothemis erythraea, Lestes dryas, L. barbarus, Aeshna mixta, Ischnura elegans*. Their exuviae were found in H11, therefore their larvae can survive increased salinity. An additional interesting discovery here is *Orthetrum cancellatum*, in a lithotelm with salinity 13‰ (Beschovski, 1964).

References

Abadjiev, S., 2001. An atlas of the distribution of the butterflies in Bulgaria (Lepidoptera: Hesperioidea & Papilionoidea). Zoocartographia Balcanica 1.

Aguesse, P., 1968. Les Odonates de l'Europe occidentale, du Nord de l'Afrique et des îles atlantiques. In: Faune de l'Europe et du bassin Mediterranéen. 4. Masson et Cie, Paris.

Askew, R.R., 1988. The Dragonflies of Europe. Harley Books, Colchester, UK.

Bellmann, H., 1993. Libellen: Beobachten, Bestimmen. Weltbild Verlag, Augsburg.

Belyshev, B.F. & A.Yu. Kharitonov, 1981. Geografiya strekoz (Odonata) Boreal'nogo faunisticheskogo tsarstva [Geography of the Dragonflies (Odonata) of the Boreal Faunistic Realm]. Nauka [Science], Novosibirsk (in Russian).

Belyshev, B.F. & A.Yu. Kharitonov, 1983. Geografiya strekoz (Odonata) Meridional'nogo faunisticheskogo tsarstva [Geography of the Dragonflies (Odonata) of the Meridional Faunistic Realm]. Nauka [Science], Novosibirsk (in Russian).

Beschovski, V., 1960. [A contribution to the order Odonata from high-mountain lakes and bogs in Bulgaria]. Izvestiya na Zoologicheskiya institut (Bulletin de l'Institut zoologique de l'Académie des sciences de Bulgarie) 9:451–453 (in Bulgarian).

Beschovski, V., 1964. [Odonata from the Bulgarian Black Sea Coast]. Izvestiya na Zoologicheskiya institut s muzei (Bulletin de l'Institut de zoologie et musée) 15:115–129 (in Bulgarian).

Beschovski, V., 1966. [The Mediterranean dragonflies in Bulgaria]. Priroda [Nature] (Sofia) 15(3):66–68 (in Bulgarian).

Beschovski, V., 1967. [Ecological survey of the larvae of Odonata in the Bulgarian rivers]. Izvestiya na Zoologicheskiya institut s muzei (Bulletin de l'Institut de zoologie et musée) 24:5–20 (in Bulgarian).

Beschovski, V., 1968. [Odonata-Larven der stehenden Gewässer in Bulgarien]. Izvestiya na Zoologicheskiya institut s muzei (Bulletin de l'Institut de zoologie et musée) 26:5–27 (in Bulgarian).

Beschovski, V., 1994a. Comparative zoogeographical review of Odonata fauna of Bulgaria (Insecta, Odonata). Acta zoologica bulgarica 47:3–15.

Beschovski, V., 1994b. [Insecta, Odonata]. In Fauna na Balgariya. 23 (Fauna bulgarica. 23). Izdatelstvo na Balgarskata akademiya na naukite (Aedibus Academiae scientiarum bulgaricae), Sofia (in Bulgarian).

Beschovski, V. & V. Gastarov, 1997. *Selysiothemis nigra* (Vander Linden, 1825) – a new genus and species for the Bulgarian fauna (Odonata: Anisoptera: Libellulidae). Entomologische Zeitschrift 197(7):309–310.

Beutler, H., 1987a. Ein Fund von *Cordulegaster heros* Theischinger in Pirin-Gebirge in Bulgarischen Mazedonien (Insecta: Odonata, Cordulegastridae). Faunistische Abhandlungen aus dem Staatlichen Museum für Tierkunde in Dresden 15(2):11–14.

Beutler, H., 1987b. Libellen aus dem Einzugsgebiet der Struma in Bulgarisch-Mazedonien (Odonata). Opuscula Zoologica Fluminensia 16:1–8.

Buchholtz, C., 1955. Eine vergleichende Ethologie der orientalischen Calopterygiden (Odonata) als Beitrag zu ihrer systematischen Deutung. Zeitschrift für Tierpsychologie 12:364–386.

Campion, H., 1918. On a small collection of dragonflies from Macedonia. Entomologist 51:128–129.

Campion, H., 1921. A supplementary note on Macedonian Dragonflies. Entomologist 54:262.

Chovanec, A. & R. Raab, 1997. Dragonflies (Insecta, Odonata) and the ecological status of newly created wetlands – examples for long-term bioindication programms. Limnologica 27(3/4):381–392.

Corbet, P.S., 1962. A Biology of Dragonflies. H.F. & G. Witherby Ltd., London.

Corbet, P.S., 1999. Dragonflies. Behaviour and Ecology of Odonata. Harley Books, Colchester, UK.

Dévai, G., B. Bodnar & P. Benedek, 1976. An analysis of the faunistic records of dragonflies (Odonata) in Hungary. In Dévai, G. (ed), Research in the Chorology and Phenology of the Dragonfly (Odonata) Fauna of Hungary. Acta Biologica Debrecina 13(suppl. 1):9–92.

Dreyer, W., 1986. Die Libellen. Das umfassende Handbuch zur Biologie und Ökologie aller mitteleuropäischen Arten mit Bestimmungsschlüsseln für Imagines und Larven. Gestenberg Verlag, Hildesheim.

Dumont, H.J. & K. Desmet, 1990. Transsahara and transmediterranean migratory activity of *Hemianax ephippiger* (Burmeister) in 1988 and 1989 (Anisoptera: Aeshnidae). Odonatologica 19(2):181–185.

Fischer, S., 1964. Cycle vital de certaines espèces du genre Lestes dans les petits bassins astatiques. Polskie Archiwum Hydrobiologii 12(25):349–382.

Galletti, P. & M. Pavesi, 1983. Su alcuni Odonati di Grecia. Giornale Italiano di Entomologia 1(5):247–260.

Galletti, P. & M. Pavesi, 1987. Considerazioni su Odonati Ellenici (Odonata). In Crucitti, P. (ed), Atti del Convegno sul tema Zoologia Ellenica, Roma, 17 maggio 1986. Società Romana di Scienze Naturali, Roma: 35–37.

Gerken, B. & K. Sternberg, 1999. Die Exuvien europäischer Libellen (Insecta, Odonata). Huxaria Druckerei, Verlag und Werbeagentur, Höxter, Jena.

Haymer, A. & H. Plattner, 1969. Beschreibung der bisher unbekannten Larve von *Agrion ornatum* aus Rumänien (Odonata Zygoptera; Agrionidae). Annales de la Société Entomologique de France (N.S.) 5(1):891–908.

Heidemann, H. & R. Seidenbusch, 1993. Libellenlarven Deutschlands und Frankreichs. Handbuch für Exuviensammler. Verlag Ema Bauer, Keltern.

Jacob, U., 1969. Untersuchungen zu den Beziehungen zwischen Ökologie und Verbreitung heimischer Libellen. Faunistische Abhandlungen aus dem Staatlichen Museum für Tierkunde in Dresden 2(24):197–239.

Marinov, M., 1995. New data on dragonflies (Odonata) of the Bulgarian high mountain lakes and marshlands in Rila and Pirin mountains. In Tsankov, G. (ed), Treta natsionalna nauchna konferentsiya po entomologiya, 18–20 septemvri 1995 g., Sofia (Third National Scientific Conference of Entomology, 18–20 September 1995, Sofia). Sayuz na uchenite v Balgariya, Balgarsko entomologichesko druzhestvo (Union of the Scientists in Bulgaria, Bulgarian Society of Entomology), Sofia: 15–17.

Marinov, M., 1996. *Anaciaeschna isosceles antehumeralis* (Schneider) on snow, at Erciyes Dagi, Central Turkey (Anisoptera: Aeshnidae). Notulae odonatologicae 4(6):107.

Marinov, M., 1999. *Chalcolestes parvidens* (Artobolevski) and *Somatochlora meridionalis* Nielsen in Bulgaria (Zygoptera: Lestidae; Anisoptera: Corduliidae). Notulae odonatologicae 5(3): 31–33.

Marinov, M., 2000. An atypical *Caliaeschna microstigma* (Schneider) habitat in Bulgaria (Anisoptera: Aeshnidae). Notulae odonatologicae 5(6):83.

Marinov, M., 2001a. Does *Coenagrion mercuriale* (Charpentier, 1840) occur in Bulgaria? Exuviae 8(1):13–19.

Marinov, M., 2001b. The *Orthetrum coerulescens* complex in Bulgaria (Odonata: Libellulidae). International Journal of Odonatology 4(1):35–40.

Marinov, M., 2001c. *Somatochlora borisi* spec. nov., a new European dragonfly species from Bulgaria (Anisoptera: Corduliidae). The International Dragonfly Fund (IDF) Report 3(1/2):9–16.

Marinov, M., 2001d. The genus *Somatochlora* Selys in Bulgaria. The International Dragonfly Fund (IDF) Report 3(1/2):46–53.

Marinov, M., 2001e. Review of *Hemianax ephippiger* (Burm.) records from Bulgaria (Anisoptera: Aeshnidae). Notulae odonatologicae 5(8):105–106.

Marinov, M. & N. Simov, 2004. *Somatochlora arctica* (Zett.) and *Leucorrhinia dubia* (Vander L.) new for the fauna of Bulgaria (Anisoptera: Corduliidae, Libellulidae). Notulae odonatologicae 6(3):34–35.

Marmels, J. de, 1975. Die Larve von *Hemianax ephippiger* (Burmeister, 1839) (Anisoptera, Aeshnidae). Odonatologica 4:259–263.

Peters, G., 1987. Die Edellibellen Europas. In: Die neue Brehm-Bücherei. 585. Ziemsen Verlag, Wittenberg Lutherstadt.

Peters, G. & H. Hackenthal, 1986. Notizen über die Libellen (Odonata) in Mazedonien. Acta Musei Macedonici Scientiarum Naturalium 18(5/151):125–158.

Petkov, P., 1921. [Beitrag zur Kenntnis die bulgarischen Odonaten]. Godishnik na Sofiiskiya universitet. Fiziko-matematicheski fakultet (Annuaire de l'Université de Sofia. Faculté physico-mathématique) 15/16:1–39 (in Bulgarian).

Russev, B., 1962. [Die Insektenfauna der Donau vor dem bulgarischen Ufer]. Izvestiya na Opitnata stantsiya po sladkovodno ribarstvo – Plovdiv (Mitteilungen der Versuchsstation für Süsswasserfischzucht – Plovdiv) 1:115–128 (in Bulgarian).

Theischinger, G., 1979. *Cordulegaster heros* sp. nov. und *Cordulegaster heros pelionensis* ssp. nov., zwei neue Taxa des *Cordulegaster boltonii* (Donovan)-Komplexes aus Europa (Anisoptera: Cordulegastridae). Odonatologica 8(1):23–38.

Verschuren, D., 1989. Revision of the larvae of West-Palearctic *Cordulegaster* Leach, 1815 (Odonata, Cordulegastridae), with a key to the considered taxa and a discussion on their affinity. Bulletin et Annales de la Société Royale Belge d'Entomologie 125:5–35.

Wildermuth, H., 1986. Zur Habitatwahl und Verbreitung von *Somatochlora arctica* (Zetterstedt) in der Schweiz (Anisoptera: Corduliidae). Odonatologica 15(2):185–202.

Wildermuth, H., 2001. Concealment in European *Somatochlora* larvae (Anisoptera: Corduliidae). Exuviae 8(1):1–12.

Yazicioglu, T., 1982. Dragonflies from the Ergene River basin, Thrace, Turkey. Notulae odonatologicae 1(9):141–156.

Zernov, S.A., 1949. Obshchaya gidrobiologiya [General Hydrobiology]. Izdatel'stvo Akademii nauk SSSR [Publishing House of the Academy of Sciences of the USSR], Moscow & Leningrad (in Russian).

Zhadin, V.I., 1950. [The life in the rivers]. In Zhadin, V.I. (ed), Zhizn' presnykh vod SSSR. III [The Live of the Fresh Waters of the USSR. III]. Izdatel'stvo Akademii nauk SSSR [Publishing House of the Academy of Sciences of the USSR], Moscow & Leningrad: 113–256 (in Russian).

8 Fauna and Zoogeography of the Orthopterid Insects (Embioptera, Dermaptera, Mantodea, Blattodea, Isoptera, and Orthoptera) in Bulgaria

Alexi Popov

National Museum of Natural History, Bulgarian Academy of Sciences, Tsar Osvoboditel Blvd. 1, 1000 Sofia, Bulgaria, e-mail: alpopov@bulinfo.net

Abstract: Species diversity of orthopterid orders in Bulgaria is analyzed and revised. Four species and one subspecies are deleted from the list of Bulgarian fauna. *Platycleis medvedevi* and *Eupholidoptera chabrieri* are added. The number of orthopterid taxa established with certainty in Bulgaria is 251 (238 species and 13 additional subspecies). Presence of other 18 taxa (13 species and five additional subspecies) is not yet confirmed. Horizontal and vertical distribution of all species is analyzed for the first time. The richest in species are the Western Rhodopes (50% of the Bulgarian taxa). Characteristic species of nine regions are listed, and the fauna of these regions is compared. Of the six altitudinal belts, the richest is the fauna of the xerothermic oak forests with 202 taxa (81 taxa in this belt only). The boundaries of significant importance are the timberline and the boundary between the mesophilous forest and the beech forest. An original scheme of zoogeographical placement of all taxa by chorotypes and by the origin is proposed. There are 49 types of chorotypes (geographical ranges). Most numerous are the Eurosiberian species (33 species, or 12.2%) but the thermophilous species (several categories) prevail in the Bulgarian fauna. Except for two oreotundral species with Arctoalpine distribution, all other taxa belong to the Arboreal type, which is divided into 23 categories according to the origin. The Holomediterranean elements are the most numerous (37 species, or 14%). There are twice as many species with the southern (Mediterranean or Balkan) origin as those with northern origin. The origin of southern species is analyzed for six levels of speciation. The most interesting zoogeographically 11 genera are analyzed in detail.

1 Introduction

The species of the order Orthoptera found in Bulgaria are well-suited for zoogeographical conclusions due to several reasons. Many taxa from the superfamily Tettigonioidea and from the family Pamphagidae have restricted ranges within the Balkan Peninsula. The degree of endemism in Orthoptera is very high compared to other insect orders. Species with the southern origin prevail in the fauna of South Europe and the Mediterranean. These areas (including Bulgaria) are characterized by the rich species diversity as a result of the intensive speciation in the Balkan Peninsula and Anatolia.

The position of Bulgaria at the border between the Eurosiberian and the Mediterranean zoogeographical subregions is the reason for its great species richness; in the

V. Fet and A. Popov (Eds.), Biogeography and Ecology of Bulgaria, 233–295.

past, the species of both northern and southern origin dispersed into this territory and mixed with the autochthonous fauna. The varied relief of Bulgaria with altitudes from the sea level up to 2925 m (Musala Peak in the Rila Mts.) is also a prerequisite for the presence of a rich faunistic diversity of such groups as Orthoptera, which inhabit all altitudinal vegetation belts from the xerothermic oak belt to the alpine belt. The orography of Bulgaria and the adjacent areas has favored isolation of certain populations and their differentiation into distinct taxa.

The faunistic knowledge on Orthoptera and related orders in Bulgaria was regarded as very good 15 years ago, both with respect to the species recorded in this country as a whole and to their distribution in separate regions, as the result of investigations by Georgi Peshev. The advances in the taxonomy in the last period, however, necessitate revision of some taxa. Regarding the zoogeographical categorization, such categories of origin as Paleoaegeidean, Paleomediterranean, Paleoeuropean, Angaridean have been used previously (Peshev and Djingova, 1974; Pešev, 1974; Pechev, 1975; Peshev and Andreeva, 1986). In general, these categories are quite clear but the results cannot be always compared in theoretical reviews with the data for other insect groups. The better knowledge of the species' ranges also leads to the revision of their zoogeographical identity. It is therefore urgent to update the faunistic data and to conduct a new zoogeographical analysis of Orthoptera, as well as related orders with lower species richness. The existing knowledge of Bulgarian fauna gives an opportunity for zoogeographical conclusions concerning this group of insects.

2 Faunistic Diversity

2·1 Species diversity

A list of all species recorded from Bulgaria, which have not been deleted in further publications, has been compiled (Table 1). Two species newly established for this country, specimens of which were found in Bulgarian collections, are also included in this list. In total, 274 taxa (255 species and 19 additional subspecies) are listed in Table 1.

We analyzed the state of taxonomic knowledge with respect to subspecific division considering the possibility that some subspecies can be in fact only individual aberrations or ecological forms. Subspecies of widespread species were not listed because of vagueness on whether they represent distinct taxa, because of the necessity of revision of subspecies, species, or groups of species (e.g. *biguttulus*-group of the genus *Chorthippus*), and because of preferred zoogeographical comparison of species instead of subspecies with insufficiently known ranges. An exception is made for some verified subspecies, e.g. *Platycleis albopunctata grisea, Eupholidoptera chabrieri schmidti, Stenonemobius bicolor ponticus, Stenobothrus stigmaticus faberi* as well as the endemic and subendemic subspecies, which are important for zoogeographical deductions.

Table 1 Species diversity, distribution in selected regions, vertical distribution and zoogeography of Orthopterida in Bulgaria

Taxa	Distribution in Bulgaria								Vegetational belts in Bulgaria						Zoogeography	
	W-STA	C-STA	RILA	W-RHO	E-RHO	BE, SL	BSEA	STRU	XER	OAK	BEE	CON	SUB	ALP	Chorotype	Origin
EMBIOPTERA																
OLIGOTOMIDAE																
Haploembia solieri (Rambur, 1842)							+	+	+						HoMe	HoMe/s
DERMAPTERA																
ANISOLABIDIDAE																
Carcinophorinae																
Anisolabis maritima (Géné, 1832)							+		+						Cos	HoMe/e
SPONGIPHORIDAE																
Labiinae																
Labia minor (Linnaeus, 1758)			+						+	+	+				Cos	?
LABIDURIDAE																
Labidurinae																
Labidura riparia (Pallas, 1773)			+	+	+		+	+	+	+					Cos	?
FORFICULIDAE																
Anechurinae																
Anechura bipunctata (Fabricius, 1781)			+	+					+			+			EuSi	Sib
Forficulinae																
Apterygida media (Hagenbach, 1822)								+		+					Eu	CE–Me
Forficula auricularia Linnaeus, 1758		+	+	+	+	SL	+	+	+	+	+	+			Cos	CEu
Forficula smyrnensis Serville, 1839							+	+	+						EMe	PoMe/s
MANTODEA																
MANTIDAE																
Amelinae																
Ameles heldreichi Brunner von Wattenwyl, 1882				+	+	BE	+	+	+	+					EMe	PoMe/s

continued

Table 1 continued

Taxa	Distribution in Bulgaria								Vegetational belts in Bulgaria						Zoogeography	
	W-STA	C-STA	RILA	W-RHO	E-RHO	BE, SL	BSEA	STRU	XER	OAK	BEE	CON	SUB	ALP	Chorotype	Origin
Mantinae																
Mantis religiosa (Linnaeus, 1758)	+	+		+	+	+	+	+	+	+	+				PT–Pa	PaTr
Iris oratoria (Linnaeus, 1758)					+			+	+						HoMe	HoMe/s
EMPUSIDAE																
Empusinae																
Empusa fasciata Brullé, 1836				+	+	+	+	+	+	+					EMe	PoMe/s
BLATTODEA																
POLYPHAGIDAE																
Polyphaginae																
Polyphaga aegyptiaca (Linnaeus, 1758)								+	+						Tu–Me	PoMe/e
BLATTIDAE																
Blattinae																
Blatta orientalis Linnaeus, 1758	synanthropic in towns and villages in the entire Bulgaria								+	+					Cos	PaTr
Periplaneta americana (Linnaeus, 1758)	synanthropic in some towns in Bulgaria						+		+	+					Cos	PaTr
BLATTELLIDAE																
Blattellinae																
Blattella germanica (Linnaeus, 1767)	synanthropic in towns and villages in the entire Bulgaria								+	+					Cos	PaTr
Loboptera decipiens (Germar, 1817)				+	+	BE	+	+	+						HoMe	HoMe/s
ECTOBIIDAE																
Ectobiinae																
Ectobius (*Ectobius*) *sylvestris* (Poda, 1761)		+	+	+			+		+		+	+			Eu	CEu
Ectobius (*Ectobius*) *vittiventris* (Costa, 1847)				+	?	+	+		+	+					CSEu	PoMe/e
Ectobius (*Ectobius*) *lapponicus* (Linnaeus, 1758)	+		+	+		SL	?	+	+	+	+	+			EuSi	Sib
Ectobius (*Ectobius*) *balcani* Ramme, 1923		+	+	+	+	+		+	+	+	+	+	+		CSEu	CEu
Ectobius (*Ectobius*) *erythronotus* (Burr, 1898)			+	+	+	+	+		+		+				CSEu	CEu
Ectobius (*Ectobius*) *punctatissimus* Ramme, 1923									+						TrAd	PoMe/e

Phyllodromica brevipennis (Fischer, 1853)			+			SL					+	+		EMe	PoMe/e
Phyllodromica marginata (Schreber, 1781)					+	SL		+	+					EMe	PoMe/e
Phyllodromica carniolica (Ramme, 1913)			+								+	+		NWBa	Din
Phyllodromica pallida (Brunner von Wattenwyl, 1882)											+	+		Ba–An	? Ana
Phyllodromica subaptera (Rambur, 1838)	needs confirmation		?					?	?	?				SEu	HoMe/s
ISOPTERA															
Kalotermitidae															
Kalotermes flavicollis (Fabricius, 1793)							+		+					HoMe	HoMe/s
Rhinotermitidae															
Heterotermitinae															
Reticulitermes lucifugus (Rossi, 1792)					+		+	+	+		+			HoMe	HoMe/e
ORTHOPTERA															
Ensifera															
TETTIGONIOIDEA															
Phaneropteridae															
Phaneroptera falcata (Poda, 1761)	+	+		+			?		+	+	+			EuSi	Sib
Phaneroptera nana Fieber, 1853	+			+	+	+	+	+	+	+				HoMe	HoMe/e
Tylopsis lilifolia (Fabricius, 1793)	+	+		+	+	+	+	+	+	+				HoMe	HoMe/s
Isophya brevipennis Brunner von Wattenwyl, 1878	needs confirmation												?	CEu	CEu
Isophya speciosa (Frivaldszky, 1867)	+	+	+	+	+		+		+	+	+	+	+	EMe	PoMe/e
Isophya rectipennis Brunner von Wattenwyl, 1878	+						+		+	+	+			Ba–An	PoMe/s
Isophya bureschi Peshev, 1959			+	+								+		Ri–P–S	Ri–Rh
Isophya gulae Peshev, 1981									+					Tun	Thr
Isophya modestior Brunner von Wattenwyl, 1882	+								+	+				CSEu	? MoBa
Isophya obtusa Brunner von Wattenwyl, 1882		+										+	+	CBa	MoBa
Isophya petkovi Peshev, 1959				+	+		+		+					Th–Bl	Thr
Isophya modesta modesta (Frivaldszky, 1867)	does not occur in Bulgaria														
Isophya modesta longicaudata Ramme, 1951							+		+					EBa	Moe
Isophya plevnensis Peshev, 1985									+					CD–CP	Moe

continued

Table 1 continued

Taxa	Distribution in Bulgaria								Vegetational belts in Bulgaria						Zoogeography	
	W-STA	C-STA	RILA	W-RHO	E-RHO	BE, SL	BSEA	STRU	XER	OAK	BEE	CON	SUB	ALP	Chorotype	Origin
Isophya kisi Peshev, 1981						SL					+	+			Ri–P–S	Ri–Rh
Isophya andreevae Peshev, 1981								+	+						Stru	Mac
Isophya pravdini pravdini Peshev, 1985		+									+				CSt	StPl
Isophya pravdini bazyluki Peshev, 1985		+									+				CSt	StPl
Isophya pravdini adamovici Peshev, 1985											+				ESt	StPl
Isophya miksici Peshev, 1985	+									+	+				WSt	StPl
Isophya rhodopensis Ramme, 1951				+						+	+	+	+		WRh	Ri–Rh
Isophya hospodar hospodar (Saussure, 1898)					+				+						Th–Bl	Thr
Isophya hospodar medimontana Nedelkov, 1908										+					SoMo	StPl
Isophya rammei Peshev, 1981									+						Stra	Thr
Barbitistes serricauda (Fabricius, 1798)	+	+		+							+	+			CSEu	CEu
Barbitistes constrictus Brunner von Wattenwyl, 1878			+									+			CEu	CEu
Ancistrura nigrovittata (Brunner von Wattenwyl, 1878)				+	+	+		+	+	+	+				EBa	? Thr
Metaplastes pulchripennis (Costa, 1863)	does not occur in Bulgaria															
Andreiniimon nuptialis (Karny, 1918)								+	+						TrAd	Mac
Leptophyes punctatissima (Bosc, 1792)					+	+	+		+	+	+				Eu	CEu
Leptophyes albovittata (Kollar, 1833)	+	+	+		+	BE	+		+	+	+				Eu	PoMe/e
Leptophyes discoidalis (Frivaldszky, 1867)	+							+	+						CSEu	CEu
Leptophyes laticauda (Frivaldszky, 1867)									+						TrAd	Din
Polysarcus denticauda (Charpentier, 1825)	+	+	+	+		+	+		+		+	+	+		CSEu	CEu
Poecilimon ornatus (Schmidt, 1849)											+				NWBa	Din
Poecilimon affinis affinis (Frivaldszky, 1867)	+			?								+			CSEu	MoBa
Poecilimon affinis ruenensis Peshev, 1980											+	+			Oso	Ri–Rh
Poecilimon affinis rilensis Peshev, 1980			+									+	+		Ri–P–S	Ri–Rh
Poecilimon affinis medimontanus Peshev, 1980											+				CSr	StPl
Poecilimon harzi Peshev, 1980												+	+		Ri–P–S	Ri–Rh

Poecilimon mistshenkoi mistshenkoi Peshev, 1980												+		Ri–P–S	Ri–Rh
Poecilimon mistshenkoi marzani Peshev, 1980						SL						+		Ri–P–S	Ri–Rh
Poecilimon mistshenkoi tinkae Peshev, 1980						BE							+	Bel	Ri–Rh
Poecilimon mistshenkoi vlahinensis Peshev, 1980											+			Vl–M–O	Ri–Rh
Poecilimon kisi Peshev, 1980						BE							+	Bel	Ri–Rh
Poecilimon schmidti (Fieber, 1853)	+	+	+	+			+		+		+			CSEu	PoMe/e
Poecilimon thoracicus (Fieber, 1853)	+	+	+	+	+	+	+	+	+	+	+	+	+	? NCBa	? MoBa
Poecilimon elegans Brunner von Wattenwyl, 1878	+		+	+					+	+	+	+		NWBa	Din
Poecilimon fussi Brunner von Wattenwyl, 1878		+	+	+			+	+	+	+	+	+		CSEu	CEu
Poecilimon zwicki Ramme, 1939					+	+		+	+	+	+	+		M–Ri–Rh	Mac
Poecilimon pechevi Andreeva, 1978													+	Vl–M–O	Ri–Rh
Poecilimon macedonicus Ramme, 1926	doubtful information for Bulgaria													Mac	Mac
Poecilimon brunneri (Frivaldszky, 1867)	+	+	+	+	+		+	+	+	+	+	+	+	? NCBa	? MoBa
Poecilimon orbelicus Pančić, 1883	?		+	+		+					+	+	+	M–Ri–Rh	Ri–Rh
Poecilimon belasicensis Popov, 1997	taxonomic status unclear					BE							+	Bel	Ri–Rh
Poecilimon marmaraensis Naskrecki, 1991										+				EBa	Thr
Poecilimon heinrichi (Ramme, 1951)									+					Stra	Thr
Poecilimon miramae Ramme, 1933									+					Ba–An	? Ana
Meconematidae															
Meconema thalassinum (De Geer, 1773)	+	+		+		+		+	+	+	+			Eu	CEu
Meconema meridionale Costa, 1860	probably introduced						+		+					CSEu	AdMe
Conocephalidae															
Conocephalinae															
Conocephalus hastatus (Charpentier, 1825)	+	+		+	+		+		+	+				EMe	PoMe/s
Conocephalus fuscus (Fabricius, 1793)	+			+			+	+	+					Pal	Si–Me
Conocephalus dorsalis (Latreille, 1804)				+			+	+	+					E–WAs	Sib
Copiphorinae															
Ruspolia nitidula (Scopoli, 1786)	+			+	+		+	+	+					AT–Pa	AfTr

continued

Table 1 continued

Taxa	Distribution in Bulgaria								Vegetational belts in Bulgaria						Zoogeography	
	W-STA	C-STA	RILA	W-RHO	E-RHO	BE, SL	BSEA	STRU	XER	OAK	BEE	CON	SUB	ALP	Chorotype	Origin
TETTIGONIIDAE																
Tettigoniinae																
Tettigonia viridissima (Linnaeus, 1758)	+	+	+	+	+	+	+	+	+	+	+	+			Pal	Si–Me
Tettigonia caudata (Charpentier, 1845)	+				+		+	+	+	+					E–WAs	PoMe/e
Tettigonia cantans (Füssli, 1775)	+		+	+							+	+			EuSi	Sib
Decticus verrucivorus (Linnaeus, 1758)	+	+	+	+	+	+	+	+	+	+	+	+	+		Pal	Si–Me
Decticus albifrons (Fabricius, 1775)					+	+	+	+	+	+					HoMe	HoMe/s
Platycleis (*Platycleis*) *albopunctata grisea* (Fabricius, 1781)	+	+	+	+	+	+	+		+	+	+	+			CSEu	CE–Me
Platycleis (*Platycleis*) *intermedia* (Serville, 1839)	+			+	+	+	+	+	+	+	+				HoMe	HoMe/s
Platycleis (*Platycleis*) *falx* (Fabricius, 1775)	does not occur in Bulgaria															
Platycleis (*Platycleis*) *affinis* Fieber, 1853	+			+	+	+	+	+	+	+	+				Tu–Me	HoMe/e
Platycleis (*Platycleis*) *escalerai* Bolivar, 1899			?				+		+	+					EMe	PoMe/s
Platycleis (*Montana*) *stricta* (Zeller, 1849)	needs confirmation		?						?			?			CSEu	AdMe
Platycleis (*Montana*) *macedonica* (Berland et Chopard, 1922)								+	+						Mac	Mac
Platycleis (*Montana*) *medvedevi* (Miram, 1927)									+						SEE–WA	Cas
Platycleis (*Modestana*) *ebneri* (Ramme, 1926)						BE			+						NWBa	Din
Platycleis (*Tessellana*) *tessellata* (Charpentier, 1825)	?	needs confirmation						?							Tu–Me	HoMe/e
Platycleis (*Tessellana*) *veyseli* Koçak, 1984	+				+				+						E–WAs	Sib
Platycleis (*Tessellana*) *nigrosignata* (Costa, 1863)					+	SL		+	+	+					EMe	PoMe/s
Platycleis (*Tessellana*) *orina* Burr, 1899	needs confirmation								?						SBa	BaMe
Platycleis (*Tessellana*) *incerta* Brunner von Wattenwyl, 1882					+	+	+	+	+	+					EMe	PoMe/s

Metrioptera (*Metrioptera*) *bicolor* (Philippi, 1830)			+	+							+	+		EuSi	Sib
Metrioptera (*Metrioptera*) *domogledi domogledi* Brunner von Wattenwyl, 1882	+	+									+			NCBa	MoBa
Metrioptera (*Metrioptera*) *domogledi arnoldi* Ramme, 1933	+	+	+	+		SL				+	+	+	+	CBa	MoBa
Metrioptera (*Metrioptera*) *roeseli* (Hagenbach, 1822)	+			+					+		+			EuSi	Sib
Metrioptera (*Metrioptera*) *fedtschenkoi ambitiosa* Uvarov, 1924	taxonomic status unclear			+			+	+	+					SBa	BaMe
Metrioptera (*Metrioptera*) *oblongicollis* Brunner von Wattenwyl, 1882		+		+	+	SL	+		+	+	+	+		SBa	BaMe
Metrioptera (*Metrioptera*) *helleri* Schmidt, 1998	+												+	WSt	StPl
Sepiana sepium (Yersin, 1854)				+	+		+		+	+				HoMe	HoMe/s
Pholidoptera aptera aptera (Fabricius, 1793)	?			?	needs confirmation						?			CEu	CEu
Pholidoptera aptera karnyi Ebner, 1908	+	+	+	+						+	+	+	+	NCBa	Din
Pholidoptera aptera bulgarica Mařan, 1953					?		+	+	+					M–R–BM	Mac
Pholidoptera brevipes Ramme, 1939									+					Ba–An	Thr
Pholidoptera littoralis (Fieber, 1853)	+		?						+	+				SEu	CEu
Pholidoptera macedonica Ramme, 1928						+					+	+	+	Mac	Mac
Pholidoptera buresi Mařan, 1957	taxonomic status unclear											+		Ri–P–S	Ri–Rh
Pholidoptera hoberlandti Mařan, 1957			+	taxonomic status unclear								+		Ri–P–S	Ri–Rh
Pholidoptera rhodopensis Mařan, 1953			+	taxonomic status unclear							+			Ri–P–S	Ri–Rh
Pholidoptera frivaldskyi (Herman, 1871)	+	+	+	+					+		+	+	+	SEEu	CEu
Pholidoptera fallax (Fischer, 1853)	+	+	+	+			+		+	+	+			CSEu	PoMe/e
Pholidoptera griseoaptera (De Geer, 1773)	+	+									+	+		Eu	CEu
Pholidoptera ganevi Harz, 1986				+							+			WRh	Ri–Rh
Eupholidoptera chabrieri schmidti (Fieber, 1861)										+				TrAd	PoMe/s
Eupholidoptera marani Peshev, 1960								+	+					Stru	Mac

continued

Table 1 continued

Taxa	Distribution in Bulgaria								Vegetational belts in Bulgaria						Zoogeography	
	W-STA	C-STA	RILA	W-RHO	E-RHO	BE, SL	BSEA	STRU	XER	OAK	BEE	CON	SUB	ALP	Chorotype	Origin
Eupholidoptera beybienkoi Peshev, 1962		+							+	+					CSt	StPl
Eupholidoptera smyrnensis (Brunner von Wattenwyl, 1882)					+			+	+						Ba–An	Ana
Parapholidoptera castaneoviridis (Brunner von Wattenwyl, 1882)					+				+	+	+				Ba–An	Ana
Bucephaloptera bucephala (Brunner von Wattenwyl, 1882)		+		+	+	BE	+	+	+	+					Ba–An	Ana
Psorodonotus fieberi (Fieber, 1853)	+	+	+	+		BE			+		+	+	+		NCBa	MoBa
Pachytrachis gracilis (Brunner von Wattenwyl, 1861)	+	+	+	+		+	+	+	+	+	+				CSEu	PoMe/e
Pachytrachis frater (Brunner von Wattenwyl, 1882)	needs confirmation			?					?	?					CSEu	Din
Anterastes serbicus Brunner von Wattenwyl, 1882	+	+	+	+		+					+	+	+	+	Ba–An	Ana
Pterolepis germanica (Herrich-Schäffer, 1840)	+	+		+	+	+	+	+	+	+	+				CSEu	PoMe/e
Gampsocleis glabra (Herbst, 1786)								?	+						E–WAs	Cas
Gampsocleis abbreviata Herman, 1874						SL			+						SBa	BaMe
Onconotinae																
Onconotus servillei Fischer de Waldheim, 1846									+						SEE–WA	Cas
Saginae																
Saga natoliae Serville, 1839	+			+	+	BE	+	+	+	+	+				EMe	PoMe/s
Saga cf. *hellenica* Kaltenbach, 1967										+					SBa	BaMe
Saga rammei Kaltenbach, 1965								+	+						Ma–Th	Mac
Saga campbelli campbelli Uvarov, 1921						+		+	+	+	+				Mac	Mac
Saga campbelli gracilis Kis, 1962					+				+	+					EBa	Thr
Saga pedo (Pallas, 1771)							+		+						E–WAs	Cas

BRADYPORIDAE														
Ephippigerinae														
Ephippiger ephippiger ephippiger (Fiebig, 1784)			+	+	+	+	+	+	+	+	+	+	CSEu	CEu
Ephippiger ephippiger balkanicus Andreeva, 1985	+	+	taxonomic status unclear					+	+	+		+	NBu	StPl
Ephippiger ephippiger varnensis Andreeva, 1985	taxonomic status unclear						+	+					NBl	Moe
Bradyporinae														
Bradyporus dasypus (Illiger, 1800)				+	+	+	+	+	+	+			EBa	? Moe
Callimenus macrogaster (Lefebvre, 1831)				+	+		+	+					Ba–An	? Moe
RHAPHIDOPHOROIDEA														
RHAPHIDOPHORIDAE														
Troglophilinae														
Troglophilus neglectus vlasinensis Mařan, 1958	+			+	+			+	+	subspecies status unclear			CBa	StPl
Rhaphidophorinae														
Tachycines asynamorus Adelung, 1902	accidentally introduced in the past; not recently distributed												Cos	SiTi
GRYLLOIDEA														
GRYLLIDAE														
Gryllinae														
Gryllus campestris Linnaeus, 1758	+	+	+	+	+	+	+	+	+	+	+		WPa	CE–Me
Gryllus bimaculatus De Geer, 1773	doubtful information for Bulgaria												AT–Pa	AfTr
Acheta domesticus (Linnaeus, 1758)	synanthropic in villages				+		+	+					Cos	HoMe/e
Melanogryllus desertus (Pallas, 1771)	+				+		+	+	+				WPa	HoMe/e
Tartarogryllus sandanski Andreeva, 1982							+	+					Stru	Mac
Eumodicogryllus bordigalensis (Latreille, 1804)				+				+					Tu–Me	HoMe/e
Modicogryllus frontalis (Fieber, 1844)	+			+	+		+	+					E–WAs	Cas
Modicogryllus truncatus (Tarbinsky, 1940)							+	+					SEEu	?
Gryllomorphinae														
Gryllomorpha dalmatina (Ocskay, 1832)					+		+	+					HoMe	HoMe/s
Gryllomorpha cf. *miramae* Medvedev, 1933					?		+	+					SEE–WA	? PoMe/e
Discoptila buresi Mařan, 1958					+		+	+					ER–Bl	Thr

continued

Table 1 continued

Taxa	Distribution in Bulgaria								Vegetational belts in Bulgaria						Zoogeography	
	W-STA	C-STA	RILA	W-RHO	E-RHO	BE, SL	BSEA	STRU	XER	OAK	BEE	CON	SUB	ALP	Chorotype	Origin
Nemobiinae																
Pteronemobius heydeni heydeni (Fischer, 1853)	+			+	+	? SL	+	?	+	+					CSEu	HoMe/e
Pteronemobius heydeni tartarus (Saussure, 1874)								+	+						SEE–WA	? Cas
Stenonemobius bicolor ponticus Gorochov, 1984									+						SEEu	PoMe/s
Oecanthinae																
Oecanthus pellucens (Scopoli, 1763)	+	+		+	+	BE	+	+	+	+					WPa	HoMe/e
Mogoplistidae																
Arachnocephalus vestitus Costa, 1855					+		+	+	+						SEu	HoMe/s
Myrmecophilidae																
Myrmecophilus acervorum (Panzer, [1799])				?			?		+						CSEu	CEu
Myrmecophilus myrmecophilus (Savi, 1819)[1]					+				+						SEu	HoMe/s
Gryllotalpidae																
Gryllotalpa gryllotalpa (Linnaeus, 1758) s. l.	+		+	+	+		+	+	+	+	+	species complex			WPa	CEu
Caelifera																
TETRIGOIDEA																
Tetrigidae																
Tetrix subulata (Linnaeus, 1758)	+	+		+	+		+	+	+	+	+				Hol	Si–Ne
Tetrix bolivari Saulcy, 1901	+			+	+		+	+	+						Tu–Me	HoMe/e
Tetrix ceperoi (Bolivar, 1887)				+			+	+	+						WPa	HoMe/e
Tetrix tuerki (Krauss, 1876)	+			+	+			+	+	+					CSEu	CEu
Tetrix bipunctata (Linnaeus, 1758)	+		+	+		BE			+	+	+				Pal	Si–Me
Tetrix tenuicornis (Sahlberg, 1893)	+	+	+	+	+	+	+		+	+	+	+			Pal	Si–Me
Uvarovitettix depressus (Brisout de Barneville, 1848)	+	+	+	+	+	+	+	+	+	+	+	+			Tu–Me	HoMe/e
Paratettix meridionalis (Rambur, 1838)			+		+		+	+	+			+			HoMe	HoMe/s
TRIDACTYLOIDEA																
Tridactylidae																
Xya variegata Latreille, 1809	?			?	+		?	?	+	?	?				PT–Pa	?
Xya pfaendleri (Harz, 1970)					+				+						PT–Pa	?

ACRIDOIDEA																
PAMPHAGIDAE																
Pamphaginae																
Paranocarodes straubei straubei (Fieber, 1853)							+		+						Ba–An	Ana
Paranocarodes chopardi Peshev, 1965					+				+						ERh	Thr
Paranocaracris bulgaricus bulgaricus (Ebner et Drenowski, 1930)				+		SL		+	+	+	+	+			M–R–BM	Ri–Rh
Akicerinae																
Glyphanus obtusus Fieber, 1853	does not occur in Bulgaria															
Asiotmethis limbatus (Charpentier, 1845)				+	+	SL	+	+	+						Ba–An	? Ana
ACRIDIDAE																
Catantopinae																
Podisma pedestris (Linnaeus, 1758)	+		+	+		+	?		?		+	+	+	+	EuSi	Sib
Bohemanella frigida (Boheman, 1846)		+				SL							+	+	H-OTu	SiTu
Miramella sp.			+								+				? CBa	? Din
Pseudopodisma fieberi (Scudder, 1897)	+		+	+							+	+			SEu	MoBa
Odontopodisma schmidti (Fieber, 1853)	+			+					+	+					CSEu	? Din
Odontopodisma rubripes (Ramme, 1931)	+	+		+					+	+	+				CSEu	CEu
Odontopodisma decipiens Ramme, 1951	+	+	+	+		+	+	+	+	+	+				SEEu	CEu
Pezotettix giornae (Rossi, 1794)	+	+		+	+	+	+	+	+	+	+	+			HoMe	HoMe/e
Calliptaminae																
Calliptamus italicus (Linnaeus, 1758)	+	+		+	+	+	+	+	+	+	+	+			Tu–Me	HoMe/e
Calliptamus barbarus (Costa, 1836)	+			+	+	BE	+	+	+	+					Tu–Me	HoMe/e
Paracaloptenus caloptenoides (Brunner von Wattenwyl, 1861)	+		+	+	+	+	+	+	+	+	+	+	+		EMe	PoMe/e
Cyrtacanthacridinae																
Anacridium aegyptium (Linnaeus, 1764)					+	BE	+	+	+						AT–Pa	HoMe/e
Acridinae																
Acrida ungarica (Herbst, 1786)	+	+		+	+	+	+	+	+	+					HoMe	HoMe/e
Oedipodinae																
Locusta migratoria (Linnaeus, 1758)					+		+	+	+	+					PT–Pa	PaTr
Oedaleus decorus (Germar, 1826)	+			+	+	+	+	+	+	+					AT–Pa	HoMe/e

continued

Table 1 continued

Taxa	Distribution in Bulgaria								Vegetational belts in Bulgaria						Zoogeography	
	W-STA	C-STA	RILA	W-RHO	E-RHO	BE, SL	BSEA	STRU	XER	OAK	BEE	CON	SUB	ALP	Chorotype	Origin
Psophus stridulus (Linnaeus, 1758)	+	+	+	+		+				+	+	+	+		EuSi	Sib
Celes variabilis (Pallas, 1771)				+	+			+	+						E–WAs	Cas
Oedipoda caerulescens caerulescens (Linnaeus, 1758)	+	+	+	+	+	+	+	+	+	+	+	+			Pal	Si–Me
Oedipoda caerulescens ganevi Harz, 1985	status of the subspecies unclear							+	+	+	+				Vl–M–O	Ri–Rh
Oedipoda germanica (Latreille, 1804)	+	+	+	+	+	+	+	+	+	+	+				CSEu	HoMe/e
Oedipoda miniata (Pallas, 1771)	+		+	+	+	BE		+	+	+	+				Tu–Me	HoMe/e
Sphingonotus caerulans (Linnaeus, 1767)	+		+	+	+	+	+	+	+	+					E–WAs	CEu
Acrotylus patruelis (Herrich-Schäffer, 1838)	+			+	+	BE	+	+	+						AT–Pa	AfTr
Acrotylus insubricus (Scopoli, 1786)	+	+		+	+	+	+	+	+	+	+				AT–Pa	AfTr
Acrotylus longipes (Charpentier, 1845)	+			+	+		+	+	+						AT–Pa	AfTr
Aiolopus thalassinus (Fabricius, 1781)	+			+	+	+	+	+	+						Cos	PaTr
Aiolopus strepens (Latreille, 1804)	+	+		+	+	+	+	+	+	+	+	+			AT–Pa	AfTr
Epacromius tergestinus (Charpentier, 1825)							+	+	+						Pal	Mon
Epacromius coerulipes (Ivanov, 1887)							+		+						EuSi	Sib
Platypygius crassus (Karny, 1907)							+		+						SEEu	Cas
Mecostethus parapleurus (Hagenbach, 1822)	+			+			+	+	+						EuSi	Sib
Stethophyma grossum (Linnaeus, 1758)	+		+	+					+		+	+			EuSi	Sib
Paracinema tricolor (Thunberg, 1815)							+	+	+						AT–Pa	AfTr
Duroniella laticornis (Krauss, 1909)					+				+						EMe	PoMe/s
Gomphocerinae																
Arcyptera (*Arcyptera*) *fusca* (Pallas, 1773)	+	+	+	+						+	+	+			EuSi	Sib
Arcyptera (*Pararcyptera*) *microptera* (Fischer de Waldheim, 1833)	+			+	+		+		+	+					EuSi	Sib
Pallasiella turcomana (Fischer de Waldheim, 1833)								+	+						SEE–WA	? PoMe/e
Chrysochraon dispar (Germar, [1834])	+	+		+						+	+				EuSi	Sib
Euthystira brachyptera (Ocskay, 1826)	+	+	+	+		+			+		+	+			EuSi	Sib

Dociostaurus maroccanus (Thunberg, 1815)	+		+	+	+	BE	+	+	+	+	+				Tu–Me	HoMe/e
Dociostaurus brevicollis (Eversmann, 1848)	+		+	+	+	+	+	+	+	+	+	+			SEE–WA	? Cas
Dociostaurus genei (Ocskay, 1833)								+	+						HoMe	HoMe/s
Dociostaurus tartarus Uvarov, 1921	needs confirmation							?	?						Tu–Ca	Ira
Dociostaurus kraussi (Ingenizkij, 1897)							+		+						SEE–WA	? Cas
Notostaurus anatolicus (Krauss, 1896)				+	+			+	+						EMe	PoMe/e
Omocestus viridulus (Linnaeus, 1758)	+	+	+	+							+	+	+	+	EuSi	Sib
Omocestus rufipes (Zetterstedt, 1821)	+	+	+	+	+	+	+	+	+	+	+	+			Pal	Si–Me
Omocestus minutus (Brullé, 1832)	+			+	+	BE	+	+	+	+					SEEu	PoMe/e
Omocestus petraeus (Brisout de Barneville, 1855)	+	+	+	+	+	BE	+	+	+	+	+				EuSi	Si–Me
Omocestus haemorrhoidalis (Charpentier, 1825)	+	+	+	+	+	+	+		+	+	+	+	+		EuSi	Sib
Stenobothrus crassipes (Charpentier, 1825)	+	+							+	+	+				CSEu	CEu
Stenobothrus lineatus (Panzer, 1796)	+	+	+	+		+			+	+	+	+	+		EuSi	Sib
Stenobothrus nigromaculatus (Herrich-Schäffer, 1840)	+	+	+	+		BE		?	+	+	+	+	+	+	EuSi	Sib
Stenobothrus fischeri (Eversmann, 1848)	+	+	+	+	+	SL	+	+	+	+	+	+			EuSi	Sib
Stenobothrus bulgaricus Ramme, 1933	taxonomic status unclear			+						+					WRh	Ri–Rh
Stenobothrus stigmaticus faberi Harz, 1975	+	+	+	+		BE			+	+	+	+	+		CSEu	CEu
Stenobothrus rubicundulus Kruseman et Jeekel, 1967	+	+	+	+		SL				+	+	+			CSEu	CEu
Gomphocerus sibiricus (Linnaeus, 1767)	+	+	+	+		+					+	+	+	+	EuSi	Sib
Gomphocerippus rufus (Linnaeus, 1758)	+	+	+	+						+	+	+	+		EuSi	Sib
Aeropedellus variegatus (Fischer de Waldheim, 1846)			+										+	+	H-OTu	SiTu
Myrmeleotettix maculatus (Thunberg, 1815)	+	+	+	+		+		?			+	+	+	+	EuSi	Sib
Stauroderus scalaris (Fischer de Waldheim, 1846)	+	+	+	+		+				+	+	+	+		EuSi	Sib
Chorthippus vagans (Eversmann, 1848)	+		?	?							+				CSEu	CEu
Chorthippus mollis mollis (Charpentier, 1825)	+	+	+	+	+	+				+	+	+			EuSi	Sib
Chorthippus mollis pechevi Karaman, 1975	status of the subspecies unclear						+		+						NBl	Moe

continued

Table 1 continued

Taxa	Distribution in Bulgaria								Vegetational belts in Bulgaria						Zoogeography	
	W-STA	C-STA	RILA	W-RHO	E-RHO	BE, SL	BSEA	STRU	XER	OAK	BEE	CON	SUB	ALP	Chorotype	Origin
Chorthippus brunneus (Thunberg, 1815)	+	+	+	+		+	+	+	+	+	+	+	+	+	EuSi	Sib
Chorthippus bornhalmi Harz, 1971					+			+	+	+					EMe	PoMe/s
Chorthippus biguttulus (Linnaeus, 1758)	+	+	+	+		+	+	+	+	+	+	+	+		Pal	Sib
Chorthippus porphyropterus euhedickei Helversen, 1989					+				+						Ba–An	BaMe
Chorthippus apricarius (Linnaeus, 1758)	+	+	+	+							+	+	+	+	EuSi	Sib
Chorthippus parallelus parallelus (Zetterstedt, 1821)	+	+	+	+	+	BE	+	+	+	+	+	+	+		EuSi	Sib
Chorthippus parallelus tenuis (Brullé, 1832)									+						Ba–An	BaMe
Chorthippus montanus (Charpentier, 1825)	+										+	+			EuSi	Sib
Chorthippus albomarginatus (De Geer, 1773)	+	+	+	+			+	+	+	+	+	+			EuSi	Sib
Chorthippus dorsatus (Zetterstedt, 1821)	+	+	+	+					+	+	+	+			EuSi	Sib
Chorthippus dichrous (Eversmann, 1859)	+				+		+	+	+	+					EuSi	Sib
Chorthippus loratus (Fischer de Waldheim, 1846)	+				+		+	+	+	+					EMe	Cas
Euchorthippus pulvinatus (Fischer de Waldheim, 1846)	+		+	+	+		+		+	+					Tu–Me	Cas
Euchorthippus declivus (Brisout de Barneville, 1848)	+	+	+	+	+	+	+	+	+	+	+				CSEu	HoMe/e

[1] During the printing of the present paper, the only specimen published as *Myrmecophilus myrmecophilus* from Bulgaria (Popov and Chobanov, 2004) was identified by Dr. A. Gorochov as *Myrmecophilus hirticaudus* (Fischer de Waldheim, 1846). The finding of more material will solve the problem to which species belongs this specimen.

Regions of Bulgaria: BE, Belasitsa Mts.; BSEA, Black Sea coast; C-STA, Central Stara Planina National Park; E-RHO, Eastern Rhodopes Mts.; RILA, Rila National Park; SL, Slavyanka Mts.; STRU, Middle Struma Valley; W-RHO, Western Rhodopes Mts.; W-STA, Western Stara Planina Mts.

Vegetational belts: ALP, alpine; BEE, beech; CON, coniferous; OAK, mesophilous oak; SUB, subalpine; XER, xerothermic oak.
Chorotypes: AT–Pa, Afrotropical–Palearctic; Ba–An, Balkan–Anatolian; Bel, Belasitsa; CBa, Central Balkan; CD–CP, Central Danubian–Central Predbalkan; CEu, Central European; Cos, Cosmopolitan; CSEu, Central and South European; CSr, Central Sredna Gora; CSt, Central Stara Planina; EBa, Eastern Balkan; EMe, Eastern Mediterranean; ER–Bl, Eastern Rhodopean–Black Sea coast; ERh, Eastern Rhodopean; ESt, Eastern Stara Planina; Eu, European; EuSi, Eurosiberian; E–WAs, European–Western Asian; Hol, Holarctic; HoMe, Holomediterranean; H-OTu, Holarctic Oreotundral; Mac, Macedonian; Ma–Th, Macedonian–Thracian; M–R–BM, Macedonian–Rhodopean–Black or Marmara Sea coast; M–Ri–Rh, Macedonian–Rila–Rhodopean; NBl, northern part of Bulgarian Black Sea coast; NBu, North Bulgarian; NCBa, Northern and Central Balkan; NWBa, Northwestern Balkan; Oso, Osogovo; Pal, Palearctic; PT–Pa, Paleotropical–Palearctic; Ri–P–S, Rila–Pirin–Slavyanka; SBa, Southern Balkan; SEEu, Southeastern European; SEE–WA, Southeastern European–Western Asian; SEu, South European; SoMo, Sofia Mountainous; Stra, Strandja; Stru, Middle Struma Valley; Th–Bl, Thracian–Black Sea coast; TrAd, Transadriatic; Tu–Ca, Turanian–Caucasian; Tu–Me, Turanian–Mediterranean; Tun, Tundja; Vl–M–O, Vlahina–Maleshevska–Ograzhden; WPa, Western Palearctic; WRh, Western Rhodopean; WSt, Western Stara Planina.
Origin: AdMe, Adriatomediterranean; AfTr, Afrotropical; Ana, Anatolian; BaMe, Balkan Mediterranean; Cas, Caspian; CE–Me, Central European–Mediterranean; CEu, Central European; Din, Dinaric; HoMe/e, Holomediterranean/expansive; HoMe/s, Holomediterranean/stationary; Ira, Iranian; Mac, Macedonian; MoBa, Montane Balkan; Moe, Moesian; Mon, Mongolian; PaTr, Paleotropical; PoMe/e, Pontomediterranean/expansive; PoMe/s, Pontomediterranean/stationary; Ri–Rh, Rila–Rhodopean; Sib, Siberian; Si–Me, Siberian–Mediterranean; Si–Ne, Siberian–Nearctic; SiTi, Sinotibetan; SiTu, Siberian Tundral; StPl, Stara Planina; Thr, Thracian.

The occurrence of each species in Bulgaria was also analyzed. The presence of material was checked in the collections of the National Museum of Natural History in Sofia (NMNHS) where the collection of G. Peshev is deposited. The correctness of interpretation of the published species records based on material which has not been preserved was estimated depending on the time of identification and the use of literature relevant from the current point of view.

Deleted taxa. Of the taxa given in Table 1, three species and one subspecies, whose occurrence in Bulgaria is out of the question, as well as one species, which had been introduced accidentally in the past but does not occur now in this country, have to be deleted from the list of Bulgarian fauna. *Metaplastes pulchripennis, Platycleis falx*, and *Isophya modesta modesta* (only as a subspecies) have been incorrectly identified. They are not found in the Balkan Peninsula, and the information for Bulgaria must be rejected for zoogeographical reasons. *Glyphanus obtusus* is an endemic species for the eastern part of Central Greece and the adjacent areas of Peloponnesus. The statement that the *terra typica* of the species is situated in Bulgaria is a result of an erroneous interpretation of the locality. The different case is *Tachycines asynamorus*, which was accidentally introduced in greenhouses in Sofia where it has occurred for 15 years. During the last 60 years this species has not been found in Bulgaria and should be deleted from the list of Bulgarian fauna. Along with the deleted taxa, the subfamily Rhaphidophorinae and the genera *Metaplastes* (for the present), *Glyphanus*, and *Tachycines* have to be deleted as well.

Added taxa. Two newly recorded species, which have not been published so far from Bulgaria, are *Platycleis* (*Montana*) *medvedevi* from North Bulgaria and *Eupholidoptera chabrieri* (with its subsp. *schmidti*) from Southwestern Bulgaria (Fig. 4).

Number of taxa. With the abovementioned corrections, the total number of taxa of Orthopterida in Bulgaria is 269, including 251 species and 18 additional subspecies. They are distributed by orders as follows: Embioptera, one species; Dermaptera, seven species; Mantodea, four species; Blattodea, 16 species; Isoptera, two species; Orthoptera, 221 species and 18 subspecies, or 239 taxa. This number includes 238 species and 13 additional subspecies, or 251 taxa altogether, which are recorded with certainty for the fauna of Bulgaria. The remaining taxa published for Bulgaria cannot be confirmed or have unclear taxonomic status. Their occurrence in Bulgaria, however, is possible; they are retained in the list until they will be confirmed or rejected, or until their status as distinct taxa will be clarified. Such taxa, for which occurrence in Bulgaria is doubtful, belong to three categories.

Species with doubtful locality data for Bulgaria. These are *Poecilimon macedonicus* and *Gryllus bimaculatus*, published for Bulgaria without exact localities and most likely reported as a speculation and not based of a concrete record.

Taxa needing confirmation: eight taxa (seven species and one subspecies). Blattodea and Caelifera are represented with one species each; the remaining taxa belong to the superfamily Tettigonioidea. The doubt in their correct identification arises due to lack of confirmation of the records dated 100 years ago (*Platycleis stricta*) or 50 years ago (*Platycleis tessellata* and *Pachytrachis frater*), from the record based only on a female (*Isophya brevipennis*), or from identification that seems uncertain from the current point of view (*Phyllodromica subaptera* from Blattodea, *Platycleis orina, Dociostaurus tartarus*, and *Pholidoptera aptera aptera* as a subspecies).

Taxa with unclear status. Possible synonyms are four species of *Pholidoptera* and *Stenobothrus* as well as four subspecies of *Ephippiger, Oedipoda*, and *Chorthippus*, all described from Bulgaria. This, however, must be proved through taxonomic investigations.

The next two categories include species which are a part of the Bulgarian fauna, though not autochthonous.

Introduced species. The only species accidentally introduced to Bulgaria is *Meconema meridionale*. A small population has been found recently by Chobanov (2003) in a Botanical Garden north of Varna. At the same place and time Chobanov (2003) found also small but stable introduced population of *Leptophyes punctatissima*; however, this species occur autochthonously in Western Bulgaria. In the same paper, MSc Dragan Chobanov proposes two hypotheses for the occurrence of *Barbitistes constrictus* in the Rila Mts. One of these considers this species as introduced accidentally in the beginning of the 20th century. I regard as more probable the alternative hypothesis about a relict origin of the Rila population. Concerning *Tachycines asynamorus*, accidentally introduced 70 years ago but extinct later, see above under *Deleted taxa*.

Synanthropic species. Three species of Blattodea and one species of Grylloidea belong to this category. Bulgaria has not been a part of the original range of these species. This is indisputable for the representatives of Blattodea and very likely for *Acheta domesticus*. It is possible that the native area of the latter is located near to Bulgaria. Nevertheless, it seems that this cricket species was also spread passively because (as well as three cockroache species) it does not occur in Bulgaria outside of the houses.

2·2 Taxonomic and faunistic problems

The advances of the orthopterid taxonomy in the last 20 years led to the status changes for some taxa. New species and subspecies were described, new synonyms established, some species split, and the status of some taxa elevated or downgraded. Nearly all published faunistic information for Bulgaria predates this period. Also, part of it had not been interpreted according to the knowledge available at the time of publication. Because of this, the accumulation of taxonomic problems concerning the species diversity in Bulgaria, which await the solution, is not surprising. These problems can be summarized as follows.

Taxonomic problems

1. *Validity of taxa described from Bulgaria.* Many species and subspecies from the genera *Isophya, Poecilimon, Eupholidoptera*, and *Ephippiger*, described by G. Peshev, E. Andreeva, and N. Nedelkov, need to be revised. Only few of them are likely to be distinct species.
2. *Probable synonyms.* New material of *Stenobothrus bulgaricus* is necessary to be found in its type locality. *Poecilimon belasicensis* needs to be compared with related species from adjacent countries.
3. *Revision of species groups.* The *aptera*-group of the genus *Pholidoptera* needs a revision. Three species described by Dr. Josef Mařan are very likely synonyms but due to the small number of known specimens it is not clear to which species they belong. The revision will show the systematic position of the taxon often recorded from South Bulgaria and reported as *Pholidoptera* sp. This is possibly one of the three subspecies of *Pholidoptera aptera* or *Ph. macedonica* found in South Bulgaria and North Greece.
4. *Verification of the taxonomic status.* It has to be checked whether *Metrioptera domogledi domogledi* and *M. d. arnoldi* occur sympatrically on Stara Planina; in this case they are distinct species. They were regarded for a long time as species with distinct differences in the subgenital plate in both sexes but have been downgraded to subspecies status because of the similarity in the song pattern. Another problem in the same genus concerns *M. roeseli* and *M. fedtschenkoi ambitiosa*. Recently, a doubt was expressed about the occurrence of the latter along with the former in the Balkan Peninsula. A question arises about what are the populations published as *M. f. ambitiosa*.
5. *Unclear status of subspecies.* The distinct subspecies status of *Troglophilus neglectus vlasinensis, Chorthippus parallelus tenuis, Ch. mollis pechevi*, and *Oedipoda caerulescens ganevi* has to be verified. The last two subspecies are most likely synonyms but this has to be proved. The subspecific division of *Saga campbelli* and *Callimenus macrogaster* also awaits revision.
6. *Unknown male.* For *Pholidoptera ganevi*, it is not possible to determine its systematic position in the genus and the species validity until the male will be described.

Faunistic problems

7. *Species diversity of genera and subgenera.* The genera *Ectobius* (Blattodea) and *Pachytrachis* and the subgenera *Montana* and *Tessellana* (both of the genus *Platycleis*) in Bulgaria need a revision; it has to be established which, and how many, species of each of these genus-group taxa occur in Bulgaria. The published species of *Ectobius* have not been identified according to the current

criteria (a glandular pit on the seventh tergite in males). Attention should be paid particularly to the complex *Ectobius erythronotus/E. burri*.

8. *Genus known for Bulgaria from females only*. The genus *Miramella* was reported for Bulgaria only once (Nedelkov, 1908) as *Miramella alpina*, the only species of the genus known at that time. Only females, which are unidentifiable under the current criteria, have been found. The species is now split in many species and the occurrence of four of them in Bulgaria is possible. The issue cannot be settled until the male is found.
9. *Unknown or unclear subspecies identity*. The material of *Pholidoptera macedonica* and *Callimenus macrogaster* has to be revised in order to show which, and how many, of the subspecies of these two species occur in Bulgaria. Three subspecies of *Pholidoptera aptera* are reported from Bulgaria. *Ph. a. karnyi* is common in the mountains (Fig. 6), and *Ph. a. bulgarica* is very likely also a distinct subspecies distributed at least in the lowlands. The record of *Ph. a. aptera* in Bulgaria seems dubious.
10. *Insufficiently known distribution in Bulgaria of closely related species*. Species of the genera *Modicogryllus* and *Xya* have been split in two species each but reported for Bulgaria on the basis of outdated taxonomic criteria. Along with *Modicogryllus frontalis* and *Xya variegata*, the closely related *M. truncatus* and *X. pfaendleri* are recorded in Bulgaria as well. The correct identification will eliminate the confusion and will outline the distribution of each species in Bulgaria.
11. *Species whose identification is not clear*. The recent records of *Saga* cf. *hellenica* and *Gryllomorpha* cf. *miramae* (Chobanov, 2003) needs verification on the basis of more abundant material from Bulgaria and comparative material from other countries.
12. *Doubtful records*. Two species, *Poecilimon macedonicus* and *Gryllus bimaculatus*, reported for Bulgaria (most likely on the basis of speculations), have to be searched both in collections and in nature.
13. *Probable incorrect identification*. Three species, *Phyllodromica subaptera* (Blattodea), *Isophya brevipennis*, and *Dociostaurus tartarus*, have to be searched in collections or in the reported localities. The report of *Ph. subaptera* is most likely an incorrect identification but the occurrence of this species in Bulgaria is possible. The last two species are not very likely to be found in Bulgaria.
14. *Stability of the population of an introduced species*. Future studies will show whether the small introduced population of *Meconema meridionale* in the Botanical Garden near Varna (Chobanov, 2003) will survive and whether it will expand from this locality (the only in Bulgaria).
15. *Sibling species distinguished genetically*. *Gryllotalpa gryllotalpa* is considered now a complex of species with a different number of chromosomes. Their taxonomic status in most cases is not clarified. Whether one or more species, and which ones exactly, occur in Bulgaria can be decided only after genetic studies.

3 Distribution of Species

3·1 Horizontal distribution

Nine regions with well, and relatively evenly, explored fauna of Orthoptera are selected to illustrate its distribution in Bulgaria. One or two reviews for each of these regions have been published in the last five years (Central Stara Planina National Park[1], Rila National Park, Kresna Gorge, Eastern Rhodopes Mts.) or 30–40 years earlier (Belasitsa Mts., Slavyanka Mts., Western Stara Planina Mts., Black Sea coast, Western Rhodopes).

These regions have been selected to cover the maximal difference in the ecological conditions which is the premise for variation in faunistic diversity. The nine selected regions house 91% of the orthopterid species in Bulgaria (Table 1). The majority of the remaining 23 species occur in Pirin and Strandja; these mountain ranges are not included since they are insufficiently studied. The selected regions include the high mountain ranges with coldest and most humid climate (Rila, Stara Planina, Western Rhodopes) and the three regions with warmest and driest climate in this country: the Middle Struma Valley with the hottest place in Bulgaria, the sea coast, and a low mountain (Eastern Rhodopes). On the other hand, the selected regions include protected (Rila, Central Stara Planina) as well as unprotected (Belasitsa, Black Sea coast) territories, and the areas where limestone and karst forms are present (Slavyanka, most part of Stara Planina) or absent (Rila, Struma Valley).

The main reasons for the selection of these regions are the following:

Western and Central Stara Planina: their position on the dispersal route of Eurosiberian and Central European species to the Balkan Peninsula from the Carpathians; one of the two main routes of this dispersal;

Rila: the best developed alpine belt, most strongly manifested mountain climate;

Western Rhodopes: the best example of xerothermic and mountain fauna represented in a mountain range;

Belasitsa and Slavyanka: the only mountains in Bulgaria located as islands in the Mediterranean Subregion;

Middle Struma, Eastern Rhodopes, Black Sea coast: the warmest regions with richly represented Mediterranean fauna.

The data on the species distribution were compiled using all available literature for Bulgaria. It includes major reviews (MR) as well as additional references (AR)

[1] The official name of the park in English is Central Balkan National Park. In the present paper, the adjective "Balkan" is used only as an attribute for the Balkan Peninsula (Balkan Mediterranean, Montane Balkan, Balkan–Anatolian, Central Balkan etc.). Balkan is the Turkish name and Stara Planina is the Bulgarian name of the mountain range dividing North and South Bulgaria. In order to avoid confusion, for the mountain range we use only the name Stara Planina, e.g. Central Stara Planina (Mts.), Eastern Stara Planina (Mts.), etc., and for the park, Central Stara Planina National Park.

published earlier and later than these reviews, which list additional species from the respective region not mentioned in the major references. Below is the list of this literature.

Western Stara Planina. MR: Peschev (1970), Pešev (1974). AR: Buresch and Peschev (1957), Popov (1958), Peshev (1980, 1985), Andreeva (1985), Schmidt (1998).

Central Stara Planina National Park. MR: Popov (2000a), Popov et al. (2000). AR: Pešev (1990), Popov (unpubl.).

Rila National Park. MR: Popov (2000b).

Western Rhodopes. MR: Pechev (1964, 1975). AR: Nedelkov (1908), Buresch (1939), Buresch and Peschev (1955, 1957, 1958), Bazyluk (1961), Popov (1970), Harz (1984, 1985b, 1986), Pešev (1990), Popov (1998).

Eastern Rhodopes. MR: Popov and Chobanov (2004).

Belasitsa. MR: Pešev (1962b). AR: Buresch and Peschev (1957), Pešev (1962a), Andreeva (1980), Peshev (1980), Peshev and Andreeva (1986), Popov (1997a), Popov (unpubl.).

Slavyanka. MR: Pešev and Mařan (1963). AR: Drenowski (1936, 1938), Buresch and Peschev (1957), Peshev (1980, 1981), Peshev and Andreeva (1986), Popov (unpubl.).

Black Sea coast. MR: Peshev and Djingova (1974). AR: Drenowski (1938), Buresch (1939), Drensky (1942), Ramme (1951), Mařan (1953), Buresch and Peschev (1957), Mařan (1958), Bazyluk (1961), Popov (1970), Harz (1975), Karaman (1975), Andreeva (1985), Pešev (1990), Chobanov (2003), Popov (unpubl.).

Middle Struma Valley. MR: Popov et al. (2001), only for Kresna Gorge. AR: Kaltenbach (1967), Harz (1975), Andreeva (1982), Popov and Ganev (1983), Harz (1985a, 1985b, 1986), Peshev and Andreeva (1986), Gorochov and Llorente (2001), Chobanov (2003).

The species reported in the major reviews, which occur next to but not inside the certain region, were excluded after an analysis. The information on species distribution is complete only for Kresna Gorge and the Eastern Rhodopes because the entire rich collection of the NMNHS from these regions was revised and identified (Popov et al., 2001; Popov and Chobanov, 2004). The faunistic lists in the oldest reviews, namely these for Belasitsa (Pešev, 1962b) and Slavyanka (Pešev and Mařan, 1963), are corrected. The species typical only for the lowlands but not occurring in the mountain proper are deleted from these lists; the number of such species is 34 for Belasitsa, and 20 for Slavyanka. We removed also five species characteristic only for Strandja from the list of the Black Sea coast species by Peshev and Djingova (1974). The species added on the basis of publications other than the main reviews are most numerous for the Western Rhodopes (40 species), Middle Struma Valley (outside the Kresna Gorge) (34 species), and Black Sea coast (26 species).

Share of the fauna of the regions compared to the fauna of Bulgaria. Exactly 50% of the Bulgarian taxa of Orthopterida occur in the Western Rhodopes. The reason for such richness is the abovementioned combination of xerothermic and mountain fauna. In all other regions, less than 50% of the Bulgarian species are recorded. Only the orders with the lowest species diversity are represented by all their Bulgarian species in some of the selected regions: Mantodea (four species) in Struma Valley and Eastern Rhodopes, Isoptera (two species) on the Black Sea coast, and Embioptera (one species) in Struma Valley and on the Black Sea coast. All three orders include only thermophilous species. Blattodea and Dermaptera are represented in each region at most by ca. 50% of their species. Their distribution in Bulgaria, however, is not yet well studied. Five small orders taken together have the highest representation in the Struma Valley and the Black Sea coast (53% of the Bulgarian taxa). Orthoptera has its highest representation in the Western Rhodopes (51% of the Bulgarian taxa).

Among the suborders, the Western Rhodopes are ranked first for the representation of Ensifera (39% of the Bulgarian taxa), as well as Caelifera (71%). This rich fauna of Orthoptera is due to the diverse habitats and the large area of the Western Rhodopes. The thermophilous and xerophilous fauna is concentrated in the valleys of the rivers in the northern part of the range (Chepinska Reka, Vacha, Chepelarska Reka), which flow from south to north. Before entering the Thracian Lowland and joining the Maritsa River, these small rivers run through mountain gorges, which provide suitable refuges against the prevailing westerly winds for such species as *Tylopsis lilifolia, Sepiana sepium, Saga natoliae* (Ensifera), *Asiotmethis limbatus, Acrotylus patruelis*, and *A. longipes* (Caelifera). The mountain fauna in the same region includes the inhabitants of the forest belts up to the coniferous belt. The subalpine zone occupies a limited area and is present only on Golyam Perelik Peak, where such mountain species occur as *Isophya rhodopensis, Polysarcus denticauda, Psorodonotus fieberi* (Fig. 6), *Anterastes serbicus* (Ensifera), *Podisma pedestris, Omocestus viridulus, Myrmeleotettix maculatus* (Fig. 2), *Gomphocerus sibiricus* (Caelifera). The considerable difference in the share of Western Rhodopean Ensifera and Caelifera compared to all Bulgarian species is due to the broader ecological plasticity of most Gomphocerinae, Oedipodinae, and Tetrigidae (Caelifera), which are widespread in Bulgaria, on one hand, and to the endemism and often sparse populations of many Phaneropteridae and Tettigoniidae (Ensifera), which are scantily represented in Bulgaria, on the other hand.

Among the families with high species richness, the same (and highest) share of Phaneropteridae (30%) and Tettigoniidae (41%) compared to their total Bulgarian fauna is established in the Western Rhodopes and Western Stara Planina. The explanation for the species richness of Western Stara Planina is the same as for the Western Rhodopes. There are refugia containing Submediterranean (expansive Mediterranean) fauna along Western Stara Planina such as the foot of this mountain range near Belogradchik and the Iskar Gorge. Thermophilous species such as *Phaneroptera nana, Leptophyes albovittata* (Phaneropteridae), *Platycleis affinis, Pholidoptera fallax* (Tettigoniidae) occur there. The peaks of Sveti Nikolska Mts. and Berkovska Mts. (parts of Western Stara Planina) are inhabited by typical mountain taxa such as *Barbitistes serricauda* (Fig. 2), *Polysarcus denticauda,*

Poecilimon affinis affinis (Phaneropteridae), *Metrioptera domogledi domogledi, M. helleri* (Fig. 6), and *Pholidoptera frivaldskyi* (Tettigoniidae).

Among the subfamilies with high species richness, the Oedipodinae are best represented in Struma Valley, and Gomphocerinae, in the Western Stara Planina by the same share (76%) of their total Bulgarian fauna. The Struma Valley is inhabited by both xerophilous (*Celes variabilis, Sphingonotus caerulans*, three species of *Oedipoda*, and three species of *Acrotylus*) and hygrophilous (*Locusta migratoria, Aiolopus thalassinus, Mecostethus parapleurus*, and *Paracinema tricolor*) Oedipodinae but all these species are thermophilous.

The lowest relative share of the Bulgarian fauna for the suborders Ensifera and Caelifera is observed in Slavyanka with 21% and 35%, respectively. For the subfamilies, this share is the lowest for Tettigoniinae in Belasitsa (27%), for Oedipodinae, in the Central Stara Planina National Park (24%), and for Gomphocerinae, in Slavyanka (31%). Slavyanka and Belasitsa, although closely located, are rather different in geological structure and vegetation. For that reason only 60% of the total number of orthopterid species found there inhabit both mountains. The low representation for the two mountains is due to their small area. The reasons for the low species diversity of Oedipodinae in the Central Stara Planina National Park are the lack of suitable habitats for hygrophilous and psammophilous species and the relatively high altitude of large parts of this park.

Share of the main systematic groups compared to the entire orthopterid fauna of the regions. As main systematic groups here we consider the suborders, families, and subfamilies with high species richness. The share of Ensifera in the regional orthopterid fauna is the lowest in the Rila National Park (37%). This is due to the small share of Grylloidea and lack of the families Meconematidae and Conocephalidae in the fauna of Rila. On one hand, only two most common species of Grylloidea, *Gryllus campestris* and *Gryllotalpa gryllotalpa* s.l., occur in this mountain range. On the other hand, Rila is the only region where Meconematidae and Conocephalidae are not found. These facts can be explained by the high altitude of the lower border of the park and mountain range, and by the absence of wetlands typical for Conocephalidae, at the foothills of Rila. In other regions, Ensifera constitute 41–45% of all orthopterids.

Caelifera have the highest share in Western Stara Planina (54% of all orthopterids), due to the higher number of Gomphocerinae species (especially of the genus *Chorthippus*) than in any other territory. *Ch. montanus* and *Ch. vagans* are known in Bulgaria only from this region, but further verification of these data is necessary.

Among the families rich in species, the differences between the regions are insignificant. The highest share of Phaneropteridae is observed in the Central Stara Planina National Park (17%), and the lowest, in the Middle Struma Valley (9%). The share of Tettigoniidae is the highest in Slavyanka (22%), and the lowest, in the Black Sea coast (16%).

More important are the differences in distribution of Oedipodinae and Gomphocerinae. The poorest representation of Oedipodinae compared to all species is in the national parks of Central Stara Planina (6%) and Rila (7%). The Struma Valley

(14%) and the Black Sea coast and Belasitsa (13% each) are at the other extreme. Gomphocerinae have their minimal share in the Black Sea coast (15%), in the Eastern Rhodopes (16%), and in the Struma Valley (17%); and the maximal share in the national parks of Rila (33%) and Central Stara Planina (32%). The share of these two subfamilies in the selected regions appears to depend not on the hygrophily of the species but on their thermophily. The ratio of hygrophilous, mesophilous, and xerophilous species does not vary considerably among the regions. According to their thermophily, however, these two subfamilies express the opposing trends. The thermophilous species prevail in Oedipodinae, and the cold-tolerant ones, in Gomphocerinae. The regional distribution of species belonging to these two subfamilies is not surprising, keeping in mind the altitude of each region.

Species characteristic for certain regions. The most important characteristic of a zoogeographical or biogeographical region is the complex of endemic (subendemic) species found only (or nearly only) in this region. In addition, it is important to analyze other species, which for some reason (relict range, localities in the periphery of their range, isolated records of very rare stenotopic species, etc.) are restricted to a certain area within this territory. Below, we list the species characteristic for nine selected regions of Bulgaria; the species found only in the particular region are listed under (A), and those found in this region and in one other region, under (B).

1. ***Western Stara Planina***.

(A): *Metrioptera helleri* (Fig. 6), *Isophya miksici* (both local endemics), and *Chorthippus montanus* (the material on this species needs to be revised).

(B): *Metrioptera domogledi domogledi* (could occur only in this region), *Isophya modestior* (also in Vitosha Mts.), *Leptophyes discoidalis* (also in Kresna Gorge), *Odontopodisma schmidti* (also in Western Rhodopes), *Poecilimon affinis affinis, Pholidoptera littoralis*, and *Chorthippus vagans* (the last three species with doubtful data for other regions).

2. ***Central Stara Planina National Park***.

(A): *Isophya obtusa* (a Balkan endemic), *I. pravdini pravdini*, and *I. p. bazyluki* (an endemic species of Stara Planina with two local endemic subspecies in this region).

(B): *Eupholidoptera beybienkoi* (currently considered a local endemic species; also in the Central Predbalkan; the taxonomic status of this species needs revision).

3. ***Rila National Park***.

(A): *Miramella* sp. (the genus only in Rila as well), *Barbitistes constrictus* (a relict population), *Poecilimon affinis rilensis* (an endemic subspecies of Rila), *Pholidoptera hoberlandti*, and *Ph. rhodopensis* (both are local endemics; the taxonomic status of the last three taxa needs revision).

(B): *Phyllodromica carniolica* (Blattodea; also in Vitosha Mts.), *Aeropedellus variegatus* (Fig. 1), and *Isophya bureschi* (both also in Pirin Mts.).

4. ***Western Rhodopes***.

(A): *Isophya rhodopensis, Pholidoptera ganevi*, and *Stenobothrus bulgaricus* (all three are local endemics; the taxonomic status of the latter is unclear).

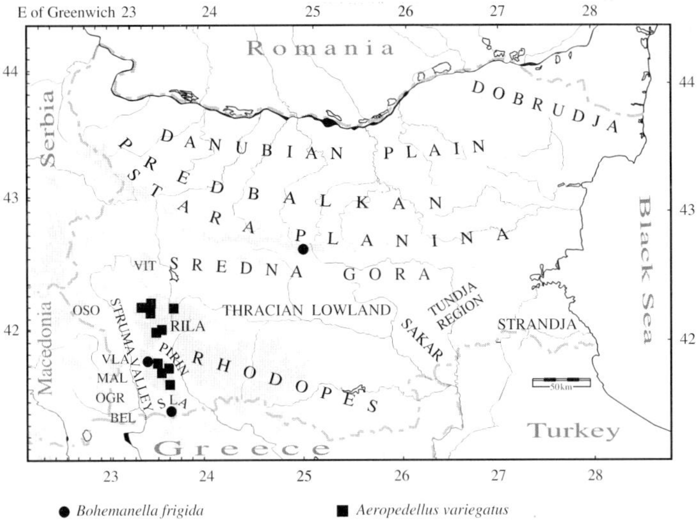

Fig. 1 Distribution in Bulgaria of the oreotundral species of Orthopterida: *Bohemanella frigida* and *Aeropedellus variegatus*. The main geographical regions are named. Abbreviations: BEL, Belasitsa Mts.; MAL, Maleshevska Mts.; OGR, Ograzhden Mts.; OSO, Osogovo Mts.; SLA, Slavyanka Mts.; VIT, Vitosha Mts.; VLA, Vlahina Mts. Areas above 1200 m altitude are shaded.

(B): *Odontopodisma schmidti* (also in Western Stara Planina), *Eumodicogryllus bordigalensis* (also in Sofia), and *Anechura bipunctata* (Dermaptera; also in Rila Mts.; the last two species are placed in this category provisorily; their distribution in Bulgaria is insufficiently known).

5. ***Eastern Rhodopes***.

(A): *Paranocarodes chopardi* (a local endemic of the Bulgarian and Greek parts of the mountain range; see Figs. 10 and 11), *Duroniella laticornis* (the only localities of the species in Europe are in this region; see Fig. 4), *Myrmecophilus myrmecophilus*, *Xya pfaendleri*, and *Chorthippus porphyropterus* (the last three species placed in this category provisorily; at least the last two species are very likely widespread in Bulgaria but their distribution is not studied).

(B): *Iris oratoria* (Mantodea) (Fig. 4), *Eupholidoptera smyrnensis* (both also in Struma Valley), *Saga campbelli gracilis* (also in Sakar Mts.; see Figs. 7 and 8), *Discoptila buresi* (also in the Black Sea coast), and *Chorthippus bornhalmi* (also in Ograzhden Mts.; the last species placed in this category provisorily; most likely widespread in Bulgaria but its distribution is not studied).

6. ***Belasitsa***.

(A): *Poecilimon kisi*, *P. belasicensis*, *P. mistshenkoi tinkae* (all currently believed to be local endemics), and *Platycleis ebneri* (a Balkan endemic).

(B): *Pholidoptera macedonica* (also in Slavyanka Mts.) and *Saga campbelli campbelli* (also in Struma Valley with the adjacent part of Slavyanka Mts.; see Fig. 8).

7. ***Slavyanka***.

(A): *Poecilimon mistshenkoi marzani* (a local endemic subspecies).

(B): *Isophya kisi* (also in Pirin Mts.), *Gampsocleis abbreviata* (also in Sakar–Tundja Region), *Saga campbelli campbelli* (Fig. 8), and *Pholidoptera macedonica* (about the last two taxa see above under Belasitsa).

8. ***Black Sea coast***.

(A): *Anisolabis maritima* (Dermaptera), *Kalotermes flavicollis* (Isoptera) (Fig. 4), *Epacromius coerulipes, Platypygius crassus, Dociostaurus kraussi*, and (introduced) *Meconema meridionale*.

(B): *Haploembia solieri* (Embioptera) (Fig. 4), *Epacromius tergestinus, Modicogryllus truncatus* (also in Struma Valley; the distribution of the latter in Bulgaria is insufficiently known), *Isophya modesta* (also in the adjacent parts of Dobrudja; see Fig. 6), *Discoptila buresi* (also in the Eastern Rhodopes), *Paranocarodes straubei* (also in the adjacent part of Strandja Mts.; an isolated population near Sliven has an uncertain taxonomic status; see Fig. 11), and *Forficula smyrnensis* (Dermaptera; also in the adjacent parts of Strandja Mts. and in Ograzhden Mts.).

9. ***Struma Valley***.

(A): *Apterygida media* (Dermaptera), *Polyphaga aegyptiaca* (Blattodea), *Isophya andreevae* (a local endemic), *Platycleis macedonica* (a Balkan endemic), *Andreiniimon nuptialis, Pteronemobius heydeni tartarus, Pallasiella turcomana, Eupholidoptera marani, Tartarogryllus sandanski* (two last species are local endemics; their taxonomic status needs revision), and *Dociostaurus tartarus* (the material on this species needs to be revised).

(B): *Dociostaurus genei* (Fig. 4), *Gryllomorpha* cf. *miramae* (both species also with doubtful data for the Rhodopes; perhaps occur only in the Struma Valley), *Leptophyes discoidalis* (also in Western Stara Planina), *Iris oratoria* (Mantodea) (Fig. 4), *Eupholidoptera smyrnensis* (both also in the Eastern Rhodopes), *Saga rammei* (also near Popovitsa in the Thracian Lowland, see Fig. 8), *Haploembia solieri* (Embioptera) (Fig. 4), *Epacromius tergestinus, Modicogryllus truncatus* (about the last three taxa see above under Black Sea coast), and *Saga campbelli campbelli* (see above under Belasitsa, Fig. 8).

3·2 Vertical distribution

Vertical distribution of species is described along the altitudinal vegetation belts, which are commonly accepted by the botanists for Bulgaria. According to the localities and ecological requirements of every species, it is estimated to which of the belts each record belongs. The vegetation belts used here are the following: *xerothermic oak forests*: Quercetum; *mesophilous oak and mixed forests*:

Quercetum, Carpinetum; *beech forests*: Fagetum; *coniferous forests*: Pinetum (*Pinus sylvestris*), Piceetum; *subalpine vegetation*: Pinetum (*Pinus mugo*), Juniperetum; and *alpine vegetation*: Caricetum, Seslerietum. The total number of taxa recorded for each belt is shown in Table 2. Not surprisingly, these numbers decrease from the lowest to the highest altitudinal belt.

All orders, suborders, superfamilies, families, and subfamilies of Orthopterida recorded for Bulgaria are found in the xerothermic oak belt (Table 1). Only 13 out of 112 Bulgarian orthopterid genera are absent from that lowest belt. One of these genera, *Apterygida*, belongs to Dermaptera and the rest to Orthoptera. Twelve of these genera are represented in the Bulgarian fauna with one species each, and only *Barbitistes*, with two species. The xerothermic belt is inhabited by 202 species and subspecies (over 75% of the Bulgarian taxa). In every other belt, less than 50% of the Bulgarian taxa are found (Table 2). This richness is due to a considerable share of southern thermophilous elements in the Bulgarian orthopterid fauna, which are found only in the lowlands, and often only in South Bulgaria or even only in its southernmost and warmest parts. This trend is illustrated by a strong decrease of the species number per area unit in the direction Greece–Bulgaria–Romania–Hungary–Slovakia–Poland. Many examples of such species can be easily found in Table 1.

Only in the xerothermic belt occur the following taxa: the order Embioptera; the families Polyphagidae, Kalotermitidae, Mogoplistidae, Myrmecophilidae, and Anisolabididae; the subfamilies Copiphorinae, Onconotinae, Gryllomorphinae, Akicerinae, and Cyrtacanthacridinae (Table 1); and 81 species and subspecies (Table 3). The abovementioned order, families, and subfamilies, however, are represented in Bulgaria only by one species each, except Gryllomorphinae (three species). Among the species-rich superfamilies of Orthoptera, the share of taxa inhabiting only the xerothermic belt as compared to all Bulgarian taxa varies from 27% (Tettigonioidea and Acridoidea) to 68% (Grylloidea).

The number of taxa in the mesophilous oak belt and the beech belt is quite close (Table 2). However, only 66% of the taxa present in the beech belt are found also in the mesophilous oak belt (78 of 119 taxa). Among the other pairs of adjoining belts, this share varies from 79% for the subalpine and coniferous belts to 86%

Table 2 Vertical distribution of Orthopterida in Bulgaria by vegetation belts

Vegetation belt	Number of taxa	Share from all Bulgarian taxa (%)
xerothermic oak forests	202	76
mesophilous oak forests	124	46
beech forests	119	45
coniferous forests	82	31
subalpine vegetation	43	16
alpine vegetation	10	4

Table 3 Upper limit of vertical distribution of Orthopterida in Bulgaria by vegetation belts

Vegetation belt	Number of taxa	Share from all Bulgarian taxa (%)
xerothermic oak forests	81	30
mesophilous oak forests	46	17
beech forests	49	18
coniferous forests	48	18
subalpine vegetation	33	12
alpine vegetation	10	4

for the mesophilous and xerothermic oak belts and even 100% for the alpine and subalpine belts, i.e. all species of the alpine belt occur also in the subalpine belt. From another perspective, the taxa found in the beech belt constitute 45% of all Bulgarian orthopterids, and the species common between the beech belt and the mesophilous oak belt constitute only 29% of Bulgarian fauna. For other pairs of adjoining belts, the difference between these shares is only 3–6%. Therefore, the boundary between the mesophilous oak belt and the beech belt has to be accepted as an important barrier for the dispersal of the orthopterid fauna. This boundary runs at 900–1000 m a.s.l. in the major part of Bulgarian mountains except the southernmost ones. Below this boundary one finds the warm and dry oak forest; above it, the cold and humid beech forest.

The maximum similarity of the faunistic diversity (number of common species) is found in the xerothermic and mesophilous oak belts as well in the beech and coniferous belts. In the first case, there are 107 common taxa (86% of 124 taxa found in the mesophilous oak belt). In the second case, there are 69 common taxa (84% of 82 species found in the coniferous belt). This similarity is due to the rather similar conditions in temperature and humidity in the xerothermic oak and mesophilous oak belts. The conditions in the beech and coniferous belts are also similar, although very different from those in the oak belts. Table 3 shows the close number (46 to 49) of species and subspecies with the upper limit of distribution in the mesophilous oak, beech, and coniferous belts.

The comparison of the number of taxa found only in a single vegetation belt (Table 4) indicates a very high number of species and subspecies characteristic for the xerothermic oak belt. This can be explained by the rich representation of Mediterranean orthopterid species in Bulgaria. The higher specificity of taxa in the beech belt compared to the other belts is due to the wide distribution of the beech forests in Stara Planina, Sredna Gora, and in the smaller Bulgarian mountains. At the same time, the subalpine and alpine belts are limited in area, and a developed coniferous belt exists only in about the half of the high mountain ranges in Bulgaria (Rila, Pirin, Western Rhodopes, Vitosha, and Slavyanka).

The timberline (the boundary between the coniferous and subalpine belts), also represents an important border for the species distribution. It runs in the Bulgarian

Table 4 Taxa of Orthopterida in Bulgaria, found only in one vegetation belt (stenotopic in vertical distribution)

Vegetation belt	Number of taxa
xerothermic oak forests	81
mesophilous oak forests	6
beech forests	12
coniferous forests	7
subalpine vegetation	6
alpine vegetation	–

mountains usually at the altitude of 1800–2000 m a.s.l.; rarely, single coniferous trees reach 2200 m. The subalpine belt is developed in the eight highest mountain ranges in Bulgaria: Rila, Pirin, Stara Planina (in its western and central parts), Vitosha, Osogovo, Belasitsa, Slavyanka, and Western Rhodopes (only around the peaks of Golyam Perelik and Syutkya). The alpine belt is found only in Rila and Pirin above 2500 m and around Botev Peak in Central Stara Planina. Eight species of Bulgarian orthopterids are found entirely above the timberline. Two of them occur in the subalpine and alpine belts: these are the only species with Arctoalpine distribution and representatives of oreotundral fauna among Bulgarian Orthoptera, *Bohemanella frigida* and *Aeropedellus variegatus* (Fig. 1). The remaining species have been found only in the subalpine belt. The timberline in Belasitsa is situated on 1700 m; *Poecilimon mistshenkoi tinkae, P. kisi*, and *P. belasicensis* occur from this altitude up to 1900 m. *P. pechevi* is found in the treeless high mountain zone of Vlahina Mts. above 1600 m, and *Metrioptera helleri*, above the timberline on Todorini Kukli Peak in Western Stara Planina at 1640 m (Fig. 6). All these taxa, characteristic only for the subalpine belt, are local endemics. *Isophya brevipennis* is recorded in the subalpine belt of Pirin (Köhler, 1988) but its occurrence in Bulgaria needs confirmation since this record is based only on a female.

All species inhabiting the alpine belt occur also in the subalpine belt. Eight of the species, distributed together above and below the timberline, occur up to the alpine belt, and 17 taxa occur in the subalpine belt and below, many of them in all belts. These are eurytopic species with broad ecological plasticity. Typical representatives reaching the alpine belt are *Stenobothrus nigromaculatus* (from 600 m up to 2600 m) and *Chorthippus brunneus* (0–2600 m) (the group of species of the latter needs revision). More species occur in all belts except the alpine one; typical examples are *Isophya speciosa, Poecilimon thoracicus, Omocestus haemorrhoidalis*, and *Chorthippus parallelus*, distributed from the seashore up to 2300–2500 m.

The maximum altitude of occurrence of the orthopterid orders in Bulgaria is shown in Table 5. For the orders found in Bulgaria also in the mountains, the maximum altitude is recorded in the south of the country (Pirin Mts. and Slavyanka Mts). These two mountain ranges are formed of limestone and marble, and therefore

Table 5 Highest altitudinal occurrence of the orthopterid orders in Bulgaria

Order	Highest altitude (m)	Region	Species
Orthoptera	2850	Pirin	*Anterastes serbicus*
Orthoptera	2850	Pirin	*Bohemanella frigida*
Orthoptera	2850	Pirin	*Gomphocerus sibiricus*
Orthoptera	2850 (2900)	Pirin (Rila)	*Aeropedellus variegatus*
Blattodea	2000	Rila, Pirin, Slavyanka	*Ectobius balcani*
Mantodea	1650	Slavyanka	*Mantis religiosa*
Dermaptera	1600	Slavyanka	*Forficula auricularia*
Isoptera	1000	Stara Planina	*Reticulitermes lucifugus*
Embioptera	200	Kresna Gorge	*Haploembia solieri*

characterized with higher temperatures. Isoptera and Embioptera, both practically entirely tropical groups, also reach high in Bulgaria in sunny and warm localities.

4 Zoogeography

Zoogeographical identity of the Bulgarian orthopterid taxa is discussed here in two aspects: the type of distributional range (chorotype) and the suggested category of origin (i.e. center of dispersal, speciation and possible origin). Placement of each species and subspecies in a certain category is a result of the original analysis of existing ranges. In many cases, categories proposed previously for zoogeographical characterization of Orthopterida or other animal groups are not used and an original scheme is provided.

4·1 Chorotypes

The species are grouped according to their distribution into a detailed system of chorotypes; only the most similar or identical ranges fall under the same chorotype category. Although it is better to use a smaller number of categories to classify geographical ranges, the high degree of endemism in Orthoptera, and the presence of other species with restricted ranges, necessitate introduction of a large number of categories in order to interpret and compare the species distribution in detail. The chorotypes are arranged in groups according to the size of their ranges. The categories proposed here and their content were not previously used in the zoogeographical literature.

Lately, the chorotypes of Vigna Taglianti et al. (1999) for the Western Palearctic have been increasingly used in the Bulgarian zoological literature. However, their scheme does not correspond well to the typification of the ranges either for the species in the Balkan Peninsula or of those in Western Palearctic. Many examples can be given but the following two are sufficient to support this statement. The typical species with Eurosiberian distribution occur in the east to Manchuria (*Tettigonia cantans, Arcyptera microptera, Gomphocerus sibiricus, Gomphocerippus rufus, Chorthippus apricarius, Ch. montanus*, and *Ch. dorsatus*), Korea (*Psophus stridulus* and *Omocestus haemorrhoidalis*), Primorye Region (*Chrysochraon dispar*), the islands of Sakhalin and Kunashir (*Chorthippus biguttulus*), and Japan (*Phaneroptera falcata* and *Mecostethus parapleurus*). In the scheme of Vigna Taglianti et al. (1999), however, such species do not correspond to the Siberian–European chorotype but instead belong to the category of Asian–European species. The other example concerns the species with Carpathian–Balkan distribution (*Pholidoptera frivaldskyi*), Southeastern European (*Odontopodisma decipiens, Platypygius crassus*, and *Omocestus minutus*), and Southeastern European–Western Asian distribution (*Onconotus servillei, Pallasiella turcomana, Dociostaurus brevicollis*, and *D. kraussi*). According to Vigna Taglianti et al. (1999), they should belong to the Turanian–European chorotype. All these species, however, do not occur in the Turanian Subregion or in the Southwestern Europe. Some of them are steppe species such as *Onconotus servillei* (Fig. 3), *Dociostaurus brevicollis, D. kraussi*, and the halophilous *Platypygius crassus*, and their placement in the Turanian–European chorotype is incorrect. It should be mentioned also that the classification of Vigna Taglianti et al. (1999) uses many chorotype names which do not seem to be correctly formed in English.

To determine the chorotypes of Bulgarian endemic species and subspecies, the categories of Hubenov (1997) for division of Bulgaria were often used. The scheme of Dr. Zdravko Hubenov, developed in order to designate the distribution of the taxa in Bulgaria, is based on the physical-geographical structures. The boundaries of the natural geographical territorial units are outlined in this scheme very precisely but the level of subdivision does not always correspond to the similarity and difference in the species diversity. On one hand, areas similar in their faunas are separated at the first level (as regions), e.g. Central Danubian Plain and Central Predbalkan, which have similar fauna and the same endemic species *Isophya plevnensis*, or the Black Sea coast and Dobrudja that share the endemic subspecies *I. modesta longicaudata* (Fig. 6). On the other hand, areas with very different faunas are separated only at the third level (as districts), e.g. Western Rhodopes and Eastern Rhodopes as well as Osogovo Mts. and Middle Struma Valley.

We grouped the Bulgarian species of Orthopterida into 49 chorological categories (chorotypes) (Table 6). In reality, the number of chorotypes could be smaller. The more complete faunistic exploration of Bulgaria and adjacent countries could lead to the removal of some regions that will be left without a single characteristic Bulgarian endemic taxon, e.g. some mountain ranges (Vlahina, Maleshevska, Ograzhden, mountains around Sofia, Eastern Stara Planina) and of the provisional regions with complicated structure and compound names (Eastern Rhodopean–Black Sea coast, Macedonian–Rhodopean–Black or Marmara Sea coast, northern part of Bulgarian Black Sea coast). In addition, some taxa could be found to be synonyms of species and subspecies from Bulgaria or adjacent countries. As a result, the number of

Table 6 Chorotypes of Orthopterida in Bulgaria

Chorotypes	Number of taxa	Share from all Bulgarian taxa (%)
Holarctic Oreotundral	2	0.7
Cosmopolitan	10	3.7
Holarctic	1	0.4
Palearctic	9	3.3
Western Palearctic	5	1.86
Eurosiberian	33	12.2
European–Western Asian	8	3.0
European	6	2.2
Central European	3	1.1
Central and South European	29	10.7
South European	5	1.86
Southeastern European	6	2.2
Southeastern European–Western Asian	7	2.6
Holomediterranean	15	5.6
Transadriatic	4	1.5
Northwestern Balkan	4	1.5
Northern and Central Balkan	5	1.86
Central Balkan	4	1.5
North Bulgarian	1	0.4
Central Danubian–Central Predbalkan	1	0.4
northern part of Bulgarian Black Sea Coast	2	0.7
Western Stara Planina	2	0.7
Central Stara Planina	3	1.1
Eastern Stara Planina	1	0.4
Sofia Mountainous	1	0.4
Central Sredna Gora	1	0.4
Osogovo	1	0.4
Vlahina–Maleshevska–Ograzhden	3	1.1
Belasitsa	3	1.1
Rila–Pirin–Slavyanka	9	3.3
Western Rhodopean	3	1.1
Eastern Rhodopean	1	0.4
Eastern Rhodopean–Black Sea Coast	1	0.4
Thracian–Black Sea Coast	2	0.7
Tundja	1	0.4
Strandja	2	0.7
Middle Struma Valley	3	1.1
Eastern Balkan	5	1.86
Macedonian	4	1.5
Macedonian–Rila–Rhodopean	2	0.7
Macedonian–Rhodopean–Black or Marmara Sea Coast	2	0.7
Macedonian–Thracian	1	0.4
Southern Balkan	5	1.86
Balkan–Anatolian	13	4.8
Eastern Mediterranean	16	5.9

continued

Table 6 continued

Chorotypes	Number of taxa	Share from all Bulgarian taxa (%)
Turanian–Mediterranean	11	4.1
Turanian–Caucasian	1	0.4
Afrotropical–Palearctic	9	3.3
Paleotropical–Palearctic	4	1.5

chorotypes as well as the number of taxa could be reduced. Furthermore, one of the categories in the list is represented by a sole species, which is very likely incorrectly identified.

Most numerous are the taxa with Eurosiberian distribution (Table 6), although the thermophilous and southern species in general prevail among the orthopterid insects in Bulgaria. That is due to the fact that the Eurosiberian species form a homogeneous group with similar ranges while the Mediterranean species are scattered among a large number of chorotypes. Ranked by the species number, the next five places after the Eurosiberian taxa are taken by chorotypes with thermophilous fauna (from Central and South European to Turanian–Mediterranean).

The ranges of 244 taxa, which are classified in 44 chorotypes, fall entirely within the borders of the Palearctic. The remaining 26 species recorded in Bulgaria now or in the past belong to five chorotypes. There are three Holarctic or Holarctic Oreotundral species whose ranges do not exceed the borders of the Holarctic. The species of Cosmopolitan, Paleotropical–Palearctic, and Afrotropical–Palearctic chorotypes are represented in Bulgaria by a total of 23 species; they are distributed in the Holarctic and beyond its borders.

The chorotypes listed in Table 6 can be arranged in three groups according to the range size. The group with wide ranges includes chorotypes reaching in the east to the Far East or Siberia, as well as those exceeding the borders of the Palearctic. The group with medium sized ranges unifies chorotypes covering the entire Europe, or Central and South Europe, and adjacent regions in the east to the Western or Central Asia. The group with small ranges consists of chorotypes including only South Europe or parts of it and Anatolia. The first group of widespread species includes only seven chorotypes (from the Cosmopolitan to the Eurosiberian) with 68 species (25.2% of all Bulgarian species). The other two groups include the same number of taxa (each has 101 species and subspecies, or 37.4% of the Bulgarian fauna) but a higher number of chorotypes. The group with medium-sized ranges consists of 10 chorotypes (from the Western Palearctic to the Central and South European), and the group with small-sized ranges unifies 32 chorotypes (65% of all 49 chorotypes in the list). This division reflects the fact that the species with restricted distribution predominate in the Bulgarian orthopterid fauna.

The Cosmopolitan chorotype includes three groups of species. Three species of Blattodea and *Acheta domesticus* are synanthropic. Four species of Dermaptera are also introduced accidentally, mostly by cargo boats, to all parts of the world. Due to their ecological plasticity, they have adapted to a great variety of conditions in all or nearly all continents. Only *Aiolopus thalassinus* has a Cosmopolitan (in fact, Subcosmopolitan) range, which is not a result of an anthropogenic introduction. It occurs almost everywhere in the East Hemisphere except the northernmost regions.

Only *Tetrix subulata* has a Holarctic distribution. The almost entire absence of Holarctic species among the Orthopterida in Bulgaria is impressive when compared to other animal groups.

The distinction between the distribution of Palearctic and Eurosiberian species is rather small, particularly in the Asian part of their ranges.

The European part of the range of the European–Western Asian chorotype includes the entire Europe, e.g. *Conocephalus dorsalis* and *Sphingonotus caerulans*, or only Central and South Europe, e.g. *Gampsocleis glabra* and *Saga pedo*.

Species found only in Europe, e.g. *Apterygida media* (Dermaptera) and *Meconema thalassinum*, as well as species found in Europe and Anatolia, e.g. *Pholidoptera griseoaptera*, are interpreted here as European. The land bridge between the Balkan Peninsula and Anatolia was severed only ca. 10,000 years ago, and the modern sea straits are not a true barrier preventing the mixing of the two faunas. Thus, there is no principal difference between species occurring in Europe plus Anatolia, or only in Europe.

As Transadriatic, we characterize the taxa found only in the Apennine part of Italy (the Apennine Peninsula proper, without North Italy) and in the Balkan Peninsula, e.g. *Andreiniimon nuptialis* and *Eupholidoptera chabrieri schmidti*. The species occurring on the Apennine Peninsula, Balkan Peninsula, and in Anatolia, e.g. *Platycleis nigrosignata*, are treated here not as Transadriatic but as Eastern Mediterranean together with the species inhabiting the Balkan Peninsula and the eastern parts of the Mediterranean, e.g. *Ameles heldreichi* (Mantodea), *Saga natoliae*, and *Duroniella laticornis*. The species found only in the Balkan Peninsula and Anatolia, e.g. *Phyllodromica pallida* (Blattodea), *Eupholidoptera smyrnensis*, and *Bucephaloptera bucephala*, are separated as the Balkan–Anatolian chorotype.

The only Turanian–Caucasian species is *Dociostaurus tartarus*. A revision of the material will most likely show that this species does not occur in Bulgaria, and this chorotype category will have to be deleted.

The Afrotropical–Palearctic and Paleotropical–Palearctic species include not only species originating from the tropical areas of the Old World. These categories include, respectively, the Holomediterranean by origin *Anacridium aegyptium* and *Oedaleus decorus* which ranges to the Afrotropical Region, and species of the genus *Xya*, for which the distribution in the Paleotropical regions is insufficiently known due to the taxonomic problems.

There are nine Balkan chorotypes (excluding those confined entirely in the territory of Bulgaria), with 32 species and subspecies (Table 6), evenly distributed among these chorotypes. If we pool the three chorotypes derived from the Macedonian chorotype in one, the seven resulting categories will include 4 to 5 taxa each. Nearly all these taxa are Balkan endemic species and subspecies. The ranges of the remaining taxa extend only slightly beyond the borders of the Balkan Peninsula.

There are 19 Bulgarian chorotypes with 41 species and subspecies (Table 6). Most of these taxa are Bulgarian endemic species and subspecies; only a few of them have ranges slightly extending beyond the territory of Bulgaria. The Rila–Pirin–Slavyanka chorotype is the most diverse, with nine endemic species and subspecies (at least three times more than any of other Bulgarian chorotypes).

4·2 Origin and dispersal

To interpret the place of species' origin, one considers not only the recent ranges but also the suggested place of origin of the genera (and for the subspecies, also the origin of the species). This analysis is initially based on the theory of zoogeographical categories proposed for the Holarctic by de Lattin (1967), in which each species can be treated as a faunal element within three biochores (high-level biomes): Arboreal, Eremial, and Oreotundral. The types of zoogeographical centers considered by de Lattin (1967) are: center of (ancestral) origin, center of differentiation, center of preservation (refugium center), and center of dispersal. Determination of the centers of origin and differentiation on the basis of distribution of a certain systematic group is quite difficult and uncertain. However, determination of the centers of preservation and dispersal could be more feasible since it is associated with the latest dynamic events (expansion or contraction), which resulted in the recent range. In some cases, the ranges have not undergone considerable changes, and are limited to the respective second-level (e.g. for the Pontomediterranean stationary species) or third-level (e.g. for the Balkan species) center of dispersal; or the route of the expansion is commonly accepted, e.g. for the Siberian species. In many cases, only the first-level center of speciation and dispersal, e.g. for the Holomediterranean species, is known.

Four of the categories of origin according to de Lattin (1967) are used here: Siberian, Mongolian, Mediterranean, and Caspian faunal elements. The content of the last category, however, is changed since we include here the steppe species with distinct type of range. Most remaining categories of origin are proposed here as subdivisions of the Mediterranean center of dispersal and of its Pontomediterranean portion. Not only the ranges but the habitats and ecological requirements are taken into consideration in this analysis. There are 24 categories of origin; 21 of them include only autochthonous (non-introduced) correctly identified species.

Eremial fauna. There are no representatives of the eremial fauna among the Bulgarian orthopterids. *Acrotylus longipes* takes a transitional position between the arboreal and eremial fauna. It is found in the Adriatomediterranean and Pontomediterranean arboreal centers and in almost the entire Afrotropical Region southward to Zambia and Namibia. The Bulgarian name of the species is Dune Grasshopper. This psammophilous species is adapted to the desert and semidesert areas in Africa as well as arid sandy and gravel seashore and river banks without vegetation in Bulgaria. The other two species of the same genus in Bulgaria, *A. insubricus* and *A. patruelis*, occur also in the entire Africa, including Sahara and Namibia, and in the desert areas of Arabia and Central Asia. They also inhabit dry sandy and rocky terrains with very scanty vegetation but according to their habitats in Bulgaria they are closer to arboreal species than *A. longipes*. Both species are among the few in Europe that hibernate as adults, which is also a consequence of their tropical origin.

Oreotundral fauna. Oreotundral fauna by origin does not exist. Oreotundral (Arctoalpine) distribution is characteristic for some strongly expansive animal species with tundral or oreal origin. Only two species, *Bohemanella frigida* and *Aeropedellus variegatus*, represent this fauna among the Orthoptera in Bulgaria (Fig. 1). The northern, continuous part of their Arctoalpine ranges prevails strongly over the southern mountain part and covers the entire extreme North of Palearctic (*B. frigida*) or its Siberian part and Finland (*Ae. variegatus*). The southern mountain part of the ranges includes the Alps, the mountains of Bulgaria, and Altai Mts. (both species), some other Central European and Balkan mountains and Southern Siberia (first species), the Apennines and Caucasus (second species). In the Nearctic, these two species occur only in Alaska (with the former also in Northwestern Canada). This indicates at least for *Aeropedellus variegatus* that its dispersal from East Asia via Bering landbridge has stopped at its initial stage. The chorotype of both species has to be identified as Holarctic Oreotundral and the origin at least of *Ae. variegatus*, as Siberian Tundral. The migration of *Bohemanella frigida* was realized most likely in the opposite direction (see below under Distribution and origin of selected genera). Several subspecies of these two species have been described (for *B. frigida* see also below under Distribution and origin of selected genera). The Bulgarian populations of *B. frigida* (Fig. 1) correspond to the subspecies *B. f. strandi* (Fruhstorfer, 1921) described from the Alps. For most Arctoalpine species, the time since separation of the northern and southern populations is not considered to be sufficient for differentiation of subspecies. Thus the question whether some populations represent distinct subspecies is left open.

Among other inhabitants of the high mountains, six species and subspecies occur only in the subalpine belt, and eight other species reach up to the alpine belt (see above under Vertical distribution). All these taxa, however, are representatives of arboreal but not of oreal fauna.

Arboreal fauna. Almost all (99.3%) of the Bulgarian orthopterid species belong to the arboreal fauna. Thamnobionts (dendrobionts) are only a small part of these taxa. Considerably more species are chortobionts, which inhabit clearings in the mountain forests but also meadows and other open areas as well as arid, treeless habitats. These open areas are secondary deforested habitats or primary steppes. Because of that, and judging from their general distribution, these chortobionts are also typical arboreal species by origin.

The origin remains unclear for *Modicogryllus truncatus* and for both species of Tridactyloidea due to taxonomic problems and lack of revision of distributional data, as well as for two species of Dermaptera because of their anthropogenic dispersal nearly all over the world. The category of origin is assigned to the remaining 265 species, but for 16 of them with some reservations. The species with uncertainly interpreted origin belong to Blattodea (one species), Tettigonioidea (seven species, including both species of Bradyporinae), Grylloidea (two species), and Acridoidea (six species), identified as Caspian, Pontomediterranean, Dinaric, Montane Balkan, Moesian, Thracian, and Anatolian.

According to their origin, the arboreal species of Orthopterida in Bulgaria are divided into 23 categories (Table 7). Three of these categories, namely Adriatomediterranean, Iranian, and Sinotibetan faunal elements, are not typical for the Bulgarian fauna and include introduced or most likely incorrectly

Table 7 Zoogeographical origin of Orthopterida in Bulgaria

Categories of origin	Number of taxa	Share from all Bulgarian taxa (%)
OREOTUNDRAL		
Siberian Tundral	2	0.8
ARBOREAL		
Northern arboreal		
Siberian	35	13.2
Siberian–Nearctic	1	0.4
Siberian–Mediterranean	8	3.0
Mongolian	1	0.4
Central European	27	10.2
Central European–Mediterranean	3	1.1
Caspian	12	4.5
Balkan		
Dinaric	9	3.4
Montane Balkan	9	3.4
Moesian	6	2.3
Stara Planina	10	3.8
Rila–Rhodopean	21	7.9
Macedonian	11	4.1
Thracian	11	4.1
Balkan Mediterranean	7	2.6
Mediterranean		
Holomediterranean	37	14.0
Pontomediterranean	30	11.3
Anatolian	8	3.0
Tropical		
Afrotropical	7	2.6
Paleotropical	6	2.3
Species introduced or unconfirmed in Bulgaria		
Adriatomediterranean	2	0.8
Iranian	1	0.4
Sinotibetan	1	0.4

identified species. The richest in species categories are Holomediterranean, Siberian, Pontomediterranean, and Central European elements (Table 7), which include 49% of all Bulgarian species.

The arboreal faunal elements in Bulgaria (Table 7) can be arranged in three groups according to the position of their centers of dispersal towards the territory of Bulgaria. The species of the categories originating from the Siberian, Mongolian, Central European, and Caspian centers of dispersal are considered to be of *northern origin*. The species originating from the Mediterranean and tropical areas are considered to be of *southern origin*. The species with *Balkan origin* are placed separately in Table 7. These three groups are almost equal in species number: northern, 87; southern, 88; and Balkan, 84 (32 to 33% of all Bulgarian taxa in each group). These categories do not include two oreotundral species of northern origin and four introduced and unconfirmed species of southern origin.

The question arises about the place of the taxa with Balkan origin. Three out of eight Balkan centers of speciation (Macedonian, Thracian, and Balkan Mediterranean) are located in the Mediterranean Subregion (Fig. 5), i.e. they fall under the categories with southern origin. Even among these faunas, however, there are three taxa with Macedonian origin which in fact also inhabit localities outside of Mediterranean Subregion. *Pholidoptera macedonica* occurs in Slavyanka above 900 m up to the highest peak of this mountain. *Saga campbelli campbelli* (Fig. 8) and *Poecilimon zwicki* (Fig. 6) are also found in Slavyanka up to 1600 and 1500 m, respectively. Other five Balkan centers of speciation are located entirely in the Eurosiberian Subregion and their faunas should be considered to have northern origin. In the theory of zoogeography, all Balkan and Bulgarian endemic taxa were earlier considered a special case of Pontomediterranean faunal elements (Gruev, 1988). Later, the same author revised his concept treating (in our opinion, correctly) the endemic taxa as faunal elements different in origin and belonging to different categories (Gruev and Kuzmanov, 1994). Here, the categories of taxa originating from the Balkan centers of speciation are separated in a special group. The great majority of these taxa is closely related to southern species and, in addition, includes representatives of genera with southern origin, e.g. *Ancistrura, Andreiniimon, Gampsocleis, Saga, Discoptila*, and *Paranocarodes*. Few are the taxa closely related to northern species, e.g. representatives of *Metrioptera* and of subfamily Catantopinae. In the genera *Isophya* and *Poecilimon*, an intensive speciation in the extra-Mediterranean areas (e.g. Bulgaria) of the ancestral taxa, which dispersed from the Mediterranean areas, has taken place. All this indicates that most of the Balkan taxa have to be treated as taxa of southern origin. Therefore, species with southern origin in Bulgaria prevail to a large degree over those with northern origin.

Division according to the origin differs among particular systematic groups. All species of Mantodea, Isoptera, and Embioptera in Bulgaria have southern origin. All these orders have tropical origin from which only a small number of representatives reaches South Europe. Southern and Balkan taxa predominate strongly in Tettigonioidea, while northern species predominate in Caelifera. The share of the

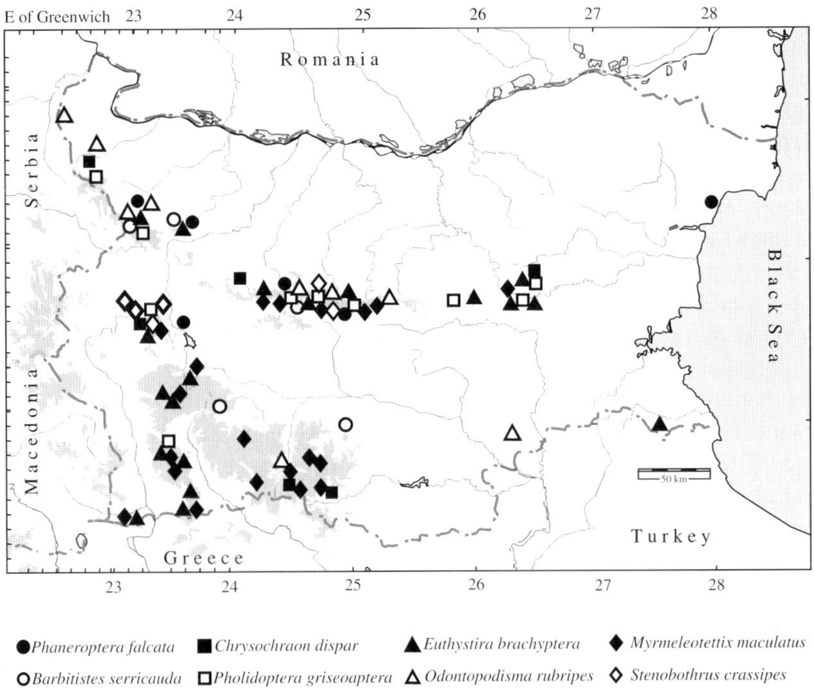

Fig. 2 Distribution in Bulgaria of some species of Orthoptera with northern arboreal origin. Siberian faunal elements: *Phaneroptera falcata, Chrysochraon dispar, Euthystira brachyptera, Myrmeleotettix maculatus*. Central European faunal elements: *Barbitistes serricauda, Pholidoptera griseoaptera, Odontopodisma rubripes, Stenobothrus crassipes.* Areas above 1200 m altitude are shaded.

taxa with northern origin compared to all Bulgarian species of each group is 22% in Tettigonioidea, and 55% in Caelifera.

The richest in species in the group of northern arboreal species are the Siberian ones (40%), followed by the Central European and Caspian faunal elements (Table 7). The Siberian elements (Fig. 2) are grouped in one category because their origin from Ussurian, Sinokorean, or Angarian subcenters of origin cannot be distinguished. They all are believed to be young species, which additionally dispersed recently from the Siberian center. Examples of species with ranges reaching eastward to Manchuria, Korea, Primorye Region, Sakhalin, and Japan are given above under the Chorotype section. The ranges of many other species do not reach the Pacific Ocean in the east, e.g. *Ectobius lapponicus* (Blattodea), *Conocephalus dorsalis, Arcyptera fusca*, and *Omocestus viridulus*. There are thermophilous Siberian elements among the orthopterids, e.g. *Anechura bipunctata* (Dermaptera), *Platycleis veyseli*, and *Mecostethus parapleurus*. Their ranges are located more southward from those of the typical Siberian species and they often do not reach the East Siberia. *Gomphocerus sibiricus* is a Siberian faunal element with Boreomontane distribution. Its range is disjunct between the lowland of North and

Central Russia, Siberia, and Kamchatka and the mountains of South Europe and Central Asia. Disjunction in this case is smaller than in the Arctoalpine ranges. The Siberian species have dispersed to the Balkan Peninsula via Dinaric Alps and via Carpathians. More cold-tolerant species are found in Bulgaria only in the mountains, e.g. *Tettigonia cantans, Metrioptera bicolor, Myrmeleotettix maculatus* (Fig. 2), and *Omocestus viridulus*. More ecologically plastic species inhabit both lowlands and mountains, e.g. *Omocestus haemorrhoidalis, Stenobothrus lineatus, S. nigromaculatus*, and *Chorthippus parallelus*, all of them from subfamily Gomphocerinae. Most thermophilous Siberian species are found in Bulgaria only in the lowlands.

The Siberian–Mediterranean species are assumed here to have an initial and an additional center of dispersal. In this scenario, they have dispersed to Europe from the initial center in Siberia during the interglacial periods. Under the influence of the cooling climate, they have retreated in two directions: to the refugia of Asia and South Europe. From there they have dispersed again both to the taiga and in the Mediterranean Subregion. The Siberian–Mediterranean species have a wide ecological plasticity; the category includes such species as *Tettigonia viridissima, Decticus verrucivorus*, and *Oedipoda caerulescens*, widespread in Palearctic as well as in Bulgaria.

The only Siberian–Nearctic species by origin is *Tetrix subulata* with Holarctic distribution.

The only species originating from the Mongolian center of dispersal is *Epacromius tergestinus*. The Asian part of its range resembles that of the Siberian species but is located more southward and does not include Siberia. The localities in the Balkan Peninsula are relict ones: Istria, Dobrudja (Dobrogea) between the Danube Delta and Varna, and Petrich in the Struma Valley. Elsewhere in Europe, the species occurs also in relict localities between Santander and Venice and between Augsburg and Rome.

Another species of the same genus is also found in Europe in the relict localities: *Epacromius coerulipes*, a Siberian faunal element with Eurosiberian distribution. It occurs also locally between Venice and Hungary, and within the Balkan Peninsula it is found only in the Black Sea coast of Romania and Bulgaria (Sozopol). The habitats inhabited by this species in Southeast Europe indicate that *E. coerulipes* represents a transitional form between the species with Siberian and Caspian (or taiga and steppe) origins. An argument for this is the fact that in Siberia this species inhabits both forests and steppes.

The Central European faunal elements (Fig. 2) are the old species, which are believed to have originated in the extra-Mediterranean part of Europe, survived the glaciations in the refugia of Central and Eastern Europe, and afterward dispersed almost over the entire Central Europe. *Barbitistes constrictus* appears to be a relict, which after its dispersal southward has survived in the Balkan Peninsula only in Rila, the refugium for its only population outside Central Europe. Other Central European species also practically do not enter South Europe. *Barbitistes serricauda* (Fig. 2) and *Stenobothrus crassipes* (Fig. 2) have dispersed only to the northernmost parts of the Balkan Peninsula and *Ectobius sylvestris* (Blattodea), *Poecilimon fussi,*

Pholidoptera littoralis, and *Stenobothrus stigmaticus faberi* are found further southward in the entire northern half of the Balkan Peninsula. More widely distributed in South Europe are *Leptophyes punctatissima* and *Meconema thalassinum*; only in the mountains are found *Polysarcus denticauda, Pholidoptera griseoaptera*, and *Stenobothrus rubicundulus*. Some Central European elements with restricted distribution actually represent Carpathian–Balkan species, e.g. *Pholidoptera frivaldskyi* and *Odontopodisma rubripes*.

The scheme of dispersal for Central European–Mediterranean species is similar to that of the Siberian–Mediterranean ones. The distinction lies in their initial center of dispersal which here is the Central European. It is assumed that, during the glaciations, the species of this category have moved to the south to occupy the refugia in the southern part of Central Europe and in the Mediterranean. After each expansion part of their populations has adapted to the conditions in the Mediterranean Subregion. *Platycleis albopunctata grisea* and *Gryllus campestris* are widespread in Bulgaria except the high mountain parts, and *Apterygida media* is a rare species but its distribution in Bulgaria is insufficiently studied.

The Caspian faunal elements include mainly steppe species which inhabit Transcaucasia and areas to the north of the Black Sea. The expansive species of this category dispersed to a varying degree into the steppe extension in the north-eastern part of the Balkan Peninsula, the eastern part of Central Europe (Hungary), and Southwestern Siberia and Kazakhstan. *Onconotus servillei* occurs in Bulgaria (Fig. 3) only in the northeastern plain part of the country (see below in Distribution

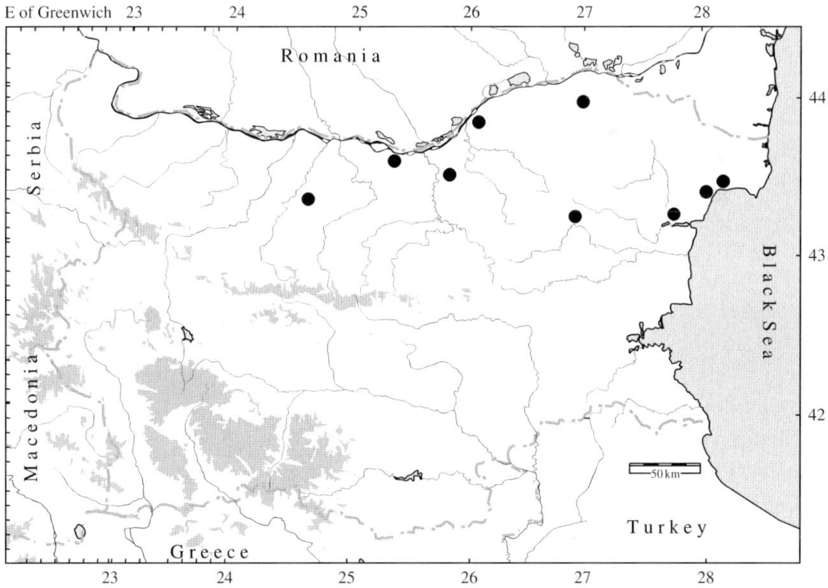

Fig. 3 Distribution in Bulgaria of *Onconotus servillei*, a typical steppe species.

and origin of selected genera). It is one of the most typical steppe species of animals in Bulgaria together with such mammals as *Sicista subtilis* (Pallas, 1773) (Rodentia: Zapodidae), *Cricetus cricetus* Linnaeus, 1758 (Rodentia: Cricetidae), and *Mustela eversmanni* Lesson, 1827 (Carnivora: Mustelidae). The similar range has the new for Bulgaria *Platycleis medvedevi*, recorded for Bulgaria from the Danube in the northernmost part of this country. More widely distributed in the lowlands of Bulgaria are *Gampsocleis glabra* and *Celes variabilis*.

In the group with the Mediterranean origin, the Pontomediterranean (including the Anatolian) and Holomediterranean faunal elements are approximately equally represented (Table 7). Nearly all Holomediterranean elements originate from the northern part of the Mediterranean Subregion and inhabit uniformly the entire territory from the Atlantomediterranean to the Pontomediterranean second-level centers of dispersal. Depending on the degree of the postglacial changes in the ranges of the Holomediterranean species, they are divided chorologically into stationary and expansive ones (de Lattin, 1967). This division is relative since gradual transitions exist between these two groups. As the limit for the stationary species, we accept the northern boundary of the range of *Tylopsis lilifolia*. Some thermophilous species do not reach this limit and inhabit in Bulgaria only Struma Valley, Eastern Rhodopes, and Black Sea coast (Fig. 4), e.g. *Iris oratoria* (Mantodea), *Kalotermes flavicollis* (Isoptera), *Haploembia solieri* (Embioptera), *Gryllomorpha dalmatina*, and *Myrmecophilus myrmecophilus*. The number of expansive Holomediterranean species is higher than that of the stationary ones (Table 8). The expansion of *Tetrix ceperoi* and *Oecanthus pellucens* reaches farthest northward.

There are no systematic groups of animals in Bulgaria which include Adriatomediterranean faunal elements[2]. Although *Meconema meridionale* and *Platycleis stricta* are expansive Adriatomediterranean species, the former is accidentally introduced in Bulgaria; and the latter is very likely incorrectly identified almost 100 years ago (it was never found again, and no preserved material exists). Therefore, the category of Adriatomediterranean species should not be included into the list.

The Pontomediterranean faunal elements are also divided into stationary and expansive ones; the expansive elements slightly predominate (Table 8). The Pontomediterranean species are also interpreted here as stationary when their ranges exceed the limits of the Pontomediterranean second-level center eastward in the habitats similar to the Mediterranean ones (e.g. on the Black Sea coast and in Transcaucasia), or when they are polycentric species with Adriatomediterranean–Pontomediterranean distribution but within the bounds of the Mediterranean Subregion, e.g. *Eupholidoptera chabrieri schmidti*. A few expansive Pontomediterranean species reach northward as far as Southern Sweden, e.g. *Leptophyes albovittata*. Two expansive species with ranges resembling those

[2] If species have been placed to this category, their finding in Bulgaria means that they are Pontomediterranean or polycentric Adriatomediterranean–Pontomediterranean but not strictly Adriatomediterranean elements.

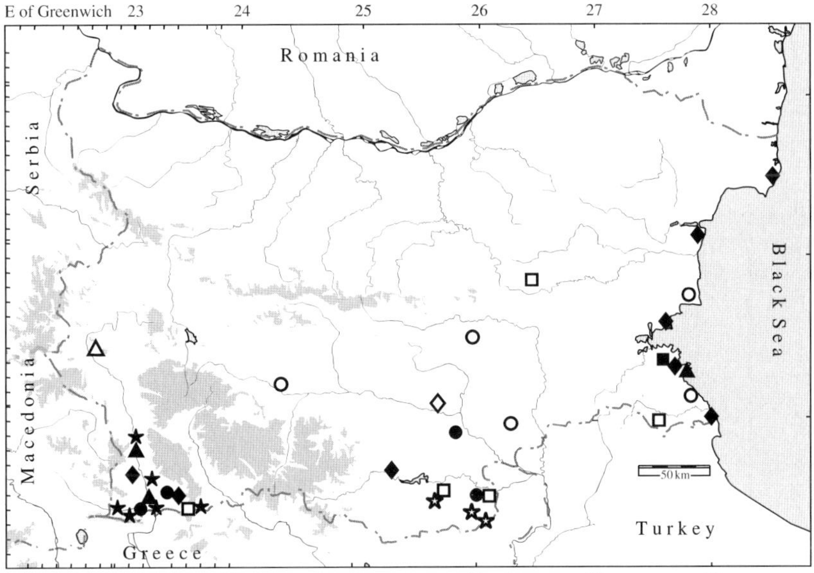

● *Iris oratoria* ■ *Kalotermes flavicollis* ▲ *Haploembia solieri* ◆ *Arachnocephalus vestitus*
★ *Dociostaurus genei* ○ *Platycleis escalerai* □ *Platycleis nigrosignata* △ *Eupholidoptera chabrieri schmidti*
◇ *Stenonemobius bicolor ponticus* ☆ *Duroniella laticornis*

Fig. 4 Distribution in Bulgaria of some stationary taxa of Orthopterida with Mediterranean origin. Holomediterranean faunal elements: *Iris oratoria, Kalotermes flavicollis, Haploembia solieri, Arachnocephalus vestitus, Dociostaurus genei*. Pontomediterranean faunal elements: *Platycleis escalerai, P. nigrosignata, Eupholidoptera chabrieri schmidti, Stenonemobius bicolor ponticus, Duroniella laticornis.*

of the Holomediterranean elements are of Pontomediterranean origin. *Pterolepis germanica* undoubtedly originated from the Pontomediterranean center where lies the main part of its range (the entire Balkan Peninsula, Anatolia, and Caucasus). It has occupied during its dispersal also the Adriatomediterranean (Italy) and Tyrrhenian (Corsica) second-level centers and has reached westward to Southern France but does not occur in the Atlantomediterranean second-level center (Iberian Peninsula). *Polyphaga aegyptiaca* (Blattodea) in its expansion has crossed non-existent now land bridges connecting the Balkan Peninsula to South Italy, Sicily, Tunisia, and Algeria (at the same time also dispersing eastward from the Pontomediterranean center).

A part of the Pontomediterranean stationary species (13 species) belong to the Balkan–Anatolian chorotype (Table 6). Eight of them are placed in the category of Anatolian origin.

Levels of centers of speciation. Analyzing the species which have evolved in the Balkan Peninsula and in the adjacent areas, i.e. the groups of the Mediterranean

Table 8 Division of the Bulgarian orthopterid taxa with Mediterranean and Balkan origin by centers of speciation. First number: the number of taxa originated in the center listed in the first column. Number in parentheses: the number of taxa originated in the given center and in parts of it, i.e. at the same level or at lower level(s)

Center of speciation	Number of taxa					
	Center of first level	Center of second level	Center of third level	Center of fourth level	Center of fifth level	Center of sixth level
Holomediterranean	39 (159)					
Holomediterranean (expansive species)	23 of 39					
Holomediterranean (stationary species)	16 of 39					
Pontomediterranean		30 (120)				
Pontomediterranean (expansive species)		17 of 30				
Pontomediterranean (stationary species)		13 of 30				
Balkan–Anatolian		0 (90)				
Anatolian			8			
Balkan			13 (82)			
Dinaric				8		
Montane Balkan				9		
Moesian				1 (4)		
Central Danubian Plain and Predbalkan					1	
Black Sea coast (northern part)					2*	
Stara Planina s.l.				1 (9)		
Stara Planina Mts.					0 (6)	
Western Stara Planina						2
Central Stara Planina						3
Eastern Stara Planina						1
Predbalkan					see under Moesian	
Sredna Gora					1	
Vitosha Group					1	

Rila–Rhodopean	1 (20)		
Osogovo–Belasitsa Group		1 (7)	
Osogovo			1
Vlahina			1
Ograzhden			1**
Belasitsa			3
Rila–Pirin Group		2 (9)	
Rila			3
Pirin			3
Slavyanka			1
Rhodopes		0 (3)	
Western Rhodopes			3
Eastern Rhodopes			see under Thracian
Macedonian	4 (7)		
Middle Struma Valley		3	
Thracian	3 (7)		
Eastern Rhodopes		1	
Tundja Region		1	
Strandja		2	
Balkan Mediterranean	5		

* Most likely synonyms
** Most likely a synonym

and Balkan origin (Table 7), we can address in detail, and at the different levels, the centers of speciation for the taxa found in Bulgaria (Table 8). In many animal groups this is not possible but the high degree of endemism in South Europe and Southwestern Asia allows that sort of analysis for Orthoptera.

The first-level center of speciation of the species of the Mediterranean and Balkan origin is the Holomediterranean one. All 159 taxa of these have originated in the Holomediterranean center or in its parts; for 120 of these taxa, we were able to identify these parts more precisely. The remaining 39 taxa can be only identified as Holomediterranean faunal elements.

The second-level center of speciation is the Pontomediterranean center with 120 taxa, 30 of which are identified as originating in this center in general, and 90, in its Balkan–Anatolian part. For some of the group of 30 species, it can be speculated with some certainty from which part of the Pontomediterranean center they originate. For instance, *Saga natoliae* and *Duroniella laticornis* are believed to be of Anatolian origin and *Pachytrachis gracilis* and *Paracaloptenus caloptenoides*, of Balkan origin. The Anatolian origin is assigned due to the range that occupies the entire Asian sector (Anatolia, Syria, Lebanon, Israel) and only a part of the European sector (the southern half of the Balkan Peninsula) of the Pontomediterranean center. Such are the ranges of *Duroniella laticornis* and *Notostaurus anatolicus*. Distribution of the species with a Pontomediterranean origin (the Pontomediterranean faunal elements) indicates that some of them are polycentric regarding their second-level center of dispersal. Their ranges cover two or more centers of second level but not the entire Atlantomediterranean center; therefore they are not Holomediterranean species. To the nine second-level centers proposed by de Lattin (1967), we add here also the Taurian center (Crimea). de Lattin (1967) considers the southern coast of Crimea a part of the Caspian primary center. Our interpretation of the Caspian faunal elements as steppe species, however, necessitates the separation of Southern Crimea as a distinct second-level center and its inclusion into the Mediterranean primary center. Here are some examples for polycentric Pontomediterranean ranges:

Atlantomediterranean–Tyrrhenian–Adriatomediterranean–Pontomediterranean: *Pholidoptera fallax* (in the Atlantomediterranean center, only in Southern France);

Atlantomediterranean–Tyrrhenian – Adriatomediterranean – Pontomediterranean–Cretan: *Pterolepis germanica* (in the Atlantomediterranean center, only in Southern France);

Tyrrhenian–Adriatomediterranean–Pontomediterranean: *Phyllodromica marginata* (Blattodea);

Adriatomediterranean–Pontomediterranean: *Platycleis nigrosignata*;

Adriatomediterranean–Pontomediterranean–Cretan–Cyprian:*Platycleis escalerai*;

Adriatomediterranean–Pontomediterranean–Cretan–Taurian: *Empusa fasciata* (Mantodea);

Pontomediterranean–Cretan: *Platycleis incerta*;

Pontomediterranean–Cyprian: *Duroniella laticornis*;

Pontomediterranean–Taurian: *Poecilimon schmidti*.

The third-level centers of speciation for the species distributed in Bulgaria are the Balkan and the Anatolian ones. This level includes all Bulgarian taxa, which are assigned to a certain part of the Pontomediterranean second-level center by their origin. The species with Balkan origin prevail in the Bulgarian orthopterid fauna (Table 8) at the third level. If the next levels (from fourth to sixth) with species distributed only in the Balkan Peninsula will be added, this predominance increases greatly. Within the framework of the entire Pontomediterranean center, however, the number of the species with Anatolian origin is quite higher than the number of the Balkan species. An example for this in the Bulgarian orthopterid fauna is found if we examine only the species of the Balkan–Anatolian chorotype (i.e. found only in the Balkan Peninsula and Anatolia): there are twice as many taxa with the Anatolian origin than with the Balkan one. At third level the origin is identified also according to the relative sizes of the Anatolian and Balkan part of the range and according to the origin of the related species. An example of the species with Anatolian origin is *Bucephaloptera bucephala* (found in the entire Anatolia and the eastern part of the Balkan Peninsula). An example of the taxon with Balkan origin is *Chorthippus porphyropterus euhedickei* (found in the southern part of the Balkan Peninsula and Northwestern Anatolia). The range of the latter subspecies is obviously not fully known since it was described 15 years ago, its correct species placement is known only for six years, and it is hard to distinguish by the stridulatory pegs from the related species of *biguttulus*-group. Nevertheless, its origin can be accepted as Balkan because nearly all representatives of *Chorthippus* and most of the species of Gomphocerinae in the Bulgarian fauna have northern origin. A particular position among the species of Anatolian origin is occupied by *Anterastes serbicus*. It is a mountain species with Montane Mediterranean distribution. Its origin is analyzed below in Distribution and origin of selected genera.

The fourth-level centers of speciation occupy specific parts of the Balkan Peninsula. There are eight such centers (Fig. 5), three of which (Dinaric, Montane Balkan, and Balkan Mediterranean) are located outside the territory of Bulgaria. The other five centers are situated partially or entirely in Bulgaria (Table 8, Fig. 6). The Rila–Rhodopean Massif is of a paramount significance for the speciation and its number of endemic species is more than two times higher than in any of the other centers. The opposite extreme is represented by the Moesian center; it is very likely that most or all taxa believed to have originated in it are not distinct species or subspecies. On the other hand, the taxa from the Balkan Mediterranean center are distinct. Approximately equal numbers of taxa (seven to nine) have assigned to the other centers at the fourth level of speciation. Altogether, 82 species and subspecies are believed to originate from the Balkan Peninsula, including Bulgaria.

The five centers present on the territory of Bulgaria are analyzed also at the fifth and sixth levels (Table 8). The Rila–Rhodopean Massif is divided into three fifth-level centers of speciation. Most species in this massif have originated in the Rila–Pirin Group and Osogovo–Belasitsa Group while the role of Rhodopes for the speciation is considerably more restricted. The formation of the fauna of Orthoptera

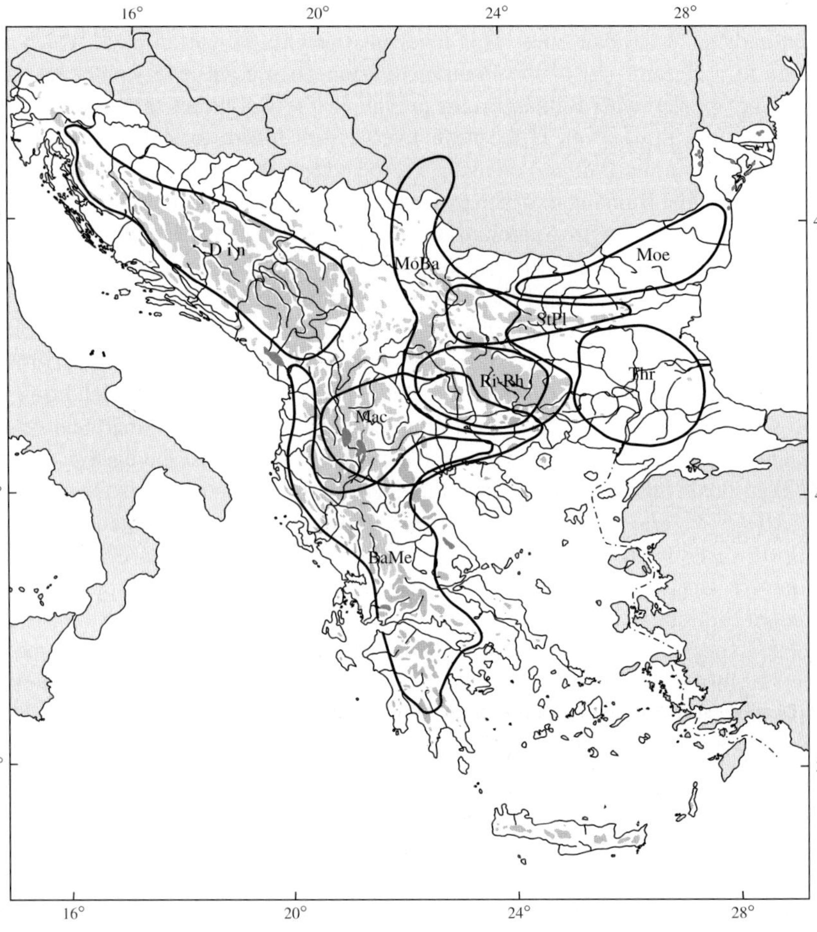

Fig. 5 Balkan centers of origin of the orthopterid species distributed in Bulgaria. Din, Dinaric; MoBa, Montane Balkan; Moe, Moesian; StPl, Stara Planina; Ri–Rh, Rila–Rhodopean; Mac, Macedonian; Thr, Thracian; BaMe, Balkan Mediterranean. Areas above 1200 m altitude are shaded.

in Stara Planina is concentrated at the fifth level while the Predbalkan, Sredna Gora, and the mountains of the Vitosha Group have a subordinate significance.

The sixth level of speciation includes parts of certain mountain ranges or groups in Bulgaria. Most numerous here are the taxa (three in each region), which have their origin in Central Stara Planina, Belasitsa, Rila, Pirin, and Western Rhodopes. The same number of species has originated in the Middle Struma Valley (fifth-level center of speciation).

The determination of the speciation centers at the fourth to sixth levels is to a great extent provisory. This is not only because it is based on the geographical division of Bulgaria. The probable synonymy of some taxa in the future could decrease the number of centers at the last two levels and the list of taxa of

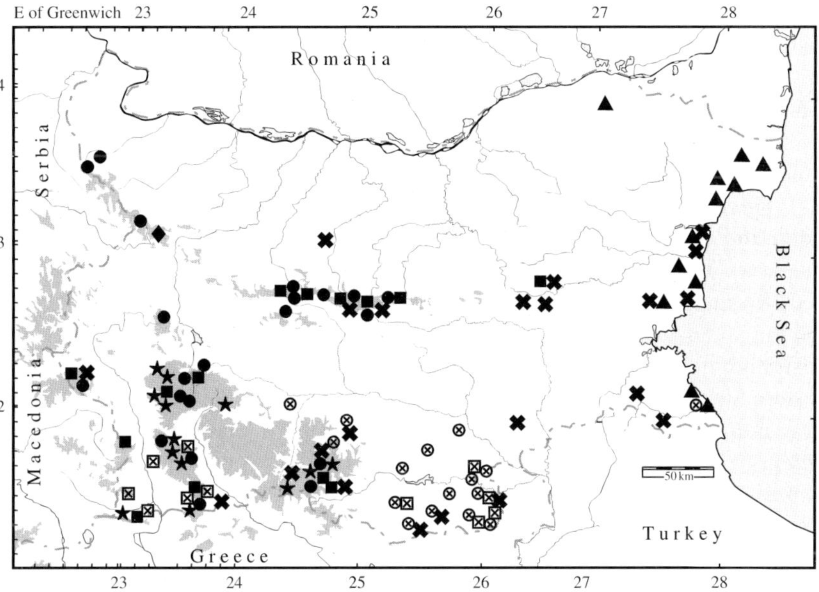

Fig. 6 Distribution in Bulgaria of some endemic taxa of Tettigonioidea with different origin. Dinaric origin: *Pholidoptera aptera karnyi*. Montane Balkan origin: *Psorodonotus fieberi*. Moesian origin: *Isophya modesta longicaudata*. Stara Planina origin: *Metrioptera helleri*. Rila–Rhodopean origin: *Poecilimon orbelicus*. Macedonian origin: *Poecilimon zwicki*. Thracian origin: *Isophya petkovi*. Balkan Mediterranean origin: *Metrioptera oblongicollis*. Areas above 1200 m altitude are shaded.

Orthoptera assigned to them. It is hard to believe that some mountain ranges such as Vlahina and Ograzhden represent centers of speciation. No local endemics are known from these mountains from other animal groups. For Orthoptera, this is possibly also true for such mountain regions as Vitosha Group, Sredna Gora, Osogovo, and Eastern Stara Planina. Four subspecies of *Poecilimon mistshenkoi* described from Pirin, Slavyanka, Belasitsa, and from Vlahina, Maleshevska, and Ograzhden as well as three subspecies of *P. affinis* described from Osogovo, Rila, and Central Sredna Gora have to be accepted with reservations. At first sight they seem to be good examples of ongoing speciation but their status is unclear.

4·3 Distribution and origin of selected genera

The most interesting in zoogeographical respect genera of Bulgarian Orthopterida are selected and analyzed below in detail.

1. *Psorodonotus* (15 taxa: ten species and five additional subspecies). Range: Balkan Peninsula (one species with three subspecies), Anatolia (four species),

Anatolia, Caucasus, and Transcaucasia (three species), Caucasus and Transcaucasia (two species and three additional subspecies). The genus consists of mountain species. Center of origin of this genus is most likely the Caucasus. *Psorodonotus fieberi* is the only species in the Balkan Peninsula. Its three subspecies (earlier considered distinct species) are clearly distinguished geographically: *P. f. illyricus* Ebner, 1923 (from Istria to Montenegro), *P. f. macedonicus* Ramme, 1931 (Montenegro, Kosovo, Macedonia, Albania, North and Central Greece southward to Oita Mts.), and *P. f. fieberi* (Serbia and Bulgaria). Transitional forms between them and probably hybrids are observed. The information on the subspecies ranges needs verification. Origin of the species and the nominate subspecies: Montane Balkan. In Bulgaria, *P. f. fieberi* is found in all high mountains (Fig. 6) from 1400 to 2200 m and in an isolated locality in Eastern Stara Planina at 600 m (Pešev, 1990). Occasionally, it occurs in unusually large numbers, e.g. in Central Stara Planina in 1996 (Popov et al., 2000).

2. *Anterastes* (ten taxa: nine species and one additional subspecies). Range: Balkan Peninsula and Anatolia (one species with two subspecies), Anatolia (nine species). The genus consists of mountain species. The origin of the genus is undoubtedly Anatolian. *Anterastes serbicus* is the only species in the Balkan Peninsula. Range of the species: Serbia, Bulgaria, Albania, North Greece southward to Ossa Mts., Western Anatolia eastward to Akşehir (*A. s. serbicus*), and Macedonia (*A. s. macedonicus* Karaman, 1961). The status of *A. s. macedonicus* needs confirmation; this subspecies most likely will be proved to be a synonym. Three species, including *A. serbicus*, occur in Northwestern Anatolia (Uludağ Mts.). This shows the Anatolian origin of *A. serbicus*. It is a relict with Montane Mediterranean (more strictly, Montane Pontomediterranean) distribution. The Montane Mediterranean species are Tertiary relicts distributed only in the mountain areas of the Mediterranean Subregion, usually in the highest parts of the mountains (Heiss and Josifov, 1990). They are representatives of genera with ranges restricted only to the Mediterranean or also to the Central Asian subregion, and are the only Mediterranean species which regularly occur and reproduce in the subalpine and alpine belts. These relicts are remainders of the ancient cold-tolerant Mediterranean fauna which inhabited the high mountain steppes of South Europe, or also of Central Asia, at the end of the Tertiary. *A. serbicus* occurs in Bulgaria in all high mountains in the subalpine, alpine, and partly in the forest belt from 1500 to 2700 m, and only on Slavyanka it is also found at 900 m (Pešev and Mařan, 1963).
3. *Onconotus* (two species). Range: from Hungary and Serbia across Ukraine, southern part of the European Russia, and North Caucasus to Siberia (two species). The genus consists of steppe species. Origin of the genus: Caspian. *O. servillei* is the only species in the Balkan Peninsula. Its range corresponds to the range of the genus and reaches eastward to West Siberia. Origin of the species: Caspian. In Bulgaria, *O. servillei* is found (Fig. 3) only in the northeastern part of the country (Dobrudja and the eastern half of the Danubian Plain), westward to Pleven (Buresch and Peschev, 1958). The other species,

O. laxmanni (Pallas, 1771), occurs from Ukraine to Siberia. The genus is separated into the distinct subfamily Onconotinae but recently an opinion was expressed that its status has to be reduced to a tribe in Tettigoniinae.

4. *Saga* (16 taxa: 14 species and two additional subspecies). Range: South Europe, southern parts of Central Europe, West Siberia (one species), Balkan Peninsula (three species, one of which with two subspecies), from the Balkan Peninsula to Syria (one species), Anatolia (four species), Eastern Mediterranean from Eastern Anatolia to Israel, Transcaucasia, and Iran (five species, one of them with two subspecies). The richest number of species is in Turkey (ten species and one additional subspecies, with one species only in European Turkey). The genus includes the largest European orthopterid species, all of them thermophilous. The origin of the genus is Pontomediterranean. The more distant ancestors of *Saga* are probably Afrotropical because, according to Kaltenbach (1967), the primitive genera of Saginae, close to *Saga*, are found in Africa. Only two taxa of those found in the Balkan Peninsula are not endemic. The widest distribution in the genus has *S. pedo*. Its origin is Caspian and it occurs in South Europe (without Greece) northward to Southern France, Switzerland, Austria, Czech Republic, Slovakia, Ukraine, Central Russia, and eastward to West Siberia. Other species of the genus have a Pontomediterranean origin in a broad sense with different centers of speciation of fourth level: *S. natoliae* (Pontomediterranean in a narrow sense; from Southern Croatia and Albania to South Bulgaria, Northeastern Greece, European Turkey, Anatolia, and Syria), *S. hellenica* (Balkan Mediterranean; Albania, Macedonia, Lyulin Mts. in Southwestern Bulgaria [Fig. 8], and Greece), *S. rammei* (Macedonian; Petrich and Popovitsa in South Bulgaria [Fig. 8], Macedonia, Greek Macedonia), *S. campbelli campbelli* (Macedonian; Kresna Gorge, Belasitsa, and Slavyanka in Southwestern Bulgaria [Fig. 8], Macedonia, Northeastern Greece), *S. c. gracilis* (Thracian; Dobrogea in Eastern Romania, Eastern Rhodopes and Sakar in South Bulgaria [Figs. 7 and 8], Northern Aegean islands, European Turkey). Another species, *S. rhodiensis* Salfi, 1929, occurs in Greece (Rhodes Island) as well as in Southwestern Anatolia but this island does not belong geographically to the Balkan Peninsula. Among other Mediterranean species, the expansive eastward species is *S. ephippigera ephippigera* Fischer de Waldheim, 1846 with the range: Israel, Syria, Eastern Anatolia, Northwestern Iran, Armenia, and Caucasus. The five Pontomediterranean taxa (Fig. 8) found in Bulgaira are distributed only in South Bulgaria; the information of Ramme (1951) on *S. natoliae* at Varna is based on the very old material; the species is probably extinct there. The Caspian element *S. pedo* is found mainly in North Bulgaria (most or all published records from South Bulgaria are very likely a result of incorrect identification).
5. *Bradyporus* (one species) and *Callimenus* (six species). These two genera are closely related. Range of both genera: Eastern Romania (Romanian Moldova and Dobrogea), Bulgaria, Serbia, Macedonia, and North Greece. Out of this territory, *Bradyporus* is found in the north in Hungary and in the southeast in European Turkey, and *Callimenus* in the northeast from Southern Romania

Fig. 7 *Saga campbelli gracilis*, an endemic species and subspecies from the eastern part of the Balkan Peninsula with Thracian origin (Photo: T. Ivanova).

and Moldova to North Caucasus and in the southeast from South Greece to Iran. Both genera consist of steppe species. The origin of the genera, or, more exactly, of their common ancestor, is probably Caspian (although *Bradyporus* does not currently inhabit this area). The genus *Callimenus* has a Pontian type of distribution and the ranges of its species cover various territories around the Black Sea: *C. multituberculatus* (Fischer de Waldheim, 1833) is found from the North Caucasian steppes to Ukraine; *C. montandoni* Burr, 1898, from Ukraine to Southern Romania; *C. oniscus* Burmeister, 1838, in Macedonia and the entire Greece (without Thrace); *C. macrogaster*, from Romanian Moldova, Bulgaria, and Serbia over the Greek Thrace to Anatolia; *C. dilatatus* Stål, 1875, in Anatolia; and *C. latipes* Stål, 1875, in Iran. Judging from their current distribution, the origin of the species occurring in Bulgaria (*Callimenus macrogaster*, see Fig. 9, and *Bradyporus dasypus*) is probably Moesian. Both species occur in Bulgaria in the lowlands but the abundance of *C. macrogaster* decreases strongly and in many localities it became extinct during the last 60 years. These two genera are separated into the distinct subfamily Bradyporinae within the family Bradyporidae, but, according to a recent opinion, the status of this group has to be reduced to the tribe Bradyporini (Tettigoniidae: Bradyporinae).

6. *Gryllomorpha* (32 taxa: 31 species and one additional subspecies). Range: the Mediterranean countries and Uzbekistan. The ranges of the species cover: the entire Mediterranean (*G. dalmatina*); the Western Mediterranean from Spain to Italy (*G. uclensis* Pantel, 1890); Portugal and Spain (two species, one of which also in North Africa); Italy (one species); Albania and continental Greece (two

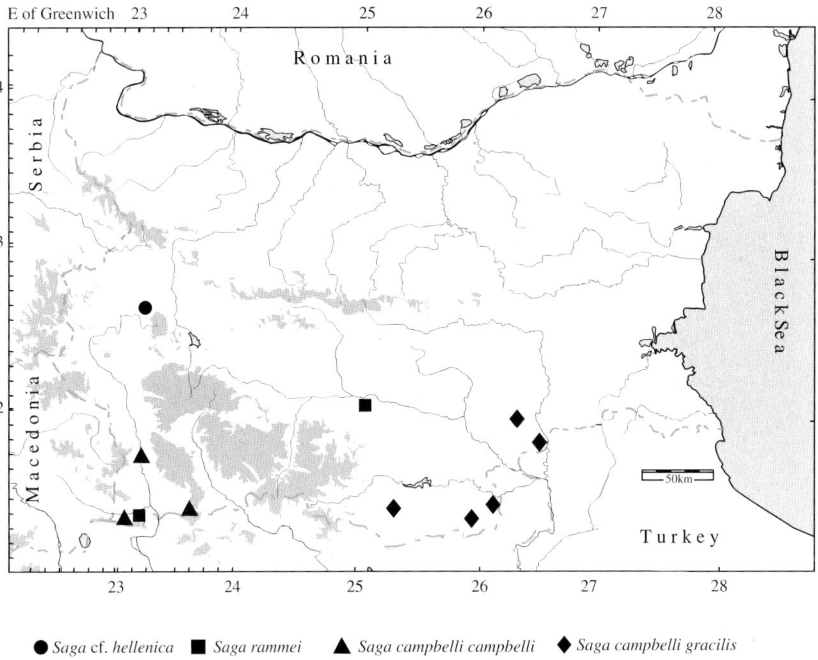

Fig. 8 Distribution in Bulgaria of the endemic species of *Saga*: *Saga* cf. *hellenica, S. rammei, S. campbelli campbelli, S. c. gracilis*. Areas above 1200 m altitude are shaded.

species); Aegean islands (one species); Crete (one species); Bulgaria, Kherson Region and Crimea in Ukraine, and Uzbekistan (*G. miramae*); Anatolia (one species); Syria (one species), Israel (three species); Egypt, Libya, and Tunisia (three species); Algeria (five species); and Morocco and Canary Islands (nine species). The highest number of species is found in Spain (four species; six with the Canary Islands) and Greece (five species). The species lead a cryptic mode of life and occur in litter, under stones, in caves and tunnels, burrows of rodents, and houses, especially in cellars. Origin of the genus: Mediterranean. Perhaps some of the Balkan species will be proved to be synonyms. In Bulgaria, *G. dalmatina* and *G. miramae* are recorded so far in the Eastern Rhodopes (the latter species uncertainly; after a female only) and in Struma Valley. Most likely the former species occurs in all hot lowland areas in Bulgaria. Origin of the species: *G. dalmatina*, Holomediterranean (stationary type of the range); *G. miramae*, probably Pontomediterranean (expansive type of the range).

7. *Discoptila* (13 taxa: 12 species and one additional subspecies). Range: the Mediterranean from Morocco and Spain to Anatolia and Crimea. The ranges of the species cover: Morocco (one species), Spain (two species, one of them also in Crimea and with a doubtful record in Greece), Italy (two species), Montenegro (one species), continental Greece (two species), Aegean islands, including Crete

Fig. 9 *Callimenus macrogaster*, a peculiar steppe Balkan-Anatolian species occurring in lowlands (Photo: B. Petrov).

(two species, one of which with two subspecies), Bulgaria (one species), and Anatolia (one species). The highest number of species is found in Greece (four species and one additional subspecies). The systematic position of the species in this genus has been discussed by Popov (1984). The species lead a cryptic mode of life and occur in caves, under stones, in old uninhabited houses, cellars, ruins, in litter. They are rare and, except *D. fragosoi* (Bolivar, 1885), all are local endemics. Most species are known from a single locality and only from the type material. The absence of this genus in the remaining part of the Mediterranean is probably due to its higher requirements for humidity compared to those of *Gryllomorpha*. An evidence for this is the higher number of the cave dwellers in *Discoptila* than in *Gryllomorpha*. Origin of the genus: North Mediterranean. *D. buresi* is an Eastern Bulgarian endemic so far known from Varna area and Eastern Rhodopes. Origin of the species: Thracian.

8. Subfamily Pamphaginae. The Bulgarian representatives of this subfamily belong to the genera *Paranocarodes* and *Paranocaracris*. The species of these two genera as well as of other related genera impress with their primitive morphology and are ancient relicts, but also remnants of a thermophilous Tertiary fauna. The taxa with similar origin, ranges, and relationships are very few in Bulgaria. The genus *Paranocarodes* includes 13 taxa (eight species and five additional subspecies). Range: Southeastern Bulgaria, Northeastern Greece, European Turkey, Aegean islands, and Western Anatolia (three species and three additional subspecies), Anatolia (four species and two additional subspecies), and Azerbaijan (one species). The genus consists of preglacial thermophilous relicts inhabiting xerothermic habitats with quite scanty herbaceous vegetation. Center of origin of the genus is Anatolia. Two species occur in the Balkan Peninsula:

Paranocarodes chopardi (Eastern Rhodopes in Bulgaria and Greece; Fig. 10) and *P. straubei*. The latter has three subspecies: *P. s. straubei* (Southeastern Bulgaria, European Turkey, Northwestern Anatolia), *P. s. serratus* Uvarov, 1949 (European Turkey and Anatolia), and *P. s. insularis* Ramme, 1951 (the islands of Lesbos and Chios). In Greece, outside the border of the Balkan Peninsula, occurs also *P. fieberi* (Brunner von Wattenwyl, 1882) with two subspecies: *P. f. fieberi* (Samos Island and Western Anatolia with a doubtful record on Crete) and *P. f. mytilenensis* Ramme, 1951 (Lesbos Island). The genus *Paranocaracris* includes 23 taxa (seven species and 16 additional subspecies). Range: Bulgaria, Greece, and European Turkey (one species with two subspecies), Anatolia (four species and 13 additional subspecies), Caucasus and Transcaucasia (four species and two additional subspecies). This genus also includes preglacial xerothermophilous relicts inhabiting rocky and stony habitats, relatively poor in herbaceous vegetation. Center of origin of this genus is Anatolia. *Paranocaracris bulgaricus* is the only species in the Balkan Peninsula. Its two subspecies are clearly distinguished geographically: *P. b. bulgaricus* (South Bulgaria, the mountains of Falakron and Pangaion in Northeastern Greece, European Turkey) and *P. b. flavotibialis* Willemse, 1974 (the mountains of Smolikas and Olympus in Northwestern Greece). The ranges of the Bulgarian species of both genera are isolated. Origin and distribution of the species occurring in Bulgaria (Fig. 11): *Paranocarodes straubei straubei*, Anatolian; the southern part of the Bulgarian Black Sea coast and adjacent areas of Strandja; *P. chopardi*, Thracian; Eastern Rhodopes (Fig. 10); *Paranocaracris bulgaricus bulgaricus*, Rila–Rhodopean; Kresna Gorge, Central Pirin, Slavyanka, Stargach Mts., Mesta Valley, Western

Fig. 10 *Paranocarodes chopardi*, a local endemic species of the Eastern Rhodopes and preglacial relict, remnant of the thermophilous Tertiary fauna (Photo: S. Beshkov).

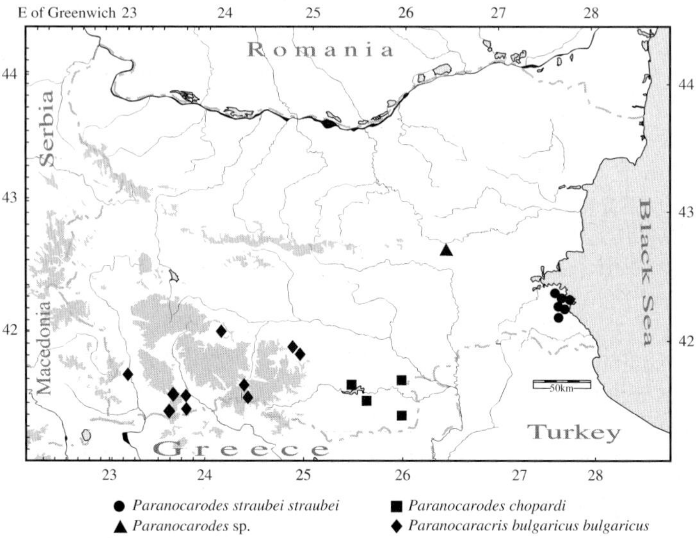

Fig. 11 Distribution in Bulgaria of the species of Pamphaginae: *Paranocarodes straubei straubei, Paranocarodes chopardi, Paranocarodes* sp., *Paranocaracris bulgaricus bulgaricus*. Areas above 1200 m altitude are shaded.

and Central Rhodopes. An isolated population at Sliven needs revision in order to establish to which genus and species it belongs. The species occur mainly at low altitudes but in the presence of suitable habitats they go high up in the mountains: *Paranocaracris b. bulgaricus* up to 1850 m in Pirin and 1800 m in Slavyanka, *P. b. flavotibialis* occurs in Smolikas Mts. (Greece) above the timberline at 1800–2300 m, *Paranocarodes straubei straubei* inhabits in Bulgaria only low altitudes but in Uludağ Mts. (Anatolia) occurs up to 2000 m. The observations in Bulgaria indicate that *Paranocaracris bulgaricus* (and perhaps other species as well) is a calciphilous species which inhabits chalky substrates only. In Pirin Mts., e.g., it occurs only in the narrow strip of Proterozoic marbles, which stretches between Slavyanka Mts. and Orelyak Peak in Central Pirin, and has not been found in the adjacent silicate areas (Popov, 1997b). Subfamily Pamphaginae includes 38 genera. Recently there have been proposals to divide it into four subfamilies: Orchaminae (four genera), Pamphaginae (25 genera), Nocarodesinae (seven genera), and Tropidaucheninae (two genera) as well as to downgrade the genus *Paranocarodes* in rank to a subgenus of the genus *Tropidauchen*. These proposals do not seem to be sufficiently justified. It should be noticed also that the formation of the Latin names of subfamilies Nocarodesinae and Tropidaucheninae has not been grammatically correct.

9. *Bohemanella* (three taxa: one species with three subspecies). Range of the monotypic genus: northern continuous part in Scandinavia, North Russia, almost entire Siberia, North Kazakhstan, Kamchatka, Alaska, and Northwestern Canada and southern isolated territories in Jura, the Alps, Sudetes Mts., Tatra Mts.,

mountains of Kosovo, Macedonia, and Bulgaria, mountains of Southern Siberia, Altai, North Mongolia, Manchuria, and Sakhalin. The three subspecies of the only species *Bohemanella frigida* cover the following parts of the range: *B. f. frigida* in the northern continuous part; *B. f. strandi* (Fruhstorfer, 1921) in the southern, isolated high mountain territories in Europe; and *B. f. kamtschatkae* (Sjöstedt, 1936) in Kamchatka. Ramme (1951) separated the genus *Bohemanella* from the taxonomically very closely related *Melanoplus*. Since then *B. frigida* sometimes has been treated as a species of *Melanoplus*, and sometimes as a representative of a distinct genus. The genus *Melanoplus* unites more than 240 species, all from North America. Therefore the origin of *Melanoplus* is Nearctic. The ancestors of *B. frigida* should have originated in Nearctic and performed a little known type of migration from America to Asia (Popov, 1997b). The same opinion was expressed by Vickery (1987) but later he considered incorrectly this species a Palearctic immigrant in North America (Vickery, 1997). *B. frigida* is a young (glacial) relict with a typical Arctoalpine type of distribution. The northern part of its disjunctive range covers the tundra and forest-tundra of Eurasia and nortwestern part of North America. The southern part includes the subalpine and alpine belts of mountains in Eurasia. The wide northern range indicates that center of origin of *Bohemanella* and *B. frigida* could be the Siberian or Alaskan tundra. The origin of *B. f. strandi* is southern, most likely in the Alps. In Bulgaria, the species occurs in three isolated localities: Botev Peak in Central Stara Planina, the crest between Kamenititsa and Vihren peaks in Northern Pirin, and on Slavyanka (Fig. 1), always above the timberline on an altitude of 2050–2850 m. Some authors divide the subfamily Catantopinae into two subfamilies (Catantopinae and Melanoplinae) and the latter into tribes Melanoplini and Podismini.

Acknowledgments

I would like to thank Dr. Stoyan Beshkov, Dr. Teodora Ivanova, and MSc Boyan Petrov for the photos of *Paranocarodes chopardi, Saga campbelli gracilis*, and *Bradyporus dasypus* kindly placed at my disposal and allowed for publication here. My gratitude is directed also to MSc Silvia Tosheva for her help in preparation of the maps.

References

Andreeva, E., 1980. *Poecilimon bulgaricus* n. sp. (Orthoptera, Tettigoniidae) – a new species from Bulgaria. Acta zoologica bulgarica 15:107–111.

Andreeva, E., 1982. A contribution to the taxonomy of Gryllidae (Orthoptera) of Bulgaria. Acta zoologica bulgarica 19:78–83.

Andreeva, E., 1985. Contribution to the taxonomical studies on genus *Ephippiger* Berth. (Orthoptera, Tettigoniidae) from Bulgaria. Acta zoologica bulgarica 29:17–24.

Bazyluk, W., 1961. Materialien zur Kenntnis von Blattodea der Paläarktis. I–IV. Annales Zoologici (Warszawa) 19(11):417–435.

Buresch, I., 1939. [Referate u. Mitteilungen]. Izvestiya na Balgarskoto entomologichno druzhestvo (Mitteilungen der Bulgarischen Entomologischen Gesellschaft in Sofia; Bulletin de la Société Entomologique de Bulgarie) 10:140–156 (in Bulgarian).

Buresch, I. & G. Peschev, 1955. [Artenbestand und Verbreitung der Geradflügler (Orthoptera) in Bulgarien unter Berücksichtigung der schädlichen Heuschrecken. I. Teil – Acridoidea]. Izvestiya na Zoologicheskiya institut (Bulletin de l'Institut zoologique de l'Académie des sciences de Bulgarie) 4–5:3–107 (in Bulgarian).

Buresch, I. & G. Peschev, 1957. [Artenbestand und Verbreitung der Geradflügler (Orthoptera) in Bulgarien. II. Teil: Blattodea, Mantodea, Gryllodea]. Izvestiya na Zoologicheskiya institut (Bulletin de l'Institut zoologique de l'Académie des sciences de Bulgarie) 6:305–356 (in Bulgarian).

Buresch, I. & G. Peschev, 1958. [Artenbestand und Verbreitung der Geradflügler (Orthopteroidea) in Bulgarien unter Berücksichtigung der schädlichen Arten. III. Teil – Tettigonioidea]. Izvestiya na Zoologicheskiya institut (Bulletin de l'Institut zoologique de l'Académie des sciences de Bulgarie) 7:3–90 (in Bulgarian).

Chobanov, D., 2003. New data on the occurrence of Orthoptera in Bulgaria. Articulata 18(2):227–246.

Drenowski, A., 1936. [Beitrag zur Insektenfauna Bulgariens u. Mazedoniens. II (Lepidoptera, Hymenoptera, Diptera und Orthoptera)]. Izvestiya na Balgarskoto entomologichno druzhestvo (Mitteilungen der Bulgarischen Entomologischen Gesellschaft in Sofia; Bulletin de la Société Entomologique de Bulgarie) 9:237–256 (in Bulgarian).

Drenowski, A., 1938. 7. Ein Beitrag zur Orthopterenfauna Bulgariens u. Mazedoniens (Neue Fundorte). In Drenowski, A. Entomologische Beiträge. Eigene Ausgabe, Sofia: 14.

Drensky, P., 1942. [Über die Insektenfauna des Küstengebietes nördlich von Warna (Ökologische Notizen)]. Izvestiya na Balgarskoto entomologichno druzhestvo (Mitteilungen der Bulgarischen Entomologischen Gesellschaft in Sofia; Bulletin de la Société Entomologique de Bulgarie) 12:15–44 (in Bulgarian).

Gorochov, A. & V. Llorente, 2001. Estudio taxonomico preliminar de los Grylloidea de España (Insecta, Orthoptera). Graellsia 57(2):95–139.

Gruev, B., 1988. Obshta biogeografiya [General Biogeography]. Nauka i izkustvo [Science and Art], Sofia (in Bulgarian).

Gruev, B. & B. Kuzmanov, 1994. Obshta biogeografiya. Vtoro osnovno preraboteno izdanie [General Biogeography. Second Basic Revised Edition]. Universitetsko izdatelstvo "Sv. Kliment Ohridski" [St. Kliment Ohridski University Publishing House], Sofia (in Bulgarian).

Harz, K., 1975. Orthopterologische Beiträge. XIV. Bericht der Naturwissenschaftlichen Gesellschaft Bayreuth 15:49–54.

Harz, K., 1984. Zur Verbreitung von Orthopteren in Bulgarien. Articulata 2(4):91.

Harz, K., 1985a. Zur Orthopterenfauna von Bulgarien. Articulata 2(6):160–161.

Harz, K., 1985b. Dermaptera in Bulgarien. Articulata 2(7):205–206.

Harz, K., 1986. Neues von Orthopteren aus Bulgarien (Saltatoria, Caelifera). Articulata 2(8):266–267.

Heiss, E. & M. Josifov, 1990. Vergleichende Untersuchung über Artenspektrum, Zoogeographie und Ökologie der Heteropteren-Fauna in Hochgebirgen Österreichs und Bulgariens. Berichte des Naturwissenschaftlich-Medizinischen Vereins in Innsbruck 77:123–161.

Hubenov, Z., 1997. Possibilities for using of a system from the really defined natural territories for the faunistic researches in Bulgaria. Acta zoologica bulgarica 49:5–9.

Kaltenbach, A., 1967. Unterlagen für eine Monographie der Saginae. I. Superrevision der Gattung Saga Charpentier (Saltatoria: Tettigoniidae). Beiträge zur Entomologie 17(1–2):3–107.

Karaman, M., 1975. A new subspecies of grasshopper from Bulgaria *Chorthippus mollis pechevi* n. ssp. Acta entomologica Jugoslavica 11(1–2):55–57.

Köhler, G., 1988. Zur Heuschreckenfauna (Saltatoria) des Pirin-Gebirges (VR Bulgarien). Entomologische Nachrichten und Berichte 32(2):77–80.

Lattin, G. de, 1967. Grundriss der Zoogeographie. VEB Gustav Fischer Verlag, Jena.

Mařan, J., 1953. Contribution to the knowledge of the genus *Pholidoptera* Wesm. Acta entomologica Musei Nationalis Pragae 28 [1952] (411):209–221.

Mařan, J., 1958. Eine neue Art der Gattung *Discoptila* Pantel aus Bulgarien. Orthoptera – Gryllidae. Acta entomologica Musei Nationalis Pragae 32(490):37–40.

Nedelkov, N., 1908. [Second contribution to the entomological fauna of Bulgaria]. Periodichesko spisanie na Balgarskoto knizhovno druzhestvo v Sofiya [Periodical Journal of the Bulgarian Literary Society in Sofia] 68/19 [1907] (5–6):411–436 (in Bulgarian).

[Peshev, G.] Pešev, G., 1962a. [Unbekannte Geradflüglerarten (Orthoptera) für die Fauna Bulgariens]. Izvestiya na Zoologicheskiya institut s muzei (Bulletin de l'Institut de zoologie et musée) 11:181–187 (in Bulgarian).

[Peshev, G.] Pešev, G., 1962b. [Composition et répartition oecologique des orthoptères de la montagne Bélassitza]. Izvestiya na Zoologicheskiya institut s muzei (Bulletin de l'Institut de zoologie et musée) 12:59–107 (in Bulgarian).

[Peshev, G.] Pechev, G., 1964. [Les orthoptères de la Thrace]. In Fauna na Trakiya. Sbornik ot statii. Chast I (Die Fauna Thrakiens. Sammelwerk. Band I). Izdatelstvo na Balgarskata akademiya na naukite (Verlag der Bulgarischen Akademie der Wissenschaften), Sofia: 107–144 (in Bulgarian).

[Peshev, G.] Peschev, G., 1970. [Die Orthopteren im Westlichen Balkangebirge]. Izvestiya na Zoologicheskiya institut s muzei (Bulletin de l'Institut de zoologie et musée) 31:173–202 (in Bulgarian).

[Peshev, G.] Pešev, G., 1974. [Les orthoptères (Orthoptera) de la Stara Planina]. Izvestiya na Zoologicheskiya institut s muzei (Bulletin de l'Institut de zoologie et musée) 41:65–93 (in Bulgarian).

[Peshev, G.] Pechev, G., 1975. [La faune orthopterologique (Orthoptera) des Rhodopes. II. Composition specifique, distribution et origine]. In Fauna na Rodopite. Materiali (La faune des Rhodopes. Materiaux). Izdatelstvo na Balgarskata akademiya na naukite (Éditions de l'Académie bulgare des sciences), Sofia: 93–120 (in Bulgarian).

Peshev, G., 1980. A contribution to the taxonomy of the genus *Poecilimon* Fisch. (Orthoptera, Tettigoniidae). Acta zoologica bulgarica 16:30–43.

Peshev, G., 1981. A contribution to the investigations on the species of the genus *Isophya* Br.-W. (Orthoptera, Tettigoniidae) from Bulgaria. Acta zoologica bulgarica 17:27–37.

Peshev, G., 1985. A contribution to the studies of the species of genus *Isophya* Br. W. (Orthoptera, Tettigoniidae) in Bulgaria. II. Acta zoologica bulgarica 29:3–16.

[Peshev, G.] Pešev, G., 1990. [Biogeographical significance of the disjunctive distribution of mountaineous Orthoptera in Bulgaria]. Acta zoologica bulgarica 39:16–24 (in Bulgarian).

Peshev, G. & E. Andreeva, 1986. [Orthoptera in Southwest Bulgaria. I. Fauna]. In Fauna na Yugozapadna Balgariya. Chast 1 (Fauna of Southwestern Bulgaria. Part 1). Izdatelstvo na Balgarskata akademiya na naukite (Publishing House of the Bulgarian Academy of Sciences), Sofia: 82–117 (in Bulgarian).

Peshev, G. & M. Djingova, 1974. [Orthoptera along the Bulgarian Black Sea coast]. Izvestiya na Zoologicheskiya institut s muzei (Bulletin de l'Institut de zoologie et musée) 40:17–46 (in Bulgarian).

[Peshev, G.] Pešev, G. & J. Mařan, 1963. [Étude sur les insects orthoptères du mont Slavianka (Alibotuš)]. Izvestiya na Zoologicheskiya institut s muzei (Bulletin de l'Institut de zoologie et musée) 14:27–69 (in Bulgarian).

Popov, A., 1958. [Cave cricket found in Bulgaria]. Priroda [Nature] (Sofia) 7(6):67 (in Bulgarian).

Popov, A., 1970. [On the study of earwigs in Bulgaria]. Priroda [Nature] (Sofia) 19(1):78–82 (in Bulgarian).

Popov, A., 1984. Le genre *Discoptila* Pant. (Orthoptera, Gryllidae). Biologia Gallo-Hellenica 11(1):65–78.

Popov, A., 1997a. *Poecilimon belasicensis* nom. nov. (Orthoptera, Tettigoniidae). Historia naturalis bulgarica 7:17–18.

Popov, A., 1997b. Orthoptera. In Sakalian, V. (ed), Endemic and Relict Insects in the Pirin National Park, Bulgaria. Pensoft, Sofia & Moscow: 12–24.

Popov, A., 1998. Insects of Bulgaria: Blattodea, Mantodea, Isoptera, Orthoptera, Dermaptera, Embioptera, Megaloptera, Raphidioptera, Neuroptera, Mecoptera. In Meine, C. (ed), Bulgaria's Biological Diversity: Conservation Status and Needs Assessment. Volumes I and II. Pensoft, Sofia & Moscow: 223–227.

Popov, A., 2000a. Entomofaunistic diversity of the Central Balkan National Park. Orthopteroidea, Neuropteroidea, Mecoptera. In Sakalian, M. (ed), Biological Diversity of the Central Balkan National Park. Pensoft, Sofia: 359–362, 540–542, 571–573.

Popov, A., 2000b. Entomofaunistic diversity of the Rila National Park. Orthopteroidea, Neuropteroidea, Mecoptera. In: Sakalian, M. (ed.). Biological Diversity of the Rila National Park. Pensoft, Sofia: 326–331, 527–531, 581–584.

Popov, A. & D. Chobanov, 2004. Dermaptera, Mantodea, Blattodea, Isoptera and Orthoptera of the Eastern Rhodopes (Bulgaria and Greece). In Beron, P. & A. Popov (eds), Biodiversity of Bulgaria. 2. Biodiversity of Eastern Rhodopes (Bulgaria and Greece). Pensoft & National Museum of Natural History, Sofia: 241-309.

Popov, A., D. Chobanov & M. Langourov, 2001. [Blattaria, Mantodea, Isoptera, Orthoptera, Dermaptera, Embioptera of the Kresna Gorge (SW Bulgaria)]. In Beron, P. (ed), Bioraznoobrazie na Kresnenskiya prolom (Biodiversity of Kresna Gorge (SW Bulgaria)). Natsionalen prirodonauchen muzei i Institut po zoologiya pri BAN (National Museum of Natural History (Sofia) & Institute of Zoology, Bulgarian Academy of Sciences), Sofia: 115–130 (in Bulgarian).

Popov, A., Ch. Deltshev, Z. Hubenov, V. Beschovski, D. Dobrev & B. Guéorguiev, 2000. Invertebrate fauna. In Popov, A. & T. Meshinev (eds), High Mountain Treeless Zone of the Central Balkan National Park. Biological Diversity and Problems of its Conservation. Pensoft, Sofia: 339–416.

Popov, A. & Yu. Ganev, 1983. [Polyphagidae (Blattodea) – a new family to the Bulgarian fauna]. Acta zoologica bulgarica 23:67–69 (in Bulgarian).

Ramme, W., 1951. Zur Systematik, Faunistik und Biologie der Orthopteren von Südost-Europa und Vorderasien. Mitteilungen aus dem Zoologischen Museum in Berlin 27:1–431.

Schmidt, G., 1998. *Metrioptera helleri* sp. n. (Ensifera: Tettigoniidae) aus Bulgarien. Articulata 13(1):11–17.

Vickery, V., 1987. The northern Nearctic Orthoptera: their origins and survival. In Baccetti, B. (ed), Evolutionary Biology of Orthopteroid Insects. Ellis Horwood Limited Publishers, Chichester: 581–591.

Vickery, V., 1997. Orthopteroid insects (Orthoptera) of the Yukon. In Danks, H. & J. Downes (eds), Insects of the Yukon. Biological Survey of Canada (Terrestrial Arthropods), Ottawa: 223–239.

Vigna Taglianti, A., P. Audisio, M. Biondi, M. Bologna, G. Carpaneto, A. De Biase, S. Fattorini, E. Piattella, R. Sindaco, A. Venchi & M. Zapparoli, 1999. A proposal for a chorotype classification of the Near East fauna, in the framework of the Western Palearctic region. Biogeographia 20:31–59.

9 Biogeography of the Endemic Carabidae (Coleoptera) in the Central and Eastern Balkan Peninsula

Borislav Guéorguiev

National Museum of Natural History, Bulgarian Academy of Sciences, Tsar Osvoboditel Blvd. 1, 1000 Sofia, Bulgaria, e-mail: bobivg@yahoo.com

Abstract: The endemic carabid faunas of 13 regions of the Central and Eastern Balkan Peninsula are compared. The analysis includes 385 species and subspecies, all of them Balkan endemics and subendemics or local endemics. The following issues are discussed: descriptive biogeography (definition of diversity and centers of diversity, chorology and specificity, taxonomic structure and macroecological habitats, and affinities of taxonomic and macrohabitat structure); causal biogeography (faunal elements and complexes, their time of formation and/or arrival, affinity of biogeographical structure, relict, orthoselective and migratory components, definition of inter- and postglacial refugia and centers of dispersal), and regional division based on taxonomic, macrohabitat, and biogeographical structure

1 Introduction

Complex faunas consisting of various biogeographical elements are formed in certain areas as a result of topography and different climates in different epochs. Could a local assemblage of species be assigned to definite faunal complex(es) (Gruev, 1990; Popov and Deltshev, 1997)? Looking for the answers to such questions, I do not envisage a simple fusion of diverse lineages, e.g. into a "Balkan complex". An analysis of a concrete fauna should address its elements using taxonomic knowledge on specific groups, faunal data, and a geographical approach in the framework of historical biogeography.

During the last 150 years, the Central and Eastern Balkans have been an attractive field for many European researchers of Carabidae. This was due to the specific geographical position, presence of mountain, temperate, Submediterranean, and Mediterranean climates, a highly complex topography, large areas occupied by limestone, and, last but not least, the specific history of this region. These conditions defined the pattern of the Central and Eastern Balkan fauna, which differs from the typical (and well-known) Central European ground beetle fauna. In addition, the core of the Central and Eastern Balkan fauna (20 to 25% of the species-group taxa) is unique and not found anywhere else. For many current carabid researchers, the region remains not sufficiently explored and still holds many secrets. I believe that the Balkan Peninsula, taken as a whole, contains the richest ground beetle diversity compared to other very diverse European areas (Turin, 1981; Vigna Taglianti, 1993; Zaballos and Jeanne, 1994; Serrano, 2003), including the Caucasus Major (Kryzhanovskij et al., 1995) as well; such statement, however, cannot be yet confirmed.

V. Fet and A. Popov (Eds.), Biogeography and Ecology of Bulgaria, 297–356.

2 Area Studied

The studied area is limited to the central and eastern parts of the Balkan Peninsula (Fig. 1). It is divided into 13 regions according to political or natural factors. Six of those regions lie west of the rivers Morava and Vardar (= Axios): *Northwestern Serbia, Southwestern Serbia, Dardanian Region* (incl. Eastern Kosovo), *Western Kosovo, Albania*, and *Western Macedonia*. Other seven areas lie east of the upper portions of these rivers: *Eastern Serbia, Northern Bulgaria* (excluding South

Fig. 1 The Balkan Peninsula and delineation of the area and regions studied. NS, Northwestern Serbia; SS, Southwestern Serbia; DR, Dardanian Region; ES, Eastern Serbia; NB, Northern Bulgaria; DO, Dobrogea; WK, Western Kosovo; AL, Albania; WM, Western Macedonia; EM, Eastern Macedonia; SB, Southern Bulgaria; NG, Northeastern Greece; ET, European Turkey.

Dobrudja), *Dobrogea* (Romanian North Dobrogea+Bulgarian South Dobrudja), *Eastern Macedonia, Southern Bulgaria, Northeastern Greece*, and *European Turkey*. The rivers Rogačica, Đetinja, and Zapadna (West) Morava limit *Northwestern Serbia* and *Southwestern Serbia*. The *Dardanian Region* is restricted by the rivers Južna (South) Morava, Zapadna (West) Morava, Ibar, Sitnica, and Nerodimka. Danube, Velika Morava, Krševicka Reka, and Pčinja (at the Prohor Pčinjski Monastery) delineate *Eastern Serbia*. *Northern Bulgaria* and *Southern Bulgaria* are separated due to geomorphology and landscape by the southern edge of the Stara Planina (= Balkan) Mts. In addition to other natural factors, the line Tutrakan–Batova River delineates *Dobrogea* in the south. *Western Macedonia* is limited in the east by the rivers of Pčinja and Vardar, and *Northeastern Greece* lies to the east of Vardar.

3 Study Goals

The major question posed here is "What are the elements composing the endemic carabid fauna in the Central and Eastern Balkans, and what are their origins"? Another, secondary goal is to test previously proposed biogeographical divisions of this region. The first step (*descriptive biogeography*) is to determine diversity, distribution (chorotype), specificity (level of endemism), taxonomic structure, and principal macrohabitats. The second step (*analytical* or *causal biogeography*) includes definition of the faunal elements and complexes; the relict, orthoselective, and migrant components; the main Pleistocene refugia and centers of postglacial dispersal; and former and present centers of speciation. We attempt also to trace the history of extant fauna and to outline the most important reasons for the current diversity and distribution. After an analysis of taxonomic, macroecological, and biogeographical structure, a biogeographical division of the Balkan Peninsula could be suggested.

4 Material and Methods

In the Balkans, the endemic Carabidae are represented by numerous species and subspecies, as well as some supraspecific taxa, with different origin in space and time. This interesting part of the carabid fauna, never in the past studied from comparative and historical biogeographical aspect, was chosen to be the object of present work, based on numerous literature as well as field and museum material. Carabidae is treated here after Lawrence and Newton (1995) and Löbl and Smetana (2003). The main boundary, which divides the Balkan fauna into western and eastern portions, is the Morava–Vardar biogeographical barrier (Jeannel, 1928; Furon, 1953; Tabacaru, 1970; Guéorguiev, 1977). The location and the size of the geographic range define the type of distribution. Taxa restricted to a certain locality (cave, karstic system, part of a mountain range, a single mountain massif, etc.)

are treated as local endemics. Some of those are listed for two or three regions when they occur in one and the same boundary massif; e.g. *Nebria attemsi* and *N. ganglbaueri ganglbaueri* both inhabit the Šar Massif and are listed for the adjacent parts of Western Kosovo, Albania, and Western Macedonia. The Balkan endemics are found either only in certain regions or in the majority of regions of the Balkan Peninsula. The Balkan subendemics are also found in smaller territories outside the Peninsula; only subendemics with well-known distribution are included here. The principal habitat is defined by the species' preference of one of three macroecological environments of the Palearctic terrestrial fauna (de Lattin, 1967): the arboreal (woodland, woodland-like, moist meadow, riverside, littoral, and other humid habitats), the eremial (steppe-like, xeric scrub, sand, and other arid habitats), and the oreal (high-altitude habitats) biochores. It is difficult to determine the specific biochore of some taxa. For example, semixeric *Lamprias* and *Paracymindis* found in open Submediterranean terrains and halophilous and halobiont species of *Dyschiriodes* known from riverside and littoral habitats are referred more or less provisionally to the arboreal biochore. Conversely, semimesic species of *Cicindela* s. str. living in open and sandy areas are treated as eremial. In addition, a gradual transition between macrohabitats is observed in several genus-group taxa. Such shifts have occurred in the Balkans from arboreal to oreal (some *Trechus* species groups, *Bembidionetolitzkya, Morphnosoma, Aphaon*) and from eremial to oreal (*Pachystus, Pelor*) habitats.

The affinity of the regional faunas is characterized through relative abundance analysis (Pesenko, 1982; Baev and Penev, 1995), which employs the BIODIV software functions "*analysis of quantitative data*" and "*association measure*", and the numerical form of the Czekanovski–Dice–Sørensen index. Following this method, we determined the taxonomic (at species and genus levels), macrohabitat, and biogeographical structure.

Division of the fauna into "relict", "orthoselective", and "migratory" categories is accepted after the Russian florist Krasilov (after Takhtajan, 1974), taking into account the combined influence of evolution, migration, and extinction on each taxon of the extant fauna. This analytical approach was recently adopted for Carabidae (Turin et al., 2003). The formula $F = f1 + f2 + f3$ is applied, where F are all extant taxa of the fauna; f1 are ancient taxa unchanged or slightly changed since the Tertiary; f2 are ancient taxa, which reflect direct results of permanent and long-term restructuring of biota; and f3 is a clearly allochthonous fauna which migrated to the area due to the recent biotic changes.

Through tracking places of probable formation and the heterochronic phases of colonization, each endemic taxon is assigned to a definite faunal element, a faunal complex, and a center of postglacial dispersal. This part of our study deals mainly with the geographical approach in the historical biogeography (Vuilleumier, 1975), which addresses the present distribution of a species and its relatives to a "center of diversity". We reviewed the available phylogenetic and ecological data for the concrete taxon as well as the respective paleoclimatic, paleogeological, and paleoecological events (Gruev and Kuzmanov, 1994; Palamarev, 1997; Vaptzarov, 1997).

We followed the methods and terminology of the European school of biogeography (Stegmann, 1938; de Lattin, 1967; Kryzhanovskij and Starobogatov, 1974; Müller, 1974; Bănărescu and Boşcaiu, 1978; Vtorov and Drozdov, 1978; Sedlag and Weinert, 1987; Lopatin, 1989). The concept of a faunal element is addressed from geographical, phyletic, and chronological perspectives. The end of Pliocene is the boundary according to which the species are assigned to be either neoendemics or paleoendemics, and either autochthonous or allochthonous. The approach to the history of the carabid fauna, its probable origin, and the directions and times of colonization follows, in general, Casale and Vigna Taglianti (1982, 1993, 1999) and Turin et al. (2003).

5 Principal Faunal Features

Family Carabidae is one of the best studied insect groups in the region; it is estimated that we know about 95% of the genera and at least 90% of the species. The same figures could be referred also to the endemic taxa, while the portion of described endemic species is estimated to be ca. 85% of their actual number. Undoubtedly, the new taxa will continue to be described in near future. The regional karst is more or less insufficiently studied, and its area is one of the largest on Earth. Most new Balkan endemic ground beetles are expected to be described from the Greek limestone habitats (Casale and Giachino, unpublished data), a region beyond the scope of our present study. Despite the good level of faunal knowledge, some systematic and chorological problems still exist. A few species such as *Tachyura krueperi, Harpalus serbicus*, and *H. serdicanus* Apfelbeck, have uncertain status (Löbl and Smetana, 2003). Information about some other species likely is based on misidentifications or inaccurate locality data (e.g. records of *Bembidion praeustum viridifluum* for Macedonia). This also refers to the records of *Deltomerus sterbai, Pterostichus macedonicus, Harpalus serdicanus, Calathus ellipticus* Reitter, *C. glabricollis*, and *Acinopus subquadratus* Brullé from Bulgaria (Guéorguiev and Guéorguiev, 1995), as well as to the record of the latter species from Albania. All these species are omitted further for the areas mentioned. *Licinus cassideus dohrnii* Fairmaire and *L. submarginatus* J.R. Sahlberg, recognized as valid taxa by Löbl and Smetana (2003), are treated here as synonyms of *L. cassideus* F.

Family Carabidae is known to a different degree in different regions: 495 species and subspecies are recorded for Albania (our unpublished data) and 766, for Bulgaria (Guéorguiev and Guéorguiev, 1995, with modifications). At least 520 taxa are known for Serbia (with Kosovo; Ćurčić, 2000), and ca. 420, for Macedonia (Hristovski, Guéorguiev, unpublished data), accounting for the species splitting by the ex-Yugoslavian specialists. The faunas of Dobrogea, Northeastern Greece, and European Turkey are inadequately explored and their total lists have still to be clarified; on the other hand, both the list and the distribution of the endemics in these areas are better known.

6 Descriptive Biogeography

6·1 Diversity and centers of diversity

Table 1 lists 385 species-group endemic taxa of Carabidae belonging to 87 genus-group taxa and to 30 tribes and subtribes, among them an undescribed species.

The highest diversity of endemic species and subspecies is recorded for Albania (161 taxa), Western Macedonia (116), Southern Bulgaria (104), and Northern Bulgaria (71). In Albania, the share of endemic carabids in the entire carabid fauna is 32.5%. This share in Western Macedonia, compared to the total Macedonian fauna, is 27.6%, and in Eastern Macedonia it is ca. 7.1% (almost four times less than in Western Macedonia). The endemic shares of Southern Bulgaria and Northern Bulgaria, both compared to the entire Bulgarian carabid fauna, are 13.58% and 9.27%, respectively. The share of carabid endemics in different regions of Serbia (with Kosovo) are: Western Kosovo, 10.58%; Eastern Serbia, 7.5%; Southwestern Serbia, 6.92%; Dardanian Region, 5.19%; and Northwestern Serbia, 3.65%. For the time being, the degree of endemism in Northeastern Greece, Dobrogea, and European Turkey is unknown. Undoubtedly, Northeastern Greece has the highest share of endemics among these three regions, due to the presence of many known endemic taxa and the important role of the Rhodopes as a focus of recent vicariant events.

The maximum diversity is located in the western portions (Albania + Western Macedonia) of the Central Balkans, due to the complex topography of this area, especially its prevailing broken mountain terrains. This type of relief defines numerous microhabitats, and represents the main reason for speciation in the Southern Dinaric center of diversity. In the north, from Albania to Western Kosovo, the richness of endemics decreases more than three times. From Western Kosovo to Southwestern Serbia and from Southwestern Serbia to Northwestern Serbia, the decrease continues: 1.5 and 1.9 times, respectively. The decrease in the richness of endemics from Western Macedonia to Dardanian Region and from Western Macedonia to Eastern Macedonia is ca. 4.3 and 3.9 times, respectively. Northward, comparing the pairs Eastern Macedonia–Eastern Serbia and Dardanian Region–Eastern Serbia, the richness of endemics increases 1.26 and 1.4 times. The increase in the Dardanian Region–Eastern Serbia is due to the direct influence of the Southern Carpathian and Stara Planina foci of speciation. Moving to the east, from Eastern Serbia to Northern Bulgaria, from Eastern Macedonia to Southern Bulgaria, and from Eastern Macedonia to Northeastern Greece, the richness of endemics increases again: 1.97, 3.2, and 1.6 times, respectively. This increase, although not so high as the western one, is due to the presence of other centres of speciation, e.g. Stara Planina and the Rhodopes. In the east (Dobrogea and European Turkey), the endemic richness considerably decreases.

The maximum diversity of endemic Carabidae in (Albania+Western Macedonia) and in (Northern Bulgaria+Southern Bulgaria) regions coincides with the mountain ranges reaching above 2000 m a.s.l., which have been both refugia and important

Table 1 Distribution of the endemic carabid species and subspecies in the Central and Eastern Balkans

Taxa	NS	SS	DR	ES	WK	AL	WM	EM	NB	SB	NG	DO	ET	DT	PH	FE
Leistus (Pogonophorus) magnicollis Motschulsky	–	–	+	–	+	+	+	–	+	+	–	–	+	BE	Arb	EU
Nebria (Alpaeus) aetolica galicica Horvatovich	–	–	–	–	–	+	+	–	–	–	–	–	–	LE	Ore	AAR
Nebria (Alpaeus) aetolica macedonica Mařan	–	–	–	–	–	–	+	–	–	–	–	–	–	LE	Ore	AAR
Nebria (Alpaeus) aetolica papingutica Horvatovich	–	–	–	–	–	+	–	–	–	–	–	–	–	LE	Ore	AAR
Nebria (Alpaeus) aetolica rambouseki Mařan	–	–	–	–	–	+	+	–	–	–	–	–	–	LE	Ore	AAR
Nebria (Alpaeus) attemsi Apfelbeck	–	–	–	–	+	+	+	–	–	–	–	–	–	LE	Ore	AAR
Nebria (Alpaeus) bosnica jezercensis Horvatovich	–	–	–	–	–	+	–	–	–	–	–	–	–	LE	Ore	AAR
Nebria (Alpaeus) bosnica kulti Mařan	–	–	–	–	–	+	–	–	–	–	–	–	–	LE	Ore	AAR
Nebria (Alpaeus) bosnica prokletiensis Horvatovich	–	–	–	–	+	+	–	–	–	–	–	–	–	LE	Ore	AAR
Nebria (Alpaeus) bosnica sturanyi Apfelbeck	–	+	–	–	+	+	–	–	–	–	–	–	–	WBE	Ore	AAR
Nebria (Alpaeus) dahlii dahlii (Duftschmid)	–	+	–	–	+	–	–	–	–	–	–	–	–	WBE	Ore	AAR
Nebria (Alpaeus) eugeniae K. Daniel	–	–	–	–	–	–	–	–	–	+	–	–	–	LE	Ore	AAR
Nebria (Alpaeus) ganglbaueri ganglbaueri Apfelbeck	–	–	–	–	+	+	+	–	–	–	–	–	–	LE	Ore	AAR
Nebria (Alpaeus) ganglbaueri gyalicaensis Horvatovich	–	–	–	–	–	+	–	–	–	–	–	–	–	LE	Ore	AAR
Nebria (Alpaeus) ganglbaueri korabensis Horvatovich	–	–	–	–	–	+	+	–	–	–	–	–	–	LE	Ore	AAR

continued

Table 1 continued

Taxa	NS	SS	DR	ES	WK	AL	WM	EM	NB	SB	NG	DO	ET	DT	PH	FE
Nebria (*Alpaeus*) *ganglbaueri ljubotensis* Horvatovich	–	–	–	–	+	–	+	–	–	–	–	–	–	LE	Ore	AAR
Nebria (*Alpaeus*) *ganglbaueri matejkai* Mařan	–	–	–	–	–	–	+	–	–	–	–	–	–	LE	Ore	AAR
Nebria (*Alpaeus*) *ganglbaueri merditana* Apfelbeck	–	–	–	–	–	+	–	–	–	–	–	–	–	LE	Ore	AAR
Nebria (*Alpaeus*) *ganglbaueri zebiaensis* Horvatovich	–	–	–	–	–	+	–	–	–	–	–	–	–	LE	Ore	AAR
Nebria (*Alpaeus*) *hybrida hybrida* Reitter	–	–	–	–	–	–	–	–	–	+	–	–	–	LE	Ore	AAR
Nebria (*Alpaeus*) *hybrida pirinensis* Horvatovich	–	–	–	–	–	–	–	–	–	+	–	–	–	LE	Ore	AAR
Nebria (*Alpaeus*) *hybrida rhodopensis* Horvatovich	–	–	–	–	–	–	–	–	–	+	–	–	–	LE	Ore	AAR
Nebria (*Alpaeus*) *peristerica peristerica* Apfelbeck	–	–	–	–	–	–	+	–	–	–	–	–	–	LE	Ore	AAR
Nebria (*Alpaeus*) *peristerica kaimakcalensis* Horvatovich	–	–	–	–	–	–	+	–	–	–	–	–	–	LE	Ore	AAR
Nebria (*Alpaeus*) *rhilensis* J. Frivaldszky	–	–	–	–	–	–	–	–	–	+	–	–	–	LE	Ore	AAR
Nebria (*Alpaeus*) *speiseri* Ganglbauer	–	–	–	–	+	–	–	–	–	–	–	–	–	LE	Ore	AAR
Nebria (*Nebria*) *kratteri kratteri* Dejean	–	–	–	–	–	+	+	–	–	–	–	–	–	WBSE	Ore	AP
Cicindela (*Cicindela*) *campestris olivieria* Brullé	–	–	–	–	–	+	–	–	–	–	+	–	+	BE	Ere	POM
Cicindela (*Cicindela*) *campestris palustris* Motschulsky	–	–	–	–	–	–	–	–	–	–	–	–	+	EBSE	Ere	POM
Cicindela (*Cicindela*) *monticola albanica* Apfelbeck	+	+	+	+	+	+	+	–	+	+	+	–	–	BE	Ere	POM
Cicindela (*Cicindela*) *monticola rumelica* Apfelbeck	–	–	–	–	–	–	–	–	+	+	–	+	+	EBSE	Ere	POM

Calosoma (*Microcallisthenes*) *pentheri* Apfelbeck	–	–	–	–	+	+	–	–	–	–	–	–	–	WBE	Ore	NA
Calosoma (*Microcallisthenes*) *relictum* Apfelbeck	–	–	–	–	+	+	+	–	–	–	–	–	–	WBE	Ore	NA
Carabus (*Morphocarabus*) *scheidleri illigeri* Dejean	+	+	+	–	–	–	–	–	–	–	–	–	–	WBE	Arb	EAS
Carabus (*Morphocarabus*) *scheidleri versicolor* I. Frivaldszky von Frivald	–	–	–	–	–	–	–	–	+	–	–	–	–	EBE	Arb	EAS
Carabus (*Morphocarabus*) *scheidleri simulator* Kraatz	–	–	–	+	–	–	–	–	–	–	–	–	–	EBE	Arb	EAS
Carabus (*Trachycarabus*) *scabriusculus bulgarus* Lapouge	+	–	+	+	–	–	+	+	+	+	–	+	+	BSE	Arb	EAS
Carabus (*Eucarabus*) *ullrichii arrogans* Schaum	–	–	–	+	–	–	–	–	–	–	–	–	–	LE	Arb	EU
Carabus (*Eucarabus*) *ullrichii rhilensis* Kraatz	+	+	–	+	–	–	–	+	+	+	–	–	–	BSE	Arb	EU
Carabus (*Tachypus*) *cancellatus alessiensis* Apfelbeck	–	–	–	–	–	+	–	–	–	–	–	–	–	WBE	Arb	EU
Carabus (*Tachypus*) *cancellatus intermedius* Dejean	+	+	+	+	–	–	–	–	+	+	–	+	–	BSE	Arb	EU
Carabus (*Archicarabus*) *montivagus vellepiticus* Hampe	–	–	–	–	–	+	–	–	–	–	–	–	–	WBE	Arb	EU
Carabus (*Oreocarabus*) *preslii preslii* Dejean	–	–	–	–	–	+	–	–	–	–	–	–	–	WBE	Arb	EU
Carabus (*Oreocarabus*) *preslii neumeyeri* (Schaum)	–	–	–	–	–	+	+	–	–	–	–	–	–	WBE	Arb	EU
Carabus (*Tomocarabus*) *convexus dilatatus* Dejean	+	+	+	+	–	+	+	+	+	+	+	+	+	BSE	Arb	EAS

continued

Table 1 continued

Taxa	NS	SS	DR	ES	WK	AL	WM	EM	NB	SB	NG	DO	ET	DT	PH	FE
Carabus (*Pachystus*) *cavernosus cavernosus* I. Frivaldszky von Frivald	+	–	+	+	–	+	+	+	+	+	+	–	–	BE	Ore	SA
Carabus (*Pachystus*) *graecus graecus* Dejean	–	–	–	–	–	+	–	–	–	–	–	–	–	BE	Ere	SA
Carabus (*Platycarabus*) *irregularis ramanus* Sokolář	–	+	–	–	–	–	–	–	–	–	–	–	–	WBSE	Arb	EU
Carabus (*Megodontus*) *caelatus dalmatinus* Duftschmid	–	–	–	–	–	+	–	–	–	–	–	–	–	WBE	Arb	EAS
Carabus (*Megodontus*) *caelatus sarajevoensis* Apfelbeck	+	+	–	–	–	+	+	–	–	–	–	–	–	WBE	Arb	EAS
Carabus (*Megodontus*) *croaticus bosnicus* Apfelbeck	–	+	+	–	+	+	+	–	–	–	–	–	–	WBE	Arb	EAS
Carabus (*Megodontus*) *violaceus azurescens* Dejean	+	+	+	+	+	+	+	+	+	+	+	–	–	BE	Arb	EAS
Carabus (*Megodontus*) *violaceus dryas* Gistl	–	–	–	–	+	+	+	–	–	+	+	–	–	BE	Arb	EAS
Carabus (*Lamprostus*) *torosus torosus* I. Frivaldszky von Frivald	–	–	–	–	–	–	–	–	+	+	–	+	+	EBE	Ere	COL
Carabus (*Procrustes*) *coriaceus excavatus* Charpentier	–	–	–	–	–	+	+	–	–	–	–	–	–	WBE	Arb	SA
Carabus (*Procrustes*) *coriaceus kindermanni* Waltl	–	–	–	–	–	–	–	–	–	+	+	+	+	EBSE	Arb	SA
Carabus (*Procerus*) *gigas gigas* Creutzer	+	+	+	+	–	+	+	–	+	+	+	+	–	BSE	Arb	SA
Carabus (*Procerus*) *scabrosus scabrosus* Olivier	–	–	–	–	–	–	–	–	+	+	–	–	+	EBE	Arb	SA
Cychrus semigranosus balcanicus Hopffgarten	–	–	–	+	–	–	–	+	+	+	+	–	–	EBE	Arb	EU
Cychrus semigranosus montenegrinus Apfelbeck	–	–	–	–	+	+	+	–	–	–	–	–	–	WBE	Arb	EU
Aptinus (*Aptinus*) *acutangulus* Chaudoir	–	–	–	–	+	+	–	–	–	–	–	–	–	WBE	Arb	PAM

Aptinus (*Aptinus*) *lugubris* Schaum	–	–	–	–	–	–	+	–	–	–	–	–	–	WBE	Arb	PAM
Aptinus (*Aptinus*) *merditanus merditanus* Apfelbeck	–	–	–	–	–	+	+	–	–	–	–	–	–	WBE	Arb	PAM
Aptinus (*Aptinus*) *merditanus orientalis* Hůrka	–	–	–	–	–	–	–	–	–	+	–	–	–	LE	Arb	PAM
Clivina subterranea Decu, Nitzu et Juberthie	–	–	–	–	–	–	–	–	–	–	–	+	–	LE	Arb	POM
Reicheadella bischoffi (Meschnigg)	–	–	–	–	–	+	–	–	–	–	–	–	–	WBE	Arb	PAM
Reicheadella xanthina Casale, Giachino, Jalžić et Vailati	–	–	–	–	–	–	–	–	–	–	+	–	–	LE	Arb	PAM
Dyschiriodes (*Dyschiriodes*) *bacillus arbensis* J. Müller	–	–	–	–	–	+	–	–	–	–	–	–	–	WBSE	Arb	POM
Dicropterus brevipennis serbicus (Ganglbauer)	–	–	–	+	–	–	–	–	–	–	–	–	–	LE	Arb	NA
Parvocaecus turcicus (Coiffait)	–	–	–	–	–	–	–	–	–	–	–	–	+	LE	Arb	SA
Winklerites lagrecai Casale, Giachino et M. Etonti	–	–	–	–	–	–	–	–	–	–	+	–	–	LE	Arb	NA
Winklerites perpusillus (Reitter)	–	–	–	–	–	–	–	–	–	–	+	–	–	LE	Arb	NA
Winklerites weiratheri (J. Müller)	–	–	–	–	–	–	–	–	–	–	+	–	–	LE	Arb	NA
Asaphidion nebulosum balcanicum Netolitzky	–	+	+	+	–	+	–	+	+	+	–	+	+	BSE	Arb	EAS
Bembidion (*Trepanes*) *maculatum serbicum* Apfelbeck	–	–	–	+	–	–	–	–	–	–	–	–	–	LE	Arb	ALT
Bembidion (*Bembidionetolitzkya*) *rhodopense* Apfelbeck	–	–	–	–	+	+	+	–	+	+	–	–	–	BE	Ore	AAR
Bembidion (*Bembidionetolitzkya*) *justinae* Meschnigg	–	–	–	–	–	+	–	–	–	–	–	–	–	LE	Arb	AAR

continued

Table 1 continued

Taxa	NS	SS	DR	ES	WK	AL	WM	EM	NB	SB	NG	DO	ET	DT	PH	FE
Bembidion (*Bembidionetolitzkya*) *meschniggi* Netolitzky	–	–	–	–	–	+	–	–	–	–	–	–	–	LE	Arb	AAR
Bembidion (*Bembidionetolitzkya*) *brunoi* Bonavita	–	–	–	–	–	+	–	–	–	–	–	–	–	LE	Arb	AAR
Bembidion (*Ocyturanes*) *balcanicum balcanicum* Apfelbeck	–	+	+	–	+	+	+	–	+	+	–	–	–	BE	Ore	PAR
Bembidion (*Ocyturanes*) *pindicum* Apfelbeck	–	–	–	–	–	+	+	+	–	+	–	–	–	BE	Ore	PAR
Bembidion (*Ocyturanes*) *stolfai* J. Müller	–	–	–	–	–	+	–	–	–	–	–	–	–	LE	Ore	PAR
Bembidion (*Peryphanes*) *grandipenne grandipenne* Schaum	–	–	–	–	–	+	+	+	–	+	–	–	–	BE	Arb	POM
Bembidion (*Peryphus*) *cruciatum albanicum* J. Müller	–	–	–	–	–	+	+	–	–	+	+	–	–	BE	Arb	ALT
Bembidion (*Ocydromus*) *aetolicum aetolicum* Apfelbeck	–	–	–	–	–	+	–	–	–	–	–	–	–	WBE	Arb	EAS
Bembidion (*Omoperyphus*) *hypocrita illyricum* Netolitzky	–	–	–	–	–	–	+	–	–	–	–	–	–	BE	Arb	AP
Bembidion (*Sinechostictus*) *kosti* Matits	–	+	–	–	–	–	–	–	–	–	–	–	–	LE	Arb	EU
Tachyura (*Tachyura*) *krueperi* (Apfelbeck)	–	–	–	–	–	–	–	–	–	+	–	–	–	LE	Arb	POM
Deltomerus (*Deltomerus*) *malissorum* Apfelbeck	–	–	–	–	–	+	–	–	–	–	–	–	–	LE	Ore	AAR
Deltomerus (*Deltomerus*) *nopcsae* Csiki (? = *parumpunctatus* Zamotajlov)	–	–	–	–	+	+	–	–	–	–	–	–	–	LE	Ore	AAR
Deltomerus (*Deltomerus*) *sterbai* (Rambousek)	–	–	–	–	–	–	+	–	–	–	–	–	–	LE	Ore	AAR
Deltomerus (*Paradeltomerus*) *paradoxus paradoxus* Apfelbeck	–	–	–	–	+	+	+	–	–	–	–	–	–	LE	Ore	AAR
Deltomerus (*Paradeltomerus*) *paradoxus korabensis* J. Müller	–	–	–	–	–	+	+	–	–	–	–	–	–	LE	Ore	AAR
Trechus (*Trechus*) *babinjensis* Jeannel	–	–	–	–	+	+	–	–	–	–	–	–	–	LE	Ore	NA

Trechus (*Trechus*) *valbonensis* Jeannel	–	–	–	–	–	+	–	–	–	–	–	–	–	LE	Ore	NA
Trechus (*Trechus*) *subnotatus subnotatus* Dejean	–	+	+	–	–	+	+	–	+	+	–	–	–	BE	Arb	SA
Trechus (*Trechus*) *subnotatus ljubetensis* Apfelbeck	–	–	–	–	+	+	+	–	–	–	–	–	–	LE	Arb	SA
Trechus (*Trechus*) *byzantinus* Apfelbeck	–	–	–	–	–	–	–	–	–	–	–	–	+	LE	Arb	NA
Trechus (*Trechus*) *cardioderus athonis* Schatzmayr	–	–	–	–	–	–	+	–	–	–	+	–	–	BE	Arb	NA
Trechus (*Trechus*) *cardioderus golesnicensis* Apfelbeck	–	–	–	–	–	–	+	–	–	+	–	–	–	BE	Arb	NA
Trechus (*Trechus*) *cardioderus transdanubiensis* Nonveiller, Pavićević et Popović	–	–	–	+	–	–	–	–	–	–	–	–	–	LE	Arb	NA
Trechus (*Trechus*) *irenis* Csiki	+	+	+	+	+	–	–	–	+	+	–	–	–	BE	Arb	NA
Trechus (*Trechus*) *rhilensis* Kaufmann	–	–	–	–	–	–	–	–	–	+	–	–	–	EBE	Ore	NA
Trechus (*Trechus*) *gulickai* Löbl	–	–	–	–	–	–	–	–	–	+	–	–	–	LE	Ore	NA
Trechus (*Trechus*) *demircapicus* P. Moravec	–	–	–	–	–	–	–	–	–	+	–	–	–	LE	Ore	NA
Trechus (*Trechus*) *orphaeus* Pawlowski	–	–	–	–	–	–	–	–	–	+	–	–	–	LE	Ore	NA
Trechus (*Trechus*) *pachycerus pachycerus* Apfelbeck	–	–	–	–	–	–	+	–	–	–	–	–	–	LE	Ore	NA
Trechus (*Trechus*) *pachycerus jacupicaensis* Roubal	–	–	–	–	–	–	+	–	–	–	–	–	–	LE	Ore	NA
Trechus (*Trechus*) *goebli goebli* Breit	–	–	–	–	–	–	+	–	–	–	–	–	–	LE	Ore	NA
Trechus (*Trechus*) *goebli matchai* Jeannel	–	–	–	–	–	–	+	–	–	–	–	–	–	LE	Ore	NA
Trechus (*Trechus*) *rambouseki* Breit	–	–	–	–	–	–	–	–	–	+	–	–	–	LE	Ore	NA
Trechus (*Trechus*) *thessalonicus* Jeannel	–	–	–	–	–	–	+	–	–	–	–	–	–	LE	Ore	NA
Trechus (*Trechus*) *centralis* Nonveiller, Pavićević et Popović	–	–	+	–	–	–	–	–	–	–	–	–	–	LE	Ore	NA

continued

Table 1 continued

Taxa	NS	SS	DR	ES	WK	AL	WM	EM	NB	SB	NG	DO	ET	DT	PH	FE
Trechus (*Trechus*) *pulchellus koprivnicensis* P. Moravec et Lompe	+	+	–	–	–	–	–	–	–	–	–	–	–	WBE	Arb	NA
Trechus (*Trechus*) *pulchellus ventricosus* Jeannel	–	–	–	+	–	–	–	–	–	–	–	–	–	LE	Arb	NA
Trechus (*Trechus*) *szujeckii* Pawlowski	–	–	–	–	–	–	–	–	–	+	–	–	–	LE	Arb	NA
Trechus (*Trechus*) *priapus priapus* K. Daniel	+	+	–	–	+	+	+	+	–	+	–	–	–	BE	Arb	NA
Trechus (*Trechus*) *priapus divergens* Meixner	–	–	–	–	–	–	–	–	+	–	–	–	–	LE	Arb	NA
Trechus (*Trechus*) *priapus medius* Meixner	–	–	–	+	–	–	–	–	–	+	–	–	–	EBE	Arb	NA
Trechus (*Trechus*) *priapus serbicus* Apfelbeck	–	–	+	–	–	–	–	–	–	–	–	–	–	LE	Arb	NA
Trechus (*Trechus*) *rhodopeius* Jeannel	–	–	–	–	–	–	–	–	+	+	–	–	–	EBE	Arb	NA
Trechus (*Trechus*) *merditanus* Apfelbeck	–	–	–	–	–	+	–	–	–	–	–	–	–	LE	Arb	NA
Trechus (*Trechus*) *hajeki* Reitter	–	–	–	–	–	–	+	–	–	–	–	–	–	LE	Ore	NA
Trechus (*Trechus*) *egregius* Jeannel	–	–	–	–	–	+	–	–	–	–	–	–	–	LE	Ore	NA
Trechus (*Trechus*) *dumitrescui* Decou	–	–	–	–	–	–	–	–	–	–	–	+	–	LE	Arb	NA
Trechus (*Trechus*) *albanicus* Apfelbeck	–	–	–	–	+	+	+	–	–	–	–	–	–	LE	Ore	NA
Trechus (*Trechus*) *kobingeri kobingeri* Apfelbeck	+	–	+	–	–	+	+	–	–	–	–	–	–	WBE	Ore	NA
Trechus (*Trechus*) *kobingeri pawlowskianus* P. Moravec et Lompe	–	–	–	–	–	–	–	+	–	+	–	–	–	EBE	Ore	NA
Trechus (*Trechus*) *merkli* Pawlowski	–	–	–	+	–	–	–	–	+	–	–	–	–	LE	Ore	NA
Trechus (*Trechus*) *pirinicus* Pawlowski	–	–	–	–	–	–	–	–	–	+	–	–	–	LE	Ore	NA
Trechus (*Trechus*) *matrismeae* Pawlowski	–	–	–	–	–	–	–	–	–	+	–	–	–	LE	Ore	NA
Trechus (*Trechus*) *bohemorum* Pawlowski	–	–	–	–	–	–	–	–	–	+	–	–	–	LE	Ore	NA
Trechus (*Trechus*) *midas* Jeannel	–	–	–	–	–	–	+	–	–	–	–	–	–	LE	Ore	NA
Trechus (*Trechus*) *ravasinianus* Lorenz (= *ravasinii* Jeannel)	–	–	–	–	–	+	–	–	–	–	–	–	–	LE	Ore	NA

Trechus (Trechus) weiratheri Jeannel	–	–	–	–	–	–	+	–	–	–	–	–	–	LE	Ore	NA
Trechus (Trechus) angusticeps Apfelbeck	–	–	–	–	–	–	–	–	–	–	–	–	+	LE	Arb	NA
Albanotrechus beroni Casale et V. Guéorguiev	–	–	–	–	–	+	–	–	–	–	–	–	–	LE	Arb	NA
Pheggomisetes buresi (Knirsch)	–	–	–	–	–	–	–	–	+	–	–	–	–	LE	Arb	NA
Pheggomisetes radevi radevi Knirsch	–	–	–	–	–	–	–	–	+	–	–	–	–	LE	Arb	NA
Pheggomisetes radevi ilcevi Knirsch	–	–	–	–	–	–	–	–	+	–	–	–	–	LE	Arb	NA
Pheggomisetes radevi tranteevi V. Guéorguiev	–	–	–	–	–	–	–	–	+	–	–	–	–	LE	Arb	NA
Pheggomisetes globiceps globiceps Buresch	–	–	–	–	–	–	–	–	+	–	–	–	–	LE	Arb	NA
Pheggomisetes globiceps mladenovi V. Guéorguiev	–	–	–	–	–	–	–	–	+	–	–	–	–	LE	Arb	NA
Pheggomisetes globiceps georgievi Z. Karaman	–	–	–	–	–	–	–	–	+	–	–	–	–	LE	Arb	NA
Pheggomisetes globiceps lakatnicensis Jeannel	–	–	–	–	–	–	–	–	+	–	–	–	–	LE	Arb	NA
Pheggomisetes globiceps cerovensis V. Guéorguiev	–	–	–	–	–	–	–	–	+	–	–	–	–	LE	Arb	NA
Pheggomisetes globiceps breiti Mandl	–	–	–	–	–	–	–	–	+	–	–	–	–	LE	Arb	NA
Pheggomisetes globiceps ilandjievi V. Guéorguiev	–	–	–	+	–	–	–	–	+	–	–	–	–	LE	Arb	NA
Pheggomisetes globiceps stoicevi V. Guéorguiev	–	–	–	–	–	–	–	–	+	–	–	–	–	LE	Arb	NA
Pheggomisetes globiceps karlukovensis Genest	–	–	–	–	–	–	–	–	+	–	–	–	–	LE	Arb	NA
Anillidius turcicus (J. Frivaldszky)	–	–	–	–	–	–	–	–	–	–	–	–	+	LE	Arb	NA
Neotrechus lonae lonae (J. Müller)	–	–	–	–	–	+	–	–	–	–	–	–	–	LE	Arb	NA
Neotrechus lonae atavus J. Müller	–	–	–	–	–	+	–	–	–	–	–	–	–	LE	Arb	NA
Neotrechus lonae shkodrensis Scheibel	–	–	–	–	–	+	–	–	–	–	–	–	–	LE	Arb	NA
Neotrechus malissorum (J. Müller)	–	–	–	–	–	+	–	–	–	–	–	–	–	LE	Arb	NA

continued

Table 1 continued

Taxa	NS	SS	DR	ES	WK	AL	WM	EM	NB	SB	NG	DO	ET	DT	PH	FE
Neotrechus suturalis pentheri Winkler	–	–	–	–	–	+	–	–	–	–	–	–	–	LE	Arb	NA
Serboduvalius dragacevensis S. Ćurčić, Pavićević et B. Ćurčić	–	+	–	–	–	–	–	–	–	–	–	–	–	LE	Arb	NA
Serboduvalius starivlahi (B. Guéorguiev, S. Ćurčić et B. Ćurčić)	–	+	–	–	–	–	–	–	–	–	–	–	–	LE	Arb	NA
Duvalius (*Paraduvalius*) *balcanicus* (J. Frivaldszky)	–	–	–	–	–	–	–	–	+	–	–	–	–	LE	Arb	NA
Duvalius (*Paraduvalius*) *bulgaricus* (Knirsch)	–	–	–	–	–	–	–	–	+	–	–	–	–	LE	Arb	NA
Duvalius (*Paraduvalius*) *beroni* V. Guéorguiev	–	–	–	–	–	–	–	–	+	–	–	–	–	LE	Arb	NA
Duvalius (*Paraduvalius*) *bureschi* Jeannel	–	–	–	–	–	–	–	–	–	+	–	–	–	LE	Arb	NA
Duvalius (*Paraduvalius*) *garevi* Casale et Genest	–	–	–	–	–	–	–	–	+	–	–	–	–	LE	Arb	NA
Duvalius (*Paraduvalius*) *germanae* Casale et V. Taglianti	–	–	–	–	–	–	–	–	–	–	+	–	–	LE	Arb	NA
Duvalius (*Paraduvalius*) *glabellus* G. Etonti et M. Etonti	–	–	–	–	–	–	–	–	–	–	+	–	–	LE	Arb	NA
Duvalius (*Paraduvalius*) *hanae* Hůrka	–	–	–	–	–	–	–	–	+	–	–	–	–	LE	Arb	NA
Duvalius (*Paraduvalius*) *joannidisi* Casale, Giachino et M. Etonti	–	–	–	–	–	–	–	–	–	–	+	–	–	LE	Arb	NA
Duvalius (*Paraduvalius*) *karelhurkai* Farkač	–	–	–	–	–	–	–	–	–	+	–	–	–	LE	Arb	NA
Duvalius (*Paraduvalius*) *kotelensis* Genest	–	–	–	–	–	–	–	–	+	–	–	–	–	LE	Arb	NA
Duvalius (*Paraduvalius*) *legrandi* (Genest)	–	–	–	–	–	–	–	–	+	–	–	–	–	LE	Arb	NA
Duvalius (*Paraduvalius*) *marani* (Knirsch)	–	–	–	–	–	–	–	–	–	+	–	–	–	LE	Arb	NA

Duvalius (*Paraduvalius*) *papasoffi* (Mandl)	–	–	–	–	–	–	–	–	+	–	–	–	–	LE	Arb	NA
Duvalius (*Paraduvalius*) *petrovi* B. Guéorguiev	–	–	–	–	–	–	–	–	–	+	–	–	–	LE	Arb	NA
Duvalius (*Paraduvalius*) *philipensis* Jeannel	–	–	–	–	–	–	–	–	–	–	+	–	–	LE	Arb	NA
Duvalius (*Paraduvalius*) *pirinensis* B. Guéorguiev	–	–	–	–	–	–	–	–	–	+	–	–	–	LE	Arb	NA
Duvalius (*Paraduvalius*) *pretneri* V. Guéorguiev	–	–	–	–	–	–	–	–	+	–	–	–	–	LE	Arb	NA
Duvalius (*Paraduvalius*) *pruinosus* Jeannel	–	–	–	–	–	–	–	–	–	–	+	–	–	LE	Arb	NA
Duvalius (*Paraduvalius*) *rajtchevi* (Genest et Juberthie)	–	–	–	–	–	–	–	–	–	+	–	–	–	LE	Arb	NA
Duvalius (*Paraduvalius*) *regisborisi* (Buresch)	–	–	–	–	–	–	–	–	+	–	–	–	–	LE	Arb	NA
Duvalius (*Paraduvalius*) *stankovitchi stankovitchi* (Jeannel)	–	–	–	+	–	–	–	–	–	–	–	–	–	LE	Arb	NA
Duvalius (*Paraduvalius*) *stankovitchi devojensis* (Jeannel)	–	–	–	+	–	–	–	–	–	–	–	–	–	LE	Arb	NA
Duvalius (*Paraduvalius*) *stankovitchi georgevitchi* (Jeannel)	–	–	–	+	–	–	–	–	–	–	–	–	–	LE	Arb	NA
Duvalius (*Paraduvalius*) *sydowi* Jeannel	–	–	–	–	–	–	–	–	–	–	+	–	–	LE	Arb	NA
Duvalius (*Paraduvalius*) *winkleri* (Jeannel)	–	–	–	+	–	–	–	–	–	–	–	–	–	LE	Arb	NA
Duvalius (*Paraduvalius*) *zivkovi* (Knirsch)	–	–	–	–	–	–	–	–	+	–	–	–	–	LE	Arb	NA
Duvalius (*Biharotrechus*) *beshkovi* Coiffait	–	–	–	–	–	–	–	–	–	+	–	–	–	LE	Arb	NA
Duvalius (*Biharotrechus*) *peristericus* (J. Müller)	–	–	–	–	–	–	+	–	–	–	–	–	–	LE	Arb	NA
Duvalius (*Platyduvalius*) *macedonicus* (J. Müller)	–	–	–	–	–	–	+	–	–	–	–	–	–	LE	Arb	NA
Duvalius (*Neoduvalius*) *cvijici cvijici* (Jeannel)	–	+	–	–	–	–	–	–	–	–	–	–	–	LE	Arb	NA

continued

Table 1 continued

Taxa	NS	SS	DR	ES	WK	AL	WM	EM	NB	SB	NG	DO	ET	DT	PH	FE
Duvalius (*Neoduvalius*) *cvijici stopicensis* (Jeannel)	–	+	–	–	–	–	–	–	–	–	–	–	–	LE	Arb	NA
Duvalius (*Duvalius*) *bischoffi* Meschnigg	–	–	–	–	–	+	–	–	–	–	–	–	–	LE	Arb	SA
Duvalius (*Duvalius*) *bolei* Pretner	–	–	–	+	–	–	–	–	–	–	–	–	–	LE	Arb	NA
Duvalius (*Duvalius*) *didonnai* Magrini, Vanni et Ingusio	–	–	–	–	–	+	–	–	–	–	–	–	–	LE	Arb	SA
Duvalius (*Duvalius*) *fodori* Scheibel	–	–	–	–	–	–	+	–	–	–	–	–	–	LE	Arb	NA
Duvalius (*Duvalius*) *fuschi* Scheibel	–	–	–	–	–	+	–	–	–	–	–	–	–	LE	Arb	SA
Duvalius (*Duvalius*) *georgi* (J. Müller)	–	–	–	–	–	+	–	–	–	–	–	–	–	LE	Arb	SA
Duvalius (*Duvalius*) *gogalai* Pretner	–	–	–	–	–	–	+	–	–	–	–	–	–	LE	Arb	NA
Duvalius (*Duvalius*) *milenae* Casale	–	–	–	–	–	–	–	–	–	–	+	–	–	LE	Arb	SA
Duvalius (*Duvalius*) *muelleri* Winkler	–	–	–	–	–	+	–	–	–	–	–	–	–	LE	Arb	SA
Duvalius (*Duvalius*) *pominii* Schatzmayr	–	–	–	–	–	+	–	–	–	–	–	–	–	LE	Arb	SA
Duvalius (*Duvalius*) *regiszogui* Meschnigg	–	–	–	–	–	+	–	–	–	–	–	–	–	LE	Arb	SA
Duvalius (*Duvalius*) *strupii* Scheibel	–	–	–	–	–	+	+	–	–	–	–	–	–	LE	Arb	NA
Duvalius (*Duvalius*) *sturanyi stenocephalus* (Apfelbeck)	–	–	–	–	–	+	–	–	–	–	–	–	–	LE	Arb	NA
Duvalius (*Duvalius*) *vignai* Casale	–	–	–	–	–	–	+	–	–	–	–	–	–	LE	Arb	SA
Stomis (*Stomis*) *rostratus rostratus* (Duftschmid)	–	–	–	–	–	+	–	–	–	–	–	–	–	WBSE	Arb	EAS
Xenion ignitum (Kraatz) s.l.	–	–	–	–	–	–	–	+	+	+	+	–	–	EBE	Arb	EU
Myas (*Myas*) *chalybaeus* (Palliardi)	–	+	–	+	–	+	+	+	+	+	+	+	–	BSE	Arb	EUA
Poecilus (*Poecilus*) *rebeli* (Apfelbeck)	–	–	–	–	–	+	–	–	–	–	–	–	–	WBE	Ere	IT
Pterostichus (*Bothriopterus*) *oblongopunctatus bosnicus* Apfelbeck	+	+	+	–	+	+	+	–	–	–	–	–	–	WBE	Arb	ALT
Pterostichus (*Morphnosoma*) *melanarius bulgaricus* (Lutshnik)	–	–	–	–	–	–	–	–	+	–	–	–	–	LE	Ore	ALT

Pterostichus (*Pseudomaseus*) *anthracinus biimpressus* (Küster)	–	–	–	–	–	+	–	–	–	–	–	–	–	BSE	Arb	ALT
Pterostichus (*Feronidius*) *melas depressus* (Dejean)	–	–	–	–	–	+	–	–	+	+	–	+	+	BE	Arb	EU
Pterostichus (*Feronidius*) *serbicus* Apfelbeck	+	+	+	+	–	–	–	–	–	+	–	–	–	BE	Arb	EU
Pterostichus (*Platypterinus*) *ottomanus ottomanus* Apfelbeck	–	–	–	–	+	+	+	–	–	–	–	–	–	LE	Arb	EU
Pterostichus (*Platypterinus*) *ottomanus kajmakcalensis* Jedlička	–	–	–	–	–	–	+	–	–	–	–	–	–	LE	Arb	EU
Pterostichus (*Aphaon*) *bischoffianus* Jedlička	–	–	–	–	–	+	–	–	–	–	–	–	–	LE	Arb	EU
Pterostichus (*Aphaon*) *brucki* Schaum	–	+	+	+	+	+	+	+	+	+	+	–	–	BE	Arb	EU
Pterostichus (*Aphaon*) *latifianus* Apfelbeck	–	–	–	–	+	+	+	–	–	–	–	–	–	LE	Ore	EU
Pterostichus (*Aphaon*) *lumensis lumensis* Apfelbeck	–	–	–	–	+	+	–	–	–	–	–	–	–	LE	Ore	EU
Pterostichus (*Aphaon*) *lumensis litae* Csiki	–	–	–	–	+	+	+	–	–	–	–	–	–	LE	Ore	EU
Pterostichus (*Aphaon*) *lumensis ljubetensis* Apfelbeck	–	–	–	–	+	–	+	–	–	–	–	–	–	LE	Ore	EU
Pterostichus (*Aphaon*) *malissorum* Apfelbeck	–	–	–	–	–	+	–	–	–	–	–	–	–	LE	Ore	EU
Pterostichus (*Aphaon*) *merklii* (J. Frivaldszky)	–	–	–	–	–	–	–	–	+	+	–	–	–	EBE	Arb	EU
Pterostichus (*Aphaon*) *pentheri* Apfelbeck	–	–	–	–	+	+	–	–	–	–	–	–	–	LE	Ore	EU
Pterostichus (*Aphaon*) *reiseri* Ganglbauer	–	–	–	–	–	+	–	–	–	–	–	–	–	WBE	Ore	EU
Pterostichus (*Aphaon*) *rhilensis rhilensis* Reitter	–	–	–	–	–	–	–	–	–	+	–	–	–	LE	Ore	EU
Pterostichus (*Aphaon*) *rhilensis kourili* Mařan	–	–	–	–	–	–	–	–	–	+	–	–	–	LE	Ore	EU
Pterostichus (*Aphaon*) *rhilensis vitosensis* Mařan	–	–	–	–	–	–	–	–	+	+	–	–	–	LE	Ore	EU

continued

Table 1 continued

Taxa	NS	SS	DR	ES	WK	AL	WM	EM	NB	SB	NG	DO	ET	DT	PH	FE
Pterostichus (*Cryobius*) *macedonicus* Apfelbeck	–	–	–	–	–	–	+	–	–	–	–	–	–	LE	Arb	EU
Pterostichus (*Cryobius*) *properans* (Chaudoir)	–	–	–	–	–	–	–	–	–	+	–	–	+	EBSE	Arb	EU
Pterostichus (*Parahaptoderus*) *brevis brevis* (Duftschmid)	+	+	+	+	–	–	+	+	–	+	–	–	–	BSE	Arb	EU
Pterostichus (*Parahaptoderus*) *vecors* (Tschitschérine)	–	–	–	–	–	–	–	+	+	+	–	–	–	EBE	Arb	EU
Pterostichus (*Parahaptoderus*) sp.	–	–	–	–	–	–	–	–	–	–	+	–	–	LE	Arb	EU
Pterostichus (*Haptoderotapinus*) *matchai* (Jedlička)	–	–	–	–	–	–	+	–	–	–	–	–	–	WBE	Arb	EU
Pterostichus (*Rambousekiella*) *ledenikensis* (Knirsch)	–	–	–	–	–	–	–	–	+	–	–	–	–	LE	Arb	NA
Tapinopterus (*Tapinopterus*) *balcanicus balcanicus* Ganglbauer	–	–	–	–	–	–	–	–	–	+	+	–	–	EBE	Arb	NA
Tapinopterus (*Tapinopterus*) *balcanicus belasicensis* Mařan	–	–	–	–	–	–	–	+	–	+	+	–	–	EBE	Arb	NA
Tapinopterus (*Tapinopterus*) *balcanicus bureschi* Mařan	–	–	–	–	–	–	–	–	–	+	+	–	–	LE	Arb	NA
Tapinopterus (*Tapinopterus*) *bartoni* Mařan	–	–	–	–	–	–	–	–	–	+	–	–	–	EBE	Arb	NA
Tapinopterus (*Tapinopterus*) *bischoffi* J. Müller	–	–	–	–	–	+	–	–	–	–	–	–	–	LE	Arb	NA
Tapinopterus (*Tapinopterus*) *comita* Jedlička	–	–	–	–	–	–	+	–	–	–	–	–	–	LE	Arb	NA
Tapinopterus (*Tapinopterus*) *detonii* Schatzmayr	–	–	–	–	–	+	–	–	–	–	–	–	–	LE	Arb	NA

Tapinopterus (*Tapinopterus*) *dochii* Apfelbeck	–	–	–	–	+	+	+	–	–	–	–	–	–	WBE	Arb	NA
Tapinopterus (*Tapinopterus*) *extensus extensoides* (Jedlička)	–	–	–	–	–	+	–	–	–	–	–	–	–	LE	Arb	SA
Tapinopterus (*Tapinopterus*) *heyrovskii* Jedlička	–	–	–	–	–	–	+	–	–	–	–	–	–	LE	Arb	NA
Tapinopterus (*Tapinopterus*) *insulicola* (Tschitschérine)	–	–	–	–	–	–	–	–	–	–	+	–	–	LE	Arb	NA
Tapinopterus (*Tapinopterus*) *kaufmanni kaufmanni* Ganglbauer	–	–	–	+	–	–	–	–	–	–	–	–	–	EBSE	Arb	NA
Tapinopterus (*Tapinopterus*) *kaufmanni kalofirensis* Mařan	–	–	–	–	–	–	–	–	+	+	–	–	–	EBE	Arb	NA
Tapinopterus (*Tapinopterus*) *kaufmanni kulti* Mařan	–	–	–	–	–	–	–	–	–	+	–	–	–	EBE	Arb	NA
Tapinopterus (*Tapinopterus*) *kaufmanni winkleri* Mandl	–	–	–	+	–	–	–	–	+	–	–	–	–	LE	Arb	NA
Tapinopterus (*Tapinopterus*) *miridita miridita* (Apfelbeck)	–	–	–	–	+	+	+	–	–	–	–	–	–	LE	Arb	NA
Tapinopterus (*Tapinopterus*) *miridita jacupicensis* Jedlička	–	–	–	–	–	–	+	–	–	–	–	–	–	LE	Arb	NA
Tapinopterus (*Tapinopterus*) *miridita jablanicensis* Mařan	–	–	–	–	–	+	+	–	–	–	–	–	–	LE	Arb	NA
Tapinopterus (*Tapinopterus*) *monastirensis monastirensis* Reitter	–	–	–	–	–	–	+	–	–	–	–	–	–	LE	Arb	NA
Tapinopterus (*Tapinopterus*) *monastirensis gestroi* J. Müller	–	–	–	–	–	+	–	–	–	–	–	–	–	LE	Arb	NA
Tapinopterus (*Tapinopterus*) *protensus aetolicus* Ganglbauer	–	–	–	–	–	+	–	–	–	–	–	–	–	WBE	Arb	SA
Tapinopterus (*Tapinopterus*) *protensus tomoricensis* J. Müller	–	–	–	–	–	+	–	–	–	–	–	–	–	LE	Arb	SA

continued

Table 1 continued

Taxa	NS	SS	DR	ES	WK	AL	WM	EM	NB	SB	NG	DO	ET	DT	PH	FE
Tapinopterus (*Tapinopterus*) *purkynei* Jedlička	–	–	–	–	–	–	+	–	–	–	–	–	–	LE	Arb	NA
Tapinopterus (*Tapinopterus*) *rambousekianus* Mařan	–	–	–	–	–	–	+	–	–	–	–	–	–	LE	Arb	NA
Tapinopterus (*Hypogium*) *albanicus* (Tschitschérine)	–	–	–	–	–	+	–	–	–	–	–	–	–	LE	Arb	NA
Tapinopterus (*Speluncarius*) *minusculus* Straneo	–	–	–	–	–	–	–	–	–	–	–	–	+	LE	Arb	NA
Tapinopterus (*Speluncarius*) *minutulus* J. Müller	–	–	–	–	–	+	–	–	–	–	–	–	–	LE	Arb	NA
Tapinopterus (*Speluncarius*) *rumelicus* (J. Müller)	–	–	–	–	–	–	–	–	–	–	+	–	–	LE	Arb	NA
Tapinopterus (*Speluncarius*) *seticeps* (J. Müller)	–	–	–	–	–	–	–	–	–	–	+	–	–	LE	Arb	NA
Molops (*Molops*) *robustus robustus* (Dejean)	–	–	–	+	–	–	–	+	+	+	–	–	–	EBSE	Arb	NA
Molops (*Molops*) *robustus parallelus* Mlynář	–	–	–	+	–	–	+	–	–	+	–	–	–	BE	Arb	NA
Molops (*Molops*) *maderi* J. Müller	–	–	–	–	–	+	–	–	–	–	–	–	–	LE	Arb	NA
Molops (*Molops*) *weiratheri* J. Müller	–	–	–	–	–	–	+	–	–	–	–	–	–	LE	Arb	NA
Molops (*Molops*) *spartanus valonensis* J. Müller	–	–	–	–	–	+	–	–	–	–	–	–	–	LE	Arb	SA
Molops (*Molops*) *simplex simplex* Chaudoir	–	–	–	–	+	–	–	–	–	–	–	–	–	WBE	Arb	NA
Molops (*Molops*) *simplex meridionalis* Mlynář	–	–	–	–	–	+	–	–	–	–	–	–	–	LE	Arb	NA
Molops (*Molops*) *dilatatus dilatatus* Chaudoir	–	–	–	–	–	–	–	+	–	+	+	–	–	EBE	Arb	NA
Molops (*Molops*) *dilatatus angulicollis* J. Müller	–	–	–	–	–	–	–	–	+	+	–	–	–	EBE	Arb	NA

Molops (*Molops*) *rhodopensis rhodopensis* Apfelbeck	–	–	–	–	–	–	–	–	–	+	+	–	–	EBE	Arb	NA
Molops (*Molops*) *rhodopensis kourili* Mařan	–	–	–	–	–	–	–	–	–	+	–	–	–	LE	Arb	NA
Molops (*Molops*) *alpestris kalofericus* Mlynář	–	–	–	–	–	–	–	–	+	+	–	–	–	EBE	Arb	NA
Molops (*Molops*) *alpestris rhilensis* Apfelbeck	–	–	–	–	–	–	–	–	–	+	–	–	–	EBE	Arb	NA
Molops (*Molops*) *parreyssi lonae* J. Müller	–	–	–	–	–	+	–	–	–	–	–	–	–	LE	Arb	NA
Molops (*Molops*) *merditanus merditanus* Apfelbeck	–	–	–	–	–	+	–	–	–	–	–	–	–	LE	Arb	NA
Molops (*Molops*) *merditanus intermedius* J. Müller	–	–	–	–	–	+	–	–	–	–	–	–	–	LE	Arb	NA
Molops (*Molops*) *curtulus cakorensis* Mlynář	–	+	–	–	+	+	–	–	–	–	–	–	–	LE	Arb	NA
Molops (*Molops*) *curtulus winklerianus* J. Müller	–	–	–	–	–	+	–	–	–	–	–	–	–	LE	Arb	NA
Molops (*Molops*) *reiseri reiseri* Apfelbeck	–	+	+	–	+	–	–	–	–	–	–	–	–	WBE	Arb	NA
Molops (*Molops*) *reiseri corallipes* J. Müller	–	+	–	–	–	–	–	–	–	–	–	–	–	LE	Arb	NA
Molops (*Molops*) *reiseri holdhausi* J. Müller	–	–	–	–	–	+	–	–	–	–	–	–	–	WBE	Arb	NA
Molops (*Molops*) *reiseri pseudoalpestris* Mlynář	–	+	–	–	–	–	–	–	–	–	–	–	–	LE	Arb	NA
Molops (*Molops*) *reiseri victoris* J. Müller	–	–	–	–	+	+	–	–	–	–	–	–	–	LE	Arb	NA
Molops (*Molops*) *albanicus* Apfelbeck	–	–	–	–	+	+	–	–	–	–	–	–	–	LE	Arb	NA
Molops (*Molops*) *plurisetosus* J. Müller	–	–	–	–	–	+	–	–	–	–	–	–	–	LE	Arb	NA
Molops (*Molops*) *osmanilis osmanilis* Apfelbeck	–	–	–	–	+	+	+	–	–	–	–	–	–	WBE	Arb	NA
Molops (*Molops*) *osmanilis tomoricensis* J. Müller	–	–	–	–	–	+	–	–	–	–	–	–	–	LE	Arb	NA

continued

Table 1 continued

Taxa	NS	SS	DR	ES	WK	AL	WM	EM	NB	SB	NG	DO	ET	DT	PH	FE
Molops (*Molops*) *rufipes rufipes* Chaudoir	–	–	–	–	+	+	+	–	–	–	–	–	–	WBE	Arb	NA
Molops (*Molops*) *rufipes belasicensis* Mlynář	–	–	–	–	–	–	+	+	–	+	+	–	–	BE	Arb	NA
Molops (*Molops*) *doderoi doderoi* Schatzmayr	–	–	–	–	–	–	–	–	–	–	+	–	–	LE	Arb	NA
Molops (*Molops*) *doderoi hlisnikovskyi* Mlynář	–	–	–	–	–	–	–	–	–	+	+	–	–	LE	Arb	NA
Molops (*Molops*) *matchai* Roubal	–	–	–	–	–	–	+	–	–	–	–	–	–	WBE	Arb	NA
Molops (*Molops*) *alpestris alpestris* Dejean	–	–	–	+	–	–	–	–	–	–	–	–	–	EBSE	Arb	NA
Molops (*Molops*) *alpestris centralis* Mlynář	–	–	–	–	–	–	–	–	–	+	–	–	–	EBE	Arb	NA
Molops (*Molops*) *alpestris imitator* J. Müller	–	–	–	–	–	+	+	–	–	–	–	–	–	WBE	Arb	NA
Molops (*Molops*) *rufipes jacupicensis* (Mařan)	–	–	–	–	–	–	+	–	–	–	–	–	–	LE	Arb	NA
Molops (*Molops*) *rufipes golobardensis* Mlynář	–	–	–	–	–	–	–	–	–	+	–	–	–	LE	Arb	NA
Molops (*Molops*) *rufipes denteletus* B. Guéorguiev	–	–	–	–	–	–	–	+	–	+	–	–	–	LE	Arb	NA
Molops (*Molops*) *rufipes klisuranus* Apfelbeck	–	–	–	–	–	–	–	–	+	+	–	–	–	LE	Arb	NA
Molops (*Molops*) *rufipes munelensis* Apfelbeck	–	–	–	–	–	+	–	–	–	–	–	–	–	LE	Arb	NA
Molops (*Molops*) *rufipes pseudoosmanilis* Mlynář	–	–	–	–	–	+	+	–	–	–	–	–	–	LE	Arb	NA
Molops (*Molops*) *rufipes steindachneri* Apfelbeck	–	–	–	–	+	+	+	–	–	–	–	–	–	WBE	Arb	NA
Molops (*Molops*) *rufipes sturanyi* Apfelbeck	–	–	–	–	–	+	+	–	–	–	–	–	–	LE	Arb	NA
Molops (*Molops*) *piceus bulgaricus* Mařan	–	–	–	–	–	–	–	+	+	+	–	–	–	EBE	Arb	NA
Molops (*Molops*) *piceus balcanicus* Mlynář	+	+	+	–	–	+	+	–	–	–	–	–	–	WBE	Arb	NA

Molops (*Molops*) *piceus byzantinus* Apfelbeck	–	–	–	–	–	–	–	–	–	+	–	–	+	EBSE	Arb	NA
Molops (*Molops*) *piceus osogovensis* B. Guéorguiev	–	–	–	–	–	–	–	+	–	+	–	–	–	LE	Arb	NA
Molops (*Molops*) *piceus vlassinensis* Matits	–	–	–	+	–	–	–	–	–	–	–	–	–	LE	Arb	NA
Molops (*Molops*) *malshentianus* Apfelbeck	–	–	–	–	–	+	–	–	–	–	–	–	–	LE	Arb	NA
Molops (*Molops*) *rufus* Matits	–	–	+	–	–	–	–	–	–	–	–	–	–	LE	Arb	NA
Molops (*Stenochoromus*) *montenegrinus cavernicola* J. Müller	–	–	–	–	–	+	–	–	–	–	–	–	–	LE	Ore	NA
Molops (*Stenochoromus*) *montenegrinus koprivnikensis* (Csiki)	–	–	–	–	+	–	–	–	–	–	–	–	–	LE	Ore	NA
Molops (*Stenochoromus*) *montenegrinus nivalis* (Apfelbeck)	–	–	–	–	–	+	–	–	–	–	–	–	–	WBE	Ore	NA
Zabrus (*Pelor*) *aegaeus* Apfelbeck	–	–	–	–	–	–	–	–	–	–	+	–	–	BE	Ere	SA
Zabrus (*Pelor*) *aetolus purkynei* Mařan	–	–	–	–	–	–	+	–	–	–	–	–	–	LE	Ere	SA
Zabrus (*Pelor*) *aetolus winkleri* J. Müller	–	–	–	–	–	+	–	–	–	–	–	–	–	LE	Ere	SA
Zabrus (*Pelor*) *albanicus albanicus* Apfelbeck	–	–	–	–	+	+	+	–	–	–	–	–	–	WBE	Ore	SA
Zabrus (*Pelor*) *albanicus jablanicensis* Mařan	–	–	–	–	–	+	+	–	–	–	–	–	–	LE	Ore	SA
Zabrus (*Pelor*) *albanicus jacupicensis* Mařan	–	–	–	–	–	–	+	–	–	–	–	–	–	LE	Ore	SA
Zabrus (*Pelor*) *albanicus latifianus* Ganglbauer	–	–	–	–	+	+	+	–	–	–	–	–	–	LE	Ore	SA
Zabrus (*Pelor*) *balcanicus balcanicus* Heyden	–	–	–	–	–	–	–	–	+	+	–	–	+	EBE	Ere	SA
Zabrus (*Pelor*) *balcanicus rhodopensis* Apfelbeck	–	–	–	+	–	+	+	+	–	+	+	–	–	BE	Ere	SA
Zabrus (*Pelor*) *boldori* Schatzmayr	–	–	–	–	–	+	–	–	–	–	–	–	–	LE	Ere	SA

continued

Table 1 continued

Taxa	NS	SS	DR	ES	WK	AL	WM	EM	NB	SB	NG	DO	ET	DT	PH	FE
Zabrus (*Pelor*) *corpulentus ponticus* Ganglbauer	–	–	–	–	–	–	–	–	+	+	–	+	+	EBSE	Ere	SA
Zabrus (*Pelor*) *ganglbaueri ganglbaueri* Apfelbeck	–	–	–	–	–	+	+	–	–	–	–	–	–	WBE	Ore	SA
Zabrus (*Pelor*) *ganglbaueri lonae* J. Müller	–	–	–	–	–	+	–	–	–	–	–	–	–	LE	Ore	SA
Zabrus (*Pelor*) *graecus subtilis* Dejean	–	–	–	–	–	–	+	–	–	+	+	–	–	BE	Ere	SA
Zabrus (*Pelor*) *incrassatus incrassatus* (Ahrens)	–	–	+	–	–	+	+	–	–	–	+	–	–	BE	Ere	SA
Zabrus (*Pelor*) *incrassatus bischoffi* J. Müller	–	–	–	–	–	+	+	+	–	–	–	–	–	BE	Ere	SA
Zabrus (*Pelor*) *incrassatus tuleschkovi* Mařan	–	–	–	–	–	–	–	–	–	+	–	–	–	EBE	Ere	SA
Zabrus (*Pelor*) *peristericus* Apfelbeck	–	–	–	–	–	–	+	–	–	–	–	–	–	LE	Ore	SA
Zabrus (*Pelor*) *spinipes insignis* J. Müller	–	–	–	–	–	+	–	–	–	–	–	–	–	WBE	Ere	SA
Harpalus (*Harpalus*) *euchlorus* Ménétriés	–	–	–	–	–	–	–	–	+	+	–	–	+	BE	Ere	POM
Harpalus (*Harpalus*) *karamani karamani* Apfelbeck	–	–	–	+	–	–	–	–	–	–	–	–	–	BE	Ere	POM
Harpalus (*Harpalus*) *rumelicus* Apfelbeck	–	–	–	–	–	–	+	+	+	–	–	+	+	BSE	Ere	IT
Harpalus (*Harpalus*) *serbicus* Matits	–	–	–	+	–	–	–	–	–	–	–	–	–	LE	Ere	IT
Harpalus (*Harpalus*) *triseriatus triseriatus* A. Fleischer	–	–	–	–	–	–	+	+	–	+	+	–	–	BE	Ere	SA
Harpalus (*Harpalus*) *triseriatus babunensis* Mlynář	–	–	–	–	–	–	+	–	–	–	–	–	–	WBE	Ere	SA
Acinopus (*Acinopus*) *baudii hellenicus* Heinz	–	–	–	–	–	+	–	–	–	–	–	–	–	BSE	Ere	IT
Pachycarus (*Mystropterus*) *atrocoeruleus atrocoeruleus* (Waltl)	–	–	–	–	–	–	+	+	–	+	+	–	–	BE	Ere	SA

Pachycarus (*Mystropterus*) *macedonicus* V. Guéorguiev et B. Guéorguiev	–	–	–	–	–	–	+	–	–	–	–	–	–	WBE	Ere	SA
Odontocarus robustus (Dejean)	–	–	–	–	–	+	–	–	–	–	–	–	–	WBE	Ere	PAM
Acupalpus (*Acupalpus*) *planicollis* (Schaum)	–	–	–	–	–	+	–	–	–	+	–	–	–	BE	Arb	POM
Anthracus pallipes (Dejean)	–	–	–	–	–	–	–	–	–	+	–	–	–	BE	Arb	IT
Licinus (*Neorescius*) *oertzeni oertzeni* Reitter	–	–	–	–	–	+	+	–	–	–	–	–	–	WBE	Arb	PAM
Licinus (*Neorescius*) *oertzeni albanicus* J. Müller	–	–	–	–	–	+	–	–	–	–	–	–	–	LE	Arb	PAM
Platynus (*Batenus*) *scrobiculatus purkynei* Obenberger	–	–	–	–	–	–	–	–	–	+	–	–	–	EBSE	Arb	EU
Platynus (*Batenus*) *scrobiculatus serbicus* (Csiki)	+	+	+	+	+	–	+	+	+	+	+	–	–	BE	Arb	EU
Platynus (*Batenus*) *proximus* (J. Frivaldszky)	–	–	–	–	–	–	–	–	+	–	–	–	–	LE	Arb	EU
Platyderus (*Platyderus*) *dalmatinus dalmatinus* L. Miller	–	–	–	–	–	+	–	–	–	–	–	–	–	WBSE	Arb	PAM
Platyderus (*Platyderus*) *graecus* (Reiche & Saulcy)	–	–	–	–	–	+	–	–	–	–	–	–	–	WBE	Arb	PAM
Platyderus (*Platyderus*) *minutus obtusangulus* J. Müller	–	–	–	–	–	+	–	–	–	–	–	–	–	LE	Arb	PAM
Calathus (*Calathus*) *bosnicus* Ganglbauer	–	–	–	–	+	+	–	–	–	–	–	–	–	WBE	Arb	PAM
Calathus (*Calathus*) *giganteus* Dejean	–	–	–	–	–	+	–	–	–	–	–	–	–	WBE	Arb	PAM
Calathus (*Calathus*) *glabricollis* Dejean	–	–	–	–	–	–	+	–	–	–	–	–	–	WBE	Arb	PAM

continued

Table 1 continued

Taxa	NS	SS	DR	ES	WK	AL	WM	EM	NB	SB	NG	DO	ET	DT	PH	FE
Calathus (*Calathus*) *korax* Reitter	–	–	–	–	–	+	+	–	–	–	–	–	–	WBE	Arb	PAM
Calathus (*Calathus*) *macedonicus* Mařan (? = *ravasinii* J. Müller)	–	–	–	–	+	+	+	–	–	–	–	–	–	WBE	Arb	PAM
Calathus (*Neocalathus*) *albanicus* Apfelbeck	–	–	–	–	+	+	+	–	–	–	–	–	–	LE	Ore	ALT
Calathus (*Neocalathus*) *melanocephalus purkynei* Mařan	–	–	–	–	–	–	+	–	–	–	–	–	–	LE	Ore	ALT
Calathus (*Neocalathus*) *metallicus aeneus* Putzeys	–	–	–	–	–	–	–	+	+	+	+	–	–	EBE	Ore	EU
Calathus (*Neocalathus*) *metallicus prokletijensis* Mařan	–	–	+	–	+	+	–	–	–	–	–	–	–	WBE	Ore	EU
Synuchidius ganglbaueri Apfelbeck	–	–	–	–	+	+	+	–	–	–	–	–	–	WBE	Ore	ALT
Laemostenus (*Antisphodrus*) *cavicola albanicus* J. Müller	–	–	–	–	–	+	–	–	–	–	–	–	–	LE	Arb	PAM
Laemostenus (*Antisphodrus*) *elongatum pentheri* J. Müller	–	–	–	–	–	+	–	–	–	–	–	–	–	LE	Arb	PAM
Laemostenus (*Actenipus*) *plasoni plasoni* (Reitter)	–	–	–	–	–	–	–	–	–	+	+	–	–	EBE	Arb	PAM
Laemostenus (*Actenipus*) *plasoni etontii* Casale	–	–	–	–	–	–	–	–	–	–	+	–	–	LE	Arb	PAM
Laemostenus (*Actenipus*) *plasoni thracicus* J. Müller	–	–	–	–	–	–	–	–	–	–	+	–	–	LE	Arb	PAM
Laemostenus (*Pristonychus*) *euxinicus* Nitzu	–	–	–	–	–	–	–	–	–	–	–	+	–	LE	Arb	POM
Laemostenus (*Pristonychus*) *stoevi* B. Guéorguiev	–	–	–	–	–	–	–	–	+	–	–	–	–	LE	Arb	POM

Laemostenus (*Pristonychus*) *derventicus* B. Guéorguiev	–	–	–	–	–	–	–	–	–	+	–	–	–	LE	Arb	POM
Laemostenus (*Pristonychus*) *cimmerius* (Fischer von Waldheim) s.l.	–	–	–	–	–	+	–	–	+	+	+	–	+	BSE	Arb	NA
Laemostenus (*Pristonychus*) *andreevi* B. Guéorguiev	–	–	–	–	–	–	–	–	–	–	+	–	–	LE	Arb	NA
Omphreus (*Omphreus*) *bischoffi* Meschnigg	–	+	–	–	+	–	–	–	–	–	–	–	–	LE	Arb	NA
Omphreus (*Omphreus*) *gracilis* Apfelbeck	–	–	–	–	+	–	+	–	–	–	–	–	–	LE	Arb	NA
Omphreus (*Omphreus*) *lonai* Winkler	–	–	–	–	–	+	–	–	–	–	–	–	–	LE	Arb	NA
Omphreus (*Omphreus*) *morio albanicus* Apfelbeck	–	–	–	–	+	+	+	–	–	–	–	–	–	LE	Arb	NA
Omphreus (*Omphreus*) *morio malissorum* Winkler	–	–	–	–	–	+	–	–	–	–	–	–	–	LE	Arb	NA
Omphreus (*Omphreus*) *morio serbicus* Winkler	–	+	–	–	–	–	–	–	–	–	–	–	–	LE	Arb	NA
Omphreus (*Omphreus*) *ravasinii* J. Müller	–	–	–	–	–	+	–	–	–	–	–	–	–	LE	Arb	NA
Lebia (*Lamprias*) *festiva krueperi* Apfelbeck	–	–	–	–	–	–	+	–	–	+	–	–	+	BSE	Arb	IT
Cymindis (*Paracymindis*) *vassili* B. Guéorguiev	–	–	–	–	–	–	–	–	–	+	–	–	–	LE	Arb	IT
TOTAL: 385	*19*	*36*	*27*	*39*	*55*	*161*	*116*	*30*	*71*	*104*	*50*	*15*	*24*			

Abbreviations above the columns (see also Fig. 1): DT, type of distribution; PH, principal macroecological habitat; FE, faunal element. Abbreviations in the columns: BE, Balkan endemic; BSE, Balkan subendemic; EBE, East Balkan endemic; EBSE, East Balkan subendemic; LE, local endemic; WBE, West Balkan endemic; WBSE, West Balkan subendemic; Arb, arboreal; Ere, eremial; Ore, oreal; AAR, Alpine–Armenian; ALT, Altaic; AP, Alpine–Pamirian; COL, Colchic; EAS, Eurasian; EU, European; EUA, European–American; IT, Iranian–Turanian; NA, North Aegeidean; PAM, Paleomediterranean; PAR, Pamirian–Armenian; POM, Pontomediterranean; SA, South Aegeidean.

foci of current speciation. In fact, those are the only places where the rich oreal fauna exists in the studied area. In three areas (Southern Dinaric zone, the Rhodopes, and Stara Planina) the processes of allogenesis and anagenesis should have started not later than in the Early Miocene. They are traceable through the whole Late Tertiary, in the Pleistocene, and continue now.

6·2 Chorology and specificity

The list of endemic carabid species-group taxa (species and subspecies) includes 28 Balkan subendemics, 124 Balkan endemics, and 242 local endemics, assigned to seven categories and four complexes (Tables 1–2).

I. Taxa more or less widely distributed in the Balkan Peninsula: 48 taxa, including 13 Balkan subendemics and 35 Balkan endemics. This group embraces taxa occurring in both Western and Eastern Balkans. The most numerous are *Carabus* (nine taxa), *Bembidion* (six), *Trechus* (five), *Pterostichus* (five), *Zabrus* (five), and *Harpalus* (four). The beetles of this complex prefer arboreal (30 taxa), eremial (14), and oreal (four) habitats at a ratio 7.5:3.5:1 (oreal is always taken as 1).

II. Taxa distributed only in the Western Balkans: 57 taxa, including five Western Balkan subendemics and 52 Western Balkan endemics. *Carabus* (ten taxa), *Molops* (ten) and *Calathus* (six) are the best represented genera. Taxa of this complex prefer arboreal (40 taxa), eremial (five), and oreal (12) habitats at a ratio 3.33:0.42:1.

III. Taxa distributed only in the Eastern Balkans: 38 taxa, including ten Eastern Balkan subendemics and 28 Eastern Balkan endemics. *Molops* (ten taxa), *Tapinopterus* (six), *Carabus* (five), *Pterostichus* (four) and *Trechus* (four) are the best represented genera. Taxa of this complex prefer arboreal (29 taxa), eremial (six), and oreal (three) habitats at a ratio 9.67:2:1.

IV. Local endemics: 242 taxa; none recorded from both the Western and the Eastern Balkans. They are numerous in the genera *Duvalius* (46 taxa), *Trechus* (32), *Molops* (31), *Nebria* (23), *Tapinopterus* (20), *Pterostichus* (16), *Pheggomisetes* (13), *Zabrus* (nine), *Laemostenus* (eight), *Omphreus* (seven), *Bembidion* (six), *Deltomerus* (five) and *Neotrechus* (five). Arboreal (168) and oreal (70) taxa are dominant, whereas there are only four eremial ones. The ratio of arboreal to eremial to oreal taxa is 2.4:0.057:1.

The specificity (level of endemism) is a component, specific to each concrete fauna, by which this fauna could be differentiated from the others (Lopatin, 1989; Turin et al., 2003). In our case, the concrete fauna is that of the endemic Balkan Carabidae, and the specificity component is the local endemism (Table 2). The highest value of specificity has Albania, 55.28%, followed by Western Macedonia (48.27%), Western Kosovo (47.27%), Northern Bulgaria (46.48%), and North-eastern Greece (42.0%). The medium levels of specificity are exhibited in Eastern Serbia (38.46%), Southern Bulgaria (32.69%), and Southwestern Serbia (27.78%). The remaining regions, in order of decreasing specificity, are European Turkey,

Table 2 Diversity and specificity of the endemic carabid species and subspecies in the Central and Eastern Balkans

Region	END	BLK	BLK %	WBLK	WBLK %	EBLK	EBLK %	LOC	LOC %
NS	19	13	68.42	6	31.58	–	–	–	–
SS	36	16	44.44	10	27.78	–	–	10	27.78
DR	27	17	62.96	7	25.93	–	–	3	11.11
ES	39	18	46.15	–	–	6	15.39	15	38.46
WK	55	10	18.18	19	34.55	–	–	26	47.27
AL	161	27	16.77	45	27.95	–	–	89	55.28
WM	116	32	27.59	28	24.14	–	–	56	48.27
EM	30	19	63.33	–	–	9	30.00	2	6.67
NB	71	21	29.58	–	–	17	23.94	33	46.48
SB	104	37	35.58	–	–	33	31.73	34	32.69
NG	50	20	40.00	–	–	9	18.00	21	42.00
DO	15	8	53.33	–	–	4	26.67	3	20.00
ET	24	10	41.67	–	–	9	37.50	5	20.83
TOTAL	*385*	*48*	*12.47*	*57*	*14.8*	*38*	*9.87*	*242*	*62.86*

Abbreviations (see also Fig. 1): END, number of species and subspecies; BLK, number of species and subspecies with Balkan endemic and subendemic chorotype; BLK %, percentage of Balkan endemic and subendemic chorotypes; WBLK, number of species and subspecies with Western Balkan chorotype; WBLK %, percentage of Western Balkan chorotype; EBLK, number of species and subspecies with Eastern Balkan chorotype; EBLK %, percentage of Eastern Balkan chorotype; LOC, number of local endemic species and subspecies; LOC %, (specificity), percentage of local endemic chorotype.

Dobrogea, Dardanian Region, and Eastern Macedonia. No local endemics are known from Northwestern Serbia.

Although the western regions occupy the combined territory three times smaller than the eastern ones, the Western Balkan endemics are 1.5 times more numerous than the eastern ones. This could be explained by more intensive former and recent speciation, and probably more favorable environmental conditions west of Vardar in the past and present. The fact that 25% of all endemics (chorotype complexes II+III) inhabit either the Western or the Eastern Balkan areas but not both of them, suggests separate centers of speciation and postglacial dispersal. For the widely distributed Balkan endemics and subendemics, their center of origin within the Balkan Peninsula is difficult to identify. Almost 63% of all endemics, however, are confined to isolated localities. Moreover, most of these local endemics are neoendemics, which suggests that important foci of speciation existed and still exist in the Central and Eastern Balkans.

6·3 The taxonomic structure and affinities of the regional faunas

Species-group taxa. The taxonomic structure at the species and subspecies level is examined from two perspectives: (a) for all species-group taxa (385 taxa), and (b) only for the local endemics (242 taxa).

For all taxa (Table 1), the regions are associated in three distinct clusters. First, the Southern Dinaric cluster [(Albania+Western Macedonia)+Western Kosovo] is separated at the 8% similarity level (Fig. 2). Next, the Subpontian cluster (Dobrogea+European Turkey), is separated at 13%. At 20% similarity level, the remaining group splits into two branches, one of them representing the Subdinaric [((Dardanian Region+Northwestern Serbia)+Southwestern Serbia)+(Eastern Macedonia+Eastern Serbia)] cluster. The second branch [(Northern Bulgaria+Southern Bulgaria)+Northeastern Greece], at 22% similarity level, is divided into Bulgarian and Northeastern Greece components. This early separation of the Rhodopean area (Southern Bulgaria, Northeastern Greece) is possibly due to several not yet confirmed endemics in Northeastern Greece. The most interesting result is the association of the transitional central zone (Eastern Serbia+Eastern Macedonia) with the Subdinaric section (Northwestern Serbia, Southwestern Serbia, Dardanian Region), and not with the Stara Planina–Rhodopean one (Northern Bulgaria, Southern Bulgaria, Northeastern Greece).

Analyzing only local endemics (Fig. 3), we see again a Southern Dinaric branch [(Albania+(Western Macedonia+Western Kosovo))+Southwestern Serbia] and an Eastern Balkan one. The Southern Dinaric branch includes two subgroups separated at a very low level, ca. 3%; obviously, these local endemic faunas are related. The Eastern Balkan branch consists also of two subgroups, separated at ca. 1%. First of these subgroups includes the pair (Eastern Serbia+Northern Bulgaria), and the next one includes the triad [(Eastern Macedonia+Southern Bulgaria)+Northeastern Greece]. Considered separately, the local endemism of Dardanian Region, Dobrogea, and European Turkey shows no resemblance to that of any other region; this is explained by the low number of local endemics in each of these three areas. The affinity of the local endemic faunas is a result rather of vicariance effect than of dispersal; apparently, it reveals the related centers of speciation. Such association could be explained by the high restriction of the local endemics to limited areas and the prevalence of neoendemics.

Genus-group taxa. The study of 87 genus-group taxa (Table 3) allows both to demonstrate the dominant groups and to track the faunal connections. There are no dominant (> 50%) genera and subgenera in the separate faunas. Significant subdominant groups in all regions (excluding Dobrogea and European Turkey) are subgenera of *Carabus, Trechus* s. str., and *Molops* s. str., as well as *Paraduvalius* (in Eastern Serbia, Northern Bulgaria, and Northeastern Greece), *Alpaeus, Aphaon*, and *Pelor* (in Western Kosovo, Albania, and Western Macedonia), and *Pheggomisetes* (in Northern Bulgaria). The only subdominant in European Turkey is *Cicindela* s. str., while in Dobrogea there is no such group. Fig. 4 shows further confirmation of regional association. The Subpontian branch (Dobrogea+European Turkey) forms again a distinct cluster separated at high level (ca. 60%). It ties with the remaining regions at the lowest level, ca. 30% similarity. Indicators for this zone are endemic taxa of *Clivina, Parvocaecus*, and *Anillidius*, but there are no local endemics at genus level. At ca. 45% similarity, three clusters are separated: Subdinaric [(Dardanian Region + Northwestern Serbia) + Southwestern

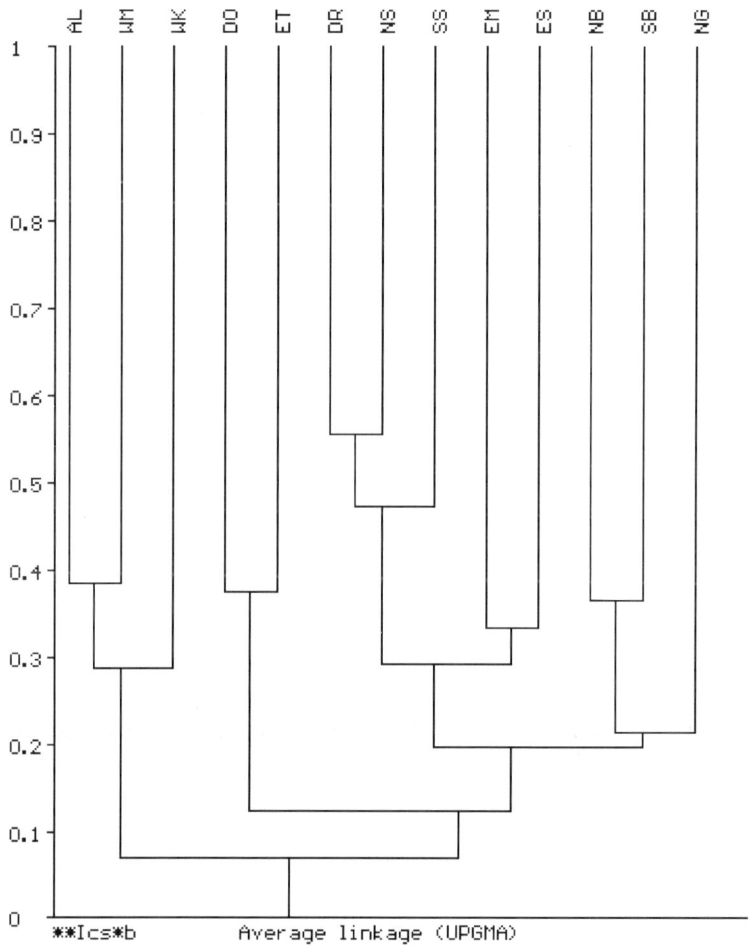

Fig. 2 Affinity of the regional faunas based on the endemic carabid species-group taxa.

Serbia], Southern Dinaric [(Albania + Western Macedonia) + Western Kosovo)], and Stara Planina–Rhodopean [(((Eastern Serbia+Northern Bulgaria) + Southern Bulgaria) + Northeastern Greece) + Eastern Macedonia]. The isolation of Eastern Macedonia in the last cluster is due to its poorly studied fauna. The indicators in the Subdinaric zone are species of *Platycarabus, Sinechostictus, Serboduvalius,* and *Neoduvalius*. The genus *Serboduvalius* is a local endemic for Southwestern Serbia. Endemics from 26 genus-group taxa are found only in the Southern Dinaric zone; of those, *Paradeltomerus, Albanotrechus, Haptoderotapinus*, and *Synuchidius* are local endemics, and *Microcallisthenes, Platyduvalius, Hypogium*, and *Stenochoromus* are subendemics. In the Stara Planina–Rhodopean zone, 13 genus-group

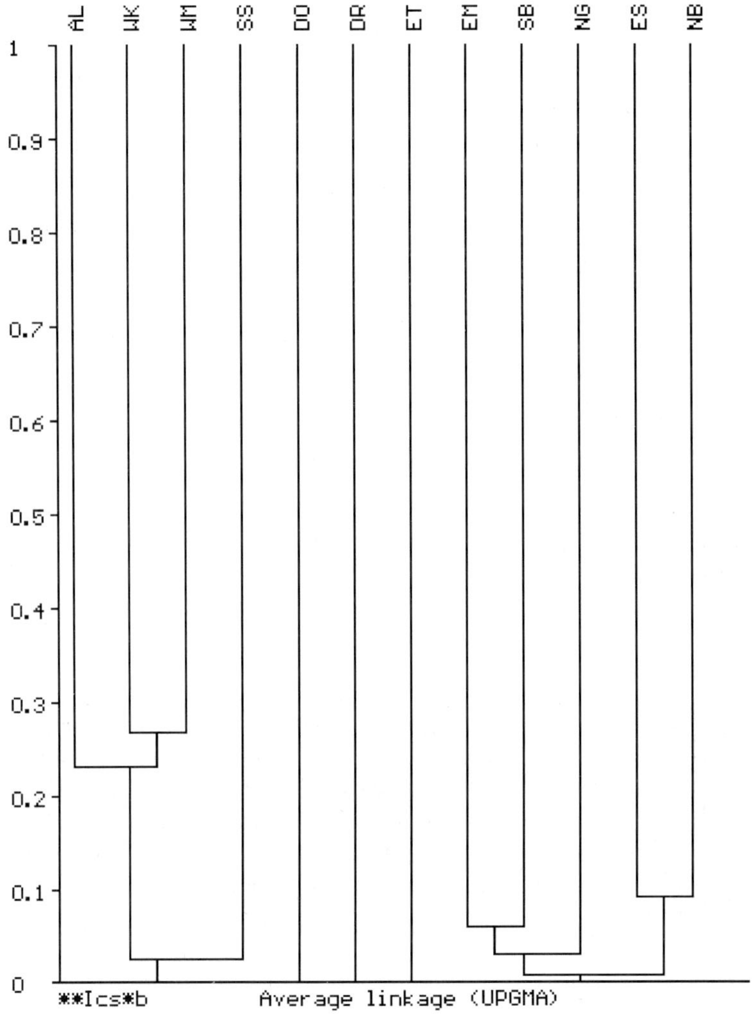

Fig. 3 Affinity of the regional faunas based on the local endemic carabid species-group taxa.

taxa are indicators; *Pheggomisetes, Paraduvalius, Xenion*, and *Rambousekiella* are endemics; and *Dicropterus* and *Winklerites* are subendemics.

The taxonomic structure at the tribe and subtribe level varied among the regions, but there are no dominant tribes or subtribes (Table 4). The most important subdominant groups are three subtribes of Pterostichini (excluding Dobrogea), Trechini (excluding Western Kosovo, Eastern Macedonia, and Dobrogea), and Carabina (excluding Western Kosovo, Albania, and Western Macedonia). The presence of Balkan subendemic tribe Omphreini is of a major biogeographical importance. For more information on this subject see also Subchapter 8.

Table 3 Taxonomic structure of endemic carabids of the Central and Eastern Balkans. All species and subspecies within genus-group taxa included. Abbreviations, see Fig. 1

Taxa	NS	SS	DR	ES	WK	AL	WM	EM	NB	SB	NG	DO	ET
Leistus (*Pogonophorus*)	–	–	3.70	–	1.82	0.62	0.86	–	1.41	0.96	–	–	4.17
Nebria (*Alpaeus*)	–	5.56	–	–	12.73	8.07	8.62	–	–	4.81	–	–	–
Nebria (*Nebria*)	–	–	–	–	–	0.62	0.86	–	–	–	–	–	–
Cicindela (*Cicindela*)	5.26	2.78	3.70	2.56	1.82	1.24	0.86	–	2.82	1.92	4.00	6.67	12.50
Calosoma (*Microcallisthenes*)	–	–	–	–	3.64	1.24	0.86	–	–	–	–	–	–
Carabus (*Morphocarabus*)	5.26	2.78	3.70	2.56	–	–	–	–	1.41	–	–	–	–
Carabus (*Trachycarabus*)	5.26	–	3.70	2.56	–	–	0.86	3.33	1.41	0.96	–	6.67	4.17
Carabus (*Eucarabus*)	5.26	2.78	–	5.13	–	–	–	3.33	1.41	0.96	–	–	–
Carabus (*Tachypus*)	5.26	2.78	3.70	2.56	–	0.62	–	–	1.41	0.96	–	6.67	–
Carabus (*Archicarabus*)	–	–	–	–	–	0.62	–	–	–	–	–	–	–
Carabus (*Oreocarabus*)	–	–	–	–	–	1.24	0.86	–	–	–	–	–	–
Carabus (*Tomocarabus*)	5.26	2.78	3.70	2.56	–	0.62	0.86	3.33	1.41	0.96	2.00	6.67	4.17
Carabus (*Pachystus*)	5.26	–	3.70	2.56	–	1.24	0.86	3.33	1.41	0.96	2.00	–	–
Carabus (*Platycarabus*)	–	2.78	–	–	–	–	–	–	–	–	–	–	–
Carabus (*Megodontus*)	10.53	8.33	7.41	2.56	5.45	3.11	3.45	3.33	1.41	1.92	4.00	–	–
Carabus (*Lamprostus*)	–	–	–	–	–	–	–	–	1.41	0.96	–	6.67	4.17
Carabus (*Procrustes*)	–	–	–	–	–	0.62	0.86	–	–	0.96	2.00	6.67	4.17
Carabus (*Procerus*)	5.26	2.78	3.70	2.56	–	0.62	0.86	–	2.82	1.92	2.00	6.67	4.17
Cychrus	–	–	–	2.56	1.82	0.62	0.86	3.33	1.41	0.96	2.00	–	–
Aptinus (*Aptinus*)	–	–	–	–	1.82	1.24	1.72	–	–	0.96	–	–	–
Clivina	–	–	–	–	–	–	–	–	–	–	–	6.67	–
Reicheadella	–	–	–	–	–	0.62	–	–	–	–	2.00	–	–
Dyschiriodes (*Dyschiriodes*)	–	–	–	–	–	0.62	–	–	–	–	–	–	–
Dicropterus	–	–	–	2.56	–	–	–	–	–	–	–	–	–
Parvocaecus	–	–	–	–	–	–	–	–	–	–	–	–	4.17
Winklerites	–	–	–	–	–	–	–	–	–	–	6.00	–	–
Asaphidion	–	2.78	3.70	2.56	–	0.62	–	3.33	1.41	0.96	–	6.67	4.17

continued

Table 3 continued

Taxa	NS	SS	DR	ES	WK	AL	WM	EM	NB	SB	NG	DO	ET
Bembidion (*Trepanes*)	–	–	–	2.56	–	–	–	–	–	–	–	–	–
Bembidion (*Bembidionetolitzkya*)	–	–	–	–	1.82	2.48	0.86	–	1.41	0.96	–	–	–
Bembidion (*Ocyturanes*)	–	2.78	3.70	–	1.82	1.86	1.72	3.33	1.41	1.92	–	–	–
Bembidion (*Peryphanes*)	–	–	–	–	–	0.62	0.86	3.33	–	0.96	–	–	–
Bembidion (*Peryphus*)	–	–	–	–	–	0.62	0.86	–	–	0.96	2.00	–	–
Bembidion (*Ocydromus*)	–	–	–	–	–	0.62	–	–	–	–	–	–	–
Bembidion (*Omoperyphus*)	–	–	–	–	–	–	0.86	–	–	–	–	–	–
Bembidion (*Sinechostictus*)	–	2.78	–	–	–	–	–	–	–	–	–	–	–
Elaphropus (*Elaphropus*)	–	–	–	–	–	–	–	–	–	0.96	–	–	–
Deltomerus (*Deltomerus*)	–	–	–	–	1.82	1.24	0.86	–	–	–	–	–	–
Deltomerus (*Paradeltomerus*)	–	–	–	–	1.82	1.24	1.72	–	–	–	–	–	–
Trechus (*Trechus*)	21.05	11.11	18.52	12.82	9.09	6.21	12.93	6.67	7.04	15.38	2.00	6.67	8.33
Albanotrechus	–	–	–	–	–	0.62	–	–	–	–	–	–	–
Pheggomisetes	–	–	–	2.56	–	–	–	–	18.31	–	–	–	–
Anillidius	–	–	–	–	–	–	–	–	–	–	–	–	4.17
Neotrechus	–	–	–	–	–	3.11	–	–	–	–	–	–	–
Serboduvalius	–	5.56	–	–	–	–	–	–	–	–	–	–	–
Duvalius (*Paraduvalius*)	–	–	–	10.26	–	–	–	–	15.49	5.77	12.00	–	–
Duvalius (*Biharotrechus*)	–	–	–	–	–	–	0.86	–	–	0.96	–	–	–
Duvalius (*Platyduvalius*)	–	–	–	–	–	–	0.86	–	–	–	–	–	–
Duvalius (*Neoduvalius*)	–	5.56	–	–	–	–	–	–	–	–	–	–	–
Duvalius (*Duvalius*)	–	–	–	2.56	–	5.59	3.45	–	–	–	2.00	–	–
Stomis (*Stomis*)	–	–	–	–	–	0.62	–	–	–	–	–	–	–
Xenion	–	–	–	–	–	–	–	3.33	1.41	0.96	2.00	–	–
Myas	–	2.78	–	2.56	–	0.62	0.86	3.33	1.41	0.96	2.00	6.67	–
Poecilus (*Poecilus*)	–	–	–	–	–	0.62	–	–	–	–	–	–	–
Pterostichus (*Cryobius*)	–	–	–	–	–	–	0.86	–	–	0.96	–	–	4.17
Pterostichus (*Parahaptoderus*)	5.26	2.78	3.70	2.56	–	–	0.86	6.67	1.41	1.92	2.00	–	–
Pterostichus (*Haptoderotapinus*)	–	–	–	–	–	–	0.86	–	–	–	–	–	–

Pterostichus (Bothriopterus)	5.26	2.78	3.70	–	1.82	0.62	0.86	–	–	–	–	–	–
Pterostichus (Morphnosoma)	–	–	–	–	–	–	–	–	1.41	–	–	–	–
Pterostichus (Pseudomaseus)	–	–	–	–	–	0.62	–	–	–	–	–	–	–
Pterostichus (Feronidius)	5.26	2.78	3.70	2.56	–	0.62	–	–	1.41	1.92	–	6.67	4.17
Pterostichus (Platypterinus)	–	–	–	–	1.82	0.62	1.72	–	–	–	–	–	–
Pterostichus (Aphaon)	–	2.78	3.70	2.56	10.91	4.97	3.45	3.33	4.23	4.81	2.00	–	–
Pterostichus (Rambousekiella)	–	–	–	–	–	–	–	–	1.41	–	–	–	–
Tapinopterus (Tapinopterus)	–	–	–	5.13	3.64	5.59	7.76	3.33	2.82	5.77	8.00	–	–
Tapinopterus (Hypogium)	–	–	–	–	–	0.62	–	–	–	–	–	–	–
Tapinopterus (Speluncarius)	–	–	–	–	–	0.62	–	–	–	–	4.00	–	4.17
Molops (Molops)	5.26	13.89	11.11	10.26	14.55	13.66	10.34	20.00	7.04	16.35	10.00	–	4.17
Molops (Stenochoromus)	–	–	–	–	1.82	1.24	–	–	–	–	–	–	–
Zabrus (Pelor)	–	–	3.70	2.56	3.64	6.83	9.48	6.67	2.82	4.81	8.00	6.67	8.33
Harpalus (Harpalus)	–	–	–	5.13	–	–	2.59	6.67	2.82	1.92	2.00	6.67	8.33
Acinopus (Acinopus)	–	–	–	–	–	0.62	–	–	–	–	–	–	–
Pachycarus (Mystropterus)	–	–	–	–	–	–	1.72	3.33	–	0.96	2.00	–	–
Odontocarus	–	–	–	–	–	0.62	–	–	–	–	–	–	–
Acupalpus (Acupalpus)	–	–	–	–	–	0.62	–	–	–	0.96	–	–	–
Anthracus	–	–	–	–	–	–	–	–	–	0.96	–	–	–
Licinus (Neorescius)	–	–	–	–	–	1.24	0.86	–	–	–	–	–	–
Platynus (Batenus)	5.26	2.78	3.70	2.56	1.82	–	0.86	3.33	2.82	1.92	2.00	–	–
Platyderus (Platyderus)	–	–	–	–	–	1.86	–	–	–	–	–	–	–
Calathus (Calathus)	–	–	–	–	3.64	2.48	2.59	–	–	–	–	–	–
Calathus (Neocalathus)	–	–	3.70	–	3.64	1.24	1.72	3.33	1.41	0.96	2.00	–	–
Synuchidius	–	–	–	–	1.82	0.62	0.86	–	–	–	–	–	–
Laemostenus (Antisphodrus)	–	–	–	–	–	1.24	–	–	–	–	–	–	–
Laemostenus (Actenipus)	–	–	–	–	–	–	–	–	–	0.96	6.00	–	–
Laemostenus (Pristonychus)	–	–	–	–	–	0.62	–	–	2.82	1.92	4.00	6.67	4.17
Omphreus (Omphreus)	–	5.56	–	–	5.45	2.48	1.72	–	–	–	–	–	–
Lebia (Lamprias)	–	–	–	–	–	–	0.86	–	–	0.96	–	–	4.17
Cymindis (Paracymindis)	–	–	–	–	–	–	–	–	–	0.96	–	–	–
TOTAL	*100.0*	*100.0*	*100.0*	*100.0*	*100.0*	*100.0*	*100.0*	*100.0*	*100.0*	*100.0*	*100.0*	*100.0*	*100.0*

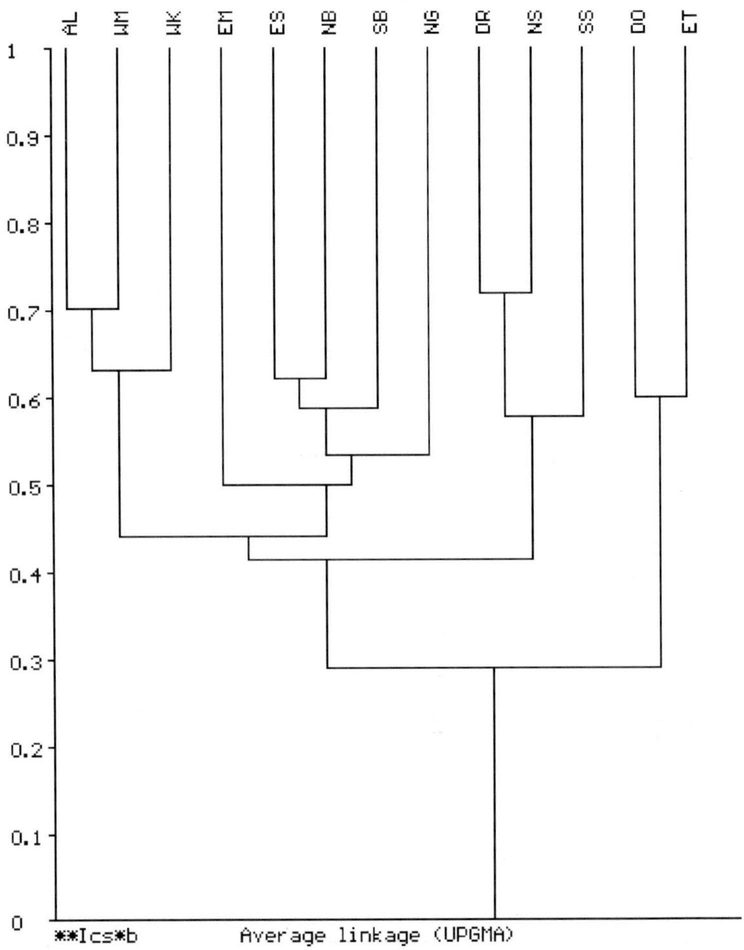

Fig. 4 Affinity of the regional faunas based on the genus-group taxa that include endemic carabid species-group taxa.

6·4 The principal macroecological habitats

The forest environment is definitely the most ancient in the Balkans (Popov and Deltshev, 1997). More or less mesic microhabitats in the woodland and woodland-like ecosystems house 267 species-group taxa of carabids, or 69.35% of all endemics. The predominant arboreal morphoecotype are cryptobionts with lower mobility: geobionts, stratobionts, and subterranean species with slow evolution rate (Matvejev, 1980). Most diverse genera are *Molops* (50 taxa), *Duvalius* (46), *Tapinopterus* (29), *Carabus* (22), *Trechus* (18), *Pterostichus* (16), *Pheggomisetes* (13), *Laemostenus* (ten), *Bembidion* (nine), and *Omphreus* (seven); predominant tribes are Pterostichini (101 taxa) and Trechini (87). The specificity of this

Table 4 Taxonomic structure of endemic carabids in the Central and Eastern Balkans at tribe and subtribe level. Abbreviations, see Fig. 1

Taxa	NS	SS	DR	ES	WK	AL	WM	EM	NB	SB	NG	DO	ET
Nebriini	–	5.56	3.70	–	14.55	9.32	10.34	–	1.41	5.77	–	–	4.17
Cicindelini	5.26	2.78	3.70	2.56	1.82	1.24	0.86	–	2.82	1.92	4.00	6.67	12.50
Calosomatina	–	–	–	–	3.64	1.24	0.86	–	–	–	–	–	–
Carabina	47.37	25.00	29.63	23.08	5.45	8.70	8.62	16.67	14.08	10.58	12.00	40.00	20.83
Cychrini	–	–	–	2.56	1.82	0.62	0.86	3.33	1.41	0.96	2.00	–	–
Brachinini	–	–	–	–	1.82	1.24	1.72	–	–	0.96	–	–	–
Clivinina	–	–	–	–	–	–	–	–	–	–	–	6.67	–
Reicheiina	–	–	–	–	–	0.62	–	–	–	–	2.00	–	–
Dyschiriini	–	–	–	–	–	0.62	–	–	–	–	–	–	–
Anillini	–	–	–	2.56	–	–	–	–	–	–	6.00	–	4.17
Bembidiina	–	8.33	7.41	5.13	3.64	6.83	5.17	10.00	4.23	5.77	2.00	6.67	4.17
Tachyina	–	–	–	–	–	–	–	–	–	0.96	–	–	–
Patrobini	–	–	–	–	3.64	2.48	2.59	–	–	–	–	–	–
Trechini	21.05	22.22	18.52	28.21	9.09	15.53	18.10	6.67	40.85	22.12	16.00	6.67	12.50
Stomiina	–	–	–	–	–	0.62	–	–	–	–	–	–	–
Myadina	–	2.78	–	2.56	–	0.62	0.86	6.67	2.82	1.92	4.00	6.67	–
Pterostichina	15.79	11.11	14.81	12.82	18.18	14.91	16.38	13.33	12.68	15.38	16.00	6.67	12.50
Molopina	5.26	13.89	11.11	10.26	16.36	14.91	10.34	20.00	7.04	16.35	10.00	–	4.17
Zabrini	–	–	3.70	2.56	3.64	16.83	9.48	6.67	2.82	4.81	8.00	6.67	8.33
Harpalina	–	–	–	5.13	–	0.62	2.59	6.67	2.82	1.92	2.00	6.67	8.33
Ditomina	–	–	–	–	–	0.62	1.72	3.33	–	0.96	2.00	–	–
Stenolophina	–	–	–	–	–	0.62	–	–	–	1.92	–	–	–
Licinini	–	–	–	–	–	1.24	0.86	–	–	–	–	–	–
Platynini	5.26	2.78	3.70	2.56	1.82	–	0.86	3.33	2.82	1.92	2.00	–	–
Atranopsina	–	–	–	–	–	1.86	–	–	–	–	–	–	–
Calathina	–	–	3.70	–	9.09	4.35	5.17	3.33	1.41	0.96	2.00	–	–
Sphodrina	–	–	–	–	–	1.86	–	–	2.82	2.88	10.00	6.67	4.17
Omphreini	–	5.56	–	–	5.45	2.48	1.72	–	–	–	–	–	–
Lebiina	–	–	–	–	–	–	0.86	–	–	0.96	–	–	4.17
Cymindidina	–	–	–	–	–	–	–	–	–	0.96	–	–	–
TOTAL	*100.0*	*100.0*	*100.0*	*100.0*	*100.0*	*100.0*	*100.0*	*100.0*	*100.0*	*100.0*	*100.0*	*100.0*	*100.0*

component is 62.92%. Among the arboreal fauna, the most ancient phyletic lineages are retained including some evident Tertiary relicts. There are the Gondwanan *Reicheadella* and Anillina (*Dicropterus, Parvocaecus*, and *Winklerites*) as well as two groups with a less certain descent: *Aptinus* s.l. and Omphreini. The current distribution of the arboreal complex does not appear to have been influenced much by the Pleistocene climatic oscillations. Presence of old arboreal lineages in certain preglacial microhabitats suggests a continuous existence of Pliocene or even more ancient environmental conditions in some localities in the Central and Eastern Balkan Peninsula.

The eremial biochore includes 29 moderately xerophilous species and subspecies, all preferring grasslands and other open dry habitats. They constitute 7.53% of all endemics and belong to the genera *Carabus* (subgenera *Lamprostus* and *Pachystus*), *Cicindela, Poecilus, Zabrus, Harpalus, Acinopus, Pachycarus*, and *Odontocarus* (tribes Cicindelini, Carabini, Pterostichini, Zabrini, and Harpalini). Last two tribes are predominant, with 12 and ten species-group taxa, respectively. The common morphoecotype here are phanerobionts: very mobile geobionts and stratobionts with evolutionary rate faster than that of the arboreal taxa. Local endemism among eremial carabids is very low (13.79%), as opposed to the arboreal and oreal groups. Local endemic eremial subspecies are *Zabrus aetolus purkynei* and *Z. aetolus winkleri*; endemic eremial species are *Z. boldori* and *Harpalus serbicus* (the last with uncertain status; Kataev et al., 2003). Most eremial species-group taxa belong to the South Aegeidean (16 taxa) and Iranian–Turanian (ten taxa) faunal elements.

The oreal complex consists of 89 subalpine and alpine microthermophilous carabids, or 23.12% of all endemics; it includes genera *Nebria* (26 taxa), *Trechus* (25), *Pterostichus* (11), *Zabrus* (seven), *Deltomerus* (five), *Bembidion* (four), *Calathus* (four), *Molops* (three), *Calosoma* (two), *Carabus* (one), and *Synuchidius* (one). Predominant tribes are Nebriini and Trechini. The common morphoecotype are cryptobionts with low mobility, which include geobionts and stratobionts but with the evolutionary rate faster than that of the arboreal taxa. The specificity of the oreal element is exceptionally high (78.65%). Such groups as *Nebria* (Kavanaugh, 1980) and *Deltomerus* are primarily adapted to cool or cold habitats, and their association with the oreal ecosystems is considered to be the oldest in the Balkans. Conversely, recent adaptive shifts from arboreal to oreal (*Bembidionetolitzkya, Trechus, Aphaon, Stenochoromus*, and *Neocalathus*) and from eremial to oreal (*Pachystus* and *Pelor*) macrohabitats have taken place within a single species group or subgenus. Such ecological changes are reflected in functional morphoadaptations. Also, several neoendemic genus-group taxa (*Aphaon, Stenochoromus*, and *Synuchidius*) result from the evolution of the highest mountain ecosystems in the region. The "taxon reorganization" among the Balkan Carabidae supports the view of Erwin (1979) and earlier authors (Jablokoff-Khnzorian, 1963) about the nature of a taxon pulse and the course of carabid evolution from ecological perspective at a regional level. The oreal environmental type appears to be the youngest of the three types in the Balkans.

Table 5 Macrohabitats of the endemic carabid species and subspecies in the Central and Eastern Balkans. Abbreviations, see Fig. 1

Macrohabitat	NS	SS	DR	ES	WK	AL	WM	EM	NB	SB	NG	DO	ET
Arboreal	84.21	88.89	74.07	84.61	50.91	63.36	55.17	70.00	80.28	68.27	80.00	73.33	66.67
Eremial	5.26	2.78	7.41	10.26	1.82	7.45	9.48	16.67	9.86	10.58	16.00	26.67	33.33
Oreal	10.53	8.33	18.52	5.13	47.27	29.19	35.35	13.33	9.86	21.15	4.00	–	–
TOTAL	*100.0*	*100.0*	*100.0*	*100.0*	*100.0*	*100.0*	*100.0*	*100.0*	*100.0*	*100.0*	*100.0*	*100.0*	*100.0*

Distribution of the endemic carabids among biochores varies in the different regions (Table 5). Distinct Southern Dinaric and Pontian types of macrohabitat distributions (Fig. 5) are distinguished. The Dinaric type [(Albania+Western Macedonia)+Western Kosovo] is characterized by the highest share of oreal fauna (29–48%), a moderate contribution of eremial fauna (2–10%), and a relatively low share of arboreal fauna. The Pontian zone [Dobrogea+European Turkey] has the highest share of eremial endemics (26–34%) and lacks the oreal ones. The remaining regions form one mixed Subdinaric–Rhodopean group [(Dardanian Region+Southern Bulgaria)+Eastern Macedonia]+[(((Northern Bulgaria + North-western Serbia)+Eastern Serbia)+Northeastern Greece)+Southwestern Serbia] with

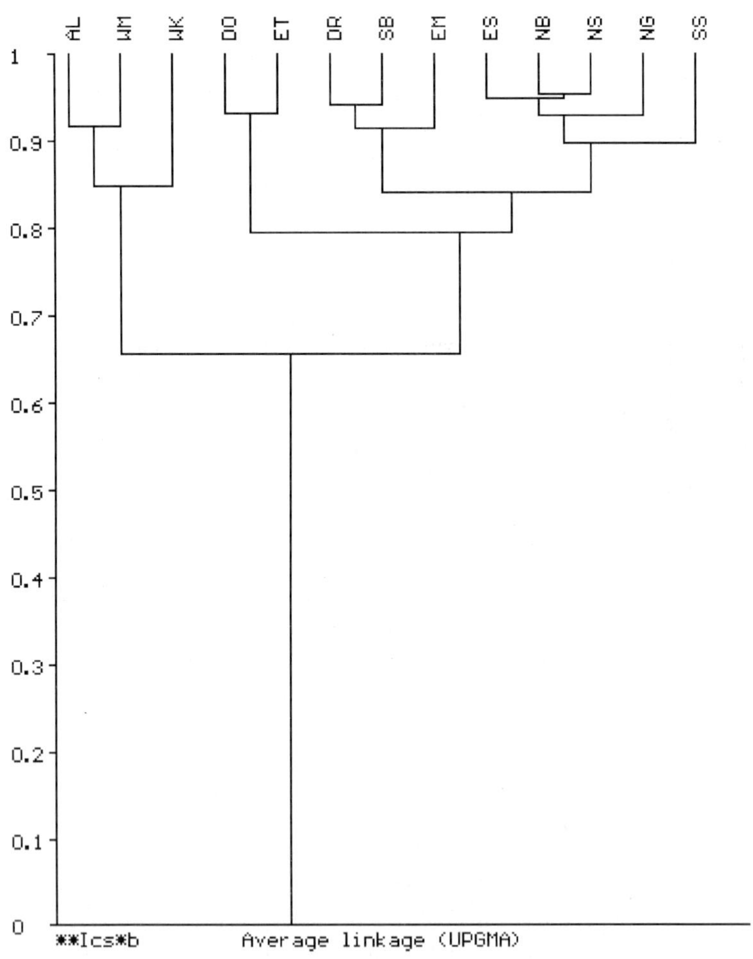

Fig. 5 Affinity of the regional faunas based on microhabitat affiliation of the endemic carabid species-group taxa.

the highest contribution of arboreal fauna (68–89%), the share of eremial fauna (2–17%) lower than in the Pontian zone; and the share of oreal fauna (4–21%) lower than in the Southern Dinaric zone.

7 Causal Biogeography

7·1 The faunal elements and complexes, their formation and arrival

Considering the origin of separate faunas, some authors grouped boreal species into a "Eurosiberian" faunal complex, and southern species into a "Mediterranean" one. Such definitions emphasize the geographical and rarely the temporal (phylogenetic) component of a faunal element. About 30 years ago, Kryzhanovskij and Starobogatov (1974) stated that each *faunal element* unites species with similar ecological preferences, common origin, and common history of dispersal. Further, each *faunal complex* is formed by species originated in a common territory and environment, and within a uniform temporal interval. Good examples of combination of these components, when the origin of a fauna is traced, are given by La Greca (1985, 1996, 1999), Gruev (1990, 2000), Casale and Vigna Taglianti (1999), and Popov (2002). Following a similar model, all endemic Balkan Carabidae are grouped into 13 *faunal elements*, each composed of taxa related chronologically, geographically, and phyletically (Tables 1 and 6). The distribution of species composing a separate monophyletic taxon is traced at superspecies, species group, subgenus, or genus levels. The faunal elements are further grouped in five *faunal complexes* (Boreal, Nemoral, Alpine–Pamirian, Mediterranean, and Eurasian xeric); each of them corresponding more or less to a distinct biome.

The *Boreal* (taiga) *faunal complex* is composed of a single Altaic (Mongolian) faunal element. Such subgenera as *Trepanes, Peryphus, Bothriopterus, Pseudomaseus*, and *Neocalathus* have center of diversity in and around the mountains of Altai, Sayan, and Tien Shan. In addition, their most widely distributed species, with Transpalearctic, West Palearctic, or Eurosiberian ranges, occur in the same area. The Altaic region is also an important center of diffusion of the Pliocene–Pleistocene cold-adapted boreal flora to the Balkans (Stojanoff, 1940). The subgenus *Morphnosoma* of the genus *Pterostichus* has a similar distribution, if *P.* (*M.*) *brevistylis* (Jeannel), a species with uncertain status from the Western Alps, is excluded. *Synuchidius* is a possible recent local derivative of *Neocalathus*. Most likely, the species of this complex were eurytopic and adapted to cool habitats already in the Early Pliocene. Later on, they diffused more or less far in Europe and Siberia, and, in some cases, circumboreally. During the interglacials, those taxa evolved under isolation in the high Balkan mountains (*Bembidion maculatum serbicum, Pterostichus melanarius bulgaricus, Calathus albanicus, C. melanocephalus purkynei*, and *Synuchidius ganglbaueri*). The Altaic faunal element includes four oreal and four arboreal taxa, most of them (six taxa) found in the Southern Dinaric zone. Western Kosovo and Northwestern Serbia are relatively well represented (Table 6), while Eastern Macedonia, Dobrogea, and European Turkey have no endemics of Boreal origin.

The *Nemoral* (temperate nemoral) *faunal complex* includes 55 species-group taxa belonging to three faunal elements: European–American (one), European (41), and Eurasian (13). All these taxa are remainders of the Arctotertiary Boreal fauna of the East Asian stock. It is possible that these remnants represent less temperate, most adapted to cool habitats lineages of this Asian fauna. Their dispersal within Mediterranean and Submediterranean Europe was due to their plasticity and favorable environmental conditions in the postglacial time. Endemics from this complex are dominant in Northwestern Serbia and subdominant in Southwestern Serbia, Dardanian Region, Eastern Macedonia, Dobrogea, and Eastern Serbia (Table 6). They are relatively poorly represented in Albania and Western Macedonia because the Mediterranean faunal elements prevail there. *Myas chalybaeus* is the only member of the European–American element. Except this species, genus *Myas* includes three southeastern Nearctic species (Allen, 1980; Bousquet and Larochelle, 1993). This type of distribution is unique in the Palearctic and is considered to be a result of Eocene–Oligocene land connections between Europe and North America (Noonan, 1986, 1988). The European element is composed of relict groups with more or less distinct Northeastern American and/or East Asian relatives. In Europe, these taxa escaped the glacial cooling in Submediterranean arboreal refugia (see below); current Submediterranean ranges of most of these endemics support this suggestion. In the interglacials, some stenothermic groups adapted to cool environment (*Aphaon* and *Calathus metallicus*) probably went up into the high forest and oreal terrains. The lineages of the Eurasian nemoral element live, as a rule, both in Europe and East Asia (*Morphocarabus, Trachycarabus, Tomocarabus, Asaphidion*, and *Stomis*). Most subgenera and genera of this group have the lowest species richness in West and Central Asia due to the Pleistocene degradation of the nemoral biome there.

The *Alpine–Pamirian* (West Palearctic high mountain and oreal) *faunal complex* includes 39 species-group taxa belonging to three close faunal elements: Alpine–Pamirian, Alpine–Armenian, and Pamirian–Armenian. Their history is obscure because of the very dynamic evolution of the biota in the Alpine–Himalayan orographic belt. However, I believe that the genesis of that complex is coherent with that of the Arctotertiary boreal fauna, and that both are derivatives of the Paleogene Angarian fauna. The existence of Western Palearctic oreal elements is possibly due to the disjunction of the more southern Arctotertiary biota. An important difference between the two complexes consists in the different routes used for their longitudinal dispersal. The Alpine–Pamirian carabids, probably more adapted to the elevated ecosystems, could have differentiated in the mountains south of the Paratethys, and used the Alpine–Himalayan chain to migrate from east to west and vice versa. The Arctotertiary boreal fauna moved north of Paratethys mainly from Asia to Europe across the lowlands. The Alpine–Pamirian element includes a single oreal (*Nebria* s. str.) and single arboreal (*Omoperyphus*) endemics.

Table 6 Biogeographic structure of the endemic carabid species and subspecies in the Central and Eastern Balkans

Faunal element	NS	SS	DR	ES	WK	AL	WM	EM	NB	SB	NG	DO	ET
ALT	5.26	2.78	3.70	2.56	5.46	3.11	4.31	–	1.41	0.96	2.00	–	–
EUA	–	2.78	–	2.57	–	0.62	0.86	3.33	1.41	0.96	2.00	6.67	–
EU	26.31	16.67	22.22	12.82	5.45	5.59	5.17	13.33	7.04	4.81	6.00	20.00	12.50
EAS	26.32	22.22	25.93	20.51	20.00	10.56	11.21	26.68	18.30	17.31	12.00	13.33	12.50
AAR	–	5.55	–	–	18.18	13.04	12.08	–	1.41	5.77	–	–	–
AP	–	–	–	–	–	0.62	1.72	–	–	–	–	–	–
PAR	–	2.78	3.70	–	1.82	1.87	1.72	3.33	1, 41	1.92	–	–	–
COL	–	–	–	–	–	–	–	–	1.41	0.96	–	6.67	4.17
NA	26.32	38.88	25.93	46.16	36.37	32.92	36.22	30.00	52.11	44.23	46.00	6.67	25.00
PAM	–	–	–	–	5.45	9.94	5.17	–	–	1.92	8.00	–	–
POM	5.26	2.78	3.70	5.13	1.82	3.11	1.72	3.33	5.63	6.73	4.00	20.00	16.67
SA	10.53	5.56	14.82	7.69	5.45	17.38	18.10	16.67	8.46	11.54	20.00	20.00	20.83
IT	–	–	–	2.56	–	1.24	1.72	3.33	1.41	2.89	–	6.66	8.33
TOTAL	*100.0*	*100.0*	*100.0*	*100.0*	*100.0*	*100.0*	*100.0*	*100.0*	*100.0*	*100.0*	*100.0*	*100.0*	*100.0*

Abbreviations in the first column: AAR, Alpine–Armenian; ALT, Altaic; AP, Alpine–Pamirian; COL, Colchic; EAS, Eurasian; EU, European; EUA, European–American; IT, Iranian–Turanian; NA, North Aegeidean; PAM, Paleomediterranean; PAR, Pamirian–Armenian; POM, Pontomediterranean; SA, South Aegeidean. Other abbreviations, see Fig. 1.

At present time, these two subgenera are widespread from the Pyrenees to the Himalayas. The Alpine–Armenian element consists of 31 oreal (*Alpaeus, Bembidionetolitzkya, Deltomerus*) and three arboreal (*Bembidionetolitzkya*) species-group taxa; of these, 31 taxa are local endemics and three have wider Balkan ranges. The Pamirian–Armenian element includes three oreal taxa of the subgenus *Ocyturanes*, a group with two distinct centers of diversification, Armenia and Pamir. Carabids from the Alpine–Pamirian complex are most numerous in Albania (25 taxa), Western Macedonia (18), Western Kosovo (11), and Southern Bulgaria (eight). The highest shares of these taxa are in the same four regions and in Southwestern Serbia (Table 6).

The *Mediterranean* (thermoxeric nemoral and subtropical forest) *faunal complex* consists of 276 species-group taxa and six faunal elements: Paleomediterranean (23 taxa), North Aegeidean (Dinaric) (193), South Aegeidean (45), Pontomediterranean (14), and Colchic (South Euxinian in other classifications) (one). Different elements reflect the heterochronic faunal climaxes in the Pontomediterranean area. The Paleomediterranean element is the successor of the local tropical Paleogene fauna – a mixture of Angarian (Laurasian) and Gondwanan lineages. It is believed that, in the Early Tertiary, this tropical fauna existed on the drifting segments of dry land between Laurasia and the African section of Gondwana. In this respect, the Paleomediterranean faunal element is the oldest and includes the most ancient Balkan lineages, like the Gondwanan *Reicheadella* (Jeannel, 1961) and *Aptinus* s.l. Other groups of probably pre-Miocene age are *Albanotrechus* (representative of the *Aphaenops* lineage), *Odontocarus, Neorescius, Calathus* s. str., *Platyderus, Antisphodrus*, and *Actenipus*, most of them from Laurasian stock. Among these, the only eremial exception is *Odontocarus*. I have little doubts that further diversification of these groups in the Mediterranean basin is a result of vicariance rather than dispersal (see e.g. La Greca, 1996). This statement is supported by the actual low to almost none dispersal potential of most species belonging to this element (except some *Aptinus, Calathus* s. str., and *Platyderus*). It is easier to trace the descendants of the Aegeidean fauna, at least those of Early Tertiary origin, due to the absence of closely related groups in the Nearctic and central and eastern parts of the Palearctic (in contrast with the Arctotertiary nemoral fauna). Both the North Aegeidean and the South Aegeidean elements are, to a large extent, successors of the Paleomediterranean fauna (La Greca, 1990). They are representatives of two main parts of the Miocene Aegeidean dry land. The modern geology and geomorphology of the Eastern Mediterranean region suggests the following possible sequence of events (see, among others, La Greca, 1996; Casale and Vigna Taglianti, 1999). The Aegeidean dry land (Aegeide), the heir of the Cretaceous North Gondwanan fragment, drifted northward in Early Tertiary and occupied a portion of the western Tethys. In Miocene–Early Pliocene, this land split into two main blocks (Apulian plate excluded), and the entire block or its components were connected with different sections of Eurasia. The lineages which further speciated in the Northern Aegeide were the older Paleomediterranean groups as well as new Gondwanan (Pantropical and Paleotropical) and

Angarian immigrants. The second group included the ancestors of *Pheggomisetes, Anillidius, Neotrechus, Serboduvalius, Duvalius, Trechus*, the subtribe Molopina of Pterostichini, *Tapinopterus* s.l., and *Rambousekiella*. The recent distribution of *Molops* s.l. and its probable relative *Cyclotrachelus* from the Southern Nearctic (Bousquet, 1999) demonstrate again the European–American faunal connection (Noonan, 1988), in this case, at the suprageneric level. The subendemic Balkan tribe Omphreini, of the Paleomediterranean stock (Casale and Vigna Taglianti, 1999), has an unclear position among the Conchifera carabids. Except the Western and Central Balkans, this tribe inhabits Taurus (southwestern and central part of South Anatolia). The species of *Omphreus* s. str. most likely originated in the Northern Aegeide. After the Upper Miocene connection of the Southern Aegeide (especially its Anatolian part) to Asia and Africa–Arabia, new immigrants entered from Central Asia and from the Arabian subcontinent, several of them good dispersers. Among the Central Asian migrants probably were *Callisthenes*, which in the Northern Aegeide gave rise to *Microcallisthenes*. The founders of the "*stussineri*"-" *conspicuus*"-"*andreevi*" and "*terricola*" lineages of *Pristonychus* perhaps also entered into the region at this time. Among the Paleotropical immigrants, representatives of Anillini, especially of these of subtribe Scotodipnina, arrived at this time; the latter differentiated further in the Northern Aegeide into *Dicropterus* and *Winklerites*, and in the Southern Aegeide into *Caecoparvus* and *Parvocaecus*. In Early Pliocene, after the complete integration of the two plates, an exchange between both Aegeidean elements became possible. In general, 50.13% of the recent Balkan endemics and subendemics belong to the North Aegeidean faunal element, i.e. it dominates over all other faunal elements. The arboreal forms among the carabids of this element are predominant, including numerous Trechini, *Molops* s. str., and *Tapinopterus* s. str. (84.46%), while the oreal taxa (*Microcallisthenes*, some *Trechus*, and *Stenochoromus*) constitute 15.54% and the eremial ones are absent. The following lineages arose in the Southern Aegeide: *Pachystus, Procrustes, Procerus, Trechus subnotatus, Duvalius* s. str., "*spartanus*" species group of *Molops* s. str., taxa related to *Tapinopterus extensus* and *T. protensus* of *Tapinopterus* s. str., *Pelor, Mystropterus*, and a few others. After the connection with the Northern Aegeide, carabids of these groups appeared in the Central and Eastern Balkans. Currently, the endemics of the South Aegeidean origin occur arboreal (44.44%), eremial (37.78%), and oreal (17.78%) macrohabitats. The beginning of the Quaternary likely marked the formation of the proper Mediterranean faunal elements. The formation of its local variant, the Pontomediterranean element, is due to the Pleistocene isolation of the Mediterranean biota, its probable extinction in other sections of the great Mediterranean refugium (de Lattin, 1967), and further speciation and subspeciation in the Pontian and Eastern Mediterranean areas. Examples of such evolution are Balkan endemics and subendemics of *Cicindela* s. str., *Clivina, Dyschiriodes, Peryphanes, Harpalus* s. str., and "*terricola*" species group of *Pristonychus*. The Pontomediterranean descent is indicated also for *Tachyura krueperi* (uncertain species status) and *Acupalpus planicollis*, two taxa belonging to Paleotropical lineages. Because of its uncertain relationship with the Nemoral and the Mediterranean complexes, the Colchic faunal element (with a single eremial *Lamprostus* member) is placed provisionally in its own group. The Balkan endemics of the Mediterranean complex form over 50% in nine regions (Table 6), with their maximum in Northeastern Greece. Their lowest shares are found in Northwestern Serbia, Southwestern Serbia, Dardanian Region, and Western Kosovo due to the higher role of the nemoral forms.

The *Eurasian xeric* (steppe) *faunal complex* includes a single Iranian–Turanian faunal element with three arboreal and four eremial species-group taxa. This fauna dispersed from the Western and Central Asian steppes. All local endemics of this group are young derivatives of lineages, which expanded their ranges during the interglacial extensions of grassland habitats. Several of these, like *Cymindis vassili*, are "xerothermic relicts" (thermophilic populations, which once had wider distribution). Most endemics of Iranian–Turanian origin live in Southern Bulgaria (three taxa), as well as in Albania, Western Macedonia, and European Turkey (two taxa each). Their share is the highest in European Turkey, Dobrogea, Eastern Macedonia, Southern Bulgaria, and Eastern Serbia, which is well explained by the three main routes for dispersal of this element to the Balkans, i.e. Pannonian steppes (to Eastern Serbia–Eastern Macedonia), Pontian steppes (to Dobrogea), and Central Anatolian steppe (to European Turkey–Southern Bulgaria).

The biogeographical association (Fig. 6) displays again an isolated Pontian group [Dobrogea + European Turkey]. The highest share, among the regions, of the triad of Pontomediterranean, South Aegeidean, and Iranian–Turanian faunal elements (Table 6) supports such an association. The remaining regions are divided in two groups. In the first one, Northeastern Greece groups closely with the distinct Southern Dinaric zone [((Albania + Western Macedonia) + Western Kosovo) + Northeastern Greece]. This grouping is based on the lowest shares of the European and Eurasian Arctotertiary elements, the highest shares of the Alpine–Armenian (excluding Northeastern Greece) and Paleomediterranean elements, and the lowest shares of the Pontomediterranean (excluding Northeastern Greece) element. The remaining regions form one, more or less mixed group [(Dardanian Region + Northwestern Serbia) + Eastern Macedonia] + [((Eastern Serbia + Northern Bulgaria) + Southern Bulgaria) + Southwestern Serbia], a result contrary to that obtained formerly.

7·2 The relict, orthoselective, and migratory elements

The "relict" species (*sensu* Krasilov, see 9.4) are characterized by the ancient age, sufficient number of plesiotypic characters, and restricted distribution. The supraspecific groups have usually a distinct phyletic and/or geographic isolation. From the ecological point of view, the relict elements are examples of little changed, or successfully preadapted, Tertiary forms, which were not seriously affected by the unfavorable Pleistocene environmental trends. Their evolution

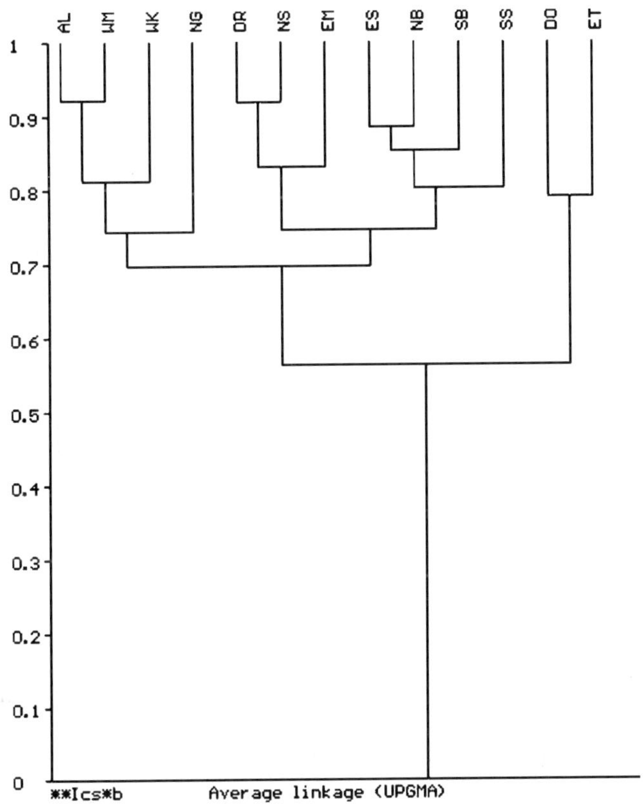

Fig. 6 Biogeographical structure based on the faunal element affiliation of the endemic carabid species-group taxa.

continued during the glaciations *in situ* without dramatic events, and their habitats remained more or less unchanged. All endemics from both Arctotertiary and Alpine–Pamirian faunal complexes belong here. From the Mediterranean faunal complex, those are most endemics belonging to the Paleomediterranean element (all *Aptinus, Odontocarus, Neorescius, Platyderus*, and *Calathus* s. str.). From both Aegeidean lineages, those are *Microcallisthenes, Carabus, Trechus, Tapinopterus, Molops, Pelor, Pachycarus*, and *Omphreus, Harpalus triseriatus* s.l., and *Laemostenus cimmerius* s.l. All relicts survived the Pleistocene climatic disturbances in refugia. In the interglacials and in the postglacial time, despite the competition with the new Boreal and steppe immigrants, the relict fauna succeeded to remain in its pre-Pleistocene microhabitats and to adapt to the new, less favorable abiotic conditions of the Holocene environement, thus avoiding significant morphoecological reorganization. As a consequence of inter- and postglacial dispersal and further cladovicariant events, the relict component contributed to the origin of many neoendemic subspecies and young species. In total, 251 endemic species and subspecies (or 65.19%) belong to the "relict" fauna.

There is no doubt that the "orthoselective" (or "transformed") fauna is also a relict in terms of its Tertiary age. Prevalence of plesiotypic characters and taxonomic split (phylogenetic isolation) between the related phyletic lineages define the relict character of this fauna. The supraspecific taxa are monotypic or oligotypic with disjunct ranges, mostly of island type. In ecological sense, the habitats of the "transformed fauna", where they once lived, were more or less naturally destroyed which led to the occupation of new niches not preferred by the competitors. As a result of their preadaptation, these carabids experienced significant morphoecological reorganization. The main portion of the "transformed fauna" was, and still is, stenobiotic in regard to humidity. Taxa with similar features can be found only in the complicated Mediterranean faunal complex. There are: *Clivina* (one taxon), *Reicheadella* (two), *Albanotrechus* (one), *Pheggomisetes* (13), *Anillidius* (one), *Neotrechus* (five), *Serboduvalius* (two), *Duvalius* (46), all Scotodipnina (five), *Rambousekiella* (one), and subgenera *Hypogium* and *Speluncarius* of *Tapinopterus* (total of five taxa). There are also nine species of *Laemostenus* (excluding *L. cimmerius*). This fauna comprises 23.64% of all endemics. Except *Reicheadella bischoffi* and single *Laemostenus* taxa, all other relict carabids are local endemics. Most representatives are blind and confined exclusively to limestone habitats; all of them are arboreal forms. The absence of oreal and eremial forms suggests the absence of ecological plasticity among the taxa of the "orthoselective" component.

The "migratory" fauna includes recent immigrants, which entered the Balkan Peninsula since the end of Pliocene, during the glacials and interglacials, and after the last glacial. Upon their arrival, the populations of separate species underwent contractions of their ranges, which led to isolation and subsequent speciation and subspeciation or extinction. Usually, these ground beetles are morphologically hardly distinct from their close, non-endemic relatives. Within the "migratory" fauna, adaptive radiation and vicariance were the evolutionary processes of equal value. Representatives of this group are all carabids of the Boreal (eight taxa) and Eurasian xeric (seven taxa) faunal complexes. From the Mediterranean complex, there is a single Colchic representative, *Carabus (Lamprostus) torosus* s.l., and 14 taxa of Pontomediterranean stock: *Cicindela monticola* s.l., *Dyschiriodes bacillus* s.l., etc. Most complicated is the distinction of the "migratory" element within the Nemoral (Arctotertiary) faunal complex. This is due to a very probable preglacial occurrence of such taxa in the Balkans. Similarly, 11 subspecies of *Carabus scabriusculus, C. ullrichii, C. cancellatus, C. convexus, C. irregularis, C. violaceus*, and *Calathus metallicus* can be regarded as recent immigrants. The larger part of the ranges of the respective species lies out of the Balkan Peninsula. Two South Aegeidean species, *Carabus (Procerus) scabrosus* and *Zabrus (Pelor) corpulentus*, also reflect the post-Tertiary disperal to the Balkan Peninsula. Altogether, 43 species-group taxa, or 11.17% of the endemics, belong to the recent migratory fauna.

7·3 Inter- and postglacial refugia, centers of dispersal, and areas of origin

According to the model of de Lattin (1967) for the Palearctic terrestrial fauna degradation in the Pleistocene, all Balkan endemics should have been distributed among a few neighboring refugia. Those could be of an "open" type as well as of a "closed" type (Lindroth, 1979). For the arboreal biota, such refugia seem to be easily found. One can connect the arboreal Balkan endemic ground beetles with the great Pontomediterranean refugium. Its northern limits passed along the western and/or southern slopes of the North Albanian Alps, the Šar Massif, the Korab–Dešat–Jablanica range, etc. In the northwestern Greece, this boundary went around the Pindus Mts., further passing along the southern slopes of the Belasitsa, the Rhodopes and the Strandja massifs. However, some European biogeographers (e.g., Bănărescu and Boşcaiu, 1978; Buchar, 1983), examining the native fauna in depth, established presence of other arboreal refugia outside the Mediterranean one. Buchar (1983) emphasized a limited distribution of many forest animals to either the Carpathians or the Alps, or to both of these massifs; he demonstrated four refugia in the Submediterranean Europe: Atlantic, Alpine, Carpathian, and Balkan. Animals with such Submediterannean/Central European distribution normally do not inhabit the Mediterranean. I support this view through the analysis of a specific, non-Mediterranean distribution of many arboreal Balkan endemic and subendemic carabids. As a whole, species with arboreal preferences survived in the glacials within the Pontomediterranean and the Submediterranean Balkan refugia. Considering the regional relief, there is little doubt that their enclaves, especially in the area of their overlap, existed together. Each endemic ground beetle species-group taxon can be associated more or less successfully with one of those arboreal refugia when analyzing its range, altitudinal distribution, macro- and microhabitat preferences, etc. The endemics that survived in the Pontomediterranean refugium are most likely inhabitants of the lowlands and low mountains in the Mediterranean climatic zone, such as *Carabus (Oreocarabus) preslii* s.l. and *C. (Procerus) scabrosus* s.l., and usually do not pass northward across the climatic border. They prefer higher winter temperatures and have an aestivation diapause. The arboreal endemics, which survived the glacials in the Submediterranean refugium, live now northward from the Mediterranean climatic zone. Examples of such taxa are *Carabus (Morphocarabus) scheidleri* s.l. and *Cychrus semigranosus*. Normally, such endemics live in higher altitudes, endure lower winter temperatures and have no aestivation diapause. Further research on this trend is necessary to elucidate details of the arboreal taxa's past and to explain their present.

The area of maximal diversity and a probable center of postglacial diffusion of the subgenus *Pelor* (genus *Zabrus*) lie in the inland Anatolia. An eremial focus of secondary importance clearly exists in Central Anatolia and the adjacent Armenia, which was not delineated by de Lattin (1967) from the close Iranoeremial and/or Syroeremial centers of dispersal. The same area was, and still seems to be, the motherland of many xeric taxa with Anatolian or Subanatolian distribution (Casale and Vigna Taglianti, 1999). I also think that a part of the ground beetle assemblage there could have a primary South Aegeidean origin (*Ischnocarabus, Pachystus, Lamprostus, Hoplodactylus, Pelor*, and *Paramystropterus*), due to the phyletic isolation of some local groups and their restriction to a part of the former Aegeide. Since its Messinian land connection with Asia, Anatolia experienced a massive influx of the competitive Iranian–Turanian eremial fauna (*Callisthenes, Amara, Pangus*, various other Harpalini, *Cymindis*, etc.). The biogeographical place of Central Anatolia remains more or less unsettled: it has been treated as a transitional area between the Iranian–Turanian Superprovince (Kryzhanovskij, 1983) in the east and the Pontomediterranean area (de Lattin, 1967) in the north, west, and south. In the southeast, a narrow forest strip in Southern Kurdistan divides the region from the Paleoeremic domain (Por, 1975). Other eremial endemics, *Pachycarus (Mystropterus)* spp., *Acinopus baudii hellenicus, Harpalus euchlorus, H. karamani* s.l., *H. triseriatus* s.l., *Odontocarus robustus*, and several species of *Zabrus (Pelor)*, cannot be associated with the Central Anatolian refugium. I suggest the existence, during some of the xerothermic interglacials, of small western, restricted enclaves of this refugium within the Pontomediterranean arboreal somewhere in the present Central or Southern Greece. The current distribution of two other endemics, *Poecilus rebeli* and *Harpalus serbicus*, cannot be explained by the southern eremial refugia. Here, the influence of a northern "turano-eremische" (de Lattin, 1967) or, more exactly, "aralo-kaspische" (Bănărescu and Boşcaiu, 1978) fauna has to be suggested. According to the latter authors, the westernmost enclave of this eremial focus has existed in the center of the Pannonian Lowland. The association of *Cicindela* s. str. with littoral and inland open semi-dry habitats contributed to their survival.

The eremial Carabidae survived in two interglacial refugia, Central Anatolian and Iranian–Turanian; the latter was of less importance for the current endemics. Such forms could enter the Balkan Peninsula from the north via Anatolia through Southern and Central Greece to Northern Greece (*Carabus graecus, Acinopus baudii, Harpalus triseriatus, Pachycarus* spp., *Odontocarus robustus*, and *Zabrus aegeus*), or across the Bosporus and Dardanelles along the Black Sea coast (*Cicindela campestris palustris, C. monticola rumelica, Carabus torosus torosus*, and *Zabrus corpulentus ponticus*). The ancestors of *Poecilus rebeli* and *Harpalus serbicus* probably dispersed from the north, via Pannonian steppes to Vojvodina, and from there to the Drina and Morava valleys. The lower share of the eremial local endemics compared to the arboreal and oreal ones could be explained by both historical and recent environmental factors. The subtropical and temperate warm preglacial climates, as well as similar current climate, have limited the expansion and distribution of the typical xeric ecosystems. Predominance of subspecies and vicariant species points at a more recent expansion and speciation of the eremial endemics (excluding *Odontocarus robustus*, a distinct Paleomediterranean faunal element), probably due to the interglacials and the postglacial xerothermic period.

All oreal ecosystems in the studied area are confined to the eastern Alpine oreal refugium (Southeastern Alps and Balkan and West Anatolian high mountains), which is of a secondary importance after de Lattin (1967). Based on the more detail data for distribution of the endemics, I infer that the oreal in the Central/Eastern Balkan area can be divided into two distinct parts. The first one is the Southern Dinaric oreal refugium: the highest landscapes of the North Albanian Alps (=Prokletije Massif), the Šar Massif, the Korab–Dešat crest, the Jakupica Massif, Baba (=Pelister) and Nidže mountains. These landscapes are restricted to Western Kosovo, Albania, and Western Macedonia. The corresponding numbers of oreal carabid endemic taxa are 26, 47, and 41, with the total 68. The second oreal refugium is located in Southern Bulgaria and is confined to the highest ecosystems of the Rila and Pirin Mts. The total number of oreal endemics there is 22. The number of oreal endemic taxa in the remaining regions is as follows: seven in Northern Bulgaria, five in Dardanian Region, four in Eastern Macedonia, three in Southwestern Serbia, and two in Northwestern Serbia, Eastern Serbia, and Northeastern Greece. These lower numbers in the last seven regions are due to the absence of mountain ranges higher than 2400 m. There are more essential distinctions between the endemic fauna of Southern Dinaric and Rhodopean oreals besides the vicariance of related species. The main distinction is the much more diverse fauna of the Southern Dinaric zone. Many supraspecific taxa found there are absent from the Rhodopean oreal: *Nebria* s. str., *Microcallisthenes, Deltomerus*, *Stenochoromus, Pelor, Neocalathus*, and *Synuchidius*, and some species groups of *Trechus*. Almost all of the endemics from the Rhodopean oreal have common origin with those of the Southern Dinaric oreal fauna, e.g. they are their geographical vicariants. Regarding the Balkan endemic fauna, the centers of postglacial distribution (in the sense of de Lattin, 1967) could be a place of origin for at least all neoendemic forms.

7·4 A scenario for the formation of the Balkan oreal ground beetle taxocoenoses

As a rule, the Tertiary paleoendemics as well as Quaternary neoendemic taxa have participated in the structure of the high mountain Mediterranean faunas (La Greca, 1966; Gruev, 1990; Heiss and Josifov, 1990). The formation of the Balkan oreal community is also composite, and not very ancient. Because of the evident phyletic relationships and recent macroecological shifts within some supraspecific taxa, a scenario for the formation of this community is proposed here (Fig. 7).

The origin of *Nebria* (*Nebria*) *krateri* is obscure since species of this subgenus are widespread from the Pyrenees to the Western Himalayas. The derivation of the species of *Nebria (Alpaeus)* and *Deltomerus* s.l., the oldest oreal inhabitants of the Balkan Peninsula, is connected to both the Alpine and the Armenian oreals. Their adaptation to the highest mountain ecosystems dates from the pre-Pleistocene times. *Bembidion* (*Bembidionetolitzkya*) *rhodopense* reveals the same origin; however, the penetration of this species into the oreal should be more recent. More obscure is the geographic-phyletic evolution of the subgenus *Ocyturanes* (genus *Bembidion*) but undoubtedly it is connected with either the Armenian or the Pamirian oreal. The original North Aegeidean (Dinaric) faunal elements in the high mountain grasslands

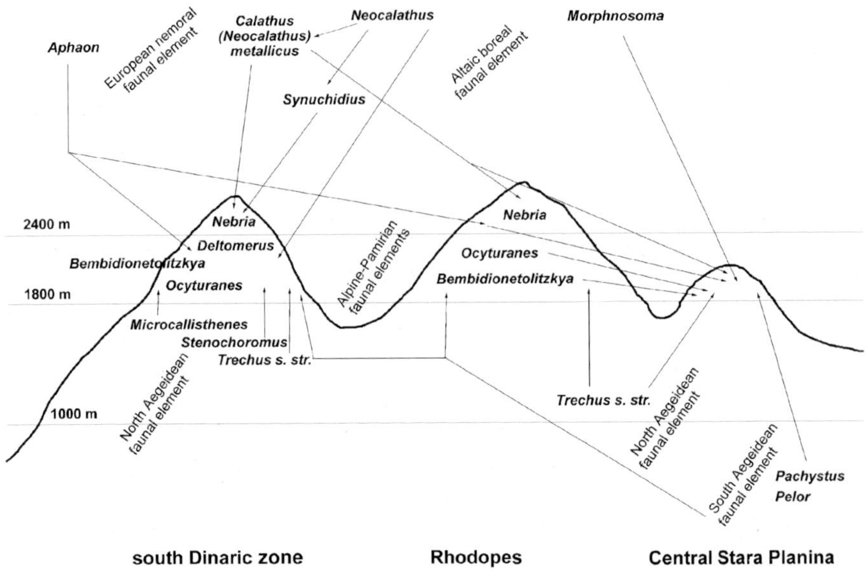

Fig. 7 A scenario for the formation of the endemic carabid oreal taxocoenoses in the Central and Eastern Balkan Peninsula.

are the taxa from five species groups of *Trechus* (*"babinjensis"*, *"rhilensis"*, *"hajeki"*, *"egregius"*, and *"obtusiusculus"*, cf. Casale and Laneyrie, 1982), the polytypic *Molops* (*Stenochoromus*) *montenegrinus* s.l., and the subgenus *Microcallisthenes* (genus *Calosoma*). La Greca (1977) indicated that the Italian oreal *Trechus* species are a result of the Pliocene and early Pleistocene speciation. *Microcallisthenes* evolved in the North Aegeidean territory as a result of a possible migration of an Anatolian lineage of *Callisthenes* during the Late Tertiary (Casale and Vigna Taglianti, 1999). Currently, this genus is the most diverse in the mountains of Central Asia, and probably has Angarian origin. The original South Aegeidean faunal elements are *Carabus* (*Pachystus*) *cavernosus* s.l. and several *Pelor* species. *C. cavernosus* is probably the most widely distributed oreal species in the region. This fact and the presence of this species in the lower open habitats (e.g., in Northwestern Serbia and Southern Bulgaria) suggest its recent adaptation to the oreal environement. Also, in the Osogovo Mts. (Southern Bulgaria), this beetle is subdominant in the orophyte belt from 1583 m to 2251 m (pers. obs.). The oreal penetration of the *Pelor* triad *"albanicus-ganglbaueri-peristericus"* is also a consequence of a recent habitat preadaptation in a single, presumably monophyletic, lineage. The shifts from eremial to oreal macrohabitat have occurred only among the descendants of the South Aegeidean faunal element, whereas shifts from arboreal to oreal environment have taken place in various other elements. I have little doubt about the origin of the complex *Aphaon*, which is very closely related to (and possibly is a synonym of) *Pterostichus* s. str. The former should have evolved from a single or a few ancestors, which in the past inhabited the high woodlands of the Western Balkans. Other route, through the Carpathians, is almost impossible (the only Carpathian *Pterostichus* s. str. species, *P. sacheri*, is perhaps a derivative of the Balkan *Aphaon*). Among the extant Balkan species of this lineage, the Eastern Balkan *Pterostichus rhilensis* s.l. is the most primitive, and unique in having more than one (from two to four) lateral setae on each side of pronotum. The formation of this species is a result of isolation in the highest parts of Rila and Pirin Mts. During some of the glacials, this species successfully dispersed to the Central Balkan subalpine oreal. A definite Carpathian origin is reflected by the range of *Calathus* (*Neocalathus*) *metallicus* s.l. This species is more or less isolated systematically from other representatives of *Neocalathus*; in the Balkans, it is confined only to the high northern mountains and to the North Albanian Alps. It is absent from the rest of high mountains in Albania and Western Macedonia (where it is replaced by *Synuchidius ganglbaueri* and two *Neocalathus* taxa; see below). *Calathus* (*N.*) *metallicus* totally dominates the most Bulgarian alpine and subalpine carabid taxocoenoses, e.g. those of the Central Stara Planina, Vitosha Mts., Osogovo Mts., and Rila Mts. (pers. obs.). Other recent immigrants of northern origin (Altaic elements) are one *Neocalathus* species, two subspecies from *Morphnosoma* and *Neocalathus*, and one *Synuchidius* species. *Pterostichus* (*Morphnosoma*) *melanarius bulgaricus* lives in the subalpine grasslands of the Central Stara Planina. This distinct morphosubspecies is a derivative of West Palearctic eurytopic species. The penetration of *Neocalathus* and the speciation of *Calathus* (*Neocalathus*) *albanicus* and *C.* (*N.*) *melanocephalus purkynei* probably reflect two different in time events. The elytral chaetotaxy of the first species (probably derivative of *C.* (*N.*) *micropterus*) is in the stage of reduction (Guéorguiev, 1998), in contrast to the triad *"cinctus-melanocephalus-mollis"*. *Synuchidius ganglbaueri*, a species and a genus with *incertae sedis* taxonomic status (Hovorka and Sciaky, 2003), seems to be a derivative of *Neocalathus*.

8 Regional Division Based on Taxonomic, Macrohabitat, and Biogeographical Structure of Endemic Balkan Carabidae

The conclusions given in the sections 9.6.3, 9.6.4, and 9.7.1, which follow from the data in Figs. 2–6 and Tables 1–6, propose a biogeographical partition of the Central and Eastern Balkan Peninsula and produce some ideas about the recent history of its biota. The majority of the endemic Balkan Carabidae is considered direct or modified descendants of an older, rich Tertiary fauna.

The highest diversity of the ancient fauna is concentrated in the Southern Dinaric zone (Western Kosovo, Albania, Western Macedonia). The prevalence of Nebriini and Zabrini over Carabini, a good representation of Calosomatina, Patrobini, Calathina, and Omphreini as well as the highest share of oreal fauna, are specific to this zone. Further characteristics are the lowest shares of European and Eurasian Arctotertiary faunal elements, the highest share of endemics of the Alpine–Armenian and Paleomediterranean origin, and the lowest share of Pontomediterranean endemics.

The lineages directly related to the ancient fauna dominate also the Balkan–Rhodopean zone (Eastern Serbia, Northern Bulgaria, Eastern Macedonia, Southern Bulgaria, and Northeastern Greece), with its highest share of the North Aegeidean faunal element (excluding Eastern Macedonia) and the very high impact of Trechini, Pterostichini, and arboreal endemics. The share of Carabini is also considerable but is lower than in the next zone. Absence of Calosomatina, Patrobini, Calathina, Omphreini, and the low share of Zabrini are additional characters for the Balkan–Rhodopean region.

The Subdinaric zone (Northwestern Serbia, Southwestern Serbia, and Dardanian Region) is more or less distinct. This zone is characterized by the highest contribution of the European Arctotertiary faunal element and in the highest share of Carabina (excluding Dobrogea). Other characters are the very high impact of arboreal forms and many endemics from Trechini, Pterostichina, and Molopina. However, a strong Southern Dinaric presence in Southwestern Serbia (Nebriini, *Neoduvalius, "curtulus"* species group of *Molops* s. str., Omphreini) and a weaker Alpine–Pannonian influence (*Platycarabus*) are also noticeable in this zone.

Finally, a Subpontian zone (Dobrogea, European Turkey) is highly evident. Part of its fauna consists of derivatives of both the Colchic (*Lamprostus*) and the South Aegeidean (*Procrustes, Procerus*, and *Pelor*) relict elements. Carabina and Trechini dominate in this zone as well, but the share of Cicindelini and Harpalina is the highest. The highest contributions of eremial fauna and of Pontomediterranean, South Aegeidean, and Iranian–Turanian faunal elements are other most distinctive features of the Subpontian zone.

The demonstrated differentiation of the carabid fauna into the western and eastern portions is due to the continentalization of the climate in latter area (Popov and Deltshev, 1997) as well as to the smaller area occupied by the mountain landscapes in the Northern and Eastern Balkan Peninsula. Another probable reason is the stronger decreasing of the temperatures in the Western Balkan Peninsula (owing to the larger area of mountain landscapes there) during the Pleistocene. Those are the most important reasons for the impoverishment of the diversity of endemic Balkan taxa in the west-to-east direction.

Acknowledgments

I am very obliged to MSc. P. Petrov (Institute of Physiology, Sofia) for technical support and to Dr. L. Penev (Central Laboratory of General Ecology, Sofia) for the literature provided and a useful discussion of the BIODIV program. Thanks are also due to Dr. A. Popov (National Museum of Natural History, Sofia) for the encouragement to complete this research and to Dr. V. Fet (Marshall University, Huntington, West Virginia) for the linguistic help. Any mistakes or misinterpretations of the data are of course my responsibility.

References

Allen, R.T., 1980. A review of the subtribe Myadi: description of a new genus and species, phylogenetic relationships, and biogeography (Coleoptera: Carabidae: Pterostichini). Coleopterists Bulletin 34:1–29.

Baev, P. & L. Penev, 1995. BIODIV. Program for Calculating Biological Diversity Parameters, Similarity, Niche Overlap, and Cluster Analysis. Version 5.1. Pensoft, Sofia & Moscow.

Bănărescu, P. & N. Boşcaiu, 1978. Biogeographie. Gustav Fischer Verlag, Jena.

Bousquet, Y., 1999. Supraspecific classification of the Nearctic Pterostichini (Coleoptera: Carabidae). Association des entomologistes amateurs du Québec, Supplement 9:1–292.

Bousquet, Y. & A. Larochelle, 1993. Catalogue of the Geoadephaga (Coleoptera: Trachypachidae, Rhysodidae, Carabidae including Cicindelini) of America north of Mexico. Memoirs of the Entomological Society of Canada 167:1–397.

Buchar, J., 1983. Zoogeografie. Státní pedagogické nakladatelství, Prague.

Casale, A. & R. Laneyrie, 1982. Trechodinae et Trechinae du monde. Tableau des sous-families, tribus, séries phylétiques, genres et catalogue général des espèces. Mémoires de Biospéologie 9:1–221.

Casale, A. & A. Vigna Taglianti, 1982. I Coleotteri Carabidi delle Alpi Liguri: composizione della fauna ed origine del popolamento. Lavori della Societá Italiana di Biogeografia, Nuova Serie, 9:567–598.

Casale, A. & A. Vigna Taglianti, 1993. I Coleotteri Carabidi delle Alpi occidentali e centro-orientali (Coleoptera, Carabidae). Biogeographia 16:331–399.

Casale, A. & A. Vigna Taglianti, 1999. Caraboid beetles (excl. Cicindelidae) of Anatolia, and their biogeographical significance (Coleoptera, Caraboidea). Biogeographia 20:277–406.

Ćurčić, S.B., 2000. On the diversity of some carabid beetles (Coleoptera: Carabidae: Carabinae and Harpalinae) in Serbia. Arhiv bioloških nauka (Archive of Biological Sciences) (Belgrade) 52(4):219–226.

Erwin, T.L., 1979. [Chapter] 3.22. Thoughts on the evolutionary history of ground beetles: hypotheses generated from comparative faunal analysis of lowland forest sites in temperate and tropical regions. In Erwin, T.L., G.E. Ball & D.R. Whitehead (eds), Carabid Beetles: Their Evolution, Natural History and Classification. W. Junk, The Hague, Boston & London: 539–592.

Furon, R., 1953. Histoire de l'Egéide. Revue générale des sciences pures et appliquées 60:79–95.

Gruev, B., 1990. [Mountain leaf beetles of Eumolpinae, Chrysomelinae and Alticinae (Coleoptera, Chrysomelidae) in Bulgaria, fauna and zoogeography]. Nauchni trudove na Plovdivskiya universitet "Paisii Hilendarski" (Travaux scientifiques de l'Université de Plovdiv "Paisii Hilendarski"), Biology, 28(6):27–61 (in Bulgarian).

Gruev, B., 2000. [Zoogeographical belonging of the leaf beetles species of Lamprosomatinae, Eumolpinae, Chrysomelinae, Alticinae, Hispinae and Cassidinae (Coleoptera, Chrysomelidae) and their distribution in the biogeographical regions of Bulgaria]. Nauchni trudove na Plovdivskiya universitet "Paisii Hilendarski" (Travaux scientifiques de l'Université de Plovdiv "Paisii Hilendarski"), Biology, 36(6):5–33 (in Bulgarian).

Gruev, B. & B. Kuzmanov, 1994. Obshta biogeografiya. Vtoro osnovno preraboteno izdanie [General Biogeography. Second Basic Revised Edition]. Universitetsko izdatelstvo "Sv. Kliment Ohridski" [St. Kliment Ohridski University Publishing House], Sofia (in Bulgarian).

Guéorguiev, B., 1998. Ground-beetles (Coleoptera: Carabidae) collected by Bulgarian zoologists in Republic of Macedonia. Historia naturalis bulgarica 9:35–51.

Guéorguiev, V., 1977. La faune troglobie terrestre de la péninsule Balkanique. Origine, formation et zoogéographie. Éditions de l'Academie bulgare de science, Sofia.

Guéorguiev, V. & B. Guéorguiev, 1995. Catalogue of the Ground-beetles of Bulgaria (Coleoptera: Carabidae). Pensoft, Sofia & Moscow.

Heiss, E. & M. Josifov, 1990. Vergleichende Untersuchung über Artenspektrum, Zoogeographie und Ökologie der Heteropteren-Fauna in Hochgebirgen Österreichs und Bulgariens. Berichte des naturwissenschaftlich-medizinischen Vereins, Innsbruck 77:123–161.

Hovorka, O. & R. Sciaky, 2003. Subtribe Calathina. In Löbl, I. & A. Smetana (eds), Catalogue of Palearctic Coleoptera. Vol. 1. Archostemata–Myxophaga–Adephaga. Apollo Books, Steenstrup: 524–529.

Jablokoff-Khnzorian, S.M., 1963. [Sur le rythme de l'evolution]. Zoologicheskiy zhurnal [Zoological Journal] 42(10):1433–1445 (in Russian).

Jeannel, R., 1928. Monographie des Trechinae. Morphologie comparé et distribution géographique d'un groupe de Coleoptères (Troisième livraison). L'Abeille 35:1–808.

Jeannel, R., 1961. La Gondwanie et le peuplement de l'Afrique. Annales du Musée royal de l'Afrique centrale, Tervuren, ser. 8, 102:1–161.

Kataev, B.M., D.W. Wrase & N. Ito, 2003. Subtribe Harpalina. In Löbl, I. & A. Smetana (eds), Catalogue of Palearctic Coleoptera. Vol. 1. Archostemata–Myxophaga–Adephaga. Apollo Books, Steenstrup: 367–397.

Kavanaugh, D.H., 1980. Insects of western Canada, with special reference to certain Carabidae (Coleoptera): Present distribution patterns and their origins. The Canadian Entomologist 112:1129–1144.

Kryzhanovskij, O.L., 1983. [The beetles of the suborder Adephaga: Families Rhysodidae, Trachipachidae, family Carabidae (an introduction and a review of the USSR fauna)]. In Fauna SSSR. Nasekomye zhestkokrylye. Tom 1, vypusk 2. [Fauna of the USSR. Coleoptera. Vol. 1, No. 2]. Nauka [Science], Leningrad (in Russian).

Kryzhanovskij, O.L., I.A. Belousov, I.I. Kabak, M.B. Kataev, K.V. Makarov & V.G. Shilenkov, 1995. A Checklist of the Ground-beetles of Russia and Adjacent Lands (Insecta, Coleoptera, Carabidae). Pensoft, Sofia & Moscow.

Kryzhanovskij, O.L. & Ya.I. Starobogatov, 1974. [Current state of the theory of the geographic range and faunal complexes, and goals for research]. Problemy dolgosrochnogo planirovaniya biologicheskikh issledovanii, Zoologiya [Problems of the Long-term Planning of Biological Research, Zoology] 1:44–52 (in Russian).

La Greca, M., 1966. Origine degli ortotteri appenninici di alta quota. Atti della Accademia nazionale italiana di entomologia, Rendiconti 13:19–33.

La Greca, M., 1977. L'evoluzione plio-pleistocenica degli insetti di alta montagna. Bollettino di zoologia 44:261–285.

La Greca, M., 1985. Origine des Orthoptères de la faune d'Italie. Mitteilungen der Deutschen Gesellschaft für allgemeine und angewandte Entomologie 4:223–225.

La Greca, M., 1990. The insect biogeography of West Mediterranean islands. Accademia Nazionale dei Lincei. Atti dei Convegni Lincei 85 (International Symposium on Biogeographical Aspects of Insularity): 469–491.

La Greca, M., 1996. Storia biogeografica degli ortotteri d'Italia: origine e distribuzione (Insecta, Orthoptera). Bollettino Museo Civico di Storia Naturale di Verona 20:1–46.

La Greca, M., 1999. Biogeography of the Palearctic Pamphagidae (Orthoptera). Memorie della Società entomologica italiana 77:123–159.

Lattin, G. de. 1967. Grundriss der Zoogeographie. VEB Gustav Fischer Verlag, Jena.

Lawrence, J.F. & A.F. Newton, Jr., 1995. Families and subfamilies of Coleoptera (with selected genera, notes, references, and data on family-group names). In Pakaluk, J. & S.A. Ślipiński (eds), Biology, Phylogeny, and Classification of Coleoptera: Papers Celebrating the 80th Birthday of Roy A. Crowson. Muzeum i Instytut Zoologii PAN, Warszawa: 779–1006.

Lindroth, C.H. 1979. [Chapter] 2.41. The theory of glacial regufia. In Erwin, T.L., G.E. Ball & D.R. Whitehead (eds), Carabid Beetles: Their Evolution, Natural History and Classification. W. Junk, The Hague, Boston & London: 385–394.

Löbl, I. & A. Smetana (eds), 2003. Catalogue of Palearctic Coleoptera. Vol. 1. Archostemata–Myxophaga–Adephaga. Apollo Books, Steenstrup.

Lopatin, I.K., 1989. Zoogeografiya [Zoogeography]. Vysheyshaya shkola [Higher Education], Minsk (in Russian).

Matvejev, S.D., 1980. [Possibilities of a common biogeographical division of eastern Yugoslavia for all faunistic groups]. Zbornik radova o faune Srbije (Proceedings on the Fauna of SR Serbia) 1:181–206 (in Serbian).

Müller, P., 1974. Aspects of Zoogeography. W. Junk, The Hague.

Noonan, G., 1986. Distribution of insects in the Northern Hemisphere: Continental drift and epicontinental seas. Bulletin of the Entomological Society of America 32(2):80–84.

Noonan, G., 1988. Faunal relationships between eastern North America and Europe as shown by insects. Memoirs of the Entomological Society of Canada 144:39–53.

Palamarev, E., 1997. [Basic stages of the development of flora and vegetation in the geologic history of Bulgarian lands]. In Yordanova, M. & D. Donchev (eds), Geografiya na Balgariya. Fizicheska geografiya. Sotsialno-ikonomicheska geografiya (Geography of Bulgaria. Physical Geography. Socio-Economic Geography). Akademichno izdatelstvo "Prof. Marin Drinov" (Prof. Marin Drinov Academic Publishing House), Sofia: 260–265 (in Bulgarian).

Pesenko, Y.A., 1982. Printsipy i metody kolichestvennogo analiza v faunisticheskikh issledovaniyakh [Principles and Methods of Quantitative Analysis in Faunistic Investigations]. Nauka [Science], Moscow (in Russian).

Popov, A., 2002. Zoogeographical analysis of Neuroptera in Bulgaria. Acta Zoologica Academiae Scientiarum Hungaricae 48(Suppl. 2):271–280.

Popov, V. & Ch. Deltshev, 1997. [Formation of the modern fauna]. In Yordanova, M. & D. Donchev, (eds), Geografiya na Balgariya. Fizicheska geografiya. Sotsialno-ikonomicheska geografiya (Geography of Bulgaria. Physical Geography. Socio-Economic Geography). Akademichno izdatelstvo "Prof. Marin Drinov" (Prof. Marin Drinov Academic Publishing House), Sofia: 310–316 (in Bulgarian).

Por, F.D., 1975. An outline of the zoogeography of the Levant. Zoologica Scripta 4:5–20.

Sedlag, U. & E. Weinert, 1987. Biogeographie, Artbildung, Evolution. VEB Gustav Fischer Verlag, Jena.

Serrano, J., 2003. Catálogo de los Carabidae (Coleoptera) de la Península Ibérica/Catalogue of the Carabidae (Coleoptera) of the Iberian Peninsula. Monografías de la Sociedad Entomológica Aragonesa (Zaragoza) 9.

Stegmann, B., 1938. [Grundzüge der ornithogeographischen Gliederung der Paläarktischen Gebietes]. In: Fauna SSSR. Ptitsy. Tom 1, vypusk 2 (Faune de l'URSS. Oiseaux. Vol. 1, No. 2). Izdatel'stvo Akademii nauk SSSR (Édition de l'Académie des sciences de l'URSS), Moscow & Leningrad (in Russian).

Stojanoff, N., 1940. Zur Frage über die Herkunft des arktischen Elements in der Flora der Balkanländer. Godishnik na Sofiiskiya universitet. Fiziko-matematicheski fakultet (Annuaire de l'Université de Sofia. Faculté physico-mathématique) 36(3 – Natural History): 195–250.

Tabacaru, I., 1970. Sur la répartitions des diplopodes cavernicoles européens. In Orghidan, T. (ed), Livre du centenaire Émile G. Racovitza 1868–1968. Éditions de l'Académie de la République socialiste de Roumanie, Bucharest: 421–443.

Takhtajan, A.L., 1974. [Floristic subdivision of land]. In Zhizn' rasteniy v shesti tomakh. Tom 1. Vvedenie. Bakterii i aktinomitsety [Life of Plants in Six Volumes. Volume 1. Introduction. Bacteria and Actinomycetes]. Prosveshchenie [Education], Moscow: 117–153 (in Russian).

Turin, H. 1981. Provisional checklist of the European ground-beetles. Monografieën van de Nederlandse Entomologische Vereniging 9.

Turin, H., L. Penev & A. Casale (eds), 2003. The Genus *Carabus* in Europe. A Synthesis. Pensoft & European Invertebrate Survey, Sofia, Moscow & Leiden.

Vaptzarov, I., 1997. [Basic stages of the geologic development and their significance for the modern appearance of the relief]. In Yordanova, M. & D. Donchev (eds), Geografiya na Balgariya. Fizicheska geografiya. Sotsialno-ikonomicheska geografiya (Geography of Bulgaria. Physical Geography. Socio-Economic Geography). Akademichno izdatelstvo "Prof. Marin Drinov" (Prof. Marin Drinov Academic Publishing House), Sofia: 33–39 (in Bulgarian).

Vigna Taglianti, A., 1993. Coleoptera Archostemata, Adephaga 1 (Carabidae). In Minelli, A., S. Ruffo & S. La Posta (eds), Checklist delle specie della fauna italiana. 44. Calderini, Bologna.

Vtorov, P.P. & N.N. Drozdov, 1978. Biogeografiya [Biogeography]. Prosveshchenie [Education], Moscow (in Russian).

Vuilleumier, F., 1975. Zoogeography. In Farner, D.S. & J.R. King (eds), Avian Biology. Vol. 5. Academic Press, New York: 421–496.

Zaballos, J.P. & C. Jeanne, 1994. Nuevo catálogo de los Carabidos (Coleoptera) de la peninsula Ibérica. Monografías de la Sociedad Entomológica Aragonesa (Zaragoza) 1.

10 Fauna and Zoogeography of Jewel Beetles (Coleoptera: Buprestidae) in Bulgaria

Vladimir Sakalian[1] and Mario Langourov[2]

[1]*Institute of Zoology, Bulgarian Academy of Sciences, Tsar Osvoboditel Blvd. 1, 1000 Sofia, Bulgaria, e-mail: sakalian@inagency.com*

[2]*Institute of Zoology, Bulgarian Academy of Sciences, Tsar Osvoboditel Blvd. 1, 1000 Sofia, Bulgaria, e-mail: langourov@gmail.com*

Abstract: The Bulgarian fauna of jewel beetles (Buprestidae) was analyzed according to area similarity and taxonomic structure, and compared to other European and Asian faunas. The Bulgarian fauna of Buprestidae belongs to the South European faunal complex as a part of East Balkan faunal type. The fauna includes 179 species and subspecies belonging to 22 zoogeographical categories, combined in ten zoogeographical complexes. The general appearance of the Bulgarian buprestid fauna is determined by the taxa of Mediterranean and European complexes with relatively narrow ranges (ca. 60 % of all species and subspecies). Affinity of Bulgarian jewel beetle fauna to the East Balkan faunal type is also demonstrated by the presence of five species and subspecies endemic for this region. The recent Bulgarian fauna has been formed mainly during Pleistocene and Holocene by autochthonous and allochthonous elements. Some of the endemics, subendemics, and East Mediterranean taxa, with their possible center of speciation in the eastern part of Balkan Peninsula, could be autochthonous for Bulgaria. Allochthonous elements colonized Bulgaria during preglacial, glacial, and postglacial periods. Our zoogeographical results demonstrate four main routes of buprestid colonization in Bulgaria: from the Ancient Mediterranean (especially East Mediterranean), Central European, Boreal, and Eurasian Steppe centers of speciation

1 Introduction

Bulgarian jewel beetle fauna is relatively well studied. First information on distribution of jewel beetles in Bulgaria is found in Frivaldszky (1835, 1837). Since 1837, more than 120 papers, containing data about buprestid species in Bulgaria, have been published. One of the most important publications is the "Catalogue raisonné des Buprestides de Bulgarie" by Obenberger (1932, 1933, 1935). Since the publication of Obenberger's work, large amount of new information has been established, summarized in the "Catalogue of the Jewel Beetles of Bulgaria (Coleoptera: Buprestidae)" by Sakalian (2003). The present paper is mainly based on this Catalogue.

Zoogeographical characteristics of the Bulgarian fauna (as a whole) can be found only in the mentioned Obenberger's catalogue, published 70 years ago. Since then many new data about buprestid distribution have been established. Sakalian (1994, 1996) published a regional zoogeographical characteristic of Sandanski–Petrich and Gotse Delchev valleys (Southwestern Bulgaria) and the Bulgarian Black Sea coast.

V. Fet and A. Popov (Eds.), Biogeography and Ecology of Bulgaria, 357–378.

Present level of investigation of buprestid distribution, in Bulgaria as well as beyond its borders, allowed conducting an updated zoogeographical analysis of Bulgarian buprestid fauna. The general goals of this paper are: to determine the specific position and connections of Bulgarian fauna with other Western Palearctic faunas; to analyze the general trends of Bulgarian buprestid fauna formation; and to present the results of regional (within Bulgaria) faunistic and zoogeographical analysis.

2 Methods

We conducted fauna similarity analysis based on Czekanowski–Dice–Sørensen similarity index. Index calculation and dendrogram construction were performed using the BIODIV software (Baev and Penev, 1995). The analysis included species lists for all investigated countries (547 species belonging to 43 genera); status of some infraspecific taxa is not clear. We also analyzed the taxonomic structure of represented faunas (grouping of species into genera). The information about species composition and species number of compared regional faunas can be find in following papers: Schaefer (1971, 1972), Mühle (1980), Bílý (1982, 1993), Cobos (1986), Lodos and Tezcan (1992), Niehuis and Tezcan (1993), Curletti (1994), Volkovich and Alexeev (1994), Niehuis (1996), Karaman and Tezcan (1998), Tezcan and Bílý (1998), Tozlu and Özbek (1998), Ulay and Tezcan (1998), Chikatunov et al. (1999), Halperin and Argaman (2000), Mühle et al. (2000), and Sakalian (2000, 2003). The regions (countries) were chosen so as to be representative for main faunal types around Bulgaria and to have moderately to well studied buprestid faunas. Unfortunately, some faunas (Albania, Bosnia and Herzegovina, Macedonia, Israel) are not studied well enough. The data from other Balkan Peninsula countries (Serbia and Montenegro, Croatia) are partial and incomplete, and therefore were not used in the fauna similarity analysis.

The zoogeographical characteristic of each taxon is based mainly on its recent geographical range. The information on distribution of Bulgarian buprestid species within and beyond Bulgaria can be found in Sakalian (2003).

3 Zoogeographical Position and Connections of the Buprestid Fauna of Bulgaria

The results of the similarity analysis and taxonomic structure (genera/species) are presented in Fig. 1 and Table 1.

Figure 1 shows that Central and South European buprestid faunas have the similarity level about 55%. The faunas of Israel and Turkmenistan have a lower level of similarity. The position of Bulgarian fauna is as follows: together with Greek (high similarity level), Macedonian, and Albanian faunas, it forms the East Balkan Peninsula faunal type. The faunas of the western part of the Balkan Peninsula (Bosnia and Herzegovina) and Central Europe (Czech Republic and Slovakia) are

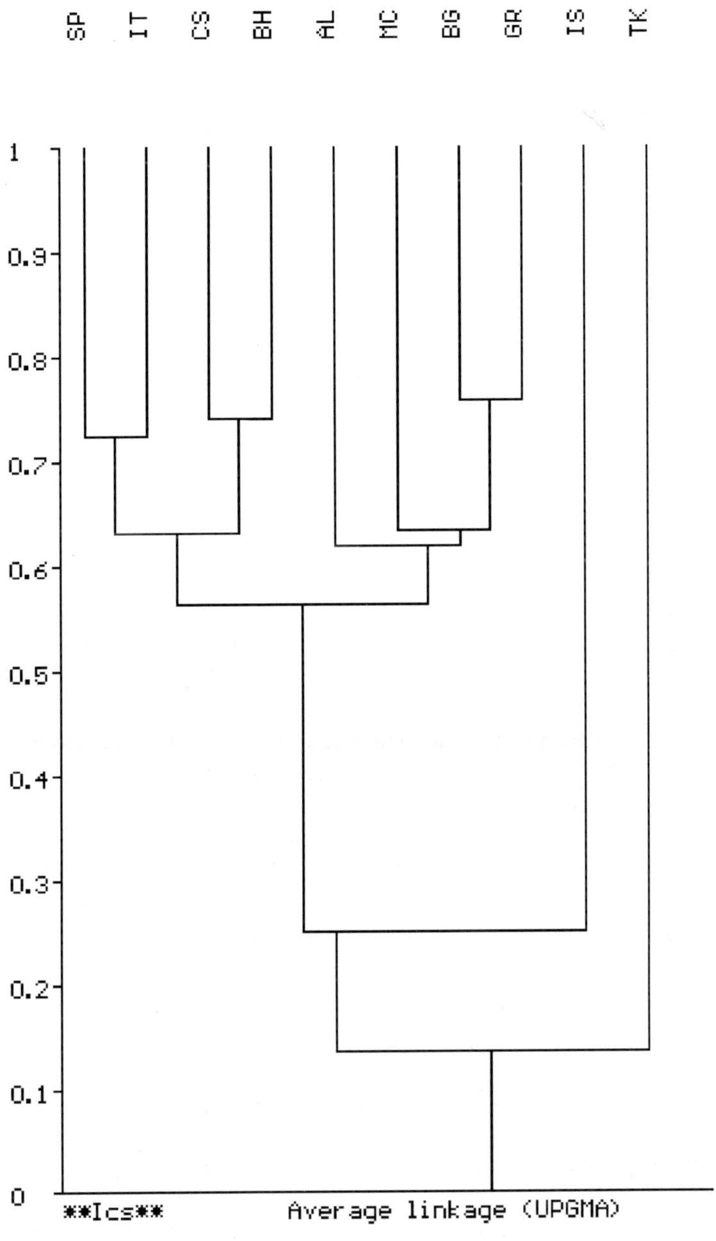

Fig. 1 Faunistic similarity between the studied buprestid faunas. SP, Spain and Portugal; IT, Italy; CS, Czech Republic and Slovakia; BH, Bosnia and Herzegovina; AL, Albania; MC, Macedonia; BG, Bulgaria; GR, Greece; IS, Israel; TK, Turkmenistan.

Table 1 Taxonomic structure (number of species) of buprestid regional faunas (doubtful species from some countries excluded)

Genera	Countries												
	TK	IS	TU	GR	BG	MC	AL	BH	IT	FR	SP	CS	FN
Julodis	3	7	11	2	2	1	1		1	1	1		
Julodella	3		3										
Ptosima		1	1	1	1	1	1	1	1	1	1	1	
Polyctesis		1	1										
Strigopteroides	1	1		2									
Thurntaxisia				1									
Acmaeodera	7	19	21	17	8	6	5	3	9	6	9	1	
Xantheremia	3	1											
Acmaeoderella	32	19	22	19	10	8	4	2	7	6	9	1	
Steraspis		1											
Chalcophora		1	4	3	1	1	3	2	4	3	1	1	1
Chalcophorella		1	4	2	2	2	2	2	1				
Psiloptera	1	1	3										
Capnodis	8	8	10	6	4	5	4	3	4	2	2	1	
Cyphosoma	2	2	3	2	1		1		2		1		
Perotis	1	1	5	3	1	1	1	1	2	1	1	1	
Latipalpis		1	3	2					1	1	1		
Sphenoptera	65	33	114	20	11	7	8	6	12	6	10	3	
Dicerca	2		2	5	4	2	4	5	4	4	4	5	4
Poecilonota	1		1	1	1			1	1	1	1	1	1
Scintillatrix	1		6	5	4	2	2	3	5	3	3	3	1
Palmar			2	3	2	2		1	1	1	1		
Buprestis	1	2	13	8	5	4	6	6	7	4	8	4	5
Eurythyrea			3	3	3		3	3	3	3	3	2	
Kisanthobia		1	1	1	1			1	1	1			

Melanophila	1	1	2	2	1	1	2	2	2	2	2	1	1
Trachypteris	1	1	1	1	1	1			1	1	1	1	
Phaenops			5	4	2		1	3	3	4	2	3	1
Chalcogenia		1											
Anthaxia	21	43	86	66	44	35	23	23	38	38	36	23	6
Chrysobothris	5	3	6	5	5	2	1	2	4	3	2	3	2
Clema	2	1											
Coroebus	1	2	6	5	3	3	2	4	4	4	4	4	
Meliboeus	6	8	10	5	6	2	2	3	6	5	4	2	
Nalanda		1	1	1	1	1	1	1	1	1	1	1	
Agrilus	13	16	37	33	36	15	22	33	49	38	33	37	20
Paracylindromorphus	4		1	1	1	1	1	1	1	1	1	1	
Cylindromorphus	2		1	1	2	1		2	4	3	2	4	
Aphanisticus	2	2	4	5	5		3	2	7	6	6	2	2
Habroloma	1				1			1	2	2	2	1	1
Trachys	1	2	10	10	10	3	4	5	10	9	9	8	3
Galbella		1	1										
Total species	191	187	404	245	179	107	107	122	198	161	161	115	48

TK, Turkmenistan; IS, Israel; TU, Turkey; GR, Greece; BG, Bulgaria; MC, Macedonia; AL, Albania; BH, Bosnia and Herzegovina; IT, Italy; FR, France; SP, Spain and Portugal; CS, Czech Republic and Slovakia; FN, Fennoscandia and Denmark.

combined with those of Apennine Peninsula (Italy) and Iberian Peninsula (Spain and Portugal). While the high similarity level between the Apennine and Iberian Peninsula faunas is expected due to their historical and recent connections, it is interesting that they are joined by the faunas of Bosnia and Herzegovina, Czech Republic, and Slovakia; this issue requires a more detailed analysis.

The analysis of taxonomic structure (Table 1) was conducted in addition to the similarity analysis; it also included some additional faunas (Turkey, France, Fennoscandia, and Denmark; Fig. 1). For Turkey, unfortunately, we have only information for the genus-level buprestid fauna but not its species composition; thus Turkey was excluded from the dendrogram.

Comparison of the buprestid faunas of Turkmenistan and Turkey with those of other countries shows a great domination of genus *Sphenoptera* (ca. 30 %), typical for the arid regions of Central and West Asia. However, while in Turkmenistan the second place by species richness belongs to *Acmaeoderella*, and third, to *Anthaxia*, in Turkey the order is different (*Anthaxia*, *Agrilus*, and *Acmaeoderella*). A relatively good representation of *Anthaxia* and *Agrilus* make Turkish fauna more similar to East Balkan faunas (especially those of Greece and Bulgaria) where representatives of these two genera, especially *Anthaxia*, dominate. The genera *Sphenoptera* and *Acmaeoderella* are relatively well represented in Bulgaria as well. The more hygrophilous genus *Agrilus* dominated over *Anthaxia* in the following faunas: Bosnia and Herzegovina, Italy, Czech Republic/Slovakia, and Fennoscandia/Denmark. The portion of *Agrilus* increases to ca. 42 % in Northern Europe. In France and Iberian Peninsula the impact of *Anthaxia* and *Agrilus* is relatively equal. The species number of such genera as *Acmaeodera*, *Acmaeoderella*, *Anthaxia*, *Capnodis*, *Chalcophorella*, *Sphenoptera*, etc. (Table 1) decreases in Europe from south to north. Some genera (*Strigopteroides* and *Thurntaxisia*) are found in Europe only in Greece. The number of representatives of Nearctic and Palearctic genera such as *Buprestis*, *Dicerca*, *Eurythyrea*, *Melanophila*, *Phaenops*, *Scintillatrix*, etc. (Table 1) is relatively similar in all European countries. The species richness decreases from Turkey and Greece to the east (Turkmenistan) and to the west and, especially, to the north (judging only from the best known faunas). The jewel beetle fauna of Israel is interesting with domination of *Anthaxia* as in East Balkan Peninsula and *Sphenoptera* as in Turkey and Turkmenistan.

It is possible to characterize the Bulgarian buprestid fauna as a part of South European fauna, or the member of a specific East Balkan faunal type, most similar to the buprestid fauna of Greece.

4 Zoogeographical Characteristics of the Bulgarian Buprestid Fauna

According to the classification of Buprestidae (Bellamy, 1985), the Bulgarian 179 species (Table 2) belong to ten subfamilies, 17 tribes, 31 genera, and 24 subgenera (Sakalian, 2003).

Table 2 Distribution and zoogeographical characteristics of the Bulgarian fauna of Buprestidae

Taxa	Regions						Zocgeographical categories	
	DP	SP	TN	TS	RR	BS	species	subspecies
Julodis ehrenbergi Laporte, 1835			+		+		EM	
Julodis pubescens (Olivier, 1790)					+		BSE	
Ptosima undecimmaculata undecimmaculata (Herbst, 1784)	+	+	+	+	+	+	SWP	SWP
Acmaeodera (*Acmaeodera*) *brevipes brevipes* Kiesenwetter, 1858				+	+		MIT	EM
Acmaeodera (*Acmaeodera*) *edmundi edmundi* Obenberger, 1940					+		EMIT	EM
Acmaeodera (*Acmaeodera*) *pilosellae pilosellae* (Bonelli, 1812)		+		+	+	+	MIT	NM
Acmaeodera (*Palaeotethya*) *bipunctata bipunctata* (Olivier, 1790)					+		ME	ME
Acmaeodera (*Acmaeotethya*) *crinita crinita* Spinola, 1838		+	+	+	+	+	TM	EM
Acmaeodera (*Acmaeotethya*) *degener degener* (Scopoli, 1763)				+	+	+	SWP	PM
Acmaeodera (*Acmaeotethya*) *ottomana ottomana* (Frivaldszky, 1837)		+	+	+	+	+	EMIT	EMIT
Acmaeodera (*Acmaeotethya*) *quadrizonata* Abeille de Perrin, 1891				+			EM	
Acmaeoderella (*Liogastria*) *chrysanthemi* (Chevrolat, 1854)		+			+		EM	
Acmaeoderella (*Acmaeoderella*) *circassica* (Reitter, 1890)	+			+	+	+	PM	
Acmaeoderella (*Acmaeoderella*) *levantina* (Obenberger, 1934)			+		+		EM	
Acmaeoderella (*Acmaeoderella*) *seminata* (Abeille de Perrin, 1895)		+		+	+	+	EM	
Acmaeoderella (*Omphalothorax*) *adspersula adspersula* (Illiger, 1803)					+		MIT	TM
Acmaeoderella (*Carininota*) *flavofasciata flavofasciata* (Piller et Mitterpacher,1783)		+	+	+	+	+	WP	SWP
Acmaeoderella (*Carininota*) *mimonti mimonti* (Boieldieu, 1865)			+	+	+	+	SWP	SWP
Acmaeoderella (*Euacmaeoderella*) *gibbulosa* (Ménétriés, 1832)					+		EMIT	
Acmaeoderella (*Euacmaeoderella*) *subcyanea* (Reitter, 1889)					+		EMIT	
Acmaeoderella (*Euacmaeoderella*) *vetusta* (Ménétriés, 1832)					+		EMIT	
Chalcophora mariana mariana (Linnaeus, 1758)	+	+	+	+	+		TP	ES
Chalcophorella (*Chalcophorella*) *stigmatica* (Schoenherr, 1817)		+	+	+	+	+	EMIT	

continued

Table 2 continued

Taxa	Regions						Zoogeographical categories	
	DP	SP	TN	TS	RR	BS	species	subspecies
Chalcophorella (*Rossiella*) *fabricii* (Rossi, 1794)			+	+	+		EM	
Capnodis cariosa (Pallas, 1776)	+	+	+	+	+	+	EMIT	
Capnodis porosa (Klug, 1829)					+		EMIT	
Capnodis tenebricosa (Olivier, 1790)		+	+	+	+	+	SWP	
Capnodis tenebrionis (Linnaeus, 1761)	+	+	+	+	+	+	SWP	
Cyphosoma euphraticum (Laporte et Gory, 1839)						+	EM	
Perotis lugubris lugubris (Fabricius, 1777)	+	+	+	+	+	+	EMIT	EM
Sphenoptera (*Sphenoptera*) *antiqua* (Illiger, 1803)		+	+	+	+	+	MIT	
Sphenoptera (*Sphenoptera*) *coracina* (Steven, 1829)					+		EMIT	
Sphenoptera (*Sphenoptera*) *cuprina* Motschulsky, 1860				+			EAS	
Sphenoptera (*Deudora*) *buresi* Obenberger, 1932					+		BE	
Sphenoptera (*Deudora*) *rauca* (Fabricius, 1787)			+		+	+	TM	
Sphenoptera (*Chilostetha*) *basalis* Morawitz, 1861			+		+		EAS	
Sphenoptera (*Chilostetha*) *jugoslavica* Obenberger, 1926		+	+	+	+	+	EM	
Sphenoptera (*Chilostetha*) *laportei* Saunders, 1871			+				NM	
Sphenoptera (*Chilostetha*) *petriceki* Obenberger, 1952		+	+		+	+	PM	
Sphenoptera (*Chilostetha*) *sceptrifera* Obenberger, 1952					+	+	BE	
Sphenoptera (*Chilostetha*) *substriata* Krynicki, 1834					+	+	EAS	
Dicerca (*Dicerca*) *aenea aenea* (Linnaeus, 1761)	+		+	+	+	+	TP	WP
Dicerca (*Dicerca*) *alni* (Fischer, 1824)	+	+	+	+	+	+	WP	
Dicerca (*Dicerca*) *berolinensis* (Herbst, 1779)	+	+	+	+	+	+	E	
Dicerca (*Argante*) *moesta* (Fabricius, 1794)	+				+		WES	
Poecilonota variolosa variolosa (Paykull, 1799)			+		+	+	TP	WES
Scintillatrix dives (Guillebeau, 1889)					+		E	
Scintillatrix gloriosa (Marseul, 1865)			+	+	+	+	EM	
Scintillatrix mirifica mirifica (Mulsant, 1855)	+	+	+	+	+	+	SWP	E

Scintillatrix rutilans (Fabricius, 1777)	+	+	+	+	+	+	E	
Palmar bella (Gory, 1840)			+	+	+		E	
Palmar festiva (Linnaeus, 1767)			+	+	+		ME	
Buprestis (*Buprestis*) *cupressi* Germar, 1817	+				+		EM	
Buprestis (*Buprestis*) *haemorrhoidalis haemorrhoidalis* Herbst, 1780	+		+	+	+	+	TP	WP
Buprestis (*Buprestis*) *novemmaculata* Linnaeus, 1767				+	+		TP	
Buprestis (*Buprestis*) *octoguttata octoguttata* Linnaeus, 1758					+		WP	WES
Buprestis (*Buprestis*) *rustica* Linnaeus, 1758			+		+		ES	
Eurythyrea aurata (Pallas, 1776)			+		+	+	PM	
Eurythyrea austriaca (Linnaeus, 1767)		+	+	+	+		E	
Eurythyrea quercus (Herbst, 1780)					+		E	
Kisanthobia ariasi (Robert, 1858)				+		+	TM	
Trachypteris picta decastigma (Fabricius, 1787)	+	+	+	+	+	+	SP	ME
Melanophila acuminata (De Geer, 1774)			+		+		H	
Phaenops cyanea (Fabricius, 1775)	+	+	+		+	+	TP	
Phaenops knoteki knoteki (Reitter, 1898)		+			+		ME	EE
Anthaxia (*Cratomerus*) *diadema diadema* (Fischer, 1823)			+	+	+	+	EMIT	PM
Anthaxia (*Cratomerus*) *hungarica hungarica* (Scopoli, 1772)	+	+	+	+	+	+	SWP	ME
Anthaxia (*Cratomerus*) *scorzonerae* (Frivaldszky, 1837)	+	+	+	+	+	+	EM	
Anthaxia (*Cratomerus*) *sponsa* Kiesenwetter, 1857				+	+		EMIT	
Anthaxia (*Haplanthaxia*) *cichorii cichorii* (Olivier, 1790)	+	+	+	+	+	+	SWP	ME
Anthaxia (*Haplanthaxia*) *kiesenwetteri* Marseul, 1865	+	+	+		+	+	EM	
Anthaxia (*Haplanthaxia*) *millefolii millefolii* (Fabricius, 1801)	+	+	+	+	+	+	ME	PM
Anthaxia (*Haplanthaxia*) *olympica olympica* Kiesenwetter, 1880	+	+	+	+	+	+	PM	PM
Anthaxia (*Haplanthaxia*) *praeclara praeclara* Mannerheim, 1837					+		TM	TM
Anthaxia (*Haplanthaxia*) *rossica* Daniel, 1903		+	+	+	+	+	PM	
Anthaxia (*Haplanthaxia*) *umbellatarum* (Fabricius, 1787)					+	+	ME	
Anthaxia (*Anthaxia*) *amasina amasina* Daniel, 1903		+					EM	EM
Anthaxia (*Anthaxia*) *anatolica anatolica* Chevrolat, 1837				+	+	+	MIT	EM
Anthaxia (*Anthaxia*) *bicolor bicolor* Falderman, 1835	+	+	+	+	+	+	EMIT	EMIT
Anthaxia (*Anthaxia*) *brevis brevis* Gory et Laporte, 1839				+	+		EMIT	PM

continued

Table 2 continued

Taxa	Regions						Zoogeographical categories	
	DP	SP	TN	TS	RR	BS	species	subspecies
Anthaxia (*Anthaxia*) *candens* (Panzer, 1789)		+	+	+	+	+	E	
Anthaxia (*Anthaxia*) *deaurata deaurata* (Gmelin, 1788)	+		+	+	+	+	ME	ME
Anthaxia (*Anthaxia*) *discicollis discicollis* Gory et Laporte, 1839				+	+		EMIT	EM
Anthaxia (*Anthaxia*) *fulgurans* (Schrank, 1789)	+	+	+	+	+	+	E	
Anthaxia (*Anthaxia*) *funerula funerula* (Illiger, 1803)	+			+	+	+	WP	WE
Anthaxia (*Anthaxia*) *hackeri* Frivaldszky, 1884					+	+	EE	
Anthaxia (*Anthaxia*) *hypomelaena* (Illiger, 1803)	+	+	+	+	+	+	MIT	
Anthaxia (*Anthaxia*) *lucens lucens* Küster, 1852					+		EM	EM
Anthaxia (*Anthaxia*) *manca* (Linnaeus, 1767)	+			+	+	+	ME	
Anthaxia (*Anthaxia*) *midas* Kiesenwetter, 1857					+		TM	
Anthaxia (*Anthaxia*) *muliebris* Obenberger, 1918						+	EMIT	
Anthaxia (*Anthaxia*) *nitidula* (Linnaeus, 1758)		+	+				EME	
Anthaxia (*Anthaxia*) *passerini* (Pecchioli, 1837)		+			+		EMIT	
Anthaxia (*Anthaxia*) *plicata plicata* Kiesenwetter, 1859					+	+	PM	PM
Anthaxia (*Anthaxia*) *podolica podolica* Mannerheim, 1837	+	+	+	+	+	+	E	E
Anthaxia (*Anthaxia*) *salicis salicis* (Fabricius, 1777)	+	+	+	+	+	+	SWP	ME
Anthaxia (*Anthaxia*) *semicuprea* Küster, 1851		+	+		+	+	E	
Anthaxia (*Anthaxia*) *signaticollis* Krynicki, 1832	+	+	+	+	+	+	PM	
Anthaxia (*Anthaxia*) *thalassophila pseudokervillei* Niehuis, 1990			+		+		NM	BSE
Anthaxia (*Anthaxia*) *tuerki* Ganglbauer, 1886		+		+	+	+	E	
Anthaxia (*Melanthaxia*) *godeti* Gory et Laporte, 1839	+	+	+	+	+	+	WP	
Anthaxia (*Melanthaxia*) *helvetica helvetica* Stierlin, 1868		+	+	+	+	+	E	E
Anthaxia (*Melanthaxia*) *istriana* Rosenhauer, 1847		+	+	+	+		NM	
Anthaxia (*Melanthaxia*) *morio* (Fabricius, 1792)					+		E	
Anthaxia (*Melanthaxia*) *muehlei* Niehuis, 1983					+		BE	
Anthaxia (*Melanthaxia*) *nigrojubata incognita* Bílý, 1974					+	+	EIT	E

Anthaxia (*Melanthaxia*) *quadripunctata quadripunctata* (Linnaeus, 1758)		+	+	+	+	+	TP	WP
Anthaxia (*Melanthaxia*) *tenella tenella* Kiesenwetter, 1858					+		EM	BE
Anthaxia (*Melanthaxia*) *thessalica* Brandl, 1981		+	+	+	+		BE	
Chrysobothris (*Chrysobothris*) *affinis affinis* (Fabricius, 1794)		+	+	+	+	+	WP	WE
Chrysobothris (*Chrysobothris*) *chrysostigma chrysostigma* (Linnaeus, 1758)	+		+				ES	ES
Chrysobothris (*Chrysobothris*) *igniventris* Reitter, 1895			+	+			E	
Chrysobothris (*Chrysobothris*) *leonhardi* Obenberger, 1916	+	+	+	+	+	+	BSE	
Chrysobothris (*Chrysobothris*) *solieri* Laporte et Gory, 1838			+		+		ME	
Coroebus elatus (Fabricius, 1787)	+	+	+	+	+	+	WP	
Coroebus florentinus (Herbst, 1801)	+	+		+	+	+	ME	
Coroebus rubi (Linnaeus, 1767)		+	+	+	+	+	EIT	
Meliboeus (*Meliboeus*) *episcopalis* (Mannerheim, 1837)				+	+	+	EM	
Meliboeus (*Meliboeus*) *graminis* (Panzer, 1799)		+	+	+		+	ME	
Meliboeus (*Meliboeus*) *graminoides* (Abeille de Perrin, 1896)					+	+	PM	
Meliboeus (*Meliboeus*) *subulatus* Morawitz, 1861					+	+	EAS	
Meliboeus (*Meliboeoides*) *violaceus* (Kiesenwetter, 1857)			+	+	+	+	EMIT	
Meliboeus (*Melixes*) *aeratus* (Mulsant et Rey, 1863)					+		TM	
Nalanda fulgidicollis (Lucas, 1846)	+	+		+	+	+	ME	
Agrilus acutangulus Thery, 1942		+	+	+	+		E	
Agrilus albogularis Gory, 1841				+	+	+	SWP	
Agrilus alexeevi alexeevi Bellamy, 1998						+	PM	PM
Agrilus angustulus (Illiger, 1803)	+	+	+	+	+	+	WP	
Agrilus antiquus Mulsant et Rey, 1863					+		ME	
Agrilus ater ater (Linnaeus, 1767)		+			+		ES	WES
Agrilus auricollis auricollis Kiesenwetter, 1857	+				+	+	E	E
Agrilus biguttatus (Fabricius, 1777)		+	+	+	+	+	WP	
Agrilus buresi Obenberger, 1935		+	+	+	+	+	EM	
Agrilus convexicollis Redtenbacher, 1849	+		+	+	+	+	E	
Agrilus croaticus Abeille de Perrin, 1897				+	+	+	EE	
Agrilus cuprescens Ménétriés, 1832			+	+	+	+	TP	
Agrilus curtulus Mulsant et Rey, 1863			+	+		+	NM	

continued

Table 2 continued

Taxa	Regions						Zoogeographical categories	
	DP	SP	TN	TS	RR	BS	species	subspecies
Agrilus cyanescens Ratzeburg, 1837			+		+		ES	
Agrilus derasofasciatus Lacordaire, 1835	+	+	+	+	+	+	SWP	
Agrilus graminis Gory et Laporte, 1837	+	+	+	+	+	+	ME	
Agrilus hastulifer Ratzeburg, 1837	+	+	+	+	+	+	SWP	
Agrilus hyperici (Creutzer, 1799)	+	+	+	+	+	+	E	
Agrilus integerrimus Ratzeburg, 1837		+			+		E	
Agrilus laticornis (Illiger, 1803)	+	+	+	+	+	+	E	
Agrilus lineola lineola Redtenbacher, 1849				+	+	+	EIT	E
Agrilus litura Kiesenwetter, 1857		+	+	+	+	+	E	
Agrilus macroderus Abeille de Perrin, 1897		+	+	+	+	+	PM	
Agrilus marozzinii Gobbi, 1974			+		+		TM	
Agrilus obscuricollis Kiesenwetter, 1857	+	+	+		+	+	E	
Agrilus olivicolor Kiesenwetter, 1857	+	+			+	+	E	
Agrilus populneus Schaefer, 1946			+		+	+	E	
Agrilus pratensis pratensis Ratzeburg, 1837			+	+	+	+	ES	WES
Agrilus ribesi Schaefer, 1946			+				ES	
Agrilus roscidus Kiesenwetter, 1857	+	+	+	+	+	+	ME	
Agrilus sericans sericans Kiesenwetter, 1857					+	+	SP	EIT
Agrilus sinuatus (Olivier, 1790)		+	+	+	+	+	E	
Agrilus vaginalis philipovi Alexeev, 1965			+		+		EMIT	PM
Agrilus viridicaerulans rubi Schaefer, 1937		+	+	+	+	+	EIT	E
Agrilus viridis (Linnaeus, 1758)	+	+	+	+	+	+	TP	
Agrilus zigzag Marseul, 1865					+	+	EAS	
Paracylindromorphus subuliformis subuliformis (Mannerheim, 1837)		+		+		+	ES	WES
Cylindromorphus bifrons Rey, 1889					+	+	EAS	
Cylindromorphus filum (Gyllenhal, 1817)		+	+	+	+	+	E	
Aphanisticus angustatus Lucas, 1846				+	+	+	TM	

Aphanisticus elongatus Villa, 1835				+	+	+	WP	
Aphanisticus emarginatus (Olivier, 1790)			+		+	+	WP	
Aphanisticus pusillus (Olivier, 1790)			+		+	+	WP	
Aphanisticus pygmaeus Lucas, 1846					+		SWP	
Habroloma geranii (Silfverberg, 1977)		+	+	+	+		WE	
Trachys fragariae Brisout, 1874	+		+	+	+	+	E	
Trachys minutus (Linnaeus, 1758)	+	+	+	+	+	+	ES	
Trachys phlyctaenoides Kolenati, 1846				+		+	PM	
Trachys problematicus Obenberger, 1918				+	+		E	
Trachys pumilus Illiger, 1803	+		+	+	+	+	TM	
Trachys puncticollis rectilineata Abeille de Perrin, 1900					+	+	EIT	E
Trachys quercicolus Marseul, 1871		+					NM	
Trachys scrobiculatus Kiesenwetter, 1857				+	+		WP	
Trachys troglodytes (Gyllenhal, 1817)			+		+	+	E	
Trachys troglodytiformis Obenberger, 1918	+	+	+	+		+	ME	

Regions: DP, Danubian Plain; SP, Stara Planina; TN, Transitional region; TS, Thracian Lowland and Tundja–Strandja region; RR, Rila–Rhodope Massif; BS, Black Sea coast. Zoogeographical categories (in alphabetical order): BE, Balkan endemic; BSE, Balkan subendemic; E, European; EAS, Eurasian Steppe; EE, East European; EIT, European–Iranian–Turanian; EM, East Mediterranean; EME, European–Mediterranean; EMIT, East Mediterranean–Iranian–Turanian; ES, Eurosiberian; H, Holarctic; ME, Mediterranean–European; MIT, Mediterranean–Iranian–Turanian; NM, North Mediterranean; PM, Pontomediterranean; SP, South Palearctic; SWP, Southwest Palearctic; TM, Transmediterranean; TP, Transpalearctic; WES, West Eurosiberian; WM, West Mediterranean; WP, West Palearctic.

There are no Bulgarian or Balkan endemic or subendemic higher taxa among Bulgarian jewel beetles. The genera, representatives of which are found in Bulgaria, are more or less widely distributed in Palearctic region and beyond.

The zoogeographical character of Bulgarian buprestid fauna is clearly expressed at the species and subspecies level. We consider two types of zoogeographical characteristics; first one is based on species distribution only; second, on the distribution of both species and subspecies. Following Sakalian's (1994) system (with some additions), all taxa are arranged in 22 zoogeographical categories and ten complexes (Table 3, Fig. 2). The new additions are based of designation of European–Iranian–Turanian and Mediterranean–Iranian–Turanian complexes. The Iranian–Turanian region is accepted after Emeljanov (1974).

The Bulgarian jewel beetle fauna is dominated by the representatives of the Mediterranean zoogeographical complex (Table 3, Fig. 2), especially those from the East Mediterranean center of recent species diversity and speciation. The largest number among those belongs to the East Mediterranean (*sensu lato*) taxa, arranged in two zoogeographical categories: Pontomediterranean (distributed in the eastern part of Mediterranean region and around Black Sea) and East Mediterranean (*sensu strictum*). This group includes 28 species (15.7% of the total species number and 49.1% of the number of Mediterranean species). The impact of these elements increases if we include subspecies. In this case, total number of the East Mediterranean taxa is 38 (21.2% of the total species number and 55.1% of the number of Mediterranean species and subspecies). This is due to the fact that some polytypic species have East Mediterranean subspecies (Table 2).

It is interesting to comment on the distribution of *Trachys quercicolus*, characterized as a West Mediterranean element. This species is distributed in the western part of the Mediterranean (from Spain to North Italy and Switzerland). The only known specimen from the East Mediterranean (Bulgaria) is deposited in the collection of the Institute of Zoology in Sofia (see details in Sakalian, 2003). The distribution of *Agrilus curtulus* is similar, but there are more known specimens from Bulgaria. In these cases, the term "West Mediterranean" does not completely describe the recent species range, but it demonstrates the probable center of speciation of analyzed taxa. Two main hypotheses about occurrence of these species in Bulgaria can be proposed. The first one includes restriction of a wider range (probably North Mediterranean) and its recent disjunction into eastern and western parts. Such range restriction is typically due to considerable environmental impacts (catastrophes, great tectonic and/or climatic changes, etc); in our case, it is quite difficult to accept this hypothesis. We prefer the second hypothesis: the west to east dispersal of "avant-garde" populations.

The second place in zoogeographical structure of Bulgarian jewel beetle fauna belongs to the taxa which mainly originated and are found in Europe (European complex) (Table 3, Fig. 2). Most of these taxa are trophically connected with the nemoral forests.

The jewel beetles distributed in Europe and West or Central Asia are grouped as the European–Iranian–Turanian complex.

Table 3 Zoogeographical complexes and categories of the Bulgarian Buprestidae

Complexes	Categories	Species		Species and subspecies	
		number	%	number	%
Holarctic		1	0.56	1	0.56
	Holarctic	1	0.56	1	0.56
Palearctic		22	12.29	16	8.94
	Transpalearctic	9	5.03	4	2.24
	Western Palearctic	13	7.26	12	6.70
Eurosiberian		10	5.59	15	8.38
	Eurosiberian	8	4.47	6	3.35
	Western Eurosiberian	2	1.12	9	5.03
European		32	17.97	38	21.23
	European	29	16.29	34	18.99
	European–Mediterranean	1	0.56	1	0.56
	East European	2	1.12	3	1.68
Eurasian Steppe		6	3.35	6	3.35
	Eurasian Steppe	6	3.35	6	3.35
South Palearctic		15	8.38	9	5.03
	South Palearctic	2	1.12	0	0
	Southwest Palearctic	13	7.26	9	5.03
European–Iranian–Turanian		5	2.79	2	1.12
	European–Iranian–Turanian	5	2.79	2	1.12
Mediterranean–Iranian–Turanian		25	13.96	15	8.38
	Mediterranean–Iranian–Turanian	6	3.35	2	1.12
	East Mediterranean–Iranian–Turanian	19	10.61	13	7.26
Mediterranean		57	31.85	69	38.55
	Transmediterranean	9	5.03	9	5.03
	Mediterranean–European	15	8.38	17	9.50
	North Mediterranean	3	1.68	3	1.68
	West Mediterranean	2	1.12	2	1.12
	Pontomediterranean	11	6.15	16	8.94
	East Mediterranean	17	9.50	22	12.29
Balkan endemic		6	3.35	8	4.47
	Balkan endemic	4	2.23	5	2.79
	Balkan subendemic	2	1.12	3	1.68

The taxa of the Mediterranean–Iranian–Turanian complex demonstrate recent connections between Mediterranean and Central Asian faunas. Similar type of distribution, and probably speciation, is exhibited by the buprestids of the South Palearctic complex. Those taxa are more widely distributed to the east and to the north than

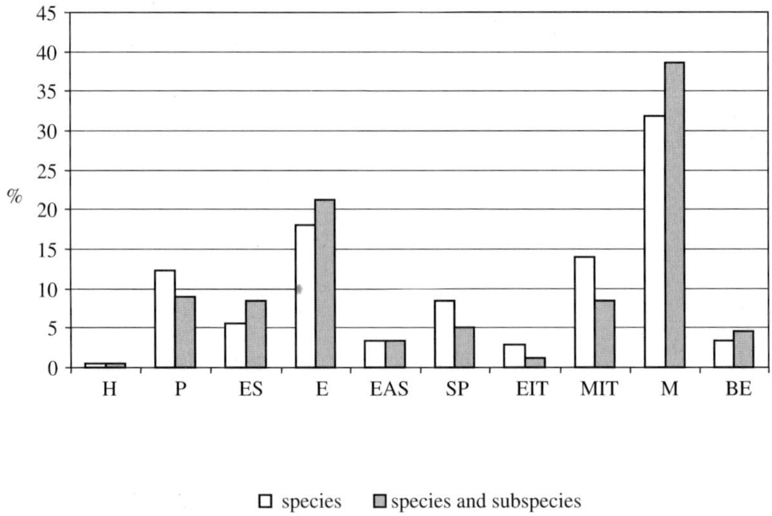

Fig. 2 Zoogeographical complexes of Buprestidae in Bulgaria. H, Holarctic; P, Palearctic; ES, Eurosiberian; E, European; EAS, Eurasian Steppe; SP, South Palearctic; EIT, European–Iranian–Turanian; MIT, Mediterranean–Iranian–Turanian; M, Mediterranean; BE, Balkan endemic.

taxa of the previous groups (Sakalian, 1994), and their distribution is limited by their thermo- and xerophilous preference. The elements, which probably originated somewhere in the Ancient Mediterranean region (Mediterranean, Mediterranean–Iranian–Turanian, and South Palearctic complexes), are the most widely distributed taxa of the Bulgarian fauna (Table 3, Fig. 2).

The representatives of the Eurasian Steppe complex are more or less widely distributed in the steppe and forest-steppe regions of Europe and Asia and trophically connected to steppe herbaceous plants. For *Sphenoptera cuprina* and *Agrilus zigzag*, Bulgaria is the southwestern boundary of their range.

The taxa with the widest ranges are combined in the Holarctic (*Melanophila acuminata*), Palearctic, and Eurosiberian complexes (ca. 20% of Bulgarian buprestid speceis; Tables 2 and 3). Most of these taxa originated in the boreal areas.

The unique character of the Bulgarian fauna is further determined by its endemic and subendemic taxa. The area of distribution and probable speciation of some of those (*Sphenoptera buresi, S. sceptrifera, Anthaxia muehlei, A. tenella tenella*, and *A. thessalica*) is the East Balkan Peninsula (Bulgaria, Greece, and Macedonia). Probably, some species and subspecies such as subendemics *Julodis pubescens, Anthaxia thalassophila pseudokervillei* and East Mediterranean *Acmaeoderella levantina* and *A. seminata* also originated in the East Balkans. The role of East Balkans as a part of East Mediterranean center of buprestid speciation becomes even more important if we take into account the existence of 12 endemic taxa in the continental Greece.

Further analysis of the distribution of Bulgarian polytypic buprestids would allow establishing some general boundaries among their subspecies. For greater part of European–Iranian–Turanian, Mediterranean–Iranian–Turanian, and East Mediterranean–Iranian–Turanian species, the range boundaries of their subspecies are located in Turkey and/or adjacent countries. One of the causes of the intraspecific divergence could be the adaptation of buprestids to a more arid climate in West and Central Asia. The occurrence of vicariant species or subspecies in the western and eastern parts of the Palearctic region could be explained by the disjunction of nemoral forest. Some other subspecies are isolated in the Northwest Africa and on the Mediterranean Sea islands.

Not only allopatric but also sympatric speciation or subspeciation in jewel beetles, on the base of trophic divergence, is quite possible but is very difficult to prove. The "classical" example is *Agrilus viridis* species group. This group contains one polyphagous species (*A. viridis*) and some oligophagous taxa (trophic forms or/and subspecies or/and species), which are adapted to feeding on certain tree genera. Morphologically, however, these forms do not differ much.

The role of the ranges of host plants in recent buprestid distribution is interesting but rather poorly studied. Some problems here include: predominant role of polyphagous and wide oligophagous taxa among jewel beetles, lack of data on trophic specialization and on detailed distribution of some taxa. In our opinion, the real potential of distribution for many species is limited mainly by their thermophily, xerophily, photophily, and species competition rather than by lack of suitable host plants.

The main character of the Bulgarian fauna of Buprestidae is determined by taxa from Mediterranean and European complexes with relatively narrow ranges, ca.60% of all species and subspecies. The endemic East Balkan species and subspecies demonstrate the position of the Bulgarian buprestid fauna as a part of the East Balkan faunal type.

5 Regional Faunistic and Zoogeographical Characteristics

We use the regional geographical division of Bulgarian territory by Hubenov (1997). The species composition of each of the regions is presented in Table 2. The characteristic of each of the areas (north to south) is as follows:

Danubian Plane. The total number of taxa in this region is 53 (or 29.6% of the total Bulgarian species number). No unique taxa of buprestids are found exclusively in the Danubian Plane. From zoogeographical perspective, species and subspecies of the Mediterranean complex (20, or 35.7% of the region's fauna) are predominant here. Of those, 15 are representatives of the more widely distributed Mediterranean–European and Pontomediterranean taxa. Only five taxa (one Transmediterranean and four East Mediterranean) are Mediterranean in a narrow sense. Taxa belonging to the European complex (12, or

22.6%) are followed by those from Palearctic (eight, or 15.1%), Eurosiberian (five, or 9.4%), South Palearctic (four, or 7.6%) and Mediterranean–Iranian–Turanian complexes (three, or 5.7%). Only one subendemic Balkan species, *Chrysobothris leonhardi*, is found in this region.

Stara Planina Mountains. The number of taxa found in this area is 80 (44.7%). Only there, *Anthaxia amasina amasina* and *Trachys quercicolus* were collected. In this region, the Mediterranean complex is represented by 28 species and subspecies (35.0%). It is followed by European (22 taxa, or 27.5%), Palearctic (eight taxa, or 10.0%), Mediterranean–Iranian–Turanian (seven taxa, or 8.8%), Eurosiberian and South Palearctic complexes (six taxa each, or 7.5%), and European–Iranian–Turanian (one species). Balkan endemics are represented by *Anthaxia thessalica*; and Balkan subendemics, by *Chrysobothris leonhardi*.

Transitional region. The total number of buprestid taxa here is 103 (57.5%). The unique taxa are *Sphenoptera laportei* and *Agrilus ribesi*. In this region also Mediterranean (36, or 35.0%) and European (24, or 23.3%) taxa are predominant. They are followed by Palearctic (13, or 12.6%), Eurosiberian (ten, or 9.7%), and South Palearctic and Mediterranean–Iranian–Turanian complexes (seven species each, or 6.8%). Holarctic, Eurasian Steppe, and European–Iranian–Turanian complexes are represented by one species each. Balkan endemics are represented by *Anthaxia thessalica*; and Balkan subendemics, by *Anthaxia thalassophila pseudokervillei* and *Chrysobothris leonhardi*.

Thracian Lowland and Tundja–Strandja region. The total of 105 (58.7%) taxa was collected there. *Acmaeodera quadrizonata* and *Sphenoptera cuprina* were found only in this region. The total number of species and subspecies belonging to Mediterranean complex is 42 (40.0%). The European complex is represented by 23 taxa (21.9%). Contribution of other complexes is as follows: Palearctic (13 taxa, or 12.4%), South Palearctic and Mediterranean–Iranian–Turanian (eight taxa each, or 7.6%), Eurosiberian (seven taxa, or 6.7%), Eurasian Steppe and European–Iranian–Turanian (one species each). Balkan endemics are represented by *Anthaxia thessalica*; and Balkan subendemics, by *Chrysobothris leonhardi*.

Rila–Rhodope Massif. This region contains the highest diversity of Buprestidae in Bulgaria, 161 species and subspecies (89.9%), 21 of them unique (Table 2). Taxa of the Mediterranean complex dominate in this region as well (58 taxa, or 36.0%). The European complex includes 36 taxa (22.4%); Palearctic, 16(9.9%), Mediterranean–Iranian–Turanian, 14 (8.7%), Eurosiberian, 12 (7.5%), South Palearctic, nine (5.6%), Eurasian Steppe, five (3.1%), European–Iranian–Turanian, two (1.2%), and Holarctic, one (0.6%). All Balkan endemics and subendemics known from Bulgaria (Table 2) are also represented in this region (eight taxa, or 5.0%).

Black Sea coast. The total number of 115 (64.2%) taxa (second place among all regions) was found there. The unique taxa are *Cyphosoma euphraticum, Anthaxia muliebris*, and *Agrilus alexeevi alexeevi*. In this region, the number of the Mediterranean taxa is 44 (38.3%), and of the European taxa, 27 (23.5%). The contribution of other complexes is as follows: Palearctic, 14 (12.2%); South Palearctic and Mediterranean–Iranian–Turanian, eight taxa each (7.0%); Eurosiberian, six

(5.2%); Eurasian Steppe, four (3.5%), and European–Iranian–Turanian, two (1.7%). A Balkan endemic species is *Sphenoptera sceptrifera*; a Balkan subendemic is *Chrysobothris leonhardi*.

The Bulgarian buprestid fauna is still incompletely studied. The investigations so far concerned mainly the southern part of Bulgaria. Nevertheless, it is possible to make conclusions about the jewel beetle distribution in Bulgaria. As one of the most thermophilous and xerophilous beetle groups, Buprestidae prefer the areas with higher temperature and middle to low humidity. They are more rare in the north, and obviously are much more easily collected in South Bulgaria. On the other hand, the diversity of natural habitats is greater in South Bulgaria, especially in the Rila–Rhodope Massif and at the Black Sea coast. The high mountains in Bulgaria coexist with the areas characterized with the highest Mediterranean influence (Struma and Mesta valleys). The Mediterranean impact at the Black Sea coast is also significant, the high humidity allowing for the distribution of mesophilous taxa normally found in the mountains. The mentioned factors also determine the highest diversity level of buprestid species recorded in these two regions.

Zoogeographical analysis shows the domination of Mediterranean elements in all regions of Bulgaria; the level of such domination is similar in different regions and Bulgaria as a whole. The Mediterranean taxa are slightly more dominant in the Thracian Lowland and Tundja–Strandja region (40.0 %), and slightly less, in the Transitional and Stara Planina regions (ca. 35 %). The European complex contribution is in the second place, from 27.5 % in Stara Planina to ca. 22 % in the Thracian Lowland and Tundja–Strandja. The contribution of widely distributed taxa (mainly of the boreal origin) such as Holarctic, Palearctic, and Eurosiberian is the highest in the Danubian Plain. The Eurasian Steppe elements and Balkan endemics and subendemics are widely represented in the lowlands of Rila–Rhodope region. The Eurasian Steppe taxa are very rare in Bulgaria; their impact is probably due to the highest level of investigation of the Rila–Rhodope region. All other zoogeographical complexes are almost equally represented in the regional faunal structure.

The Bulgarian regions clearly differ in the number of species belonging to the main zoogeographical complexes. In general, however, it depends on the total species number found in each region; their contribution (percentage of taxa belonging to different zoogeographical complexes) does not exhibit any significant regional differentiation.

6 Trends in the Recent Formation of the Bulgarian Buprestid Fauna

The recent species composition of the Bulgarian fauna is determined by the geographical position of the country, historical processes of faunogenesis, recent climatic factors, orographic and river systems, and the level of anthropogenic impact on natural habitats. Combination of all these factors allows a relatively high diversity level of different zoogeographical elements with mainly autochthonous or allochthonous origin.

The recent Bulgarian fauna was formed during glacial and, especially, postglacial periods (Pleistocene and Holocene). Some of the endemics, subendemics, and East Mediterranean species and subspecies (see section 13.4), whose possible territory of speciation is the eastern part of Balkan Peninsula, could be autochthonous for Bulgaria. Other species with wider recent ranges could have also originated in East Balkans, including the territory of Bulgaria. Due to the absence of fossil buprestid evidence from Bulgaria, it is very difficult to prove that.

Allochthonous elements could colonize Bulgaria during preglacial, glacial, and postglacial periods. The finding of such preglacial relicts as *Buprestis splendens* in Bulgaria is very possible (it is found in Greece and Albania) but so far it has not been collected. In our opinion, there were four main routes of buprestid colonization of the Bulgarian territory. First route was the migration from the centers of speciation in the Ancient Mediterranean region. The recent representatives of these taxa are grouped in Mediterranean, Mediterranean–Iranian–Turanian, and South Palearctic complexes as well as some other with wider recent ranges. As mentioned above (section 12. 4), taxa which originated in East Mediterranean center of speciation (Pontomediterranean, East Mediterranean and, probably, the majority of East Mediterranean–Iranian–Turanian elements) have the greatest contribution in the Bulgarian fauna.

The second route was represented by the European taxa, which speciated in Central and East Europe. The third trend, characteristic mainly for glacial periods, is migration of the Boreal elements, which originated in the North Eurasian regions (Holarctic and main part of Eurosiberian and Palearctic taxa). The last route of faunal formation is dispersal from the northeastern part of Bulgaria of taxa found in the steppe and forest-steppe regions of Europe and Asia (Eurasian Steppe complex).

References

Baev, P. & L. Penev, 1995. BIODIV. Program for Calculating Biological Diversity Parameters, Similarity, Niche Overlap and Cluster Analysis. Version 5.1. Pensoft, Sofia & Moscow.

Bellamy, C., 1985. A catalogue of the higher taxa of the family Buprestidae (Coleoptera). Navorsinge van die Nasionale Museum Bloemfontein, Natural Science 4(15):405–472.

Bílý, S., 1982. The Buprestidae (Coleoptera) of Fennoscandia and Denmark. Fauna Entomologica Scandinavica, 10.

Bílý, S., 1993. Buprestidae. In: J. Jelinek (ed), Check-list of Czechoslovak Insects IV (Coleoptera). Folia Heyrovskiana, Supplementum 1:73–75.

Chikatunov, V., T. Pavlicek & E. Nevo, 1999. Coleoptera of "Evolution Canyon". Pensoft, Sofia & Moscow.

Cobos, A., 1986. Fauna Iberica de Coleopteros Buprestidae. Consejo Superior de Investigaciones Cientificas, Madrid.

Curletti, G., 1994. I Buprestidi d'Italia. Monografie di "Natura Bresciana" 19.

Emeljanov, A., 1974. [Proposals on the classification and nomenclature of areals]. Entomologicheskoe obozrenie (Revue d'entomologie de l'URSS) 53(3):497–522 (in Russian).

Frivaldszky, I., 1835. Á Balkány vidékén tett természettudomanyi utazásrol. Magyar tudós Társasag Évkönyvei 2(2):235–276.

Frivaldszky, I., 1837. Termeszettudomany-osztalyi ertekezes. VIII. Balkány vidéki természettudomanyi utázas. Magyar tudós Társasag Évkönyvei 3(3):156–184.

Halperin, J. & Q. Argaman, 2000. Annotated list of Buprestidae (Col.) and their host plants of Israel. Zoology in the Middle East 20:99–116.

Hubenov, Z., 1997. Possibilities for using of a system from the really defined natural territories for the faunistic researches in Bulgaria. Acta zoologica bulgarica 49:5–9.

Karaman, S. & S. Tezcan, 1998. Contribution to the study of the genus *Anthaxia* (Subgenus *Anthaxia* s. str.) Eschscholtz, 1829 (Coleoptera, Buprestidae) of Turkey. Türkiye Entomoloji Dergisi 22(1):19–35.

Lodos, N. a& S. Tezcan, 1992. Türkiye Buprestidae (Coleoptera) faunasinin genel görünümü ve zoocografi yönden degerlendirilmesi. The Journal of Agricultural Faculty of Ege University 29(1):15–22.

Mühle, H., 1980. Ergebnisse der Albanien-Expedition 1961 des Deutshen Entomologischen Institutes. 96. Beitrag. Coleoptera: Buprestidae. Beiträge zur Entomologie 30:369–383.

Mühle, H., P. Brandl & M. Niehuis, 2000. Catalogus Faunae Graeciae: Coleoptera, Buprestidae. Selbstverlag, Augsburg.

Niehuis, M., 1996. Prachtkäferfunde aus Israel mit Beschreibung von *Anthaxia martinhauseri* n. sp. (Coleoptera: Buprestidae) Mitteilungen des Internationalen Entomologischen Vereins 21(3/4):131–151.

Niehuis, M. & S. Tezcan, 1993. Beitrag zur Kenntnis der *Agrilus*-Arten der Türkei (Coleoptera: Buprestidae). Mitteilungen des Internationalen Entomologischen Vereins 18(1/2):1–74.

Obenberger, J., 1932. Catalogue raisonné des Buprestides de Bulgarie. Izvestiya na Tsarskite prirodonauchni instituti v Sofia (Mitteilungen aus den Königlichen naturwissenschaftlichen Instituten in Sofia – Bulgarien; Bulletin des Institutions royales d'histoire naturelle à Sofia – Bulgarie) 5:15–66.

Obenberger, J., 1933. Catalogue raisonné des Buprestides de Bulgarie. II. partie. Izvestiya na Tsarskite prirodonauchni instituti v Sofia (Mitteilungen aus den Königlichen naturwissenschaftlichen Instituten in Sofia – Bulgarien; Bulletin des Institutions royales d'histoire naturelle à Sofia – Bulgarie) 6:49–115.

Obenberger, J., 1935. Catalogue raisonné des Buprestides de Bulgarie. III. partie. Izvestiya na Tsarskite prirodonauchni instituti v Sofia (Mitteilungen aus den Königlichen naturwissenschaftlichen Instituten in Sofia – Bulgarien; Bulletin des Institutions royales d'histoire naturelle à Sofia – Bulgarie) 8:23–96.

Sakalian, V., 1994. Studies on Buprestidae (Coleoptera) in the Sandanski–Petrič and Goce Delčev valleys – Southwest Bulgaria. III. Zoogeographical characteristic. Acta zoologica bulgarica 47:35–42.

Sakalian, V., 1996. Jewel beetles (Coleoptera: Buprestidae) from the Bulgarian Black Sea coast. Quaderni della Stazione del Ecologia del Museo Civico di Storia Naturale di Ferrara 10:151–163.

Sakalian, V., 2000. Contribution to the knowledge of jewel beetles of Albania (Coleoptera: Buprestidae). Beiträge zur Entomologie 50(1):91–101.

Sakalian, V., 2003. A Catalogue of the Jewel Beetles of Bulgaria (Coleoptera: Buprestidae). Zoocartographia Balcanica 2.

Schaefer, L., 1971. Catalogue des Coléoptères Buprestides de France. Bulletin mensuel de la Société Linnéenne de Lyon 40(9):275–284.

Schaefer, L., 1972. Catalogue des Coléoptères Buprestides de France. Bulletin mensuel de la Société Linnéenne de Lyon 41(8):155–164.

Tezcan, S. & S. Bílý, 1998. Contribution to the study of the genus *Anthaxia* Eschscholtz, 1829 (subgenera *Cratomerus* Solier, 1833 and *Melanthaxia* Richter, 1944) (Coleoptera, Buprestidae) of Turkey. Türkiye Entomoloji Dergisi 22(3):171–186.

Tozlu, G. & H. Özbek, 1998. *Sphenoptera (Chrysoblemma) artemisiae* Reitter, 1889 (Coleoptera: Buprestidae), a new record for the Turkish fauna. Zoology in the Middle East 16:107–110.

Ulay, M. & S. Tezcan, 1998. Contribution to the study of the genus *Anthaxia* Eschscholtz, 1829 (subgenus *Haplanthaxia* Reitter, 1911) (Coleoptera, Buprestidae) of Turkey. Türkiye Entomoloji Dergisi 22(2):109–121.

Volkovich, M. & A. Alexeev, 1994. Buprestid beetles (Coleoptera: Buprestidae) from Kopetdagh and the adjacent regions of Southern Turkmenistan. In Fet, V. & K.I. Atamuradov (eds), Biogeography and Ecology of Turkmenistan. Kluwer Academic Publishers, Dordrecht, Boston & London: 419–449.

11 Fauna and Zoogeography of Myriapoda in Bulgaria

Pavel Stoev
National Museum of Natural History, Bulgarian Academy of Sciences, Tsar Osvoboditel Blvd. 1, 1000 Sofia, Bulgaria, e-mail: stoev@nmnh.bas.bg

Abstract: The paper summarizes all information on taxonomic diversity of subphylum Myriapoda in Bulgaria, and provides a preliminary faunal and zoogeographical analysis. At present, 240 species and subspecies from 31 families and 70 genera are known from Bulgaria; further 28 taxa have uncertain status. About half of Bulgarian myriapods (45%) belongs to the endemic (Balkan) complex, while the rest originate from outside the Balkan Peninsula. A substantial part of the endemics are cave inhabitants. The work is supplemented by a complete bibliography

1 Introduction

Although Bulgarian myriapods have been an object of investigation for about 110 years, and no less than 100 publications have been devoted to different aspects of their systematics, distribution, and biology, no attempt at reviewing all that information has ever been made. This obvious gap in our current knowledge leads to a number of difficulties when some more general issues of myriapod zoogeography and ecology are concerned.

This paper reviews all literature dealing with Bulgarian myriapods and provides an annotated list of the species known to date (2005), considering also some aspects of their zoogeography and ecology. While centipeds (Chilopoda) have already been an object of a more recent review (e.g. Stoev, 2002), the information concerning the millipeds (Diplopoda) and pauropods (Pauropoda) is still scattered among scores of old papers and needs a critical treatment. This hiatus is partly filled by the annotated checklist presented here. The class Symphyla remains entirely unexplored in Bulgaria. Since no zoogeographical and ecological analysis of Bulgarian myriapods has ever been made, the distribution patterns (chorotypes) of myriapods are defined and briefly considered here. For determining the species' chorotypes, the paper of Vigna Taglianti et al. (1999) was mainly used, with some additions. A complete bibliography supplements the work.

2 History of Bulgarian Myriapod Studies

The studies of Myriapoda in Bulgaria began in 1892 with Georgi Hristovich (Hristovich, 1892), who, among other invertebrates in his general faunistic reports, mentioned also a few myriapods from the surroundings of Sofia. Twelve years later

V. Fet and A. Popov (Eds.), Biogeography and Ecology of Bulgaria, 379–404.

Stefan Jurinich, the real father of the Bulgarian myriapodology, published his observations, very serious for their time, based on a large material collected by himself and his collaborators throughout the country (Jurinich, 1904). Written in Bulgarian, this remarkable scientific work remained almost unknown to the European systematists. Since then, scores of publications has been devoted to the subject. From the pleiad of the European zoologists dealing with the issue, the most significant contributions were made by Verhoeff (1925, 1926a, 1926b, 1926c, 1926d, 1926e, 1928a, 1928b, 1937), Remy (1945), Strasser (1960, 1962a, 1962b, 1966, 1969, 1973, 1975), Matic (1964, 1967, 1973a, 1973b), Matic and Golemansky (1964, 1965, 1967a, 1967b, 1967c, 1969), and Kaczmarek (1969a, 1969b, 1970, 1972, 1973, 1975). Other papers dedicated, entirely or in part, to the Bulgarian myriapod fauna are those of Attems (1904, 1929a, 1929b, 1959), Anonymous (1907), Schubart (1934), Lang (1935, 1958), Folkmanova (1936), Jawlowski (1938), Krestewa (1940), Demange (1961), Negrea (1965, 1971), Gulička (1967a, 1967b), Tabacaru (1967, 1983), Turk (1970), Ceuca (1973), Würmli (1973), Matic and Darabanţu (1974), Dobroruka (1977), Golovatch and Kondeva (1992), and Golovatch et Stoev (2004).

Myriapods have also been mentioned in some general works, like those of Guéorguiev and Beron (1962), Beron and Guéorguiev (1967), Popov (1969), Beron (1972, 1978, 1994, 2001), Guéorguiev et al. (1994), and Beron et al. (2004) on the cave fauna, those of Beron (1999) on the mountain fauna, and of Deltshev et al. (1998) on the biodiversity of Bulgaria. The parasites of myriapods have been studied by Golemansky (1973a, 1973b), Golemansky and Taschev (1973), and Golemansky and Lipa (1991). The publications of Eason (1970, 1983), Scheller (1977, 1981), Enghoff (1985, 2004), Stoev (1997b, 2005), Ćurčić and Makarov (2000), Djursvoll et al.(2000) considered some taxonomical and morphological aspects, and included species from Bulgaria.

The latest achievements are connected with the work of Georgi Ribarov on centipeds (Ribarov, 1984, 1985, 1986a, 1986b, 1987a, 1987b, 1987c, 1989a, 1989b, 1990, 1992, 1996) and Yasen Christov (Christow, 1983, 1984b; Christov, 1984a, 1985, 1986a, 1986b) and Emilia Kondeva (Kondeva, 1996, 1998, 2000, 2002, 2004) on millipeds.

Finally, I began my study on this group in 1993, and so far only a small part of the myriapods in the collection of the National Museum of Natural History in Sofia has been examined and published (Stoev and Ribarov, 1995; Stoev, 1997a, 2001a, 2001b, 2002, 2003, 2004a, 2004b, 2004c; Stoev in Deltshev et al., 2000a, 2000b; Stoev and Beron, 2001; Stoev et Enghoff, 2003; Stoev and Lapeva-Gjonova, 2005).

3 Fauna of Myriapoda

Four recent classes, namely, Chilopoda, Symphyla, Diplopoda, and Pauropoda compose the subphylum Myriapoda. All of them are well represented in the fauna of Bulgaria, although the Bulgarian Symphyla have never been an object of a taxonomic study.

The list of Bulgarian myriapods (Table 1) currently includes 240 species and subspecies belonging to 70 genera and 31 families. Additional 28 species and subspecies, six genera and one family (Mecistocephalidae) have been recorded but their occurrence should be reconfirmed (Table 2). Some of these records are probably based on misidentifications, while others could belong to species with uncertain identity. Bulgarian milliped list encompasses 120 species and subspecies, with Julidae without any doubt the largest family (48 taxa), followed by Polydesmidae (23 taxa) and Anthroleucosomatidae (12 taxa). Among centipeds, 102 species and subspecies have been recorded from Bulgaria; this number could increase slightly with the future research, but will probably not surpass 130 taxa. The largest share among Chilopoda belongs to Lithobiidae (66 taxa), followed by Geophilidae (11 taxa). Lithobiids

Table 1 List of Bulgarian myriapods and their chorotypes

Taxa	Chorotypes	Caves
CHILOPODA		
SCUTIGERIDAE		
Scutigera coleoptrata (Linnaeus, 1758)	MED	
LITHOBIIDAE		
Eupolybothrus (*Eupolybothrus*) *andreevi* Matic, 1964	SPE	*
Eupolybothrus (*Eupolybothrus*) *fasciatus* (Newport, 1845)	n/a	
Eupolybothrus (*Eupolybothrus*) *gloriastygis* (Absolon, 1916)	n/a	*
Eupolybothrus (*Eupolybothrus*) *litoralis* (L. Koch, 1867)	EME	
Eupolybothrus (*Mesobothrus*) *transsylvanicus* (Latzel, 1882)	CPBK	
Eupolybothrus (*Leptopolybothrus*) *tridentinus* (Fanzago,1874)	EEU	
Eupolybothrus valkanovi (Kaczmarek, 1973)	n/a	
Lithobius (*Lithobius*) *agilis* C.L. Koch, 1847	CEU	
Lithobius (*Lithobius*) *borealis* Meinert, 1872	EU	
Lithobius (*Lithobius*) *borisi* Verhoeff, 1928	EBK	
Lithobius (*Lithobius*) *dentatus* C.L. Koch, 1844	CEU	
Lithobius (*Lithobius*) *electron* Verhoeff, 1928	n/a	
Lithobius (*Lithobius*) *ergus* (Chamberlin, 1952)	ANBK	
Lithobius (*Lithobius*) *erythrocephalus* C.L. Koch, 1847	EU	
Lithobius (*Lithobius*) *forficatus* (Linnaeus, 1758)	EU	
Lithobius (*Lithobius*) *glaciei* Verhoeff, 1928	n/a	
Lithobius (*Lithobius*) *lakatnicensis* Verhoeff, 1926	EBK	*
Lithobius (*Lithobius*) *latro* Meinert, 1872	CEU	
Lithobius (*Lithobius*) *lucifugus* L. Koch, 1862	CEU	
Lithobius (*Lithobius*) *mutabilis* L. Koch, 1862	CEU	
Lithobius (*Lithobius*) *muticus* C.L. Koch, 1847	CEU	
Lithobius (*Lithobius*) *nigripalpis* L. Koch, 1867	EME	
Lithobius (*Lithobius*) *oglednicus* Ribarov, 1987	STRE	
Lithobius (*Lithobius*) *parietum* Verhoeff, 1899	CPBK	
Lithobius (*Lithobius*) *peggauensis* Verhoeff, 1937	CEU	
Lithobius (*Lithobius*) *pelidnus* Haase, 1880	EU	
Lithobius (*Lithobius*) *peregrinus* Latzel, 1880	EME	
Lithobius (*Lithobius*) *piceus* L. Koch, 1862	CEU	

continued

Table 1 continued

Taxa	Chorotypes	Caves
Lithobius (*Lithobius*) *rushovensis* Matic, 1967	ANBK	*
Lithobius (*Lithobius*) *schuleri* Verhoeff, 1925	CEU	
Lithobius (*Lithobius*) *stygius* Latzel, 1880	BK	*
Lithobius (*Lithobius*) *tenebrosus* Meinert, 1872	EU	
Lithobius (*Lithobius*) *thracicus* Matic et Golemansky, 1967	n/a	
Lithobius (*Lithobius*) *tiasnatensis* Matic, 1973	EBK	*
Lithobius (*Lithobius*) *tricuspis* Meinert, 1872	CEU	
Lithobius (*Lithobius*) *uniunguis* Matic et Golemansky, 1967	BG	
Lithobius (*Lithobius*) *viriatus* Sselivanoff, 1878	EME	
Lithobius (*Monotarsobius*) *aeruginosus* L. Koch, 1862	EU	
Lithobius (*Monotarsobius*) *catascaphius* (Verhoeff, 1937)	EME	
Lithobius (*Monotarsobius*) *crassipes* L. Koch, 1862	ESI	
Lithobius (*Monotarsobius*) *curtipes* C.L. Koch, 1847	CAE	
Lithobius (*Monotarsobius*) *dobrogicus* Matic, 1962	EBK	
Lithobius (*Sigibius*) cf. *apfelbecki* Verhoeff, 1900	n/a	
Lithobius (*Sigibius*) *beroni* Negrea, 1965	EBK	
Lithobius (*Sigibius*) *bifidus* (Matic, 1973)	STRE	*
Lithobius (*Sigibius*) *burzenlandicus* Verhoeff, 1931	CPBK	
Lithobius (*Sigibius*) *jurinici* Matic et Golemansky, 1965	ANBK	
Lithobius (*Sigibius*) *micropodus* (Matic, 1980)	MED	
Lithobius (*Sigibius*) *microps* Meinert, 1868	EU	
Lithobius (*Sigibius*) *proximus* Matic et Golemansky, 1967	n/a	
Lithobius (*Sigibius*) *pustulatus* Matic, 1964	CPBK	
Lithobius (*Sigibius*) *strandzanicus* (Ribarov, 1987)	STRE	
Lithobius (*Sigibius*) *totevi* Kaczmarek, 1975	n/a	
Lithobius (*Sigibius*) *vizicae* (Ribarov, 1987)	EBK	
Lithobius (*Sigibius*) *wardaranus* (Verhoeff, 1937)	BK	
Lithobius (*Sigibius*) *zelazovae* Kaczmarek, 1975	n/a	
Pleurolithobius patriarchalis (Berlese, 1894)	EME	
Harpolithobius anodus anodus (Latzel, 1880)	EEU	
Harpolithobius anodus dentatus Matic, 1957	CPBK	
Harpolithobius aseni Kaczmarek, 1975	RHE	
Harpolithobius banaticus rhodopensis Kaczmarek, 1975	RHE	
Harpolithobius folkmanovae Kaczmarek, 1975	STRE	
Harpolithobius hemusi Kaczmarek, 1975	SPE	
Harpolithobius cf. *dollfusi* (Verhoeff, 1901)	n/a	
Harpolithobius maculipes (Folkmanova, 1936)	n/a	
Harpolithobius radui (Matic, 1955)	CPBK	
HENICOPIDAE		
Lamyctes emarginatus (Newport, 1844)	COS	
SCOLOPENDRIDAE		
Scolopendra cingulata Latreille, 1829	MED	
CRYPTOPIDAE		
Cryptops anomalans Newport, 1844	EU	
Cryptops croaticus Verhoeff, 1931	EEU	
Cryptops hortensis (Donovan, 1810)	EMT	
Cryptops parisi Brolemann, 1920	CEU	
Cryptops rucneri Matic, 1967	EEU	

continued

Table 1 continued

Taxa	Chorotypes	Caves
Himantariidae		
Himantarium gabrielis (Linnaeus, 1767)	MED	
Bothriogaster signata (Kessler, 1874)	TUM	
Thracophilus beroni Matic and Darabanţu, 1974	RHE	*
Thracophilus bulgaricus Verhoeff, 1926	BK	
Schendylidae		
Schendyla delicatula Kaczmarek, 1969	PO	
Schendyla montana Attems, 1895	SEU	
Schendyla nemorensis (C.L. Koch, 1836)	EU	
Schendyla varnensis (Kaczmarek, 1969)	PO	
Schendyla walachica Verhoeff, 1900	SEU	
Geophilidae		
Geophilus carpophagus Leach, 1815	EU	
Geophilus electricus (Linnaeus, 1758)	EU	
Geophilus flavus (De Geer, 1778)	ESI	
Geophilus proximus C.L. Koch, 1847	ESI	
Geophilus linearis C.L. Koch, 1835	EU	
Geophilus rhodopensis Kaczmarek, 1970	BG	
Geophilus "*strictus*" Latzel, 1880	n/a	
Clinopodes flavidus C.L. Koch, 1847	EMT	
Clinopodes polytrichus (Attems, 1903)	n/a	
Clinopodes trebevicensis (Verhoeff, 1898)	BK	
Pachymerium ferrugineum (C.L. Koch, 1835)	WPA	
Dignathodontidae		
Dignathodon microcephalus (Lucas, 1846)	MED	
Henia (*Henia*) *illyrica* (Meinert, 1870)	EEU	
Henia (*Meinertia*) *bicarinata* (Meinert, 1870)	MED	
Henia (*Turkophilus*) *porosa* (Verhoeff, 1941)	ANBK	
Linotaeniidae		
Strigamia acuminata (Leach, 1815)	EU	
Strigamia crassipes (C.L. Koch, 1835)	EU	
Strigamia engadina (Verhoeff, 1935)	CEU	
Strigamia transsilvanica (Verhoeff, 1928)	CPBK	
DIPLOPODA		
Polyxenidae		
Polyxenus lagurus (Linnaeus, 1758)	HO	
Lophoproctidae		
Lophoproctus cf. *lucidus* (Chalande, 1888)	n/a	
Glomeridae		
Glomeris balcanica Verhoeff, 1906	EBK	
Glomeris hexasticha Brandt, 1833	CEU	
Glomeris pustulata Latreille, 1804	CEU	
Doderiidae		
Trachysphaera costata (Waga, 1857)	CEU	
Trachysphaera orghidani (Tabacaru, 1958)	CPBK	*
Trachysphaera rotundata (Lignau, 1911)	EME	

continued

Table 1 continued

Taxa	Chorotypes	Caves
Polydesmidae		
Polydesmus bureschi Verhoeff, 1928	PBE	
Polydesmus complanatus (Linnaeus, 1761)	CEU	
Polydesmus denticulatus C.L. Koch, 1847	EU	
Polydesmus herzogowinensis Verhoeff, 1897	BK	
Polydesmus jawlowskii Strasser, 1966	EBK	
Polydesmus mediterraneus oertzeni Verhoeff, 1901	EME	
Polydesmus planinensis Gulička, 1967	SPE	
Polydesmus renschi Shubart, 1934	CPBK	
Polydesmus rodopensis Gulička, 1967	RHE	
Polydesmus rubellus Attems, 1902	Crete–Dobrudja	
Polydesmus zonkovi Verhoeff, 1937	BG	
Brachydesmus cristofer Strasser, 1966	VE	
Brachydesmus dadayi dadayi Verhoeff, 1895	CPBK	
Brachydesmus dadayi brusenicus Gulička, 1967	PBE	
Brachydesmus herzogowinensis confinis Strasser, 1969	SPE	
Brachydesmus herzogowinensis reflexus Strasser, 1975	SPE	
Brachydesmus herzogowinensis trifidus Strasser, 1973	OE	
Brachydesmus peristerensis Verhoeff, 1932	EBK	
Brachydesmus polydesmoides Verhoeff, 1895	BK	
Brachydesmus radewi Verhoeff, 1926	PBE	*
Brachydesmus superus superus Latzel, 1884	EU	
Brachydesmus superus bulgaricus Verhoeff, 1928	CPBK	
Brachydesmus tetevensis Strasser, 1973	SPE	
Paradoxosomatidae		
Strongylosoma stigmatosum (Eichwald, 1830)	EEU	
Strongylosoma jaqueti Verhoeff, 1898	CPBK	
Oxidus gracilis (C.L. Koch, 1847)	INTR	
Metonomastus pomak Golovatch et Stoev, 2004	RHE	
Trichopolydesmidae		
Bacillidesmus bulgaricus bulgaricus Strasser, 1962	SPE	*
Bacillidesmus bulgaricus dentatus Strasser, 1966	PBE	*
Anthroleucosomatidae		
Bulgarosoma bureschi Verhoeff, 1926	SPE	*
Bulgarosoma superficiei Strasser, 1975	VE	
Rhodoposoma rhodopinum (Strasser, 1966)	RHE	*
Troglodicus meridionale (Tabacaru, 1967)	RHE	*
Troglodicus tridentifer Gulička, 1967	RHE	*
Stygiosoma beroni Gulička, 1967	RHE	*
Anamastigona alba (Strasser, 1960)	SPE	*
Anamastigona bilselii (Verhoeff, 1940)	EBK	
Anamastigona delcevi (Strasser, 1973)	SLE	*
Anamastigona falcata (Gulička, 1967)	SPE	*
Anamastigona lepenicae (Strasser, 1975)	RHE	*
Bulgardicus tranteevi Strasser, 1960	PBE	*
Haaseidae		
Haasea vidinensis (Strasser, 1973)	DPE	

continued

Table 1 continued

Taxa	Chorotypes	Caves
MASTIGOPHOROPHYLLIDAE		
Mastigophorophyllon bulgaricum bulgaricum Schubart, 1934	BG	
Mastigophorophyllon bulgaricum pirinicum Gulička 1967	PNE	
Mastigona bosniensis (Verhoeff, 1897)	BK	
Mastigona transsylvanica (Verhoeff, 1897)	CPBK	
CHORDEUMATIDAE		
Melogona broelemanni (Verhoeff, 1897)	CPBK	
CRASPEDOSOMATIDAE		
Craspedosoma transsylvanicum Verhoeff, 1897	EEU	
SCHIZOPETALIDAE		
Balkanopetalum armatum Verhoeff, 1926	SPE	
Balkanopetalum beskovi Strasser, 1973	RHE	
Balkanopetalum bulgaricum Stoev et Enghoff, 2003	SL–PNE	
Balkanopetalum petrovi Stoev et Enghoff, 2003	RHE	
Balkanopetalum rhodopinum Verhoeff, 1937	RHE	
Eurygyrus ochraceus C.L. Koch, 1847	INTR	
Callipodella fasciata (Latzel, 1882)	BK	
DORYPETALIDAE		
Dorypetalum bulgaricum Strasser, 1966	RHE	
BLANIULIDAE		
Nopoiulus kochii (Gervais, 1847)	EUAN	
Blaniulus guttulatus (Fabricius, 1798)	EU	
Archiboreoiulus pallidus (Brade-Birks, 1920)	n/a	
Cibiniulus phlepsii (Verhoeff, 1897)	EEU	
Proteroiulus fuscus (Am Stein, 1857)	EU	
NEMASOMATIDAE		
Nemasoma varicorne C.L. Koch, 1847	EU	
JULIDAE		
Leptoiulus borisi Verhoeff, 1926	BG	
Leptoiulus sarajevensis Verhoeff, 1898	BK	
Leptoiulus trilineatus (C.L. Koch, 1847)	EEU	
Xestoiulus fontisherculis (Verhoeff, 1899)	CPBK	
Xestoiulus rebeli (Attems, 1904)	SPE	
Xestoiulus pirinicus (Gulička, 1967)	PNE	
Typhloiulus bureschi Verhoeff, 1926	SPE	*
Typhloiulus georgievi Strasser, 1962	SPE	*
Typhloiulus kotelensis Jawlowski, 1938	SPE	
Typhloiulus longipes Strasser, 1973	SPE	*
Typhloiulus staregai Strasser, 1973	SPE	*
Typhloiulus strictus (Latzel, 1882)	CPBK	
Serboiulus spelaeophilus Gulička, 1967	SPE	*
Cylindroiulus abaligetanus Verhoeff, 1901	ANBK	
Cylindroiulus arborum Verhoeff, 1928	CEU	
Cylindroiulus bellus (Lignau, 1903)	PO	
Cylindroiulus boleti (C.L. Koch, 1847)	EEU	
Cylindroiulus horvathi (Verhoeff, 1897)	CEU	
Cylindroiulus luridus (C.L. Koch, 1847)	CEU	
Enantiulus nanus (Latzel, 1884)	CEU	

continued

Table 1 continued

Taxa	Chorotypes	Caves
Unciger transsilvanicus (Verhoeff, 1899)	CEU	
Brachyiulus apfelbecki Verhoeff, 1898	BK	
Brachyiulus bagnalli (Brolemann, 1924)	CPBK	
Brachyiulus lusitanus Verhoeff, 1898	BK	
Megaphyllum anatolicum denticulatum (Strasser, 1969)	PO	
Megaphyllum beroni (Strasser, 1973)	RHE	
Megaphyllum bosniense (Verhoeff, 1897)	BK	
Megaphyllum dentatum (Verhoeff, 1898)	BK	
Megaphyllum glossulifer (Schubart, 1934)	BG	
Megaphyllum hercules (Verhoeff, 1901)	BK	
Megaphyllum lictor (Attems, 1904)	EBK	
Megaphyllum rhodopinum (Verhoeff, 1928)	EBK	
Megaphyllum rossicum strandschanum (Verhoeff, 1937)	EBK	
Megaphyllum transsylvanicum (Verhoeff, 1897)	EEU	
Megaphyllum unilineatum (C.L. Koch, 1838)	EEU	
Balkanophoenix borisi Verhoeff, 1937	BG	
Pachyiulus cattarensis (Latzel, 1884)	BK	
Pachyiulus cattarensis (Latzel, 1884)	BK	
Pachyiulus hungaricus (Karsch, 1881)	CPBK	
Pachyiulus varius (Fabricius, 1781)	MED	
Apfelbeckiella bulgarica Verhoeff, 1926	STRE	
Apfelbeckiella golemanskyi Ceuca, 1973	RHE	
Apfelbeckiella trnowensis trnowensis (Verhoeff, 1928)	BG	
Apfelbeckiella trnowensis deliormana (Strasser, 1966)	DPE	
Apfelbeckiella trnowensis rhodopina (Strasser, 1966)	RHE	
Rhodopiella beroni beroni Strasser, 1966	EBK	
Rhodopiella beroni saetigera Strasser, 1973	STRE	
Rossiulus vilnensis (Jawlowski, 1925)	EEU	
Ommatoiulus sabulosus (Linnaeus, 1758)	EU	
POLYZONIDAE		
Polyzonium germanicum Brandt, 1837	EU	
HIRUDISOMATIDAE		
Hirudisoma hirsutum (Verhoeff, 1901)	EBK	
PAUROPODA		
EURYPAUROPODIDAE		
Trachypauropus cf. *glomerioides* Tömösváry, 1882	n/a	
PAUROPODIDAE		
Stylopauropus pedunculatus (Lubbock, 1867)	HO	
Pauropus furcifer Silvestri, 1902	WPA	
Pauropus huxleyi Lubbock, 1867	EU	
Allopauropus (*Allopauropus*) *brevisetus* Silvestri, 1902	HO	
Allopauropus (*Allopauropus*) *danicus* (Hansen, 1902)	COS	
Allopauropus (*Allopauropus*) *furcula* Silvestri, 1902	EU	
Allopauropus (*Allopauropus*) *humilis* Remy, 1945	EBK	
Allopauropus (*Allopauropus*) *krestewaae* Remy, 1945	RHE	
Allopauropus (*Allopauropus*) *longisetus* Remy, 1945	n/a	
Allopauropus (*Decapauropus*) *aristatus* Remy, 1936	CEU	

continued

Table 1 continued

Taxa	Chorotypes	Caves
Allopauropus (*Decapauropus*) *corsicus* Remy, 1940	MED	
Allopauropus (*Decapauropus*) *florisimilis* Krestewa, 1940	BG	
Allopauropus (*Decapauropus*) *gracilis* (Hansen, 1902)	COS	
Allopauropus (*Decapauropus*) *monachorum* Remy, 1945	SPE	
Allopauropus (*Decapauropus*) *rostratus* Krestewa, 1940	MED	
Allopauropus (*Decapauropus*) *tripartitus* Krestewa, 1940	EEU	
Total: 240 species and subspecies		

HO, Holarctic; WPA, West Palearctic; EMT, European–Mediterranean–Turanian; ESI, Eurosiberian; CAE, Central Asian–European; EUAN, European–Anatolian; EU, European; CEU, Central European; EEU, East European; CPBK, Carpathian–Balkan; SEU, South European; TUM, Turanian–Mediterranean; MED, Holomediterranean; EME, East Mediterranean; ANBK, Anatolian–Balkan; PO, Pontian; BK, Holobalkan; EBK, East Balkan; BG, Bulgarian; SPE, Stara Planina Mts. endemic; RHE, Rhodope Mts. endemic; PNE, Pirin Mts. endemic; SLE, Slavyanka Mts. endemic; VE, Vitosha Mts. endemic; OE, Osogovo Mts. endemic; STRE, Strandja Mts. endemic; PBE, Predbalkan endemic; DPE, Danubian Plain endemic; INTR, Introduced species; COS, Cosmopolitans; *, strict cave inhabitant (=troglobite); n/a, not applicable.

are the most diverse myriapod family in Bulgaria (Table 3). Members of only two families of Pauropoda, Pauropodidae and Eurypauropodidae, have been recorded. While Pauropodidae comprises at least 17 taxa, and is the third richest in species family of Myriapoda, Eurypauropodidae is represented by a single, tentatively identified species (Turk, 1970).

Provisional estimates show that among Chilopoda probably no less than 80% of the Bulgarian fauna have already been recorded. In Diplopoda, this number does not exceed 70%. Pauropods are very scarcely known; only three outdated papers deal with their taxonomy and distribution (Krestewa, 1940; Remy, 1945; Turk, 1970).

4 Zoogeographic Aspects

Inasmuch as the species list is still to be completed with taxa new either to Bulgaria or to the science, only some aspects of the zoogeography of the group are discussed here. This analysis should be considered provisional, showing only some of the patterns of distribution and origin of Bulgarian myriapods.

Out of 240 species and subspecies, and another 28 taxa probably incorrectly recorded from Bulgaria, only 222 were used for this study. Being very poorly known, the remaining taxa were omitted from the analysis. According to their geographic ranges, Bulgarian myriapods can be grouped into three zoogeographical complexes: Northern complex (89 taxa; ca. 40% of the considered species and subspecies), Southern complex (29 taxa; 13%), and species restricted to the Balkans (99 taxa; 45%). The Cosmopolitans (three taxa, ca. 1%) and the introduced species (two taxa, ca. 1%) are placed separately.

Table 2 Myriapods with uncertain identity or doubtful records from Bulgaria

Taxa	References
CHILOPODA	
Lithobius (*Lithobius*) *castaneus* Newport, 1844	Jurinich, 1904
Lithobius (*Sigibius*) *trebinjanus* Verhoeff, 1900	Attems, 1929a
Dicellophilus carniolensis (C.L. Koch, 1847)	Attems, 1929a
DIPLOPODA	
Glomeris conspersa (C.L. Koch, 1847)	Jurinich, 1904; Anonymous, 1907
Glomeris marginata (Villers, 1789) (sub *limbata*)	Jurinich, 1904
Glomeris pulchra C.L. Koch, 1847	Jurinich, 1904
Trachysphaera multiclavigera (Verhoeff, 1898)	Strasser, 1973
Polydesmus edentulus C.L. Koch, 1847	Jurinich, 1904
Polydesmus subscabratus Latzel, 1884	Jurinich, 1904
Strongylosoma stigmatosum balcanica Schubart, 1934	Schubart, 1934
Acanthopetalum carinatum (Brandt, 1840)	Ceuca, 1973
Dischizopetalum illyricum (Latzel, 1884)	Lang, 1958
Eurygyrus rufolineatus C.L. Koch, 1847	Jurinich, 1904
Haasea flavescens (Latzel, 1884)	Lang, 1958
Mastigophorophyllon saxonicum Verhoeff, 1916	Lang, 1958
Choneiulus palmatus (Nemec, 1895)	Lang, 1958
Leptoiulus trilineatus bureschi Verhoeff, 1928	Verhoeff, 1928; Verhoeff, 1937
Leptoiulus macedonicus (Attems, 1927)	Lang, 1935
Leptoiulus proximus (Nemec, 1896)	Lang, 1958
Cylindroiulus meinerti (Verhoeff, 1891)	Lang, 1958
Julus scandinavius Latzel, 1884	Lang, 1958
Julus terrestris Linnaeus, 1758	Christovich, 1892; Lang, 1958
Unciger foetidus (C.L. Koch, 1838)	Jurinich, 1904; Lang, 1958
Ophyiulus pilosus (Newport, 1842)	Jurinich, 1904; Anonymous, 1907; Lang, 1958
Megaphyllum austriacum (Latzel, 1884)	Jurinich, 1904
Megaphyllum platyurum (Latzel, 1884)	Jurinich, 1904
Megaphyllum projectus Verhoeff, 1894	Lang, 1958
Megaphyllum sjaelandicum (Meinert, 1868)	Lang, 1958

The Northern complex consists of species widespread in the Holarctic or Palearctic with all possible extensions toward Central Africa and adjacent parts of Asia. This complex includes species showing the following distribution patterns: Holarctic (HO), West Palearctic (WPA), Eurosiberian (ESI), European–Mediterranean–Turanian (EMT), Central Asian–European (CAE), European–Anatolian (EUAN), European (EU), Central European (CEU), East European (EEU), and Carpathian–Balkan (CPBK).

The Southern complex consists of species widespread in the Mediterranean region with some extensions towards the Central European, Turanian, and Pontian regions. This complex includes species showing the following distribution patterns: South European (SEU), Turanian–Mediterranean (TUM), Holomediterranean (MED), East Mediterranean (EME), Anatolian–Balkan (ANBK), and Pontian (PO).

Table 3 Species number in Bulgarian myriapod families

Families	Species number
CHILOPODA	102
Scutigeridae	1
Lithobiidae	66
Henicopidae	1
Scolopendridae	1
Cryptopidae	5
Himantariidae	4
Schendylidae	5
Geophilidae	11
Dignathodontidae	4
Linotaeniidae	4
DIPLOPODA	120
Polyxenidae	1
Lophoproctidae	1
Glomeridae	3
Doderiidae	3
Polydesmidae	23
Paradoxosomatidae	4
Trichopolydesmidae	2
Anthroleucosomatidae	12
Haaseidae	1
Mastigophorophyllidae	4
Chordeumatidae	1
Craspedosomatidae	1
Schizopetalidae	7
Dorypetalidae	1
Blaniulidae	5
Nemasomatidae	1
Julidae	48
Polyzonidae	1
Hirudisomatidae	1
PAUROPODA	18
Eurypauropodidae	1
Pauropodidae	17

The group of taxa with a Balkan distribution is very heterogeneous. It consists of several distribution types but depending on their origin, each of the taxa could also be assigned to either the Northern or Southern complex. The following main chorotypes are found among Balkan myriapods: Holobalkan (BK), East Balkan (EBK), Bulgarian (BG), endemics of Stara Planina (SPE), endemics of Rhodopes (RHE), and endemics of Strandja Mts. (STRE). Less numerous are the endemics of the Slavyanka (SLE), Pirin (PNE), Vitosha (VE) and Osogovo (OE) Mts., the Predbalkan (PBE), and the Danubian Plain (DPE). Another interesting pattern is found in *Polydesmus rubellus* which is currently known only from the island of Crete and Dobrudja.

Each chorotype and the myriapods assigned to it are briefly discussed below. Table 4 demonstrates the number of species assigned to each distribution pattern. The European chorotype (23 taxa) is the richest in species number, followed by the Central European (22 taxa), the endemics of Stara Planina and Rhodopes (each with 19), and the Carpathian–Balkan taxa (18). As a whole, the Balkan (s.l.) forms dominate in Bulgarian fauna, followed by those with generally northern and Mediterranean (s.l.) distribution. The introduced species and those with a (Sub)Cosmopolitan distribution have the smallest share (Table 5).

4·1 Holarctic species

Myriapods with this distribution pattern appear to be very rare in the Bulgarian fauna. Two members of Pauropoda, *Stylopauropus pedunculatus* and *Allopauropus brevisetus*, belong here. Both are rare in Bulgaria with no records since Krestewa (1940) and Remy (1945). The only milliped with a Holarctic distribution is *Polyxenus lagurus*.

4·2 Palearctic species

Here, we place all myriapods with West Palearctic, European–Mediterranean–Turanian, Eurosiberian, Central Asian–European and European–Anatolian distribution. Only two species, *Pachymerium ferrugineum* and *Pauropus furcifer*, are considered West Palearctic by distribution. The former is quite common in Bulgaria, while the latter has only few records. *Cryptops hortensis* and *Clinopodes flavidus* are widespread in Europe, with their ranges extending eastward and reaching the Turanian region. *Geophilus flavus, G. proximus*, and *Lithobius crassipes* occur in Siberia and Europe. The European–Anatolian chorotype is represented only by *Nopoiulus kochii*, and the Central Asian–European, by *Lithobius curtipes*.

4·3 European species

This group consists of 23 taxa, found throughout Europe. Some of them, like *Lithobius forficatus, L. microps*, and *Schendyla nemorensis*, are inclined to synanthropization and could be found in human-made habitats. In this way, they sometimes cross the continent's boundaries, although being undoubtedly of European origin. *Lithobius borealis* has been recorded only once in the Rhodope Mts. *Pauropus huxleyi* and *Allopauropus furcula* are also insufficiently known in Bulgaria.

Table 4 Chorotypes of Bulgarian myriapods

Chorotype	Number of species
Holarctic	3
West Palearctic	2
European–Mediterranean–Turanian	2
Eurosiberian	3
Central Asian–European	1
European–Anatolian	1
European	23
Central European	22
East European	14
Carpathian–Balkan	18
South European	2
Turanian–Mediterranean	1
Holomediterranean	9
East Mediterranean	8
Anatolian–Balkan	5
Pontian	4
Holobalkan	15
East Balkan	16
Bulgarian	9
Stara Planina endemic	19
Rhodope endemic	19
Pirin and Slavyanka endemic	4
Vitosha endemic	2
Osogovo endemic	1
Strandja endemic	6
Predbalkan endemic	5
Danubian Plain endemic	2
Crete–Dobrudja	1
Introduced species	2
Cosmopolitans	3

Table 5 Main zoogeographical complexes of Bulgarian myriapods

Complex	Number of species
Northern complex	89
Southern complex	29
Balkan (s.l.) complex	99
Cosmopolitans	3
Introduced species	2
Total	*222*

4·4 Central European species

Their origin is linked to the nemoral broad-leaved forest of Central Europe. During the Quaternary glaciations most of these taxa are believed to have reached southern Europe, where they have survived in the mesophilous broad-leaved forest of *Fagus, Betula, Alnus, Acer, Carpinus*, etc. They are mostly confined to the middle belts of the Bulgarian mountains with extensions toward the orophyte zone and the lowlands. Currently, 22 species with Central European distribution are known from Bulgaria. Four of them, *Lithobius dentatus, L. peggauensis, L. tricuspis*, and *Strigamia engadina*, have so far been recorded in Bulgaria only once. The rest are more or less widespread.

4·5 East European species

Fourteen species of Bulgarian myriapods are found only in East Europe. Some of them have wider ranges, including the Caucasus, Central Asia, or Central Europe. While most of them are widespread, *Cryptops rucneri* and *Rossiulus vilnensis* have been recorded in Bulgaria only occasionally.

4·6 Carpathian–Balkan species

Eighteen species are distributed only in the Carpathian Mts. and the Balkan Peninsula. Some of them, like *Lithobius pustulatus, Harpolithobius radui, Trachysphaera orghidani*, and *Typhloiulus strictus*, are found only in the Carpathians and the mountains of Bulgaria, while others (*Lithobius parietum, Eupolybothrus transsylvanicus, Melogona broelemanni*, and *Pachyiulus hungaricus*) are more or less widespread in the Balkans. *Leptoiulus trilineatus* and *Cylindroiulus boleti* have wider ranges beyond the Carpathians, reaching the Alps in the west.

4·7 Mediterranean species

The group includes all species with Holomediterranean, Turanian–Mediterranean, and East Mediterranean distribution. The largest share belongs to the Holomediterranean forms (nine taxa), followed by East Mediterranean (eight) and Turanian–Mediterranean (one). *Eupolybothrus litoralis, Pleurolithobius patriarchalis, Dignathodon microcephalus*, and *Bothriogaster signata* are known only from southern Bulgaria, not surpassing the southern slopes of the Stara

Planina Mts. *Lithobius peregrinus, L. catascaphius, L. micropodus, Henia bicarinata, Trachysphaera rotundata, Polydesmus mediterraneus oertzeni, Pachyiulus varius, Allopauropus aristatus*, and *A. corsicus* are poorly known in Bulgaria. *Scutigera coleoptrata* is a common synanthrope, which also inhabits natural habitats such as caves, rock crevices, and is found under stones.

4·8 Anatolian–Balkan species

Some forms originating in Anatolia extend their ranges toward the southeastern Balkan Peninsula (Strandja Mts., Thrace) and to some of the Aegean islands. Out of five species belonging here, only *Henia porosa* and *Cylindroiulus abaligetanus* are taxonomically better known. The systematics and distribution of *Lithobius ergus, L. rushovensis*, and *L. jurinici* require further investigation.

4·9 Balkan species

These species originate from the Balkan Peninsula, rarely extending their ranges beyond its boundaries. The Holobalkan distribution is demonstrated by 15 taxa, most of them very common in Bulgaria. *Lithobius stygius* and *Thracophilus bulgaricus* are known only from Bulgaria and the western Balkans, with no records from the central part of the Balkan Peninsula.

4·10 East Balkan species

This category includes all species that occur in Bulgaria and in at least one of the neighboring countries: Greece, Turkey, Republic of Macedonia, Serbia, and Romania. No less than 16 taxa are considered East Balkan; six of them belong to *Lithobius* (s.l.). Some are certainly Bulgarian by origin. *Lithobius borisi, L. dobrogicus, L. beroni, Allopauropus humilis*, and *A. rostratus* are still insufficiently known.

4·11 Species found only in Bulgaria

Nine species belong here. Their origin is most probably connected with some of the main mountain massifs in the country (Stara Planina or Rila–Rhodopes), from where they have later dispersed to other regions. Some of them, like *Geophilus rhodopensis, Mastigophorophyllon bulgaricum, Leptoiulus borisi*, and *Apfelbeckiella trnowensis*, are more or less widespread. Further research may reveal their presence in other Balkan countries as well.

4·12 Stara Planina endemics

Stara Planina is one of the main mountain ranges in Bulgaria. It stretches across the country from the western border to the Black Sea coast. The origin of 19 myriapods is linked with this mountain range, thus proving its unique position and significance as a center of active speciation. Among these, Julidae are represented by seven taxa, five of which belong to *Typhloiulus* (s.l.), and Polydesmidae, by four. The endemics of the Predbalkan (5), a region located close to Stara Planina and having several natural junctions with it, should also be added to this chorotype. Among those, special attention should be paid to *Bulgardicus*, a monotypic genus restricted to a cave near Karlukovo. About 70–80% of all Stara Planina endemics are species restricted to, and originating from, the caves of Western and Central Stara Planina and the karst of the Predbalkan.

4·13 Rhodopean endemics

The Rhodope Mts. stretch between Bulgaria and Greece, forming one of the largest and well-defined mountain massifs in the eastern part of the Balkan Peninsula. It is surrounded by, and demonstrates geological connections with, Rila, Pirin, and Slavyanka Mts. and the region of Kraishte. The list of myriapods with a Rhodopean distribution and origin comprises at least 19 species. The genera *Stygiosoma*, *Rhodoposoma*, and *Troglodicus* (Anthroleucosomatidae) characterize the myriapod fauna of this mountain range. The genus *Balkanopetalum* (Schizopetalidae) also seems to originate from the Rhodopes; only two out of six species of this genus are known outside this massif. Almost half of all endemic species are cave-dwellers restricted to the karst territories of the Rhodopes. The regions of Dobrostan, Peshtera–Velingrad, and Trigrad are the richest in endemics. The pauropods *Allopauropus krestewaae* and *A. moesiacus* have been described from the Rhodopes, but their taxonomic status needs reevaluation.

4·14 Pirin and Slavyanka endemics

Situated in southwestern Bulgaria, Pirin and Slavyanka Mts. are linked by the Paril Pass. *Xestoiulus pirinicus* and *Mastigophorophyllon bulgaricum pirinicum* are the only endemic myriapods discovered in Pirin. *Anamastigona delcevi* is a species restricted to Slavyanka (Orvilos). *Balkanopetalum bulgaricum* is known only from a cave in southern Pirin and superficially in Slavyanka. Since this region is still unexplored, many more interesting species are to be expected in the future.

4·15 Vitosha endemics

Vitosha Mts. are one of the highest mountain massifs in Bulgaria. Despite that, only two endemic myriapods, *Brachydesmus cristofer* and *Bulgarosoma superficiei*, hitherto have been described from there. *Lithobius totevi*, a species with an uncertain taxonomic status (Stoev, 2002), should also be added to that group.

4·16 Osogovo endemics

Despite their large territory, the Osogovo Mts. do not seem to support a high number of endemics. So far, only *Brachydesmus herzogowinensis trifidus* has been described from this massif and was not found outside it. *Lithobius glaciei* is another local endemic, but its taxonomic status is not clear.

4·17 Strandja endemics

Six endemic myriapods are known from Strandja Mts., which stretch between Bulgaria and Turkey. Two of them, *Lithobius oglednicus* and *L. strandzanicus*, are still poorly known, and further studies may lead to their synonymization. *L. bifidus* and *Apfelbeckiella bulgarica* are cave-dwellers; the latter reported from epigean habitats as well (Kondeva, 1996).

4·18 Other distribution patterns

Only two species, *Haasea vidinensis* and *Apfelbeckiella trnowensis deliormana*, are considered endemics of the Danubian Plain, where both inhabit caves. *Schendyla delicatula, S. varnensis*, and *Megaphyllum anatolicum denticulatum* have been described from the Bulgarian Black Sea coast with no further records outside this region. *Cylindroiulus bellus* is another species whose origin is probably connected with a territory around the Black Sea and could be categorized as a Pontian element. The South European species *Schendyla walachica* and *S. montana* are widespread in the Mediterranean region, being found also in countries like Romania and Austria. A very unusual distribution pattern has *Polydesmus rubellus*, which was described from Crete and later found in Bulgarian Dobrudja (Strasser, 1973).

4·19 (Sub)Cosmopolitan species

Only three species could be assigned to this category: *Lamyctes emarginatus, Allopauropus danicus*, and *A. gracilis*. Their distribution and biology in Bulgaria are not sufficiently known; only few records exist for each species.

4·20 Introduced species

This group includes only *Oxidus gracilis* and *Eurygyrus ochraceus*. The former was recently discovered in the Botanical Garden of the University of Sofia (Stoev,

2004b); the latter is known from the gardens of Euxinograd (Former Royal Residence), where it was probably introduced with plants or soil from outside Bulgaria. The genus *Eurygyrus* includes about a dozen species distributed in Anatolia, some of the Greek islands, and Peloponnesus.

5 Ecological Aspects

5·1 Cave species

Bulgarian cave fauna is one of the richest in the Balkan Peninsula, surpassed only by that of the Dalmatian karst. About 100 terrestrial troglobite species have been recorded from Bulgaria (Beron, 1994), and this number is still expected to increase very fast due to the recent intensive biospeleological investigations. About 68 taxa of millipeds, 43 centipeds, and one pauropod are currently known from the Bulgarian caves (Beron, 1994; Stoev, 2001b, 2004a). Out of these 112 cave-inhabiting myriapod taxa, 27 (ca. 24%) are considered strict cave-dwellers (= troglobites). This accounts for approximately 11% of all Bulgarian myriapods.

5·2 High-altitude species

Several species have been recorded from the subalpine and alpine zones of the Bulgarian mountains, but with almost no exception all were found in the lower, forest belts as well. Only *Lithobius electron, L. glaciei*, and *L. borisi* have been described from the altitudes above 2000 m a.s.l. in the Rhodopes, Osogovo, and Pirin Mts. respectively, and were not collected again. Since their taxonomical status is yet to be established (Stoev, 2002), at this moment they are not characterized as true high-altitude species. As a result, no real subalpine or alpine species have been found so far among the Bulgarian myriapods. Nevertheless, such elements could probably be discovered with more intensive faunistic research.

6 Main Gaps in Knowledge

The lack of taxonomists dealing with myriapod systematics is the main reason for an insufficient knowledge of the Bulgarian myriapod fauna. However, compared to the neighboring Republic of Macedonia, Albania, Montenegro, and Turkey, Bulgarian myriapods are much better studied. The Danubian Plain, Ludogorie, the Black Sea coast, Maleshevska Mts., Vlahina Mts., Ograzhden, Slavyanka, Osogovo, and Belasitsa are less explored areas of the country. Only Sredna Gora, Western Rhodopes, and Strandja–Sakar region are better studied. Due to the efforts of three generations of biospeleologists, the cave-dwelling myriapods are faunistically and systematically better known than other ecological types of Myriapoda. Meanwhile, the high diversity of caves and other karst phenomena are still not sufficiently explored and future investigations will certainly increase the number of myriapods from these habitats. A large collection of still unidentified specimens of Chilopoda, Diplopoda, and Symphyla is currently preserved in the collection of the National Museum of Natural History in Sofia and awaits taxonomical treatment.

Acknowledgments

I thank Prof. Alessandro Minelli (Padova, Italy) and Prof. Henrik Enghoff (Copenhagen, Denmark) for their valuable comments on the earlier stages of preparation of the manuscript.

References

Anonymous, 1907. Myriapoda. In: Anon., Collections du Musée d'histoire naturelle de Son Altesse Royale Ferdinand I, Prince de Bulgarie. Imprimerie de L'État, Sofia: 397.

Attems, C., 1904. Neue paläarktische Myriopoden nebst Beiträgen zur Kenntnis einiger alter Arten. Archiv für Naturgeschichte (Berlin) 1(2):179–196.

Attems, C., 1929a. Die Myriopodenfauna von Albanien und Jugoslavien. Zoologische Jahrbücher, Abteilung für Systematik 56:270–306.

Attems, C., 1929b. Myriapoda 1. Geophilomorpha. In: Das Tierreich. Eine Zusammenstellung und Kennzeichnung der rezenten Tierformen. 52. Walter de Gruyter & Co., Berlin & Leipzig.

Attems, C., 1959. Die Myriopoden der Höhlen der Balkanhalbinsel. Annalen des naturhistorischen Museums in Wien 63:281–406.

Beron, P., 1972. Essai sur la faune cavernicole de Bulgarie. III. Résultats des recherches biospéologiques de 1966 à 1970. International Journal of Speleology 4:285–349.

Beron, P., 1978. Aperçu sur la composition, l'origine et la formation de la faune cavernicole de la Stara Planina occidentale (Bulgarie). International Journal of Speleology 9:197–220.

Beron, P., 1994. Résultats des recherches biospéléologiques en Bulgarie de 1971 à 1994 et liste des animaux cavernicoles bulgares. Tranteeva (Sofia) 1:1–137.

Beron, P., 1999. Biodiversity of the high mountain terrestrial fauna in Bulgaria. Historia naturalis bulgarica 10:13–33.

Beron, P., 2001. Etude comparative des faunes cavernicoles de la Bulgarie et de la Grèce. Historia naturalis bulgarica 13:53–68.

Beron, P. & V. Guéorguiev, 1967. Essai sur la faune cavernicole de Bulgarie. II. Résultats des recherches biospéologiques de 1961 à 1965. Izvestiya na Zoologicheskiya institut s muzei (Bulletin de l'Institut de zoologie et musée) 24:151–212.

Beron, P., B. Petrov & P. Stoev, 2004. The invertebrate cave fauna of the Eastern Rhodopes (Bulgaria and Greece). In Beron, P. & A. Popov (eds), Biodiversity of Bulgaria. 2. Biodiversity of Eastern Rhodopes (Bulgaria and Greece). Pensoft & National Museum of Natural History, Sofia: 791–822.

Ceuca, T., 1973. Contribution à la connaissance des diplopodes (Myriapoda, Diplopoda) de Bulgarie. Izvestiya na Zoologicheskiya institut s muzei (Bulletin de l'Institut de zoologie et musée) 38:241–247.

[Christov, J.] Christow, J., 1983. [Über die Fauna der Diplopoden im Zentralen Mittelwaldgebirge]. Nauchni trudove na Plovdivskiya universitet "Paisii Hilendarski" (Travaux scientifiques de l'Université de Plovdiv "Paisii Hilendarski") 21(4):39–43 (in Bulgarian).

Christov, J., 1984a. [Beitrag zur Fauna der Diplopoda (Myriapoda) im Zentralen Mittelwaldgebirge]. Nauchni trudove na Plovdivskiya universitet "Paisii Hilendarski" (Travaux scientifiques de l'Université de Plovdiv "Paisii Hilendarski") 22(1):91–94 (in Bulgarian).

[Christov, J.] Christow, J., 1984b. [Die Verteilung der Diplopoda (Myriapoda) in einer Formation vom Querceta delechampii in Zentralen Sredna Gora Gebirge]. Nauchni trudove na Plovdivskiya universitet "Paisii Hilendarski" (Travaux scientifiques de l'Université de Plovdiv "Paisii Hilendarski") 22(1):95–101 (in Bulgarian).

Christov, J., 1985. [Senkrechte Verteilung und Saisondynamik der Diplopoden von der Formation Fageta silvaticae im mittleren Teil des Sredna Gora Gebirges]. Nauchni trudove na Plovdivskiya universitet "Paisii Hilendarski" (Travaux scientifiques de l'Université de Plovdiv "Paisii Hilendarski") 23(1):89–96 (in Bulgarian).

Christov, J., 1986a. [Zusammensetzung der Arten, Zahl, Biomasse und Raumverteilung der Diplopoden von der Formation Fageta silvaticae im mittleren Teil des Sredna Gora Gebirges]. Nauchni trudove na Plovdivskiya universitet "Paisii Hilendarski" (Travaux scientifiques de l'Université de Plovdiv "Paisii Hilendarski") 24(1):161–168 (in Bulgarian).

Christov, J., 1986b. Vidov sastav i razprostranenie na mnogonozhkite ot klas Diplopoda (Myriapoda) v Sredna gora s tsenologichni izsledvaniya v Sashtinska Sredna gora. [Species composition and distribution of the millipeds of class Diplopoda (Myriapoda) in the Sredna Gora Mountains with coenological investigations in the Sashtinska Sredna Gora Mountains]. Doktorska disertatsiya [PhD Thesis]. Plovdivski universitet "Paisii Hilendarski" [Paisii Hilendarski University of Plovdiv], Plovdiv (in Bulgarian).

Ćurčić, B.P.M. & S. Makarov, 2000. A revision of some species of *Bulgarosoma* Verhoeff (Antroleucosomatidae, Myriapoda) from the Balkan peninsula. Arhiv bioloških nauka [Archive of Biological Sciences] (Belgrade) 52(3):173–178.

Deltshev, Ch., S. Andreev, G. Blagoev, V. Golemansky, G. Milojkova, V. Peneva, D. Dobrev, M. Todorov, M. & Z. Hubenov, 1998. Invertebrates (non-Insecta) in Bulgaria. In Meine, C. (ed), Bulgaria's Biological Diversity: Conservation Status and Needs Assessment. Volumes I and II. Pensoft, Sofia & Moscow: 109–161.

Deltshev, Ch., P. Beron, G. Blagoev, V. Golemansky, V. Peneva, P. Stoev, P., M. Todorov & Z. Hubenov, 2000a. Faunistic diversity of invertebrates (non Insecta) of the Rila National Park. In Sakalian, M. (ed.). Biological Diversity of the Rila National Park. Pensoft, Sofia: 249–284, 521–522.

Deltshev, Ch., P. Beron, G. Blagoev, V. Golemansky, V. Peneva, P. Stoev, P., M. Todorov & Z. Hubenov, 2000b. Faunistic diversity of invertebrates (non Insecta) in Central Balkan National Park. In Sakalian, M. (ed), Biological Diversity of the Central Balkan National Park. Pensoft, Sofia: 289–317, 535–536.

Demange, J.-M., 1961. Un nouveau *Lithobius* cavernicole de Roumanie. Annales de Spéléologie 16(2):179–182.

Djursvoll, P., S. Golovatch, K. Johanson & B. Meidell, 2000. Phylogenetic relationships within *Polydesmus* sensu lato (Diplopoda: Polydesmidae). Fragmenta Faunistica (Warszawa) 43(Suppl.): 37–59.

Dobroruka, L., 1977. Bulgarische Chilopoden in der Sammlung von Nationalmuseum Praha. Věstnik Československé společnosti zoologické 41(1):5–7.

Eason, E.H., 1970. On certain characters used to separate species and subspecies in the Lithobiidae. Bulletin du Muséum national d'histoire naturelle, 2-e sér., 41(Suppl. 2):58–60.

Eason, E.H., 1983. The identity of the European and Mediterranean species of Lithobiidae (Chilopoda) described by K.W. Verhoeff and now represented by material preserved in the British Museum (Natural History). Zoological Journal of the Linnean Society 77:111–144.

Enghoff, H., 1985. Modified mouthparts in hydrophilous cave millipedes (Diplopoda). Bijdragen tot de Dierkunde 55(1):66–77.

Enghoff, H., 2004. Myriapoda. In: Fauna Europaea. Version 1.1. http://www.faunaeur.org.

Folkmanova, B., 1936. Über einige von Dr. Jaroslav Štorkan in Bulgarien gesammelte Chilopoden. Izvestiya na Tsarskite prirodonauchni instituti v Sofia (Mitteilungen aus den Königlichen naturwissenschaftlichen Instituten in Sofia – Bulgarien; Bulletin des Institutions royales d'histoire naturelle à Sofia – Bulgarie) 9:93–97.

Golemansky, V., 1973a. *Stenophora beroni* n.sp. (Gregarinida: Stenophoridae) – une nouvelle eugrégarine du diplopode troglophile *Balkanopetalum armatum* Verhoeff (Diplopoda: Lysiopetalidae). Zoologischer Anzeiger 191(1/2):151–156.

Golemansky, V., 1973b. *Stenophora bulgarosomae* n.sp. – une nouvelle grégarine (Gregarinida: Stenophoridae) de diplopode troglobie *Bulgarosoma bureschi* Verhoeff. Zoologischer Anzeiger 190(3/4):159–166.

Golemansky, V. & J. Lipa, 1991. Gregarines (Apicomplexa: Eugregarinida) from cave and terrestrial arthropods in Bulgaria. Acta Protozoologica 30:201–208.

Golemansky, V. & V. Taschev, 1973. Recherches sur les grégarines (Gregarinida: Stenophoridae) du diplopode troglobie *Typhloiulus bureschi* Verhoeff en Bulgarie. International Journal of Speleology 5:87–94.

Golovatch, S. & E. Kondeva, 1992. Contribution to the millipede fauna of Bulgaria (Diplopoda). Acta zoologica bulgarica 44:19–26.

Golovatch, S. & P. Stoev, 2004. A new species of the millipede genus *Metonomastus* Attems, 1937 from a cave in the Eastern Rhodopes (Bulgaria), with some remarks on its congeners (Diplopoda: Paradoxosomatidae). In Beron, P. & A. Popov (eds), Biodiversity of Bulgaria. 2. Biodiversity of Eastern Rhodopes (Bulgaria and Greece). Pensoft & National Museum of Natural History, Sofia: 199–205.

Guéorguiev, V. & P. Beron, 1962. Essai sur la faune cavernicole de Bulgarie. Annales de Spéléologie 17(2/3):285–441.

Guéorguiev, V., Ch. Deltshev & V. Golemansky, 1994. Bulgarie. In Juberthie, C. & V. Decu (eds), Encyclopaedia Biospeologica. Vol. 1. Société de Biospéologie, Moulis & Bucharest: 619–629.

Gulička, J., 1967a. Neue und interessante Diplopoden aus Bulgarien. Annotationes zoologicae et botanicae (Bratislava) 39:1–9.

Gulička, J., 1967b. Neue Höhlen-Diplopoden aus Bulgarien. Annotationes zoologicae et botanicae (Bratislava) 42:1–5.

Hristovich, G., 1892. [Materials for the exploration of the Bulgarian fauna]. Sbornik za narodni umotvoreniya, nauka i knizhnina [A Collection of Folklore, Science and Literary Works] 8:337–346 (in Bulgarian).

Jawlowski, H., 1938. Über einige interessantere Diplopoden-Arten aus Bulgarien. Annales Musei Zoologici Polonici 13(14):167–173.

Jurinich, S., 1904. [A contribution to the Bulgarian fauna of Myriapoda]. Sbornik za narodni umotvoreniya, nauka i knizhnina [A Collection of Folklore, Science and Literary Works] 20:1–44 (in Bulgarian).

Kaczmarek, J., 1969a. Beiträge zur Kenntnis bulgarischer Chilopoden. Teil I. Bulletin de la Société des amis des sciences et des lettres de Poznan, Sér. D, 9:263–277.

Kaczmarek, J., 1969b. Beiträge zur Kenntnis bulgarischer Chilopoden. Teil II. Die Familie Schendylidae. Bulletin de la Société des amis des sciences et des lettres de Poznan, Sér. D, 10:99–109.

Kaczmarek, J., 1970. Beiträge zur Kenntnis bulgarischer Chilopoden. Teil III. Bulletin de la Société des amis des sciences et des lettres de Poznan, Sér. D, 11:81–89.

Kaczmarek, J., 1972. Beiträge zur Kenntnis bulgarischer Chilopoden. Teil IV. Bulletin de la Société des amis des sciences et des lettres de Poznan, Sér. D, 12/13:261–264.

Kaczmarek, J., 1973. Beiträge zur Kenntnis bulgarischer Chilopoden. Teil V. Das Genus *Polybothrus*. Bulletin de la Société des amis des sciences et des lettres de Poznan, Sér. D, 14:181–191.

Kaczmarek, J., 1975. Beiträge zur Kenntnis bulgarischer Chilopoden. Teil VI (Lithobiomorpha: *Harpolithobius, Pleurolithobius, Monotarsobius*). Annales Zoologici (Warszawa) 33(4):47–66.

Kondeva, E., 1996. Millipedes (Diplopoda) in mixed *Quercus–Carpinus* and *Fagus–Rhododendron* forests in the northern parts of Strandja Mountain (Bulgaria). In Proceedings. Scientific Papers. Second Balkan Scientific Conference on Study, Conservation and Utilisation of Forest Resources (3–5 June 1996), Sofia. Volume 2. PSSA, Sofia: 128–131 (in Bulgarian).

Kondeva, E., 1998. Consumption and assimilation of beech leaves *Fagus sylvatica* L. by millipedes *Pachyiulus hungaricus* (Karsch, 1881) (Diplopoda: Julidae). Acta zoologica bulgarica 50(1):89–92.

Kondeva, E., 2000. Phenology of litter-inhabiting millipedes (Diplopoda) in monitoring mixed oak and oak–hornbeam forests near Bourgas (Bulgaria). Acta zoologica bulgarica 52(1):89–94.

Kondeva, E., 2002. Millipedes (Diplopoda) in litter of xerothermic oak forests in Bulgaria. Journal of Balkan Ecology, 5(4):417–421.

Kondeva, E., 2004. Millipedes on Lozen Mountain. Seasonal activity and migrations in typical plant communities of Bulgaria. Journal of Balkan Ecology, 7(3):305–308.

Krestewa, P., 1940. [Pauropoda in Bulgaria]. Izvestiya na Balgarskoto entomologichno druzhestvo (Mitteilungen der Bulgarischen Entomologischen Gesellschaft in Sofia; Bulletin de la Société Entomologique de Bulgarie) 11:161–202 (in Bulgarian).

Lang, J., 1935. Über einige von Doz. Dr. Jar. Štorkán in Bulgarien gesammelte Diplopoden. Izvestiya na Tsarskite prirodonauchni instituti v Sofia (Mitteilungen aus den Königlichen naturwissenschaftlichen Instituten in Sofia – Bulgarien; Bulletin des Institutions royales d'histoire naturelle à Sofia – Bulgarie) 8:178–180.

Lang, J., 1958. Diplopoda z bulharských sberu Dr. M. Kunsta. Věstník Československé společnosti zoologické 22(1):34–44.

Matic, Z., 1964. Description d'un nouveau Lithobiidae cavernicole de Bulgarie (Chilopoda). Annales de Spéléologie, 19(3):507–510.

Matic, Z., 1967. Contributions à la connaissance des chilopodes cavernicoles de Bulgarie. Izvestiya na Zoologicheskiya institut s muzei (Bulletin de l'Institut de zoologie et musée) 23:235–242.

Matic, Z., 1973a. *Lithobius balcanicus* n. sp. – une nouvelle lithobiide (Chilopoda, Lithobiomorpha) de Bulgarie. Izvestiya na Zoologicheskiya institut s muzei (Bulletin de l'Institut de zoologie et musée) 38:249–251.

Matic, Z., 1973b. Nouvelles contributions à la connaissance des chilopodes cavernicoles de Bulgarie. Izvestiya na Zoologicheskiya institut s muzei (Bulletin de l'Institut de zoologie et musée) 38:253–263.

Matic, Z. & C. Darabanţu, 1974. Contributions à la connaissance du genre *Thracophilus* Verhoeff (Geophilomorpha, Himantariidae). Izvestiya na Zoologicheskiya institut s muzei (Bulletin de l'Institut de zoologie et musée) 40:93–98.

Matic, Z. & V. Golemansky, 1964. Contribution à la connaissance des lithobiidés (Chilopoda, Lithobiomorpha) en Bulgarie. Godishnik na Sofiiskiya universitet "Kliment Ohridski". Biologicheski fakultet (Annuaire de l'Université de Sofia "Kliment Ohridski", Faculté de Biologie) 57(1 – Zoology):99–106.

Matic, Z. & V. Golemansky, 1965. Nouvelle contribution à la connaissance des lithobiidés (Chilopoda, Lithobiomorpha) en Bulgarie. Godishnik na Sofiiskiya universitet "Kliment Ohridski". Biologicheski fakultet (Annuaire de l'Université de Sofia "Kliment Ohridski", Faculté de Biologie) 58(1 – Zoology):13–28.

Matic, Z. & V. Golemansky, 1967a. Le sous-genre *Monotarsobius* (Chilopoda, *Lithobius*) en Bulgarie. Izvestiya na Zoologicheskiya institut s muzei (Bulletin de l'Institut de zoologie et musée) 24:39–50.

Matic, Z. & V. Golemansky, 1967b. Recherches sur les espèces et l'écologie des lithobiidés (Chilopoda, Lithobiomorpha) en Bulgarie. Izvestiya na Zoologicheskiya institut s muzei (Bulletin de l'Institut de zoologie et musée) 24:121–132.

Matic, Z. & V. Golemansky, 1967c. Matériaux sur la faune des lithobiidés (Chilopoda, Lithobiomorpha) en Bulgarie. Izvestiya na Zoologicheskiya institut s muzei (Bulletin de l'Institut de zoologie et musée) 25:17–21.

Matic, Z. & V. Golemansky, 1969. Scolopendromorphes (Chilopoda, Scolopendromorpha) de la faune de Bulgarie. Izvestiya na Zoologicheskiya institut s muzei (Bulletin de l'Institut de zoologie et musée) 29:167–173.

Negrea, S., 1965. Contributions à l'étude des Lithobiidae (Chilopoda) de Bulgarie. Fragmenta balcanica 5(14):91–104.

Negrea, S., 1971. *Lithobius* (*Monotarsobius*) *taschevi* Matic & Golemansky est un synonyme de *Lithobius* (*Monotarsobius*) *beroni* Negrea (Chilopoda, Lithobiomorpha). Travaux de l'Institut de Spéologie "Emile Racovitza" 10:231–233.

Popov, A., 1969. [La faune cavernicole dans la région du v. Gabare, distr. Vraca]. Godishnik po speleologiya i peshterno delo (Annuel de Spéléologie) (Sofia), 1 [1968], 36–40:43 (in Bulgarian).

Remy, P., 1945. Révision des pauropodes récoltés en Bulgarie par Mlle P. Krestewa. Archives de zoologie experimentale et generale 84, N et R:78–99.

Ribarov, G., 1984. [Est-ce qu'on peut rencontrer en Bulgarie *Scolopendra morsitans* (L.)?] Izvestiya na muzeite ot Yugoiztochna Balgariya (Bulletin des musées de la Bulgarie du sud-est) 7:273–276 (in Bulgarian).

Ribarov, G., 1985. Vidov sastav, ekologiya, razprostranenie i znachenie na mnogonozhkite ot klas Chilopoda v Yugoiztochna Balgariya [Species Composition, Ecology, Distribution and Significance of the Centipeds (Chilopoda) in Southeastern Bulgaria]. Doktorska disertatsiya [PhD Thesis].

Plovdivski universitet "Paisii Hilendarski" [Paisii Hilendarski University of Plovdiv], Plovdiv (in Bulgarian).

Ribarov, G., 1986a. [Contribution à la connaissance de Lithobiomorpha (Chilopoda) en Bulgarie sud-est]. Izvestiya na muzeite ot Yugoiztochna Balgariya (Bulletin des musées de la Bulgarie du sud-est) 8:217–226 (in Bulgarian).

Ribarov, G., 1986b. [A contribution to the study of Geophilomorpha (Chilopoda) from southeastern Bulgaria]. Nauchni trudove na Plovdivskiya universitet "Paisii Hilendarski" (Travaux scientifiques de l'Université de Plovdiv "Paisii Hilendarski") 24(1):155–160 (in Bulgarian).

Ribarov, G., 1987a. [Species list, ecology, distribution and importance of Lithobiomorpha (Chilopoda) from the Strandja and Sakar Mountains]. Sbornik Strandja–Sakar [Strandja–Sakar Collection] 4:279–287 (in Bulgarian).

Ribarov, G., 1987b. Five new species of Lithobiomorpha (Chilopoda) from southeastern Bulgaria. Acta zoologica bulgarica 34:45–52.

Ribarov, G., 1987c. *Henia angelovi* sp. n. (Chilopoda, Geophilidae) – eine neue*Henia*-Art aus Südostbulgarien. Acta zoologica bulgarica 35:86–89.

Ribarov, G., 1989a. On some Chilopoda species (Myriapoda) published by S. Jurinich and K. Verhoeff. Historia naturalis bulgarica 1:34–35.

Ribarov, G., 1989b. [A contribution to the study of centipedes (Chilopoda) from Sredna Gora Mountain]. Nauchni trudove na Plovdivskiya universitet "Paisii Hilendarski" (Travaux scientifiques de l'Université de Plovdiv "Paisii Hilendarski") 27(6):165–184 (in Bulgarian).

Ribarov, G., 1990. Chilopoda new to the science or to the Bulgarian fauna. In Minelli, A. (ed), Proceedings of 7th International Congress of Myriapodology. Brill, Leiden: 429.

Ribarov, G., 1992. The distribution of the centipedes of the genera *Harpolithobius* and *Eupolybothrus* in Bulgaria (Chilopoda, Lithobiomorpha). Berichte des naturwissenschaftlich-medizinischen Vereins, Innsbruck 10(Suppl.):361–365.

Ribarov, G., 1996. Check-list, distribution and habitat in Bulgarian centipedes. In Geoffroy, J.-J., J.P. Mauriès & M. Nguyen Duy-Jaquemin (eds), Acta Myriapodologica. Mémoires du Museum national d'histoire naturelle 169:235–241.

Scheller, U., 1977. The Pauropoda and Symphyla of the Geneva Museum IV. A basic list of the Pauropoda of Greece. Revue suisse de Zoologie 84(2):361–408.

Scheller, U., 1981. The Pauropoda and Symphyla of the Geneva Museum VIII. Pauropoda collected in Greece by Dr. Bernd Hauser. Revue suisse de Zoologie 88(1):101–107.

Schubart, O., 1934. Über einige von Dr. Rensch in Bulgarien gesammelte Diplopoden. Izvestiya na Tsarskite prirodonauchni instituti v Sofia (Mitteilungen aus den Königlichen naturwissenschaftlichen Instituten in Sofia – Bulgarien; Bulletin des Institutions royales d'histoire naturelle à Sofia – Bulgarie) 7:36–50.

Stoev, P., 1997a. *Lithobius christovici* Matic et Golemansky, 1964 – a junior synonym of *Lithobius ergus* (Chamberlin, 1952) (Chilopoda: Lithobiomorpha). Godishnik na Sofiiskiya universitet "Sv. Kliment Ohridski". Biologicheski fakultet (Annuaire de l'Université de Sofia "St. Kliment Ohridski". Faculté de Biologie) 88(4):111–113.

Stoev, P., 1997b. A check-list of the centipedes of the Balkan Peninsula with some taxonomic notes and a complete bibliography (Chilopoda). Entomologica scandinavica 51(Suppl.):87–105.

Stoev, P., 2001a. [Myriapoda (Chilopoda, Diplopoda) of the Kresna Gorge]. In Beron, P. (ed), Bioraznoobrazie na Kresnenskiya prolom (Biodiversity of Kresna Gorge (SW Bulgaria)). Natsionalen prirodonauchen muzei i Institut po zoologiya pri BAN (National Museum of Natural History (Sofia) & Institute of Zoology, Bulgarian Academy of Sciences), Sofia: 103–107 (in Bulgarian).

Stoev, P., 2001b. A synopsis of Bulgarian cave centipedes (Chilopoda). Arthropoda Selecta 10(1):31–54.

Stoev, P., 2002. A Catalogue and Key to the Centipedes (Chilopoda) of Bulgaria. Pensoft, Sofia & Moscow, Series Faunistica No 25.

Stoev, P., 2003. [Myriapods (Chilopoda, Diplopoda) of the Rila Monastery Natural Park]. In Peev, D. (ed), Barza ekologichna otsenka na Priroden park "Rilski manastir" (A Rapid Ecological Assessment of Rila Monastery Nature Park). Amerikanska agentsiya za mezhdunarodno razvitie (United States Agency for International Development), Sofia: 134–138 (in Bulgarian).

Stoev, P., 2004a. New distributional records of millipedes from Bulgarian caves (Myriapoda: Diplopoda). Acta zoologica bulgarica 56(2): 145–154.

Stoev, P., 2004b. Myriapoda (Chilopoda, Diplopoda) in urban environments in the City of Sofia. In Penev, L., D.J. Niemelä, D. Kotze & N. Chipev (eds), Ecology of the City of Sofia. Species and Communities in Urban Environment. Pensoft, Sofia: 299–306.

Stoev, P., 2004c. The myriapods (Chilopoda, Diplopoda) of the Eastern Rhodopes (Bulgaria and Greece). In Beron, P. & A. Popov (eds), Biodiversity of Bulgaria. 2. Biodiversity of Eastern Rhodopes (Bulgaria and Greece). Pensoft & National Museum of Natural History, Sofia: 207–220.

Stoev, P., 2005. On the identity of some poorly known lithobiid centipedes described by Karl Verhoeff (Chilopoda: Lithobiomorpha). Zootaxa 796:1–12.

Stoev, P. & P. Beron, 2001. On the type specimens of Myriapoda (Chilopoda, Diplopoda) in the collection of the National Museum of Natural History in Sofia. Arthropoda Selecta 9(2) [2000]:91–102.

Stoev, P. & H. Enghoff, 2003. Systematics, phylogeny and biogeography of genus *Balkanopetalum* Verhoeff, 1926 (Diplopoda: Callipodida: Schizopetalidae). Zootaxa 272:1–26.

Stoev, P. & A. Lapeva-Gjonova, 2005. Myriapods from ant nests in Bulgaria (Chilopoda, Diplopoda). Peckiana 4:131–142.

Stoev, P. & G. Ribarov, 1995. Contribution to the study of cave Myriapoda from Bulgaria. Acta zoologica bulgarica 48:89–96.

Strasser K., 1960. Diplopoden aus Alpen-, Apenninen- und Balkanländern. Fragmenta entomologica 3(6):95–140.

Strasser K., 1962a. Die Typhloiulini (Diplopoda Symphyognatha). Atti del Museo Civico di Storia Naturale di Trieste 23, I(1):1–77.

Strasser K., 1962b. Diplopoden aus Bulgarien und Jugoslawien. Senckenbergiana biologica 43(6):437–470.

Strasser K., 1966. Über Diplopoden Bulgariens, I. Annales Zoologici (Warszawa) 23(12):325–385.

Strasser K., 1969. Über Diplopoden Bulgariens, II. Annales Zoologici (Warszawa) 27(7):133–168.

Strasser K., 1973. Über Diplopoden Bulgariens, III. Annales Zoologici (Warszawa) 30(15):411–469.

Strasser K., 1975. Weiterer Beitrag zur Kenntnis der bulgarischen Diplopoden. Acta zoologica bulgarica 3, 70–77.

Tabacaru, I., 1967. Beiträge zur Kenntnis der cavernicolen Antroleucosomiden (Diplopoda, Ascospermophora). International Journal of Speleology 3:1–31.

Tabacaru, I., 1983. Recherches zoogéographiques sur les diplopodes du sud-est de l'Europe et du Proche-Orient (II suite). Travaux de l'Institut de Spéologie "Emile Racovitza" 22:9–20.

Turk, F., 1970. Some notes on the Acari and Myriapods collected by Dr. P. Browne and Mr. W.G.R. Maxwell in Bulgarian caves. Transactions of the Cave Research Group of Great Britain 12(1):39–41.

Verhoeff, K., 1925. Beiträge zur Kenntnis der Steinläufer, Lithobiiden. Archiv für Naturgeschichte, A, 91(9):124–158.

Verhoeff, K., 1926a. Zwei neue Höhlen-Myriapoden aus Bulgarien. Zoologischer Anzeiger 65(11/12):294–296.

Verhoeff, K., 1926b. Chilognathen-Beiträge (103. Diplopoden-Aufsatz). Zoologischer Anzeiger 68(1/2):57–71.

Verhoeff, K., 1926c. Zwei neue Geophilomorphen-Gattungen aus Tracien und Mexico. Zoologischer Anzeiger 69(3/4):97–105.

Verhoeff, K., 1926d. Über einige von Dr. I. Buresch in Bulgarien gesammelte Diplopoden. 1. Aufsatz. Trudove na Balgarskoto prirodoizpitatelno druzhestvo (Travaux de la Société bulgare des sciences naturelles) 12:67–78.

Verhoeff, K., 1926e. Über einige von Dr. I. Buresch in Bulgarien gesammelte Diplopoden. 2. Aufsatz. Izvestiya na Balgarskoto entomologichno druzhestvo (Mitteilungen der Bulgarischen Entomologischen Gesellschaft in Sofia; Bulletin de la Société Entomologique de Bulgarie) 3:193–210.

Verhoeff, K., 1928a. Über Chilopoden aus Bulgarien gesammelt von Herrn Dr. I. Buresch (I. Aufsatz). Izvestiya na Balgarskoto entomologichno druzhestvo (Mitteilungen der Bulgarischen Entomologischen Gesellschaft in Sofia; Bulletin de la Société Entomologique de Bulgarie) 4:115–124.

Verhoeff, K., 1928b. Über Diplopoden aus Bulgarien, gesammelt von Herrn Dr. I. Buresch (III. Aufsatz). Izvestiya na Tsarskite prirodonauchni instituti v Sofia (Mitteilungen aus den Königlichen naturwissenschaftlichen Instituten in Sofia – Bulgarien; Bulletin des Institutions royales d'histoire naturelle à Sofia – Bulgarie) 1:28–44.

Verhoeff, K., 1937. Über Diplopoden aus Bulgarien, gesammelt von Dr. I. Buresch und seinen Mitarbeitern. IV. Aufsatz. Izvestiya na Tsarskite prirodonauchni instituti v Sofia (Mitteilungen aus den Königlichen naturwissenschaftlichen Instituten in Sofia – Bulgarien; Bulletin des Institutions royales d'histoire naturelle à Sofia – Bulgarie) 10:93–120.

Vigna Taglianti, A., P. Audisio, M. Biondi, M. Bologna, G. Carpaneto, A. De Biase, S. Fattorini, E. Piattella, R. Sindaco, A. Venchi & M. Zapparoli, 1999. A proposal for a chorotype classification of the Near East fauna, in the framework of the Western Palearctic region. Biogeographia 20:31–59.

Würmli, M., 1973. Zur Systematik der Scutigeriden Europas und Kleinasiens (Chilopoda, Scutigeromorpha). Annalen des naturhistorischen Museums in Wien 77:399–408.

12 Fauna and Zoogeography of Scorpions (Arachnida: Scorpiones) in Bulgaria

Victor Fet[1] and Michael E. Soleglad[2]

[1] *Department of Biological Sciences, Ma;shall University, Huntington, West Virginia 25755-2510, USA, e-mail: fet@marshall.edu*

[2] *P.O. Box 250, Borrego Springs, California 92004, USA, e-mail: soleglad@la.znet.com*

Abstract: The fauna and zoogeography of scorpions in Bulgaria have not been well researched. At least four species are recorded currently for this country: the Balkan–Anatolian *Mesobuthus gibbosus* (Buthidae), known only from Pirin Mts., and several taxa of the genus *Euscorpius* Thorell (Euscorpiidae). The Balkan species *E. hadzii* inhabits the southwest of Bulgaria, and an undetermined species of the Balkan–Anatolian "*Euscorpius mingrelicus*" complex is known only from Pirin Mts. Remaining Bulgarian populations of *Euscorpius*, widely ranging from north to south (mainly in the mountain areas, up to 1850 m a.s.l.), belong to "*Euscorpius carpathicus*" complex. Their true taxonomic identity is not yet known, but these populations are not homogeneous. A clearly separated cluster of Rhodope populations could represent a different taxon from the Stara Planina (=Balkan) ones (mainly known from the Western Stara Planina) and those from the southwest of Bulgaria. Taxonomic affinities of Bulgarian scorpion taxa are discussed, as well as their geographic and altitudinal distribution

1 Introduction

The scorpion fauna of Bulgaria has been a subject of a very few studies, and the information on this fauna has been rather brief and sometimes unclear. One of the earliest Bulgarian zoologists Stefan Jurinich (1855–1947) was the first to record the scorpion species *Euscorpius carpathicus* (Linnaeus, 1767) from Bulgaria (Jurinich, 1905). Fet (1993) mentioned for Bulgaria the subspecies *Euscorpius germanus croaticus* Caporiacco, 1950, which was described from Croatia. Kovařík (1998) mentioned *Mesobuthus gibbosus* (Brullé, 1832) (Buthidae) and *Euscorpius carpathicus* for Bulgaria but without precise locality records, and a year later he (Kovařík, 1999) added with doubt *Euscorpius germanus* (C.L. Koch, 1837), again without precise locality records. Fet and Lowe (2000) did not list *M. gibbosus* from Bulgaria. Fet and Sissom (2000) recorded the following taxa of *Euscorpius* for this country: *E. carpathicus* (not assigned to subspecies), *E. germanus* (under question, as its enigmatic subspecies *Euscorpius germanus croaticus* Caporiacco, 1950) and also with doubt *Euscorpius mingrelicus* (Kessler, 1874). In a detailed paper on the Balkan scorpions, Fet (2000) listed as confirmed for Bulgaria only scorpions of the "*carpathicus*" complex (or *Euscorpius carpathicus* (Linnaeus, 1767) *sensu lato*), which he divided in three "groups" on the basis of trichobothrial pattern, metasomal carination, and coloration. In a recent paper dealing with the

V. Fet and A. Popov (Eds.), Biogeography and Ecology of Bulgaria, 405–422.

taxonomy of the "*carpathicus*" complex, Fet and Soleglad (2002) restricted the scope of *E. carpathicus* exclusively to the populations inhabiting southwestern Romania and revalidated as full species its "old" junior synonyms *Euscorpius hadzii* Caporiacco, 1950, *Euscorpius koschewnikowi* Birula, 1900 and *Euscorpius tergestinus* (C.L. Koch, 1837). Among these, *E. hadzii* was recorded from southwestern Bulgaria and corresponded to the "Group B" previously defined by Fet (2000). Most recently, in a note by Teruel et al. (2004) on scorpions collected in Pirin Mts., two additional taxa were recorded for Bulgaria: *Mesobuthus gibbosus* (for the first time confirmed with the exact locality data) and an undetermined *Euscorpius* sp. belonging to the "*mingrelicus*" complex of this genus (Fet, 1993; Fet and Sissom, 2000).

2 Material and Methods

Most of our information on scorpions of Bulgaria is original, and is derived from the analysis of the magnificent collection of the National Museum of Natural History at Sofia, Bulgaria (NMNHS). Spanning over 600 specimens from Bulgaria, this collection was accumulated over last 100 years. As in any large zoological museum, we can see that, being exotic and dangerous animals, scorpions attracted the attention of many professional zoologists and amateur naturalists alike. The NMNHS labels include many famous names of classical Bulgarian zoologists such as Ivan Buresch and Pencho Drenski (1930s); modern zoologists of all generations, especially Petar Beron; local amateur naturalists such as the late Dimitar Raichev of Chepelare who first collected unusual Western Rhodope *Euscorpius* in early 1980s; and even members of the royal family, Tsar Ferdinand I (1901) and Prince Kiril (1939–1940). The fact that important discoveries still are possible is illustrated by the recent findings of Teruel et al. (2004) who reported two new species of scorpions for Bulgaria from Pirin Mts.

Trichobothrial designations follow Fet and Soleglad (2002). Detailed label data are available from the authors.

3 Fauna and Zoogeography

Family Euscorpiidae Laurie, 1896
Genus *Euscorpius* Thorell, 1876
Subgenus *Euscorpius* Thorell, 1876
***Euscorpius hadzii* Caporiacco, 1950**

In Bulgaria, *E. hadzii* is found only in the southwestern corner of the country (Fig. 1). We analyzed 93 specimens (29 males and 64 females).

E. hadzii is common in the southwest (along valleys of Struma and Mesta to the Rila Mountains). It is found along the Macedonian border from Belasitsa in the southwest to Ograzhden Mts. and Maleshevska Mts. and as far toward

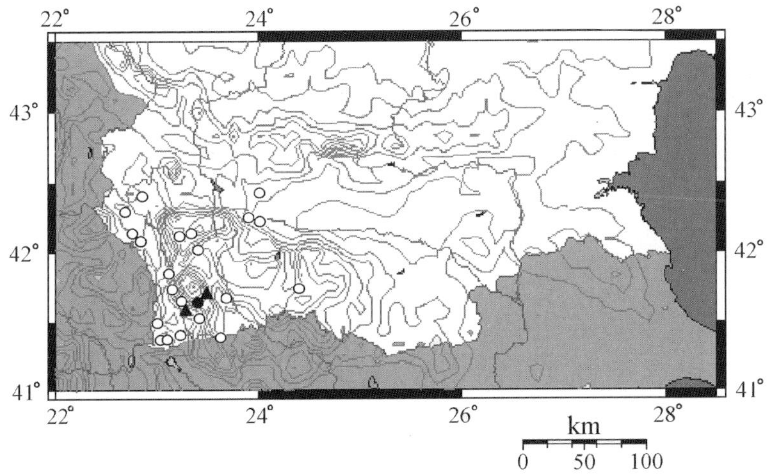

Fig. 1 Distribution of *Euscorpius hadzii* (circle), *Euscorpius* (*Alpiscorpius*) sp. (closed circle), and *Mesobuthus gibbosus* (triangle) in Bulgaria.

north as Konyavska Mts. (Baikalsko) and Osogovo Mts. The Kyustendil record of *E. hadzii* is the westernmost for scorpions in Bulgaria. In the southeastern direction it reaches Slavyanka (=Alibotush) Mts. but only a single record comes from the Western Rhodopes (Devin). Three other disjunct records show that *E. hadzii* is also found in Maritsa Valley. The sole record from Pirin is undoubtedly due to poor representation of this area in collections (Teruel et al., 2004). Outside of Bulgaria, *E. hadzii* is found in Albania, Bosnia and Herzegovina, Croatia, Greece, Serbia and Montenegro, and Macedonia.

The highest altitudes at which *E. hadzii* is found in Bulgaria (1600 m, Western Pirin; 1500 m, Ograzhden; 1450 m, Rila) match its record in Albania (Boga, Maya Tchardakut) at the 1400–1600 m (Fet, 2000, as "Group B, Subgroup B2" of *E. carpathicus*). This, however, is not a record altitude for scorpions in Bulgaria (see below).

For detailed description and discussion of this species see Fet and Soleglad (2002: 24–30) who redescribed *E. hadzii* and designated a neotype from Albania (Prokletije). A typical trichobothrial pattern of *E. hadzii* from Bulgaria is shown in Fig. 2.

Our statistical analysis of morphology of this species in Bulgaria uncovered, when compared to the populations from the Western Balkans, differences in the numbers of accessory trichobothria on the pedipalp, and also some subtle differences in key morphometric ratios.

Trichobothria: We see a *reduction* in all variable patellar trichobothria series in the Bulgarian population of *E. hadzii*, specifically the ventral and external series eb_a, *em*, and *et* (see Table 1). The populations exhibiting the largest counts in these series center in Croatia and Serbia and Montenegro where we see reductions in

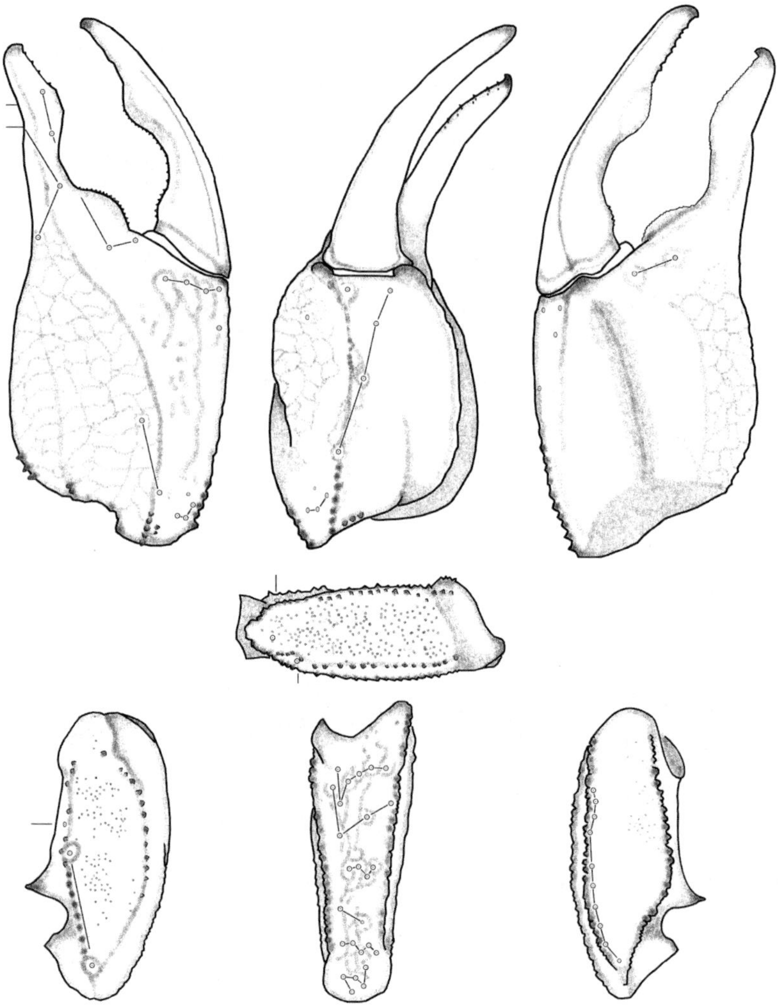

Fig. 2 Trichobothrial pattern of *Euscorpius hadzii*, male from Rila Mts., Bulgaria.

these counts as the species radiates southward to Albania and eastward to Bulgaria (Fig. 3 histogram). The Bulgarian populations show statistically the lowest counts in these trichobothrial series. Interestingly, in the eb_a series we see a reduction of over 1.5 trichobothria, averaging a little over five in this series, where the other populations exhibit counts averaging seven (see Figs. 4–9 showing diagrams of the external aspect of pedipalp patella). In the other series, we find that the Albanian population is intermediate between the Bulgarian and the Croatia and Serbia and Montenegro populations.

Table 1 Statistical distribution of diagnostic pedipalp patella trichobothria series for *Euscorpius hadzii*. The data for each group of populations are presented as: **min–max(mean)(±SD)[n]{cmin–cmax}**, where min = minimum value, max = maximum value, SD = standard deviation, n = number of samples, cmin = corrected minimum (mean-SD), cmax = corrected maximum (mean+SD)

	Ventral	eb_a
Bulgaria	8–11 (9.567) (±0.540) [178] {9.027–10.108}	4–8 (5.449) (±0.621) [176] {4.823–6.070}
Albania	9–13 (10.617) (±0.764) [094] {9.854–11.381}	5–8 (6.860) (±0.457) [093] {6.404–7.317}
Croatia	10–13 (11.923)(±0.845) [026] {11.078–12.768}	6–8 (6.962) (±0.720) [026] {6.241–7.682}
Serbia & Montenegro	10–13 (11.686) (±0.678)[051] {11.008–12.364}	6–8 (7.000) (±0.566) [051] {6.434–7.566}
All populations	8–13 (10.335) (±1.096) [349] {9.240–11.431}	4–8 (6.171)(±0.937) [346] {5.233–7.108}
	em	*et*
Bulgaria	3–5 (4.000) (±0.150) [178] {3.850–4.150}	6–8 (6.949) (±0.514)[177] {6.435–7.463}
Albania	3–5 (4.073) (±0.332) [096] {3.741–4.405}	6–9 (7.432) (±0.630)[095] {6.802–8.062}
Croatia	4–5 (4.885) (±0.326) [026] {4.559–5.210}	6–9 (7.500) (±0.906)[026] {6.594–8.406}
Serbia & Montenegro	4–5 (4.902) (±0.300) [051] {4.602–5.202}	6–9 (7.725) (±0.750)[051] {6.975–8.476}
All populations	3–5 (4.217) (±0.439) [351] {3.777–4.656}	6–9 (7.235) (±0.688) [349] {6.547–7.923}

Note: Bulgarian populations exhibit the lowest counts for each series, and Croatian and Serbian/Montenegrin populations have the highest counts. Compare this table to the histograms presented in Fig. 3. See Material and Methods section for definition of statistical data group. eb_a = external basal-a series; *em* = external median series; *et* = external terminal series.

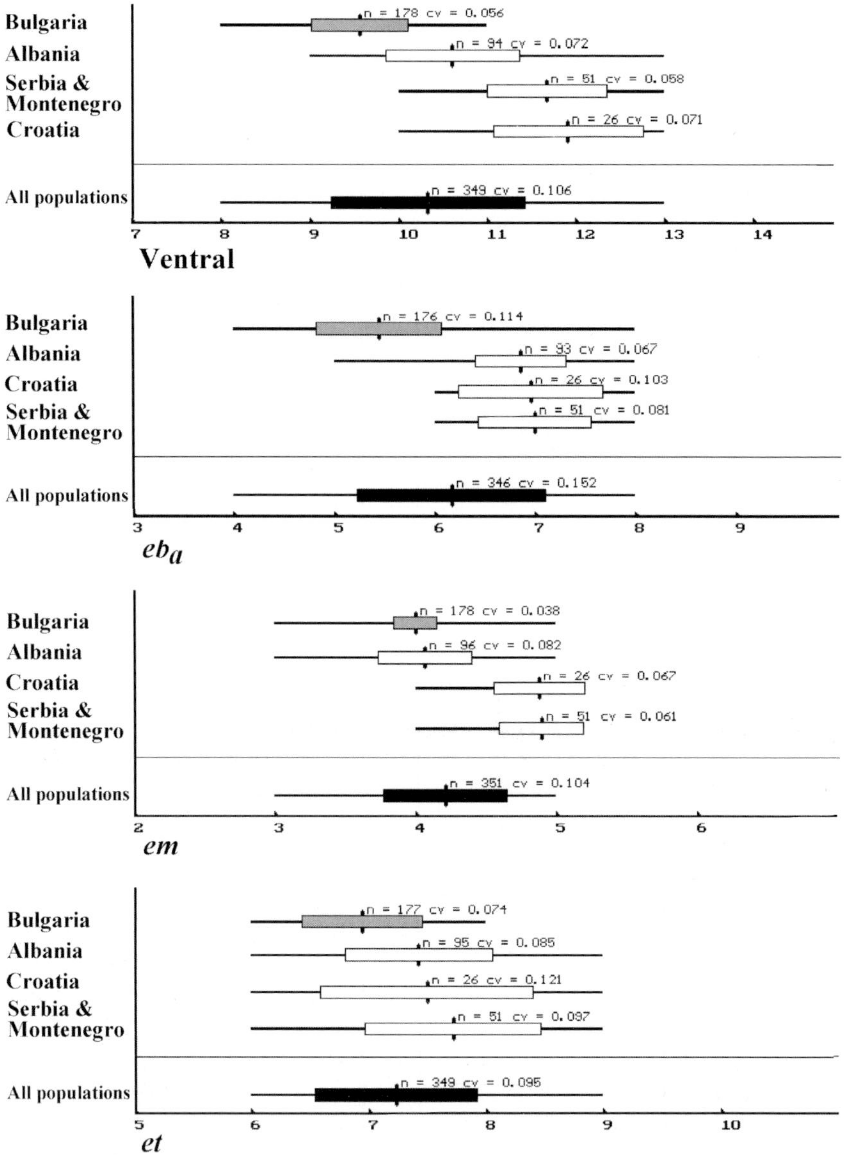

Fig. 3 Trichobothria counts for *Euscorpius hadzii* distributed by populations. Bulgarian populations depicted with grey bar, all populations depicted with black bar.

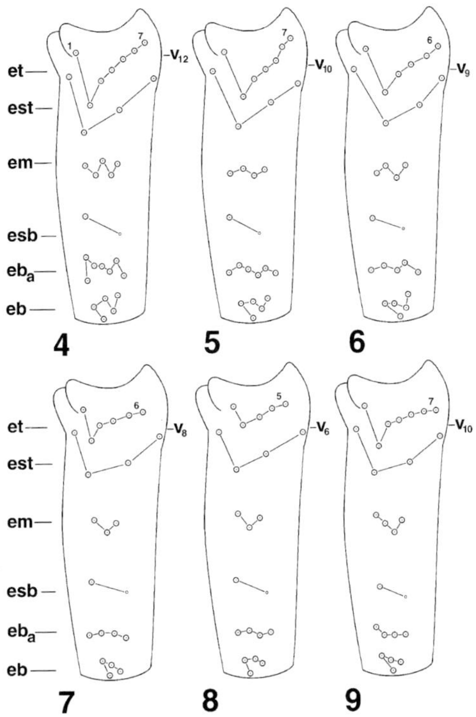

Fig. 4–9 Diagrammatic trichobothrial pattern of external aspect of pedipalp patella; 4, *Euscorpius hadzii*, Croatia; 5 and 6, *E. hadzii*, Bulgaria; 7, *E. carpathicus*, Romania; 8 and 9, *Euscorpius* sp. ("*carpathicus*" complex), Bulgaria (8, Trigrad; 9, Plovdiv District).

Morphometric ratios: Fet and Soleglad (2002) established several ratios that provided good diagnostic characters for separation of several species of the "*carpathicus*" complex. We investigated two of these ratio, the Dorsal Patellar Spur (DPS) ratio and metasoma (length/width) ratios. In Fig. 10, we contrast the DPS ratio (see Fet and Soleglad, 2002: Fig. 69, for details on this ratio) for Bulgarian *E. hadzii* populations with other species described in Fet and Soleglad (2002). The species *E. mingrelicus* (subgenus *Alpiscorpius*) is also included in this comparison to illustrate its considerably blunted DPS, typical of this subgenus. As indicated in Fig. 10, we see that the Bulgarian population DPS is slightly less elongated than the other *E. hadzii* populations, but the SD ranges overlap considerably. In Figs. 11–12, which show histograms for metasomal segments I, III, and V (female and male), we see, except for metasomal segment I of females, that the Bulgarian population SD range considerably overlaps with other *E. hadzii* populations. In these figures we also see that *E. koschewnikowi* exhibits by far the thinnest metasoma and that *E. mingrelicus* and *E. carpathicus* have more stout metasomas. In general, subgenus *Alpiscorpius* species have the stoutest metasoma in *Euscorpius*, whereas *E. carpathicus* has the stoutest metasoma in the "*carpathicus*" complex.

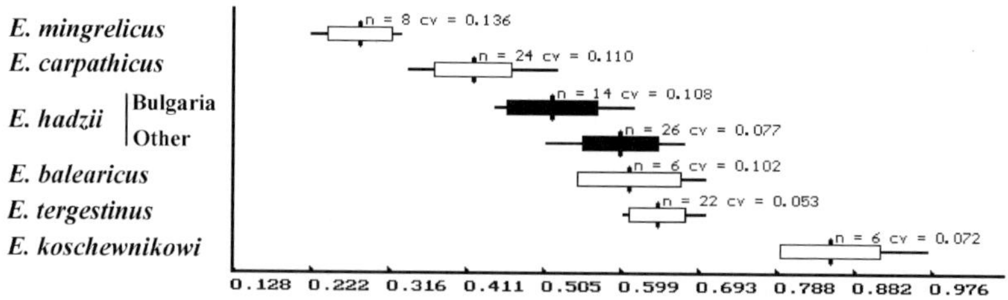

Fig. 10 Morphometric ratio for Dorsal Patellar Spur (DPS) contrasting Bulgarian and other populations of *E. hadzii* (black bars) with other *Euscorpius* species.

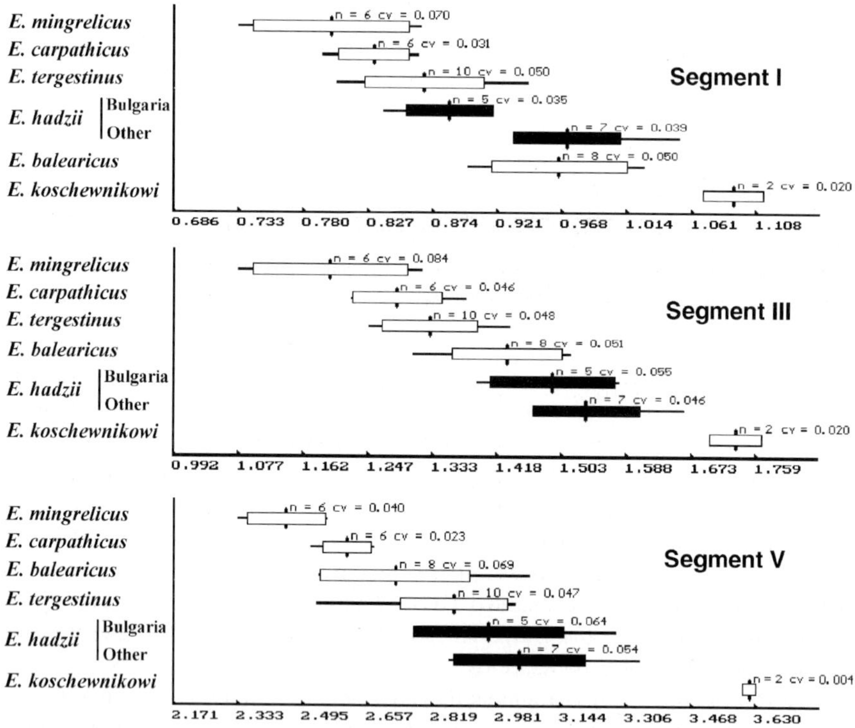

Fig. 11 Morphometric ratio for metasomal segments I, III, and V (length/width) in females, contrasting Bulgarian and other populations of *E. hadzii* (black bars) with other *Euscorpius* species.

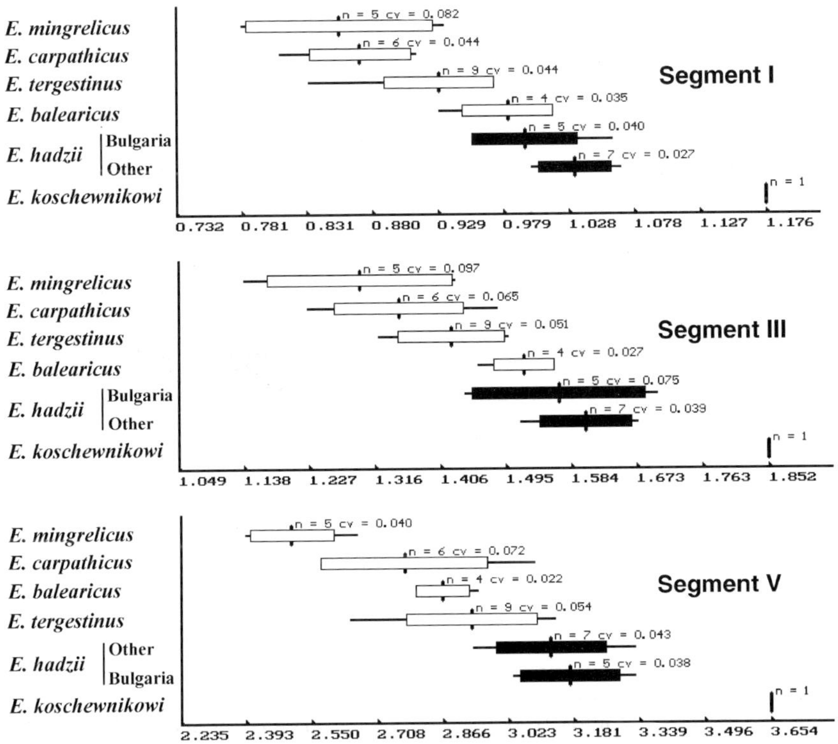

Fig. 12 Morphometric ratio for metasomal segments I, III, and V (length/width) in males, contrasting Bulgarian and other populations of *E. hadzii* (black bars) with other *Euscorpius* species.

Euscorpius sp. (*"carpathicus"* complex)

The populations, which are listed here as belonging to "*carpathicus*" complex (but which exclude *E. hadzii*), are the most widespread in Bulgaria. We analyzed 510 specimens (155 males and 355 females) from various localities, mostly from Western Stara Planina and South Bulgaria. Scorpions of this complex were clearly more commonly collected in the mountains than in the lowlands (Fig. 13). However, collection localities are not evenly distributed and could be biased toward better studied, popular mountain sites (e.g. valleys of Iskar and Struma).

A preliminary analysis of morphology (first of all trichobothrial numbers on external and ventral aspects of a pedipalp, as well as some morphometric ratios; Figs. 7–9) allows us to separate at least three groups (clusters) of populations, which geographically roughly represent North Bulgaria, Southwest Bulgaria, and the Rhodope Mountains.

(1) The "northern" group of populations is widespread across the Stara Planina (=Balkan) mountain system and some adjacent areas (Fig. 13). It is found

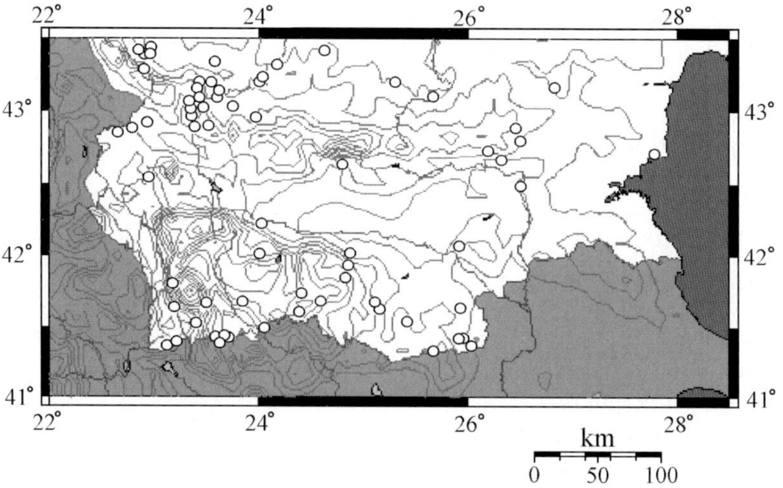

Fig. 13 Distribution of "*Euscorpius carpathicus*" complex populations in Bulgaria.

in the northwestern corner of the country in the mountains bordering Serbia (the northernmost record of scorpions for Bulgaria is in Belogradchik) as far southwest as Tran in Kraishte, and in numerous localities of Western Stara Planina (especially Iskar Gorge). Data from Central North Bulgaria (Pleven, Veliko Tarnovo) and Central Stara Planina (Karlovo, Sliven) are scarce, most likely due to insufficient collecting.

Scorpions were not collected on the lowlands of Danubian Plain, thus there is no evidence that the current Bulgarian range of *Euscorpius* comes close to that of the Romanian *E. carpathicus*. Very few scorpions were recorded from the east and northeast of Bulgaria: populations near Sliven, Yambol, and Shumen are known. A single record exists from the Black Sea Coast (Sveti Vlas near Cape Emine, the easternmost record of scorpions for Bulgaria). In our analysis, data on all these populations were pooled with those of the Western Stara Planina. We also include in the "northern" populations those from the northern foothills of Central Rhodopes (Bachkovo, Asenovgrad) since they show high morphological similarity to Stara Planina specimens but not to ones from the inner valleys of the Rhodope Mts. The highest altitude at which the "northern" group was found is 1400 m at Mt. Izdremets (above Bov village, Sofia District) in the Iskar watershed.

(2) The "southwestern" group of "*carpathicus*" complex populations is geographically separated from the "northern" group, and concentrated around Struma River valley, flanked in the south by Belasitsa and Slavyanka (=Alibotush) mountain massifs along the Bulgarian–Greek border (Fig. 13). The Belasitsa record is the southernmost for scorpions in Bulgaria.

In its morphology, this group is intermediate between "northern" populations and the Rhodope ones. It is unclear if the range of "southwestern" populations is continuous with the "northern" populations. In the east, its range borders that of

Rhodope populations. Groups (1) and (2) were listed by Fet (2000) as Bulgarian populations of Group A. The highest altitude at which the "southwestern" group was found is 1850 m (Tsarev Vrah summit, Slavyanka Mts.). This is so far the highest altitude for scorpions in Bulgaria (compare *E. hadzii* above). We should notice here that the highest altitudinal records for the subgenus *Euscorpius* is from Crete (Lefka Ori Mt., 2200 m, P. Beron coll.; Fet, 2000).

(3) The Rhodope populations of "*carpathicus*" complex in this treatment are limited by Pirin Mts. in the west, and Arda River in the east. In the "Rhodope" cluster we include all populations found east from Mesta River valley in the Western Rhodopes (Kovachevitsa, Devin, Trigrad, Smolyan, Rozhen, Shiroka Laka) and also a few known populations from the Eastern Rhodopes (Kardjali, Krumovgrad, and Ivailovgrad Districts). Morphology of "Rhodope" populations is clearly different from the "northern" ones (Figs. 4–9, 14). Statistical analysis shows that ventral and external median series of patellar trichobothria (Fig. 14) do not overlap in Rhodope and "northern" group within SD ranges. Part of group (3) was listed by Fet (2000) as Group C. The highest altitude at which the "Rhodope" group was found is 1500 m (Rozhen Pass, Smolyan District).

Interestingly, by both of these characters the Rhodope group falls in the same range with a form from Western Croatia (Mali Halam, Velebit Mts.) currently known as *E. carpathicus croaticus* Caporiacco, 1950. This Croatian form is known from

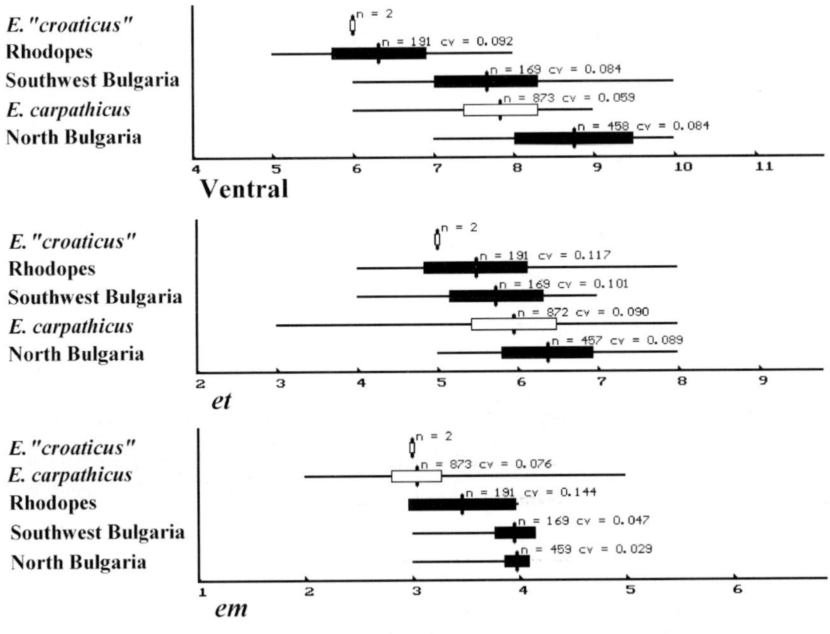

Fig. 14 Trichobothria counts for Bulgarian "*carpathicus*" complex populations (black bars). *E. carpathicus* (Romania) and type specimen of *E. "croaticus"* (Croatia) (white bars) are shown for comparison.

a single specimen (analyzed by the authors; deposited in the Museo Zoologico "La Specola" dell'Universita de Firenze, Florence, Italy); its validity and status are currently unclear as populations of the "*carpathicus*" complex from inland Croatia and neighboring areas of the western Balkans are not studied. In Fig. 14, this form, shown for comparison to the Bulgarian populations, is designated as *E. "croaticus"*. A certain similarity of the western Croatian form to the Rhodope populations (and especially those from Trigrad area which exhibit $em = 3$) is the reason for Fet (1993) mentioning this subspecies (as *Euscorpius germanus croaticus* Caporiacco, 1950) for Bulgaria. Fet and Braunwalder (2000) commented that this taxon does not belong to *Euscorpius germanus*, and Gantenbein et al. (2000a) transferred it to the "*carpathicus*" complex (=subgenus *Euscorpius*). It does not, however, belong to *E. carpathicus* (L., 1767), which is restricted to Romania by Fet and Soleglad (2002), although the Romanian species also exhibits $em = 3$ character.

Fig. 15 shows the composite index of trichobothria (external terminal series et+ventral series v) plotted for all Bulgarian populations of "*carpathicus*" complex (excluding *E. hadzii*). It is clear that a heterogeneity is observed: the Rhodope group does not overlap with the "northern" group at all. This division has to be analyzed further; at this moment, we refrain from any taxonomic decisions. It is possible that some of the observed clusters will be either described as new species or assigned to the existing Balkan species.

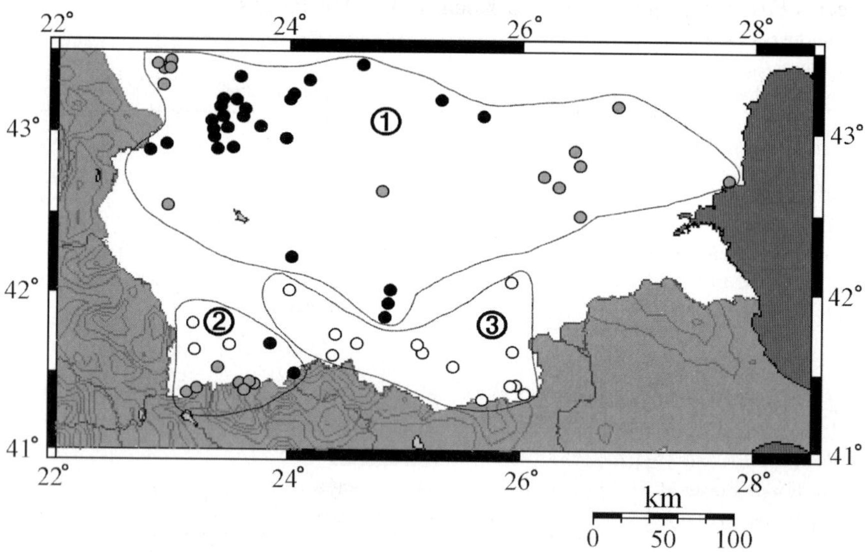

Fig. 15 Distribution of trichobothria counts for Bulgarian "*carpathicus*" complex populations. Index ($et + v$) (total sum of patellar trichobothria in external terminal series and ventral series) was scored as follows: below 13, white circles; 13 to 15, grey circles; 15 and higher, black circles. Populations: 1, northern; 2, southwestern; 3, Rhodope.

Currently, Bulgarian populations of *Euscorpius* belonging to "*carpathicus*" complex do not have a taxonomic name; no scorpion taxa have been ever named and described from Bulgarian territory. The name *Euscorpius carpathicus* (L., 1767) has been restricted to Romanian populations (Fet and Soleglad, 2002), and we see no reason to expand it including geographically adjacent but rather morphologically different populations from Bulgaria.

Subgenus *Alpiscorpius* Gantenbein, Fet, Largiadèr et Scholl, 1999 *Euscorpius* sp. ("*mingrelicus*" complex)

Only one specimen, a unique subadult male of this unidentified species, is known from Pirin Mountains in Bulgaria (Teruel et al., 2004), collected at 1200–1300 m a.s.l. This specimen clearly showed all characters defining the "*mingrelicus*" complex of the genus (obsolescence of metasomal carination, *est-et/dsb-est* ratio ≥ 1.5).

Members of this complex (which is a part of subgenus *Alpiscorpius* Gantenbein, Fet, Largiadèr et Scholl, 1999) are widely distributed over the western part of the Balkan Peninsula (Fet, 2000; Fet and Sissom, 2000). Fet (1993) reviewed all known distribution and taxonomic composition of the species (*sensu lato*) *Euscorpius mingrelicus* (Kessler, 1874) which was originally described from Georgia (Caucasus) but later redefined (Bonacina, 1980) as a part of the former species *Euscorpius germanus* (C.L. Koch, 1837). The latter species is in fact limited to the Alpine zone of Europe (Gantenbein et al., 2000a). Treated earlier as one species widespread from the Alps to Caucasus (Bonacina, 1980; see Fet and Sissom, 2000 for the taxonomic history), "*E. mingrelicus*" complex appears to be a group of species, which are currently under revision using both morphological and molecular techniques (Fet, 2000; Gantenbein et al., 1999, 2000a; Scherabon et al., 2000; Fet et al., 2003). Currently, three species are recognized in this complex, with *E. mingrelicus* (Kessler, 1874) *sensu stricto* ranging from Bosnia to Caucasus (Fet and Sissom, 2000); however, this division is clearly not satisfactory. Fet (2000) described *E. beroni* from the high mountains of Albania (Prokletije Mts.), and Scherabon et al. (2000) demonstrated a separate status of *E. gamma* Caporiacco from the northeastern part of the geographic range of this complex (Northeastern Italy, Slovenia, Croatia, and Austria). For ecology and distribution of *E. gamma* in Slovenia and Austria, see also Scherabon (1987), Fet et al. (2001), and Komposch et al. (2001). The Pirin specimen falls into the range of "*mingrelicus*" complex for which taxonomic identity of populations is not yet determined. It is an important biogeographic find—the first specimen of this complex from Bulgaria, and most likely a glacial relict. We should notice here that the highest altitudinal records for the subgenus *Alpiscorpius* is from Anatolia (type series of *E. ciliciensis* Birula, 1898; Bulghar Dagh, 2600 m, M. Holtz coll., Fet, 1986).

Family Buthidae C. L. Koch, 1837
Genus *Mesobuthus* Vachon, 1950
***Mesobuthus gibbosus* (Brullé, 1832)**

Only one specimen, a juvenile female of this species is known from Pirin Mountains, collected at 1200–1300 m a.s.l. This is the first precise record of *M. gibbosus* from Bulgaria (Teruel et al., 2004). This Bulgarian specimen was identical in morphology to other examined specimens of the same species from Greece and European Turkey (Teruel et al., 2004).

M. gibbosus was described from the Peloponnesus, Greece; it is found in Albania, Bulgaria, Greece, Macedonia, Montenegro, and Turkey (both European and Asian). Its formerly reported (Fet and Lowe, 2000: 177) populations from Cyprus, Israel, Lebanon, and Syria belong to different species: *M. cyprius* Gantenbein et Kropf, 2000 on Cyprus (Gantenbein et al., 2000b) and *M. nigrocinctus* Ehrenberg, 1828 in Israel, Lebanon, and Syria (Fet et al., 2000). The genus *Mesobuthus* has an Asian center of diversity, and most likely the Asian origin, and *M. gibbosus* is its westernmost species (Gantenbein et al., 2003). It is very possible that *M. gibbosus* could reach Pirin Mountains directly from the south via the Struma or Mesta river valleys, as mentioned above for *E. hadzii*. These two biogeographic routes are common avenues of penetration of Submediterranean elements into Bulgaria (Fet, 2000). Such a dispersal could be a recent postglacial event, or alternatively could happen during the Pleistocene interglacial periods. Mountain habitats are normally occupied by various species of *Mesobuthus* from Greece to Central Asia; the highest recorded altitude was that for *M. eupeus* (C. L. Koch, 1838) at southeastern Caucasus (Marayurt, Zuvant) at 8033 ft (ca. 2680 m) (Birula, 1917).

4 Further Research

Taxonomy (and, consequently, zoogeography) of Bulgarian and, in general, Balkan *Euscorpius* species at this moment is not settled, for both subgenera *Euscorpius* and *Alpiscorpius*.

Traditionally treated as one species widespread in Europe (from Baleares to Crimea; Hadži, 1930, 1931; Caporiacco, 1950; Ćurčič, 1972; Valle, 1975; Vachon, 1975; Vachon and Jaques, 1977; Fet, 1986; Crucitti, 1993; etc.; see Fet and Sissom, 2000 for the detailed if convoluted taxonomic history), "*E. carpathicus*" complex is a complicated group of species currently under revision using both morphological and molecular techniques (Fet, 2003; Fet and Soleglad, 2002; Fet et al., 2003b; Gantenbein et al., 2001, 2003). Kinzelbach (1975) divided *E. carpathicus* into two species, designating the second one as "*E. mesotrichus* Hadži, 1929". This name, listed by Fet and Sissom (2000) as *E. tergestinus* (C.L. Koch, 1837) was later demonstrated to belong to the species (or possibly a complex of species) *E. sicanus* (C.L. Koch, 1837), which has a relict distribution in Italy, Malta, and Greece (Fet et al., 2003b). Other morphologically distinct species, recently separated from "*carpathicus*" complex are *E. balearicus* Caporiacco, 1950 (Balearic

Islands), *E. concinnus* (C.L. Koch, 1837) (France, Italy), *E. hadzii* Caporiacco, 1950 (Balkans), *E. koschewnikowi* Birula, 1900 (Northeastern Greece), *E. tauricus* (C.L. Koch, 1837) (Crimea), and *E. tergestinus* (C.L. Koch, 1837) (Italy, Slovenia, Croatia, Austria) (Fet and Soleglad, 2002; Fet, 2003; Fet et al., 2004; Vignoli et al., 2005). *E. carpathicus* (L., 1767) *sensu stricto* is currently restricted to Southwestern Romania (Fet and Soleglad, 2002). However, within Bulgaria, as well as for the large portion of the geographic range of "*carpathicus*" complex in the Balkans, the taxonomy is not defined or is defined only partially. The status of the following five taxonomic entities remains to be clarified: *E. carpathicus croaticus* Caporiacco, 1950 (transferred from *E. germanus* subspecies by Gantenbein et al., 2000a; type locality: Velebit Mts., Croatia); *E. c. scaber* Birula, 1900 (type locality: Mt. Athos, Northeastern Greece); *E. c. candiota* Birula, 1903 (type locality: Crete); *E. c. ossae* Caporiacco, 1950 (type locality: Mt. Ossa, Thessaly, Greece); and *E. c. aegaeus* Caporiacco, 1950 (type locality: Antiparos, Aegean Sea). For large territory of Croatia (except Dalmatian coast), Bosnia and Herzegovina, Serbia, Montenegro, Greece (including Aegean islands), Bulgaria, and Turkey, the taxonomy of the "*E. carpathicus*" complex remains to be established.

For the large part of the "*mingrelicus*" complex of the subgenus *Alpiscorpius*, the taxonomy is also currrently not defined or is defined partially. There are formally seven valid subspecies of *E. mingrelicus* from Balkans to Anatolia (Fet and Sissom, 2000), and status of these forms is still unclear. The neotype for *E. mingrelicus* (Kessler, 1874) was designated from Georgia by Gantenbein et al. (2000a). The status of the following eight taxonomic entities remains to be clarified: *Scorpius bosnensis* von Möllendorf, 1873 (type locality: Bosnia); *E. m. ciliciensis* Birula, 1898 (Cilician Taurus, Anatolia, Turkey); *E. m. dinaricus* Caporiacco, 1950 (type locality: Bosnia); *E. m. caporiaccoi* Bonacina, 1980 (type locality: Bosnia); and four subspecies described from NW Anatolia, Turkey: *E. m. phrygius* Bonacina, 1980; *E. m. legrandi* Lacroix, 1995; *E. m. ollivieri* Lacroix, 1995; and *E. m. uludagensis* Lacroix, 1995. The high genetic diversity of populations within Turkey (Fet et al., 2003a) indicates a possibility of cryptic species as recently discovered in the related Alpine "*germanus*" complex (Gantenbein et al., 2000a).

Further investigation of both "*carpathicus*" complex (=subgenus *Euscorpius*) and the "*mingrelicus*" complex of the subgenus *Alpiscorpius* from Bulgaria, Greece, and other Balkan areas (Fet and Soleglad, in progress) will shed more light at the species structure of these groups. It is likely that a number of new species will be described in future to accommodate a considerable diversity of this complex in the Balkans.

Acknowledgments

We thank Dr. Petar Beron (National Museum of Natural History, Sofia, Bulgaria) for providing the unique opportunity to examine the entire NMNHS collection of scorpions, spanning 100 years and over 600 specimens. V.F.'s travel to

Bulgaria in 1999 was supported by the COBASE (Cooperation in Basic Science and Engineering) grant from the National Research Council, Washington, DC, USA. More extended V.F.'s travel to Bulgaria in 2005 was supported by the Fulbright Scholar Award 04-11-08 from CIES (Council of International Exchange of Scholars), Washington, DC, USA. Help, hospitality, and friendship of many Bulgarian colleagues, especially Christo Deltshev, Alexi Popov, Petar Beron, Vladimir Sakalian, Milen Vassilev, Dobrin Dobrev, Gergin Blagoev, and Ivan Pandourski made the 1999 and 2005 visits productive and enjoyable.

References

[Birula, A.] Byalynitsky-Birula, A.A., 1917. [Arachnoidea Arthrogastra Caucasica. Pars I. Scorpiones]. Zapiski Kazvazskogo Muzeya (Mémoires du Musée du Caucase) (Tiflis) A(5) (in Russian). English translation: 1964. Arthrogastric Arachnids of Caucasia. 1. Scorpions. Jerusalem: Israel Program for Scientific Translations.

Bonacina, A., 1980. Sistematica specifica e sottospecifica del complesso "*Euscorpius germanus*" (Scorpiones, Chactidae). Rivista del Museo Civico di Scienze Naturali "Enrico Caffi" (Bergamo) 2:47–100.

Caporiacco, L. di., 1950. Le specie e sottospecie del genere "*Euscorpius*" viventi in Italia ed in alcune zone confinanti. Memorie. Accademia nazionale dei Lincei (ser. 8) 2:159–230.

Crucitti, P., 1993. Distribution and diversity of Italian scorpions. Redia 76(2):281–300.

Ćurčić, B.P.M., 1972. Considerations upon the geographic distribution and origin of some populations in the genus *Euscorpius* Thorell (Chactidae, Scorpiones). Rapport de la Commission International de la Mer Méditerranée (Monaco) 21(3):83–88.

Fet, V., 1986. Notes on some *Euscorpius* (Scorpiones: Chactidae) from Greece and Turkey. Rivista del Museo Civico di Scienze Naturali "Enrico Caffi" (Bergamo) 9 [1985]:3–11.

Fet, V., 1993. Notes on *Euscorpius mingrelicus* (Kessler, 1874) from the Caucasus. Rivista del Museo Civico di Scienze Naturali "Enrico Caffi" (Bergamo) 16:1–8.

Fet, V., 2000. Scorpions (Arachnida, Scorpiones) from the Balkan Peninsula in the collections of the National Museum of Natural History, Sofia. Historia naturalis bulgarica 11:47–60.

Fet, V., 2003. The Crimean scorpion, *Euscorpius tauricus* (C.L. Koch, 1837) (Scorpiones: Euscorpiidae): an endemic species supported by mitochondrial DNA evidence. Arthropoda Selecta (Moscow) 11(4):271–276.

Fet, V. & M.E. Braunwalder, 2000. The scorpions (Arachnida, Scorpiones) of the Aegean area: current problems in taxonomy and biogeography. Belgian Journal of Zoology 130(Suppl. 1):17–22.

Fet, V., B.E. Hendrixson, W.D. Sissom & G. Levy, 2000. First record for the genus *Mesobuthus* Vachon, 1950 in Israel: *Mesobuthus nigrocinctus* (Ehrenberg, 1828), comb. n. (Scorpiones: Buthidae) from Mt. Hermon. Israel Journal of Zoology 46:287–295.

Fet, V., Ay. Karataş, E.V. Fet & A. Karataş, 2003a. [First data on the molecular phylogeny of *Euscorpius* (Scorpiones: Euscorpiidae) from Turkey]. Zoologicheskiy zhurnal [Zoological Journal] 82(12):1518–1521 (in Russian); English translation (2003): Entomological Review 83 (Suppl. 2): 249–252.

Fet, V., M. Kuntner & B. Sket, 2001. Scorpions of Slovenia: a faunistic and biogeographical survey. In Fet, V. & P.A. Selden (eds), Scorpions 2001. In Memoriam Gary A. Polis. British Arachnological Society, Burnham Beeches, Bucks: 255–256.

Fet, V. & G. Lowe, 2000. Family Buthidae. In Fet, V., Sissom, W.D., Lowe, G. and Braunwalder, M.E. Catalog of the Scorpions of the World (1758–1998). New York Entomological Society, New York: 54–286.

Fet, V. & W.D. Sissom, 2000. Family Euscorpiidae. In: Fet, V., Sissom, W.D., Lowe, G. and Braunwalder, M.E. Catalog of the Scorpions of the World (1758–1998). New York Entomological Society, New York: 355–381.

Fet, V. & M.E. Soleglad, 2002. Morphology analysis supports presence of more than one species in the "*Euscorpius carpathicus*" complex (Scorpiones: Euscorpiidae). Euscorpius 3:1–51.

Fet, V., M.E. Soleglad & B. Gantenbein, 2004. The Euroscorpion: genus *Euscorpius* (Scorpiones: Euscorpiidae). Proceedings of the 3d Scorpiology Symposium, American Arachnological Society 28th Annual Meeting, Norman, Oklahoma, 23–27 June 2004. Euscorpius 17:47–59.

Fet, V., M.E. Soleglad, B. Gantenbein, V. Vignoli, N. Salomone, E.V. Fet & P.J. Schembri, 2003b. New molecular and morphological data on the "*Euscorpius carpathicus*" species complex (Scorpiones: Euscorpiidae) from Italy, Malta, and Greece justify the elevation of *E. c. sicanus* (C.L. Koch, 1837) to the species level. Revue suisse de Zoologie 110(2):355–379.

Gantenbein, B., V. Fet, M. Barker & A. Scholl, 2000a. Nuclear and mitochondrial markers reveal the existence of two parapatric scorpion species in the Alps: *Euscorpius germanus* (C.L. Koch, 1837) and *E. alpha* Caporiacco, 1950, stat. nov. (Scorpiones, Euscorpiidae). Revue suisse de Zoologie 107(4): 843–869.

Gantenbein, B., V. Fet & A.V. Gromov, 2003. The first DNA phylogeny of four species of *Mesobuthus* Vachon, 1950 (Scorpiones: Buthidae) from Eurasia. Journal of Arachnology 31(3):412–420.

Gantenbein, B., V. Fet, C.R. Largiadèr & A. Scholl, 1999. First DNA phylogeny of *Euscorpius* Thorell, 1876 (Scorpiones: Euscorpiidae) and its bearing on taxonomy and biogeography of this genus. Biogeographica (Paris) 75(2):49–65.

Gantenbein, B., C. Kropf, C.R. Largiadèr & A. Scholl, 2000b. Molecular and morphological evidence for the presence of a new buthid taxon (Scorpiones: Buthidae) on the Island of Cyprus. Revue suisse de Zoologie 107:213–232.

Gantenbein, B., M.E. Soleglad & V. Fet, 2001. *Euscorpius balearicus* Caporiacco, 1950, stat. nov. (Scorpiones: Euscorpiidae): molecular (allozymes and mtDNA) and morphological evidence for an endemic Balearic Islands species. Organisms, Diversity and Evolution 1(4):301–320.

Hadži, J., 1930. Die europäischen Skorpione des Polnischen Zoologischen Staatsmuseums in Warszawa. Annales Musei Zoologici Polonici 9(4):29–38.

Hadži, J., 1931. Der Artbildungsprozess in der Gattung *Euscorpius* Thor. Archivio Zoologico Italiano 16(1–2):356–362 (IX Congrès international de zoologie).

Jurinich, S., 1905. [The genus *Euscorpius* in Bulgaria]. Periodichesko spisanie na Balgarskoto knizhovno druzhestvo v Sofiya [Periodical Journal of the Bulgarian Literary Society in Sofia] 65(1–2):134–140 (in Bulgarian).

Kinzelbach, R., 1975. Die Skorpione der Ägäis. Beiträge zur Systematik, Phylogenie und Biogeographie. Zoologische Jahrbücher, Abteilung für Systematik 102(1):12–50.

Komposch, C., B. Scherabon & V. Fet, 2001. Scorpions of Austria. In Fet, V. & P.A. Selden (eds.), Scorpions 2001. In Memoriam Gary A. Polis. British Arachnological Society, Burnham Beeches, Bucks: 267–272.

Kovařík, F., 1998. Štiří [Scorpions]. Madagaskar, Jihlava.

Kovařík, F., 1999. Revision of European scorpions, with a key to species. Serket 6(2):38–44.

Scherabon, B., 1987. Die Skorpione Österreichs in vergleichender Sicht unter besonderer Berücksichtigung Kärntens. Carinthia II./Naturwiss. Beiträge zur Heimatkunde Kärntens/Mitteilungen des Naturwissenschaftlichen Vereins für Kärnten 45:78–158.

Scherabon, B., B. Gantenbein, V. Fet, M. Barker, M. Kuntner, C. Kropf & D. Huber, 2000. A new species of scorpion from Austria, Italy, Slovenia and Croatia: *Euscorpius gamma* Caporiacco, 1950, stat. nov. (Scorpiones: Euscorpiidae). In Gajdoš, P. & S. Pekár (eds), Proceedings of the 18th European Colloquium of Arachnology, Stará Lesná, 1999. Ekológia (Bratislava) 19(Suppl. 3): 253–262.

Teruel, R., V. Fet, V. & L.F. de Armas, 2004. A note on the scorpions from the Pirin Mountains, southwestern Bulgaria (Scorpiones: Buthidae, Euscorpiidae). Euscorpius 14:1–11.

Vachon, M., 1975. Recherches sur les Scorpions appartenant ou déposés au Muséum d'Histoire naturalle de Genève. I. Contribution à une meilleure connaissance des espèces et des sous-espèces de Scorpions du genre "*Euscorpius*" Thorell, 1876 (Fam. des Chactidae). Revue suisse de Zoologie 82(3):629–645.

Vachon, M. & M. Jaques, 1977. Recherches sur les Scorpions appartenant ou déposés au Muséum d'Histoire naturelle de Genève. 2. Contribution à la connaissance de l'ancienne espèce *Scorpius banaticus* C.L. Koch 1841, actuellement considérée comme synonyme de *Euscorpius carpathicus* (Linné 1767) (Fam. des Chactidae). Revue suisse de Zoologie 84(2):409–436.

Valle, A., 1975. Considerazioni intorno alle sottospecie di *Euscorpius carpathicus* (L.) (Scorpiones, Chactidae). L'Ateneo Parmense, Acta Naturalia 11(1):209–234.

Vignoli, V., N. Salomone, T. Caruso & F. Bernini, 2005. The *Euscorpius tergestinus* (C.L.Koch, 1837) complex in Italy: biometrics of sympatric hidden species (Scorpiones: Euscorpiidae). Zoologischer Anzeiger 244(2):97–114.

13 Spatial Niches of Opiliones (Arachnida) from Vitosha Mountains, Bulgaria

Plamen Mitov

Department of Zoology and Anthropology, Faculty of Biology, St. Kliment Ohridski University of Sofia, Dragan Tsankov Blvd. 8, 1164 Sofia, Bulgaria, e-mail: pl_mitov@yahoo.com

Abstract: The spatial niche breadths of 23 opilionid species (*Paranemastoma radewi, P. aurigerum ryla, Pyza bosnica, Carinostoma ornatum, Mitostoma chrysomelas, Trogulus tricarinatus, T. closanicus, Phalangium opilio, Opilio parietinus, O. saxatilis, O. dinaricus, O. ruzickai, Rilaena balcanica, R.* cf. *serbica, Lophopilio palpinalis, Zachaeus crista, Z. anatolicus, Lacinius horridus, L. ephippiatus, L. dentiger, Odiellus lendli, Mitopus morio, Leiobunum rumelicum*) were estimated using Hill's index (N_2). Harvestmen were sampled from 54 habitats (grouped in 21 habitat types) in Vitosha Mountains, Bulgaria. A significant correlation between the number of occupied habitats and niche breadth was established ($R^2 = 0.3055$, $p = 0.003676$). The harvestmen species were classified in two groups: (1) species with "narrower" niches ($1 \leq N_2 \leq 6$) and (2) species with "wider" ecological niches ($N_2 > 6$). The first group contains harvestmen that are relatively more dependent on the biotic and abiotic environmental factors. The second group includes species occurring mostly in forest habitats or both in forest and open grass habitats. A comparative analysis of niche dimensions revealed specific guilds of harvestmen species. The closely related species have strongly different niche dimensions, which might be an important feature for avoiding or reducing the interspecific competition. Various ranked-abundance distribution models were fitted to the dataset. The relevant species-abundance model for the harvestmen assemblages studied is the Preston's Log-normal Model

1 Introduction

The order Opiliones is one of the most species-rich orders of the class Arachnida, with ca. 6000 species (Harvey, 2002). The greatest species diversity of harvestmen was recorded in tropical and subtropical regions; members of the order are absent only in Antarctic. The European fauna of Opiliones has 310 species (69 genera, ten families) while this of the Balkan Peninsula contains 130 species (44 genera, nine families) (Karaman, 1995). Out of these, 61 species (31 genera, six families) have been recorded in Bulgaria (Mitov 2003; Mitov and Stoyanov, 2004; Deltshev et al., 2005).

Several aspects of the Opiliones are of special interest: an ancient animal group (fossils known since the Devonian; Dunlop et al., 2003, 2004), the harvestmen are very sensitive to the environmental conditions, especially to the moisture and the temperature regime (Hillyard and Sankey, 1989). Most of them prefer specific habitats and plant communities; therefore, the group is of great indicative value in ecological and zoogeographical studies (Guéorguiev, 1992). Each habitat is characterized by a specific range of harvestman species reflecting its microclimate.

V. Fet and A. Popov (Eds.), Biogeography and Ecology of Bulgaria, 423–446.

Their occurrence at a certain region can be informative about its climatic history, thus providing data useful for paleoecologists. The Opiliones, as epigeobionts, have an indicative value, which is used by zoologists and urban ecologists for the purposes of the analysis of the synanthropization and environmental conditions (Schaefer, 1973; Weiss and Sarbu, 1977; Hammelbacher and Mühlenberg, 1986; Klausnitzer, 1990; Novak et al., 2002; Mitov and Stoyanov, 2004). The interest in the ecology of the Opiliones increases in terms of evaluating their impact on the community structure and their interrelationships with the components of the abiotic and biotic environment. Species of the order are important for the dynamics of the natural and human-modified ecosystems. Harvestmen are predators on some pests as snails, mites and insects, and may have an important role for the regulation of their populations (Bristowe, 1949; Todd, 1950; Ashby, 1974; Butcher, 1986; Dixon and McKinlay, 1989; Drummond et al., 1990).

The Opiliones feed on some of the most important composers of the litter, i.e. Collembola and Acari (Mitov, 1988, 2000). The range of their prey in nature includes more than 100 invertebrate species (Mitov, 2000). They are known to be a prey for about 300 species of predators (Cokendolpher and Mitov, in press). In view of this, the harvestmen attract attention as concentrators of chemical elements and indicators of radioactive pollution (Pokarzhevskiy, 1985). The Opiliones are also important from parasitological point of view. There are about 100 reported species of pathogens and parasites, and about two dozen of phoretic, endozoic or epizoic associations recorded (Cokendolpher and Mitov, in press). Harvestmen attracted the attention of biochemists as a source of new and rare substances (Fieser and Ardao, 1956; Ekpa et al., 1984, 1985; all cited after Cokendolpher, 1993) and bioengineers as a basis for technical ideas and principles (Lange, 1984; Garmash, 1985; Mitov, 1990, 1991).

Studies on the Opiliones in Bulgaria have been mostly faunistic and taxonomic (Staręga, 1976; Martens, 1978). Data about the biology, ecology, and zoogeography of the group are scarce (Mitov, 2000). In view of this, the present paper is a continuation of the series of ecological studies on Bulgarian harvestmen (Mitov, 1995, 1996, 1997; Mitov and Stoyanov, 2004, 2005). The aim of the present work is to characterize the spatial (habitat) niches of the species of the Opiliones in Vitosha Mountains, the region with the most fully studied opilionid assemblages in Bulgaria, and the type of their relationships (Mitov, 2000).

As a whole, the niches of the Opiliones are not studied. The only papers containing some data about the dimensions of the spatial niches (NB) of harvestmen are those by Weiss (1984) and Simon (1995), both using the index of Colwell and Futuyma (1971). Weiss (1984) studied 16 species in Romania: *Trogulus nepaeformis* (Scopoli, 1763) (NB = 1.00), *Opilio saxatilis* (NB = 1.18), *Lacinius ephippiatus* (NB = 1.36), *Rilaena triangularis* (Herbst, 1799) (NB = 1.60), *Lacinius dentiger* (NB = 1.72), *Opilio ruzickai* (NB = 1.81), *Paranemastoma silli* (Herman, 1871) (NB = 3.05), *Carinostoma elegans* (Sørensen, 1894) (NB = 3.66), *Lacinius horridus* (NB = 3.77), *Egaenus convexus* (C.L. Koch, 1835) (NB = 3.92), *Lophopilio palpinalis* (NB = 4.00), *Mitostoma chrysomelas* (NB = 4.79), *Trogulus tricarinatus* (NB = 5.36), *Dicranolasma scabrum* (Herbst, 1799) (NB = 5.41), *Zachaeus crista* (NB = 5.62), and *Phalangium opilio* (NB = 5.66). Almost all of these (excluding *Paranemastoma silli*) occur in Bulgaria. Simon (1995) reported the spatial niches of two species from Germany: *Rilaena triangularis* (NB = 2.41) and *Paroligolophus agrestis* (Meade, 1855) (NB = 2.50–3.04). The latter species does not occur in Bulgaria.

The lack of information on the spatial niches of harvestmen species in Bulgaria provoked the present study. Hopefully, these results will also contribute to the autecological characteristics of the species and, to some extent, to the knowledge on their interspecific interactions.

2 Material and Methods

The present study was based on the examination of 31,644 specimens (8315 males, 14,865 females, 8464 juv.), belonging to 23 species, collected by the author from 28 February 1987 to 28 April 1989 from Vitosha Mts. In total, 653 pitfall traps filled with 4% formalin solution were used. They were placed at 54 sampling localities (habitats), grouped in 21 habitat types. The localities were situated at an altitude between 750 and 2290 m (the latter is the maximum altitude for these mountains), at intervals between 200 and 500 m. The studied localities are representative for all the parts of Vitosha Mts., for all four vegetation belts and for slopes of various expositions. They are located in the low-mountain zone (up to 1450 m), middle-mountain zone (between 1450 and 1850 m), and high-mountain zone (above 1850 m) (classification of Hubenov, 1990). Samples were collected monthly. For further details on the sampling scheme see Mitov (1996). The specimens are preserved in the author's collection.

Localities and habitat types studied

Each studied habitat (Figs. 7–9) is marked by a code consisting of capital letters indicating the name of the locality, followed by the altitude and lowercase letters showing the type of the habitat. The codes of the names of the localities are: **B**, Bosnek; **CK**, Cherniya Kos site; **CP**, above Chuipetlovo; **CV**, Cherni Vrah summit; **D**, above Dragalevtsi; **K**, above Kladnitsa; **R**, Rudartsi; **SH**, Shumako site; **Z**, above Zheleznitsa.

The codes of the habitat types are as follows: *Open habitats*: **g**, glades in forests of beech (*Fagus sylvatica* L.) and hornbeam (*Carpinus betulus* L.); **m1**, oligotrophic and xerophytic meadows; **m2**, dry meadows in deciduous forests; **m3**, fresh to moist meadows in deciduous forests; **m4**, fresh meadows in coniferous forests; **m5**, fresh subalpine meadows; **m6**, moist to fresh subalpine meadows; **sb**, subalpine peat moss bogs; *Forest habitats*: **f1**, xeromesophytic Turkey oak forests (*Quercus cerris* L.); **f2**, hornbeam–beech forests; **f3**, beech–hornbeam forests; **f4**, beech forests; **f5**, spruce forests (*Picea abies* (L.) Karst); **f6**, artificial Macedonian pine forests (*Pinus* *peuce* Griseb.); **f7**, artificial Austrian pine forests (*Pinus nigra* Arn.); *Forest rivulet-bank habitats*: **r1**, rivulet-bank in f2; **r2**, rivulet-bank in f2 (islet); **r3**, rivulet-bank in f3; **r4**, rivulet-bank in f4; **r5**, rivulet-bank in f5; *Open rivulet-bank habitats*: **om**, rivulet-bank through meadows.

The breadths of spatial niches of the studied 23 species were estimated on the basis of the Hill's index (N_2) (Bengtson, 1980; Pesenko, 1982; Arndt and Pellmann, 1996):

$$(13.1) \quad N_2 = (\textstyle\sum_i p_i^2)^{-1}, (i = 1, 2, 3, \ldots, \mathrm{S}), \; 0 = N_2 = \mathrm{S},$$

where S is number of species in the j-th site (habitat) (j-th column of the data matrix); p_i is percentage of the i-th species. The software package BIODIV (Baev and Penev, 1993) was used for analysis.

For visualization of the competition level among the species, a UPGMA dendrogram presenting the overlap of the spatial niches was built based on Pianka's index (Ip) using the package BIODIV (Baev and Penev, 1993):

$$(13.2) \quad I_p = \frac{\sum_i p_{ij} p_{ik}}{\sqrt{\sum_i p_{ij}^2 \sum_j p_{ik}^2}},$$

where p_{ij} (p_{ik}) is a proportion of the i-th species in the $j(k)$-th site (habitat) of the total abundance of the same species in all samples (in terms of the niche theory, p_{ij} is defined as a proportion of resource j utilized by the i-th consumer).

The niche overlap was ranked from very high (> 85%) to high (85–55%), moderate (54–35%), and low (< 35%). This scale was modified from Zlotin's scale for faunistic similarity (1975).

The following nonparametric correlation analyses (Kruskal–Wallis rank sum test) were conducted: between the niche breadths and the number of habitats in which a certain species occurs; between the number of specimens of each species and its spatial niche breadth; between the average body length of both sexes (Table 1) and the spatial niche breadths. Estimation of the average body size of *Opilio parietinus* was based mostly on samples from the city of Sofia (collected in parks and walls of the buildings); a singe male (L: 5.0 mm) and 3 females (L: 3.8; 4.6; 6.0 mm) from Vitosha Mts. were included.

The species-abundance curves (Whittaker, 1965, 1972, cited after Odum,, 1986), termed also "rank-abundance curves" (Begon et al., 1989), of the harvestman assemblages of Vitosha Mts. (Mitov, 2000) were built and used for revealing the type of relationships among the spatial niches of the studied harvestman species. They were analyzed by the Kolmogorov–Smirnov test. All analyses were performed in the *R* language environment (Ihaka and Gentleman, 1996). Of course, these curves of the ranked distribution of relative abundance have to be considered an abstract reflection of the complex community structure (Begon et al., 1989). Figs. 7–9 present a collection of rank-abundance distribution examples from each habitat type.

Table 1 Harvestman species from Vitosha Mts. and their average body lengths

Species	Average body length (mm)	
	Male	Female
Paranemastoma radewi (Roewer, 1926)	$\bar{x} = 3.81 \pm 0.056$ (3.00–5.50), n = 101	$\bar{x} = 4.75 \pm 0.05$ (3 31–6.00), n = 113
Paranemastoma aurigerum ryla (Roewer, 1951)	$\bar{x} = 3.41 \pm 0.05$ (3.00–4.61), n = 38	$\bar{x} = 4.21 \pm 0.07$ (3 60–5.00), n = 36
Pyza bosnica (Roewer, 1919)	$\bar{x} = 3.42 \pm 0.03$ (2.50–4.50), n = 140	$\bar{x} = 3.81 \pm 0.03$ (3 00–5.38), n = 148
Carinostoma ornatum (Hadži, 1940)	$\bar{x} = 1.83 \pm 0.12$ (1.40–2.20), n = 7	$\bar{x} = 1.95 \pm 0.11$ (1 10–2.50), n = 13
Mitostoma chrysomelas (Hermann, 1804)	$\bar{x} = 1.87 \pm 0.16$ (1.60–2.20), n = 4	$\bar{x} = 2.45 \pm 0.06$ (2 20–2.69), n = 9
Trogulus tricarinatus (Linnaeus, 1758)	$\bar{x} = 4.99 \pm 0.14$ (4.10–6.30), n = 13	$\bar{x} = 5.648 \pm 0.045$ (4.50–6.46), n = 65
Trogulus closanicus Avram, 1971	$\bar{x} = 7.59 \pm 0.033$ (6.85–8.61), n = 96	$\bar{x} = 8.66 \pm 0.09$ (7 20–9.46), n = 36
Phalangium opilio Linnaeus, 1758	$\bar{x} = 4.81 \pm 0.07$ (2.70–7.50), n = 135	$\bar{x} = 6.12 \pm 0.08$ (3 50–10.00), n = 229
Opilio parietinus (De Geer, 1778)	$\bar{x} = 4.51 \pm 0.17$ (2.25–6.00), n = 25	$\bar{x} = 6.14 \pm 0.193$ (3.20–8.60), n = 55
Opilio saxatilis C.L. Koch, 1839	$\bar{x} = 3.01 \pm 0.11$ (2.00–4.23), n = 27	$\bar{x} = 4.20 \pm 0.125$ (2.54–6.00), n = 46
Opilio dinaricus Šilhavý, 1938	$\bar{x} = 3.43 \pm 0.077$ (2.20–5.00), n = 66	$\bar{x} = 4.81 \pm 0.186$ (2.50–7.31), n = 49
Opilio ruzickai Šilhavý, 1938	$\bar{x} = 4.51 \pm 0.17$ (3.00–6.00), n = 21	$\bar{x} = 5.76 \pm 0.23$ (4 34–7.30), n = 19
Rilaena balcanica Šilhavý, 1965	$\bar{x} = 3.14 \pm 0.07$ (2.20–4.50), n = 48	$\bar{x} = 4.61 \pm 0.08$ (3 00–6.15), n = 100
Rilaena cf. *serbica* Karaman, 1992	–	$\bar{x} = 6.50 \pm 0.297$ (3.50–8.50), n = 59
Lophopilio palpinalis (Herbst, 1799)	$\bar{x} = 2.87 \pm 0.03$ (2.10–4.00), n = 183	$\bar{x} = 3.87 \pm 0.04$ (2 69–5.00), n = 193
Zachaeus crista (Brullé, 1832)	$\bar{x} = 6.64 \pm 0.16$ (4.50–12.00), n = 69	$\bar{x} = 7.53 \pm 0.17$ (4 70–13.00), n = 91
Zachaeus anatolicus (Kulczyński, 1903)	$\bar{x} = 7.75 \pm 0.87$ (6.20–9.50), n = 4	$\bar{x} = 9.47 \pm 0.61$ (6 80–12.60), n = 10
Lacinius horridus (Panzer, 1794)	$\bar{x} = 4.46 \pm 0.04$ (2.46–8.00), n = 535	$\bar{x} = 6.36 \pm 0.05$ (3 00–10.00), n = 757
Lacinius ephippiatus (C.L. Koch, 1835)	$\bar{x} = 3.51 \pm 0.08$ (2.30–5.00), n = 42	$\bar{x} = 4.61 \pm 0.15$ (3 00–6.92), n = 75
Lacinius dentiger (C.L. Koch, 1847)	$\bar{x} = 4.65 \pm 0.13$ (3.46–8.00), n = 51	$\bar{x} = 6.57 \pm 0.09$ (4 00–10.00), n = 195
Odiellus lendli (Sørensen, 1894)	$\bar{x} = 4.45 \pm 0.08$ (3.92–5.00), n = 17	$\bar{x} = 6.02 \pm 0.10$ (5 00–7.08), n = 31
Mitopus morio (Fabricius, 1779)	$\bar{x} = 4.63 \pm 0.065$ (2.90–7.20), n = 182	$\bar{x} = 6.567 \pm 0.078$ (4.00–10.30), n = 260
Leiobunum rumelicum Šilhavý, 1965	$\bar{x} = 3.46 \pm 0.12$ (2.00–4.50), n = 24	$\bar{x} = 4.74 \pm 0.29$ (2 80–6.20), n = 14

The expression "phenological peak" is used to designate a maximum of the abundance of a particular phenological stage connected with mass hatching of juveniles, copulation, or egg laying.

Symbols and abbreviations used: $\bar{x}$ = mean; ± = standard error of the mean; C.V. = coefficient of variation; L = body length; ML = maximum likelihood estimate; moment = moment estimate; n = sample size; N = numbers of the specimens; NB$\bar{x}$ = average niche breadth; R^2 = coefficient of correlation; S.D. = standard deviation.

3 Results and Discussion

The niche concept is useful for quantitative evaluation of the differences among species (or among populations of a species) on the basis of either a single character or several principal characters (Odum,, 1986). The major parameters used for the niche characterization are niche overlap and niche breadth (Odum,, 1986).

3·1 Spatial niche breadths

Species evolve by specializing toward various zones of a certain resource gradient; this reduces interspecific competition (Whittaker, 1980). Broader niches correspond to weaker levels of competition, greater diversity of habitats and food, and easier access to resources (Odum, 1975, 1986; Giller, 1988). When there is a trend for a limited distribution of a certain species in several habitats out of all habitats available for the community, then the values of niche breadth index (in our case, N_2) are low; on the contrary, the values are high when each species exhibits a trend toward an indifferent occurrence in many or all habitats available (Pielou, 1977).

On the basis of the Hill's index (N_2) (Fig. 1), the harvestman species from all habitats studied in Vitosha Mts. may be tentatively classified into two groups: (1) species with spatial niche breadths $1 \leq N_2 \leq 6$, and (2) species with spatial niche breadths $N_2 > 6$.

The first group includes harvestmen, which seem to be adapted to a limited range of environmental conditions, i.e. stenotopic species. These species are rare for Vitosha Mts.: *Opilio parietinus, Odiellus lendli, Trogulus closanicus, Zachaeus anatolicus, Mitostoma chrysomelas, Lacinius ephippiatus, Carinostoma ornatum*, and *Opilio saxatilis*. Most of them occur in small numbers (Fig. 3), and their distribution is restricted to a few habitats (Fig. 2) where they are probably in competitive relationships with numerous other harvestman species. Some of these species are thermophilous and, therefore, their distribution is strongly limited by the temperature regime. Spatial niches of *Opilio saxatilis* and *Lacinius ephippiatus* in Romania are also narrow (Weiss, 1984; see Introduction). Other species, such as *Carinostoma ornatum* and *Trogulus closanicus*, are also characterized by low vagility, and their abilities to colonize new habitats strongly depend on the available

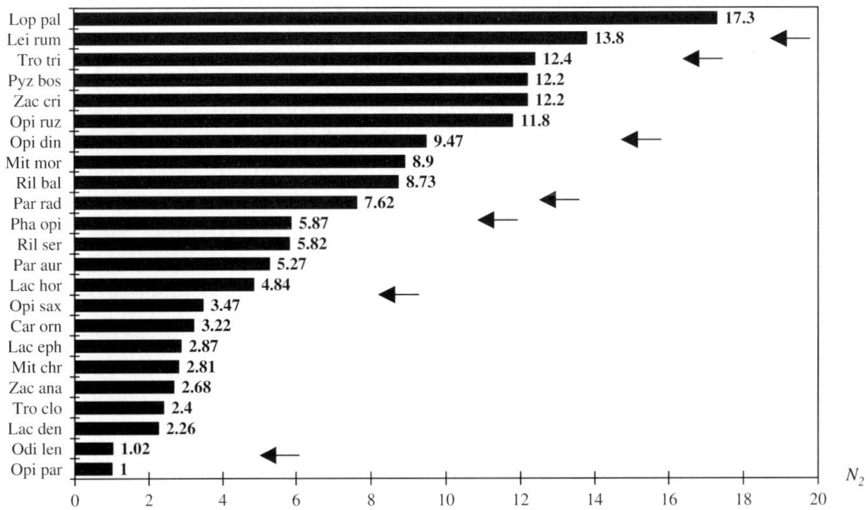

Fig. 1 Spatial niche dimensions (Hill's index, N_2) of harvestmen in Vitosha Mts. based on all the studied habitats. Arrows show the rapid decreases of the N_2 values.

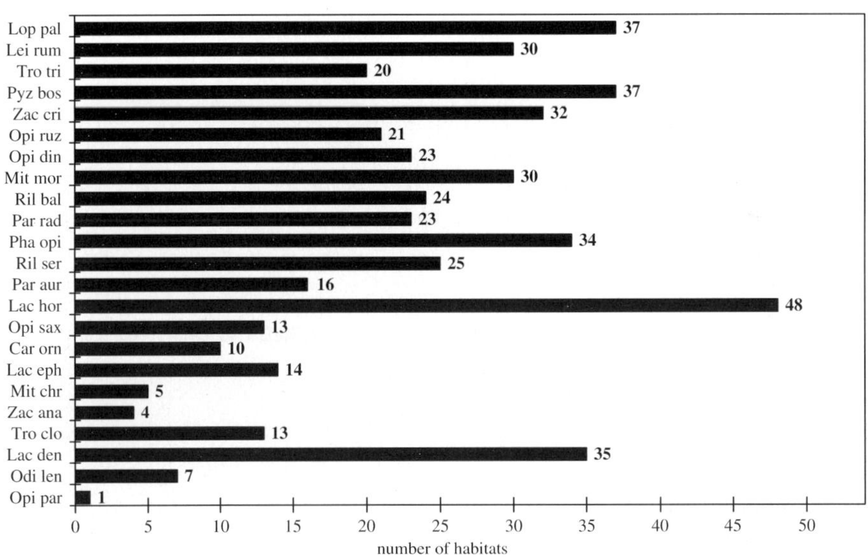

Fig. 2 Number of habitats occupied by each of the harvestman species in Vitosha Mts.

food resources (e.g. *Trogulus closanicus* probably feeds on certain terrestrial snail species, similarly to other congeneric species; see Pabst, 1953).

The first group includes also *Lacinius horridus, L. dentiger, Rilaena* cf. *serbica, Phalangium opilio*, and *Paranemastoma aurigerum ryla*. The former two are among the most abundant harvestman species in Vitosha Mts. They also occur in a large number of habitats (48 and 35, respectively) (Figs. 2, 3). In spite of this, their spatial niches are narrow. The probable restricting factors are low temperatures (these species are thermophilous), food deficiency, predators, strong interspecific and intraspecific competition (Fig. 4). According to Weiss (1984), the spatial niche breadth of *Lacinius horridus* from Romania is also not large; in contrast, the niche of *Phalangium opilio* is larger compared to niches of the remaining species studied in Romania.

Almost equal niche breadth of *Phalangium opilio* ($N_2 = 5.87$) and *Rilaena* cf. *serbica* ($N_2 = 5.82$) (Fig. 1) are of a great interest. *Phalangium opilio* is a photophilous and eurytopic species (Martens, 1978; Weiss, 1996) (Table 2) while *Rilaena* cf. *serbica* is a parthenogenetic (thelytokous) form of mountain origin (Mitov and Stoyanov, 2004) and, as a result, exhibits higher plasticity. In Vitosha Mts., both species are found in large numbers in open grass habitats (mostly in moist to fresh subalpine or forest meadows), which contain more uniform microhabitats compared to forest habitats. This could be the main reason for their relatively narrow niches ($N_2 < 6$). It could be suggested that these species reduce

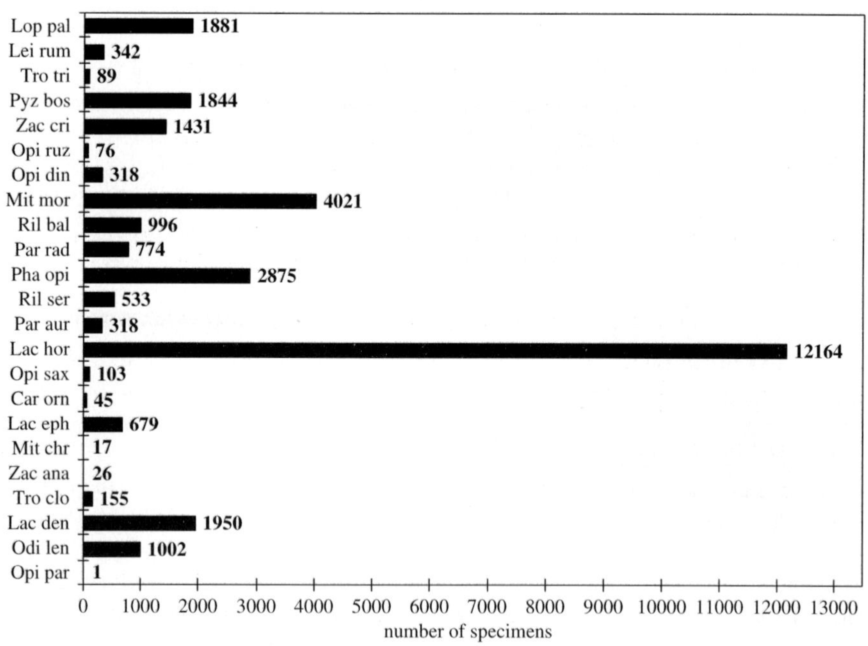

Fig. 3 Number of harvestman specimens captured by pitfall traps in Vitosha Mts.

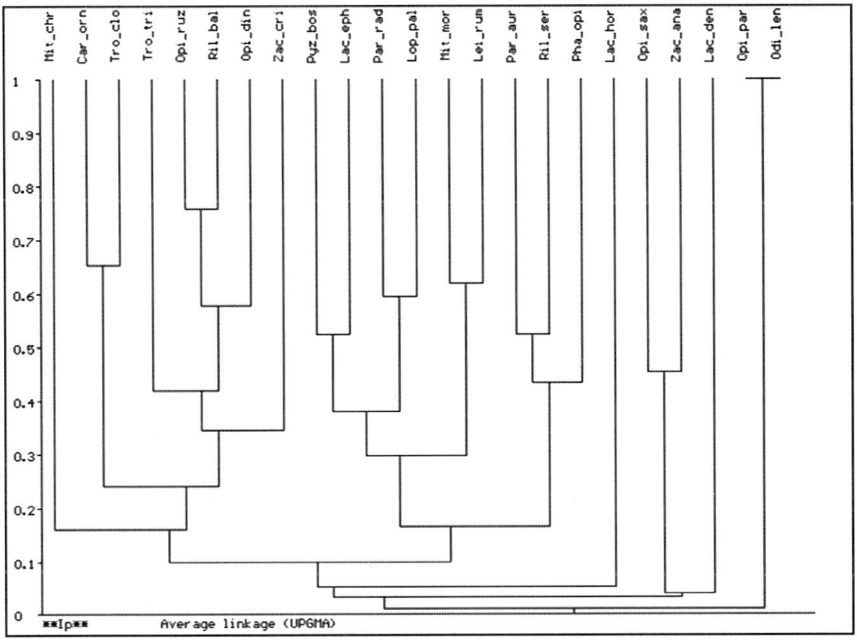

Fig. 4 Niche overlap of harvestman species in Vitosha Mts. based on all studied habitats.

their competitive interaction in habitats where they co-occur (Fig. 4) due to their phenological (their phenological peaks do not coincide), microhabitat-selective and behavioral strategies (differences in their diurnal locomotory activities).

Paranemastoma aurigerum ryla, also belonging to the same group, exhibits a niche breadth ($N_2 = 5.27$) very similar to those of the previous two species. It occurs also in open grass habitats but, being a hygrophilous, ombrophilous, psychrophilous, and frigostable species, has a distribution much more dependent on abiotic and biotic factors (Fig. 4).

The second group of species ($N_2 > 6$) includes harvestmen preferring mostly forest habitats, i.e. *Rilaena balcanica, Paranemastoma radewi, Opilio dinaricus, O. ruzickai, Leiobunum rumelicum*, or species occurring in both forest and open grass habitats, i.e. *Mitopus morio, Zachaeus crista, Pyza bosnica, Trogulus tricarinatus*, and *Lophopilio palpinalis*. Habitat preferences of these species feature greater abundance and diversity of microhabitats and nutrients. Furthermore, some of these species (*Paranemastoma radewi, Pyza bosnica, Lophopilio palpinalis, Mitopus morio*, and *Leiobunum rumelicum*) are hygrophilous, ombrophilous, psychrophilous, and frigostable, and occur in habitats characterized by a small number of harvestman species sharing limited resources. *Trogulus tricarinatus, Lophopilio palpinalis*, and *Zachaeus crista*, similarly to their populations in Romania (Weiss, 1984), have broad spatial niches in Vitosha Mts. as well.

Mitopus morio is the second species among the harvestmen of Vitosha Mts. by its abundance (over 4000 specimens collected). It occurs in more than half of the habitats situated above 1000 m (Figs. 2, 3). Nevertheless, its niche ($N_2 = 8.9$) is not among the broadest ones, which could be a result of its tolerance to severe environmental conditions. This species is a good illustration of the first biocenotic principle of Thienemann (see Dajoz, 1975, p. 270) stating that in unfavourable conditions, the number of species is small but they are represented by large numbers of individuals.

Opilio dinaricus is rare and occurs in isolated localities throughout its geographical range; according to Rafalski (1962), this species has a regressive distribution. In spite of this, the breadth of its spatial niche ($N_2 = 9.47$) in Vitosha Mts. is the fourth by size among the niches of the group considered. Probably, the deciduous forests, which are preferred by this species, provide a greater diversity of microhabitats and feeding resources. Furthermore, it competes only with a few harvestman species (Fig. 4), probably avoiding competitive interactions due to its microhabitat-selective (juveniles live in soil, adults inhabit trees and shrubs; Martens, 1978) and phenological strategies (its phenological peaks not coinciding with those of the remaining species).

Leiobunum rumelicum is characterized by a very broad spatial niche ($N_2 = 13.8$), though it has low abundance (342 specimens collected) (Figs. 1, 3). In Vitosha Mts., it occurs in numerous forest and some grassland habitats (Fig. 2, Table 2), which are a prerequisite for diverse microhabitats, with low levels of competition with other harvestmen (Fig. 4).

Among all the harvestmen in Vitosha Mts., *Lophopilio palpinalis* has the broadest spatial niche ($N_2 = 17.3$). This species occurs in 37 habitats. Influence of various factors determining the distribution of this species seems weaker, and its populations are not strongly limited by the environmental resources. Possibly, the small body size of this species is an advantage allowing its occurrence in spatially restricted soil and litter microhabitats where it is not readily discernable by predators. In these microhabitats, *L. palpinalis* is well protected against desiccation or freezing because its phenological peaks are in the hottest and the coolest periods during the year. Furthermore, presence of abundant food resources and trophic specialization of *L. palpinalis*, especially the feeding on much smaller animals, reduce its competition with other harvestman species (Fig. 4).

A comparison between *Lophopilio palpinalis* and *Pyza bosnica* shows that these species occur in equal numbers of habitats (37) and have similar abundance (Figs. 2, 3). However, *Pyza bosnica* has a narrower spatial niche ($N_2 = 12.2$). This might be a result of the severe conditions tolerated by this species and its preference of habitats characterized by low temperatures.

Spatial niches of closely related species have very different dimensions (Fig. 1). This might be a feature for avoiding or reducing the interspecific competition, illustrated by the following pairs or groups of closely related species: *Paranemastoma aurigerum ryla* ($N_2 = 5.27$) and *P. radewi* ($N_2 = 7.62$); *Trogulus closanicus* ($N_2 = 2.4$) and *T. tricarinatus* ($N_2 = 12.4$); *Lacinius dentiger*

($N_2 =$ 2.26), *L. ephippiatus* ($N_2 = 2.87$), and *L. horridus* ($N_2 = 4.84$); *Zachaeus anatolicus* ($N_2 = 2.68$) and *Z. crista* ($N_2 = 12.2$); *Opilio parietinus* ($N_2 = 1.0$), *O. saxatilis* ($N_2 = 3.47$), *O. dinaricus* ($N_2 = 9.47$), and *O. ruzickai* ($N_2 = 11.8$); *Rilaena* cf. *serbica* ($N_2 = 5.82$) and *R. balcanica* ($N_2 = 8.73$).

The general appearance of the dataset containing the dimensions of the spatial niches of all studied species (Fig. 1) is stairway-like. Seven "stairs", i.e. rapid decreases of the N_2 values between neighboring species, can be observed (see arrows in Fig. 1). These can be explained by the "unsaturated" condition of the habitats studied in relation to the species composition of harvestmen. Therefore, it could be expected that these habitats are potentially capable of accepting species-invaders. Moreover, three species recorded from Vitosha Mts. are not involved in the present analysis. These are *Dicranolasma scabrum, Histricostoma drenskii* Kratochvíl, 1958 (Mitov and Stoyanov, 2004), and *Rafalskia olympica* (Kulczyński, 1903) (see Staręga, 1976). The former two species are absent from the pitfall trap samples, and the latter species has not been recorded during the present study.

The number of habitats used by a certain species depends on the dimensions of its spatial niche (Pielou, 1977; Odum,, 1986). The regression analysis revealed significant correlation ($R^2 = 0.3055$, $p = 0.003676$) between the number of habitats utilized by the species and the breadths of their niches (N_2) (Fig. 5). No significant correlation was found between the numbers of the specimens ($\log N + 1$) of each species and the dimensions of their niches ($R^2 = 0.06586$, $p = 0.1252$). Number of studies suggests that there is a correlation between the body size and the niche dimensions (Bengtson, 1980; Whittaker, 1980; Odum,, 1986; Giller, 1988). However, application of regression analysis to the present dataset showed that there was no significant correlation between the mean body length of both sexes and the niche dimensions (males: $R^2 = 0.01748$, $p = 0.2550$; females: $R^2 = -0.00892$, $p = 0.3796$).

The harvestman species from Vitosha Mts. and their spatial niche dimensions in various types of habitats are presented in Table 2. As it can be seen, the highest number of species occurs in open grass habitats. Among them, *Phalangium opilio* has the broadest spatial niche, followed by *Zachaeus crista, Trogulus tricarinatus*, and *Lophopilio palpinalis*. This group includes photophilous and thermophilous species as well as species, which prefer small-sized microhabitats in order to be reliably protected. In the rivulet-bank habitats in grasslands, hygrophilous and frigostable *Lophopilio palpinalis* and *Pyza bosnica* have the broadest spatial niches. In forest habitats, *Leiobunum rumelicum* has the broadest niche, followed by *Opilio ruzickai, Rilaena* cf. *serbica, Opilio dinaricus, Pyza bosnica*, and *Lophopilio palpinalis*. These species are mostly ombrophilous, mesoombrophilous, mesohygrophilous, and frigostable, preferring forest habitats. In the forest rivulet-bank habitats, *Lophopilio palpinalis* possesses the broadest niche again, followed by *Paranemastoma radewi*. The latest species is ombrophilous, hygrophilous, and psychrophilous.

It should be mentioned that the average spatial niche breadths ($NB\bar{x}$) of harvestmen in forest habitats and that in the open habitats have very similar sizes

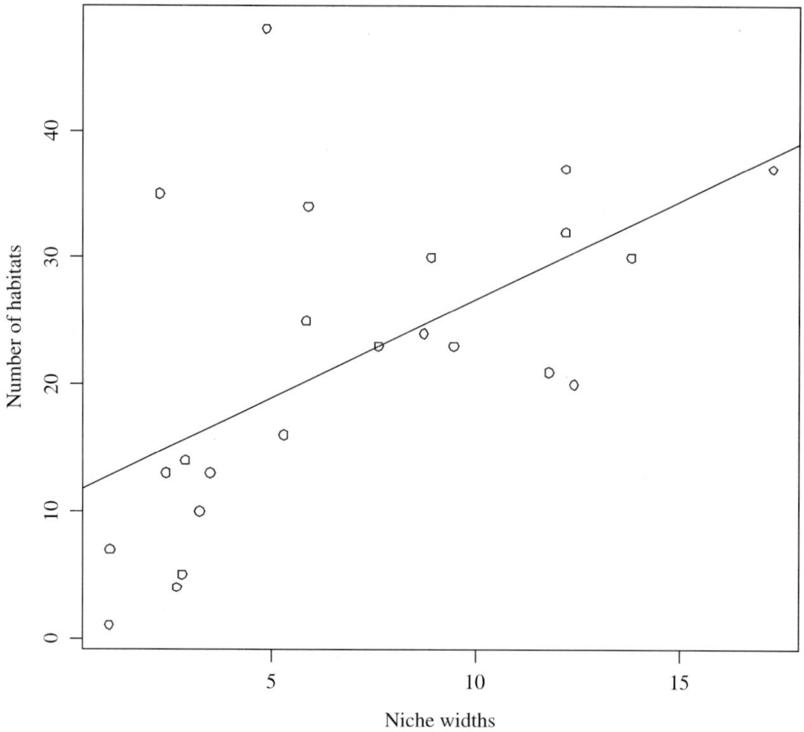

Fig. 5 Relationship between the numbers of habitats occupied by each species and the niche dimensions.

(Table 2). However, a comparison of niche dimensions among habitats showed that there was a significant difference (Kruskal–Wallis statistic = 15.8481, p = 0.001218): the niches of Opiliones inhabiting open rivulet-bank habitats were on average lower than those in the other habitat types (Fig. 6).

3·2 Guilds of harvestmen

One way to describe the functional relationships among species is by using the concept of ecological guilds (Giller, 1988). In view of the definition of a guild (Odum, 1986; Giller, 1988; see also Uetz et al., 1999), and because the space is a resource of primary importance, the comparison between the dimensions of spatial niches of the species studied (Fig. 1) reveals that *Zachaeus crista* ($N_2 = 12.2$) and *Pyza bosnica* ($N_2 = 12.2$) having equal niche dimensions are members of the same guild. Two further species can probably be added to these because the similar dimensions of their niches: *Trogulus tricarinatus* ($N_2 = 12.4$) and *Opilio ruzickai* ($N_2 = 11.8$). Several further guilds can be detected on the same basis. These are the guilds of: *Odiellus lendli* ($N_2 = 1.02$) plus *Opilio parietinus* ($N_2 = 1.0$); *Trogulus*

Table 2 Harvestman species from Vitosha Mts. and their spatial niche dimensions in various types of habitats

Open habitats ($n = 20$)		Forest habitats ($n = 20$)		Forest rivulet-bank habitats ($n = 8$)		Open rivulet-bank habitats ($n = 6$)	
Pha_opi	7.93	Lei_rum	8.32	Lop_pal	5.78	Lop_pal	3.06
Zac_cri	7.07	Opi_ruz	7.44	Par_rad	5.70	Pyz_bos	3.00
Tro_tri	6.35	Ril_ser	6.76	Lei_rum	4.59	Lac_den	2.46
Lop_pal	5.55	Opi_din	6.68	Opi_din	3.60	Ril_bal	2.27
Pyz_bos	4.97	Pyz_bos	6.09	Lac_den	3.21	Par_aur	2.25
Mit_mor	4.85	Lop_pal	5.25	Lac_hor	2.83	Ril_ser	2.00
Ril_bal	4.28	Mit_mor	4.23	Mit_mor	2.68	Mit_mor	1.88
Ril_ser	4.15	Tro_tri	4.17	Ril_bal	2.50	Lac_hor	1.72
Lac_hor	3.47	Ril_bal	4.10	Zac_cri	2.42	Tro_tri	1.68
Lac_den	3.25	Zac_cri	3.71	Par_aur	2.31	Pha_opi	1.63
Lei_rum	3.00	Car_orn	3.00	Car_orn	2.27	Par_rad	1.48
Opi_sax	2.75	Par_rad	2.85	Opi_ruz	2.27	Opi_ruz	1.38
Zac_ana	2.68	Opi_sax	2.81	Pyz_bos	2.15	Zac_cri	1.25
Opi_ruz	2.67	Pha_opi	2.70	Ril_ser	2.09	Lac_eph	1.0
Par_aur	2.53	Tro_clo	2.10	Tro_tri	2.00	Lei_rum	1.0
Car_orn	2.23	Lac_den	1.97	Lac_eph	1.92	Mit_chr	1.0
Lac_eph	2.18	Odi_len	1.81	Mit_chr	1.74	Odi_len	1.0
Par_rad	1.67	Lac_hor	1.66	Tro_clo	1.46	Opi_din	1.0
Opi_din	1.29	Par_aur	1.17	Pha_opi	1.38	Opi_sax	1.0
Tro_clo	1.15	Lac_eph	1.08				
Mit_chr	1.0	Mit_chr	1.00				
Odi_len	1.0						
Opi_par	1.0						
23 species		21 species		19 species		19 species	
NB$\bar{x} = 3.35 \pm 0.42$		NB $\bar{x} = 3.76 \pm 0.48$		NB $\bar{x} = 2.78 \pm 0.29$		NB $\bar{x} = 1.69 \pm 0.15$	
S.D. = 2.00		S.D. = 2.23		S.D. = 1.28		S.D. = 0.67	
C.V. = 0.598		C.V. = 0.595		C.V. = 0.46		C.V. = 0.40	

closanicus ($N_2 = 2.4$) plus *Lacinius dentiger* ($N_2 = 2.26$); *Lacinius ephippiatus* ($N_2 = 2.87$) plus *Mitostoma chrysomelas* ($N_2 = 2.81$); *Opilio saxatilis* ($N_2 = 3.47$) plus *Carinostoma ornatum* ($N_2 = 3.22$); *Mitopus morio* ($N_2 = 8.9$) plus *Rilaena balcanica* ($N_2 = 8.73$); *Phalangium opilio* ($N_2 = 5.87$) plus *Rilaena* cf. *serbica* ($N_2 = 5.82$), with a further probable member *Paranemastoma aurigerum ryla* ($N_2 = 5.27$).

Zachaeus anatolicus ($N_2 = 2.68$) might be related either to the guild of *Lacinius dentiger* and *Trogulus closanicus* or in the guild of *Mitostoma chrysomelas* and *Lacinius ephippiatus*. However, it is not placed in these guilds because it shares habitats with a single member of each of them.

To the guild of *Mitopus morio* plus *Rilaena balcanica, Opilio dinaricus* ($N_2 = 9.47$) and *Paranemastoma radewi* ($N_2 = 7.62$) might be affiliated. In addition, *Lacinius horridus* ($N_2 = 4.84$) might be considered a member either of the guild of

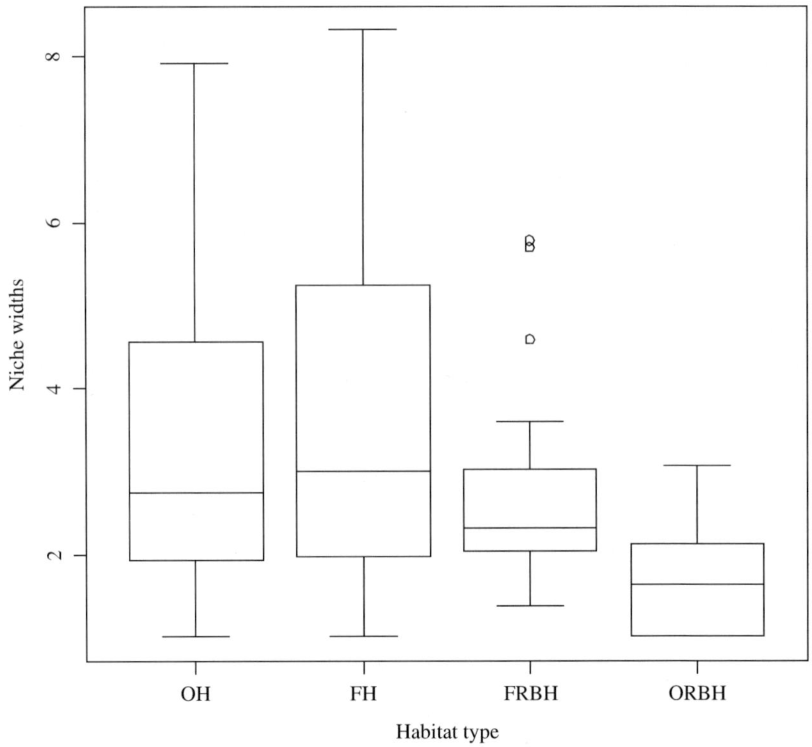

Fig. 6 Boxplot of the relationship of opilionid niche dimensions among habitat types. OH, Open habitats; FH, Forest habitats; FRBH, Forest rivulet-bank habitats; ORBH, Open rivulet-bank habitats.

Opilio saxatilis plus *Carinostoma ornatum*, or that of *Phalangium opilio, Rilaena* cf. *serbica* plus *Paranemastoma aurigerum ryla.*

The competition among the species within these guilds is mitigated. This is due to the slight overlap in both their spatial niches (Fig. 4) and phenological peaks. Furthermore, body sizes of species belonging to the same guild are distinctly different (Table 1), which might be a prerequisite for substantial differences in the feeding ranges relative to the size of the prey. It should be mentioned that the ratio between the average body lengths of species belonging to the same guild (calculated for both sexes) exhibits similar values ranging between 0.51 and 0.98.

Some species belonging to the same guild are characterized by similar body sizes, strongly overlapping spatial niches, and coinciding phenological peaks; most probably, they avoid strong interspecific competition due to their microhabitat selective strategies. A good example for this are *Opilio parietinus* occurring in shrubs, on trunks of trees, and on the walls, and *Odiellus lendli* living under stones in open grasslands. Vice versa, species belonging to the same guild and characterized by similar environmental requirements and occurring in similar microhabitats may avoid the feeding and spatial competition by only slightly overlapping their phenological peaks or by the great differences in their body size (Table 1). Examples for this are the guild of *Lacinius ephippiatus* and *Mitostoma chrysomelas* preferring shady and moist habitats, and the guild of *Opilio saxatilis* and *Carinostoma ornatum* inhabiting mostly open warm habitats.

Species belonging to the same guild and having overlapping phenological peaks may be clearly separated not only by their microhabitats but also by their feeding, e.g. two thermophilous species, *Trogulus closanicus* and *Lacinius dentiger*, feeding on different prey. The former occurs under stones and the latter is a hylobiont. Similarly, microhabitat and feeding isolation can be observed relative to the comparatively more plastic species *Trogulus tricarinatus* and *Opilio ruzickai*, which form a guild together with *Pyza bosnica* and *Zachaeus crista*. The latter two species and these of the guild of *Mitopus morio* plus *Rilaena balcanica* compete only in a small

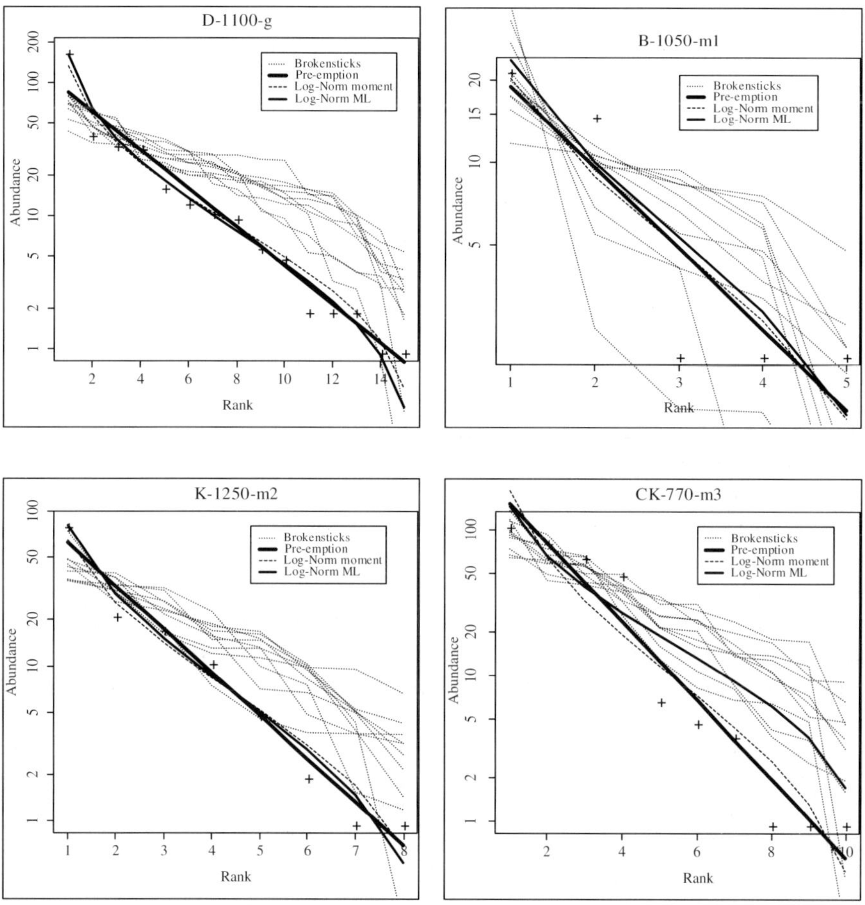

Fig. 7 Models of harvestman species abundance in the open habitats of Vitosha Mts.

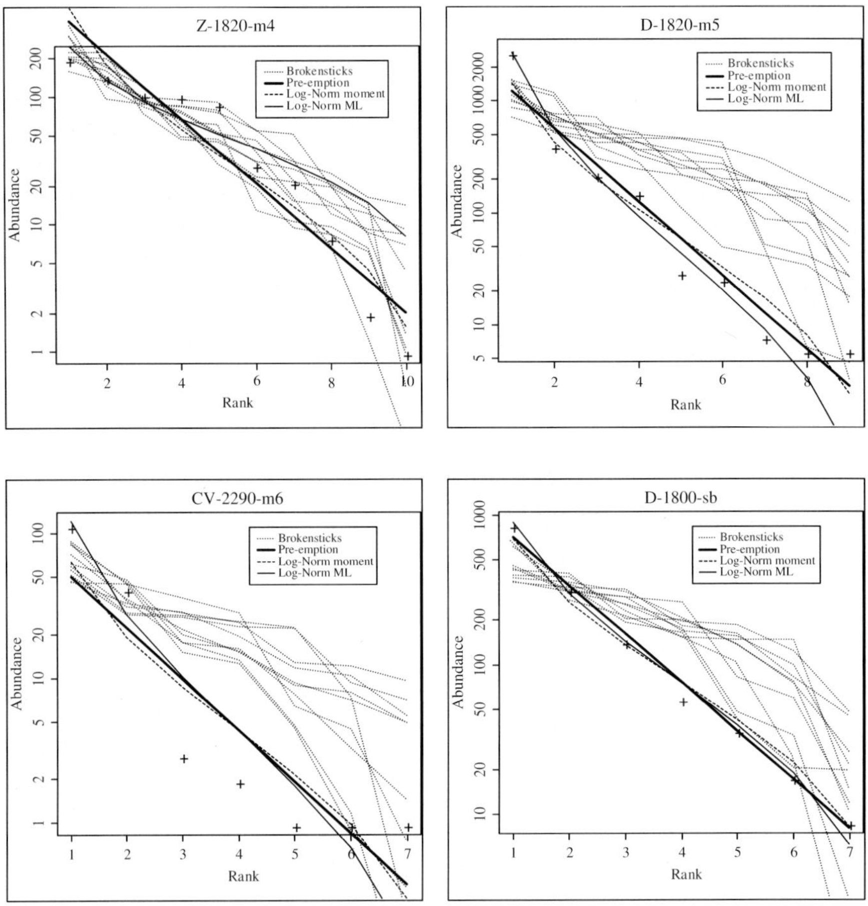

Fig. 7 continued

number of the common habitats of these pairs of species. The possible reason is that *Zachaeus crista* and *Rilaena balcanica* are thermophilous while *Pyza bosnica* and *Mitopus morio* are frigostable. In the relevant habitats, the interspecific competition between them is reduced due to the patchy distribution along the gradients of temperature, humidity and light.

In the guild of *Phalangium opilio, Rilaena* cf. *serbica* plus *Paranemastoma aurigerum ryla*, the latter two species are in a stronger spatial competition compared to the remaining species (Fig. 4). Their phenological peaks are overlapping but the average size of females of these two species is clearly different (Table 1). The average body lengths of the females of *Phalangium opilio* and *Rilaena* cf. *serbica* are almost equal (Table 1) but they use other strategies to reduce their interspecific competition (see above).

3·3 Types of niche relationships of harvestmen

As all other organisms, harvestmen exist in the conditions of indirect competition. Presence of numerous adaptations allows the niche differentiation without mutual elimination from the habitat. The way used by the species to share the niche volume, or the "hypervolume", substantially affects the type of abundance distribution (Odum, 1975). Vice versa, the abundance-distribution model, characteristic for a certain group, can be a source of information about the niche structure of the species co-occurring in a certain habitat. Three types of niche relationships were recognized: random, discrete, and overlapping (MacArthur, 1957, 1960, cited after Odum, 1975; Giller, 1988).

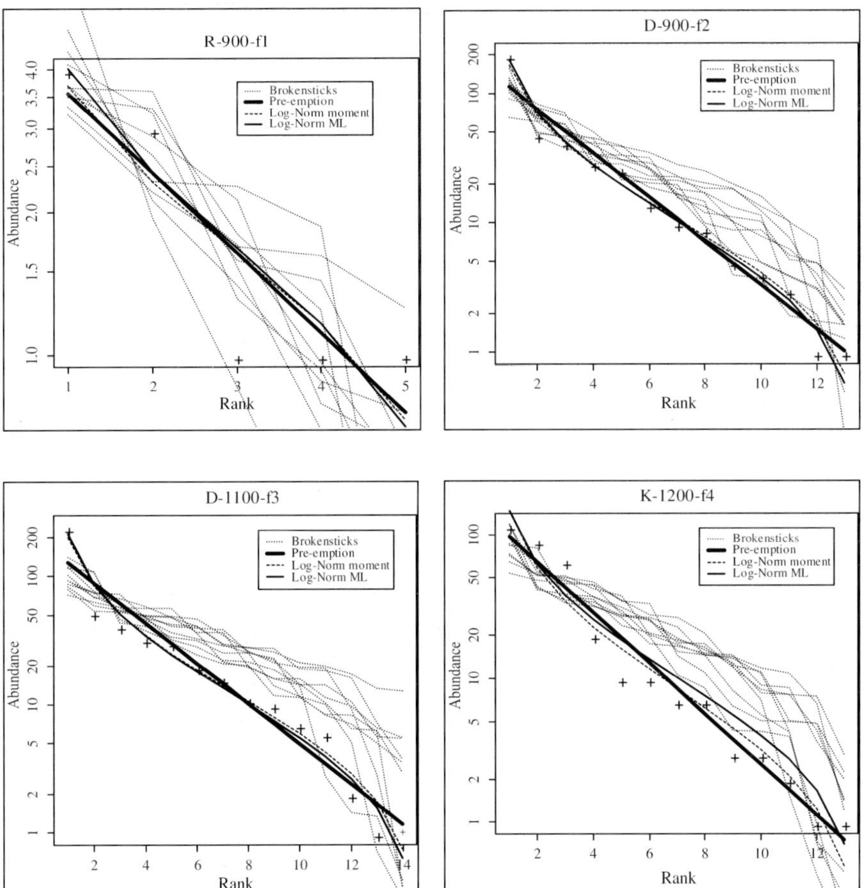

Fig. 8 Models of harvestman species abundance in the forest habitats of Vitosha Mts.

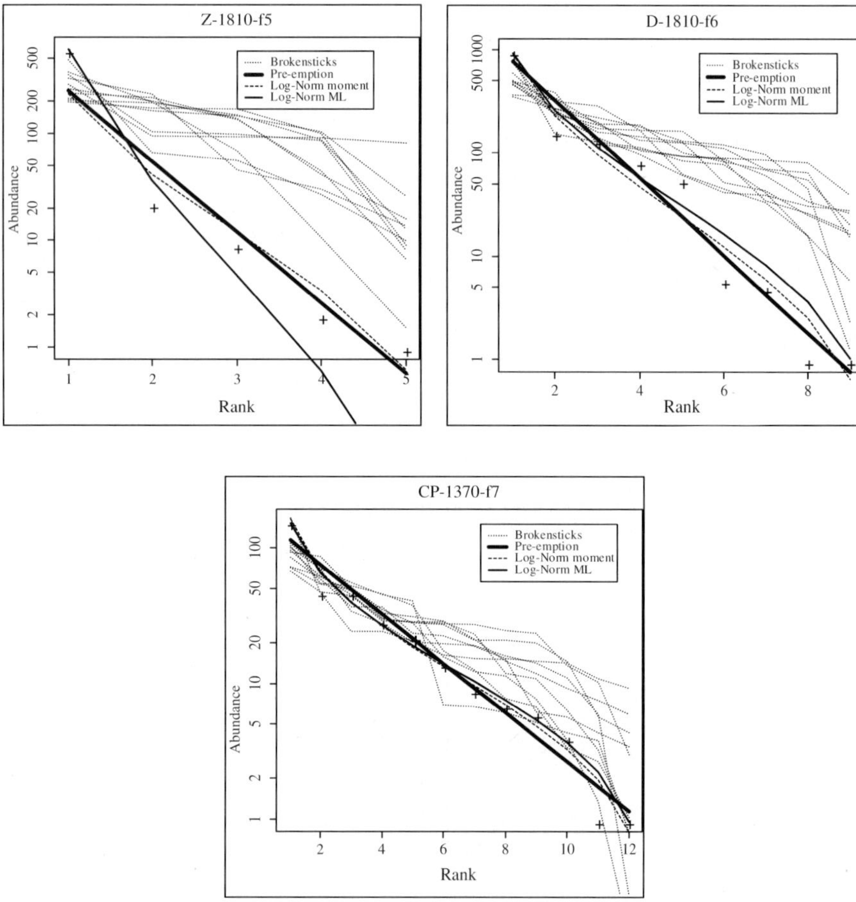

Fig. 8 continued

In order to understand how harvestmen share space, the model describing distribution of their relative abundance has to be found. In connection with this, we determined relationships between the relative abundance and the position of a certain species in the abundance-based rank hierarchy for all the habitats, expressed as the species-abundance curves (see Material and Methods). These curves were interpreted according to the hypotheses for random borders between niches (MacArthur, 1957), for geometrical row (preferential occupation of niches) (Whittaker, 1965), and for log-normal distribution (Preston, 1948, cited after Whittaker, 1980; Giller, 1988; Begon et al., 1989).

As a result of analysis of the curves using the Kolmogorov–Smirnov test, we found that these fit to the log-normal distribution (Figs. 7–9). On this basis, we can conclude that neither MacArthur's random ("broken-stick") model nor

Whittaker's model (preferential niche occupation) are consistent with the present results. These results fit better to a more complicated Preston's model of niche differentiation and overlapping. Therefore, the dimensions of space occupied by a certain harvestman species and quantity of the remaining resources are determined by several independent factors of primary importance, or by a large number of independent factors influencing the relative success of its competition with remaining species (Whittaker, 1980; Giller, 1988). Although the interpretation of the log-normal distribution is vague, it can be supposed that it reflects the shared space utilization into realized niches among various guilds (Giller, 1988).

According to Giller (1988), the log-normal distribution is also characteristic: (a) for species composition of mature communities influenced by multiple ecological factors (especially in undisturbed communities; Odum,, 1986); (b) when the species samples originate from numerous diverse habitats and communities (as it is in

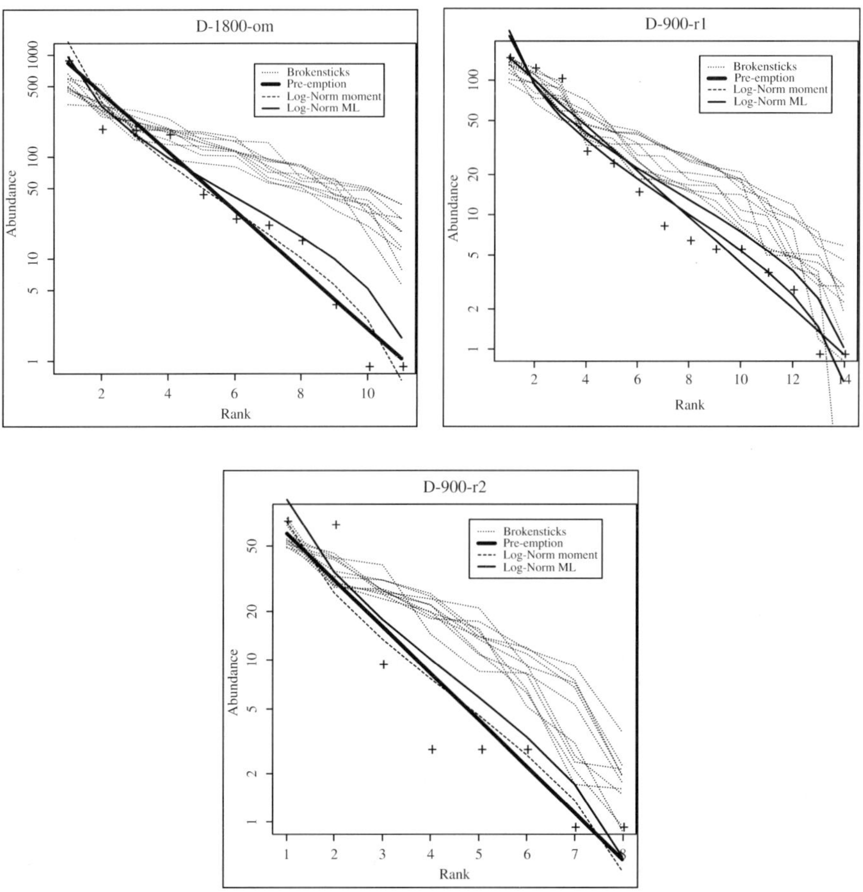

Fig. 9 Models of harvestman species abundance in open and forest rivulet-bank habitats of Vitosha Mts.

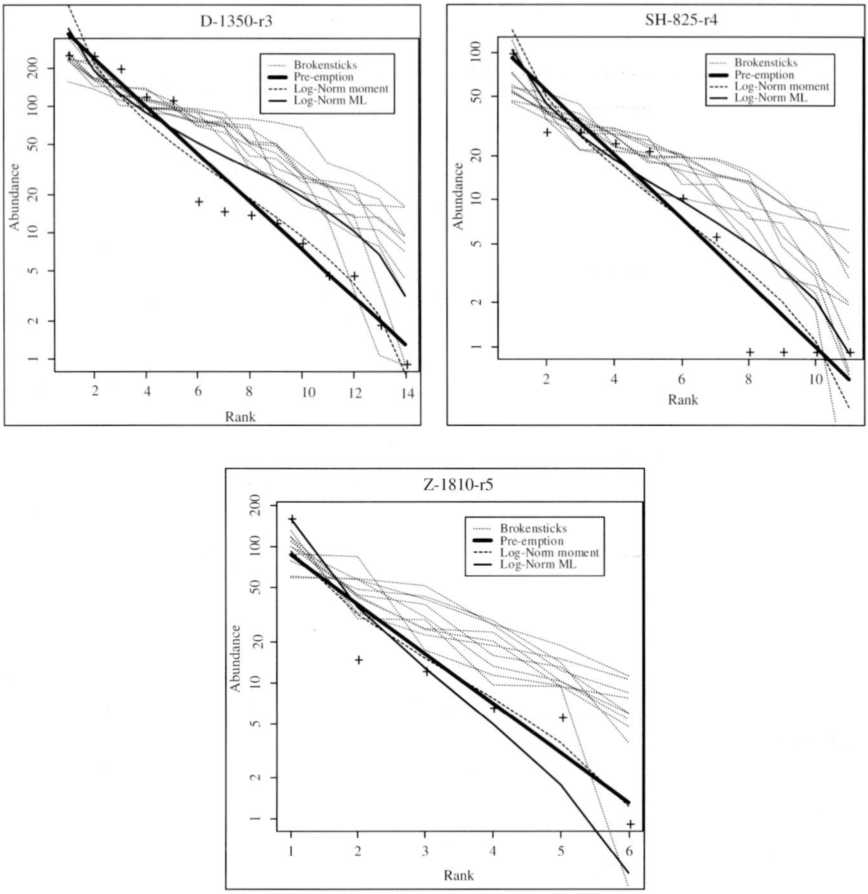

Fig. 9 continued

the present study); (c) when a small portion of species simultaneously have high prevalence and abundance; in addition, they might be characterized by restricted prevalence and low abundance, or may reach high abundance in a few communities. Finally, it can reflect the low abundance of these species in numerous communities (Giller, 1988), as in the present study.

Acknowledgments

I am grateful to Dr. Boyko B. Georgiev (Central Laboratory of General Ecology, Sofia) for reading the manuscript. I would also like to thank Ivailo Stoyanov (Sofia) for his help with statistical data analysis.

References

Arndt, E. & H. Pellmann, 1996. Ökologische Charakterisierung von Biotopen im urbanen Raum am Beispiel von Modelltiergruppen. UFZ-Bericht (Leipzig & Halle) 6, Stadtökologische Forschungen 2:1–119.

Ashby, J.W., 1974. A study of arthropod predation of *Pieris rapae* L. using serological and exclusion techniques. Journal of Applied Ecology 11: 419–425.

Baev, P. & L. Penev, 1993. BIODIV. Programme for Calculating Biological Diversity Parameters, Similarity, Niche Overlap, and Cluster Analysis. Version 4.1. Pensoft, Sofia.

Begon, M., J. Harper & C. Townsend. 1989. Ekologiya osobi, populyatsii i soobshchestva. Tom 2 [Ecology of Individuals, Populations and Communities. Vol. 2]. Mir [World], Moscow (in Russian).

Bengtson, S.-A., 1980. Species assemblages and coexistence of Faroe Island ground beetles (Coleoptera: Carabidae). Entomologia Generalis 6(2/4):251–266.

Bristowe, W.S., 1949. The distribution of harvestmen (Phalangida) in Great Britain and Ireland, with notes on their names, enemies and food. Journal of Animal Ecology 18(1):100–114.

Butcher, M.R., 1986. Aspects of the ecology of a strawberry crop with special reference to twospotted spider mite (Tetranychidae: Acari). PhD Thesis. University of Canterbury, Lincoln College, Christchurch, New Zealand.

Cokendolpher, J.C., 1993. Pathogenes and parasites of Opiliones (Arthropoda: Arachnida). Journal of Arachnology 21, 120–146.

Cokendolpher, J.C. & P.G. Mitov, in press. Enemies, endozoism, and epizoism. In Pinto da Rocha, R., G. Machado & G. Giribet (eds), The Harvestmen: The Biology of Opiliones. Harvard University Press, Boston.

Colwell, R.K. & D.J. Futuyma, 1971. On the measurement of niche breadth and overlap. Ecology 52:567–576.

Dajoz, R., 1975. Osnovy ekologii [Fundamentals of Ecology]. Progress, Moscow (in Russian).

Deltshev, Ch., B. Petrov & P. Mitov, 2005. [Faunistic diversity of Class Arachnida (non Acari) in Bulgaria – present state, importance and perspectives]. In Petrova, A. (ed), Savremenno sastoyanie na bioraznoobrazieto v Balgariya – problemi i perspektivi (Current State of Bulgarian Biodiversity – Problems and Perspectives). Balgarska platforma za bioraznoobrazie (Bulgarian Biodiversity Platform), Sofia: 129–151 (in Bulgarian).

Dixon, P.L. & R.G. McKinlay, 1989. Aphid predation by harvestmen in potato fields in Scotland. Journal of Arachnology 17(2):253–255.

Drummond, F., Y. Suhaya & E. Groden, 1990. Predation on the Colorado potato beetle (Coleoptera: Chrysomelidae) by *Phalangium opilio* (Opiliones: Phalangidae). Journal of Economic Entomology 83:772–778.

Dunlop, J., L.I. Anderson, H. Kerp & H. Hass, 2003. Preserved organs of Devonian harvestmen. Nature 425:916.

Dunlop, J.A, L.I. Anderson, H. Kerp & H. Hass, 2004. A harvestman (Arachnida: Opiliones) from the Early Devonian Rhynie cherts, Aberdeenshire, Scotland. Transactions of the Royal Society of Edinburgh: Earth Sciences 94:341–354.

Garmash, I.I., 1985. Tayny bioniki [Secrets of Bionics]. Radyanska shkola [Soviet School], Kiev (in Russian).

Giller, P.S., 1988. Struktura soobshchestv i ekologicheskaya nisha [Community Structure and the Niche]. Mir [World], Moscow (in Russian).

Guéorguiev, V., 1992. [Caractéristique zoogéographique de l'ordre Opilionida (Arachnida) en Bulgarie]. Acta zoologica bulgarica 43:3–60 (in Bulgarian).

Hammelbacher, K. & M. Mühlenberg, 1986. Laufkäfer- (Carabidae) und Weberknechtarten (Opiliones) als Bioindikatoren für Skibelastung auf Almflächen. Natur und Landschaft 61(12):463–466.

Harvey, M.S., 2002. The neglected cousins: what do we know about the smaller arachnid orders? Journal of Arachnology 30(2):357–372.

Hillyard, P.D. & J.H.P. Sankey, 1989. Harvestmen: Keys and notes for the identification of the species. 2nd revised edition. Synopses of the British Fauna. New Series 4.

Hubenov, Z., 1990. [Physico-geographical characteristics of Vitoša]. In Fauna na Yugozapadna Balgariya. Chast 3 (Fauna of Southwestern Bulgaria. Part 3). Izdatelstvo na Balgarskata akademiya na naukite (Publishing House of the Bulgarian Academy of Sciences), Sofia: 7–18 (in Bulgarian).

Ihaka, R. & R. Gentleman, 1996. R: a language for data analysis and graphics. Journal of Computational and Graphical Statistics 5(3): 299–314.

Karaman, I., 1995. Diverzitet faune kosača (Opiliones, Arachnida) Jugoslavije sa pregledom vrsta od međunarodnog značaja. In Stevanović, V. & V. Vasić, V. (eds), Biodiverzitet Jugoslavije sa pregledom vrsta od međunarodnog značaja. Biološki fakultet i Ecolibri, Belgrade: 329–335.

Klausnitzer, B., 1990. Ekologiya gorodskoy fauny [Urban Ecology]. Mir [World], Moscow (in Russian).

Lange, A.B., 1984. [Order Harvestmen (Opiliones, or Phalangida)]. In: Gilyarov, M.S. & F.N. Pravdin (eds), Zhizn' zhivotnykh. Tom 3. Tip Chlenistonogie (Arthropoda) [Animal Life. Vol. 3. Phylum Arthropods (Arthropoda)]. Prosveshchenie [Education], Moscow: 39–44 (in Russian).

MacArthur, R.H., 1957. On the relative abundance of bird species. Proceedings of the National Academy of Science USA, 43:293–295.

Martens, J., 1978. Spinnentiere, Arachnida: Weberknechte, Opiliones. In Die Tierwelt Deutschlands. 64. Teil. G. Fischer, Jena.

Mitov, P., 1988. [Contribution to the study of the food spectrum of Opiliones]. Nauchni trudove na Plovdivskiya universitet "Paisii Hilendarski", Biologiya (Travaux scientifiques de l'Université de Plovdiv "Paisii Hilendarski", Biologie) 26(6):483–488 (in Bulgarian).

Mitov, P., 1990. [Harvestmen]. Priroda i znanie [Nature and Knowledge] (Sofia) 52(10):23–26 (in Bulgarian).

Mitov, P., 1991. [Harvestmen, animals on stilts]. Priroda [Nature] (Sofia) 30(1):43–45 (in Bulgarian).

Mitov, P., 1995. Artenbestand und Tagesaktivität von Opiliones (Arachnida) aus einigen Geröllhalden im Vitoscha-Gebirge, Bulgarien. Arachnologisches Magazin 3(12):1–6.

Mitov, P., 1996. Vorläufige Beobachtungen zur Tag- und Nacht-Bewegungsaktivität von epigäischen Opiliones (Arachnida) aus verschiedenartigen Habitaten im Vitoscha-Gebirge (Nordteil), SW-Bulgarien. Revue suisse de Zoologie, vol. hors série: 479–489.

Mitov, P., 1997. Preliminary observation on the spatial distribution of the harvestmen (Opiliones, Arachnida) from Vitosha Mt. (SW Bulgaria). In: Żabka, M. (ed), Proceedings of the 16th European Colloquium of Arachnology. Wyższa Szkoła Rolniczo-Pedagogiczna, Siedlce, Poland: 249–258.

Mitov, P., 2000. Faunistichni, biologichni i ekologichni izsledvaniya varhu Opiliones (Arachnida) ot Vitosha planina s nyakoi zoogeografski belezhki [Faunistic, Biological, and Ecological Investigations on Opiliones from Vitosha Mt. with Some Zoogeographical Notes]. Doktorska disertatsiya [PhD Thesis]. Plovdivski universitet "Paisii Hilendarski" [Paisii Hilendarski University of Plovdiv], Plovdiv (in Bulgarian).

Mitov, P., 2003. Rare and endemic harvestmen (Opiliones, Arachnida) species from the Balkan Peninsula. II. Three new for the Bulgarian fauna Opiliones (Arachnida) with zoogeographical notes. Linzer Biologische Beiträge 35(1):273–288.

Mitov, P. & I. Stoyanov, 2004. The harvestmen fauna (Opiliones, Arachnida) of the city of Sofia (Bulgaria) and its adjacent regions. In: Penev, L., J. Niemelä, J., D.J. Kotze & N. Chipev (eds), Ecology of the City of Sofia. Species and Communities in an Urban Environment. Pensoft, Sofia & Moscow: 319–354.

Mitov, P. & I. Stoyanov, 2005. Ecological profiles of harvestmen (Arachnida, Opiliones) from Vitosha Mountain (Bulgaria): A mixed modeling approach using GAMS. The Journal of Arachnology 33(2):256–268.

Novak, T., Slana, L., Červek, N., Mlakar, M., Žmaher, N. & J. Gruber, 2002. Harvestmen (Opiliones) in human setttlements of Slovenia. Acta Entomologica Slovenica 10(2):131–154.

Odum, E., 1975. Osnovy ekologii [Fundamentals of Ecology]. Mir [World], Moscow (in Russian).

Odum, E., 1986. Ekologiya. Tom 2 [Ecology. Vol. 2]. Mir [World], Moscow (in Russian).

Pabst, W., 1953. Zur Biologie der mitteleuropäischen Troguliden. Zoologische Jahrbücher, Abteilung für Systematik und Ökologie 82(1/2):1–46.

Pesenko, Yu.A., 1982. Printsipy i metody kolichestvennogo analiza v faunisticheskikh issledovaniyakh [Principles and Methods of Quantitative Analyses in the Faunistic Investigations]. Nauka [Science], Moscow (in Russian).

Pielou, E.C., 1977. Mathematical Ecology. Wiley–Interscience, New York, London, Sydney & Toronto.

Pokarzhevskiy, A.D., 1985. [Accumulation of isotopic and non-isotopic carriers of artificial radionucleotides by soil animals]. In Krivolutskiy, D.A. (ed), Radioekologiya pochvennykh zhivotnykh [Radioecology of Soil Animals]. Nauka [Science], Moscow: 80–96 (in Russian).

Rafalski, J., 1962. *Opilio dinaricus* Šilhavý a little known species of harvestmen (Opiliones). Studia Societatis scientiarum Torunensis E6:1–12.

Schaefer, M., 1973. Welche Faktoren beeinflussen die Existenzmöglichkeit von Arthropoden eines Stadtparks – untersucht am Beispiel der Spinnen (Araneida) und Weberknechte (Opilionida)? Faunistisch-Ökologische Mitteilungen 4:305–318.

Simon, U., 1995. Untersuchung der Stratozönosen von Spinnen und Weberknechten (Arachn.: Araneae, Opilionida) an der Waldkiefer (*Pinus sylvestris* L.). Wissenschaft & Technik Verlag, Berlin.

Staręga, W., 1976. Die Weberknechte (Opiliones, excl. Sironidae) Bulgariens. Annales Zoologici (Warszawa) 33(18):287–433.

Todd, V., 1950. Prey of harvestmen (Arachnida, Opiliones). Entomologist's Monthly Magazine 86:252–254.

Uetz, G.W., J. Halaj & A.B. Cady, 1999. Guild structure of spiders in major crops. Journal of Arachnology 27:270–280.

Weiss, I., 1984. Ökofaunistische Untersuchung der Spinnen und Weberknechte eines Hangprofils bei Seica Mare im siebenbürgischen Hügelland. Studii şi Comunicari, Stiinţe naturale, Muzeul National Brukenthal 26:243–277.

Weiss, I. 1996. Die Weberknechtfauna Siebenbürgens (Arachnida: Opiliones). Stapfia 45:259–280.

Weiss, I. & S. Sarbu. 1977. Zur Kenntnis der Spinnen und Weberknechte des Botanischen Gartens Iaşi. Studii şi Comunicari, Stiinţe naturale, Muzeul National Brukenthal 21:225–243.

Whittaker, R. 1965. Dominance and diversity in land plant communities. Science 147:250–260.

Whittaker, R. 1980. Soobshchestva i ekosistemy [Communities and Ecosystems]. Progress, Moscow (in Russian).

14 Fauna and Zoogeography of Spiders of the Family Linyphiidae (Araneae) in Bulgaria

Christo Deltshev

Institute of Zoology, Bulgarian Academy of Sciences, Tsar Osvoboditel Blvd. 1, 1000 Sofia, Bulgaria, e-mail: cdeltshev@zoology.bas.bg

Abstract: The family Linyphiidae is represented in Bulgaria by 218 species belonging to 92 genera. Linyphiidae are distributed in all zoogeographical districts in Bulgaria, occurring in lowlands, forest, mountains, and caves. According to their current distribution, the Bulgarian linyphiid spiders can be divided into 19 zoogeographical categories, grouped into five chorological complexes (Cosmopolitan, widely distributed in Holarctic, European, endemic, and Mediterranean). The largest number of species belongs to the complex of widely distributed Holarctic species. The most characteristic for Bulgaria are the endemics; their number (28 species) is high and reflects the local character of the linyphiid fauna. This phenomenon can be attributed to the relative isolation of the mountains compared with the lowlands, in the context of paleoenvironmental changes since the Pliocene

1 Introduction

The fauna of the family Linyphiidae in Bulgaria is comparatively well studied. Drensky (1936) in his "Katalog der echten Spinnen (Araneae) der Balkanhalbinsel" reported 114 species from 46 genera. Numerous recent publications (Deltshev, 1972, 1973a, 1973b, 1974, 1975a, 1975b, 1980, 1983a, 1983b, 1983c, 1984, 1985, 1987, 1988a, 1988b, 1990, 1992, 1995, 1996, 1998, 1999, 2000; Helsdingen et al., 1977, 2001; Deltshev and Blagoev, 1992, 1997, 2001; Thaler et al., 1994; Deltshev and Ćurčić, 1997, 2002; Lazarov et al., 2001; Blagoev et al., 2002; Deltshev et al., 2003) are the result of intensive faunistic research conducted after 1966. The accumulation of new data therefore makes possible a critical taxonomic and faunistic review, together with a zoogeographical analysis.

2 Study Area and Materials

The material treated herein is based on the original collections obtained during 1965–1992 field surveys covering most of the districts in Bulgaria (Fig. 1) as well as on the critical incorporation of all available literature records about the distribution of linyphiid spiders in Bulgaria.

V. Fet and A. Popov (Eds.), Biogeography and Ecology of Bulgaria, 447–467.

Fig. 1 The geographical regions and districts of Bulgaria (according to Hubenov, 1997). Abbreviations, see text.

The geographical regions of Bulgaria and their abbreviations according to Hubenov (1997), used in the text are as follows:

DW, Western Danubian Plain; DM, Middle Danubian Plain; DE, Eastern Danubian Plain; DEL, Ludogorie–Dobrudja District; DEP, Popovo–Provadiya District; SPW, Western Predbalkan; SPM, Middle Predbalkan; SPE, Eastern Predbalkan; SBW, Western Stara Planina Mts.; SBM, Middle Stara Planina Mts.; SBE, Eastern Stara Planina Mts.; PKR, Rui Mts.; PKG, Golo Bardo Mts.; PKV, Verila Mts.; PKK, Kraishte; PKZ, Zemenska Planina Mts.; PKQ, Konyavska Planina Mts.; PVS, Sofia Basin; PVL, Lyulin Mts.; PVV, Vitosha Mts.; PVP, Plana Mts.; PSP, Podbalkan Basins; PSI, Ihtimanska Sredna Gora Mts.; PSL, Lozenska Planina Mts.; PSC, Sashtinska Sredna Gora Mts.; PSA, Sarnena Gora Mts.; PT, Thracian Lowland; PBT, Sakar–Tundja District; PBC, Sakar Mts.; PBB, Bakadjik–Burgas District; PBD, Strandja–Dervent District; PBS, Strandja Mts.; ROO, Osogovo Mts.; ROV, Vlahina Planina Mts.; ROM, Maleshevska Planina Mts.; ROG, Ograzhden Mts.; ROB, Belasitsa Mts.; ROT, Boboshevo–Simitli Valley; ROP, Krupnik–Sandansky–Petrich Valley; RPR, Rila Mts.; RPP, Pirin Mts.; RPS, Slavyanka Mts.; RPT, Stargach Mts.; RPM, Mesta Valley; RRW, Western Rhodope Mts.; RRE, Eastern Rhodope Mts.; BN, Northern Black Sea coast; BS, Southern Black Sea coast.

Table 1 Comparison of area and spider species richness of some European countries

Country	Area(km^2)	Number of species	Sources
Austria	83, 858	364	Blick et al. (2004)
Bulgaria	110, 993	218	Deltshev & Blagoev (2001)
Czech Republic	77, 280	301	Buchar and Růžička (2002)
Greece	128, 900	65	Bosmans & Chatzaki (2005)
Hungary	92, 340	191	Samu & Szinetár (1999)
Portugal	91, 500	75	Cardoso (1999)
Serbia	102, 000	154	Deltshev et al. (2003)

3 Species Composition

The family Linyphiidae is represented in Bulgaria by 218 species belonging to 92 genera (Table 2). The number of species is high compared with the number of spiders recorded from other countries of Europe with similar territories (Table 1). This richness, however, depends not only on the size of the regions, but also on the degree of exploration by araneologists. Best represented in Bulgaria are the genera *Centromerus* (16 species, or 7.3%), *Walckenaeria* (14 species, or 6.4%), *Tenuiphantes* (11 species, or 5%), and *Diplocephalus* (nine species, or 4.1%).

4 Zoogeographical Analysis

According to their current distribution, Bulgarian linyphiid spiders can be split into 19 zoogeographical chorotypes, grouped into five chorological complexes (I, Cosmopolitan; II, widely distributed in Holarctic; III, European; IV, endemic; V, Mediterranean). The data concerning general distribution of spiders are taken from Mikhailov (1997), Marusik et al. (2000), Platnick (2004), and Vigna Taglianti et al. (1999).

Cosmopolitan species complex (COS + SCO) includes five species (2.3%) widespread in Bulgaria (Table 3). *Prinerigone vagans* is well represented in the lowlands, woodlands, and high altitude zones of mountains. This complex also includes the Holarctic species *Microctenonyx subitaneus*, introduced to other continents.

Complex of species widely distributed in the Holarctic Region (HOL + PAL + WPA + ECA + EEC + EUCA) is best represented and comprises 111 (50.9%) linyphiid species widespread in Bulgaria (Table 3). Palearctic species (*sensu lato*) are dominant (46 species, 41.4%), followed by West Palearctic (35 species, 31.5%) and Holarctic linyphiids (22 species, 19.8%). The remaining chorotypes are represented by a single species each.

Table 2 Species diversity and distribution of linyphiid spiders in Bulgaria

Taxa	Districts	A	B	C	D	E	F	Chorotype
Abacoproeces saltuum (L. Koch, 1872)	DEL, PSC, ROO, ROP, RPP	+	+	+				PAL
Acartauchenius scurrilis (O. P.-Cambridge, 1872)	ROO, RPP, PVS		+	+				WPA
Agyneta cauta (O. P.-Cambridge, 1902)	SPM, ROO				+			PAL
Anguliphantes angulipalpis (Westring, 1851)	SBW			+				WPA
Antrohyphantes balcanicus (Drensky, 1931)	SPM			+				BULG
Antrohyphantes rhodopensis (Drensky, 1931)	RPR, RPP, RRW				+	+	+	BULG
Antrohyphantes sophianus (Drensky, 1931)	SBW			+				BULG
Araeoncus anguineus (L. Koch, 1869)	SPM, PVV, ROO, RPR, RPP, PBS			+	+	+	+	EUR
Araeoncus clivifrons Deltshev, 1987	RPR, RPP					+	+	BULG
Araeoncus crassiceps (Westring, 1861)	SBM			+				PAL
Araeoncus humilis (Blackwall, 1841)	SBW, PVV, RPP, ROP			+	+	+		WPA
Archaraeoncus prospiciens (Thorell, 1875)	PBS		+					EEC
Bathyphantes gracilis (Blackwall, 1841)	SBM, RPR, RPP			+	+			HOL
Bathyphantes nigrinus (Westring, 1851)	SBM, PKZ, RPP		+	+	+			WPA
Bolyphantes alticeps (Sundevall, 1833)	SBM, PBS, PVV, ROO, RPR, RPP			+	+			PAL
Bolyphantes kolosvaryi (Caporiacco, 1936)	RPR, RPP			+	+			SEU
Bolyphantes luteolus (Blackwall, 1833)	SBM, PVV, PSC, RPR, RPP			+	+	+	+	WPA
Caviphantes dobrogicus (Dumitrescu et Miller, 1962)	RRE		+					EUR
Centromerita bicolor (Blackwall, 1833)	RPP				+	+		EUR
Centromerus acutidentatus Deltshev, 2002	ROV, RPS			+	+			BALK

Centromerus brevivulvatus Dahl, 1912	PSC		+				EUR
Centromerus bulgarianus (Drensky, 1931)	SBW		+				BULG
Centromerus capucinus (Simon, 1884)	PSC	+					EUR
Centromerus cavernarum (L. Koch, 1872)	SBW, SBM, PSC, PBS		+	+	+		EUR
Centromerus incilium (L. Koch, 1881)	SBM, PKZ		+	+	+		EUR
Centromerus lakatnikensis (Drensky, 1931)	SBW, PSC, RPP, RRW		+	+	+		BALK
Centromerus milleri Deltshev, 1974	RRE	+	+				BULG
Centromerus pabulator (O. P.-Cambridge, 1875)	PKK		+				EUR
Centromerus prudens (O. P.-Cambridge, 1873)	DEL	+					EUR
Centromerus semiater (L. Koch, 1879)	PSC		+	+	+		WPA
Centromerus serratus (O. P.-Cambridge, 1875)	RPP			+	+		EUR
Centromerus silvicola (Kulczyński, 1887)	SBM, PSC			+	+		EUR
Centromerus sylvaticus (Blackwall, 1841)	PKZ, PVV, PSC, BN		+	+	+		HOL
Centromerus sylvaticus paucidentatus Deltshev, 1983	RPP				+	+	BULG
Centromerus valkanovi Deltshev, 1983	BS	+	+				BULG
Ceratinella brevipes (Westring, 1851)	SBM, PSC			+	+		EUR
Ceratinella brevis (Wider, 1834)	PSC, RPP, BN			+	+		PAL
Ceratinella major Kulczyński, 1894	PKZ, PSC		+				PAL
Ceratinella scabrosa (O. P.-Cambridge, 1871)	PKZ, PSC		+	+			WPA
Cinetata gradata (Simon, 1881)	RPP				+	+	EUR
Cresmatoneta mutinensis (Canestrini, 1868)	ROP	+					MED
Dicymbium nigrum (Blackwall, 1834)	PKZ, RPR		+	+	+		WPA
Dicymbium tibiale (Blackwall, 1836)	SBM, PKZ, PSC		+	+	+		WPA

continued

Table 2 continued

Taxa	Districts	A	B	C	D	E	F	Chorotype
Diplocephalus altimontanus Deltshev, 1984	RPP						+	BULG
Diplocephalus crassilobus (Simon, 1884)	DEL	+						EUR
Diplocephalus cristatus (Blackwall, 1833)	DEL, SBW, SBM		+	+	+			SCO
Diplocephalus foraminifer (O. P.-Cambridge, 1875)	SBM, PSC, RPR, EPP				+	+	+	EUR
Diplocephalus graecus (O. P.-Cambridge, 1872)	ROP	+						MED
Diplocephalus latifrons (O. P.-Cambridge, 1863)	PSI, RPP		+	+	+			EUCA
Diplocephalus permixtus (O. P.-Cambridge, 1871)	BN	+						EUR
Diplocephalus picinus (Blackwall, 1841)	DEL, SBM, PKZ, PSC, RPR		+	+	+			WPA
Diplocephalus protuberans (O. P.-Cambridge, 1875)	PKZ		+	+				EUR
Diplostyla concolor (Wider, 1834)	SBW, SBM, PKZ, PVV, PBS, PSC, RPR, RPP, BN, BS	+	+	+	+			HOL
Dismodicus bifrons (Blackwall, 1841)	SBM, PKZ, RPR			+	+			PAL
Dismodicus elevatus (C. L. Koch, 1838)	PKK, PVV, PSC		+					WPA
Donacochara speciosa (Thorell, 1875)	DEL, BN	+						ECA
Drapetisca socialis (Sundevall, 1833)	PKZ, PVV, PSC, RPR, RPP		+	+	+			PAL
Drepanotylus pirinicus Deltshev, 1992	RPP					+	+	BULG
Entelecara acuminata (Wider, 1834)	DW	+						HOL
Entelecara flavipes (Blackwall, 1834)	PKK		+					EUR
Entelecara media Kulczyński, 1887	RPP						+	HOL

Erigone atra Blackwall, 1833	ROO, BN	+	+					HOL
Erigone dentipalpis (Wider, 1834)	SBW, SBM, PKZ, PVV, PBS, PSC, RPR, RPP, BN, BS		+	+	+	+		HOL
Erigone longipalpis pirini Deltshev, 1983	SBM, RPR, RPP					+	+	BULG
Erigonella hiemalis (Blackwall, 1841)	PVV					+		WPA
Evansia merens O. P.-Cambridge, 1900	SBM, RPR, RPP					+	+	EUR
Frontinellina frutetorum (C. L. Koch, 1834)	DEL, SBW, SBM, PKZ, RPR, RPP, RRW, RRE		+	+	+			WPA
Gnathonarium dentatum (Wider, 1834)	PVV			+				EUR
Gonatium hilare (Thorell, 1875)	PSC, RPS		+	+				PAL
Gonatium nemorivagum (O. P.-Cambridge, 1875)	PKZ	+	+					SEU
Gonatium orientale Fage, 1931	SBM, RPR, RPP				+	+		BCAR
Gonatium paradoxum (L. Koch, 1869)	PKZ, PSC		+	+				PAL
Gonatium rubellum (Blackwall, 1841)	RPP			+	+			PAL
Gonatium rubens (Blackwall, 1833)	PSC, RPR, RPP			+	+	+		PAL
Gongylidiellum latebricola (O. P.-Cambridge, 1871)	PSC, RPP				+	+		WPA
Gongylidiellum murcidum Simon, 1884	BS	+						WPA
Gongylidium rufipes (Linnaeus, 1758)	DEL, PVV, BN	+	+					PAL
Halorates reprobus (O. P.-Cambridge, 1879)	SBW		+	+	+			EUR
Hilaira excisa (O. P.-Cambridge, 1871)	SBM			+	+			EUR
Hilaira montigena (L. Koch, 1872)	PVV		+					PAL
Hylyphantes graminicola (Sundevall, 1830)	BN	+						PAL
Hypomma cornutum (Blackwall, 1833)	PKZ	+						PAL
Improphantes decolor (Westring, 1861)	PSC, RPR			+	+			WPA

continued

Table 2 continued

Taxa	Districts	A	B	C	D	E	F	Chorotype
Improphantes improbulus (Simon, 1929)	SBM, RPR, RPP					+	+	EUR
Incestophantes annulatus (Kulczyński, 1882)	RPR						+	BCAR
Incestophantes crucifer (Menge, 1866)	PSC, BN	+	+	+				WPA
Lepthyphantes centromeroides Kulczyński, 1914	SBW, SBM, RPP, RRW				+	+	+	BALK
Lepthyphantes keyserlingi (Ausserer, 1867)	PKK		+					WPA
Lepthyphantes leprosus (Ohlert, 1865)	DEL, SBW, SBM, SBE, PKZ, PVV, PSC, PBS, ROO, RPR, RPP, RRW			+	+	+		HOL
Lepthyphantes melanotus Drensky, 1921	RPP					+		BULG
Lepthyphantes notabilis Kulczyński, 1887	PKK		+					CSEE
Lepthyphantes quadrimaculatus Kulczyński, 1898	RPP						+	EUR
Leptothrix hardyi (Blackwall, 1850)	RPR			+	+			WPA
Lessertia dentichelis (Simon, 1884)	SPM			+				SCO
Linyphia hortensis Sundevall, 1830	PVV, PSC, RPR, RPP			+	+	+		PAL
Linyphia triangularis (Clerck, 1757)	SBW, SBM, PSC, RPR, BN		+	+	+			PAL
Macrargus carpenteri (O. P.-Cambridge, 1894)	SBM, RPR, PSC			+	+			WPA
Macrargus rufus (Wider, 1834)	SBM, PSC, RPP			+	+			WPA
Mansuphantes mansuetus (Thorell, 1875)	PKZ, PVV, SBM, PSC, RPR, RPP		+	+	+			WPA

Mansuphantes rectilamellus (Deltshev, 1988)	PSC, RPR, RPP				+	+		BULG
Maso gallicus Simon, 1894	RPR, RPP, BN				+	+		EUR
Maso sundevalli (Westring, 1851)	PKZ, PVV, SBM, PBS		+	+	+			HOL
Mecopisthes peusi Wunderlich, 1972	PSC, PBS	+						EUR
Mecynargus paetulus (O. P.-Cambridge, 1875)	SBM, RPP					+	+	HOL
Megalepthyphantes collinus (L. Koch, 1872)	PKZ, SBM, PSC, RPP			+				WPA
Megalepthyphantes nebulosus (Sundevall, 1830)	PVV			+	+			HOL
Meioneta equestris (L. Koch, 1881)	PSC	+						CEE
Meioneta fuscipalpa (C. L. Koch, 1836)	RPP					+		PAL
Meioneta gulosa (L. Koch, 1869)	RPP					+		PAL
Meioneta rurestris (C. L. Koch, 1836)	PVV, SBW, SBM, RPR, RPP	+	+	+	+	+	+	PAL
Meioneta simplicitarsis (Simon, 1884)	PKZ		+					EUR
Metopobactrus orbelicus Deltshev, 1985	RPP						+	BULG
Micrargus herbigradus (Blackwall, 1854)	PSC, RPR, RPP				+	+		PAL
Micrargus subaequalis (Westring, 1851)	SBM, RPP				+	+		PAL
Microctenonyx subitaneus (O. P.-Cambridge 1875)	SBW, SBM, PSC, RPP			+	+			SCO
Microlinyphia pusilla (Sundevall, 1830)	PSC, RPR, RPP			+	+	+		HOL
Microneta viaria (Blackwall, 1841)	PKZ, SBW, PSC, PBS, RPR, RPP		+	+	+			HOL
Minyriolus pusillus (Wider, 1834)	PSC			+	+			PAL
Moebelia penicillata (Westring, 1851)	PSC	+	+					WPA
Mughiphantes lithoclasicola (Deltshev, 1983)	RPR, RPP						+	BULG

continued

Table 2 continued

Taxa	Districts	A	B	C	D	E	F	Chorotype
Mughiphantes pulcher (Kulczyński, 1881)	RPR, RPP					+	+	CSEE
Nematogmus sanguinolentus (Walckenaer, 1842)	SBM			+	+			PAL
Neriene clathrata (Sundevall, 1830)	PKZ, SBM, PSC, PBS	+	+					HOL
Neriene emphana (Walckenaer, 1842)	PSC, RPR			+	+			PAL
Neriene furtiva (O. P.-Cambridge, 1871)	SBM, PSC		+					EUR
Neriene montana (Clerck, 1757)	PKZ, PSC, PBS, RPR		+	+				HOL
Neriene peltata (Wider, 1834)	PKZ, PSC, PBS		+	+	+			WPA
Neriene radiata (Walckenaer, 1842)	PKZ, SBM, PSC, PBS, RPR, RPP		+	+	+			HOL
Obscuriphantes obscurus (Blackwall, 1841)	SBM, RPP			+	+			EUR
Oedothorax agrestis (Blackwall, 1853)	SBM, PSC, RPR, RPP				+	+		PAL
Oedothorax apicatus (Blackwall, 1850)	PKZ, PVV, SBM, PSC, RPR, RPP				+	+		PAL
Oedothorax fuscus (Blackwall, 1834)	PVV, SBW, RPP				+	+		WPA
Oedothorax gibbifer (Kulczyński, 1882)	RPP			+	+			EUR
Oedothorax gibbosus (Blackwall, 1841)	DW, PVV		+	+				PAL
Oedothorax retusus (Westring, 1851)	PKZ, PSC, BN		+	+				PAL
Oreonetides glacialis (L. Koch, 1872)	RPR						+	EUR
Ostearius melanopygius (O. P.-Cambridge, 1879)	PSC, RPP	+	+	+	+	+		COS
Palliduphantes alutacius (Simon, 1884)	SBW, SBW, PSC, RPR			+				EUR
Palliduphantes byzantinus (Fage, 1931)	DEP, SBE, PBB	+	+					BALK
Palliduphantes insignis (O. P.-Cambridge, 1913)	PKZ	+						EUR

Palliduphantes istrianus (Kulczyński, 1914)	PKZ, SBW, SBM, SBE, PSC, PBS, RRW		+	+				EEU
Palliduphantes pallidus (O. P.-Cambridge, 1871)	SBE, RPP			+	+			PAL
Palliduphantes pillichi (Kulczyński, 1915)	SBW			+				CSEE
Palliduphantes spelaeorum (Kulczyński, 1914)	RPS, RRW		+	+	+			BALK
Palliduphantes trnovensis (Drensky, 1931)	SBW, SBM		+	+				BALK
Panamomops inconspicuus (Miller et Valesova, 1964)	PSC			+				EUR
Panamomops sulcifrons (Wider, 1834)	PSC			+	+			EUR
Pelecopsis elongata (Wider, 1834)	PKZ, SBM, RPR, RPP		+	+	+	+		EUR
Pelecopsis krausi Wunderlich, 1980	PKZ		+					BCAU
Pelecopsis parallela (Wider, 1834)	PVV			+				PAL
Piniphantes pinicola (Simon, 1884)	RPP				+	+		EUR
Pityohyphantes phrygianus (C. L. Koch, 1836)	RPR, RPP				+	+		PAL
Pocadicnemis juncea Locket et Millidge, 1953	SBM, PSC			+				EUR
Pocadicnemis pumila (Blackwall, 1841)	PVV			+				HOL
Poeciloneta variegata (Blackwall, 1841)	PKZ, PSC, RPP		+	+	+	+	+	HOL
Porrhomma convexum (Westring, 1851)	DW, DM, DEP, SBW, SBM, SBE, PBS, RPR, RRW	+	+	+	+	+		PAL
Porrhomma lativelum Tretzel, 1956	PBS	+						EUCA
Porrhomma microphthalmum (O. P.-Cambridge, 1871)	SBM, PSC			+	+			PAL
Porrhomma microps (Roewer, 1931)	SBE			+				EUR
Porrhomma pygmaeum (Blackwall, 1834)	PVV			+				PAL

continued

Table 2 continued

Taxa	Districts	A	B	C	D	E	F	Chorotype
Prinerigone vagans (Audouin, 1826)	SBM, PBS, RPR, RRW			+	+	+		SCO
Sauron rayi (Simon, 1881)	PSC			+				EUR
Scotargus pilosus Simon, 1913	DEL, PSC, RPR			+	+			WPA
Scotinotylus alpigena (L. Koch, 1869)	RPR						+	PAL
Silometopus bonessi Casemir, 1970	PSC			+	+			CSE
Silometopus reussi (Thorell, 1871)	SBM, PSC	+						PAL
Sintula retroversus (O. P.-Cambridge, 1875)	PKZ, SBM, PSC, RRW	+	+					EUCA
Sintula spiniger (Balogh, 1935)	PKZ, SBM, PSC, RRW	+	+					EEU
Stemonyphantes lineatus (Linnaeus, 1758)	PKZ, SBM, PSC, RRW		+	+	+			WPA
Styloctetor romanus (O. P.-Cambridge, 1872)	SBW, SBM, PSC		+	+	+			PAL
Tallusia experta (O. P.-Cambridge, 1871)	DW, PKZ, SBM	+	+	+				PAL
Tallusia vindobonensis (Kulczyński, 1898)	PKZ	+	+					CEE
Tapinocyba biscissa (O. P.-Cambridge, 1872)	PSC		+	+				EUR
Tapinocyba insecta (L. Koch, 1869)	PVV			+				EUR
Tapinocyba mitis (O. P.-Cambridge, 1882)	PBS		+					CSEE
Tapinocyba pallens (O. P.-Cambridge, 1872)	SBM, PSC			+				EUCA
Tapinopa longidens (Wider, 1834)	SBM, PSC			+				PAL
Tenuiphantes alacris (Blackwall, 1853)	SBM, RPR, RPP				+	+		EUR
Tenuiphantes cristatus (Menge, 1866)	PVV, SBW, SBM			+	+			PAL
Tenuiphantes drenskyi (van Helsdingen, 1977)	PVV, RPR				+	+		BULG
Tenuiphantes flavipes (Blackwall, 1854)	PKZ, SBM, PSC, PBS, RPR, RPP, RRW		+	+	+	+		EUCA

Tenuiphantes floriana (van Helsdingen, 1977)	SBW, PSC, PBS, RRW		+				BCAR
Tenuiphantes jacksoni (Schenkel, 1925)	PBS		+				CSEE
Tenuiphantes jacksonoides (van Helsdingen, 1977)	RRW		+	+			CSEE
Tenuiphantes mengei (Kulczyński, 1887)	PKZ, PSC, RPR, RPP	+	+	+			PAL
Tenuiphantes tenebricola (Wider, 1834)	SBM, PSC, RPR, RPP				+	+	EUR
Tenuiphantes tenuis (Blackwall, 1852)	PKZ, SBW, SBM, PSC, PBS, RPR, RPP		+	+	+	+	WPA
Tenuiphantes zimmermanni (Bertkau, 1890)	PKZ, PSC, PBS, RRW		+	+	+		EUR
Theonina kratochvili Miller et Weiss, 1979	PSC, PBS	+					EUR
Thyreosthenius biovatus (O. P.-Cambridge, 1875)	PSC	+	+				PAL
Thyreosthenius parasiticus (Westring, 1851)	SBW, SBM, PSC		+	+			HOL
Tiso aestivus (L. Koch, 1872)	RPR					+	HOL
Tiso vagans (Blackwall, 1834)	SBM, PSC, RPP			+	+	+	EUR
Trichoncoides piscator (Simon, 1884)	ROP	+	+				WPA
Trichoncus affinis Kulczyński, 1894	SBW, PSC, ROO		+				EUR
Trichoncus auritus (L. Koch, 1869)	PSC		+				EUR
Trichoncus hackmani Millidge, 1956	PSC			+			EUR
Trichoncus saxicola (O. P.-Cambridge, 1861)	PKZ, PSC		+	+			EUR
Trichopterna cito (O. P.-Cambridge, 1872)	PKZ, PSC		+	+			WPA
Troglohyphantes bureschianus Deltshev, 1975	RRW			+	+		BULG
Troglohyphantes drenskii Deltshev, 1973	RRW			+	+		BULG

continued

Table 2 continued

Taxa	Districts	A	B	C	D	E	F	Chorotype
Troxochrus cirrifrons (O. P.-Cambridge, 1871)	ROP	+	+					EUR
Troxochrus scabriculus (Westring, 1851)	PKZ, SBW, PSC	+	+	+				WPA
Typhochrestus digitatus (O. P.-Cambridge, 1872)	PSC	+						EUR
Walckenaeria acuminata Blackwall, 1833	PKZ, SBM		+	+				EUR
Walckenaeria alticeps (Denis, 1952)	PKZ, PSC, PSB, RPR		+	+				EUR
Walckenaeria antica (Wider, 1834)	SBW, PSC, PSB		+	+				WPA
Walckenaeria capito (Westring, 1861)	SBW, PT, RPP			+	+			HOL
Walckenaeria corniculans (O. P.-Cambridge, 1875)	ROO	+	+					EUR
Walckenaeria cucullata (C. L. Koch, 1836)	PKZ, PSC		+	+				WPA
Walckenaeria dysderoides (Wider, 1834)	PSC	+						EUR
Walckenaeria furcillata (Menge, 1869)	PKZ, PSC		+	+				WPA
Walckenaeria mitrata (Menge, 1868)	SBM			+				PAL
Walckenaeria monoceros (Wider, 1834)	SBM		+	+				ECA
Walckenaeria obtusa Blackwall, 1836	ROO			+				PAL
Walckenaeria simplex Chyzer, 1894	SBW, PSC			+				EEU
Walckenaeria stylifrons (O. P.-Cambridge, 1875)	PSC	+						EUR
Walckenaeria vigilax (Blackwall, 1853)	PKZ, SBM, RPR, RPP		+	+	+			HOL

District abbreviations, see Fig. 1. Altitudinal belts: A, Submediterranean; B, Submontane; C, Montane deciduous; D, Montane coniferous; E, Subalpine; F, Alpine. Zoogeographical chorotypes: COS, Cosmopolitan; SCO, Subcosmopolitan; HOL, Holarctic; PAL, Palearctic; WPA, West Palearctic; ECA, European–Central Asian; EEC, East European–Central Asian; EUCA, European–Caucasian; EUR, European; CEE, Central and East European; CSE, Central and South European; CSEE, Central and Southeast European; SEU, South European; EEU, East European; BULG, Bulgarian endemic; BALK, Balkan endemic; BCAR, Balkan–Carpathian endemic; BCAU, Balkan–Caucasian endemic; MED, Mediterranean

Table 3 Zoogeographical characteristics of Bulgarian linyphiid spiders. Abbreviations, see text and Table 2

Complexes of chorotypes	Chorotypes	Total number of species	Altitudinal belts					
			A	B	C	D	E	F
I	COS	1	1	1	1	1	1	–
	SCO	4	–	1	4	3	1	–
II	HOL	22	4	12	16	13	6	3
	PAL	46	10	16	30	23	11	2
	WPA	35	5	22	27	19	6	1
	ECA	2	1	1	1	–	–	–
	EEC	1	–	1	–	–	–	–
	EUCA	5	2	3	3	2	1	–
III	EUR	58	12	20	25	21	12	6
	CEE	2	2	1	–	–	–	–
	CSE	1	–	–	1	1	–	–
	CSEE	6	–	4	2	–	1	1
	SEU	2	1	1	1	1	–	–
	EEU	3	1	2	2	–	–	–
IV	BULG	18	2	3	4	5	8	7
	BALK	6	1	4	4	4	1	1
	BCAR	3	–	1	–	1	1	1
	BCAU	1	–	1	–	–	–	–
V	MED	2	2	–	–	–	–	–
Total		*218*	*44*	*94*	*121*	*94*	*49*	*22*

This complex includes especially widespread species associated with lowlands, woodlands, and high altitude zones of the mountains. Most of these species are well represented in the mountains. Characteristic mountain elements are *Bolyphantes alticeps, B. luteolus, Frontinellina frutetorum, Gonatium rubens, Pityohyphantes phrygianus, Tenuiphantes alacris*, and *T. tenebricola*. High altitude elements are *Entelecara media* and *Mecynargus paetulus*, which are not found in the forest belt. Some xenotopic species (Thaler, 1988) are widely distributed in the mountains and reach the highest summits as aeronauts. These are *Dicymbium nigrum, Diplostyla concolor, Meioneta rurestris, Oedothorax agrestis, O. apicatus*, and *O. fuscus*, which inhabit the mountain zone in dense populations (Deltshev, 1990, 1995).

European chorological complex (EUR + CEE + CSE + CSEE + SEU + EEU) includes 72 species (3%), widespread in Europe and Bulgaria (Table 3). European species *sensu lato* are dominant (58 species, 80.5%); followed by Central and Southeast European (six species, 8.3%) and East European linyphiids (three species, 4.2%). The remaining chorotypes are represented by a single species each. This complex includes widespread species which inhabit both lowlands and mountains. The interesting group of European mountain species is best represented in the forest, subalpine, and alpine belts. Characteristic mountain elements are *Araeoncus anguineus, Bolyphantes kolosvaryi, Cinetata gradata, Diplocephalus foraminifer, Improphantes improbulus, Maso gallicus, Mughiphantes pulcher, Oreonetides glacialis*, and *Tiso vagans*. Species *Palliduphantes istrianus, Centromerus capucinus, C. cavernarum, Porrhomma lativelum*, and *P. microps* are characteristic for the caves.

Complex of endemics (BALK + BULG + BCAU + BCAR) includes 28 species (12.8%) found in Bulgaria, the Balkan Peninsula, and neighboring territories of Asia Minor, Caucasus, and the Carpathians (Table 3). The recorded number is high and reflects the local character of Bulgarian linyphiid fauna. The issue of the status and distribution of endemic spiders found in Bulgaria is complicated. Some of the endemics are found only in restricted areas, while others show wider distributions, sometimes even over the entire peninsula.

According to their origin, endemics form two groups. Some of the species can be regarded as probable remnants of ancient Mediterranean mountain fauna (paleoendemics), while others likely came from the northern parts of Europe during the glacials and evolved under isolation in the mountains during the interglacials (neoendemics). The curious distribution of *Antrohyphantes rhodopensis*, found only in the high altitude zone and in caves, suggests a relict of ancient Mediterranean mountain fauna (Deltshev, 1990, 1996). Concerning the formation of cave fauna, Deeleman-Reinhold (1976) wrote that "many European cave spiders are probably relics of populations of moist Tertiary forests". Due to the lack of knowledge, it is difficult to determine with certainty which of the cave spider endemics of Bulgaria are Tertiary, and which are Quaternary elements. Nevertheless, other species of the endemic genus *Antrohyphantes* (*A. balcanicus, A. sophianus*) together with the blind species *Centromerus bulgarianus, Troglohyphantes drenskii* and *Troglohyphates bureschianus*, species with primitive three-branched paracymbium, also can be regarded as probable paleoendemics (Deltshev, 1996).

The following species: *Araeoncus clivifrons, Diplocephalus altimontanus, Drepanotylus pirinicus, Erigone longipalpis pirini, Incestophantes annulatus, Mughiphantes lithoclasicola*, and *Metopobactrus orbelicus*, known only from the high alpine parts of Pirin and Rila Mts., are high alpine elements. This group also includes *Tenuiphantes drenskyi*, which occurs in the high altitude belts of Pirin, Rila, Central Stara Planina and Vitosha Mts. These species are regarded as derivatives of the respective North or Central European related species (*Diplocephalus picinus, Drepanotylus borealis, Erigone longipalpis, Lepthyphantes armatus, Metopobactrus prominulus*), due to the disjunction of ranges during the glacial and interglacial (Deltshev, 1995, 1996, 1998). The largest portion of endemics was encountered mainly in caves and high altitude zones.

Mediterranean species complex (MED) includes only two species (0.45%). *Diplocephalus graecus*, known from South Europe and North Africa (Table 3), is one of the commonest erigonines of the Mediterranean region, often occurring in human-made or human-influenced habitats (Bosmans, 1996). *Cresmatoneta mutinensis* is also a characteristic Mediterranean element, known in Bulgaria only from Black Sea coast.

5 Vertical Distribution

The vertical belts are very well represented in Bulgaria, due to the relief and climate, characterized by specific vegetation and fauna.

Submediterranean belt includes 45 species (20.6%) (Table 3), with European (26.6%) and Palearctic species (24.4%) dominant. The endemics are represented by *Centromerus milleri, C. valkanovi* and *Palliduphantes byzantinus.*

Submontane belt has 96 species recorded (44%) (Table 3). Characteristic for this belt are the cavernicolous elements, most of which are endemics: *Centromerus bulgarianus, C. lakatnikensis, C. milleri, Palliduphantes byzantinus*, and *P. trnovensis*. The endemics *Centromerus valkanovi* and *Pelecopsis krausi* have restricted distribution in Bulgaria. They are characteristic for the oak forest of the southern Black Sea coast (*C. valkanovi*) and Struma Valley in southwestern Bulgaria (*P. krausi*).

Montane deciduous belt. In this belt, the linyphiid fauna is represented best (121 species, or 55.5%) (Table 3). The West Palearctic species have a dominant position here (25.6%). *Bolyphantes alticeps, B. luteolus, Ceratinella brevis, Diplocephalus picinus*, and *Gonatium rubellum* are mountain elements. The European complex is also well represented (21.5%) and *Araeoncus anguineus, Diplocephalus latifrons, Hilaira excisa, Palliduphantes alutacius, Oedothorax gibbifer*, and *Tiso vagans* can be regarded as mountain elements. *Palliduphantes spelaeorum* and *Porrhomma microps* are cavernicolous elements. The endemics, *Antrohyphantes sophianus, Centromerus acutidentatus, C. lakatnikensis, Troglohyphantes drenskii* and *Troglohyphantes bureschianus* must be regarded also as relict cavernicolous elements.

Montane coniferous belt has 94 species (43.1%) (Table 3). Best represented are the West Palearctic and European species, where the mountain elements arc 61%. Characteristic moutain elements are *Bolyphantes alticeps, B. luteolus, Cinetata gradata, Frontinellina frutetorum, Mansuphantes mansuetus, Piniphantes pinicola, Micrargus herbigradus, M. subaequalis, Tenuiphantes alacris.* The endemics *Centromerus sylvaticus paucidentatus, Gonatium orientale, Mansuphantes rectilamellus* and *Tenuiphantes drenskyi* can be regarded as mountain relicts. Cavernicolous elements are *Centromerus cavernarum, Porrhomma microphthalmum*, and *P. convexum.* The endemics *Antrohyphantes rhodopensis, Centromerus lakatnikensis, Lepthyphantes centromeroides, Troglohyphantes drenskii* and *T. bureschianus* can be regarded as cavernicolous relicts.

Subalpine belt is inhabited by 49 species (22.5%), some of which occur also in the forest zone (Table 3). The Palearctic species include mountain elements, such as *Ceratinella brevis, Bolyphantes alticeps, B. luteolus*, and *Meioneta gulosa*. The European species, which are numerous in this belt, include mainly forms typical for the high mountains in Central Europe: *Araeoncus anguineus, Cinetata gradata, Diplocephalus foraminifer, Evansia merens, Improphantes improbulus*, and *Mughiphantes pulcher*. The endemics include mountain (*Centromerus sylvaticus paucidentatus, Gonatium orientale, Mansuphantes rectilamellus, Tenuiphantes drenskyi*) and high mountain elements (*Araeoncus clivifrons,*

Erigone longipalpis pirini). The distribution of *Antrohyphantes rhodopensis* is striking, as this species occurs at high altitudes and in caves (Deltshev, 1990, 1995, 1996, 1999, 2000).

Alpine belt is populated by 22 species (10.1%), including mainly high altitude elements (Table 3). Most characteristic are endemic high altitude elements. *Mughiphantes lithoclasicola* and *Metopobactrus orbelicus* are known only in the high alpine belt of Rila and Pirin; *Incestophantes annulatus* is found only in the high altitude zone of Rila; and *Diplocephalus altimontanus* and *Drepanotylus pirinicus* are recorded only from the high alpine belt of Pirin. To this group also belong *Araeoncus clivifrons* and *Erigone longipalpis pirini*, which occur in the subalpine and alpine belts of both mountain ranges (Deltshev, 1995, 1996). The Holarctic species *Entelecara media, Mecynargus paetulus*, the Palearctic *Scotynotilus alpigena*, and European *Improphantes improbulus* and *Oreonetides glacialis* are high altitude elements, which can be regarded as ancient (Arctoalpine, Boreoalpine) species.

6 Discussion and Conclusions

The faunistic diversity of the 218 linyphiid spider species shows that Bulgaria is a territory of considerable species richness. This conclusion is supported also by the existence of 28 endemic species, which reflect the local character of the linyphiid fauna.

The vertical belts are very well represented in Bulgaria, due to its relief and climate, characterized by specific vegetation and fauna. The linyphiid fauna is best represented in the montane deciduous belt, followed by submontane and montane coniferous belts. The high mountain elements are best represented in the subalpine and alpine belts.

In zoogeographical respect, the widely distributed spiders of the Holarctic Region are dominant, but the most characteristic faunal elements are the endemics. They reflect quite well the history of the region and include paleoendemics (mainly in caves) and neoendemics (mainly at the high altitudes and in mountains).

According to their origin, Bulgarian endemics belong to two principal faunistic complexes: Mediterranean and European. This phenomenon can be explained by the relative isolation of the mountains compared with the lowlands, in the context of paleoenvironmental changes that have occurred since the Pliocene.

Acknowledgments

I am especially indebted to my colleagues Gergin Blagoev, Stoyan Lazarov, Alexi Popov, Zdravko Hubenov, and Dobrin Dobrev for the discussion and helpful assistance.

References

Blagoev, G., Ch. Deltshev & S. Lazarov, 2002. The Spiders (Araneae) of Bulgaria. http://cl.bas.bg/bulgarianspiders/.

Blick, T., R. Bosmans, J. Buchar, P. Gajdoš, A. Hänggi, P. Van Helsdingen, V. Růžička, W. Staręga & K. Thaler, 2004. Checkliste der Spinnen Mitteleuropas. Checklist of the spiders of Central Europe. (Arachnida: Araneae). Version 1. Dezember 2004. http://www.arages.de/checklist.html#2004_Araneae.

Bosmans, R., 1996. The genera *Araeoncus* Simon, *Delorripis* Simon and *Diplocephalus* Bertkau in North Africa (Araneae: Linyphiidae: Erigoninae). Studies on North African Linyphiidae VII. Belgian Journal of Zoology 126:123–151.

Bosmans, R. & M. Chatzaki, 2005. A catalogue of the spiders of Greece. A critical review of all spider species cited from Greece with their localities. Newsletter of the Belgian Arachnological Society 20 (Suppl. 2):1–124.

Buchar, J. & V. Růžička, 2002. Catalogue of Spiders of the Czech Republic. Praha Press, Prague.

Cardoso, P., 1990. Portuguese Spiders Checklist. http://www.geocites.com/RainForest/Vines/15197/checklist.html.

Deeleman-Reinhold, C., 1976. Distribution patterns in European cave spiders. In: First International Symposium on Cave Biology and Cave Paleontology, Cape Town: 25–35.

Deltshev, Ch., 1972. A contribution to the study of spiders (Araneae) from the caves in Bulgaria. II. Genus *Lepthyphantes* in Bulgarian caves. Izvestiya na Zoologicheskiya institut s muzei (Bulletin de l'Institut de zoologie et musée) 36:137–147.

Deltshev, Ch., 1973a. A new *Troglohyphantes* from Bulgarian caves (Araneae, Linyphiidae). International Journal of Speleology 5:103–109.

Deltshev, Ch., 1973b. Redescription of *Centromerus bulgarianus* (Drensky, 1931) and *Centromerus lakatnikensis* (Drensky, 1931) (Araneae, Linyphiidae). International Journal of Speleology 5:117–126.

Deltshev, Ch., 1974. A new *Centromerus* from Bulgarian caves (Araneae, Linyphiidae). International Journal of Speleology 6:81–86.

Deltshev, Ch., 1975a. A new species (*Troglohyphantes bureschianus* n. sp., Araneae, Linyphiidae) from Bulgarian caves. Acta zoologica bulgarica 3, 99–104.

Deltshev, Ch., 1975b. The genus *Lepthyphantes* in Bulgarian caves. In: Proceedings of the 6th International Arachnological Congress, 19–20 March 1974, Amsterdam. Nederlandse Entomologische Vereniging, Amsterdam: 210–213.

Deltshev, Ch., 1980. A contribution to the taxonomical study of *pallidus* group of genus *Lepthyphantes* Menge (Araneae, Linyphiidae) in Bulgaria. Acta zoologica bulgarica 16:44–56.

Deltshev, Ch., 1983a. A contribution to the study of *sylvaticus* group of genus *Centromerus* F. Dahl (Araneae, Linyphiidae) in Bulgaria. Acta zoologica bulgarica 21:53–58.

Deltshev, Ch., 1983b. Notes on the spiders of genus *Erigone* Audouin (Araneae, Erigonidae) in Bulgaria. Acta zoologica bulgarica 22:71–75.

Deltshev, Ch., 1983c. A contribution to the taxonomical and faunistic study of genus *Lepthyphantes* Menge (Araneae, Linyphiidae) from Pirin Mountain. Acta zoologica bulgarica 23:25–32.

Deltshev, Ch., 1984. A new *Diplocephalus* species from Bulgarian mountains (Arachnida, Araneae, Erigonidae). Reichenbachia 22:91–93.

Deltshev, Ch., 1985. A contribution to the study of the family Erigonidae (Araneae) from Pirin Mountain, Bulgaria, with a description of a new species (*Metopobactrus orbelicus* sp. n.). Bulletin of the British Arachnological Society 6:359–366.

Deltshev, Ch., 1987. A critical review of genus *Araeoncus* Simon in Bulgaria, with description of a new species (*Araeoncus clivifrons* sp. n.) (Arachnida, Araneae, Erigonidae). Reichenbachia 25:97–102.

Deltshev, Ch., 1988a. A contribution to the study of genus *Lepthyphantes* Menge (Araneae, Linyphiidae) from the Pirin Mountain with a description of a new species (*Lepthyphantes rectilamellus* sp. n.). Acta zoologica bulgarica 36:52–55.

Deltshev, Ch., 1988b. The genus *Fageiella* Kratochvil and the genus *Antrohyphantes* Dumitresco (Araneae, Linyphiidae, Lepthyphanteae) in the caves of Balkan peninsula. In Haupt, J. (ed), TUB-Dokument. Kongress Tagung, Berlin 38:293–299.

Deltshev, Ch., 1990. The high-altitude spiders (Araneae) in the Pirin Mountains, Bulgaria. Acta Zoologica Fennica 190:1–115.

Deltshev, Ch., 1992. *Drepanotylus pirinicus* n. sp. from Pirin Mountain (Bulgaria), with comparative remarks on the other species of the genus (Arachnida, Araneae, Linyphiidae). Berichte des naturwissenschaftlich-medizinischen Vereins, Innsbruck 79:173–176.

Deltshev, Ch., 1995. Spiders (Araneae) from the high altitude zone of Rila Mountain (Bulgaria). Berichte des naturwissenschaftlich-medizinischen Vereins, Innsbruck 82:217–225.

Deltshev, Ch., 1996. The origin, formation and zoogeography of endemic spiders of Bulgaria (Araneae). Revue suisse de Zoologie hors serie 1:141–151.

Deltshev, Ch., 1998. Spiders from the high altitude zone of Central Stara Planina Mountain (Bulgaria) (Araneae). Berichte des naturwissenschaftlich-medizinischen Vereins, Innsbruck 85:213–221.

Deltshev, Ch., 1999. A faunistic and zoogeographical review of spiders (Araneae) of the Balkan Peninsula. Journal of Arachnology 27:255–261.

Deltshev, Ch., 2000. The endemic spiders (Araneae) of the Balkan Peninsula. In Gajdoš, P. and Pekár, S. (eds.). Proceedings of the 18th European Colloquium of Arachnology, Stará Lesná, 1999. Ekológia (Bratislava) 19(Suppl. 3):59–65.

Deltshev, Ch. & G. Blagoev, 1992. A faunistic and zoogeographic analysis of spiders (Araneae) in Zemen Gorge (Southwestern Bulgaria). Acta zoologica bulgarica 45:26–35.

Deltshev, Ch. & G. Blagoev, 1997. The spiders of Pirin Mountain (Bulgaria). Taxonomic, faunistic and zoogeographical analysis (Araneae). Berichte des naturwissenschaftlich-medizinischen Vereins, Innsbruck 84:269–286.

Deltshev, Ch. & G. Blagoev, 2001. A critical check list of Bulgarian spiders (Araneae). Bulletin of the British Arachnological Society 12:110–138.

Deltshev, Ch. & B.P.M. Ćurčić, 1997. Contribution to the knowledge of the group *europaeus* of *Centromerus* Dahl (Linyphiidae, Araneae) in the Balkan Peninsula. Revue suisse de Zoologie 104:49–55.

Deltshev, Ch. & B.P.M. Ćurčić, 2002. A contribution to the study of the genus *Centromerus* Dahl (Araneae, Linyphiidae) in the caves of the Balkan Peninsula. Revue suisse de Zoologie 109:1–10.

Deltshev, Ch., B.P.M. Ćurčić & G. Blagoev, 2003. The Spiders of Serbia. Committee for Karst and Speleology, Serbian Academy of Sciences and Arts; Institute of Zoology, Bulgarian Academy of Sciences; Institute of Zoology, Faculty of Biology, University of Belgrade; Siniša Stanković Institute for Biological Research. Belgrade & Sofia.

Drensky, P., 1936. Katalog der echten Spinnen (Araneae) der Balkanhalbinsel. Sbornik na Balgarskata akademiya na naukite. Klon prirodo-matematichen [Collected Works of the Bulgarian Academy of Sciences. Natural-Mathematical Series] 32(15):1–223.

Helsdingen, P. van, K. Thaler & Ch. Deltshev, 1977. The *tenuis* group of *Lepthyphantes* Menge (Araneae, Linyphiidae). Tijdschrift voor Entomologie 120:1–54.

Helsdingen, P. van, K. Thaler & Ch. Deltshev, 2001. The European species of *Bolyphantes* with an attempt of a phylogenetic analysis (Araneae, Linyphiidae). Memorie della Società Entomologica italiana 80:3–35.

Hubenov, Z., 1997. Possibilities for using of a system from the really defined natural territories for the faunistic researches in Bulgaria. Acta zoologica bulgarica 49:5–9.

Lazarov, S., Ch. Deltshev & G. Blagoev, 2001. Spiders (Araneae) of Sushtinska Sredna Gora Mountain, Bulgaria. Acta zoologica bulgarica 53:3–28.

Marusik, Yu.M., D.V. Logunov & S. Koponen, 2000. Spiders of Tuva, South Siberia. IBPN FEB RAS, Magadan.

Mikhailov, K.G., 1997. Catalogue of the Spiders of the Territories of the Former Soviet Union (Arachnida, Aranei). Zoological Museum of the Moscow State University, Moscow.

Platnick, N., 2006. The World Spider Catalog, Version 7.0. http://research.amnh.org/entomology/spiders/catalog/index.html.

Samu, F. & C. Szinetár, 1999. Bibliographic check list of the Hungarian spider fauna. Bulletin of the British Arachnological Society 11:161–184.

Thaler, K., 1988. Arealformen in der nivalen Spinnenfauna der Ostalpen (Arachnida: Araneae). Zoologischer Anzeiger 220:233–244.

Thaler, K., P. van Helsdingen & Ch. Deltshev, 1994. Vikariante Verbreitung im Artenkomplex von Lepthyphantes annulatus in Europa und ihre Deutung (Araneae, Linyphiidae). Zoologischer Anzeiger 232:111–127.

Vigna Taglianti, A., P. Audisio, M. Biondi, M. Bologna, G. Carpaneto, A. De Biase, S. Fattorini, E. Piattella, R. Sindaco, A. Venchi & M. Zapparoli, 1999. A proposal for a chorotype classification of the Near East fauna, in the framework of the Western Palearctic region. Biogeographia 20:31–59.

15 Fauna and Zoogeography of Wolf Spiders (Araneae: Lycosidae) in Bulgaria

Gergin Blagoev
Institute of Zoology, Bulgarian Academy of Sciences, Tsar Osvoboditel Blvd. 1, 1000 Sofia, Bulgaria, e-mail: gerginblagoev@yahoo.ca

Abstract: The spider family Lycosidae is represented in Bulgaria by 81 species belonging to 11 genera. Wolf spiders occur in all regions of Bulgaria, inhabiting lowlands, forests and mountains. According to their current distribution, the Bulgarian wolf spiders are classified in 18 zoogeographical categories and four chorological complexes (widely distributed Holarctic species, European, endemic, and Mediterranean species). The largest group is that of the widely distributed Holarctic species. *Arctosa alpigena lamperti* and *Pirata insularis* are new records for the Bulgarian lycosid fauna

1 Introduction

Drensky (1936), in his summary work on the spiders of the Balkan Peninsula, reported 62 species of the family Lycosidae in Bulgaria. Later, Buchar (1968) recorded nine new species for the Bulgarian lycosid fauna. Further, several authors contributed to the study of the wolf spiders reporting eight more species (Deltshev, 1967, 1972a, 1972b, 1973, 1976a, 1976b, 1980a, 1980b, 1997, 1998; Blagoev and Deltshev, 1989; Deltshev and Blagoev, 1995, 2001; Dimitrov, 1999; Popov et al., 2000; Lazarov et al., 2001; Blagoev et al., 2002).

The goal of the present study was to make a critical faunistic and zoogeographical review of the wolf spiders in Bulgaria based both on published information and the original field survey data.

2 Materials and Methods

The material treated herein concerns original collections obtained in 1984–2003 during field surveys covering most of the districts in Bulgaria and the critical incorporation of all available literature records on the distribution of the wolf spiders in Bulgaria. The taxonomy of Lycosidae is given according to Platnick (2004).

The geographical regions of Bulgaria and their abbreviations used in the text are given according to Hubenov (1997), with original modifications. They are as follows (Fig. 1):

D00, DANUBIAN PLAIN; **D10**, Western Danubian Plain; **D20**, Middle Danubian Plain; **D30**, Eastern Danubian Plain; **D31**, Eastern Danubian Plain: Popovo–Provadiya District; **D32**, Eastern Danubian Plain: Ludogorie–Dobrudja District.

V. Fet and A. Popov (Eds.), Biogeography and Ecology of Bulgaria, 469–480.

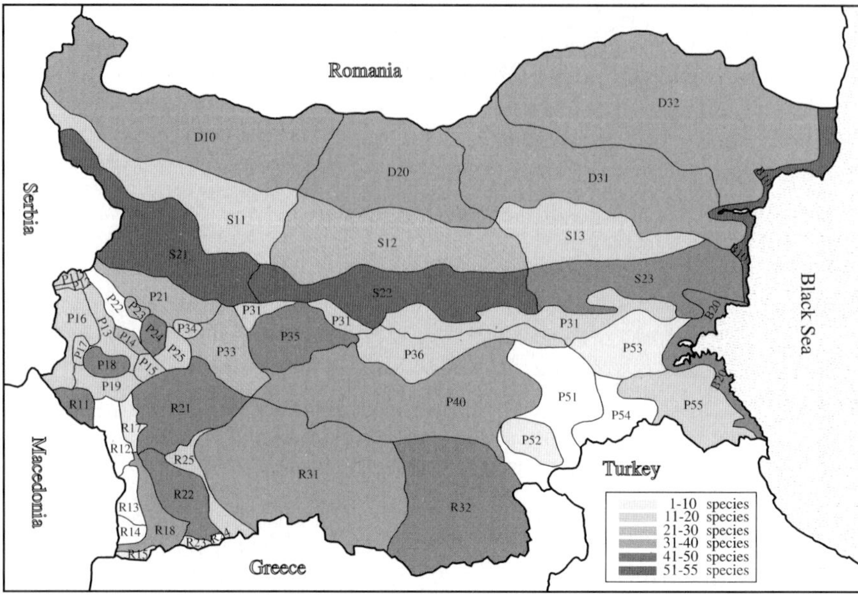

Fig. 1 Number of recorded species of Lycosidae in Bulgaria (geographical regions according to Hubenov, 1997, with modifications). Abbreviations, see text.

S00, STARA PLANINA RANGE SYSTEM; **S10**, Predbalkan Mts.; **S11**, Western Predbalkan Mts.; **S12**, Middle Predbalkan Mts.; **S13**, Eastern Predbalkan Mts.; **S20**, Stara Planina Mts. (Balkan); **S21**, Western Stara Planina Mts.; **S22**, Middle Stara Planina Mts.; **S23**, Eastern Stara Planina Mts.

P00, TRANSITIONAL REGION; **P10**, Kraishte–Konyavo District; **P11**, Rui Mts.; **P12**, Tran Basin (Znepole); **P13**, Strazha–Cherna Gora–Rudini Mts.; **P14**, Golo Bardo Mts.; **P15**, Verila Mts.; **P16**, Kraishte; **P17**, Zemenska Planina Mts.; **P18**, Konyavska Planina Mts.; **P19**, Kyustendil Basin; **P20**, Vitosha District; **P21**, Sofia Basin; **P22**, Zavalska–Viskyar Mts.; **P23**, Lyulin Mts.; **P24**, Vitosha Mts.; **P25**, Plana Mts.; **P30**, Srednogorie–Podbalkan Subregion; **P31**, Podbalkan Basins; **P32**, Sredna Gora Mts.; **P33**, Ihtimanska Sredna Gora Mts.; **P34**, Sredna Gora Mts.: Lozenska Planina Mts.; **P35**, Sashtinska Sredna Gora Mts.; **P36**, Sarnena Gora Mts.; **P40**, Thracian Lowland; **P50**, Tundja–Strandja Subregion; **P51**, Sakar–Tundja District; **P52**, Sakar Mts.; **P53**, Bakadjik–Burgas District; **P54**, Strandja–Dervent District; **P55**, Strandja Mts.

R00, RILA–RHODOPE MASSIF; **R10**, Osogovo–Belasitsa Group; **R11**, Osogovska Planina Mts.; **R12**, Vlahina Planina Mts.; **R13**, Maleshevska Planina Mts.; **R14**, Ograzhden Mts.; **R15**, Belasitsa Mts.; **R16**, Srednostrumska Valley; **R17**, Boboshevo–Simitli Srednostrumska Valley; **R18**, Krupnik–Sandanski–Petrich Srednostrumska Valley; **R20**, Rila–Pirin Group; **R21**, Rila Mts.; **R22**,

Pirin Mts.; **R23**, Slavyanka Mts.; **R24**, Stargach Mts.; **R25**, Mesta Valley; **R30**, Rhodope Mts.; **R31**, Western Rhodope Mts.; **R32**, Eastern Rhodope Mts.

B00, BLACK SEA COAST; **B10**, Northern Black Sea coast; **B20**, Southern Black Sea coast.

The data concerning general distribution of spiders are taken from Mikhailov (1997), Marusik et al. (2000), and Platnick (2004). Zoogeographic units follow Vigna Taglianti et al. (1999).

The quantitative index of Czekanowski–Dice–Sørensen (Baev and Penev, 1993) was used for comparisons concerning the lycosid fauna.

3 Species Composition and Zoogeography

3·1 Species composition

A total of 81 species of the family Lycosidae, belonging to 11 genera, have been recorded in Bulgaria so far (Table 1). Lycosids have been found in all regions of Bulgaria, including lowlands, forests, and mountains. The most diverse genera are *Pardosa* (35 species, or 42%), *Alopecosa* (19 species, or 23%), *Arctosa* (10 species, or 12%) and *Pirata* (7 species, or 8%) (Table 1). *Arctosa alpigena lamperti* Dahl, 1908 and *Pirata insularis* Emerton, 1885 (marked with an asterisk in Table 1) are new records for the Bulgarian lycosid fauna.

3·2 Zoogeographical analysis

According to their current distribution, Bulgarian wolf spiders are classified in 18 zoogeographical chorotypes and four chorological complexes (I, widely distributed Holarctic species; II, European species; III, endemic species; IV, Mediterranean species) (Table 1).

The complex of the species widely distributed in the Holarctic Region (HOL, PAT, PAL, ECA, EEC, SEC, SECA, MCA, and NECA) is the best represented one, and comprises 60 lycosid species (74%) widespread in Bulgaria. Among this complex, dominant are the Palearctic species (*sensu lato*) (43%), followed by the European–Central Asian species (14%), the Holarctic species (7%), and the Mediterranean–Central Asian species (4%). All other chorotypes are represented by a single species each.

This complex includes mainly widespread species associated with the lowlands, woodlands, and high altitude zones of the mountains. Most of the species are well represented in the lowlands. Frequently found there are *Alopecosa fabrilis, A. schmidti, A. sulzeri, Arctosa leopardus, A. stigmosa, Hogna radiata, Lycosa praegrandis, L. singoriensis, Pardosa italica, P. luctinosa*, and *Pirata tenuitarsis*. Frequently occurring in the mountains are *Alopecosa accentuata, A. inquilina,*

Table 1 Species diversity and distribution of Lycosidae in Bulgaria

Taxa	Chorotype	Districts
Alopecosa accentuata (Latreille, 1817)	PAL	D31, S20, P14, P18, P21, P23, P24, P32, P33, P35, P40, P55, R11, R21, R22, R30, B10
Alopecosa aculeata (Clerck, 1757)	HOL	D31, S00, P21, P24, P32, R11, R21, B00
Alopecosa albofasciata (Brullé, 1832)	MCA	D00, S00, P18, P31, P35, P40, P55, R11, R17, R21, R22, R25, R30, B00
Alopecosa cuneata (Clerck, 1757)	PAL	D32, S12, S20, P18, P23, P24, P32, P33, P35, P40, R11, R21, R22, R31, R32, B00
Alopecosa cursor (Hahn, 1831)	PAL	S20, P14, P18, P21, P25, P31, P33, P40, R21, R22, R32, B00
Alopecosa etrusca Lugetti et Tongiorgi, 1969	NME	S20, P18, R30
Alopecosa fabrilis (Clerck, 1757)	PAL	S12, S20, P14, P21, P40
Alopecosa inquilina (Clerck, 1757)	PAL	D00, S00, P24, P35, R11, R21, R22, R30, B10
Alopecosa pentheri (Nosek, 1905)	NEM	S21, P18, P32, P33, P40, P55, R22, B00
Alopecosa pinetorum (Thorell, 1856)	PAL	S21, P24, P35, P40, R11, B10
Alopecosa pulverulenta (Clerck, 1757)	PAL	S12, S20, P18, P21, P23, P24, P31, P40, P55, R11, R21, R22, R32, B00
Alopecosa schmidti (Hahn, 1835)	PAL	S21, B10
Alopecosa solitaria (Herman, 1879)	ECA	S20, P55
Alopecosa striatipes (C. L. Koch, 1839)	ECA	P10, R11, R21, B00
Alopecosa sulzeri (Pavesi, 1873)	PAL	D00, S21, P18, P23, P35, R11, B00
Alopecosa taeniata (C. L. Koch, 1835)	PAL	B00
Alopecosa taeniopus (Kulczyński, 1895)	ECA	D00, S20, P18, P21, P24, P35, P40, R21, B00
Alopecosa trabalis (Clerck, 1757)	ECA	S12, S20, P10, P21, P23, P24, P35, P55, R11, R21, R22, R30
* *Arctosa alpigena lamperti* Dahl, 1908	CEE	P23
Arctosa cinerea (Fabricius, 1777)	PAT	S20, P18, P23, P24, P35, P55, R17, R22, R25, R30, B00
Arctosa figurata (Simon, 1876)	EUR	D00, S21, P23, P35, R11
Arctosa leopardus (Sundevall, 1833)	PAL	D00, S20, P10, P40, R11, R17, B00
Arctosa lutetiana (Simon, 1876)	EUR	D00, P18, P35, R11, R32, B00
Arctosa maculata (Hahn, 1822)	EUR	D00, S12, S20, P18, P21, P24, P31, P32, P36, P40, R17, R21, R22, R25, R30, B20
Arctosa perita (Latreille, 1799)	HOL	D00, P18, P40, R21, R30
Arctosa stigmosa (Thorell, 1875)	PAL	D00, S20, P18, P40, R17, R32

Arctosa tbilisiensis Mcheidze, 1946	BCAU	R17
Arctosa variana C. L. Koch, 1847	MCA	R17
Aulonia albimana (Walckenaer, 1805)	PAL	S20, P10, P21, P23, P24, P25, P33, P35, P40, P55, R11, R17, R21, R22, R30, B00
Geolycosa vultuosa (C. L. Koch, 1838)	NEM	D00, S12, S21, P21, P32, P33, P40, P55, R11, R17, R22, R25, R30, B00
Hogna radiata (Latreille, 1817)	MCA	D20, S00, P18, P21, P23, P25, P31, P32, P33, P35, P40, P52, P53, P55, R11, R17, R21, R22, R30, B00
Lycosa praegrandis C. L. Koch, 1836	NECA	P40, P55, R17
Lycosa singoriensis (Laxmann, 1770)	PAL	D00, S20, P40, R30, B00
Pardosa agrestis (Westring, 1861)	PAL	D00, S12, S20, P14, P18, P21, P24, P25, P31, P33, P35, P40, R11, R15, R17, R21, R22, R23, R30, B00
Pardosa agricola (Thorell, 1856)	ECA	D00, S20, P10, P21, P23, P24, P32, P35, P40, R11, R15, R17, R21, R22, R23, R30, B00
Pardosa alacris (C. L. Koch, 1833)	EUR	D00, S22, P18, P21, P23, P24, P35, P40, P55, R11, R21, R32, B00
Pardosa albatula (Roewer, 1951)	EUR	S00, P24, P32, P35, R11, R21, R22, R30
Pardosa amentata (Clerck, 1757)	EUR	D00, S20, P18, P21, P23, P24, P35, R11, R15, R21, R22, R23, R30, B00
Pardosa atomaria (C. L. Koch, 1847)	NEM	P24
Pardosa bifasciata (C. L. Koch, 1834)	PAL	D00, S20, P23, P24, P25, P33, P35, R11, R21
Pardosa blanda (C. L. Koch, 1833)	PAL	D00, S22, P24, P35, R11, R21, R22
Pardosa cribrata Simon, 1876	MED	D00, S11, S20, P10, P21, P24, R17, R23, B00
Pardosa drenskii Buchar, 1968	BALK	S22, P24, R11, R21, R22
Pardosa ferruginea (L. Koch, 1870)	PAL	D00, S11, R21, R22
Pardosa hortensis (Thorell, 1872)	PAL	D00, S20, P14, P18, P21, P23, P24, P25, P31, P32, P33, P35, P40, P55, R11, R17, R21, R22, R25, R30, B00
Pardosa incerta Nosek, 1905	NEM	R21, R22, R31
Pardosa italica Tongiorgi, 1966	SEC	D00, P25, B20
Pardosa luctinosa Simon, 1876	PAL	D00, P21, B00
Pardosa lugubris (Walckenaer, 1802)	PAL	D00, S00, P10, P21, P23, P24, P31, P32, P35, P36, P40, P55, R11, R17, R21, R22, R22, R25, R30, B00
Pardosa mixta (Kulczyński, 1887)	ECA	R21, R22

continued

Table 1 continued

Taxa	Chorotype	Districts
Pardosa monticola (Clerck, 1757)	PAL	D00, S00, P18, P23, P24, P32, P35, P40, R11, R15, R21, R22, R23, R30, B00
Pardosa morosa (L. Koch, 1870)	ECA	S00, P14, P18, P23, P24, P35, P40, R21, R22, R30
Pardosa nebulosa (Thorell, 1872)	PAL	S20, R17, B10
Pardosa nigra (C. L. Koch, 1834)	PAL	R21, R22
Pardosa nigriceps (Thorell, 1856)	EUR	D00, S11, P21
Pardosa paludicola (Clerck, 1757)	PAL	S00, P10, B00
Pardosa palustris (Linnaeus, 1758)	HOL	S12, S20, P18, P21, P23, P24, P25, P31, P35, R11, R15, R21, R22, R23, R25, R30, B00
Pardosa pontica (Thorell, 1875)	EEC	B00
Pardosa prativaga (L. Koch, 1870)	ECA	D32, S12, S20, P10, P21, P23, P24, P25, P33, P40, R17, R21, R22, R31, B00
Pardosa proxima (C. L. Koch, 1847)	PAL	D00, S12, S20, P14, P18, P21, P23, P24, P25, P35, P40, P55, R11, R17, R21, R22, R30, B00
Pardosa pullata (Clerck, 1757)	ECA	S20, P10, P24, R11, R21, R22
Pardosa riparia (C. L. Koch, 1833)	PAL	S00, P24, R21, R22, R30, B10
Pardosa roscai (Roewer, 1951)	BALK	D00, P25, R17, B00
Pardosa saltuaria (L. Koch, 1870)	ECA	R21
Pardosa tasevi Buchar, 1968	SECA	S21, P23, P24, P35, R11, R21
Pardosa tatarica (Thorell, 1875)	PAL	D00, S20, P10, P24, P35, P40, P52, R11, R17, R22, R25, R30, B00
Pardosa vittata (Keyserling, 1863)	EUR	P18, P23, P25, P33, P55, R17, R22, R32, B00
Pirata hygrophilus Thorell, 1872	PAL	D00, S20, P18, P21, P23, P24, P31, P35, R11, R17, R21, R22, R31, B00
* *Pirata insularis* Emerton, 1885	HOL	P55

Pirata knorri (Scopoli, 1763)	PAL	S20, P10, P21, P24, P31, R11, R21, R25, R30, B20
Pirata latitans (Blackwall, 1841)	EUR	D00, S00, P18, P21, P23, P24, P25, P40, R17, R21, R22, R25, R32, B00
Pirata piraticus (Clerck, 1757)	HOL	S00, P35, P40, R11, R17, R21, R31, B00
Pirata piscatorius (Clerck, 1757)	PAL	S20, P10, P24, P25, P32, P33, R21, R25, R30, B00
Pirata tenuitarsis Simon, 1876	ECA	D00, S22, R30, B00
Trabea paradoxa Simon, 1876	SEU	B20
Trochosa hispanica Simon, 1870	MED	D00, P18, P23, P35, P52, P55, R11, R17, R32, B00
Trochosa robusta (Simon, 1876)	PAL	S20, P32, P33, P40, R21, B00
Trochosa ruricola (De Geer, 1778)	PAL	D32, S20, P18, P21, P23, P24, P25, P32, P33, P40, R11, R17, R22, R30, B00
Trochosa terricola Thorell, 1856	HOL	D00, S20, P18, P21, P23, P24, P32, P35, P40, R11, R17, R21, R22, R25, R30, B00
Xerolycosa miniata (C. L. Koch, 1834)	PAL	D00, S00, P10, P25, P32, P33, P35, P40, R11, R17, R21, R22, R30, B10
Xerolycosa nemoralis (Westring, 1861)	PAL	D00, S00, P14, P18, P21, P24, P25, P31, P33, P35, P40, R11, R17, R21, R22, R23, R30, B10

Zoogeographical chorotypes: (a) widely distributed Holarctic species complex: HOL, Holarctic; PAT, Palearctic–Afrotropical; PAL, Palearctic; ECA, European–Central Asian; EEC, East European–Central Asian; SEC, South European–Central Asian; SECA, Southeast European–Central Asian; MCA, Mediterranean–Central Asian; NECA, Northeast Mediterranean–Central Asian; (b) European complex: EUR, European; CEE, Central and East European; SEU, South European; BCAU, Balkan–Caucasian; (c) endemic complex: BALK, Balkan endemic; (d) Mediterranean complex: MED, Mediterranean; NME, North Mediterranean; NEM, Northeast Mediterranean

Pardosa blanda, and *P. ferruginea*. Specific high altitude elements are *Pardosa mixta* and *P. nigra*; they are found only in the alpine belts of the Rila and Pirin Mts.

The European chorological complex (EUR, CEE, SEU, and BCAU) comprises 12 species (15%) widespread in Europe and Bulgaria. The European species (*sensu lato*) are dominant (9%). All other chorotypes are represented by a single species each. This complex includes widespread species, mainly in the lowlands, and in the submontane and mountain belts. Most frequently found in the lowlands are *Arctosa figurata, A. lutetiana, A. maculata, Pardosa alacris, P. amentata, P. vittata*, and *Trabea paradoxa*. The species *Pardosa albatula*, widespread in most of the Bulgarian mountains, is a characteristic mountain element.

The complex of endemic species (BALK) includes two species (2%) recorded in Bulgaria and the neighboring territories. *Pardosa drenskii* is a representative of the high altitude complex of the Bulgarian wolf spiders and is not found below the timberline. The second endemic species, *Pardosa roscai*, inhabits localities along the Bulgarian and Romanian Black Sea coasts as well as localities with Mediterranean influence (Struma Valley).

The Mediterranean species complex (MED, NME, and NEM) includes seven species (9%), found mainly in the North Mediterranean region. This complex includes species associated mainly with lowlands, but also those reaching the alpine belt. Most frequently occurring in the lowlands are *Alopecosa etrusca, Pardosa atomaria*, and *Trochosa hispanica*. Specific Mediterranean mountain element is *Pardosa incerta* which is found in the high altitude zone of the mountains in Bulgaria, Turkey, and Caucasus. This species can be considered a remnant of an ancient mountain Mediterranean fauna in the region. In this connection, the vertical distribution of *Pardosa atomaria* and *Alopecosa pentheri* is interesting. These species are found in the lowlands as well as in the mountains.

4 Discussion

The number of the lycosid species recorded so far in Bulgaria (81) is high compared with the number of the lycosid species recorded in other regions of the Balkan Peninsula: Albania, 29 (Blagoev, 2005); Bosnia, 6; Croatia, 52; Greece, 55 (Bosmans & Chatzaki, 2005); Macedonia, 72 (Blagoev, 2002); Montenegro, 2; Romania, 54; Serbia, 63 (Deltshev et al., 2003); Slovenia, 48; Turkey, 27. This different species richness is due not only to the size of the region studied, but also to the degree of faunistic research. As a whole, the wolf spiders in the Balkan Peninsula are not very well studied (Fig. 2).

The faunistic complexes of wolf spiders in different regions in the Balkan Peninsula show interesting similarity trends (Fig. 3). The greatest similarity is observed between the faunistic complexes of Bulgaria, Macedonia, and Serbia. Close to these is the lycosid fauna of the Romanian Dobrudja. Second group of similarity is formed by the faunistic complexes of Croatia, Turkey, and Greece.

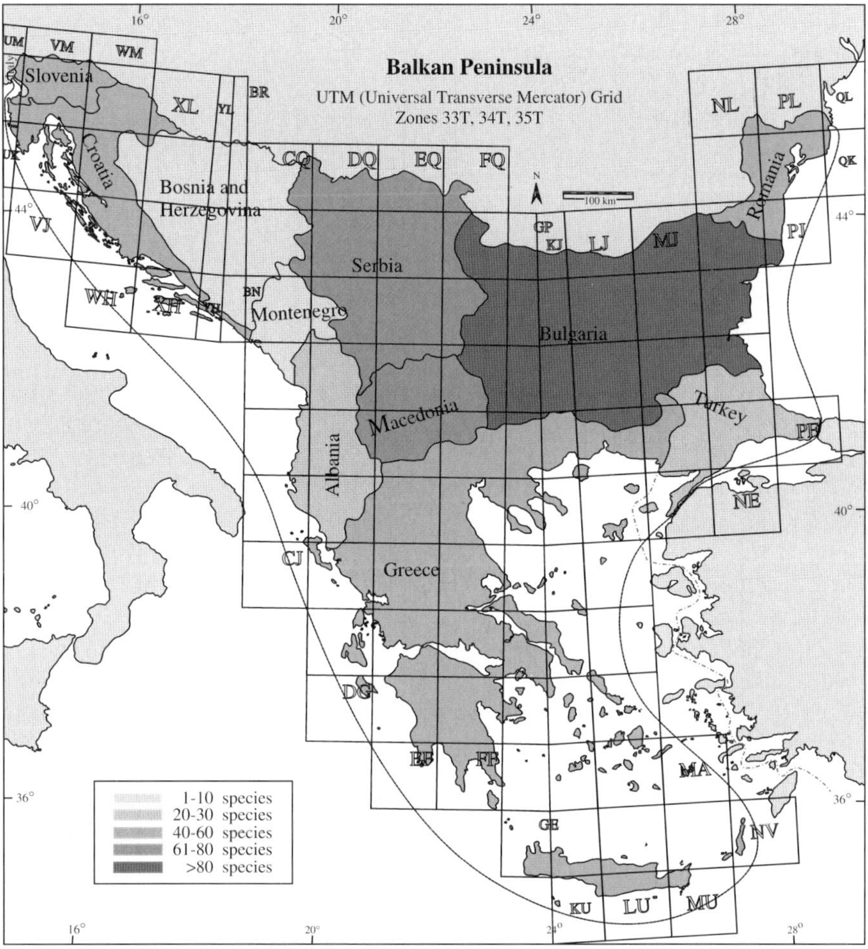

Fig. 2 Number of recorded species of Lycosidae in the Balkan Peninsula.

Close to these is the lycosid fauna of Albania. Most distant from other regions is the lycosid fauna of Slovenia. The wolf spiders in other countries of the Balkan Peninsula have been not studied so far.

Although Bulgaria has the richest lycosid fauna in the Balkan Peninsula its territory has not been evenly studied. Special attention was paid to the high mountain fauna. As a result, there is a lack of data on the occurrence of the wolf spiders in some large regions such as P22, R12, R13, R14, R24, P51, and P54 (Fig. 1). Some other regions such as Predbalkan Mts., Podbalkan Basins (P31), Kraishte (P11, P12, P13, P15, P16, P17, P19), and Sakar Mts. (P52) also are not very well studied.

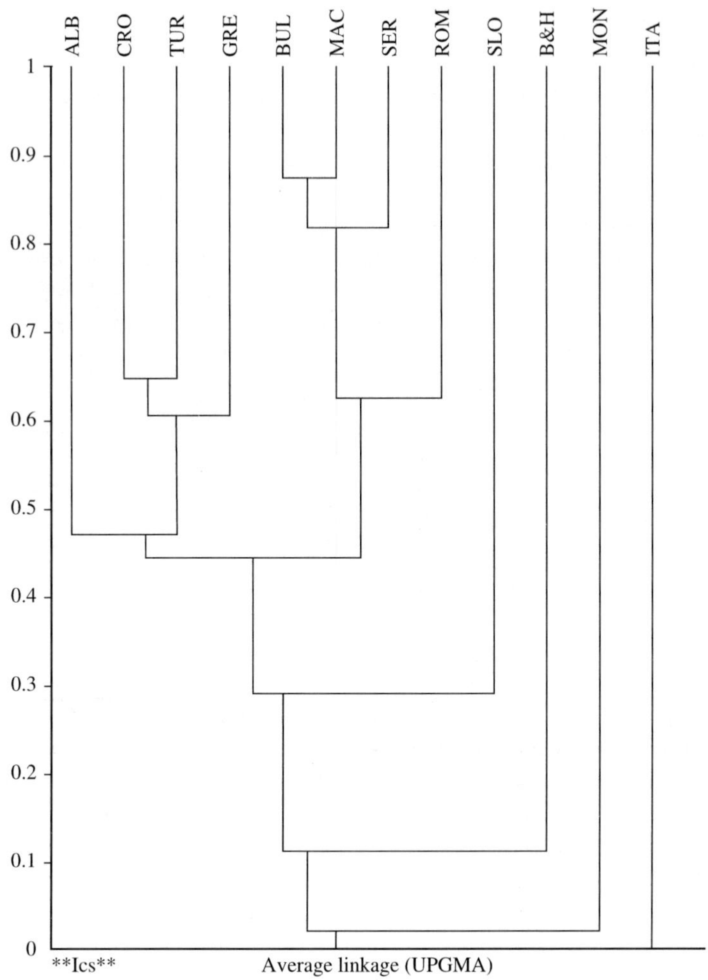

Fig. 3 Similarity of the regional lycosid faunas in the Balkan Peninsula.

The following regions are characterized with the highest species richness: Western Stara Planina Mts. (S21), Middle Stara Planina Mts. (S22) and Northern Black Sea coast (B10).

In terms of zoogeography, dominant are the species with wide geographic ranges (i.e., Palearctic and Holarctic).

Acknowledgments

I am especially indebted to Christo Deltshev, Zdravko Hubenov, and Alexi Popov for the discussion and helpful assistance.

References

Baev, P. & L. Penev, 1993. BIODIV: a Program for Calculating Biological Diversity Parameters, Similarity, Niche Overlap, and Cluster Analysis. Exeter Software, New York.

Blagoev, G., 2002. Check list of Macedonian spiders (Araneae). Acta zoologica bulgarica 54(3):9–36.

Blagoev, G. 2005. Contribution to the knowledge of the wolf spiders (Araneae, Lycosidae) of Albania. Acta zoologica bulgarica. 57(2):139–144.

Blagoev, G. & Ch. Deltshev, 1989. [Biotopical distribution of wolf-spiders (Araneae, Lycosidae) in the Zemen Gorge, Southwestern Bulgaria]. Ekologiya (Ecology) (Sofia) 22:73–80 (in Bulgarian).

Blagoev, G., Ch. Deltshev & S. Lazarov, 2002. The Spiders (Araneae) of Bulgaria. http://cl.bas.bg/bulgarianspiders/.

Bosmans, R. & M. Chatzaki, 2005. A catalogue of the spiders of Greece. A critical review of all spider species cited from Greece with their localities. Newsletter of the Belgian Arachnological Society 20 (Suppl. 2):1–124.

Buchar, J., 1968. Zur Lycosidenfauna Bulgariens (Arachn., Araneae). Časopis Československé společnosti entomologické 32:116–130.

Deltshev, Ch., 1967. [On the studies of spiders (Araneae) in the Vitosha Mountain]. Izvestiya na Zoologicheskiya institut s muzei (Bulletin de l'Institut de zoologie et musée) 24:51–56 (in Bulgarian).

Deltshev, Ch., 1972a. A review of spiders (Araneae) from Bulgarian caves. In Folk, C. (ed), Proceedings of the 5th International Congress of Arachnology, Brno: 99–104.

Deltshev, Ch., 1972b. A contribution to the study of spiders (Araneae) from the caves in Bulgaria. II. Genus *Lepthyphantes* in Bulgarian Caves. Izvestiya na Zoologicheskiya institut s muzei (Bulletin de l'Institut de zoologie et musée) 36:137–147.

Deltshev, Ch., 1973. [A contribution to the study of spiders (Araneae) from the caves of Bulgaria. III. Ecological notes on spiders (Araneae) from the entrance parts of the caves]. Izvestiya na Zoologicheskiya institut s muzei (Bulletin de l'Institut de zoologie et musée) 38:39–47 (in Bulgarian).

Deltshev, Ch., 1976a. [Spiders (Araneae) along the Bulgarian Black Sea coast]. In Suhozemna fauna na Balgariya. Materiali (Terrestrial Fauna of Bulgaria. Materials). Izdatelstvo na Balgarskata akademiya na naukite (Publishing House of the Bulgarian Academy of Sciences), Sofia: 83–96 (in Bulgarian).

Deltshev, Ch., 1976b. [On the spiders (Araneae) in the Ludogorie]. In Suhozemna fauna na Balgariya. Materiali (Terrestrial Fauna of Bulgaria. Materials). Izdatelstvo na Balgarskata akademiya na naukite (Publishing House of the Bulgarian Academy of Sciences), Sofia: 251–260 (in Bulgarian).

Deltshev, Ch., 1980a. On the high altitude spiders (Araneae) in Bulgaria. In Gruber, J. (ed), Verhandlungen. 8. Internationaler Arachnologen-Kongress abgehalten an der Universität für Bodenkultur Wien, 7–12 Juli, 1980. H. Egermannn, Vienna: 405–409.

Deltshev, Ch., 1980b. [Spiders (Araneae) from the high altitude belt of the Vitosha Mountain]. Acta zoologica bulgarica 15:78–92 (in Bulgarian).

Deltshev, Ch., 1997. Spiders (Araneae) from the coastal habitats of Shabla–Ezerets Lake, Bulgaria. Acta zoologica bulgarica 49:58–63.

Deltshev, Ch., 1998. Spiders from the high altitude zone of Central Stara Planina Mountain (Bulgaria) (Araneae). Berichte des naturwissenschaftlich-medizinischen Vereins, Innsbruck 85:213–221.

Deltshev, Ch. & G. Blagoev, 1995. A critical review of family Lycosidae (Araneae) in Bulgaria. Revue arachnologique 10:171–198.

Deltshev, Ch. & G. Blagoev, 2001. A critical check list of Bulgarian spiders (Araneae). Bulletin of the British Arachnological Society 12:110–138.

Deltshev, Ch., B.P.M. Ćurčić & G. Blagoev, 2003. The Spiders of Serbia. Committee for Karst and Speleology, Serbian Academy of Sciences and Arts; Institute of Zoology, Bulgarian Academy of Sciences; Institute of Zoology, Faculty of Biology, University of Belgrade; Siniša Stanković Institute for Biological Research. Belgrade & Sofia.

Dimitrov, D., 1999. The spider fauna of the Strandzha Mountain (South-East Bulgaria). I. Faunistic data and taxonomic remarks (Arachnida: Araneae). Acta zoologica bulgarica 51(2/3):15–26.

Drensky, P., 1936. Katalog der echten Spinnen (Araneae) der Balkanhalbinsel. Sbornik na Balgarskata akademiya na naukite. Klon prirodo-matematichen [Collected Works of the Bulgarian Academy of Sciences. Natural-Mathematical Series] 32(15):1–223.

Hubenov, Z., 1997. Possibilities for using of a system from the really defined natural territories for the faunistic researches in Bulgaria. Acta zoologica bulgarica 49:5–9.

Lazarov, S., Ch. Deltshev & G. Blagoev, 2001. Spiders (Araneae) of Sushtinska Sredna Gora Mountain, Bulgaria. Acta zoologica bulgarica 53:3–28.

Marusik, Yu.M., D.V. Logunov & S. Koponen, 2000. Spiders of Tuva, South Siberia. IBPN FEB RAS, Magadan.

Mikhailov, K.G., 1997. Catalogue of the Spiders of the Territories of the Former Soviet Union (Arachnida, Aranei). Zoological Museum of the Moscow State University, Moscow.

Platnick, N., 2006. The World Spider Catalog, Version 7.0. http://research.amnh.org/entomology/spiders/catalog/index.html.

Popov, V., Ch. Deltshev, G. Blagoev, I. Krusteva & D. Deltshev, 2000. Epigeobiont animal assemblages from two landscapes of the Bulgarian Black Sea coast: Relationship to habitat type, assemblage structure and biodiversity. II. Spiders (Araneae). Acta zoologica bulgarica 52(1):51–88.

Vigna Taglianti, A., P. Audisio, M. Biondi, M. Bologna, G. Carpaneto, A. De Biase, S. Fattorini, E. Piattella, R. Sindaco, A. Venchi & M. Zapparoli, 1999. A proposal for a chorotype classification of the Near East fauna, in the framework of the Western Palearctic region. Biogeographia 20:31–59.

16 Haplogyne Spiders (Araneae) in Bulgaria: Faunistic and Zoogeographical Analysis

Stoyan Lazarov

Institute of Zoology, Bulgarian Academy of Sciences, Tsar Osvoboditel Blvd. 1, 1000 Sofia, Bulgaria, e-mail: slazarov@zoology.bas.bg

Abstract: The spider group Haplogynae is represented in Bulgaria by 44 species belonging to six families. The Bulgarian haplogyne spiders are classified into 17 zoogeographical categories, which are grouped into five chorological complexes: Cosmopolitan, Holarctic (widely distributed), European, endemic, and Mediterranean. Most diverse are the widely distributed European–Caucasian species but endemics (25 %) are the most characteristic. Analysis of vertical distribution shows that the haplogyne spiders are distributed in all natural habitats of Bulgaria but prefer lowlands and the lower mountain belts, mainly inhabiting xerothermic oak forest

1 Introduction

The fauna of Haplogynae in Bulgaria is comparatively well studied. Drensky (1911, 1913, 1915, 1921, 1932, 1936, 1938, 1955) and Jurinitsch and Drensky (1917) reported 26 species. More recent publications (Gueorguiev and Beron, 1962; Deltshev, 1967, 1972, 1976, 1977, 1996, 1998; Deeleman-Reinhold and Deeleman, 1988; Deltshev and Blagoev, 1992, 1997; Beron, 1994; Dimitrov, 1997, 1999; Deltshev et al., 1998, 2000, 2003b; Lazarov, 1998; Dimitrov and Lazarov, 1999, 2002; Popov et al., 2000; Blagoev et al., 2001; Lazarov et al., 2001; Tzonev and Lazarov, 2001) are the result of the intensive faunistic research since 1967.

2 Study Area and Materials

The original material includes the spider collections made in 1993–2003 during the field surveys of Sredna Gora, Osogovo, Eastern Rhodopes, Chepan, and Strandja Mts. in Bulgaria. In addition, we critically reviewed all available literature. The geographical regions of Bulgaria (Fig. 1) and their abbreviations are according to Hubenov (1997); abbreviations used in the text are as follows: DW, Western Danubian Plain; DM, Middle Danubian Plain; DE, Eastern Danubian Plain; DEL, Ludogorie–Dobrudja District; DEP, Popovo–Provadiya District; SPW, Western Predbalkan; SPM, Middle Predbalkan; SPE, Eastern Predbalkan; SBW, Western Stara Planina Mts.; SBM, Middle Stara Planina Mts.; SBE, Eastern Stara Planina Mts.; PKR, Rui Mts.; PKG, Golo Bardo Mts.; PKV, Verila Mts.;

V. Fet and A. Popov (Eds.), Biogeography and Ecology of Bulgaria, 481–492.

Fig. 1 Geographical regions and districts of Bulgaria (according to Hubenov, 1997). Abbreviations, see text.

PKK, Kraishte; PKZ, Zemenska Planina Mts.; PKQ, Konyavska Planina Mts.; PVS, Sofia Basin; PVL, Lyulin Mts.; PVV, Vitosha Mts.; PVP, Plana Mts.; PSP, Podbalkan Basins; PSI, Ihtimanska Sredna Gora Mts.; PSL, Lozenska Planina Mts.; PSC, Sashtinska Sredna Gora Mts.; PSA, Sarnena Gora Mts.; PT, Thracian Lowland; PBT, Sakar–Tundja District; PBC, Sakar Mts.; PBB, Bakadjik–Burgas District; PBD, Strandja–Dervent District; PBS, Strandja Mts.; ROO, Osogovo Mts.; ROV, Vlahina Planina Mts.; ROM, Maleshevska Planina Mts.; ROG, Ograzhden Mts.; ROB, Belasitsa Mts.; ROT, Boboshevo–Simitli Valley; ROP, Krupnik–Sandanski–Petrich Valley; RPR, Rila Mts.; RPP, Pirin Mts.; RPS, Slavyanka Mts.; RPT, Stargach Mts.; RPM, Mesta Valley; RRW, Western Rhodope Mts.; RRE, Eastern Rhodope Mts.; BN, Northern Black Sea coast; BS, Southern Black Sea coast.

3 Species Composition

The haplogyne spiders are represented in Bulgaria by 44 species belonging to six families (Scytodidae, one species; Leptonetidae, two; Pholcidae, seven; Segestriidae, three; Dysderidae, 28; Oonopidae, three). The following species are known only from single localities: *Hoplopholcus labyrinthi, Pholcus ponticus, Segestria florentina, Dasumia amoena, D. canestrinii, Dysdera ninnii, D. punctata, Dysderocrates egregius, Harpactea abantia, H. deltshevi, H. strandjica, Silhouettella loricatula, Oonops pulcher*, and *Tapinesthis inermis*.

Among interesting faunistic records are:

Hoplopholcus labyrinthi, known only from Crete and Bulgaria, which inhabits mainly caves and buildings (Lazarov et al., 2001).

Tapinesthis inermis, known from Europe and introduced in the USA; Bulgaria is the single locality in the Balkan Peninsula.

Dysdera enguriensis, known only from Bulgaria and Turkey. Bulgarian records are from the Strandja Mts.

Dysdera punctata, a characteristic Mediterranean element, found in Bulgaria, Greece (including Crete), Cyprus, and Albania. Bulgarian localities of the species at the lower southern parts of the Osogovo Mts. indicate a substantial Mediterranean influence over this area.

Harpactea apolinea, known only from Bulgaria and Greece. Bulgarian records are from the Strandja Mts.

Harpactea pr. *doblikae*. The range of *H. doblikae* is not entirely clarified. This species was described from the Crimean Peninsula, and the only data outside of this territory are for the southern coast of the Bulgarian Black Sea (Popov et al., 2000). A review of Ukrainian comparative material shows that the Bulgarian specimens are similar to the Ukrainian ones, but some differences exist; it could be a closely related, undescribed new species.

Most of the haplogyne species in Bulgaria inhabit southern lowlands and lower parts of the mountains, mostly the xerothermic oak forest. The greatest number of haplogyne species is recorded from the Strandja Mts., the Rhodopes, and the Bulgarian Black Sea coast. The number of species in Bulgaria is high compared with that from other countries of Europe having the territories of similar size. The species richness of haplogyne spiders in Europe increases from north to south (Table 1).

Table 1 Comparison of area and haplogyne spider species richness in some European countries

Country	Area (sq. km)	Number of species	Sources
Austria	83, 858	20	Blick et al. (2004)
Bulgaria	110, 550	44	Deltshev & Blagoev (2001)
Czech Republic	77, 280	13	Buchar & Růžička (2002)
Greece	128, 900	95	Bosmans & Chatzaki (2005)
Hungary	92, 340	19	Samu & Szinetár (1999)
Portugal	91, 500	43	Cardoso (1999)
Serbia	102, 000	26	Deltshev et al. (2003a)

4 Zoogeographical Analysis

The Bulgarian haplogyne spiders can be grouped into 17 zoogeographical chorotypes, grouped into five chorological complexes: Cosmopolitan, widely distributed in the Holarctic, European, endemic, and Mediterranean. The data concerning general distribution of spiders are taken from Mikhailov (1997), Marusik et al. (2000), Platnick (2004), and Vigna Taglianti et al. (1999).

The complex of species widely distributed in the Holarctic Region (HOL + PAL + WPA + EUCA + SCA + EUT + EET) is best represented and comprises 13 (29.6%) species widespread in Bulgaria. European–Caucasian species are dominant (five species, or 11.4%), followed by Holarctic species (three species, or 6.8%). The remaining chorotypes are represented by single species each. This complex includes mainly widespread species associated whit lowlands and woodlands. Most of the species are well represented in low mountains. Characteristic elements are *Scytodes thoracica, Pholcus opilionoides, Segestria bavarica, Dysdera erythrina, Dysdera hungarica*, and *Harpactea rubicunda*. The single high altitude element of this complex is *Segestria senoculata*, inhabithing the mountains up to 2200 m (Deltshev and Blagoev, 1997).

The complex of endemics (BALK + BULG) includes 11 species (25.0%) found in Bulgaria and Balkan Peninsula. The established number of endemics is high and reflects the local character of the fauna. The issue about the status and distribution of endemic spiders found in Bulgaria is complicated. Some of them occur only in restricted areas, while others show wider distribution, sometimes even across the whole Balkan Peninsula. According to their origin, the endemics form two groups. Some of the species can be regarded as probable remnants of ancient Mediterranean mountain fauna (paleoendemics), and others as elements that came from the northern parts of Europe during the glacials and evolved under isolation on mountains during the interglacials (neoendemics) (Deltshev, 1996).

The group of Balkan endemics includes six species (13.6%); best represented in Bulgaria is *Dasumia kusceri*. The species *Harpactea abantia* and *H. apollinea* are known only from Greece.

The group of Bulgarian endemics includes five species (11.4%). *Harpactea srednogora, H. deltshevi* and *H. strandjica* are Bulgarian endemics with local distribution. *Harpactea srednogora* and *H. deltshevi* have been found only in Sushtinska Sredna Gora Mountains, while *H. strandjica* so far is known only from Strandja Mountains. *Protoleptoneta beroni* and *P. bulgarica* are characteristic for the caves.

The high percentage of endemic species shows that the local speciation is considerable. All endemic haplogyne spiders recorded in Bulgaria are paleoendemics.

The European species complex (EUR + CSEE + SEU + EEU) includes ten species (22.7%), widespread in Europe and Bulgaria. East European species are dominant (four species, or 9.1%), followed by European (three species, or 6.8 %), South European (2 species, or 4.6%), and Central and Southeast European (one species, or 2.3%). The complex comprises widespread species which inhabit lowland

Table 2 Species list, distribution, and zoogeographical characteristic of haplogyne spiders in Bulgaria

Taxa	Districts	A	B	C	D	E	Chorotype
SCYTODIDAE							
Scytodes thoracica (Latreille, 1802)	PSC, RPP, RPR, ROB, BS, ROO, RR	+	+	+	+		HOL
LEPTONETIDAE							
Protoleptoneta beroni Deltshev, 1977	SBW	+					BULG
Protoleptoneta bulgarica Deltshev, 1972	SBW	+					BULG
PHOLCIDAE							
Holocnemus pluchei (Scopoli, 1763)	PSC, PBD, BS, PT, PBT	+	+	+	+		HMED
Hoplopholcus forskali (Thorell, 1871)	PSC, PV, DEP, DEL, SBW, DM, DEP, DW, PKK	+	+				EET
Hoplopholcus labyrinthi (Kulczyński 1903)	PSC	+					BALK
Pholcus opilionoides (Schrank, 1781)	PSC, PV, SBW, SPM, RRW, RPP, RPR, SBE	+	+	+	+		HOL
Pholcus phalangioides (Fuesslin, 1775)	PSC, PV, SBW, SPM, RRW, RPP, RPR, SBE, PK	+	+	+	+		COS
Pholcus ponticus Thorell, 1875	BN, BS	+					SCA
Spermophora senoculata (Dugès, 1836)	PV, ROT	+	+				HOL

continued

Table 2 continued

Taxa	Districts	A	B	C	D	E	Chorotype
SEGESTRIIDAE							
Segestria bavarica C. L. Koch, 1843	PSC, SBW, ROO	+	+				EUT
Segestria florentina (Rossi, 1790)	RPS			+			EUCA
Segestria senoculata (Linnaeus, 1758)	PSC, SBM, RPR, RPP, RRW, BN	+	+	+	+	+	PAL
DYSDERIDAE							
Dasumia amoena (Kulczyński, 1897)	BN	+					EEU
Dasumia canestrinii (L. Koch, 1876)	PVL	+	+				SEU
Dasumia kusceri (Kratochvil, 1935)	ROO, SBW	+	+				BALK
Dysdera argaeica Nosek, 1905	PBS	+					NEM
Dysdera crocota C. L. Koch, 1838	BN, PVV, BS, SBW, RRW, RRE	+	+	+	+		COS
Dysdera enguriensis Deeleman-Reinhold, 1988	PBS	+					NEM
Dysdera erythrina (Walckenaer, 1802)	RPP, RRW, PKQ, SBM, RPR, BN, PT	+	+	+	+		EUCA
Dysdera hungarica Kulczyński, 1897	BN, BS, PKQ, RPR, SBM, RPP	+					EUCA
Dysdera longirostris Doblika, 1853	BN, BS, PSC, SBM, ROO, RRW, RPR, RPP, PBS	+	+	+	+		EEU

Dysdera ninnii Canestrini, 1868	PKQ, PVL		+	+		SEU
Dysdera pectinata Deeleman-Reinhold, 1988	ROO	+	+			BALK
Dysdera punctata C. L. Koch, 1838	ROO	+				NEM
Dysdera richteri Charitonov, 1956	BS, DEL	+				EUCA
Dysdera taurica Charitonov, 1956	PSC, BS, DEL	+				EEU
Dysdera westringi O. P.-Cambridge, 1872	BN, BS, SBM, ROB	+	+			NEM
Dysderocrates egregius (Kulczyński, 1897)	PVL		+			CSEE
Harpactea abantia (Simon, 1884)	BS	+				BALK
Harpactea apollinea Brignoli, 1979	PBS	+				BALK
Harpactea babori (Nosek, 1905)	PSC, BS, BN, SBM, PBS, ROT	+	+	+	+	NEM
Harpactea deltshevi Dimitrov et Lazarov, 1999	PSC				+	BULG
Harpactea pr. *doblikae* (Thorell, 1875)	BS	+				NEM
Harpactea hombergi (Scopoli, 1763)	RPP, ROO, RRW	+	+	+	+	EUR

continued

Table 2 continued

Taxa	Districts	A	B	C	D	E	Chorotype
Harpactea lepida (C. L. Koch, 1838)	BN, BS, RPP, SBM, PT, RRW, RPR, PVS	+	+	+	+		EUR
Harpactea rubicunda (C. L. Koch, 1838)	BN, BS, SBM, PBS, ROO, PSC, PKK	+	+				EUCA
Harpactea saeva (Herman, 1879)	PSC, BN, BS, ROO, PKK, ROT	+	+	+	+		EEU
Harpactea srednogora Dimitrov et Lazarov, 1999	PSC	+	+				BULG
Harpactea strandjica Dimitrov, 1997	PBS	+					BULG
Harpactea sturanyi (Nosek, 1905)	PBS	+					BALK
OONOPIDAE							
Silhouettella loricatula (Roewer, 1942)	RPS			+			WPA
Oonops pulcher Templeton, 1835	SBM		+				SCO
Tapinesthis inermis (Simon, 1882)	PSC	+					EUR

Altitudinal belts: A, Submediterranean; B, Submontane; C, Montane deciduous; D, Montane coniferous; E, Subalpine. Zoogeographical chorotypes: COS, Cosmopolitan; SCO, Subcosmopolitan; HOL, Holarctic; PAL, Palearctic; WPA, West Palearctic; EUT, European–Turanian; EET, East European–Turanian; SCA, Southeast European–Central Asian; EUCA, European–Caucasian; EUR, European; CSEE, Central and Southeast European; SEU, South European; EEU, East European; BULG, Bulgarian endemic; BALK, Balkan endemic; HMED, Holomediterranean; NEM, Northeast Mediterranean and mountains, found under stones and in litter. Widespread in Bulgaria are *Dysdera longirostris, D. taurica, Harpactea hombergi, H. lepida*, and *H. saeva*. Rare species in Bulgaria from this complex are *Dasumia amoena, D. canestrinii*, and *Tapinesthis inermis* (which was recorded from a single locality).

The Mediterranean species complex (MED + NEM) includes seven species (15.9%). The best represented are Northeast Mediterranean elements with six species (13.6%), widespread in the lowland, xerothermic oak forest. Characteristic species for Bulgarian fauna are *Dysdera westringi* and *Harpactea babori*. Four species (*Dysdera argaeica, D. enguriensis, D. punctata* and *Harpactea* pr. *doblikae*) are recorded only from single locality. The single Holomediterranean (HMED) species is *Holocnemus pluchei*, which inhabits both houses and caves.

The Cosmopolitan species complex (COS + SCO) includes three species (6.8%). Widespread in Bulgaria are *Pholcus phalangioides* and *Dysdera crocota* (Table 2). The Subcosmopolitan species *Oonops pulcher* is very rare in Bulgaria and was not recorded since it was found by Drensky.

5 Vertical Distribution

The vertical belts are very well represented in Bulgaria, due to its relief and climate, and are characterized by specific vegetation and fauna.

Submediterranean belt has the best represented haplogyne fauna, represented by 38 species (86.4%). Characteristic species are *Hoplopholcus labyrinthi, Dasumia amoena, Dysdera argaeica, D. enguriensis, D. punctata, D. richteri, D. taurica, Harpactea abantia, H. apollinea, H.* pr. *doblikae, H. strandjica, H. sturanyi*, and *Tapinesthis inermis*. Here also occur cavernicolous species: *Protoleptoneta beroni, P. bulgarica*, and *Hoplopholcus forskali*.

Submontane belt contains 24 species of haplogynes (54.6%) and includes forms characteristic for oak forest of South Bulgaria (*Scytodes thoracica, Spermophora senoculata, Dysdera ninnii, Dysderocrates egregius, Harpactea rubicunda*, and *Oonops pulcher*). The endemics *Harpactea srednogora* and *Dasumia kusceri* are rare, known in Bulgaria only from restricted areas.

Montane deciduous belt is inhabited by 15 species (34.1%). The inhabitants of this belt can be regarded as mountain elements. Characteristic species are *Segestria florentina, Dysdera crocota, D. erythrina, Harpactea hombergi*, and *Silhouettella loricatula*.

Montane coniferous belt has 13 species (29.6%), known from single specimens. The Bulgarian endemic *Harpactea deltshevi* is described from this belt at Sashtinska Sredna Gora Mts. (above Koprivshtitsa), and has only a local distribution. Characteristic species of this belt are also *Dysdera longirostris, Harpactea lepida*, and *H. saeva*, all three being typical mountain elements widespread in Europe.

Subalpine belt is inhabited only by *Segestria senoculata* (2.3%), which occur also in the forest zone. The species reaches to 2500 m in the mountains and can be found under stones and bark.

Acknowledgments

I would like to express my thanks to Ch. Deltshev, G. Blagoev, and D. Dobrev for their helpful discussions.

References

Beron, P., 1994. Résultats des recherches biospéléologiques en Bulgarie de 1971 à 1994 et liste des animaux cavernicoles bulgares. Tranteeva (Sofia) 1:1–137.

Blagoev, G., Ch. Deltshev & S. Lazarov, 2002. The Spiders (Araneae) of Bulgaria. http://cl.bas.bg/bulgarianspiders/.

Blagoev, G., S. Lazarov & Ch. Deltshev, 2001. [Spiders (Araneae) in Kresna Gorge (Southwestern Bulgaria)]. In Beron, P. (ed), Bioraznoobrazie na Kresnenskiya prolom (Biodiversity of Kresna Gorge (SW Bulgaria)). Natsionalen prirodonauchen muzei i Institut po zoologiya pri BAN (National Museum of Natural History (Sofia) & Institute of Zoology, Bulgarian Academy of Sciences), Sofia: 85–102 (in Bulgarian).

Blick, T., R. Bosmans, J. Buchar, P. Gajdoš, A. Hänggi, P. Van Helsdingen, V. Růžička, W. Staręga & K. Thaler, 2004. Checkliste der Spinnen Mitteleuropas. Checklist of the spiders of Central Europe. (Arachnida: Araneae). Version 1. Dezember 2004. http://www.arages.de/checklist.html#2004_Araneae.

Bosmans, R. & M. Chatzaki, 2005. A catalogue of the spiders of Greece. A critical review of all spider species cited from Greece with their localities. Newsletter of the Belgian Arachnological Society 20 (Suppl. 2):1–124.

Buchar, J. & V. Růžička, 2002. Catalogue of Spiders of the Czech Republic. Praha Press, Prague.

Cardoso, P., 1990. Portuguese Spiders Checklist. http://www.geocites.com/RainForest/Vines/15197/checklist.html.

Deeleman-Reinhold, C. & P. Deeleman, 1988. Revision des Dysderinae (Araneae, Dysderidae), les espèces mediterranéennes occidentales exceptées. Tijdschrift voor Entomologie 131:141–269.

Deltshev, Ch., 1967. [On the studies of spiders (Araneae) in the Vitosha Mountain]. Izvestiya na Zoologicheskiya institut s muzei (Bulletin de l'Institut de zoologie et musée) 24:51–56 (in Bulgarian).

Deltshev, Ch., 1972. A new genus of Bulgarian cave spiders (*Protoleptoneta hulgarica*, n. g., n. sp., Leptonetidae). International Journal of Speleology 4:275–283.

Deltshev, Ch., 1976. [Spiders (Araneae) along the Bulgarian Black Sea coast]. In Suhozemna fauna na Balgariya. Materiali (Terrestrial Fauna of Bulgaria. Materials). Izdatelstvo na Balgarskata akademiya na naukite (Publishing House of the Bulgarian Academy of Sciences), Sofia: 83–96 (in Bulgarian).

Deltshev, Ch., 1977. A new *Protoleptoneta* from caves (Araneae, Leptonetidae). Acta zoologica bulgarica 7:3–8.

Deltshev, Ch., 1996. The origin, formation and zoogeography of endemic spiders of Bulgaria (Araneae). Revue suisse de Zoologie hors serie 1:141–151.

Deltshev, Ch., 1998. Spiders from the high altitude zone of Central Stara Planina Mountain (Bulgaria) (Araneae). Berichte des naturwissenschaftlich-medizinischen Vereins, Innsbruck 85:213–221.

Deltshev, Ch., S. Andreev, G. Blagoev, V. Golemansky, G. Milojkova, V. Peneva, D. Dobrev, M. Todorov & Z. Hubenov, 1998. Invertebrates (non-Insecta) in Bulgaria. In Meine, C. (ed), Bulgaria's Biological Diversity: Conservation Status and Needs Assessment. Volumes I and II. Pensoft, Sofia & Moscow: 109–161.

Deltshev, Ch., P. Beron, G. Blagoev, V. Golemansky, V. Peneva, P. Stoev, P., M. Todorov & Z. Hubenov, 2000. Faunistic diversity of invertebrates (non Insecta) of the Rila National Park. In Sakalian, M. (ed), Biological Diversity of the Rila National Park. Pensoft, Sofia: 249–284.

Deltshev, Ch. & G. Blagoev, 1992. A faunistic and zoogeographic analysis of spiders (Araneae) in Zemen Gorge (Southwestern Bulgaria). Acta zoologica bulgarica 45:26–35.

Deltshev, Ch. & G. Blagoev, 1997. The spiders of Pirin Mountain (Bulgaria). Taxonomic, faunistic and zoogeographical analysis (Araneae). Berichte des naturwissenschaftlich-medizinischen Vereins, Innsbruck 84:269–286.

Deltshev, Ch. & G. Blagoev, 2001. A critical check list of Bulgarian spiders (Araneae). Bulletin of the British Arachnological Society 12:110–138.

Deltshev, Ch., B.P.M. Ćurčić & G. Blagoev, 2003. The Spiders of Serbia. Committee for Karst and Speleology, Serbian Academy of Sciences and Arts; Institute of Zoology, Bulgarian Academy of Sciences; Institute of Zoology, Faculty of Biology, University of Belgrade; Siniša Stanković Institute for Biological Research. Belgrade & Sofia.

Deltshev, Ch., S. Lazarov, & B. Petrov, 2003. A contribution to the study of the spiders (Araneae) from the caves of Bulgaria. Acta zoologica bulgarica 56:155–166.

Dimitrov, D., 1997. Description of *Harpactea strandjica*, sp. n. from Bulgaria (Araneae, Dysderidae). Bulletin of the British Arachnological Society 10:322.

Dimitrov, D., 1999. The spider fauna of the Strandzha Mountain (South-East Bulgaria). I. Faunistic data and taxonomic remarks (Arachnida: Araneae). Acta zoologica bulgarica 51(2/3):15–26.

Dimitrov, D. & S. Lazarov, 1999. Two new species *of Harpactea* from Bulgaria (Araneae, Dysderidae). Berichte des naturwissenschaftlich-medizinischen Vereins, Innsbruck 86:127–129.

Dimitrov, D. & S. Lazarov, 2002. A contribution to the study of the spiders (Araneae) in Chepun Mountain and Dragoman Swampland (NW Bulgaria). Acta zoologica bulgarica 54(2):47–53.

Drensky, P., 1911. Über die Spinnen-Fauna des Trojan-Balkans und seiner nördlichen Abhänge bis der Stadt Lowetsch. Entomologische Rundschau 28, 41–43.

Drensky, P., 1913. [Über die Spinnenfauna Bulgariens]. Spisanie na Balgarskata akademiya na naukite [Journal of the Bulgarian Academy of Sciences] 2, 1–146 (in Bulgarian).

Drensky, P., 1915. [Araneides nouveaux ou peu connus de Bulgarie]. Spisanie na Balgarskata akademiya na naukite. Klon prirodo-matematichen [Journal of the Bulgarian Academy of Sciences. Natural-Mathematical Series] 7(2):141–176 (in Bulgarian).

Drensky, P., 1921. [Contribution à l'étude des araignées de la Macédoine orientale et de Pirine planina]. Spisanie na Balgarskata akademiya na naukite. Klon prirodo-matematichen [Journal of the Bulgarian Academy of Sciences. Natural-Mathematical Series] 23(11):1–50 (in Bulgarian).

Drensky, P., 1932. [Zur Kenntnis der Spinnenfauna Tscham-Kuria im Rilagebirge (Bulgarien)]. Trudove na Balgarskoto prirodoizpitatelno druzhestvo (Travaux de la Société bulgare des sciences naturelles) 15/16:326–332 (in Bulgarian).

Drensky, P., 1936. Katalog der echten Spinnen (Araneae) der Balkanhalbinsel. Sbornik na Balgarskata akademiya na naukite. Klon prirodo-matematichen [Collected Works of the Bulgarian Academy of Sciences. Natural-Mathematical Series] 32(15):1–223.

Drensky, P., 1938. [Die Spinnenfauna Bulgariens. II. Unterordnung Arachnomorphae, I Gruppe Tetrastica, Familie: Filistatidae, Dysderidae und Oonopidae]. Izvestiya na Tsarskite prirodonauchni instituti v Sofia (Mitteilungen aus den Königlichen naturwissenschaftlichen Instituten in Sofia – Bulgarien; Bulletin des Institutions royales d'histoire naturelle à Sofia – Bulgarie) 11:81–113 (in Bulgarian).

Drensky, P., 1955. [Fliegen und Arachnoideen aus den Wäldern und den Feldschutzwaldstreifen der Süd-Dobrudscha]. In: Stojanov, N. & B. Kitanov (eds), Sbornik na ekspeditsiyata po polezashtitnite poyasi v Dobrudja prez 1952 godina [Materials of Expedition on Field-protection Belts in Dobrudja in 1952]. Izdanie na Balgarskata akademiya na naukite [Edition of the Bulgarian Academy of Sciences], Sofia: 385–397 (in Bulgarian).

Guéorguiev, V. & P. Beron, 1962. Essai sur la faune cavernicole de Bulgarie. Annales de Spéléologie 17(2/3):285–441.

Hubenov, Z., 1997. Possibilities for using of a system from the really defined natural territories for the faunistic researches in Bulgaria. Acta zoologica bulgarica 49:5–9.

Jurinitch, S. & P. Drensky, 1917. [Contribution à l'étude des araignées de Bulgarie]. Spisanie na Balgarskata akademiya na naukite. Klon prirodo-matematichen [Journal of the Bulgarian Academy of Sciences. Natural-Mathematical Series] 15(7):109–136 (in Bulgarian).

Lazarov, S., 1998. A contribution to the study of the spiders (Araneae) in Sushtinska Sredna Gora Mountain, Bulgaria. Historia naturalis bulgarica 9:27–34.

Lazarov, S., Ch. Deltshev & G. Blagoev, 2001. Spiders (Araneae) of Sushtinska Sredna Gora Mountain, Bulgaria. Acta zoologica bulgarica 53:3–28.

Marusik, Yu.M., D.V. Logunov & S. Koponen, 2000. Spiders of Tuva, South Siberia. IBPN FEB RAS, Magadan.

Mikhailov, K.G., 1997. Catalogue of the Spiders of the Territories of the Former Soviet Union (Arachnida, Aranei). Zoological Museum of the Moscow State University, Moscow.

Platnick, N., 2006. The World Spider Catalog, Version 7.0. http://research.amnh.org/entomology/spiders/catalog/index.html.

Popov, V., Ch. Deltshev, G. Blagoev, I. Krusteva & D. Deltshev, 2000. Epigeobiont animal assemblages from two landscapes of the Bulgarian Black Sea coast: Relationship to habitat type, assemblage structure and biodiversity. II. Spiders (Araneae). Acta zoologica bulgarica 52(1):51–88.

Samu, F. & C. Szinetár, 1999. Bibliographic check list of the Hungarian spider fauna. Bulletin of the British Arachnological Society 11:161–184.

Tzonev, G. & S. Lazarov, 2001. A contribution to the study of spiders (Araneae) in Osogovo Mountain, South-West Bulgaria. Acta zoologica bulgarica 53(2):67–78.

Vigna Taglianti, A., P. Audisio, M. Biondi, M. Bologna, G. Carpaneto, A. De Biase, S. Fattorini, E. Piattella, R. Sindaco, A. Venchi & M. Zapparoli, 1999. A proposal for a chorotype classification of the Near East fauna, in the framework of the Western Palearctic region. Biogeographia 20:31–59.

17 Terrestrial Cave Animals in Bulgaria

Petar Beron
National Museum of Natural History, Bulgarian Academy of Sciences, Tsar Osvoboditel Blvd. 1, 1000 Sofia, Bulgaria, e-mail: beron@mail.bg

Abstract: Karstic landscape cover about 22.7% of the Bulgarian territory, with more than 5000 caves and potholes. Regular exploration of the cave fauna of Bulgaria started in 1922, and by the end of 2005 more than 800 animal species have been reported from 780 caves. Troglobite and stygobite species are 167, most of them Bulgarian endemics. The most important groups are Diplopoda (54 spp, 17 troglobites), Coleoptera (123 spp, 41 troglobites), and others. Monograph on the cave fauna of Bulgaria is being prepared, containing all available data and bibliography of over 500 titles. Main centers of biospeleological studies are the National Museum of Natural History and the Institute of Zoology (Bulgarian Academy of Sciences)

1 Introduction

Bulgaria is situated in one of the richest regions housing cave animals, especially the terrestrial ones: the eastern part of the Balkan Peninsula. Bulgarian cave fauna has been explored very actively, mostly by the Bulgarian biospeleologists. This is due mainly to the solid tradition left by the school of Dr. Ivan Buresch but also to the well-developed speleology, traditionally centered on the scientific research and led by specialists from Bulgarian Academy of Sciences. Already for more than 80 years, Bulgarian biospeleologists explore the rich troglobite fauna of their country, its karst containing almost 5000 caves. The results have been put together by several authors and now form a very solid basis for zoogeographical analysis. Such analysis was undertaken by V. Guéorguiev (1977, 1992) and other specialists and continues further with the accumulation of new data.

2 Karst and Caves in Bulgaria

The following review is given according to the monograph "Caves and Speleology in Bulgaria" (Beron et al., 2006; based on the subdivision of Popov, 1970).

Karstic features are developed on 22.7% of Bulgarian territory (Popov, 1970), or $25,171\,km^2$ (Fig. 1). According to Boyadjiev (1964), the karst areas occupy $15,778\,km^2$; however, as Popov (1970) stated, this figure did not include the buried karst. Together with this category, most of the karstified area belongs to Danubian Plain (66% of karst in Bulgaria). The least in karst representation is the Transitional

V. Fet and A. Popov (Eds.), Biogeography and Ecology of Bulgaria, 493–526.

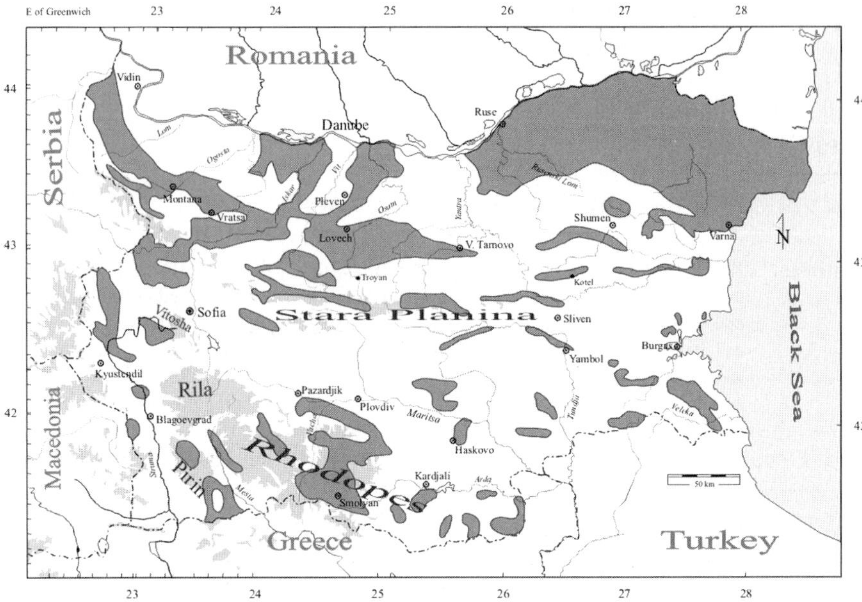

Fig. 1 Karstic areas in Bulgaria.

Geomorphological Region (6%). After many years of research, Vladimir Popov in 1968–1978 subdivided Bulgaria into four regions and 50 districts.

The Danubian Plain is subdivided into eight districts. Out of its extensive karst, 70% is buried karst (Sarmatian sediments). Several important caves are situated between the Serbian–Bulgarian border and Pleven area: Varkan near Druzhba Village (807 m), Bashovishki Pech near Oreshets Railway Station (over four km long), Sedlarkata near Rakita Village (1040 m), Gininata Peshtera near Sadovets Village (501 m), Kirov Vartop near Bohot Village (776 m), Aladjanskata Peshtera (1083 m), and Haidushkata Peshtera (438 m) near Gortalovo Village. The region of the valley of Rusenski Lom River is of particular interest. There, on the steep slopes of the river valleys near the villages Basarbovo, Ivanovo, Tabachka, Pepelina, and others many caves are formed, including Orlova Chuka (13,437 m), second in length in Bulgaria. Tamnata Dupka near Shirokovo is 878 m; Kulina Dupka near Krivnya, 326 m long.

In Ludogorie and Dobrudja there are many (over 350) relatively small caves. More important are Stoyanova Dupka near Ruino (357 m) and Dogulite near Topchii (305 m). From the abrasive caves between Kaliakra and Shabla the longest is Tyulenovata Peshtera (107 m).

The karst of Shumensko Plateau harbors several caves including Zandana (length 2220 m) and the pothole Tainite Ponori (−115 m deep and 1283 m long).

The Karstic Region of Stara Planina (the Predbalkan and the Stara Planina range) is subdivided into 19 districts. The karst of this region covers 4980 km^2, or 19.2% of its total area (Popov, 1970); this region is the richest in caves in Bulgaria. We find there 19 of the 53 Bulgarian potholes deeper than 100 m, and 46 of the 62 Bulgarian caves longer than 1000 m. In all parts of the region there are large caves and potholes.

In Belogradchik District, built of the Mesozoic limestone, are formed such caves as Magura (length 2500 m), Vodni Pech near Dolni Lom Village (1300 m), Mishin Kamik near Gorna Luka (695 m), as well as Pleshovska Dupka near Prevala Village (−102 m deep).

In Salash District, a noticeable cave is Rushkovitsa near Stakevtsi Village (450 m).

A classic karst area, and one among the richest in Bulgaria, is Vratsa District. More than 500 caves and potholes have been discovered in its thick Jurassic and Cretaceous limestone. In the higher parts of Vrachanska Planina have been explored potholes Barkite 14 (denivelation −356 m, length 2600 m), Belyar (−282 m, length 2560 m), Barkite 18 (−178 m), Pukoya near Pavolche Village (−178 m), Yavorets (−147 m) and Panchovi Gramadi (−104 m) near Zverino, and Haidushkata near Bistrets (−108 m). The water caves near Chiren (Ponora, length 3172 m; Mladenovata Propast, length 1732 m) are among the cavers' favorites. Other caves over 500 m in length are Toshova Dupka near Stoyanovo (1302 m), Mizhishnitsa (885 m), Sokolskata Dupka near Lyutadjik (815 m), and Gardyuva Dupka near Zgorigrad (510 m). The longest of the 130 little caves near Cherepish is Studenata Dupka (623 m).

The caves near Lakatnik also belong to Vratsa Region. Here we find Temnata Dupka (4500 m), Kozarskata Peshtera (709 m), Razhishkata Dupka (316 m), and Svinskata Dupka (300 m).

More than 140 caves and potholes have been recorded from the Ponor Karstic Region. Among them are the caves near Gintsi, Komshtitsa, Gubesh, Iskrets, Zimevitsa, and Tserovo. Some of these are well-known to the Sofia cavers: Dushnika (876 m), Katsite (−220 m, length 2560 m), Golemata Temnota near Drenovo (−106 m, length 2000 m), Golyamata Balabanova Dupka (4800 m), Malkata Balabanova Dupka (−125 m), Radolova Yama (−88 m), Krivata Pesht (1462 m), Tizoin (−320 m, length 3599 m), and Saguaroto (−135 m, length 2217 m). Some of the longest caves in West Bulgaria are the caves near Tserovo Railway Station (Vodnata Peshtera, 3264 m, and Mayanitsa, 1419 m).

At the border with Serbia the cave Temnata Dupka near Kalotina (Berende) marks the end of the present Bulgarian territory. Beyond the border, imposed by the Neuilly-sur-Seine Treaty of 1919, remain the former Bulgarian caves near Odorovtsi and Vetrena Dupka near Vlasi.

Another classical karstic area is Kameno Pole–Karlukovo. In this majestic karst has been built the National Caver's House "Petar Tranteev". In the valley of Iskar have been discovered more than 600 caves, including Bayov Komin (2169 m) and Drashanskata Peshtera (693 m) near Drashan; Starata Prodanka

(559 m) and Popovata Peshtera near Gabare; Golyamata Voda (−104 m, length 612 m), Zadanenka (1150 m), Bankovitsa (689 m), Stublenska Yama (−72 m, length 562 m); EC-20 (−94 m), and Tipchenitsa (−78 m), all near Karlukovo. A prominent cave in the majestic karstic landscape of Karlukovo also is Prohodna, the lower entrance of which is 42.5 m high.

In the thick (up to 450 m) limestone in the valleys of Vit, Panega, Batulska Reka, and Yablanishka Reka, are formed the caves of the Panega Region: Morovitsa near Glozhene (−105 m, length 3250 m), Vodnata Pesht near Lipnitsa (1010 m), the potholes Partizanskata (−107 m), Bezdanniyat Pchelin (−105 m), Yasenski Oblik (−104 m), Nanovitsa (−101 m), and others. The second largest karstic source in Bulgaria, Glava Panega (mean outflow 4.6 m^3/sec), was explored by divers up to −52 m depth and 230 m length but it is reasonable to presume that amazing discoveries are yet to come in this huge underground system.

Large water caves are formed in Dragana–Bezhanovo Region of the Predbalkan: Parnitsite (2500 m), Gergitsovata (408 m), Sedlarkata (1070 m), and Skoka (723 m).

The caves of Teteven and Troyan regions are among the most interesting in Bulgaria. They belong to Vasilyov, Cherni Vit, and Cherni Osam areas. Some are particularly important, such as Raichova Dupka (−377 m, the deepest in Bulgaria, length 3333 m), Malkata Yama (−232 m, length 1101 m), Borova Dupka (−156 m), Ptichata Dupka (−108 m, length 652 m), Kumanitsa (−104 m, length 1656 m), Golyamata Garlovina (−100 m), and Varlata (1110 m), all in the area of Cherni Osam Village. The water cave Trona (Duhaloto) near Apriltsi is 1040 m long.

Many caves are known in Lovech area, including Golyamata Peshtera (1921 m) and Malkata Peshtera (1295 m) near Mikre, and Sopotskata Peshtera near Sopot (1225 m).

The subterranean karst in Devetaki Plateau is forming a special cave area. It includes the famous Devetashka Peshtera, 2442 m long and containing the largest cave chamber in Bulgaria, and also the caves near Krushuna: Popskata, or Boninskata Peshtera (4530 m), Vodopada (1995 m), Urushka Maara (1600 m), Gornik (1074 m); near Gorsko Slivovo (Chernata Pesht, 741 m), Karpachevo (Futyovskata Peshtera, 700 m), Alexandrovo (Brashlyanskata Peshtera, 608 m), and Chavdartsi (Mandrata, 450 m). In this area are also the potholes Kanchova Varpina (−100 m, length 425 m) and Blagova Yama near Etropole (−130 m).

Some of the remarkable Bulgarian caves are developed in the Aptian–Urgonian limestone of the plateau of Arbanasi and Belyakovets. The cavers are proud with the new discoveries near Emen: Ruse (−100 m, length 3306 m), Troana (2750 m), Bambalovata (2923 m), as well as with the extension of the "old" Emenska Peshtera up to 3113 m. Other important caves are Genchova Dupka (740 m) near Malak Chiflik and the water cave Musinskata Peshtera (382 m) near Musina.

Large caves have been found in Strazha–Debelets District such as Machanov Trap near Zdravkovets (1921 m) and Izvora near Yantra (1400 m). The longest cave in Debeli Dyal Subdistrict is Marina Dupka near Genchevtsi (2393 m). Favorite meeting point of cavers are the big caves near Dryanovski Monastery, Bacho Kiro (3500 m) and Andaka (over 4000 m). In the same area, in the small cave Polichki have been conducted some of the first archeological explorations in Bulgarian caves, as early as the end of 19th century.

The relatively small caves high on Shipchenska Planina (Stoletovskata Peshtera and others) contain interesting cave fauna. In the mountains of Elena–Tvarditsa only some 30 caves and potholes have been recorded but they include the recently discovered pothole Magliviya Snyag (−146 m, length 2568 m) near Tvarditsa, and the cave Dolnata Maaza near Byala (280 m), known for a long time.

In the Cretaceous limestone of Kotlenska Planina there are many caves and potholes but the cave fauna is rather poor. The pothole Golyamata Yama near Kipilovo (−350 m, third deepest in Bulgaria) is an outstanding one. Important caves are known in the area of Zelenich (Prikazna, 4728 m, with the richest fauna in the area; Lutsifer, −130 m, length 3200 m; Karvavata Lokva, −140 m; Bilernika, −80 m; and Golyamata Humba, −94 m). Already from the time of Georgi Rakovski (1821–1867), the caves in the area of Zlosten were known: potholes Lednika (−242 m, length 1367 m), Maglivata (−220 m), Uzhasat na Imanyarite (−158 m), Akademik (−140 m), and the cave-sink Subatta (460 m). There are caves also near Medven Village (including the ice cave Lednitsata). The source cave in Kotel has an outflow up to 2.5 m^3/sec.

In the Preslavska Mts. the most important is the cave Prolazkata (Derventskata) Peshtera (446 m).

The Transitional Karstic Region comprises a long strip stretching from the border with Serbia to the border with Turkey, between the regions of Stara Planina and the Rhodopes. In the mountains around Tran, Zemen, and Treklyano the caves are small but in the Triassic dolomite and limestone on the western slope of Vitosha near Bosnek Village have been formed some of the most important caves in Bulgaria: Duhlata (17,600 m, the longest cave in Bulgaria), Vreloto (5280 m), and PPD (1020 m). In this belt from Sofia to Sakar Mts. there are no caves of some importance but the source cave Chirpan Bunar near Belozem (220 m, 30–40 l/sec) has to be noticed.

In the districts of Sakar and Dervent Hills are known several dozens of caves in the Triassic marble-limestone. They include Bozkite (324 m) and Dranchi Dupka (145 m) near Mramor, and Kirechnitsata (225 m) and Dranchi Dupka (−25 m) near Melnitsa.

In the Strandja District 77 relatively small caves and potholes have been recorded, among which are Bratanovata (384 m), Kaleto (301 m), Bazat 1 (208 m), Bazat 2 (201 m), Kerechnitsata (224 m), and Stoyanovata Peshtera near Kosti (150 m). The deepest among the few potholes in Strandja are Golyamata Vapa near Stoilovo (−125 m, length 381 m) and Tangarachkata Dupka near Bogdanovo (−74 m).

Rila–Rhodopean Region contains 13 cave districts in the mountains Pirin and Rhodope, developed mainly in the Proterozoic marble and (less) in limestone.

In Pirin are known mostly potholes, especially in the Vihren Subdistrict. Right under the summit of Vihren the pothole Vihrenska Propast was descended down to −170 m; in the Bayuvi Dupki circus there is the pothole Chelyustnitsa (−104 m); in Banski Suhodol, "20 Godini Akademik" (−118 m), Propast

No. 9 (−230 m), and Propast No.14 (−103 m). Prominent among the caves in Sinanitsa Subregion are Aleko (−130 m, length 600 m), Sharaliiskata (470 m), Rimskata (293 m), and Ruikovata (120 m). Near Razlog is the source cave Spropadnaloto (605 m).

The caves in South Pirin, Slavyanka, and Stargach are few but interesting. The deepest pothole is Garvanitsa (−60 m) near Gotsev Vrah; some interesting cave animals have been found in Rupite near Paril and Starshelitsa near Goleshovo. In the marble of Dabrash part of the Rhodopes the cave Manailovata (Manoilovskata) Peshtera near Ribnovo is prominent (−115 m, length 2119 m).

In the Velingrad District between Velingrad and Rakitovo the cave Lepenitsa is situated; it was one of the first to be opened as "show cave" but was vandalized many times. Near the town of Peshtera are the caves Vodnata Peshtera (1000 m), Novata Peshtera (825 m), Yubileyna (814 m), and the show cave Snezhanka (276 m).

In Dobrostan District, the largest karst district in the Western Rhodopes, we know over 200 caves and potholes. Some important among them are Garvanitsa near Kosovo (925 m), Topchika (727 m), Hralupa (311 m), and Gargina Dupka near Mostovo (534 m). Particularly interesting are the potholes Lisek (−164 m) and Kutelska Yama (−88 m) near Dryanovo, Druzhba (−130 m) and Ivanova Voda (−112 m) near Dobrostan. The stone bridges Erkyupriya (Chudnite Mostove) near Zabardo and Mostovo are to be noted.

The majestic gorges of the Trigrad District, cut in the marble (up to 1600 m thick), have many caves. Among them are Yagodinskata Peshtera (or Imamova Dupka, 8501 m, third longest in Bulgaria) and Sanchova Dupka (8880 m) near Yagodina; Izvora (2480 m) and Eminova Dupka (592 m) near Borino; the potholes Drangaleshkata (−225 m, length 1142 m) and Kambankite (−158 m) near Mugla; and Ledenitsata near Gela (−108 m, length 1419 m). In Chepelare District, the longest cave is Samurskata Dupka (634 m).

Borikovskata Peshtera (440 m) and Goloboitsa (362 m) are caves in the marble south and southeast of Smolyan. The deepest pothole in the area is Kladeto near Polkovnik Serafimovo (−147 m).

In the Ardino District, 44 caves and potholes are known, including Vodnata Peshtera near Nedelino (203 m), Gyaurhambar near Stomantsi (112 m), and Karagug near Tyutyunche (105 m).

Among the small caves is Maazata near Madretsi (114 m). In the easternmost part of the Rhodopes, 35 caves have been explored, including Karangil near Shiroko Pole (490 m), Samara (327 m), Ogledalnata Peshtera (157 m) near Ribino, and Belopolyanskata Peshtera near Belopolyane.

The volcanic caves (caves in non-carbonate rocks) in Bulgaria have been studied mainly by Boris Kolev. Only on the territory of Eastern Rhodopes, he has recorded 77 caves. They are short but spacious (Golyamata Peshtera near Byal Kladenets, 51 m; Ehtyashtata, 30 m; and Prilepnata, 18 m).

Among the caves in sandstone, conglomerate and gneiss, Lepenishki Pech near Prolaznitsa, Belogradchik area (35 m), Uske near Chetirtsi (88 m), and others are to be noticed.

3 History of Biospeleological Studies of Bulgarian Cave Animals

The first to find cave animals in Bulgaria was the Hungarian entomologist Eduard Merkl. In 1878 (the year of founding of the independent Bulgarian state), he visited some caves near the summits of Kurudja and Sveti Nikola (now Stoletov) in Central Stara Planina. There, Merkl discovered three blind beetle species, described in 1879 by Johan Frivaldzsky as *Trechus* (*Anophthalmus*) *balcanicus, Pholeuon merkli*, and *Ph. merkli* var. *simile*. This discovery remained largely unknown for a long time as it was not mentioned in the descriptions that the beetles were found in caves. The next "biospeleological" trip to Bulgaria took place more than 30 years after Merkl's.

In 1909, the Austrian entomologist Fritz Netolitzky visited the cave Zmeyovi Dupki near Tryavna and found another blind beetle, described later by Josef Müller from Trieste as *Netolitzkya maneki*. The second discovery also did not attract the attention of biospeleologists, who concentrated on the fauna of the French and West Balkan caves.

The curiosity of Bulgarian zoologists for studying cave animals was aroused by a geologist, V. Arnaudov, who in 1922 indicated to Dr. Ivan Buresch the existence of strange yellowish insects in the cave Ledenika near Vratsa. The experienced Director of the Royal Museum of Natural History in Sofia immediately realized how important this observation was. Together with his fellow entomologists from the Museum, Dr. Buresch undertook a series of visits to the caves near Vratsa and in Iskar Gorge. In several years (1923–1926), dozens of troglobites from different groups have been discovered and described. Prominent foreign specialists visited Bulgaria or identified the material collected (we can mention K. Verhoeff, E. Knirsch, E. Handschin, V. Redikorzev, A. Wagner, L. Fage, C. F. Roewer). Their publications increased notably the knowledge on groups like Isopoda, Myriapoda, Pseudoscorpiones, Opiliones, Collembola, Coleoptera, and others in Bulgaria. Many visits of Dr. Buresch and his associates to the caves, by 1927 brought the number of known cave organisms to 44; among them were some remarkable troglobites. A general list of this fauna was presented by Dr. Buresch in September 1927 to the International Zoological Congress in Budapest (Buresch, 1929). Meanwhile, he published two important papers on the cave fauna of Bulgaria (Buresch, 1924, 1926). Nenko Radev started publishing Catalogue of Bulgarian caves, of which two parts were published (in 1926 and 1928). On 18 March 1929, the Bulgarian Speleological Society was founded. New visits to the caves followed, and a series of new publications contributed to the study of Araneae (Drensky), Diptera (Czerny), Thysanura (Silvestri), Coleoptera (Buresch, Jeannel, Mandl), Pseudoscorpiones (Hadži), Oligochaeta (Černosvitov), Nycteribiidae (Z. Karaman), Crustacea (Klie), Isopoda (Strouhal), Gastropoda (A. Wagner), Isopoda, Diplopoda, and Chilopoda (Verhoeff), and other groups. A third general paper on Bulgarian biospeleology was published by Buresch (1936).

After the World War II, again Dr. Buresch organized two "caving brigades" within Bulgarian Academy of Sciences. They studied some caves in Northeastern Bulgaria and collected many new cave animals, arranged in the collection "Fauna cavernicola bulgarica" in the Institute of Zoology in Sofia. Part of this material was identified by J. Kratochvil (Opiliones), A. Angelov (Gastropoda), C. Attems (Myriapoda), I. Buresch and V. Guéorguiev (*Stenasellus*), C. F. Roewer (Opiliones), Z. Karaman (Coleoptera), E. Pretner (Coleoptera), F. Miller (Araneae), J. Lang (Diplopoda), S. and G. Karaman (Amphipoda), and others.

These papers concern, together with the older material of Dr. Buresch and his associates, some new animals (like *Paralola buresi* and *Tranteeva paradoxa*) collected in the 1950s by Petar Tranteev and the group of young biospeleologists formed around him (V. Guéorguiev, P. Beron, V. Beshkov, T. Michev, M. Kwartirnikov, A. Popov, S. Andreev, Ch. Deltshev).

After the resurrection of the organized caving in Bulgaria in 1958 the Caving Commission at Bulgarian Tourist's Union formed many caving clubs and a much more intensive research in caves took place everywhere in the country, even abroad. Considerable new collections have been accumulated and studied by the new generation of Bulgarian and some foreign specialists: Protozoa (D. Tashev, V. Golemansky), Hirudinea (A. Angelov), Mollusca (A. Riedel, especially Zonitidae), Isopoda (A. Vandel, S. Andreev), Chilopoda (J.-M. Demange, J. Gulička, Z. Matic, V. Golemansky), Diplopoda (K. Strasser), Opiliones (V. Šilhavý, W. Staręga), Araneae (Ch. Deltshev), Orthoptera (A. Popov), Collembola (J. Rusek, M. M. da Gama), Diplura (J. Rusek), Homoptera Aphidodea (D. Tashev), Coleoptera (V. Guéorguiev, Z. Karaman, L.Genest, L. Zerche), Trichoptera (K. Kumanski), Diptera and Siphonaptera parasites of bats (K. Hůrka), Diptera (V. Beschovski). More specialized studies on the nervous system and ecology of *Pheggomisetes* and other cave beetles have been started (M. Kwartirnikov). Specializations in cave biology in the Laboratoire Souterrain de Moulis (France) took place (P. Beron in 1967; V. Guéorguiev in 1968), and the first dissertation on cave animals was presented by P. Beron (1976).

For a long time the studies on bats (Chiroptera) in Bulgarian caves have been carried out without nets, which resulted in very limited results (I. Buresch, G. Heinrich, P. Beron, V. Beshkov, M. Kwartirnikov). The visit of the first Czech theriologists with nets (Horaček et al., 1974) contributed new species to the Bulgarian fauna. Another visit by British cavers brought new data, published by M. Hazelton (1970) and F. Turk (1970). Some other visits or publications by foreign cave biologists also contributed to better understanding of the Bulgarian fauna (H. Coiffait, J.-M. Thibaud, D. Dancau, L. Botosaneanu, V. Decu, A. Riedel, L. Genest). Most data on the cave fauna of Bulgaria, however, were collected after 1960 by the Bulgarian scientists. The latest bibliography of Bulgarian cave fauna (Beron, 1994) contains 409 titles, of which 230 belong entirely or partly to the Bulgarians.

Numerous data obtained have been compiled by Guéorguiev and Beron (1962), then by Beron and Guéorguiev (1967), Beron (1972), and finally by Beron (1994) who published a complete list of 704 animal species from 646 Bulgarian caves.

Since the beginning of the 1990s, a considerable "new wave" of biospeleologists marked another leap forward in the knowledge of the cave and underground fauna of Bulgaria. Several young researchers (P. Stoev, B. Petrov, S. Beshkov, T. Ivanova, I. Pandourski) carried further the efforts of the older workers. Meanwhile, some of the explorers of Bulgarian cave animals died (P. Tranteev in 1979, I. Buresch in 1980, V. Guéorguiev in 1994) or retired (V. Beshkov, 1995). The young specialists work in the National Museum of Natural History (NMNH) and in the Institute of Zoology, Bulgarian Academy of Sciences; they study selected groups of cave fauna (Myriapoda, P. Stoev; Pseudoscorpiones, B. Petrov; Copepoda, I. Pandourski; Lepidoptera, S. Beshkov). Very active is also the group for study and protection of bats, based in the NMNH (T. Ivanova and others). Bulgarian biospeleologists work actively also in the Bulgarian Federation of Speleology (President, P. Beron) and in the international organizations.

The compilation of a new "Cave Fauna of Bulgaria" by P. Beron, B. Petrov, and P. Stoev is in progress. The updated bibliography in this book contains over 500 titles.

4 Survey of Terrestrial Cave Animals in Bulgaria

By the end of 2006, from Bulgarian caves have been recorded over 800 animal species. Among them more than 115 (including several under study) are considered to be troglobites. Table 1 lists the terrestrial troglobites in Bulgaria.

Protozoa. The free-living protozoans in Bulgarian caves are not well known. Some epibiont *Vorticella* and *Tokophrya* (Ciliata), living on the cave crustaceans, have been recorded. Four endoparasitic gregarines (Sporozoa) of the genera *Lepismatophila* (Stylocephalidae) and *Stenophora* (Stenophoridae) have been described. They live in *Plusiocampa bureschi* (Diplura) and in three species of millipeds.

Nemathelminthes. Almost nothing is known about the multitude of tiny nematods inhabiting cave substratum.

Annelida. Eight species of Lumbricidae have been recorded so far in Bulgarian caves but they are not of special interest as the same species live also outside the caves.

Mollusca. Snails (Gastropoda) are many and include troglobites as well as typical troglophiles. Thanks to the Polish malacologist A. Riedel, the family Zonitidae is well known in Bulgaria and in the Balkans. One probable troglobite belonging to this family has been described from the Ptichata Dupka pothole, the tiny white *Spinophallus* (*Lindbergia* auct.) *uminskii*. Many species of the closely related genus *Lindbergia* live in the caves of Greece but *S. uminskii* is so far the sole species from Bulgaria. Other interesting taxa of cave snails should be expected in Bulgaria, taking into account the rich fauna of the neighboring countries. So far, 16 terrestrial snail species have been recorded from Bulgarian caves. Besides *Spinophallus uminskii*, particularly interesting is the troglophile *Balcanodiscus frivaldskyanus* living in the caves of Eastern Rhodopes.

Crustacea. Terrestrial crustaceans are represented by the order Isopoda. One crab (Decapoda) has also been found in a Rhodopean cave outside water. Isopoda

Table 1 Troglobites in Bulgarian caves

Taxa	Cave(s)	Endemic Status
Phylum MOLLUSCA		
Class GASTROPODA		
Order BASOMMATOPHORA		
Family Zonitidae		
Spinophallus uminskii (Riedel)	Ptichata Dupka (Cherni Osam); troglobite or troglophile?	Central Stara Planina
Phylum ARTHROPODA		
Class CRUSTACEA		
Order ISOPODA		
Suborder ONISCIDEA		
Family Styloniscidae		
Cordioniscus bulgaricus Andreev	Boichovata Peshtera (Logodash)	Vlahina Mts.
Cordioniscus schmalfussi Andreev	Zmiin Burun (Mostovo), Shepran Dupka (Belitsa)	Western Rhodopes
Family Trichoniscidae		
Bureschia bulgarica Verhoeff	Vodnata Peshtera (Tserovo Railway Station), Temnata Dupka (Lakatnik Railway Station), Belyar (Vratsa)	Western Stara Planina
Balkanoniscus corniculatus Verhoeff	Haidushkata Dupka (Karlukovo), Rushovata Peshtera and Djebin trap (Gradeshnitsa), Cave Yamata (Tserovo Railway Station), Razrushenata Peshtera (Kunino Railway Station)	Western Stara Planina and Central Predbalkan
Balkanoniscus minimus Vandel	Parnitsite (Bezhanovo)	Central Predbalkan
Rhodopioniscus beroni (Vandel)	Topchika and Yamata (Dobrostan)	Western Rhodopes
Trichoniscus bulgaricus Andreev	Yulen Ere (Hristo Danovo)	Central Stara Planina
Trichoniscus rhodopiense Vandel	12 caves in the Rhodopes, mostly Kardjali District	Rhodopes and Thassos Island, Greece
Trichoniscus valkanovi Andreev	Sarpiiskata Peshtera and Kirechnitsata (Kosti)	Strandja
Trichoniscus anophthalmus Vandel	Studenata Dupka and Ezeroto (Cherepish Railway Station), Mishin Kamik, Vreloto, and Aiduchkata Dupka (Prevala), Tatarska Dupka (Replyana), Parasinskata Propast (Belimel), Toshova Dupka (Stoyanovo)	Western Stara Planina

Trichoniscus tranteevi Andreev (= *T. anophthalmus intermedius* Vandel, praeoccup.)	Tamnata Peshtera (Tabachka)	Ludogorie
Trichoniscus petrovi Andreev	Shepran Dupka (Belitsa)	Rhodopes
Trichoniscus tenebrarum Verhoeff	Andaka (Dryanovski Manastir), Golyama Podlistsa (Veliko Tarnovo), Devetashkata Peshtera (Devetaki)	Central Predbalkan
Trichoniscus stoevi Andreev	Stoletovskata Peshtera (Shipka)	Central Stara Planina
Trichoniscus bononiensis Vandel	Varkan (Tsar Petrovo), Suhi Pech, Propast, and Parnak (Oreshets), Redaka 1, Prelaz, and Mechata Dupka (Salash), Yame 2 and Yame 3 (Targovishte), Krachimirskoto Vrelo (Stakevtsi), Zmiiskata Propast (Krachimir), Tsankinoto Vrelo (Granitovo), Neprivetlivata (= Gornata Propast) and Haidushkata Propast (Belogradchik), Tamni Pech and Golemi Pech (Varbovo), Pech (Gorna Luka), Yankulova Dupka (Prolaznitsa)	Western Danubian Plain, Western Predbalkan, and Western Stara Planina
Trichoniscus garevi Andreev	Sinyoto Ezero (Dragana), Alchashkata Peshtera (Bezhanovo), Popskata Peshtera (Krushuna), Tanyova Peshtera and Voditsata (Aglen)	Central Predbalkan
Trichoniscus beroni Andreev	Bratanovskata peshtera (Kosti)	Strandja
Beroniscus capreolus Vandel	Parnitsite (Bezhanovo)	Central Predbalkan
Cyphoniscellus (= *Bulgaroniscus*) *gueorguievi* (Vandel)	Desni Suhi Pech (Dolni Lom), Zhivkova Dupka (Gorna Luka), Vreloto, Aiduchkata Dupka, and Mitsina Dupka (Prevala), Neprivetlivata (Belogradchik), Yame 2 (Targovishte), Tatarska Dupka (Replyana), Parnak (Oreshets)	Western Predbalkan and Western Stara Planina
Bulgaronethes haplophthalmoides Vandel	Novata Peshtera (Peshtera)	Western Rhodopes
Vandeloniscellus (= *Cyphoniscellus*) *bulgaricus* (Vandel)	Toshova Dupka (Stoyanovo)	Western Stara Planina
Tricyphoniscus bureschi Verhoeff	Haidushka Dupka, Bezimenna 22, and Cherdjenitsa (Karlukovo), Vodnata Pesht, Kozarnika, and Bankovets (Lipnitsa)	Western and Central Predbalkan

continued

Table 1 continued

Taxa	Cave(s)	Endemic Status
Hyloniscus flammula Vandel	Magura, Magura 2, and Varnitsata (Rabisha)	Western Predbalkan
Class ARACHNIDA		
Order OPILIONES		
Suborder CYPHOPHTHALMI		
Family SIRONIDAE		
Siro beschkovi Mitov	Haidushkata peshtera (Deventsi)	Central Predbalkan
Tranteeva paradoxa Kratochvil	Rushovata Peshtera (Gradeshnitsa), Toplya (Golyama Zhelyazna)	Central Predbalkan
Suborder LANIATORES		
Family PHALANGODIDAE		
Paralola buresi Kratochvil	Kozarskata Peshtera, Svinskata Dupka, Zidanka, and Temnata Dupka (Lakatnik Railway Station)	Western Stara Planina
Suborder PALPATORES		
Family NEMASTOMATIDAE		
Paranemastoma (*Buresiolla*) *bureschi* (Roewer)	31 caves	Western Stara Planina
Order PSEUDOSCORPIONES		
Family CHTHONIIDAE		
Chthonius troglodites Redikorzev	Kasapnitsite and Haidushka Dupka (Karlukovo Railway Station), Haidushkata Peshtera (Deventsi)	Central Predbalkan
Family NEOBISIIDAE		
Neobisium (*Heoblothrus*) *bulgaricum* (Redikorzev) (= *Obisium subterraneum* Redikorzev)	Yalovitsa (Golyama Zhelyazna)	Central Predbalkan
Neobisium (*Heoblothrus*) *beroni* Beier	Svinskata Dupka (Lakatnik Railway Station)	Western Stara Planina
Neobisium (*Blothrus*) *kwartirnikovi* Mahnert	Duhlata (Bosnek)	Vitosha
Balkanoroncus bureschi (Redikorzev) (= *Obisium bureschi* Redikorzev, = *B. praeceps* Ćurčić)	Saeva Dupka (Brestnitsa), Djebin Trap (Gradeshnitsa)	Central Predbalkan
Balkanoroncus hadzii Harvey (= *Roncus bureschi* Hadži)	Razhishka Peshtera (Lakatnik Railway Station)	Western Stara Planina
Roncus mahnerti Ćurčić et Beron	Vodnata Dupka (Botunya)	Western Stara Planina

Order ARANEAE		
Family Linyphiidae		
Troglohyphantes drenskii Deltshev	Suhata Dupka (Velingrad)	Western Rhodopes
Centromerus bulgarianus (Drensky)	Zidanka and Razhishka Peshtera (Lakatnik Railway Station), Sokolskata Dupka (Lyutadjik), Belyar (Vratsa)	Western Stara Planina
Porrhomma microps (Roewer)	Kulina Dupka (Krivnya), Samar Daala (Voden), Prikazna (Kotel)	Ludogorie and Eastern Stara Planina
Family Leptonetidae		
Protoleptoneta beroni Deltshev	Belimelskata Peshtera (Belimel)	Western Stara Planina
Protoleptoneta bulgarica Deltshev	Mecha Dupka (Lepitsa), Djurdjina Dupka (Erden)	Western Predbalkan
Family Nesticidae		
Nesticus beroni Deltshev	Dupkata (Mostovo)	Western Rhodopes
MYRIAPODA		
Class CHILOPODA		
Family Lithobiidae		
Lithobius tiasnatensis Matic (= *L. popovi* Matic)	Devetashkata Peshtera (Devetaki), Tyasnata Propast (Mramor)	Central Predbalkan and Sakar
Lithobius stygius Latzel	Lepenitsa and Suhata Peshtera (Velingrad), Yubileina (Peshtera Town), Modarskata Peshtera (Modar), Lednitsata (Gela)	Western Rhodopes and western part of the Balkan Peninsula
Lithobius rushovensis Matic (= *L. beschkovi* Matic et Golemansky)	Rushovata Peshtera (Gradeshnitsa)	Central Predbalkan
Lithobius lakatnicensis Verhoeff	many caves	Western Stara Planina and Western Rhodopes
Lithobius bifidus (Matic)	Izvornata Peshtera (Mladezhko)	Strandja
Eupolybothrus andreevi Matic	Vodnata Peshtera (Tserovo Railway Station)	Western Stara Planina

continued

Table 1 continued

Taxa	Cave(s)	Endemic Status
Class DIPLOPODA		
Order GLOMERIDA		
Family DODERIIDAE		
Trachysphaera orghidani lakatnicensis Tabacaru	Kozarskata Peshtera and Sedmovratitsa (Lakatnik Railway Station), Tamna Dupka (Targovishte)	Western Stara Planina
Order POLYDESMIDA		
Family POLYDESMIDAE		
Brachydesmus radewi Verhoeff	Promakinyalo (Dolna Beshovitsa)	Western Predbalkan
Family TRICHOPOLYDESMIDAE (= BACILLIDESMIDAE)		
Bacillidesmus bulgaricus bulgaricus Strasser	Mishin Kamik (Gorna Luka), Mladenovata Peshtera (Chiren)	Western Predbalkan and Western Stara Planina
Bacillidesmus bulgaricus dentatus Strasser	Drashanskata Peshtera (Drashan)	Western Predbalkan
Order CHORDEUMATIDA		
Family ANTHROLEUCOSOMATIDAE		
Bulgarosoma bureschi Verhoeff	Ledenika, Belyar, Malkata Mecha Dupka, and Reznyovete (Vratsa), Kitova Kukla (Druzhevo)	Western Stara Planina
Troglodicus meridionale (Tabacaru)	Imamova Dupka (Yagodina)	Western Rhodopes
Troglodicus tridentifer Gulička	Chelechkata (Choveshkata) Peshtera (Orehovo)	Western Rhodopes
Stygiosoma beroni Gulička	Manailovata Peshtera (Ribnovo)	Western Rhodopes
Anamastigona falcatus Gulička	Kipilovskata Peshtera (Kipilovo); troglobite or troglophile?	Eastern Stara Planina
Anamastigona lepenicae (Strasser)	Lepenitsa (Velingrad)	Western Rhodopes
Anamastigona delcevi (Strasser)	Artificial galleries and caves near Paril; troglobite or troglophile?	Slavyanka
Anamastigona (*Balkandicus*) *alba* (Strasser)	Ptichata Dupka and Kumanitsa (Cherni Ossam)	Central Stara Planina
Bulgardicus tranteevi Strasser	Bankovitsa (Karlukovo)	Central Predbalkan

Order JULIDA		
Family JULIDAE		
Typhloiulus bureschi Verhoeff	31 caves	Western and Central Predbalkan, Western Stara Planina
Typhloiulus georgievi Verhoeff	Popskata Peshtera (Krushuna); troglobite or troglophile?	Central Predbalkan
Serboiulus spelaeophilus Gulička	13 caves	Western Stara Planina
Typhloiulus (*Inversotyphlus*) *longipes* Strasser	Belyar (Vratsa)	Western Stara Planina
Typhloiulus staregai Strasser	Prelaz (Salash)	Western Stara Planina
Class INSECTA		
Order COLLEMBOLA		
Family ENTOMOBRYIDAE		
Pseudosinella bulgarica Gama	Stoletovskata Peshtera (Shipka), Prikazna (Kotel)	Central and Eastern Stara Planina
Pseudosinella duodecimocellata Handschin	13 caves	Western Stara Planina, Pirin, and Western Rhodopes
Pseudosinella kwartirnikovi Gama	Gurlyova Dupka and Tavancheto (Krushuna), Stoletovskata Peshtera (Shipka), Ovnarkata (Karlukovo), Kozarnika (Lipnitsa)	Western and Central Predbalkan, Central Stara Planina
Family ONYCHIURIDAE		
Onychiurus sensitivus Handschin	Ledenika (Vratsa)	Western Stara Planina
Onychiurus vornatscheri Stach	Tsankinoto Vrelo (Granitovo)	Western Predbalkan
Protaphorura beroni (Gruia)	Venetsa, Propast (Oreshets)	Western Predbalkan
Order DIPLURA		
Family CAMPODEIDAE		
Plusiocampa bulgarica Silvestri	Magura (Rabisha), Yavoretskata Peshtera (Lakatnik), Lepenitsa (Velingrad), Hvoinenskata Peshtera (Hvoina), Sbirkovata Peshtera (Progled), Haramiiskata Dupka (Trigrad), Dupkite (Chepelare), Imamova Dupka (Yagodina)	Western Predbalkan, Western Stara Planina, and Western Rhodopes

continued

Table 1 continued

Taxa	Cave(s)	Endemic Status
Plusiocampa bureschi Silvestri (= *P. rauseri* Rusek)	Temnata Dupka and Razhishkata Peshtera (Lakatnik Railway Station), Vodnata Peshtera (Tserovo Railway Station)	Western Stara Planina
Plusiocampa beroni Bareth et Condé	Magura (Rabisha), Varkan (Tsar Petrovo)	Western Danubian Plain and Western Predbalkan
Plusiocampa cf. *beroni* 1	Novata Peshtera (Peshtera Town)	Western Rhodopes
Plusiocampa cf. *beroni* 2	Padezh (Breze)	Western Stara Planina
Plusiocampa gueorguievi Bareth et Condé	Toplya (Golyama Zhelyazna)	Central Predbalkan
Plusiocampa arbanasiensis Bareth et Condé	Lyaskovskata Peshtera (Arbanasi)	Central Predbalkan
Plusiocampa vodniensis Bareth et Condé	Vodni Pech (Dolni Lom)	Western Stara Planina
Order COLEOPTERA		
Family CARABIDAE		
Duvalius (*Paraduvalius*) *balcanicus* (Frivaldszky)	Cave (?) in Shipchenska Planina	Central Stara Planina
Duvalius (*Paraduvalius*) *beroni* Guéorguiev	Toshova Dupka (Stoyanovo)	Western Stara Planina
Duvalius (*Paraduvalius*) *bulgaricus* Knirsch	Zmeyovi Dupki (Hitrevtsi), Kumincheto (Genchevtsi)	Central Predbalkan
Duvalius (*Paraduvalius*) *bureschi* Jeannel	Lepenitsa (Velingrad)	Western Rhodopes
Duvalius (*Paraduvalius*) *garevi* Casale et Genest	Sinyoto Ezero (Dragana)	Central Predbalkan
Duvalius (*Paraduvalius*) *joakimovi* B. Guéorguiev	Stoikova Dupka 1 (*ex errore* Golyama Stoikovitsa)	Slavyanka
Duvalius (*Paraduvalius*) *kotelensis* Genest	Prikazna (Kotel)	Eastern Stara Planina
Duvalius (*Paraduvalius*) *papasoffi* Mandl	Temnata Dupka, Zidanka, and Pyasachnata Dupka (Lakatnik Railway Station)	Western Stara Planina
Duvalius (*Paraduvalius*) *petrovi* B. Guéorguiev	Zandana (Dolno Cherkovishte), ? Hasarskata Peshtera (Gorna Snezhinka)	Eastern Rhodopes
Duvalius (*Paraduvalius*) *pirinensis* B. Guéorguiev	Sharaliiskata Peshtera (Pirin)	Pirin
Duvalius (*Paraduvalius*) *pretneri* Guéorguiev	Mechata Dupka (Bov)	Western Stara Planina
Duvalius (*Paraduvalius*) *regisborisi* Buresch	Yalovitsa (Golyama Zhelyazna)	Central Predbalkan
Duvalius (*Paraduvalius*) *zivkovi* Knirsch (= *D. zivkovi deltshevi* Guéorguiev)	Ledenika, Malkata Mecha Dupka, 25 Godini Akademik, and Malkata Nevestina Propast (Vratsa)	Western Stara Planina
Duvalius (*Paraduvalius*) *legrandi* Genest	Kalugerova Dupka (Arbanasi)	Central Predbalkan

Duvalius (Paraduvalius) karelhurkai Farkac	Erkyupriya (Mostovo)	Western Rhodopes
Duvalius (Biharotrechus) beshkovi Coiffait	Mecha (Lisicha) Dupka (Stradalovo)	Osogovo
Pheggomisetes buresi buresi Knirsch	Ledenika and 25 Godini Akademik (Vratsa)	Western Stara Planina
Pheggomisetes buresi medenikensis Knirsch	Medenik (Eliseina Railway Station)	Western Stara Planina
Pheggomisetes globiceps globiceps Buresch	Dushnika (Iskrets)	Western Stara Planina
Pheggomisetes globiceps breiti Mandl	Dinevata Pesht, Svetata Voda, and Krivata Pesht (Gintsi)	Western Stara Planina
Pheggomisetes globiceps georgievi Z. Karaman	Propast 30 (Karlukovo)	Central Predbalkan
Pheggomisetes globiceps lakatnicensis Jeannel	Temnata Dupka and Zidanka (Lakatnik Railway Station), Radyova Propast (Milanovo), Kolkina Dupka (Zimevitsa), Golemata Mecha Dupka (Vratsa), Govedarskata Dupka (Chiren)	Western Predbalkan and Western Stara Planina
Pheggomisetes globiceps stoicevi Guéorguiev	Elata (Zimevitsa), Nevestina Propast (Vratsa)	Western Stara Planina
Pheggomisetes globiceps karlukovensis Genest	Bezimenna 84 (Karlukovo)	Central Predbalkan
Pheggomisetes globiceps ilandjievi Guéorguiev	Balabanova Dupka, Malata Balabanova Dupka, and Granicharskata Propast (Komshtitsa)	Western Stara Planina
Pheggomisetes globiceps mladenovi Guéorguiev	Malkata Mecha Dupka and 25 Godini Akademik (Vratsa)	Western Stara Planina
Pheggomisetes globiceps cerovensis Guéorguiev	Yamata, Peshterata, and Propastta (Tserovo Railway Station)	Western Stara Planina
Pheggomisetes radevi radevi Knirsch	Ledenika (Vratsa)	Western Stara Planina
Pheggomisetes radevi ilcevi Knirsch	Medenik (Eliseina Railway Station)	Western Stara Planina
Pheggomisetes radevi tranteevi Guéorguiev	Suhata Yama (Druzhevo)	Western Stara Planina
Pterostichus ledenikensis Knirsch	Ledenika (Vratsa), Grebenyo (Gorno Ozirovo)	Western Stara Planina
Family CHOLEVIDAE (= CATOPIDAE)		
Beronia micevi Guéorguiev	Neprivetlivata (Gornata Propast) and Haidushkata Propast (Belogradchik)	Western Predbalkan
Beskovia bulgarica Guéorguiev	Studenata Dupka and Serapionovata Peshtera (Cherepish Railway Station)	Western Stara Planina
Hexaurus merkli (Frivaldszky)	Cave under Kurudja Summit	Central Stara Planina

continued

Table 1 continued

Taxa	Cave(s)	Endemic Status
Hexaurus schipkaensis Zerche	Stoletovskata Peshtera (Shipka)	Central Stara Planina
Hexaurus similis (Frivaldszky)	a cave in Shipchenska Planina	Central Stara Planina
Hexaurus paradisi Zerche	Han Maara (Rai Hut)	Central Stara Planina
Netolitzkya jeanneli jeanneli Buresch	Bacho Kiro (Dryanovski Manastir)	Central Predbalkan
Netolitzkya jeanneli matroffi Jeannel	Lyaskovska Peshtera and Kalugerova Dupka (Arbanasi)	Central Predbalkan
Netolitzkya maneki maneki J. Müller	Zmeyovi Dupki (Hitrevtsi), Kumincheto (Genchevtsi)	Central Predbalkan
Netolitzkya maneki iltschewi Jeannel	Golyama Podlistsa (Belyakovets)	Central Predbalkan
Radevia hanusi Knirsch	Ledenika, Bezimenna, Zmeyova Dupka I, Zmeyova Dupka III, Propast 13, Bulina Dupka, Golemata Mecha Dupka, and Radyova Propast (Vratsa)	Western Stara Planina
Rhodopiola cavicola Guéorguiev	Sipeya (Bachkovski Manastir)	Western Rhodopes
Vratzaniola pandurskii Dupré	25 Godini Akademik and Barkite 9 (Vratsa)	Western Stara Planina
Tranteeviella bulgarica Pretner	Rushovata Peshtera (Gradeshnitsa), Varovit (Malka Zhelyazna)	Central Predbalkan
Bureschiana drenskii Guéorguiev	Tilkiini (Ostrovitsa), Hasarskata Peshtera (Gorna Snezhinka)	Eastern Rhodopes
Balcanobius etropolensis Guéorguiev	Bezimennata Peshtera I, Zahlupena Dupka, and Neikova Dupka (Etropole)	Central Stara Planina
Genestiellina gueorguievi (Giachino)	Yalovitsa and Toplya (Golyama Zhelyazna)	Central Predbalkan
Beroniella tetevensis Giachino et Guéorguiev	Dyado Draganovata Peshtera (Teteven)	Central Predbalkan
Bathyscia raitchevi Casale, Giachino et Etonti	Imamova Dupka (Yagodina)	Western Rhodopes
Family CURCULIONIDAE		
Troglorrhynchus beroni Angelov	Inkaya (Tsvyatovo)	Eastern Rhodopes
Troglorrhynchus gueorguievi Angelov	Yalovitsa (Golyama Zhelyazna)	Central Predbalkan
Troglorrhynchus angelovi Gueorguiev et Petrov	Zandana (Dolno Cherkovishte)	Eastern Rhodopes

are represented in Bulgarian caves by three suborders, Asellota, Flabellifera, and Oniscidea; only the latter is terrestrial. Terrestrial isopods include some of the most numerous and interesting troglobites in Bulgarian caves. They live on rotten wood, on clay, and on stalagmite surface. One of the most remarkable species (*Bureschia bulgarica*) is amphibious and occurs in large numbers around and in the lakes in the cave Temnata Dupka near Lakatnik. In total, 50 species of terrestrial isopods have been recorded from Bulgarian caves; at least 23 of them could be considered troglobites (Table 1). Almost all of them are endemic. They live in a number of caves (ca. 90 Bulgarian caves are known to contain troglobitic isopods), mostly in Stara Planina and the Rhodopes, but also in isolated localities in Ruse and Blagoevgrad areas, and in Strandja. The troglobitic Isopoda belong mostly to the family Trichoniscidae (*Hyloniscus, Trichoniscus, Balkanoniscus, Rhodopioniscus, Bureschia, Bulgaronethes, Cyphoniscellus* = *Bulgaroniscus, Tricyphoniscus, Beroniscus, Vandeloniscellus, Alpioniscus* = *Illyrionethes*; in total not less than 21 troglobites). Only one of the remaining seven families of terrestrial isopods, found in Bulgarian caves, contains troglobites: Styloniscidae, with two species, the northernmost record for this family.

Arachnida. The arachnids are represented in Bulgarian caves by seven orders: Palpigradi, Scorpiones, Pseudoscorpiones, Opiliones, Araneae, and two orders of Acari. The Palpigradi are tiny, white, archaic animals. In Bulgaria they have been so far recorded from three caves in Western Stara Planina: Samuilitsa near Kunino, Mecha Dupka near Salash, and the pothole Randjolova Tarsha near Prevala (Montana District). They are unknown in Bulgaria outside the caves. Scorpions (*Euscorpius* spp., Euscorpiidae) are rarely found in the Balkan caves, mainly near the entrances; they are trogloxenes. Among the pseudoscorpions, on the contrary, there are many troglobites; 15 species have been recorded from Bulgarian caves, including seven troglobites of the genera *Chthonius, Neobisium, Balkanoroncus*, and *Roncus* (Table 1). Pseudoscorpions found in many Bulgarian caves are deposited in NMNH; this abundant material collected is still under study.

One of the most numerous and important groups of animals in Bulgarian caves are spiders (Araneae). They are well known in Bulgaria, thanks to Pencho Drensky (before the World War II) and now to Christo Deltshev and his team. About 80 species of spiders have been recorded from the Bulgarian caves but only six are troglobites (*Porrhomma microps, Centromerus bulgarianus*, and *Troglohyphantes drenskii*, Linyphiidae; *Protoleptoneta bulgarica* and *P. beroni*, Leptonetidae; and *Nesticus beroni*, Nesticidae). Of the 17 families of spiders in Bulgarian caves the most numerous is Linyphiidae, containing ca. 30 species, including three troglobites and many common troglophiles from the genera *Centromerus, Troglohyphantes, Lepthyphantes, Porrhomma*, and others. They are small spiders, often found among stones on the cave floor and on rotten wood. The larger troglophilic spiders are found most often near the entrances. They belong to genera *Tegenaria* (Agelenidae), *Meta* (Metidae), and *Nesticus* (Nesticidae). Spiders play an important role in the food chain of cave communities.

Out of 61 species of harvestmen (Opiliones) known in Bulgaria, 22 have been recorded from Bulgarian caves; four of them are troglobites. *Paralola buresi* from the caves near Lakatnik is the only representative of the family Phalangodidae and of the entire suborder Laniatores in Bulgarian caves. *Paranemastoma (Buresiolla) bureschi* is known from caves in Western Stara Planina. There are two members of the short-legged suborder Cyphophthalmi: *Tranteeva paradoxa* from the caves Rushovata near Gradeshnitsa and Toplya near Golyama Zhelyazna, and *Siro beschkovi* from Haidushkata Peshtera near Deventsi. The most common troglophile is *Paranemastoma radewi* (Palpatores: Nemastomatidae). The remaining species are occasional visitors to the caves.

The large group of mites and ticks (Acari) is not well represented in caves. Many mites (Acariformes and Parasitiformes) live in the guano but there are no troglobites among them. The only mites which to some extent meet the requirements for the "troglobite" category are some Trombidiidae from the Dinaric Karst and some species of Rhagidiidae (white mites, running fast on the cave clay). Often on the cave walls are seen flat, long-legged ticks. They are the males of a specific parasite on bats (host-specific, not infesting humans), *Ixodes vespertilionis* (Ixodidae). The bats are hosts also of other parasitic mites but they are not true cave dwellers.

Myriapoda. The four classes forming the large group of Myriapoda are unevenly represented in caves. The small Symphyla and Pauropoda are almost unknown there but the centipeds (Chilopoda) and even more the millipeds (Diplopoda) are among the most important cave animals. Chilopoda are represented by 28 species from Bulgarian caves (two of the order Geophilomorpha, two of Scolopendromorpha, one of Scutigeromorpha, and 23 of Lithobiomorpha, all from the family Lithobiidae). Five *Lithobius* are troglobites (Table 1); particularly widespread is *L. lakatnicensis*, known from 16 caves in Stara Planina and the Rhodopes. In Vodnata Peshtera near Tserovo lives the large troglobite *Eupolybothrus andreevi*.

The millipeds (Diplopoda) are among the most interesting cave animals (along with the beetles, the isopods, and the pseudoscorpions). So far, 54 species have been recorded from Bulgarian caves (about half of the species of this class living in Bulgaria), 17 of them being troglobites. Of the small, white, rolling in a ball, representatives of Glomerida only *Trachysphaera orghidani lakatnicensis* from the caves near Lakatnik and from Tamnata Dupka near Targovishte (Vidin District) is considered a troglobite; however *T. dobrogica*, described from Northern Dobrudja in Romania, could be expected in Southern Dobrudja in Bulgaria as well. Species of the order Polydesmida are often found in Bulgarian caves (13 species and four subspecies). Of those, troglobites are *Brachydesmus radewi* of Polydesmidae and two endemic subspecies of the family Trichopolydesmidae, *Bacillidesmus bulgaricus bulgaricus* from the caves near Tsar Petrovo and Dolni Lom and *B. b. dentatus* from Drashanskata Peshtera. Several interesting troglobites are known among the Anthroleucosomatidae (order Chordeumatida, or Ascospermophora). Such are the inhabitants of the caves of Stara Planina (*Bulgarosoma bureschi* in Ledenika Cave and other species) and of Rhodopean caves (*Troglodicus meridionale, T. tridentifer*), the species of *Anamastigona* (= *Prodicus*) in the caves in Troyan area, Stoletovskata Peshtera in Central Stara Planina, Lepenitsa in the Western Rhodopes, caves near Paril and others. The monotypic endemic genera *Stygiosoma* (*S. beroni* from Manailovata Peshtera) and *Bulgardicus* (*B. tranteevi* from Bankovitsa near Karlukovo) are also troglobites. The order Callipodida does not contain troglobites in Bulgarian caves but here belong four very large and typical troglophiles: *Balkanopetalum armatum* in the caves of Western Stara Planina, and three species in the Rhodopes. *Typhloiulus* species, particularly abundant in Temnata Dupka near Lakatnik and in Vodnata Pesht near Lipnitsa, belong to order Julida (family Julidae). Of six species from genus *Typhloiulus* in Bulgarian caves, four are considered troglobites. Similar to *Typhloiulus* is the genus *Serboiulus* with one species *S. spelaeophilus* living in the caves of Northwestern Bulgaria. Often troglophiles of *Apfelbeckiella, Troglodicus*, as well as *Nopoiulus kochii* (= *N. venustus, N. pulchellus*), are found, sometimes abundant, in bat guano. Other interesting troglobitic diplopods are currently under study.

Insecta. Many families and orders of insects avoid the cave environment. Among 15 orders of insects inhabiting Bulgarian caves, troglobites are found only among the beetles (Coleoptera: Carabidae, Cholevidae, and Curculionidae), Collembola, and Diplura (Campodedidae). The remaining 12 orders (Thysanura, Ephemeroptera, Plecoptera, Orthoptera, Psocoptera, Homoptera, Heteroptera, Hymenoptera, Siphonaptera, Diptera, Trichoptera, and Lepidoptera) contain only troglophiles, trogloxenes, and parasites. Of more than 200 species of Collembola in Bulgaria, 41 are known from caves, and seven species are considered troglobites (Table 1). We should proceed with collecting Collembola in the Bulgarian caves, as in the neighboring Romania their number is twice as large. More spectacular are the Diplura, frail white insects with long antennae and two cerci. Only seven species of this order have been recorded (including six troglobites of genus *Plusiocampa*) but others are under study by the French zoologists Bareth and Condé. Of the other dipluran family (Japygidae) troglobites in Bulgaria so far have not been found but in the show cave Kutuki near Athens has been found the most extreme troglobite of this group.

The cave beetles (Coleoptera) have an honor to be the first terrestrial troglobites to be ever described (from Slovenia); following the growth of biospeleology, they also were the first troglobites to be described from Greece, Bulgaria, and other countries. In Bulgarian caves, the family Carabidae is represented by 16 species of genus *Duvalius*, three species and many subspecies of *Pheggomisetes* and one species of *Pterostichus* (*P. ledenikensis*), altogether 16 troglobite species (Table 1), as well as many other troglophiles and trogloxenes. Another remarkable family is Cholevidae (formerly Catopidae), represented so far by 17 troglobite species of 13 genera of subfamily Leptodirinae (formerly Bathysciinae) (Table 1), as well as by eight "subtroglophile" species of the genera *Choleva, Catops, Nargus*, and *Sciodrepoides*. Certain cholevids (*Beskovia* in caves near Cherepish, *Netolitzkya* in caves near Arbanasi, *Hexaurus* in Stoletovskata Peshtera) are sometimes very numerous. Others are more rare and scarce in number. Typical cases are within the Carabidae. *Pheggomisetes, Duvalius, Rambousekiella*, and other cave beetles should not be collected excessively, as their number could decline very fast and their populations recover slowly. Most species are represented by few specimens, scattered and difficult to find without traps or bait. Cave beetles are important for zoogeography; their biology is completely unknown in Bulgaria, and is a vast field for further studies.

In the guano and litter in caves are found also other Coleoptera, of which best known are Staphylinidae. The larger black staphylinids belong usually to the genus *Quedius*, the tiny ones, to *Atheta*. In total, 27 species of Staphylinidae are recorded from Bulgarian caves but none of these are troglobites. Other families of Coleoptera found in Bulgarian caves are Dytiscidae, Hydrophilidae, Histeridae, Pselaphidae, Colydiidae, Endomychidae, Cryptophagidae, Curculionidae, and Ptinidae. Of particular biospeleological interest are Pselaphidae and Curculionidae. We could expect troglobites among pselaphids as such are known from Greece and former Yugoslavia. On plant roots in caves a careful observer may notice tiny weevils *Troglorrhynchus* (Curculionidae), three species of which are considered troglobites (Table 1).

The flying insects in caves are of limited interest; they are not troglobitic and the collected material rarely contains new species. Among the stalactites, sometime hide the large, wasplike Hymenoptera of genus *Diphyus* (= *Amblyteles*) (Ichneumonidae). Often, on the ceiling or the walls, our attention is attracted by the mothlike caddisflies (Trichoptera), 16 species of which live in Bulgarian caves. Particular attention deserve three genera of Limnephilidae: *Micropterna* (six species), *Stenophylax* (three species) and *Mesophylax* (one species), which are regular trogloxenes. The remaining six species are occasional trogloxenes. Some moths (Lepidoptera) also regularly occur in caves; of 17 lepidopteran species such are *Triphosa dubitata* and *T. sabaudiata* (Geometridae); some Noctuidae such as *Apopestes, Scoliopteryx* and others; representatives of Alucitidae (Orneodidae); and Tineidae. The latter often live in cave guano and are poorly known in Bulgaria.

Among flies (Diptera), of which 56 species are known from Bulgarian caves, typical cave dwellers belong to Muscidae, Helomyzidae, Mycetophilidae, as well as the long-legged Limnobiidae, often seen on cave walls. Wingless parasitic flies of the family Nycteribiidae can be easily seen on bat fur. Recently, *Brachytarsina flavipennis*, a species from another family of bat parasites (Streblidae), was added to the fauna of Bulgaria, thanks to the efforts of Bulgarian cave biologists.

Vertebrates are not very numerous in Bulgarian caves; rarely, frogs, toads or salamander larvae are found there. At the entrances one finds nests of pigeons, swallows, in some deep potholes nests the alpine chough (*Pyrrhocorax graculus*). This interesting bird is protected by law and its colonies are to be preserved, especially during the breeding season. Deep in caves enter mice, dormice, foxes, and other mammals.

The most interesting cave mammals are the bats (Chiroptera), of which 16 species have been recorded in Bulgarian caves (five belonging to Rhinolophidae and 11, to Vespertilionidae). Large colonies are formed by Schreibers's long-fingered bat (*Miniopterus schreibersi*) and by both large species of genus *Myotis* (*M. myotis* and *M. blythi*). Several protected winter and summer roosts have been observed, with colonies of *Myotis capaccinii* and horseshoe bats (Rhinolophidae) of European importance. Bulgaria enjoys almost the full list of European bats and some very numerous colonies are worth protection. Visits, especially of larger groups, in bat caves like Parnitsite near Bezhanovo, are very improper. Disturbing of breeding bat colonies or winter roosts could deprive Bulgaria of this component of its biodiversity, and some Central European countries already face such a threat.

5 Zoogeographical Analysis of the Terrestrial Cave Fauna in Bulgaria

The intensive research on the cave and underground living animals in Bulgaria since 1922 accumulated a rather complete information on the composition and distribution of most of the groups of underground living animals. Time has come to formulate hypotheses about the origin and zoogeography of this fauna. Such hypotheses are due mainly to V. Guéorguiev, who analyzed the terrestrial troglobites known from Bulgarian caves. His series of papers (Guéorguiev, 1966, 1973a, 1973b) was crowned by his monograph on the origin, the formation and the zoogeography of the terrestrial troglobites of Balkan Peninsula (Guéorguiev, 1977). This remarkable book was followed by his speleozoogeographical subdivision of Bulgaria (Guéorguiev, 1992). Other attempts to analyze the distribution of the terrestrial cave fauna in Bulgaria find place in the papers of Beron (1976, 1978) and in some articles on different groups of cave animals (e.g. Deltshev, 1978).

All these papers do not deal with the stygobites. Some of the stygobites have been found in caves but most are living in the stygal hyporheic water or other parts of the underground ecosystems. The distribution and origin of the stygobites have been analyzed by several qualified Bulgarian stygobiologists (the late L. Tzvetkov, A. Petrova, and the very active explorer of cave Copepoda I. Pandourski).

In his monograph, Guéorguiev (1977) subdivided the Balkan Peninsula into four provinces (Dinaric, Aegean, Stara Planina, and Rhodopean). This subdivision remains valid but needs to be refined, especially concerning Bulgaria, and the list of the species-indicators are now much longer. According to Guéorguiev (1977), Bulgarian territory falls into two of these provinces: the Stara Planina Province (with two zones, Western and Eastern) and Rhodopean Province (also with two zones, Western and Eastern). The Stara Planina Province included also Serbia east of Morava River, and the Rhodopean Province included also Northern Greece. Beron (1976) subdivided the Stara Planina Province into seven regions, some of them provisional: Eastern Serbia, Ogosta, Iskar, Osam, Ruse, Kamchiya, and Dobrudja. Guéorguiev (1977) also delineated some special regions within his zones. In Bulgaria, these regions (indicated on Fig. 1) are:

In the Western zone of Stara Planina:

- Vrachanska Planina (the richest region in troglobites in the eastern part of Balkan Peninsula): by 1977, there were 29 terrestrial troglobites in this region, including

17 indicators: "*Cyphoniscellus*" (now *Vandeloniscellus*) *bulgaricus, Bulgarosoma bureschi, Typhloiulus longipes, Centromerus bulgarianus, Neobisium beroni*, "*Microcreagris bureschi*" (now *Balkanoroncus hadzii*), *Onychiurus sensitivus, Plusiocampa bureschi, Pheggomisetes radevi, Ph. r. ilcevi, Ph. r. tranteevi, Ph. globiceps mladenovi, Duvalius beroni*, "*D. deltschevi*" (now synonym of *D. zivkovi*), *D. papasoffi, D. zivkovi, Radevia hanusi*;
- Ponor Planina: 12 troglobites, including three indicators: *Eupolybothrus andreevi, Pheggomisetes globiceps globiceps, Ph. g. cerovensis*;
- Golyama Planina: seven troglobites, including one indicator (*Duvalius pretneri*);
- Karlukovo Region: five troglobites, one indicator (*Tricyphoniscus bureschi*);
- Vasilyovska Planina: eight troglobites, six indicators: *Tranteeva paradoxa, Neobisium bulgaricum, N. subterraneum* (now synonym of *N. bulgaricum*), *Plusiocampa gueorguievi, Duvalius regisborisi, Tranteeviella bulgarica*;
- Troyanska Planina: three troglobites, all considered indicators: "*Lindbergia*" (now *Spinophallus*) *uminskii, Trichoniscus bulgaricus*, "*Prodicus*" *albus* (now *Anamastigona alba*); "*Lindbergia*" *uminskii* now is no more considered troglobite.

In the Eastern zone of Stara Planina:
- Shipchenska Planina: three troglobites, all indicators: *Duvalius balcanicus, Hexaurus merkli, H. similis*; new: *Hexaurus paradisi, H. shipkaensis*;
- Trevnenska Planina: three troglobites, all indicators: *Duvalius bulgaricus, Netolitzkya jeanneli, N. maneki*;
- as a new region, we can add now Kotlenska Planina, with the troglobite indicators *Duvalius kotelensis* and *Porrhomma microps*.

In Western Rhodopes:
- Batashki Rid: four troglobites, indicators: *Troglohyphantes drenskii, Duvalius bureschi*;
- Ravnogor: two troglobites, one indicator (*Bulgaronethes haplophthalmoides*).

In Eastern Rhodopes:
- Chernatitsa (geographically speaking, part of Western Rhodopes): two troglobites, one indicator (*Troglodicus tridentifer*);
- Dobrostan Massif (also geographically part of Western Rhodopes): three troglobites, indicators: "*Balkanoniscus*" (now *Rhodopioniscus*) *beroni, Rhodopiola cavicola*; new: *Nesticus beroni*;
- Trigrad Plateau (also geographically part of Western Rhodopes): indicator: "*Bulgarosoma*" (now *Troglodicus*) *meridionale*.

Several species have been added to the list 30 years since the publishing of the book of Guéorguiev (1977).

Almost no genera and species in common exist between the Eastern and the Dinaric parts of the Balkan Peninsula delineated by Guéorguiev. The two provinces in the Eastern part of the Peninsula (Stara Planina and Rhodopes) have very few indicator genera and species in common. Such is *Lithobius lakatnicensis* (Chilopoda; Rhodopes and Stara Planina). Some genera, previously believed to inhabit both

Stara Planina and the Rhodopes, proved to be separate, and the Rhodopean species have been placed in new genera (*Rhodopioniscus, Rhodoposoma*).

Guéorguiev (1992) considered how his general zoogeographical subdivision of Bulgaria of 1982 fits with the biospeleological subdivision. In addition to two zoogeographical zones, Bulgaria was subdivided into seven regions, and their troglobitic fauna has been tested by the index of Czekanowski–Sørensen. Since the paper of Guéorguiev (1992) is in Bulgarian, it is worth to summarize its main conclusions. Moreover, this paper was submitted in 1989, and meanwhile several new troglobites and interesting troglophiles were added to the Bulgarian cave fauna. The zoogeographical regions are characterized as follows:

1. *Danubian Region*. The cave fauna consists of 26 troglophiles (trogloxenes not included) and only two troglobites: *Trichoniscus tranteevi* (Isopoda) and *Plusiocampa beroni* (Diplura) (Table 1). No similarity with the troglobites of any other region.
2. *Stara Planina Region*. The richest cave fauna in Bulgaria (total 191 species, including 103 troglophiles and 88 troglobites). Also the richest region in endemics. Closest similarity with Rila–Rhodopean Region. The troglobitic fauna exhibits an insignificant similarity with the Rila–Rhodopean Region, the troglobites in common being only *Lithobius lakatnicensis* (Chilopoda), *Plusiocampa bulgarica* (Diplura), and *Pseudosinella duodecimocellata* (Collembola).
3. *Rila–Rhodopean Region*. The second richest region in Bulgaria for cave fauna. According to Guéorguiev (1992), 46 troglophiles and 19 troglobites were recorded here. Since 1989, some other species have been added but now (end of 2006) the number of troglobite taxa in the Rila–Rhodopean Region is still only 20: *Cordioniscus bulgaricus, C. schmalfussi, Trichoniscus rhodopiense, Rhodopioniscus beroni, Bulgaronethes haplophthalmoides, Alpioniscus* (= *Illyrionethes*) sp., *Nesticus beroni, Troglohyphantes drenskii, Lithobius lakatnicensis, Troglodicus meridionale, T. tridentifer, Stygiosoma beroni, Anamastigona lepenicae, A. delcevi, Plusiocampa bulgarica, Pseudosinella duodecimocellata, Duvalius bureschi, D. petrovi, D. karelhurkai, Rhodopiola cavicola*. Among the troglobites the most numerous are the Isopoda (six species), the Diplopoda (five species), and the Coleoptera (four species). The endemism of this fauna also places the Rila–Rhodopean Region at second place as it has 19 endemic troglobites (Balkan, Bulgarian, and local). The only troglobite in common with the Thracian Region is *Trichoniscus rhodopiense* (Isopoda), and in common with the Stara Planina are *Lithobius lakatnicensis, Pseudosinella duodecimocellata*, and *Plusiocampa bulgarica*.
4. *Struma–Mesta Region*. Very few caves, mostly in Slavyanka Mts. Three cave animals are quoted by Guéorguiev: the troglobites "*Tranteevonethes gueorguievi*" (Isopoda) (*nomen nudum*, now *Alpioniscus* sp. n.) and *Anamastigona delcevi* (Diplopoda) and the troglophile "*Lepthyphantes gueorguievi*" (Araneae, now *L. spelaeorum*). We can add to them now some remarkable troglobites from the cave Stoikova Dupka 1 in Slavyanka (still

under study), including *Duvalius joakimovi* and a new species of *Troglohyphantes* (Araneae), also *Duvalius pirinensis* from Pirin.

5. *Thracian Region*. Guéorguiev (1982) included here also the Eastern Rhodopes. The intensive recent research in this area (Beron et al., 2004) added new species to the list of the troglophiles and troglobites of Eastern Rhodopes, the area little studied previously. The actual list of troglobites in the Eastern Rhodopes is as follows: *Trichoniscus rhodopiense* (Isopoda), *Lithobius tiasnatensis* (Chilopoda), *Duvalius petrovi, Bureschiana drenskii, Troglorrhynchus angelovi, T. beroni* (Coleoptera). To this list we should add troglobites found in the caves of the Greek part of Eastern Rhodopes: *Alpioniscus thracicus* (Isopoda), *Maroniella beroni* (Coleoptera). Some interesting troglophiles in the region are *Balkanopetalum petrovi, Apfelbeckiella trnowensis rhodopina, Rhodopiella beroni* (Diplopoda), *Balcanodiscus frivaldskyanus, B. cerberus* (Mollusca).
6. *Pontian Region* (in Guéorguiev, Black Sea Coast Region). Very few caves, only one troglophile: the spider *Meta bourneti*, common with Strandja and Thrace.
7. *Strandja Region*. According to Guéorguiev (1992), the cave fauna consists of 21 troglophiles (6 of them endemics) and four troglobites: *Trichoniscus valkanovi, T. beroni* and "*T. tashevi*" (nomen nudum) (Isopoda) and *Lithobius* (*Monotarsobius*) *bifidus* (Chilopoda). All troglobites are Bulgarian endemics.

6 Endemics in the Cave Fauna of Bulgaria

Mollusca. The most important group of terrestrial Gastropoda in Bulgarian caves are Zonitidae; of ca. 33 species found in Bulgaria, one-third (11 species) live in caves. Four of them can be considered endemic (Table 1). Two slugs also seem to be endemic: *Deroceras bureschi* (H. Wagner) and *Tandonia kusceri* (H. Wagner), but they are not troglobites.

Isopoda: *Oniscidea*. Zoogeographically, and from the point of view of cave evolution, the terrestrial Isopoda are among the most important and interesting groups in Bulgarian cave fauna. Of 24 genera and 50 species (including 26 troglobites) of cave isopods in Bulgaria, seven genera (*Balkanoniscus, Rhodopioniscus, Bureschia, Bulgaronethes, Bulgaroniscus, Vandeloniscellus, Tricyphoniscus*, all belonging to Trichoniscidae) and 31 species are endemic for Bulgaria. With 32 species in caves (including 24 out of 26 isopod troglobites), Trichoniscidae is the family by far the most important among Bulgarian cave isopods. Of these 32 species, 26 are Bulgarian endemics (see Table 1 for endemic troglobites). The only two other troglobites (*Cordioniscus bulgaricus* Andreev and *C. schmalfussi* Andreev) belongs to Styloniscidae. Both are Bulgarian endemics (Table 1), respectively for Boichovata Peshtera on the border with Republic of Macedonia and for the Rhodopes. Of the 32 species of Trichoniscidae, 26 are endemic for Bulgaria (four species of *Hyloniscus*, 14 *Trichoniscus, Alpioniscus* sp., *Balkanoniscus corniculatus* Verhoeff, *B. minimus* Vandel, *Rhodopioniscus beroni* (Vandel), *Bureschia bulgarica* Verhoeff, *Bulgaronethes haplophthalmoides* Vandel,

Cyphoniscellus gueorguievi (Vandel), *Vandeloniscellus bulgaricus* (Vandel), *Tricyphoniscus bureschi* Verhoeff, *Beroniscus capreolus* Vandel, and *Monocyphoniscus bulgaricus* Strouhal), and three are Balkan endemics (*Trichoniscus stankovici* Pljakic, *T. semigranulatus* Buturovic, *T. rhodopiense* Vandel).

Some other families of Oniscidea also contain endemics (trogloxenes): the Balkan endemic *Ligidium herzegowinense* Verhoeff (Ligidiidae), Bulgarian endemics *Trachelipus bulgaricus bulgaricus* Verhoeff and *T. b. bureschi* Verhoeff (Trachelipidae), *Chaetophiloscia hastata* Verhoeff (Philosciidae), *Porcellium balkanicum* Verhoeff (Porcellionidae), and *Armadillidium elysii* Verhoeff (Armadillidiidae).

Pseudoscorpiones. The Bulgarian endemics include seven troglobite species (Table 1), with a subgenus *Heoblothrus* endemic to Stara Planina, and also *Neobisium* (*N.*) *intermedium* Mahnert from Eastern Predbalkan (Prolazkata Peshtera). A Balkan endemic is *Roncus parablothroides* Hadži.

Opiliones. Of 22 species of harvestmen known from Bulgarian caves, six are Bulgarian endemics, including all four troglobites (Table 1), and also *Dicranolasma thracium* Staręga and *Histricostoma drenskii* Kratochvil. The species *Leiobunum rumelicum* Šilhavý, *Pyza bosnica* (Roewer), and *Rafalskia olympica* (Kulczyński) are Balkan endemics. The troglophile *Paranemastoma radewi* (Roewer) is a Balkan endemic widespread in Bulgaria but also known from Bosnia and Northern Greece.

The four troglobites (Table 1) are confined to Stara Planina and the Predbalkan west of Troyan and are among the most remarkable cave animals in Bulgaria. The monotypic troglobite genera *Paralola* and *Tranteeva* are endemic for Bulgaria. *Paranemastoma* (*Buresiolla*) *bureschi* (Roewer) lives in many caves in Western Stara Planina and, most certainly, in Western Confines (now Serbia), as we found it at the very border. *Siro beschkovi* is a member of a cyphophthalmid genus with many species in the Balkans, and is known so far only from one cave in the Eastern Predbalkan.

Araneae. Of 80 species of cave spiders in Bulgaria, 20 are considered here to be local, Bulgarian, or Balkan endemics. According to Deltshev (1996), the linyphiid genus *Antrohyphantes* is related to *Fageiella* from the western part of the Balkan Peninsula and is therefore an ancient (paleoendemic) element. Also, according to the same author, "the group of cave endemic spiders has similar presence, 17 (39.4%) as the group of high altitude spiders". Six troglobite species belong to Leptonetidae, Linyphiidae, and Nesticidae, and their endemic status are listed in Table 1. In addition to the troglobites, we list below other endemic spiders found in Bulgarian caves. The most diverse are Linyphiidae, which include: *Centromerus lakatnikensis* (Drensky), a Balkan endemic (Western Stara Planina, Republic of Macedonia); *C. milleri* Deltshev, a Balkan endemic (Eastern Rhodopes); *C. acutidentatus* Deltshev, a Bulgarian endemic (South Pirin); *Palliduphantes byzantinus* (Fage), a Balkan endemic; *P. istrianus* (Kulczyński) (= *slivnensis* Drensky), a Balkan endemic; *P. trnovensis* (Drensky), a Balkan endemic (Stara Planina, Republic of Macedonia); *Tenuiphantes jacksonoides* (van Helsdingen), a Bulgarian endemic (Western Rhodopes); *Antrohyphantes sophianus* (Drensky) (= *Lepthyphantes tranteevi* Miller), a Bulgarian endemic (Western Stara Planina); *A. balcanicus* (Drensky), a Balkan endemic (Central Stara Planina, Republic of Macedonia); *A. rhodopensis* (Drensky) (= *A. rodopicus* Dumitrescu), a Bulgarian endemic (Western Rhodopes); and *Troglohyphantes bureschianus* Deltshev, a Bulgarian endemic (Western Rhodopes). Also recorded are three Bulgarian endemic species of Agelenidae: *Histopona tranteevi* Deltchev, Western Rhodopes; *Coelotes drenskii* Deltshev, Eastern Stara Planina; *C. jurinitschi* (Drensky), Western Stara Planina; one Balkan endemic species of Amaurobiidae, *Calobius balcanicus* (Drensky); and one Balkan endemic subspecies of Liocranidae, *Mesiotelus cyprius scopensis* Drensky.

Diplopoda. Four genera (*Rhodoposoma, Troglodicus*, and *Stygiosoma* from Western Rhodopes and *Bulgardicus* from the caves near Karlukovo in the Predbalkan) and all 16 species of troglobitic Diplopoda in Bulgaria (Table 1) are (so far) Bulgarian endemics; this list includes also two endemic diplopod subgenera, *Balkandicus* (genus *Anamastigona*) and *Inversotyphlus* (genus *Typhloiulus*). The following non-troglobitic diplopod taxa seem also to be endemics: order Glomerida, Glomeridae: *Glomeris balcanica* Verhoeff (= *G. bureschi* Verhoeff, = *G. latemarginata* Attems), a Balkan endemic; order Polydesmida, Polydesmidae: *Polydesmus herzogowinensis* Verhoeff, Balkan endemic; *P. renschi* Schubart, *P. zonkovi* Verhoeff, *P. bureschi* Verhoeff, and *Brachydesmus cristofer* Strasser, Bulgarian endemics; *B. herzogowinensis reflexus* Strasser, *B. h. trifidus* Strasser, *B. h. confinis* Strasser, and *B. dadayi brusenicus* Gulička, Bulgarian endemic subspecies; Paradoxosomatidae, *Metonomastus pomak* Golovatch et Stoev, a Bulgarian endemic (Eastern Rhodopes); order Chordeumatida, Mastigophorophillidae: *Mastigona bosniensis* (Verhoeff), a Balkan endemic; Haaseidae: *Haasea* (*Histrosoma*) *vidinensis* (Strasser), Bulgarian endemic species and subgenus; Anthroleucosomatidae: *Rhodoposoma rhodopinum* (Strasser), Bulgarian endemic genus and species (Rhodopes); order Julida, Julidae: *Leptoiulus borisi* Verhoeff, a Bulgarian endemic; *Typhloiulus kotelensis* Jawlowski, Eastern Stara Planina; *Megaphyllum rhodopinum* (Verhoeff) and *M. beroni* (Strasser), Rhodopes; *Balkanophoenix borisi* Verhoeff, Western Stara Planina; *Pachyiulus hungaricus gracilis* Verhoeff, Rhodopes, endemic subspecies; *P. cattarensis* (Latzel), a Balkan endemic; *Apfelbeckiella trnowensis* (Verhoeff) and its subspecies *A. t. rhodopina* Strasser and *A. t. deliormana* Strasser, Bulgarian endemics; *A. byzantina* Verhoeff, a Balkan endemic; *Rhodopiella beroni* Strasser, a Bulgarian endemic genus and species; order Callipodida, Schizopetalidae, five troglophile species of *Balkanopetalum* (Table 1).

Chilopoda. Of 44 species of this class found in Bulgarian caves, 14 are endemics (nine Bulgarian and five Balkan). Six troglobite endemic taxa are listed in Table 1. Other species are: order Lithobiomorpha, Lithobiidae: three Balkan endemics, *Lithobius beroni* Negrea (Bulgaria and Greece), *L. stygius* Latzel, and *L. wardaranus* (Verhoeff); *Harpolithobius folkmanovae* Kaczmarek, a Bulgarian endemic (Southeastern Bulgaria); *H. banaticus rhodopensis* Kaczmarek, an endemic subspecies (Rhodopes); and *Eupolybothrus gloriastygis* (Absolon), a Balkan endemic (Bulgaria and Bosnia; known from two caves in Western Stara Planina); order Geophilomorpha, Himantariidae: *Thracophilus beroni* Matic et Darabanþu, Western Rhodopes; *Th. bulgaricus* Verhoeff (Western Rhodopes and Sredna Gora); and order Scolopendromorpha, Cryptopidae: *Cryptops croaticus* Verhoeff, a Balkan endemic.

Insecta. Among springtails (Collembola), there are eight Bulgarian endemics in the caves of Bulgaria, of which six are troglobites: three species of Entomobryidae (genus*Pseudosinella*), and three species of Onychiuridae (Table 1). In addition to troglobites, the list of endemic springtails found in caves includes *Onychiurus bureschi* Handschin and *O. subgranulosus* Gama (Onychiuridae) from Western Stara Planina, and *Bilobella digitata* Cassagnau (Neanuridae) from Eastern Stara Planina.

The order Diplura includes so far six Bulgarian endemic troglobites, all belonging to the genus *Plusiocampa* (Campodeidae) (Table 1).

The order Orthoptera is represented by some trogloxenes and a single troglophile species, the Balkan subendemic *Troglophilus neglectus* Krauss (Rhaphidophoridae).

Among beetles (order Coleoptera), the family Carabidae has 44 species of ground beetles recorded from the Bulgarian caves. Two genera are Bulgarian endemics, both in Western Stara Planina: *Rambousekiella* and *Pheggomisetes*. *Pheggomisetes* is practically a Bulgarian endemic since it was found also in one cave in Eastern Serbia. It is represented by several endemic troglobite species (Table 1). Another important genus is *Duvalius*, with 16 troglobite species endemic for Bulgaria (Table 1). Three other species of *Duvalius* (*Paraduvalius*) have been described from the MSS (Milieu souterrain superficiel) and other localities outside the caves: *D. hanae* Hůrka from Central Balkan, *D. marani* (Knirsch) from Slavyanka, and *D. rajtchevi* (Genest et Juberthie) from Western Rhodopes.

Among Staphylinidae, only two species (*Quedius troglophilus* and *Qu. gueorguievi*), described by H. Coiffait from Bulgaria, are so far endemic for this country. They are not typical cave animals. Pselaphidae include only one (trogloxenic) species, *Bryaxis beroni* Z. Karaman, which seems to be a Bulgarian endemic (Western Bulgaria). Among weevils (Curculionidae), all three species of hypogeous weevils reported from Bulgaria are so far Bulgarian endemics (Table 1).

The highest endemism at the generic level in Bulgarian cave fauna is represented in the beetle family Cholevidae (= Leiodidae). Far from being completely known in Bulgarian caves, this family is represented there by 29 species belonging to 19 genera. Of these, 12 genera and 19 species are Bulgarian endemics (Table 1), including monotypic *Beronia, Radevia, Vratzaniola, Balcanobius, Beroniella, Tranteeviella* (= *Bulgariella*), and *Genestiellina*; *Hexaurus* is represented by four species, and *Netolitzkya*, by two species and four subspecies. All 12 genera are very narrow endemics, found in certain areas within Bulgaria (Table 1). All these genera belong to the subfamily Leptodirinae (= Bathysciinae).

Of the order Siphonaptera, only subspecies (so far endemic for Bulgaria) of two species of fleas have been described from Bulgarian cave bats: *Nycteridopsylla ancyluris johanae* Hůrka and *N. trigona balcanica* Hůrka.

7 Relicts in the Cave Fauna of Bulgaria

7·1 The issue of relicts

The issues of the relicts and the ancient age of troglobites are still open. According to Jeannel (1944, 1960), followed by Vandel (1964) and Guéorguiev (1977), there was no doubt that the paleotroglobites are very ancient and have no relatives among the recent animals living outside the caves. On the contrary, the neotroglobites still have relatives outside the caves and their evolutionary connection can be traced via intermediary forms. According to Leleup (1965), the main lineages of South European troglobites have their origin in the orophilic pre-Pleistocene fauna, very ancient and very rich, living in land habitats which formed in remote geological periods. Vandel (1964) stated that "terrestrial troglobites are mostly descendents of a tropical fauna populating Europe and North America in the first half of the Tertiary"; we have to keep in mind that the Paleogene (the first part of the Tertiary) occupied the time between 67 and 25 Mya!

Brignoli (1979) opposed this assertion, taken for a long time as axiomatic. This early deceased, prominent Italian specialist wrote: "...It is not true at all (or, at least, it is not for sure) that the troglobites are ancient". And, further: "...The term of "relict" (or even of "living fossil"), so often applied to the troglobites, is for me completely meaningless". However, other prominent specialists did support the opinion about the ancient nature of the troglobites. According to Beier (1969), "...the troglobite species show a high degree of specialization, and are without doubt to be considered relicts from the Tertiary".

Guéorguiev (1977) was a firm supporter of the theories of Jeannel and Vandel and his classification of the troglobites according to their origin will be resumed here, completed with some new data. The present author also thinks that the assertions of Brignoli are exaggerated, and that relicts do exist. Which troglobite is ancient and which is more recent is of course matter of analysis.

7·2 Descendents of Gondwanan phyletic lineages

Among the most ancient Balkan relicts are members of the genus *Cordioniscus* (Isopoda: Styloniscidae). Two troglobite species found in Bulgaria (*Cordioniscus bulgaricus* Andreev and *C. schmalfussi* Andreev), mark the northernmost localities of this family in Europe. The endemic troglobite genus and species *Tranteeva paradoxa* Kratochvil (Opiliones: Cyphophthalmi: Sironidae) from the caves of the Predbalkan is also considered a Gondwanan relict.

7·3 Descendents of Laurasian phyletic lineages

Guéorguiev (1977) considered as Laurasian relicts some spiders of the genus *Nesticus* and some Collembola (*Acherontides spelaeus* Ionesco, considered now a troglophile). Bulgarian *Nesticus* species are also troglophiles. A typical Laurasian element, according to Guéorguiev (1977), would be also the endemic genus and species *Bulgaronethes haplophthalmoides* Vandel (Isopoda: Trichoniscidae) from Western Rhodopes.

7·4 Descendents of Mesogeidean (Paleomediterranean) phyletic lineages

Isopods of the endemic genera *Balkanoniscus, Beroniscus*, and *Bureschia* (all in Stara Planina and the Predbalkan) are considered to be the descendents of a phyletic lineage, which inhabited in the Paleocene and early Eocene the land stretching from Cantabric Mountains to Caucasus and called Mesogeide (Guéorguiev, 1977). We should add here the genus *Rhodopioniscus*, described by Tabacaru in 1993 for the Rhodopean "*Balkanoniscus" beroni* Vandel. Among Diplopoda, such Mesogeidean relicts would be the glomerids *Trachysphaera orghidani* Tabacaru and *T. dobrogica* Tabacaru (not yet recorded but certainly living in Bulgarian Dobrudja). Among spiders (Araneae), such descendents are species of *Troglohyphantes* (in Bulgaria, the troglobite *T. drenskii* Deltshev and the troglophile *T. bureschianus* Deltshev, both in Western Rhodopes).

Mesogeidean origin is presumed also for the troglobitic Opiliones *Paralola buresi* Kratochvil and *Paranemastoma* (*Buresiolla*) *bureschi* Roewer. The endemic genus *Paralola* (Phalangodidae) and its only species *P. buresi* from the caves near Lakatnik in Western Stara Planina represent in Bulgaria the (mostly tropical) suborder Laniatores. Among terrestrial troglobites, this strange creature gives impression of something really very ancient and alien to the present European fauna. However, Martens (1972) wrote that the Laniatores "should not be considered any more as Tertiary relicts in the European fauna as they are widespread in the areas remaining outside the Pleistocene glaciation". Nevertheless, *Paralola* is beyond doubt a relict; its age is to be considered further. As for *Buresiolla*, it is no more considered a separate genus but a subgenus of *Paranemastoma*. The only known species in Bulgaria is also an endemic of Western Stara Planina.

Pseudoscorpions of the subgenera *Blothrus* (genus *Neobisium*) and *Parablothrus* (genus *Roncus*) also belong to the category of Mesogeidean relicts. In Bulgaria, we know *Neobisium (Blothrus) kwartirnikovi* Mahnert from the cave Duhlata near Bosnek (Vitosha).

According to Guéorguiev (1977), about one-third of all endemic troglobite genera of Cholevidae (Coleoptera) on the Balkan Peninsula are of Mesogeidean origin. In Bulgaria, these are the genera *Netolitzkya, Hexaurus, Radevia, Rhodopiola, Beronia*, and *Bureschiana*.

7·5 Paleoaegeidean (Protoaegeidean) relicts

This group includes: isopod genera *Cyphoniscellus* (species *C. gueorguievi* Vandel) and *Tricyphoniscus* (species *T. bureschi* Verhoeff), both genera from

Trichoniscidae: Haplophthalminae; here also might belong the troglobitic species of the subgenus *Stygiocampa* (Diplura: Campodeidae: *Plusiocampa*), found in Bulgaria in Stara Planina and the Rhodopes.

7·6 North Aegeidean relicts

Most terrestrial troglobites of Balkan Peninsula belong to this category, due to the fact that major parts of former Yugoslavia, of Bulgaria and of Greece were located for long periods during the Tertiary within the landmass of Northern Aegeide (Guéorguiev, 1977).

Many genera of different groups which are listed by Guéorguiev (1977) in the categories of the descendents of Gondwanan phyletic lineages (*Tranteeva*), Laurasian phyletic lineages (*Bulgaronethes*) or of the descendents of Mesogeidean (Paleomediterranean) phyletic lineages (*Balkanoniscus, Bureschia, Buresiolla, Paralola*) or the Paleoaegeidean (Protoaegeidean) relicts (*Tricyphoniscus*) are considered by Guéorguiev (1977) also the North Aegeidean relicts. This assertion needs clarification. Further, as North Aegeidean relicts are considered also the following genera: Isopoda (*Hyloniscus*), Diplopoda (the anthroleucosomatids *Bulgarosoma, Stygiosoma* and the subgenus *Balkandicus*, and the julids *Typhloiulus, Serboiulus*, and *Apfelbeckiella*), and Opiliones (*Siro*). The endemic pseudoscorpion subgenus *Heoblothrus* (genus *Neobisium*) from Stara Planina can be also placed in this group, as well as the species of genus *Balkanoroncus*, not known to Guéorguiev by 1977. These are two species living in the caves of Stara Planina and the Predbalkan, *B. bureschi* and *B. hadzii*. Among beetles, here belong the species of genus *Pheggomisetes* (Carabidae) and the genera *Beskovia, Tranteeviella, Balcanobius* (Cholevidae). We may add the newly described cholevid genera *Beroniella* and *Vratzaniola*; the remarkable genus *Genestiellina* is also a North Aegeidean relict.

7·7 South Aegeidean relicts

The only Bulgarian "*Lindbergia*" ("*L.*" *uminskii* Riedel, a gastropod from Central Stara Planina) is now considered to represent a separate genus *Spinophallus* but still of South Aegeidean origin.

References

Beier, M., 1969. Reliktformen in der Pseudoscorpioniden-Fauna Europas. Memorie della Società Entomologica italiana 48:317–323.

Beron, P., 1972. Essai sur la faune cavernicole de Bulgarie. III. Résultats des recherches biospéologiques de 1966 à 1970. International Journal of Speleology 4:285–349.

Beron, P., 1976. Subdivision zoogéographique de la Stara Planina occidentale (Bulgarie) d'après sa faune cavernicole terrestre. Acta zoologica bulgarica 4:30–37.

Beron, P., 1978. Aperçu sur la composition, l'origine et la formation de la faune cavernicole de la Stara Planina occidentale (Bulgarie). International Journal of Speleology 9 [1977/78]:197–220.

Beron, P., 1994. Résultats des recherches biospéléologiques en Bulgarie de 1971 à 1994 et liste des animaux cavernicoles bulgares. Tranteeva (Sofia) 1:1–137.

Beron, P., T. Daaliev & A. Jalov, 2006. Caves and Speleology in Bulgaria. Pensoft, Sofia.

Beron, P. & V. Guéorguiev, 1967. Essai sur la faune cavernicole de Bulgarie. II. Résultats des recherches biospéologiques de 1961 à 1965. Izvestiya na Zoologicheskiya institut s muzei (Bulletin de l'Institut de zoologie et musée) 24:151–212.

Beron, P., B. Petrov, P. Stoev, 2004. The invertebrate cave fauna of the Eastern Rhodopes (Bulgaria and Greece). In Beron, P. & A. Popov (eds), Biodiversity of Bulgaria. 2. Biodiversity of Eastern Rhodopes (Bulgaria and Greece). Pensoft & National Museum of Natural History, Sofia: 791–822.

Boyadjiev, N., 1964. [Karst basins in Bulgaria and their underground waters]. Izvestiya na Instituta po hidrologiya i meteorologiya (Proceedings of the Institute of Hydrobiology and Meteorology) 2:45–96 (in Bulgarian).

Brignoli, P.M., 1979. Considérations zoogéographiques sur les Araignées cavernicoles de Grèce. Biologia Gallo-Hellenica 8:223–236.

Buresch, I., 1924. [Die Höhlenfauna Bulgariens]. Trudove na Balgarskoto prirodoizpitatelno druzhestvo (Travaux de la Société bulgare des sciences naturelles) 11:143–166 (in Bulgarian).

Buresch, I., 1926. [Untersuchungen über die Höhlenfauna Bulgariens. II]. Trudove na Balgarskoto prirodoizpitatelno druzhestvo (Travaux de la Société bulgare des sciences naturelles) 12:17–56 (in Bulgarian).

Buresch, I., 1929. Die Höhlenfauna Bulgariens. Eine kurze Übersicht der Erforschungen und Zusammensetzung der Höhlenfauna Bulgariens und der darauf bezughabenden Literatur. In Csiki, E. (ed), Xe Congrès International de Zoologie, tenue à Budapest du 4 au 10 septembre 1927. Deuxième partie. Imprimérie Stephaneum S. A., Budapest: 1427–1437.

Buresch, I., 1936. [Übersicht der bisherigen Kenntnisse und Erforschungen der rezenten Höhlen-Fauna Bulgariens). Izvestiya na Balgarskoto peshterno druzhestvo (Mitteilungen der Bulgarischen Speläologischen Gesellschaft in Sofia; Bulletin de la Société Spéléologique de Bulgarie) 1:13–41 (in Bulgarian).

Deltshev, Ch., 1978. The origin, formation and zoogeography of troglobitic spiders of the Balkan Peninsula. Symposium of the Zoological Society London 42:345–351.

Deltshev, Ch., 1996. The origin, formation and zoogeography of endemic spiders of Bulgaria (Araneae). Revue suisse de Zoologie hors serie 1:141–151.

Giachino, P.M., 1992. *Genestiella gueorguievi* n. gen. n. sp. della Bulgaria e considerazioni sistematiche e zoogeographiche sulla linea filetica di *Balcanobius* Guéorguiev, 1965 (Coleoptera, Cholevidae, Bathysciinae). Fragmenta Entomologica 23(2):315–325.

Guéorguiev, V., 1966. Aperçu sur la faune cavernicole de Bulgarie. Izvestiya na Zoologicheskiya institut s muzei (Bulletin de l'Institut de zoologie et musée) 21:157–184 (in Bulgarian).

Guéorguiev, V., 1973a. L'Egéide et la formation de la faune troglobie terrestre en Europe, en Afrique du Nord et en Asie Occidentale. Comptes rendus de l'Académie bulgare des Sciences 26(4):533–535.

Guéorguiev, V., 1973b. Sur la formation de la faune troglobie terrestre dans la péninsule Balkanique durant le Tertiaire. Comptes rendus de l'Académie bulgare des Sciences 26(9):1231–1234.

Guéorguiev, V., 1977. La faune troglobie terrestre de la péninsule Balkanique. Origine, formation et zoogéographie. Éditions de l'Academie bulgare de science, Sofia.

Guéorguiev, V., 1982. [Zoogeographical regions based on terrestrial fauna]. In Galabov, Zh. (ed), Geografiya na Balgariya. Fizicheska geografiya. Prirodni usloviya i resursi (Geography of Bulgaria. Physical Geography. Natural Conditions and Resources). Izdatelstvo na Balgarskata akademiya na naukite (Publishing House of the Bulgarian Academy of Sciences), Sofia: 472–477 (in Bulgarian).

Guéorguiev, V., 1992. [Subdivision zoogéographique de la Bulgarie d'après sa faune cavernicole terrestre]. Acta zoologica bulgarica 43:3–12 (in Bulgarian).

Guéorguiev, V. & P. Beron, 1962. Essai sur la faune cavernicole de Bulgarie. Annales de Spéléologie 17(2/3):285–441.

Hazelton, M., 1970. Fauna from some caves in Bulgaria and one in Yugoslavia. Transactions of the Cave Research Group of Great Britain (Ledbury) 12(1):33–37.

Horaček, I., J. Červeny, A. Taušl & D. Vitek, 1974. Notes on the mammal fauna of Bulgaria (Insectivora, Chiroptera, Rodentia). Vestnik Československé společnosti zoologické 38:19–31.

Jeannel, R., 1944. Les fossiles vivants des cavernes. Ed. Gallimard, Paris.

Jeannel, R., 1960. Situation géographique et peuplement des cavernes. Annales de Spéléologie 14(3/4):333–338.

Leleup, N., 1965. La faune entomologique cryptique de l'Afrique intertropicale. Annales de Musée Royal de l'Afrique Centrale, Sér. Science Zoologique 141:1–186.

Martens, J., 1972. *Ausobskya athos*, der erste Krallenweberknecht aus Griechenland (Opiliones, Phalangodidae). Senckenbergiana biologica 53(5/6):431–440.

Popov, V., 1970. [Distribution of karst in Bulgaria and its features]. Izvestiya na Geografskiya institut (Proceedings of the Geographic Institute) 13:5–19 (in Bulgarian).

Turk, F.A., 1970. Some notes on the Acari and Myriapoda collected by Dr P. Brown and Mr W.G.R. Maxwell in Bulgarian caves. Transactions of the Cave Research Group of Great Britain (Ledbury) 12(1):39–41.

Vandel, A., 1964. Biospéologie. La Biologie des animaux cavernicoles. Ed. Gautier-Villars, Paris.

18 Stygofauna of the Fresh Waters in Bulgaria

Ivan Pandourski
Institute of Zoology, Bulgarian Academy of Sciences, Tsar Osvoboditel Blvd. 1, 1000 Sofia, Bulgaria, e-mail: stygologia@gmail.com

Abstract: The species diversity and distribution of freshwater invertebrate stygofauna in Bulgaria, including 194 species belonging to 78 genera, are reviewed. Typical endemic species and genera for each taxonomic group are listed. Main paleogeographic and paleoclimatic events in Bulgaria that have had an important role in the formation of the stygobiotic faunistic complex are discussed

1 Introduction

The great diversity of karstic regions located from the sea level to the highest peaks of the Pirin Mts. (2914 m a.s.l.), as well as the paleogeographic and paleoclimatic development of Bulgarian landscapes, are the main factors which formed the relatively high biodiversity of underground water ecosystems in Bulgaria. Stygobiotic crustaceans and water mites are groups with the highest and the second highest species diversity, respectively. However, some large taxonomic groups of invertebrates, which inhabit ground waters, are poorly studied. These are, for example, Nematoda, Oligochaeta, and Ostracoda. Studies on the genus *Niphargus* belonging to the higher crustaceans are still in their initial stages. The possible number of species of these crustaceans, which inhabit caves or other underground habitats, could be several times greater than the number of recently discovered species.

At the same time, the large hydrogeological regions of Bulgaria remain totally unstudied (for example, parts of Western Rhodopes, almost the entire Eastern Rhodopes, as well as Slavyanka, Pirin, Belasitsa, Osogovo, Maleshevska and Vlahina Mts.). In other cases, the investigations were of an incidental nature. Examples of an ecological approach (Cvetkov et al., 1982) are rare. Quantitative data concerning stygofauna are also rare. Below, we concentrate on the composition, distribution, and origin of the known taxonomic groups inhabiting the ground waters of Bulgaria.

2 Species Diversity and Distribution

Protozoa: Ciliata. Free living in ground waters, unicellular Protozoa have not been studied in Bulgaria. Only two undetermined epibiotic representatives of ciliates from the genera *Vorticella* and *Tokophrya* have been found on stygobiotic higher crustaceans (Golemansky, 1983).

V. Fet and A. Popov (Eds.), Biogeography and Ecology of Bulgaria, 527–536.

Platyhelmintes: Turbellaria. Depigmented planaria have been found in many caves in Bulgaria, but their genus and species have not yet been identified.

Oligochaeta. Our studies show that Oligochaeta are found in almost all studied karstic and phreatic habitats, but the collected material has not yet been determined. *Haplotaxis bureschi* (Michaelsen) is frequently found in the karstic waters of Western Stara Planina.

Hirudinea. Leeches living in the underground waters of Bulgaria are also poorly studied. The two known species of the genus *Dina* are found mainly in the karstic waters of Western Stara Planina. The classification of *Trocheta bykowskii* Gedroys as a stygobiont is debatable since it also inhabits mountain springs; this species forms isolated populations in some caves in Bulgaria.

Crustacea: Ostracoda. Ostracods inhabit ground waters in almost every region in Bulgaria, but they are still poorly studied. The only determined species inhabiting karstic underground habitats is *Pseudocandona eremia* (Vejdovsky). *Kovalevskiella elegans* Danielopol et Cvetkov is widely distributed in Bulgaria in phreatic lowland waters and also in Strandja Mts. The two species of *Kovalevskiella* are endemic for Bulgaria and inhabit phreatic waters.

Crustacea: Copepoda. The cyclopoid copepods are frequently found in diverse underground habitats within almost every type of interstitial and karstic waters. Two copepod genera, *Acanthocyclops* and *Diacyclops*, have the greatest number of stygobiotic representatives (seven species each); those from the species group "*kieferi*" of the genus *Acanthocyclops* inhabit mainly karstic waters. The species from the group "*languidoides*" of the genus *Diacyclops* have been most frequently found in hyporheic habitats and wells (Pandourski, 1997). *Graeteriella unisetigera* (Graeter) is the only representative of this genus; it mainly inhabits the interstitial habitats. The species from the genus *Speocyclops* are typical for the karstic regions of Western Stara Planina and Predbalkan. *Speocyclops rhodopensis* Pandourski is the only species of this genus found in South Bulgaria and it inhabits karstic crevice waters in the Western Rhodope Mts.

The harpacticoid copepods are the most numerous group of stygobiotic crustaceans found in Bulgaria (48 species) (Apostolov, in press). They have colonized almost all kinds of underground habitats. Mihailova-Neikova (1986) have found 24 harpacticoid species in the karstic waters of Strandja. Mihailova-Neikova (1975) identified 26 taxa from the karstic waters near the village of Bankya, Tran District, in western Bulgaria. The endemic species from the genera *Elaphoidella, Stygoelaphoidella*, and *Neoelaphoidella* are typical for the karstic habitats as well as for the interstitial ones.

Crustacea: Malacostraca. The higher crustaceans are common in the biota of karstic water ecosystems in Bulgaria. Most numerous are representatives of the amphipod genus *Niphargus*, which are found in almost every karstic region. Sixteen species of *Niphargus* and one species of *Niphargopsis* have been described so far. The amphipod *Bogidiella skopljensis* (Karaman) has been found in the karstic waters of Dobrudja, and *B. albertimagni* Hertzog, in the phreatic waters of lowlands surrounding Sofia.

Representatives of the amphipod genus *Gammarus* are commonly found in ground waters. The endemic form *G. pulex cognominus* Karaman et Pinkster has been recorded from the Devetashko Karstic Plateau.

Isopods are also represented in underground ecosystems. Their different genera and families are usually adapted to different underground habitats. Microisopods from the genera *Microcharon* and *Microcerberus* are adapted to life in interstitial conditions and wells, and they are rarely found in karst. Conversely, Stenasellidae and Cirolanidae of large size (above 1 cm) inhabit mainly karstic habitats.

The representatives of Syncarida, which are exclusively subterranean inhabitants, are mainly adapted to life in interstitial conditions. We discovered previously unknown populations in the Duhlata Cave (Vitosha Mts.), in caves of Ponor Mts. (Western Stara Planina), and in the Ponora Cave, near the village of Chiren in Vratsa District. *Bathynella natans natans* Vejdovsky and *Parabathynella stygia* Chappuis are frequently and widely found in Bulgaria, and five species of the genus *Hexabathynella* are typical for the phreatic waters of South Bulgaria.

Acari. Underground water mites are a comparatively well-studied group due to the efforts of Petrova (1968, 1984). They comprise the larger part of the stygobiotic complex in Bulgaria (55 species). The generic diversity is very high (32 genera); 23 genera are represented only by one species each. The fauna of stygobiotic mites in the hyporheic and phreatic waters of Veleka River in Strandja Mts. (Petrova, 1968) has been well studied. Numerous new and endemic species have been described throughout Bulgaria.

Mollusca. Twelve stygobiotic mollusk species are known from Bulgaria. They belong to eight genera (Table 1). This group of invertebrates is very common in the karstic waters of Western Stara Planina, where they form numerous populations with high density.

3 Endemism

In Bulgarian stygofauna we find Carpathian–Balkan, Bulgarian, and local endemics. The specific features of the underground water ecosystems (hydrographic isolation with very restricted possibilities for the migration of stygobionts, etc.) are factors influencing the high level of endemism among the underground water fauna.

The recently discovered underground species of Oligochaeta, *Haplotaxis bureschi* (Michaelsen), is a Balkan endemic species known from the wells near Skopje in Macedonia and the karstic waters of Western Stara Planina. The leech subspecies *Dina lineata arndti* Augener is a regional endemic taxon for Western Stara Planina.

Three species of underground ostracods are endemic: *Kovalevskiella bulgarica* (Danielopol), a regional endemic of western Bulgaria; *K. cvetkovi* (Danielopol), a Balkan endemic; and *Mixtacandona elegans* Danielopol et Cvetkov, a Bulgarian endemic species.

Table 1 The generic composition of the Bulgarian stygofauna

Taxonomic groups	Genera	Number of stygobiotic species
PROTOZOA	*Vorticella* Linnaeus	1*
	Tokophrya Bütschli	1*
OLIGOCHAETA	*Haplotaxis* Hoffmeister	1
HIRUDINEA	*Trocheta* Dutrochet	1
	Dina Blanchard	2
OSTRACODA	*Pseudocandona* Kaufmann	2
	Mixtacandona Klie	1
	Kovalevskiella Klein	2
COPEPODA:	*Eucyclops* Claus	1
CYCLOPOIDA	*Acanthocyclops* Kiefer	7
	Diacyclops Kiefer	7
	Speocyclops Kiefer	4
	Graeteriella Brehm	1
COPEPODA:	*Nitocrella* Chappuis	4
HARPACTI-COIDA	*Nitocrellopsis* Petkovski	1
	Elaphoidella Chappuis	20
	Stygoelaphoidella Apostolov	7
	Neoelaphoidella Apostolov	1
	Parastenocaris Kessler	8
	Ceuthonectes Chappuis	2
	Maraenobiotus Scott	5
SYNCARIDA	*Bathynella* Vejdovsky	1
	Parabathynella Chappuis	1
	Hexabathynella Schminke	5
ISOPODA: Cirolanidae	*Sphaeromides* Dollfus	2
ISOPODA: ASELLOTA	*Asellus* E.L. Geoffroy	1
	Proasellus Dudich	1
	Balkanostenasellus Cvetkov	1
	Stenasellus Dollfus	1
	Protelsonia Mehely	2
	Microcharon Karaman	9
ISOPODA: Microcerberidae	*Microcerberus* Karaman	2
ISOPODA: ONISCIDEA	*Bureschia* Verhoeff	1
AMPHIPODA: Gammaridae	*Gammarus* Fabricius	1
AMPHIPODA:	*Niphargus* Schiodte	16
Niphargidae	*Niphargopsis* Chevreux	1
AMPHIPODA: Bogidiellidae	*Bogidiella* Hertzog	2
AMPHIPODA: Ingolfiellidae	*Ingolfiella* Hansen	1

continued

Table 1 continued

Taxonomic groups	Genera	Number of stygobiotic species
ACARI	*Halacarellus* Viets	3
	Lobohalacarus Viets	1
	Soldanellonyx Walter	3
	Parasoldanellonyx Viets	1
	Porolohmanella Viets	2
	Bandakia Thor	1
	Kawamuracarus Uchida	1
	Atractides Koch	11
	Neumania Lebert	1
	Feltria Koenike	1
	Barbaxonella Lundblad	1
	Albaxona Szalay	2
	Axonopsis Piersig	4
	Erebaxonopsis Motaş et Tanasachi	1
	Ljania Thor	1
	Lethaxona Viets	1
	Kongsbergia Thor	4
	Stygomomonia Szalay	1
	Momonisia Petrova	1
	Mideopsis Neuman	1
	Neoacarus Halbert	1
	Chappuisides Szalay	1
	Stygohydracarus Viets	1
	Phreatohydracarus Tanasachi et Orghidan	1
	Hungarohydracarus Szalay	1
	Balcanohydracarus Motaş et Tanasachi	1
	Arrenurus Duges	2
	Charonothrombium Motaş et Tanasachi	1
	Wandesia Schechtel	1
	Frontipodopsis Walter	1
	Torrenticola Piersig	1
	Aturus Kramer	1
MOLLUSCA	*Belgrandiella* A.J. Wagner	4
	Cavernista Radoman	1
	Hauffenia Pollonera	1
	Iglica A.J. Wagner	1
	Insignia Angelov	1
	Plagigeyeria Tomlin	1
	Pontobelgrandiella Radoman	1
	Saxurinator Schütt	2

* epibiotic species on stygobiotic higher crustaceans

High percentage of endemism is found among Copepoda. All endemic species of the order Cyclopoida belong to three genera: *Acanthocyclops, Diacyclops*, and *Speocyclops*. In karstic waters we frequently find different local and regional endemics from the species group "*kieferi*" of the genus *Acanthocyclops*. These are *A. chappuisi* (Naidenow et Pandourski) from the caves of Zemen Gorge in Struma Valley, *A. radevi* Pandourski from the karst of Ponor Mts. in Western Stara Planina, and *A. strimonis* (Pandourski) from the karst of Bosnek in Vitosha Mts. and Lepenitsa Cave near Velingrad in Western Rhodopes. The Carpathian–Balkan endemic taxa are *A. iskrecensis* Pandourski from Ponor Mts. and Lazareva Pečina Cave in Kučaj Mts. (eastern Serbia) and *A. propinquus* (Plesa) from western Bulgaria and Romania. A Bulgarian endemic from the species group "*crassicaudis*" is *A. fontinalis* (Naidenow), but its taxonomic status is unclear (Pandourski, 1997). A local endemic subspecies from the genus *Diacyclops* is *D. pelagonicus saetosus* Pandourski, found so far only in the Dushnika Cave near the village of Iskrets in Western Stara Planina. *Speocyclops rhodopensis* Pandourski is the only local endemic species of its genus inhabiting Sbirkovata Cave and Lednitsata Cave in the Rhodopes. Widely distributed in the karst of Western Stara Planina is the Balkan endemic species *S. lindbergi* Damian, originally described from Romania.

The harpacticoid copepods from the genera *Elaphoidella, Stygoelaphoidella*, and *Parastenocaris* show the highest percentage of endemism in the karstic waters of Bulgaria. Particularly rich in endemic species is the karst of Ponor Mts. The local endemic taxa *Parastenocaris bulgarica* Apostolov, *Elaphoidella iskrecensis* Apostolov, *Stygoelaphoidella elegans* Apostolov, and *Maraenobiotus parainsignipes* Apostolov live in the zone of the karstic springs near Iskrets. The local endemic *Elaphoidella pandurskyi* Apostolov and the Balkan endemic *Parastenocaris jeanneli* Chappuis have been found in the karstic area near the village of Bosnek. Many local endemic species are known only from one or more caves or springs: *Neoelaphoidella intermedia* Apostolov from Mishin Kamak Cave in Western Stara Planina (Apostolov, 1999), *Stygoelaphoidella stygia* Apostolov from Vasilyovska Mts. (Central Stara Planina), *S. elegans* Apostolov from Ponor Mts., *Elaphoidella balkanica* Apostolov from Vrachanska Mts. (Western Stara Planina), and *E. cavernicola* Apostolov from Ponor Mts. and the karstic region near Karlukovo (Apostolov, 1992).

Higher crustaceans (Malacostraca) also show a high percentage of endemics among the stygobiotic groups. Five species of the syncarid genus *Hexabathynella* are regional endemics from the phreatic waters of South Bulgaria. In the caves of Duhlata and Marina Dupka (Ponor Mts.) we found some representatives of the order Bathynellacea but the species were impossible to determine.

Isopods, particularly the group Asellota, are poorly studied in Bulgaria. Their species number, including the endemic taxa, is probably higher than that actually known. The Bulgarian endemic species *Balkanostenasellus rumelicus* (Cvetkov) from South Bulgaria and the two local endemics *Protelsonia lakatnicensis* (Buresch et Guéorguiev) and *P. bureschi* (Racovitza) from western Bulgaria belong to this group. Isopods of the family Cirolanidae are the largest in size of all cave water crustaceans known in Bulgaria. They are also relicts with a supposed ancient origin dating from the end of the Mesozoic. These are *Sphaeromides bureschi* Strouhal from Western Stara Planina and *S. polateni* Angelov from a karstic spring near Teteven in Central Stara Planina.

Both genera *Microcharon* and *Microcerberus* from the microisopods include regional endemic taxa, distributed mainly in interstitial habitats. They are sometimes found in caves and karstic springs (Ponora Cave near Chiren Village, Zemen Gorge, and the areas of Beglezh and Sadovets).

The only local endemic subspecies of the family Gammaridae is *Gammarus pulex cognominus* Karaman et Pinkster inhabiting the caves of Devetashko Plateau.

The amphipod genus *Niphargus* is typical for the karstic water ecosystems of Bulgaria. Unfortunately, in spite of the efforts of Dr. Stoitze Andreev from the National Museum of Natural History, Sofia (Andreev, 1972; Pandourski, 1998), the species of this genus are still poorly studied. Sixteen species of *Niphargus* have been determined so far, and five of them are local endemic taxa. The representatives of the genera *Bogidiella* and *Ingolfiella* have only one Balkan endemic species each; *B. skopljensis* (Karaman) inhabits the karstic waters of Dobrudja.

The high level of endemism is also observed among the stygobiotic water mites. Bulgarian endemic species are *Mideopsis motasi* Petrova, which is found throughout Bulgaria; *Soldanellonyx chappuisi thracicus* Petrova from the Thracian Lowland and Strandja Mts.; and *Halacarellus phreaticus* Petrova from the phreatic waters near to the Black Sea coast.

Regional endemic taxa for Strandja are *Atractides asticae* Petrova, *A. longiporus* Petrova, *Axonopsis orghidani* Petrova, *A. bureschi* Petrova, *Pontarachna valkanovi* Petrova, and *Momonisia phreatica* Petrova. Petrova suggests that the only species of the genus *Momonisia* from Strandja is a relict form descended from the tropical fauna, which inhabited the Bulgarian territory in the Tertiary.

All recently discovered underground mollusks from Bulgaria are local endemic species. Four gastropod species, two from the genus *Belgrandiella* and one each from the genera *Insignia* and *Pontobelgrandiella*, are known from the karstic waters near Teteven. The remaining species are known from caves or karstic springs in Western Stara Planina (the areas of Lakatnik, Iskrets, and Opitsvet).

4 Origin

No analysis or discussion has been published concerning the origin of the stygobiotic Mollusca, Oligochaeta, Hirudinea, and Ostracoda in Bulgaria. The origin of the stygobiotic crustaceans has been discussed by Cvetkov (1967, 1972, 1975, 1982), Pandourski and Breskovski (1995), Pandourski (1997), and Pandourski and Ognjanova (2001). In terms of their origin, the stygobiotic crustaceans are divided in two main groups: species with a marine origin and those with a freshwater origin.

To the first group belong mainly the higher crustaceans. The colonization of underground karstic waters by Cirolanidae and Stenasellidae is related to the paleogeographic situation at the time when Stara Planina (the Balkanid arch) was formed during the Upper Cretaceous. During this period, direct contact between the paleokarstic systems and the regressing marine basins occurred. Both Bulgarian species of the genus *Sphaeromides* and both Bulgarian species of the genus *Protelsonia* are relicts, and their ancestors should have inhabited parts of the northern Tethys (Pandourski and Breskovski, 1995).

The formation of the system of Neogene lake basins along the valleys of Struma, Mesta, and Maritsa (Galassi et al., 1995; Pandourski and Ognjanova, 2001) has been important for the origin of the relict endemic species of *Hexabathynella* and *Microcharon* from South Bulgaria. At some stage of their development, these lakes have been affected by marine waters. The brackish Miocene basins have influenced the region of Strandja, located near to the recent south Bulgarian Black sea coast (Pandourski and Ognjanova, 2001).

Two stages of paleogeographic and paleoclimatic development of Bulgarian territories were important for the species with freshwater origin, especially for the lower crustaceans such as copepods. The first stage is connected with the cool climate in Europe in the end of the Tertiary and with the beginning of glacial periods during the late Pliocene and early Quaternary. At that time, the surface populations of many boreal groups of species and genera are believed to have disappeared. The colonization and isolation of part of these populations is a factor in the origin of vicariant species and their divergence (Pandourski, 1997). Such Tertiary paleoendemics are found in the species group "*kieferi*" of the genus *Acanthocyclops*, in the genera *Speocyclops, Elaphoidella, Stygoelaphoidella*, etc.

The second stage is related to the retreat of the Würm Glacial Period, accompanied by climate warming. The populations of the cold-preferring species remained isolated in the ground waters. The neoendemics from the genus *Diacyclops* probably have such an origin.

The origin of part of the species and genera of underground water invertebrates is still being debated. For example, there are various opinions about the marine or freshwater origin of the amphipod genus *Niphargus*. Petrova (1968) examines the origin of the stygobiotic water mites of Strandja Mts., which is the most thoroughly studied region in Bulgaria. She concludes that the most ancient group of "paleophreatic" mites originates from the Paleogene, when Strandja had already emerged as land surrounded by sea. After the formation of underground populations they became isolated. As a result, the entire subterranean families and subfamilies evolved: *Kawamuracarus* (Limnesiidae), Hungarohydracaridae, and Chappuisididae.

For the so-called "neophreatic" water mites, Petrova (1984) suggests that they originated during the post-Tertiary stage, when the climate was cooler. Part of the populations still exhibits similarities to their surface ancestors. Many transitional species ranging from surface to strictly underground species are typical for these water mites. An example is the genus *Atractides*, for which 22 species are known from Strandja, including 11 stygobionts.

5 Conclusions

At this moment, 194 species of stygobiotic animals belonging to 78 genera are known from the underground continental waters of Bulgaria (Table 1). The crustaceans have the highest species diversity with 121 species, followed by the water mites with 55 species. The most ancient relict representatives are the crustaceans whose ancestors are believed to have inhabited the North Tethys during the Upper Cretaceous. Climatic changes leading to a general cooling and mountain formation (orogenesis) in Bulgarian territory during the Tertiary had an important role in the origin of the paleoendemics with freshwater origin. After the retreat of the Würm Glacial Period and warming of the climate many populations of cold-preferring species remained isolated in the underground water habitats and gave rise to many neoendemics.

The Western Stara Planina is characterized by the presence of a relatively high percentage of endemic stygobionts. Particularly rich is the endemic acarofauna of Strandja. Less number of endemics is found in the Rhodopes; however, we have to consider here the fact that large regions of Bulgaria are poorly studied.

The study of biodiversity in the ground waters of Bulgaria is still in its early stages, and is limited to the available taxonomists. Importance of the stygofauna as a part of the natural heritage of Bulgaria has still not been evaluated, and this fauna is excluded unreasonably from the National Action Plan for the Protection of Biodiversity (2000).

Acknowledgments

I am grateful to Rumyana and Martin Whitcher for their help in translating this article into English.

References

Andreev, S., 1972. Diffusion du genre *Niphargus* en Bulgarie et notes taxonomiques sur *Niphargus bureschi*. Actes du I^{-er} Colloque international sur le genre Niphargus, Verona, 15–19 aprile 1969; Museo Civico di Storia Naturale di Verona, Memorie fuori Serie 5:61–66.

Apostolov, A., 1992. Copépodes harpacticoïdes cavernicoles de Bulgarie. I. Trois nouveaux représentants du genre Elaphoidella Chappuis, 1929. Annales de Limnologie 28(2):121–130.

Apostolov, A., 1999. Copépodes harpacticoïdes cavernicoles de Bulgarie. 7. *Neoelaphoidella intermedia* n. sp., un nouveau représentant du genre *Neoelaphoidella* Apostolov, 1985. Bollettino Museo Civico di Storia Naturale di Verona 23:211–219.

Apostolov, A., in press. [Copepoda, Harpacticoida (Harpacticoids of fresh waters)]. In Fauna na Balgariya. 29 (Fauna bulgarica. 29). Akademichno izdatelstvo "Marin Drinov" (Editio academica "Marin Drinov"), Sofia (in Bulgarian).

Cvetkov, L., 1967. [Au sujet de la faune des *Stenasellus* de la péninsule Balkanique et sur l'origine des *Stenasellus*]. *Izvestiya na Zoologicheskiya institut s muzei* (*Bulletin de l'Institut de zoologie et musée)* 23:139–165 (in Bulgarian).

Cvetkov, L., 1972. [Crustacés des réseaux hydrographiques des rivières Strouma, Mesta et Maritza]. Izvestiya na Zoologicheskiya institut s muzei (Bulletin de l'Institut de zoologie et musée) 36:5–22 (in Bulgarian).

Cvetkov, L., 1975. [Composition, position zoogéographique et origine des crustacés thalassophréatiques en Bulgarie]. Hidrobiologiya (Hydrobiology) (Sofia) 1:35–53 (in Bulgarian).

Cvetkov, L., 1982. Podzemnata hidrobiosfera i biologichnata evolyutsiya (Underground Hydrobiosphere and Biological Evolution). Izdatelstvo na Balgarskata akademiya na naukite (Publishing House of the Bulgarian Academy of Sciences), Sofia (in Bulgarian).

Cvetkov, L., Gruncharova, T., Petrova, A. & M. Mihailova-Neikova, 1982. [The structure and dynamics of underground water communities. II. Phreatic fauna of Dobrudja and stygobiological studies on ground karst waters]. Hidrobiologiya (Hydrobiology) (Sofia) 16:3–19 (in Bulgarian).

Galassi P., G.L. Pesce & P. de Laurentis, 1995. The genus *Microcharon* Karaman (Isopoda Janiroidea) in the Balkan peninsula. Biogeography of the *major* and the *profundalis* groups. Mémoires de biospéologie 22:25–33.

Golemansky, V., 1983. Etat actuel et perspectives des études des protozoaires cavernicoles. In Dinev, L. (ed), European Regional Conference on Speleology, Sofia – Bulgaria, September 22–28, 1980. Proceedings. Volume I. Bulgarian Tourist Union, Bulgarian Federation of Speleology, Union of the Scientific Workers in Bulgaria, Sofia: 128–131.

Mihailova-Neikova, M., 1975. Contribution to the study of Harpacticoida (Copepoda, Crustacea) of karstic sourses near the village of Bankya, Trun district (West Bulgaria). Hidrobiologiya (Hydrobiology) (Sofia) 2:66–79.

Mihailova-Neikova, M. & A. Apostolov, 1986. Harpacticoida (Crustacea, Copepoda) des eaux montagneuses de Bulgarie. IV. Les eaux souterraines de la montagne Strandža. Acta zoologica bulgarica 32:36–47.

National Action Plan for the Protection of Biodiversity, 2000. Ministry of Environment and Waters, Sofia.

Pandourski, I., 1997. Composition, origine et formation de la faune cyclopidienne stygobie de Bulgarie. Définition du groupe d'espèces "*kieferi-stygius*" du genre *Acanthocyclops* (Crustacea, Copepoda, Cyclopoida). Bollettino del Museo regionale di Scienze naturali – Torino 15(2):279–297.

Pandourski, I, 1998. Catalogue of stygobiontic higher crustaceans (Crustacea, Malacostraca) from Bulgaria. Acta zoologica bulgarica 50(1):23–33.

Pandourski, I. & S. Breskovski, 1995. Origine des Isopodes Cirolanides et Stenasellides dans les eaux souterraines karstiques de la montagne de Stara Planina occidentale et moyenne (Bulgarie) en relation avec l'histoire paléogéographique de la région. Mémoires de biospéologie 22:115–119.

Pandourski, I. & N. Ognjanova, 2001. Le genre *Hexabathynella* Schminke, 1972 (Crustacea, Syncarida, Bathynellacea) dans les eaux souterraines de la péninsule Balkanique: distribution et remarques paléozoogéographiques. Historia naturalis bulgarica 13:69–77.

Petrova, A., 1968. Hydracariens souterrains de Bulgarie (Hydrachnellae, Acari). III. La région de Strandja montagne. Izvestiya na Zoologicheskiya institut s muzei (Bulletin de l'Institut de zoologie et musée) 28:47–93.

Petrova, A., 1984. Origine et formation des Acariens stygobiontes. Hidrobiologiya (Hydrobiology) (Sofia) 22:3–24.

19 Biodiversity and Ecology of the Bulgarian Black Sea Invertebrates

Vassil Golemansky

Institute of Zoology, Bulgarian Academy of Sciences, Tsar Osvoboditel Blvd. 1, 1000 Sofia, Bulgaria, e-mail: golemansky@zoology.bas.bg

Abstract: A brief morphological and hydrochemical characteristic of the Bulgarian Black Sea coast is given. A brief historical survey of the investigations on the animal biodiversity of the Bulgarian Black Sea coast is followed by the data about the present status of the Bulgarian Black Sea fauna. So far, more than 2250 species and varieties of animals and Protozoa have been recorded. Among them, protozoans include over 670 species and varieties; arthropods, ca. 746; mollusks, more than 140; annelids, ca. 140; chordates, 166, etc. Additional data about the density, biomass, and distribution of the widespread Black Sea inhabitants are given. The impact of the invasive animals on the Black Sea ecosystem, and some ecological problems and threats to the Black Sea biodiversity are also discussed

1 Introduction

The Black Sea is the largest intercontinental sea in the world. Its surface is about 413,500km^2; the maximum depth, 2122 m; and the average depth, 1240 m. The Black Sea is situated in the semiarid climatic zone between 46°32′ N and 40°55′ N, and 27°27′ E and 41°32′ E (Mishev and Popov, 1978). Due to this geographic position, the annual water evaporation (332–392 km^3/year) exceeds the average annual rainfall in the sea (225–300 km^3/year). Two basic inflows enter the Black Sea: fresh water from the large rivers flowing into the northwestern part of the sea (Danube, Dniester, Bug, Dnieper, Don) and sea water with the salinity typical to oceans, which penetrates through Bosporus from the Aegean and Marmara seas. Different salinity and density of these two flows and absence of the vertical currents in the sea are the reasons of weak vertical mixing of the water masses and the creation of the unique vertical structure of the Black Sea. Two water layers are formed: a massive deep layer with salinity of 22–22.5‰, and a lighter surface one with salinity 15–18‰ (Rozhdestvenskiy, 1986) (Fig. 1).

One of the specific ecological characteristics of the Black Sea is the hydrogen sulphide formation in the deep waters of the sea (from 120–150 m to the bottom). Because of this anoxic zone, a huge part of the water volume of the Black Sea is uninhabitable, and living organisms are found only in 12–13% of it. There are many hypotheses about the origin of the toxic gas, among which the hypotheses of Andrusov and Lebedintsev are the most popular (Valkanov and Rozhdestvenskiy, 1978). According to Andrusov, the hydrogen sulphide in the Black Sea and swamps and lakes connected with it has been formed through the decomposition of the

V. Fet and A. Popov (Eds.), Biogeography and Ecology of Bulgaria, 537–553.

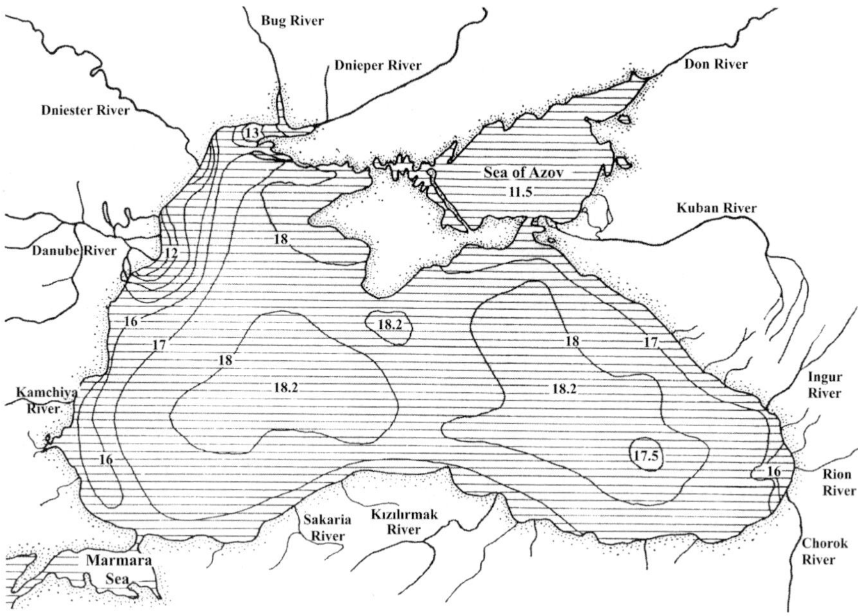

Fig. 1 The largest rivers flowing into the Black Sea and the salinity of the surface waters (after Valkanov and Rozhdestvenskiy, 1978).

organic gradients of the freshwater flora and fauna which died after the salinization and formation of the sea. According to the Lebedintsev's hypothesis, the origin of H_2S is inorganic, through the bacterial reduction of sulphates by microorganisms, mostly by *Microspira desulfuricans* (= *Desulfovibrio desulfuricans*). Formation of the hydrogen sulphide took place during the last 9,000–10,000 years of the Black Sea history. During the last 30 to 40 years a negative trend of the sulphide zone rising to the depth of 90–120 m has been established in some coastal areas. This is a serious threat for the biodiversity and sustainability of the labile Black Sea ecosystem.

Because of the lower salinity of the Black Sea, some marine taxonomic groups are absent (e.g. Cephalopoda), or only few species of them were found (e.g. Radiolaria, Echinodermata).

An important ecological peculiarity, which impacts negatively on the ecology of the Black Sea basin and the normal functioning of its ecosystem, is presence of the great human population in its watershed. There are 16 countries with 160 million inhabitants, which introduce an enormous quantity of organic substances and industrial wastes in the Black Sea through the rivers flowing into it (Fig. 2). The pollution of the Black Sea ecosystem became especially active after the World War II and it is still a central problem now, at the beginning of the 21st century. For example, before 1980 about 60,000 tons of phosphorus and over 300,000 tons of nitrogen had been discharged annually into the sea.

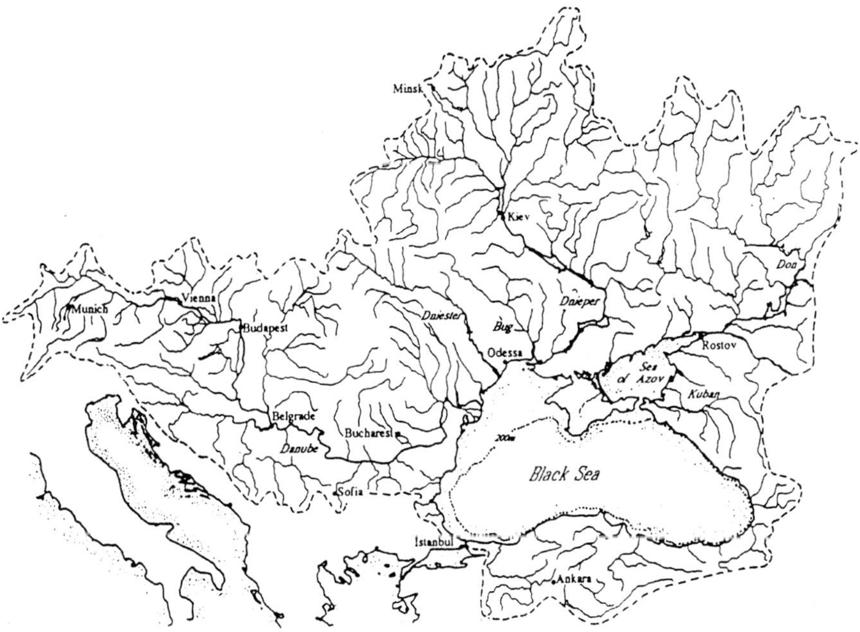

Fig. 2 The watershed of the Black Sea (after Mee, 1992).

In the 1990s these quantities doubled, reaching 740,000 tons per year (Mee, 1992; Leppäkoski and Mihnea, 1996).

2 Bulgarian Black Sea Coast

Bulgaria is situated in the western part of the Black Sea coast. The Bulgarian coast lies between the country's borders with Romania in the north and with Turkey in the south. Its total length is 371 km. Old and young cliffs form 53% of the coastal line, 34% of it is occupied by large and small beach strips, and the remaining 13% are abrasion landslide coasts (Mishev and Popov, 1978). The beach strips are formed mainly of quartz-carbonate sands. About 60 rivers and streams discharge their waters in the Bulgarian part of the Black Sea: some flow directly to the sea, and others, to the coastal lakes. Kamchiya, Hadjiiska, Veleka, and Rezovska rivers are the main ones entering the Black Sea from the Bulgarian aquatory (Fig. 3). There are 18 lakes and swamps at the Bulgarian coast. The Bulgarian coastal lakes have firth or lagoon origin. Some of them are constantly connected with the sea (Varnensko, Burgasko, Atanasovsko lakes) and the other are isolated temporarily or constantly by sand banks (Alepu). The small covered lakes like Ezerets, Shabla, Durankulak, etc. are almost freshwater, and their waters are very alkaline because of the karst spring supply. Some of the lakes connected

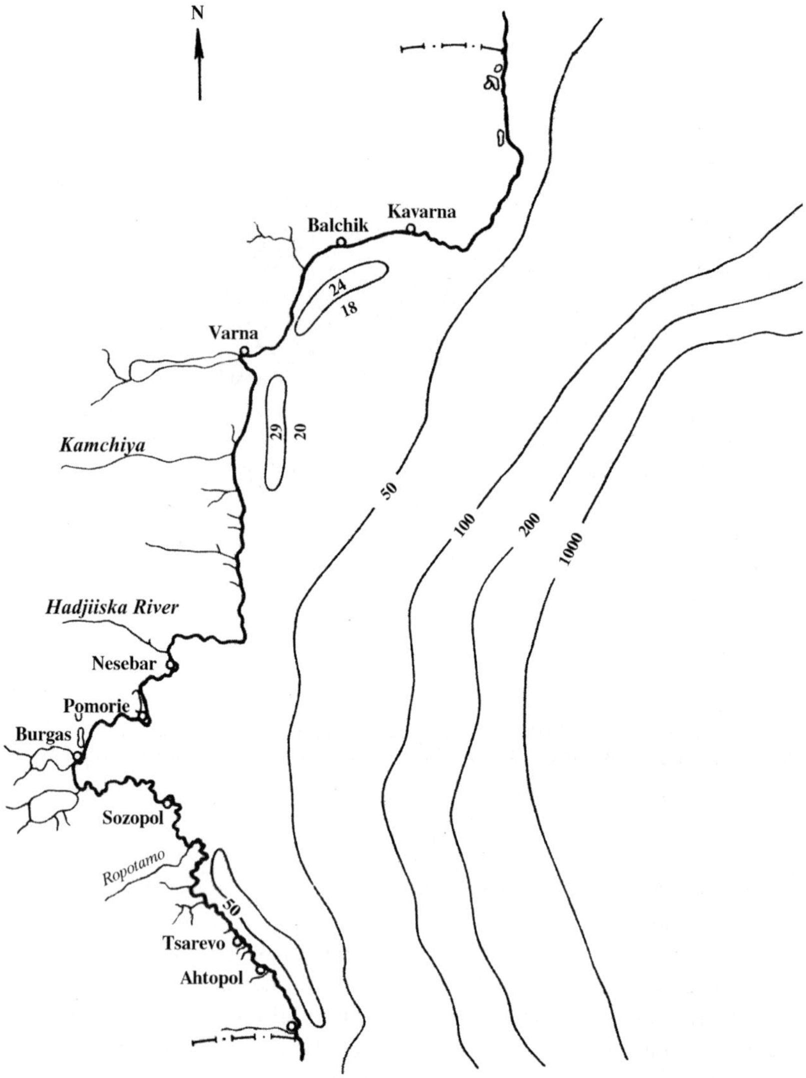

Fig. 3 The Bulgarian Black Sea coast: main rivers and settlements (after Rozhdestvenskiy, 1986).

with the sea (Pomoriisko and Atanasovsko) are hypersaline with the salinity over 100‰.

The average annual temperature of the northern part of the coast (the region of Cape Shabla) is ca. 11.5 °C, and of the southern part (the region of Tsarevo), is ca. 13 °C. The coldest month is January with an average temperature about 0.2 °C in the north (Cape Shabla), and 3.2 °C in the south (Tsarevo). The warmest month is July, with average temperatures respectively 22.5 °C and 23.5 °C (Rozhdestvenskiy,

1986). According to the last author, the average annual rainfalls in the northern part of the Bulgarian Black Sea coast (Balchik) are 443 mm, and 682 mm in the southern part (Tsarevo). The sea currents along the Bulgarian coast are changeable and depend on the wind strength. Freezing of the sea is a rare phenomenon; in the course of the 20th century it has been registered only three times: in 1929, 1942, and 1954.

3 Animal Biodiversity

3·1 Brief history of the studies

The history of the Bulgarian studies of the Black Sea animal biodiversity is comparatively brief. The first scientific report about finding of 12 halacarid species (Arachnida: Acari) in the Bulgarian coast was published by Chichkoff (1907). A few years later, Chichkoff (1912) generalized his investigations on the Bulgarian Black Sea fauna, and reported a total of 149 species and varieties of worms, mollusks, crustaceans, fishes, etc.

Founding of the Marine Biological Station in Varna in 1932 was of the great importance for the development of the studies of the Black Sea fauna. It became a basic research center, which investigated the flora and fauna of the Black Sea, the hydrobiology and hydrochemistry of the sea coast and the lakes connected with the sea. At the same time, the second research center was established in Sozopol, the Experimental Ichthyological Station, the main goal of which was studying the species diversity, distribution and biology of the fish fauna in the Bulgarian part of the Black Sea. The names of the Bulgarian zoologists G. Paspalev, A. Valkanov, A. Nechaev, K. Bulgurkov, A. Zhelyazkova-Paspaleva, etc. were closely related to the investigations on the biodiversity of the Black Sea coast at that time. The period of faunistic studies of different Black Sea animal groups lasted up to the World War II. In 1957, Valkanov summarized the existing results of these faunistic studies in the "Catalogue of our Black Sea fauna". The catalogue comprises data about 1375 species and varieties of animals and Protozoa inhabiting the open sea aquatory, as well as in the coastal brackish basins (lakes, swamps, and river mouths) closely connected with the Black Sea.

The research on the Black Sea and its Bulgarian coast fauna by Bulgarian and foreign specialists became especially intensive after the World War II. In 1954, a new scientific research institute was established on the basis of the Marine Biological Station in Varna. The main aim of this institute was the study of the flora and fauna of the Black Sea, its biological resources, and possibilities of their exploitation. Later, this institute developed into the Institute of Fisheries, which investigated the fish resources of the Black Sea and the World Ocean. In 1983, an Institute of Oceanology of the Bulgarian Academy of Sciences was founded, its main task being the research on geomorphology, hydrology, hydrochemistry, biodiversity, and ecology of the Black Sea, and conservation of the Black Sea ecosystems. At present, the specialists of the Institute of Zoology of the Bulgarian Academy of Sciences, Sofia University, Plovdiv University, and Burgas University also participate in the faunal studies of the Black Sea and Bulgarian Black Sea coast.

3·2 Present status of species diversity

The data on animal and Protozoa diversity of the Bulgarian Black Sea and its coast (studied up to 1985) were summarized by Marinov and Golemansky (1989). The numbers of species and varieties for major taxonomic groups according to Valkanov (1957), Valkanov and Marinov (1964), and Marinov and Golemansky (1989) are presented in Table 1. This table, as well as its sources, incorporates not only strictly marine species but also brackish ones. Ten years later, Konsulov (1998) published generalized data on biodiversity for the Bulgarian Black Sea coast, and discussed the ecological status of the main communities and ecosystems of the Bulgarian shelf.

So far, over 670 species and varieties of Protozoa have been recorded for this area. A large number of these taxa have been described for the first time such as the families Labyrinthodyctiidae Valkanov, 1969 (Labyrinthomorpha) and Psammonobiotidae Golemansky, 1974 (Sarcomastigophora), as well as over 90 new genera and species from different taxonomic groups.

A specific rhizopod community inhabiting the coastal underground waters of the sandy beaches has been found for the first time in the Black Sea. Later, it has been

Table 1 Number of species, subspecies, and varieties of animals and Protozoa from the Bulgarian Black Sea and its coast recorded by 1985

Taxonomic groups	Valkanov (1957)	Added by Valkanov & Marinov (1964)	Added by Marinov & Golemansky (1989)	Total
Protozoa	348	35	292	675
Porifera	29	–	–	29
Coelenterata	39	1	1	41
Platyhelminthes	36	5	17	58
Nemathelminthes	33	–	83	116
Annelida	107	16	17	140
Rotatoria, Nemertini, Acanthocephala, Gastrotricha, Kinorhyncha	77	22	19	118
Bryozoa	12	–	8	20
Mollusca	122	17	3	142
Arthropoda	415	43	288	746
Echinodermata	3	1	–	4
Chaetognatha	3	–	–	3
Chordata	151	12	3	166
Total	1375	152	731	2258

Note. The total numbers also include the species identified only by genus discovered in other regions of the World Ocean (Valkanov, 1970; Golemansky, 1970a, 1970b, 1970c, 1971, 1974, etc.). About 30 species of the class Foraminifera have been identified along the Bulgarian Black Sea coast, including the dominant species of the genera *Ammonia, Elphidium*, and *Quinqueloculina*.

Over 90 species of the order Dinoflagellida have been also registered in the Black Sea, forming the main part of the plankton. Some of them (for example, *Exuviella cordata*) are the cause of the protozoan "blooms" of the sea in front of the Bulgarian coast in the last decades. As a consequence of an intensified eutrophication of the sea, a significant increase of the biomass of *Noctiluca scintillans* has been measured after 1980. This species consumes zooplankton and negatively affects the food productivity of the marine ecosystem.

About 300 species of ciliates (Ciliata) have been observed in the Bulgarian Black Sea sector. Most of them are benthic, especially psammal, inhabitants. Among Tintinnidae (mainly of the genera *Tintinnidium* and *Tintinnopsis*), 22 species have been recorded for the Bulgarian coast. They have important role as a trophic base for the plankton-consuming fishes.

In the Bulgarian area of the Black Sea, only 22 species of Porifera have been found. They were studied mainly by Arndt (1947) and Caspers (1951). The freshwater species *Ephidatia fluviatilis*, which can withstand the salinity of 9‰, has been found in some brackish coastal lakes.

The phylum Cnidaria is represented predominantly by Hydrozoa. The main part of those is fouling organisms, found on the underwater substrates and hydrotechnical constructions (*Corine tubulosa, Aglaophenia pluma, Campanularia johnstoni*, etc.). Only three species of scyphozoans have been found. One of those, *Aurelia aurita*, reaches large quantities. The largest Black Sea jellyfish *Rhizostoma pulmo* (=*Pilema pulmo*) also occurs frequently (Fig. 4). Four species of the Anthozoa are found at the Bulgarian coast, of which most common is *Actinia equina* (Fig. 5). *Pachycerianthus solitarius* is a typical inhabitant of the community of the Phaseolina silt at the Bulgarian Sea shelf (Valkanov and Marinov, 1978).

About 450 species of various worm groups (mainly free living) have been found in the Bulgarian aquatory of the Black Sea. The annelids, especially polychaetes (Polychaeta), are best studied. They are a dominant component of the benthic community and take an important place in the food chain of the Black Sea benthic fish; 104 species of Polychaeta have been recorded along the Bulgarian coast (Fig. 6). Their distribution, density, biomass and importance in the Black Sea ecosystem have been well studied. According to Marinov (1977, 1990), about 35% of the marine polychaetes are Mediterranean immigrants and are well adapted to the Black Sea ecosystem. About 20 polychaete species are common, and form the basic benthic biomass (*Melinna palmata, Phyllodoce mucosa, Polydora ciliata, Nereis zonata, N. diversicolor, Nephthys hombergii*, etc.).

Arthropods (Arthropoda) in the Bulgarian Black Sea fauna are represented by 750 taxa, majority of which is benthic inhabitants. Crustaceans are dominant among arthropods in both biomass and density. The order Harpacticoida (ca. 200 species) is the richest in species diversity among the benthic crustaceans. Their

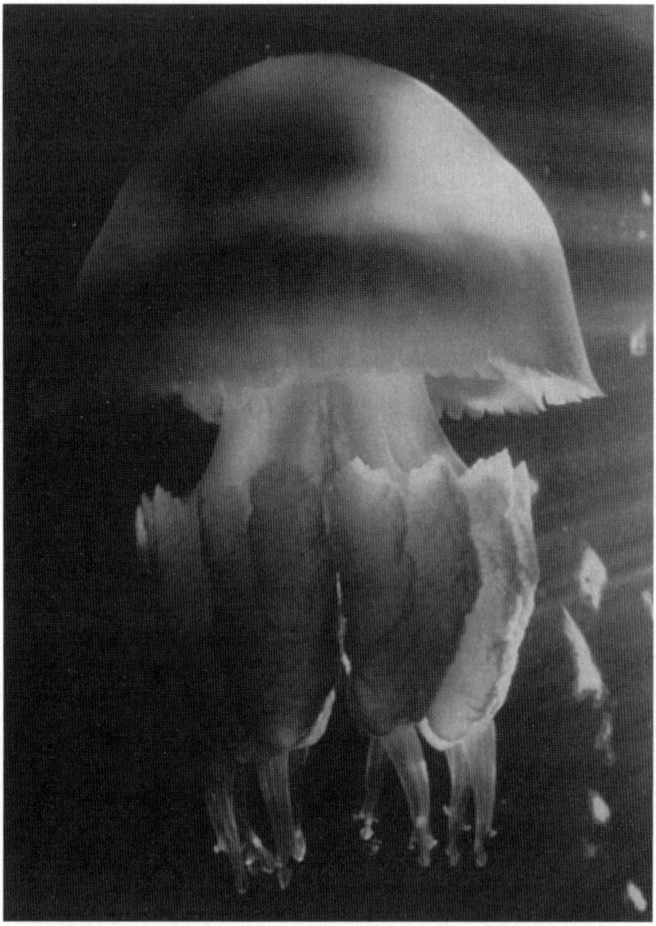

Fig. 4 *Rhizostoma pulmo* (= *Pilema pulmo*) (Scyphozoa), a common jellyfish in the Black Sea. Photo by L. Klissurov.

density and biomass are especially high in the habitats of the *Cystoseira* fouling community (up to 2000–3000 specimens/m^2). The species of the barnacle genus *Balanus* (Cirripedia) are widespread and inhabit stones, rocks, and various artificial surfaces (ship hulls, harbor constructions, etc.). According to Marinov (1990), the biomass of *B. improvisus* reached 3 kg/m^2 in some regions. The species of the orders Cladocera and Copepoda dominate among the pelagic crustaceans. They form the main mass of zooplankton, and have a significant role as the food base for the pelagic fish. The isopods, amphipods, and decapods are best represented among the Malacostraca. The main mass of isopods is formed by the species of the genus *Idotea*, which grow in fouling communities of algae, and the species of *Sphaeroma*, which inhabit the rock coastal zones. The amphipod fauna is the richest

Fig. 5 *Actinia equina* (Anthozoa). Photo by L. Klissurov.

in the zones of fouling algae. The density of amphipods in these zones is about 3800 specimens/m^2 and their biomass is about 1.4 g/m^2. Other frequently occurring amphipod genera are *Pleonexes, Stenothoe, Amphithoe*, etc. (Kaneva-Abadjieva, 1965, 1973). The total number of decapod species along the Bulgarian coast is 30. The most common are *Pachygrapsus marmoratus, Macropipus holsatus, Crangon crangon, Diogenes pugilator*, etc. (Marinov, 1990). The most interesting species from the practical point of view are the prawns *Palaemon elegans* and *P. adspersus*, and the crayfish *Astacus leptodactylus*, which live in some of the coastal lakes. Until 1960, the catch of these species reached tens of tons per year, but later their number decreased. In the last ten years, the density of the mentioned species has increased again, and *A. leptodactylus* now is a subject of farming in Bulgaria. After

Fig. 6 *Terrebellides stroemi* (Polychaeta), a species inhabiting the Mytilus and Phaseolina silt. Photo by T. Marinov.

the World War II, some new decapods have been recorded in the Bulgarian part of the Black Sea (e. g. *Callinectes sapidus, Alpheus dentipes*, and *Rhithropanopeus harrisi*), which probably have penetrated through the Mediterranean Sea. Some specimens of *Homarus vulgaris* have been caught rarely and it is supposed to be a Marmara Sea immigrant, which has entered the Black Sea through the Bosporus.

The mollusks (Mollusca) are represented by 145 species and varieties of the classes Gastropoda, Bivalvia, and Polyplacophora. The frequently occurring gastropods in the coastal rock zone, covered with fouling *Cystoseira barbata*, are *Gibbula divaricata, Tricolia pullus, Bittium reticulatum*, and *Rissoa splendida*. The snail *Rapana venosa* have been introduced in the Black Sea after the World War II, and in 1958 it has been recorded for the first time near the Bulgarian seashore (Kaneva-Abadjieva, 1958). Only in a few years the density of the species has increased and its predation pressured the population of the blue mussel *Mytilus galloprovincialis*. As a result the natural resources of *M. galloprovincialis* in the Black Sea have considerably declined. *Rapana venosa* occurs on both rock and silt bottom to the depths of 25–30 m. However, in the recent ten years the population of *Rapana venosa* has decreased because of the commercial interest in this species as a food item and the reduction of the blue mussel resources. Another widespread Black Sea gastropod is *Nassarius reticulatus,* which inhabits mostly sandy bottom along the beaches, and rarely could be found in the deep zones with a silt bottom.

The mussels (Bivalvia) are represented by 50 species. A common bivalve is the small black mussel *Mytilaster lineatus*. It is widely distributed and develops in large quantities on the algae *Cystoseira barbata. Chamelea gallina* (Fig. 7) also has high levels of occurrence on the sand and silt-sandy bottoms along the Bulgarian

Fig. 7 *Chamelea gallina* (Bivalvia) has high density in the sand and silt-sand regions of the Bulgarian Black Sea coast. Photo by T. Marinov.

coast. In some regions it has high density and biomass, for example, in Burgas Bay it reaches up to 2700 specimens/m^2 and $17,870\,g/m^2$ (Marinov, 1990). *Spisula subtruncata* is a common species found at the depth from 19 to 50 m near the Bulgarian seashore. Its density is 175 specimens/m^2 and the biomass is $85\,g/m^2$. The blue mussel (*Mytilus galloprovincialis*) is common in the open sea at the depth of 60 m and its stocks reach about 300,000 tons. *Modiolula phaseolina* develops in large quantities in the community of the Phaseolina silt at the depth from 50 to 120–130 m. Its density reaches 12,000 specimens/m^2 and the biomass, $710\,g/m^2$ (Marinov, 1990).

Due to the low salinity of the Black Sea, the phylum Echinodermata is represented only by four species. The most common is *Amphiura stepanowi* (Ophiura). It inhabits the silt bottom at the depth below 50–60 m, where its density reaches 400 specimens/m^2.

Among other phyla and classes of invertebrates found along the Bulgarian coast are: Ctenophora (three species), Nemertini (six), Bryozoa (20), and Chaetognatha (three). Most of them inhabit restricted habitats and have low density.

The parasitic fauna of the Black Sea inhabitants is poorly studied. There are data only on some groups of ectoparasitic Ciliata, Trematoda (eight species), Cestoda (six), Nematoda (four), and Acanthocephala (three), but their number probably is higher.

Among chordates, 166 species have been recorded along the Bulgarian Black Sea coast, of which fish represent over 140 species. According to "Red Data Book of the People's Republic of Bulgaria" (Botev and Peshev, 1985) and Konsulov (1998), *Knipowitschia longicaudata* (Pisces) has completely disappeared from the Bulgarian Black Sea fauna. Quite recently it has inhabited Beloslavsko Lake and Provadiiska and Devnenska rivers. It is considered to be a Pontian relict of the Caspian Sea. There are 13 species of endangered vertebrates in the Black Sea and coastal basins, of which 11 are fish, and two are mammals (the monk seal *Monachus monachus* and the porpoise *Phocaena phocaena*).

3·3 Ecological problems and threats

In the recent four decades, the Black Sea has endured a severe ecological crisis. The Romanian oceanologist Gomoiu (1985) called it the "ecosystem pathology". Another scientist talked about "a grave case of ecosystem distress". Mee (1992) wrote: "The Black Sea is the first one suffering a collapse due to the unscrupulous commercial exploitation of its bottom flora and fauna".

The eutrophication of the Black Sea basin is the main reason for its ecological crisis. One of the unfavorable consequences of the eutrophication are the phytoplankton "blooms". In the last two decades such kind of "blooms" occurred repeatedly every year along the west coast of the Black Sea, including the Bulgarian littoral. According to Mihnea (1997), in 1960–1970 the biomass of the phytoplankton increased five- to tenfold compared to the years prior to 1960. A negative tendency for decrease of the diatom algae at the expense of cyanobacteria, dinoflagellids, and chrysomonads has been registered. A similar trend has been observed for the density of the unicellular organism *Noctiluca scintillans* (Dinoflagellida). According to Leppäkoski and Mihnea (1996) the biomass of *Noctiluca* in the northwest region of the Black Sea, estimated at 260,000 tons in the 1950s and 1960s, increased to 17 million tons in the 1980–1985. *Noctiluca* is one of the most serious reducers of the useful zooplankton in the Black Sea ecosystem. The cited authors provided similar data about dynamics of the growth of another predator and zooplankton consumer, the jellyfish *Aurelia aurita* (Scyphozoa). In the 1950s and 1960s the biomass of *Aurelia* was estimated at 0.1–16 million tons, while in the 1980s it reached 300–500 million tons. *Aurelia aurita* causes serious problems for fisheries and coastal tourism.

The aggravation of ecological conditions in the Black Sea impacts the benthic fauna as well. The decline of populations of many benthic invertebrates (Crustacea, Mollusca, Polychaeta), which play a significant role in the food chain of the benthos-consuming fish, has been established in the last few decades. An exception is *Mercierella enigmatica* (= *Ficopomatus enigmaticus*) (Polychaeta), whose density has increased; however, this species grows on coastal substrates and is inaccessible for the benthos consuming fish. According to Tiganus (1992), of 79 benthic species recorded in 1960, only 16 were found in the 1980s. Zaitsev (1992) reported that the biomass of the blue mussel *Mytilus galloprovincialis* from 25 million tons in the 1960s decreased to 7 million tons in the 1980s.

The changes in the Black Sea ecosystem, especially the frequently occurring "blooms" of the unicellular organisms and the appearance of the invasive species such as *Rapana* and *Mnemiopsis* (see below), have negative influence on the fish fauna of the Bulgarian Black Sea coast. Populations of many pelagic and benthic fish species, which have been a subject of the industrial fishery, have decreased considerably and some of them (e.g. the turbot *Psetta maxima maeotica*) are under special protection. After the appearance and mass development of *Mnemiopsis leidyi* at the Bulgarian Black Sea coast, the fish stocks of some small pelagic species, such as the anchovy (*Engraulis encrasicholus ponticus*), sprat (*Sprattus sprattus sulinus*), Boyer's sand smelt (*Atherina boyeri*), etc., have declined. This has resulted in the decrease of the populations of high-value predators such as dolphin, mackerel, and bluefish, which have been replaced by small omnivorous pelagic species (Konsulov, 1998).

Some of the most serious ecological problems having a negative impact on the biological diversity of the Black Sea are caused by the invasive species of animals.

3·4 Invasive animals in the Black Sea

Invasive species are the organisms (protozoans, plants, and animals), which are introduced accidentally into new habitats, settle there, find favorable conditions for their development, and have harmful influence on the native ecosystems. The processes of globalization and especially the intensive development of the world market, tourism, and transport facilitate spreading of a large number of organisms in new habitats.

A great number of non-indigenous species, which provoked major changes in the native ecosystems, have been established in the Black Sea, including the Bulgarian coast (Zaitsev and Oztürk, 2001). Some of the first settlers, which at present are widespread along the Bulgarian coast, are the ivory barnacle (*Balanus eburneus*) and the acorn barnacle (*B. improvisus*). They were introduced in the late 19th century from the west coast of the North America through the ballast waters or hull fouling. In 1925, Valkanov registered the jellyfish *Blackfordia virginica* for the first time in the Bulgarian coast; it was introduced from the Atlantic coast of the North America. Ten years later Valkanov (1936) found the polychaete *Mercierella enigmatica*, which originates from India, where the species inhabits the brackish lakes. *M. enigmatica* develops in large numbers along the Bulgarian coast, sometimes overgrowing ship hulls, harbor, quay, and underwater constructions.

After the World War II, the introductions of invasive invertebrates in the Black Sea increased. In 1946, the gastropod *Rapana venosa*, native to the Sea of Japan, was discovered in the Novorossiysk Bay (Russian North Caucasus). After 1957 it became a widespread species along the Bulgarian Black Sea coast. Another recent intruder in the North Bulgarian coast is the softshell clam *Mya arenaria*. It was introduced from the Atlantic Ocean and was registered in Bulgaria for the first time in 1974 (Kaneva-Abadjieva, 1974). *M. arenaria* prefers areas with lower salinity and it replaces the dominant species *Lentidium mediterraneum* in the coastal sandy strips. In this way it affects negatively the biodiversity of the Black Sea ecosystem. On the other hand, the high biomass of *M. arenaria* provides food for the benthic fish and coastal birds. In the recent two decades, the mussel *Anadara inaequivalvis* has been registered in Varna Bay. Its original range is the Pacific coast of the Philippines. *A. inaequivalvis* inhabits silt and sandy-silt zones along the Bulgarian coast. According to Marinov (1990), at the depth of 40 m its density reaches 100 specimens/m^2, and the biomass, over 1 kg/m^2. *A. inaequivalvis* can withstand the low summer oxygen content in bottom layers easier than the native, autochthonous species of mussels. This has resulted in an increase of its biomass and replacement of the dominant native species *Chamelea gallina*.

Another recent invasive species is the ctenophore *Mnemiopsis leidyi*. Near the Bulgarian coast, this invader has been found by Konsulov (1990). It consumes zooplankton, and is a food competitor of the planktonic fish. In ten years, the population of *M. leidyi* has increased in large scale and reached abundance in 1990, with its biomass was up to one kg/m^3 in the open sea and five kg/m^3 in the coastal waters (Arkhipov and Malyshev, 1990). After 1990, the development of *M. leidyi* stabilized, and even a decrease of its biomass has been registered (Vinogradov et al., 2000). In the end of the 1990s, another ctenophore invader has been found in the Black Sea, *Beroe ovata* (Fig. 8), which feeds on *M. leidyi*. It could be expected

Fig. 8 *Beroe ovata* (Ctenophora), a new invader in the Black Sea. Photo by L. Klissurov.

that the increase of the *Beroe* population will result in decrease of *Mnemiopsis*, followed by an increase of the zooplankton and pelagic fish. In this way, a more stable ecological balance could be achieved in the Black Sea basin (Konsulov and Kamburska, 1998).

Other unwanted intruders in the Black Sea affecting its autochthonous fauna are the shipworm (*Teredo navalis*), the Chinese mitten crab (*Eriocheir sinensis*), the white-fingered mud crab (*Rhitropanopeus harrisi tridentale*), the blue crab (*Callinectes sapidus*), the copepod crustacean *Acartia tonsa*, etc. There are not enough authentic data and observations of the ecological impact of some of the mentioned species; it is necessary to conduct a regular monitoring of their density, horizontal and vertical distribution, ecological impact on the autochthonous communities, etc.

References

Arkhipov, A. & V. Malyshev, 1990. [Distribution and biomass of *Mnemiopsis* in the western part of the Black Sea]. Rybnoe khozyaistvo [Fish Industry] 10:30–31 (in Russian).

Arndt, W., 1947. Verzeichnis der bisher von der Schwarzen Meer-Küste Bulgariens und Rumäniens angegebenen Schwämme. Trudove na Morskata biologichna stantsiya vav Varna (Arbeiten aus der Biologischen Meeresstation in Varna, Bulgarien) 13:1–29.

Botev, B. & Tz. Peshev (eds), 1985. Chervena kniga na NR Balgariya. Tom 2. Zhivotni (Red Data Book of the People's Republic of Bulgaria. Volume 2. Animals). Izdatelstvo na Balgarskata akademiya na naukite (Publishing House of the Bulgarian Academy of Sciences), Sofia (in Bulgarian).

Caspers, H., 1951. Quantitative Untersuchungen über die Bodentierwelt des Schwarzen Meeres im bulgarischen Küstenbereich. Archiv für Hydrobiologie 45, 1–192.

Chichkoff, G., 1907. Contribution à l'étude de la faune de la Mer Noire. Archives de zoologie experimentale et génerale, 4e Ser., 7(6):247–268.

Chichkoff, G., 1912. Contribution à l'étude de la faune de la Mer Noire. Animaux recoltés sur les côtes bulgares. Archives de zoologie experimentale et génerale, 5e Ser., 10(2):29–39.

Golemansky, V., 1970a. Rhizopodes nouveaux du psammon littoral de la Mer Noire. Protistologica (Paris) 6(4):365–371.

Golemansky, V., 1970b. Contribution à la connaissance de thécamoebiens (Rhizopoda: Testacea) des eaux souterraines littorales du golf de Gdansk (Pologne). Izvestiya na Zoologicheskiya institut s muzei (Bulletin de l'Institut de zoologie et musée) 3:77–87.

Golemansky, V., 1970c. Thécamoebiens (Rhizopoda: Testacea) des eaux souterraines de quelques plages de Cuba. Izvestiya na Zoologicheskiya institut s muzei (Bulletin de l'Institut de zoologie et musée) 32:151–158.

Golemansky, V., 1971. Taxonomische und zoogeographische Notizen über die thekamöbe Fauna (Rhizopoda, Testacea) der Küstengrundgewässer der sowjetischen Fernostküste (Japanisches Meer) und der Westküste Kanadas (Stiller Ozean). Archiv für Protistenkunde 113(3):235–249.

Golemansky, V., 1974a. Psammonobiotidae fam. nova – une nouvelle famille de thécamoebiens (Rhizopoda: Testacea) du psammal supralittoral des mers. Acta Protozoologica 13(2):137–141.

Gomoiu, T., 1985. Problems concernant l'eutrophisation marine. Cercetâri Marine 18:59–95.

Kaneva-Abadjieva, V., 1958. [A new harmful snail from the Bulgarian Black Sea coast]. Priroda [Nature] (Sofia) 3:89–91 (in Bulgarian).

Kaneva-Abadjieva, V., 1965. [Distribution of the amphipods in the Bulgarian Black Sea sector]. Izvestiya na Nauchnoizsledovatelskiya institut za ribno stopanstvo i okeanografiya – Varna (Proceedings of the Research Institute of Fisheries and Oceanography – Varna) 6:75–86 (in Bulgarian).

Kaneva-Abadjieva, V., 1973. [The amphipod fauna of the biocenosis in algal encrustment off the Bulgarian Black Sea coast]. Izvestiya na Instituta po okeanografiya i ribno stopanstvo – Varna (Proceedings of the Institute of Oceanography and Fisheries – Varna) 12:87–95 (in Bulgarian).

Kaneva-Abadjieva, V., 1974. [A new mussel from the Bulgarian Black Sea coast]. Priroda [Nature] (Sofia) 2:64–65 (in Bulgarian).

Konsulov, A., 1990. *Leucothea multicornis* Eschholtz – a new species for the Black Sea. Okeanologiya (Oceanology) (Varna) 19:98–99.

Konsulov, A. (ed), 1998. Black Sea Biological Diversity. Bulgaria. U.N. Publications, New York.

Konsulov, A. & L.T. Kamburska, 1998. Ecological determination of the new Ctenophora – *Beroe ovata* invasion in the Black Sea. Trudove na Instituta po okeanologiya (Proceedings of Institute of Oceanology) (Varna):195–198.

Leppäkoski, E. & P.E. Mihnea, 1996. Enclosed seas under man-induced change: a comparison between the Baltic and Black Seas. Ambio 25:380–389.

Marinov, T., 1977. [Polychaeta]. In: Fauna na Balgariya. 6 (Fauna bulgarica. 6). Izdatelstvo na Balgarskata akademiya na naukite (Aedibus Academiae scientiarum bulgaricae), Sofia (in Bulgarian).

Marinov, T., 1990. Zoobentosat ot balgarskiya sektor na Cherno more [The Zoobenthos from the Bulgarian Sector of the Black Sea]. Izdatelstvo na Balgarskata akademiya na naukite (Publishing House of the Bulgarian Academy of Sciences), Sofia (in Bulgarian).

Marinov T. & V. Golemansky, 1989. [Second supplement to the Catalogue of the Bulgarian Black Sea fauna]. Acta zoologica bulgarica 37:3–33 (in Bulgarian).

Mee, L.D., 1992. The Black Sea in crisis: a need for concerted international action. Ambio 21(4):278–286.

Mihnea, P.E., 1999. Major shifts in the phytoplankton community (1980–1994) in the Romanian Black Sea. Oceanologica Acta 20(1):119–129.

Mishev, K. & V. Popov (1978) [Morphometric characteristics of the Black Sea basin]. In: Cherno more [The Black Sea]. G. Bakalov, Varna: 36–45 (in Bulgarian).

Rozhdestvenskiy, A., 1986. Hidrohimiya na balgarskiya sektor na Cherno more [Hydrochemistry of the Bulgarian Sector of the Black Sea]. Izdatelstvo na Balgarskata akademiya na naukite (Publishing House of the Bulgarian Academy of Sciences), Sofia (in Bulgarian).

Tiganus, V., 1992. Brief notes. Rapport du CIESM 33:54.

Valkanov, A., 1936. [Notizen über die Brackwässer Bulgariens. 2. Versuch einer hydrographischen und biologischen Erforschung derselben]. Godishnik na Sofiiskiya Universitet. Fiziko-matematicheski fakultet (Annuaire de l'Université de Sofia. Faculté physico-mathématique) 32(3):209–341 (in Bulgarian).

Valkanov, A., 1957. [Katalog unserer Schwarzmeerfauna]. Trudove na Morskata biologichna stantsiya vav Varna (Arbeiten aus der Biologischen Meeresstation in Varna, Bulgarien) 19:1–62 (in Bulgarian).

Valkanov, A., 1969. *Labyrinthodiction magnificum*, a new rhizoplasmodial unicellular organism. In: Strelkov, A.A., Sukhanova, K.M. and Raikov, I.B. (eds.). Progress in Protozoology. Proceedings of the 3rd International Conference on Protozoology, July 2–10, 1969, Leningrad. Nauka, Leningrad, 3:73–374.

Valkanov, A., 1970. Beitrag zur Kenntnis der Protozoen des Schwarzen Meeres. Zoologisher Anzeiger 184(3/4):241–290.

Valkanov, A. & T. Marinov, 1964. [Nachtrag zum Katalog der bulgarischen Schwarzmeerfauna]. Izvestiya na Zoologicheskiya institut s muzei (Bulletin de l'Institut de zoologie et musée) 17:51–59 (in Bulgarian).

Valkanov, A. & T. Marinov, 1978. Invertebrata. In Cherno more [The Black Sea], G. Bakalov, Varna: 107–124 (in Bulgarian).

Valkanov, A. & A. Rozhdestvenskiy, 1978. [Hydrobiology and hydrochemistry]. In: Cherno more [The Black Sea]. G. Bakalov, Varna: 66–85 (in Bulgarian).

Vinogradov, M., S. Vostokov, E. Arashkevich, A. Drits, E. Mussaeva, L. Anokhina & E. Shushkina, 2000. [Peculiarity of the biology of exotic ctenophores and their role in the Black Sea ecosystem]. In Vidy-vselentsy v evropeyskikh moryakh Rossii [Exotic Species in the European Seas of Russia]. Apatity: 91–113 (in Russian).

Zaitsev, Y.P., 1992. Recent changes in the trophic structure of the Black Sea. Fisheries Oceanography 1(2):180–189.

Zaitsev, Y. & B. Oztürk (eds), 2001. Exotic Species in the Aegean, Marmara, Black, Azov and Caspian Seas. Turkish Marine Research Foundation, Istanbul, Publ. 8.

20 Testate Amoebas and Monothalamous Foraminifera (Protozoa) from the Bulgarian Black Sea Coast

Vassil Golemansky
Institute of Zoology, Bulgarian Academy of Sciences, Tsar Osvoboditel Blvd. 1, 1000 Sofia, Bulgaria, e-mail: golemansky@zoology.bas.bg

Abstract: The studies on the marine interstitial testate amoebas are briefly reviewed. A characteristic of the marine supralittoral as a specific habitat is given. So far, 144 species of testate amoebas and 13 monothalamous foraminifera have been reported in the marine interstitial habitat of the World Ocean. A total of 47 interstitial testate amoebas have been reported in the Bulgarian Black Sea coast. Of these, 32 are psammobiotic, 11 are psammophilic, and 4 are psammoxenic. Eleven new genera and over 20 unknown species of psammobiotic marine interstitial testate amoebas and foraminifera have been described for the first time in the Bulgarian Black Sea coast. The morphological adaptations and some ecological preferences of the interstitial marine testate amoebas and foraminifera are discussed

1 Introduction

About 35 years ago, a new taxocoenosis of testate amoebas inhabiting the underground water of the marine sand supralittoral was observed and described in the Black Sea (Golemansky, 1969). Until then the testate amoebas were known mainly from freshwater habitats (lakes, rivers, and marshes), mosses, humid soils, etc. Earlier, Schulze (1874), Gruber (1884), Möbius (1889), Buchinski (1895, 1897), Sovinski (1902), Awerintzew (1903) and others published some records on testate amoebas observed in the marine sand littoral. However, all of the cited authors considered them to be freshwater inhabitants, which were imported occasionally into the sea through fresh waters.

Wailes (1927) was the first researcher who conducted a special study of the testate amoebas in the sand supralittoral in the region of British Columbia (the Pacific coast of Canada). He found testate amoebas in the underground waters of the sandy beach on Camp Island. One of these species he described as a new taxon, *Corythion acutum*. Wailes did not suppose that he had observed a new marine psammobiotic testate amoeba and concluded that: "Although found on the sea-shore, as described below, it is in all probably a freshwater or moss-inhabiting species". Wailes identified other observed amoebas as *Cyphoderia ampulla* (Ehrenberg) and

V. Fet and A. Popov (Eds.), Biogeography and Ecology of Bulgaria, 555–570.

Euglypha laevis Perty. He suggested that they were also carried in the mosses by the continental fresh waters.

Twenty-five years later, Valkanov (1954) observed *Corythion acutum* for the second time in the underground water of some sand beaches of the Bulgarian Black Sea coast at a salinity of 15‰. Ten years later, Valkanov & Marinov (1964) found a second testate amoeba in the same habitat, *Trinema enchelis* Penard. Valkanova (1964) reported the observation of the monothalamous foraminifera *Rhabdogromia flexilis* (= *Micatuba flexilis*) there.

Our investigations on the protists of the Black Sea have started in 1966 and allowed the description of many new and unknown testate amoebas inhabiting the underground waters of the marine sand beaches. These studies revealed the presence of a rich taxocoenosis of testate amoebas and monothalamous foraminifera, which is specific for the marine sandy supralittoral, and especially of the mixing zone of the sea water and the continental underground fresh water (Golemansky, 1969). In the following years, this specific taxocoenosis was studied in many other seas and oceans and its Cosmopolitan distribution was proved (Golemansky, 1970a, 1970b, 1970c, 1971, 1976a, 1976b; Chardez, 1972, 1977; Sudzuki, 1977, 1979, etc.).

2 The Marine Supralittoral as a Habitat

The marine supralittoral is a zone of the boundary between the sea and the continent situated above the seawater level but periodically flooded by the waves. According to Pérès and Picard (1958), this zone is also addressed as mediolittoral or eulittoral. Its inhabitants are mainly of marine origin and possess different morphological and biological adaptations to survive the specific ecological conditions. Swedmark (1964) considered this zone to be a zone of transition between the submarine littoral sand and the continental underground freshwater.

The supralittoral sand strip is relatively constant on the seashores where tidal ebb and flow are weakly expressed. There, permanent mixing of the underground continental freshwater and seawater takes place (Fig. 1). The salinity of the water in the mixing zone varies at a large scale: from 30–35‰ in the sea littoral water to 1.04‰ in the outlying sand beach strips (Fig. 2). The quantity of the oxygen also varies considerably, with a general tendency of decreasing from the sea to the continent and from the surface to the deeper zones (Pennak, 1951; Jansson, 1966). In the Bulgarian Black Sea coast, oxygen content in the supralittoral zone varies from 4.50–5.50 mg/l (at the distance of 1 m from the sea and at the depth of 0.25 m) to 0.20–0.40 mg/l (at the distance of 15 m from the sea and at the depth of 1.20 m).

The marine supralittoral psammal is a specific habitat with variable ecological conditions. It is a labile biotope with permanent fluctuations of the main ecological factors (salinity, O_2, pH, temperature, water movement, granulometry of sand, etc.), which influence its interstitial inhabitants.

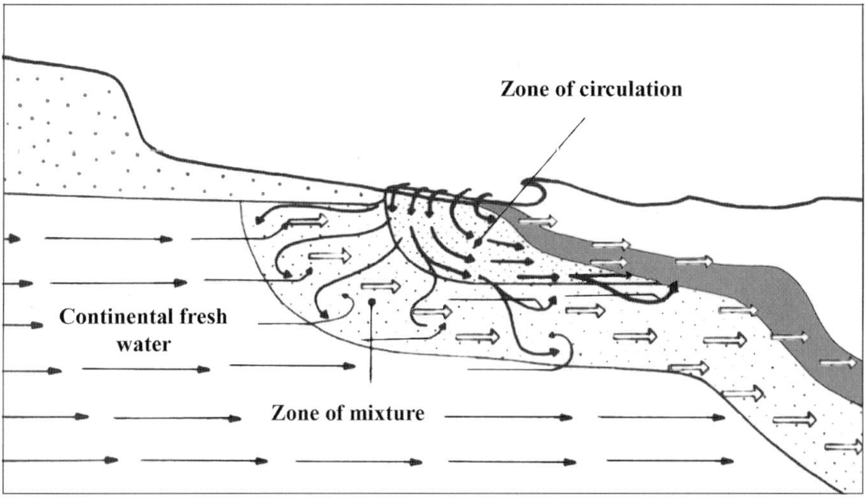

Fig. 1 Scheme of circulation and mixture of underground water in a sea without tides (after Delamare-Debouteville, 1960).

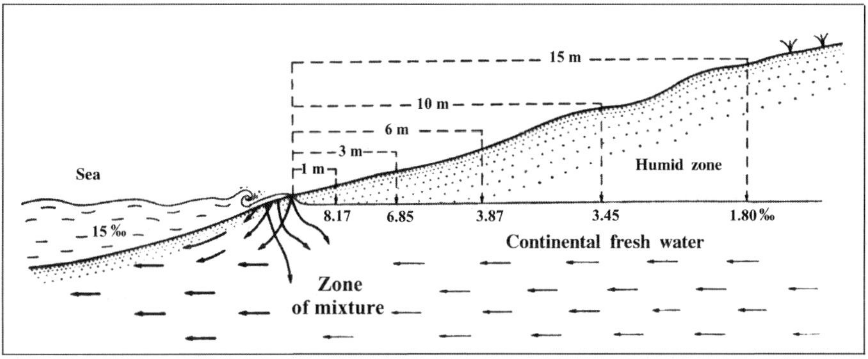

Fig. 2 Profile of Slanchev Bryag sand beach with the variation of the salinity of underground water from the sea (left) to the continent (right) (after Golemansky, 1978).

3 Biodiversity of the Marine Interstitial Testate Amoebas and Foraminifera

In the last two decades, taxonomy of Protozoa was a subject of active discussions, and some new classifications were proposed, such as those of Lee et al. (1985, 2000), Margulis et al. (1990), Cavalier-Smith (1998), Hausmann et al. (2003), etc. In the present study, we accept the basic classification of Levine et al. (1980), excluding the taxonomic position of the Foraminifera; they are considered a separate superclass of the phylum Sarcomastigophora, like the superclass Rhizopodea.

So far, a total of 141 species and varieties of testate amoebas and 13 species of foraminifera have been recorded in the marine interstitial underground waters of the sand supralittoral in the World Ocean. A great part of these (98 taxa) are strict psammobionts, which until now were observed only in the marine supralittoral psammal. The smaller portion (16 species) is considered to be psammophilic; and 22 species are freshwater testate amoebas, occasionally introduced from the continental habitats, and considered to be psammoxenic.

A total of 47 interstitial testate amoebas have been recorded in the Bulgarian Black Sea coast. Of these, 32 are psammobiotic, 11 are psammophilic, and 4 are psammoxenic. The number of the monothalamous foraminifera living in the Black Sea is relatively small (4 species) because of the sea's low salinity (ca. 16–17‰ in the coastal water). It is important to note that 11 new genera and more than 20 species of the psammobiotic marine testate amoebas and foraminifera have been found and described for the first time from the Black Sea coast. Most of the filose psammobiotic testate amoebas belong to a new family, Psammonobiotidae (Golemansky, 1974a). The testate amoebas and foraminifera known up to date in the Black Sea are listed in Table 1. Representative testate amoebas are pictured in Fig. 3.

4 Morphological Adaptations to the Marine Interstitial Habitat

As stated above, the marine interstitial supralittoral is a specific habitat, characterized by the presence of a zone of mixing of sea water and continental fresh water. There is a large variability of the salinity, dissolved oxygen, temperature, water mobility, pH, etc. Another important ecological factor is the granulometry of the sand grains.

The psammobiotic marine interstitial testate amoebas and foraminifera have various morphological and biological adaptations for this specific habitat. The most important of those were discussed by Golemansky (1969, 1978, 1980) and Chardez (1972), and can be generalized as follows:

- Presence of a large circular collar around the aperture (Fig. 4). This enlargement is formed by a proteinaceous organic membrane, covered with different idiosomes or xenosomes. The distal border of the enlargement is thin, transparent, and flexible. The border flexibility facilitates better adherence to sand grains in an unstable environment; this makes easier the movement of these organisms and their feeding on bacteria, algae, fungi, etc.
- The test dimensions of psammobiotic testate amoebas are relatively smaller in comparison with the freshwater ones. Major part of the psammobiotic testate amoebas (about 75% of all known species) has a test which varies in length from 15–60 μm (Golemansky, 1978). The list of testate amoebas and foraminifera, in

Table 1 Checklist of the marine interstitial testate amoebas and foraminifera from the Bulgarian Black Sea coast

Taxa	References	Ecological categories		
		PB	PP	PX
Phylum SARCOMASTIGOPHORA				
Subphylum SARCODINA				
Superclass Rhizopodea				
Class Lobosea				
Order Arcellinida				
Arcella hemisphaerica Perty	Golemansky (1980)			+
Microchlamys patella Clap et Lachmann	Golemansky (1980)		+	
Centropyxis aerophila Deflandre	Golemansky (1980)			+
Centropyxis constricta (Ehrenberg)	Golemansky (1980)			+
Centropyxiella arenaria Valkanov	Valkanov (1970), Golemansky (1974b)	+		
Centropyxiella elegans Valkanov	Valkanov (1970)			+
Centropyxiella gibbula Valkanov	Valkanov (1970)			+
Centropyxiella lucida Golemansky	Golemansky (1974b)	+		
Centropyxiella platystoma Golemansky	Golemansky (1981)	+		
Micropsammella retorta Golemansky	Golemansky (1970b)	+		
Alepiella tricornuta Golemansky	Golemansky (1970d)	+		
Difflugia lucida Penard	Golemansky (1974b, 1980)	+		
Difflugia subterranea Golemansky	Golemansky (1980)	+		
Pomoriella valkanovi Golemansky	Golemansky (1969, 1970a, 1974b)	+		
Hyalosphaenia cuneata Stein	Golemansky (1974b, 1980)		+	
Cryptodifflugia lanceolata Golemansky	Golemansky (1969, 1974b, 1980)	+		
Cryptodifflugia paludosa Golemansky	Golemansky (1981)	+		
Difflugiella psammophila Golemansky	Golemansky (1974b)		+	

continued

Table 1 continued

Taxa	References	Ecological categories		
		PB	PP	PX
Class FILOSEA				
Order Gromida				
Psammonobiotus balticus Golemansky	Golemansky (1974b, 1980)	+		
Psammonobiotus communis Golemansky	Golemansky (1969, 1970a, 1974b, 1980)	+		
Psammonobiotus linearis Golemansky	Golemansky (1970d, 1980)	+		
Psammonobiotus minutus Golemansky	Golemansky (1969, 1970a, 1980)	+		
Pseudocorythion acutum (Wailes)	Valkanov (1954, 1970), Golemansky (1969, 1974b, 1980)	+		
Pseudocorythion wailesi Golemansky	Golemansky (1969, 1974b, 1980)	+		
Messemvriella filosa Golemansky	Golemansky (1973, 1980)	+		
Chardezia caudata Golemansky	Golemansky (1970e, 1980)	+		
Corythionella acolla Golemansky	Golemansky (1969, 1974b, 1980)	+		
Corythionella minima Golemansky	Golemansky (1970a, 1974b, 1980)	+		
Corythionella pontica Golemansky	Golemansky (1970a, 1980)	+		
Pseudodifflugia sp.	Golemansky (1974b, 1980)		+	
Micramphora pontica Golemansky	Valkanov (1970), Golemansky (1974b)	+		
Ogdeniella elegans Golemansky	Golemansky (1969, 1970a, 1974b, 1980)	+		
Ogdeniella lucida Golemansky	Golemansky (1970d, 1974b, 1980)	+		
Ogdeniella maxima Golemansky	Golemansky (1970d, 1980)	+		
Ogdeniella taschevi Golemansky	Golemansky (1969, 1980)	+		
Euglypha compressa Carter	Golemansky (1974b, 1980)			+
Euglypha laevis Perty	Golemansky (1980)		+	
Euglypha rotunda Wailes	Golemansky (1974b, 1980)		+	
Trinema complanatum Penard	Golemansky (1974b, 1980)		+	
Trinema enchelis Ehrenberg	Valkanov & Marinov (1964), Golemansky (1974b, 1980)		+	
Trinema lineare Penard	Golemansky (1974b, 1980)		+	

Cyphoderia ampulla Ehrenberg	Golemansky (1974b, 1980)			+
Cyphoderia compressa Golemansky	Golemansky (1979)	+		
Cyphoderia littoralis Golemansky	Golemansky (1969, 1974b, 1980)	+		
Campascus interstitialis Golemansky	Golemansky (1981)	+		
Campascus sp.	Golemansky (1974b)			+
Paramphitrema pontica Golemansky	Valkanov (1970), Golemansky (1999a)	+		
Superclass Foraminifera				
Order Monothalamida				
Lagynis pontica Golemansky	Golemansky (1999)	+		
Lagenidiopsis valkanovi Golemansky	Golemansky (1974)	+		
Lagenidiopsis elegans Golemansky	Golemansky (1978, 1980)	+		
Rhumbleriella filosa Golemansky	Golemansky (1970)	+		
Micatuba (= *Rhabdogromia*) *flexilis* Höglund	Valkanova (1964), Golemansky (1980)	+		

PB, psammobiont; PP, psammophile; PX, psammoxene

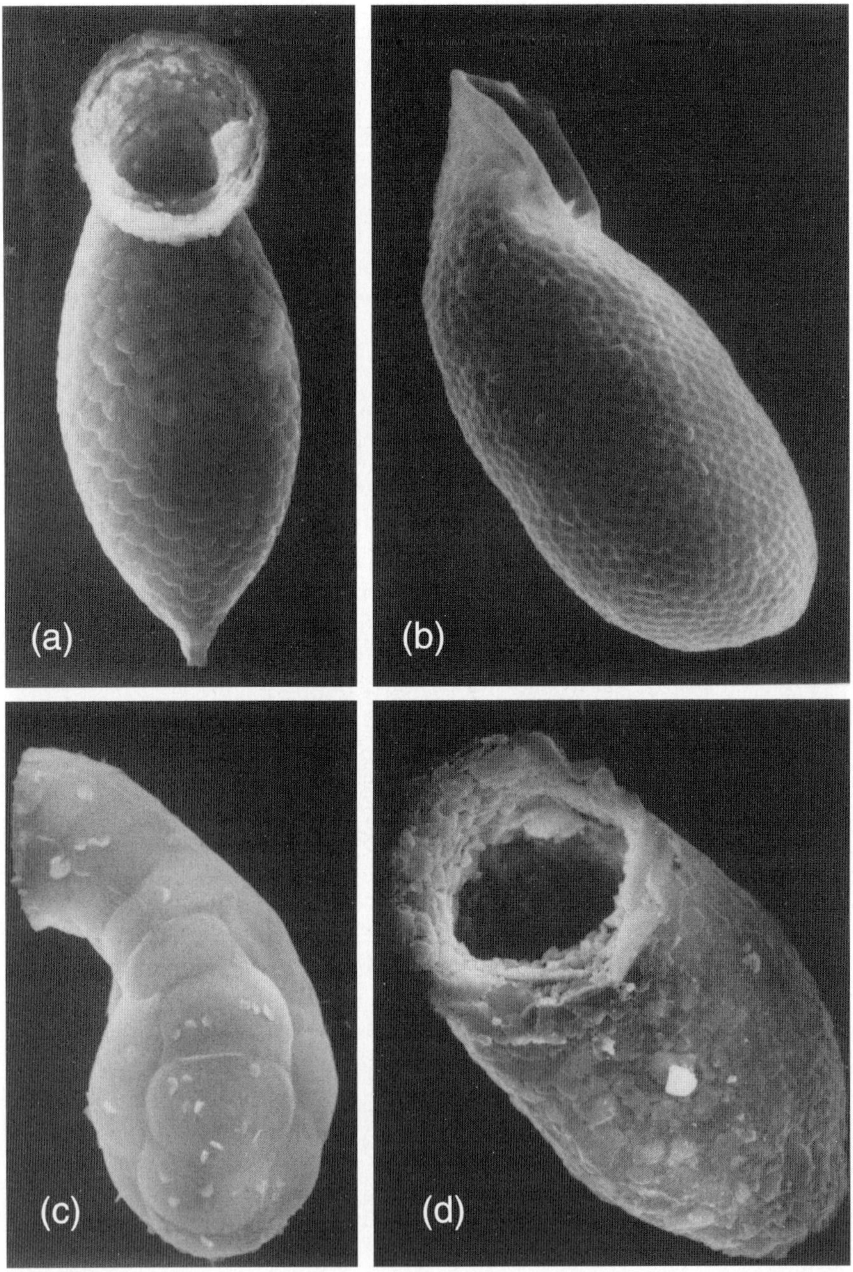

Fig. 3 Common psammobiotic testate amoebas from the Bulgarian Black Sea coast. (a) *Pseudocorythion acutum*, (b) *Messemvriella filosa*, (c) *Pomoriella valkanovi*, (d) *Centropyxiella arenaria*.

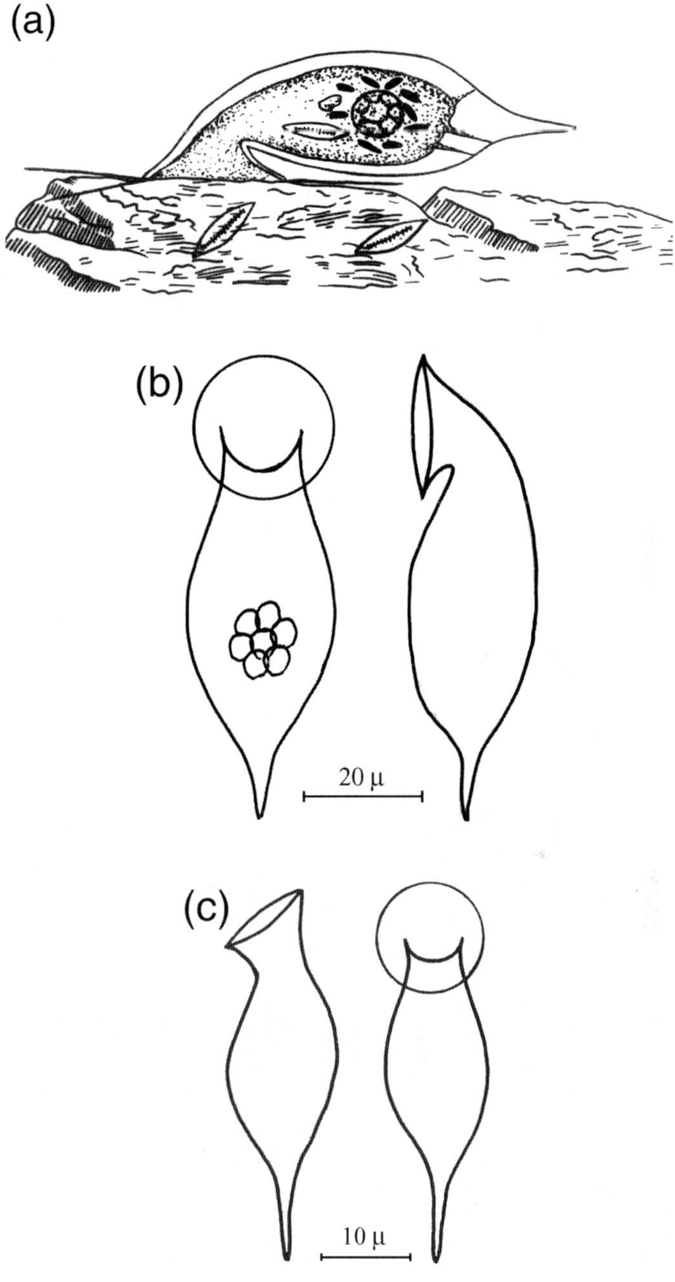

Fig. 4 Psammobiotic testate amoebas with a large circular collar around the aperture of the test (after Golemansky, 1986). (a–b) *Pseudocorythion acutum*: moving specimen (a), ventral and lateral view (b), (c) *Chardezia caudata,* (d) *Messemvriella filosa,* (e) *Pseudocorythion wailesi.*

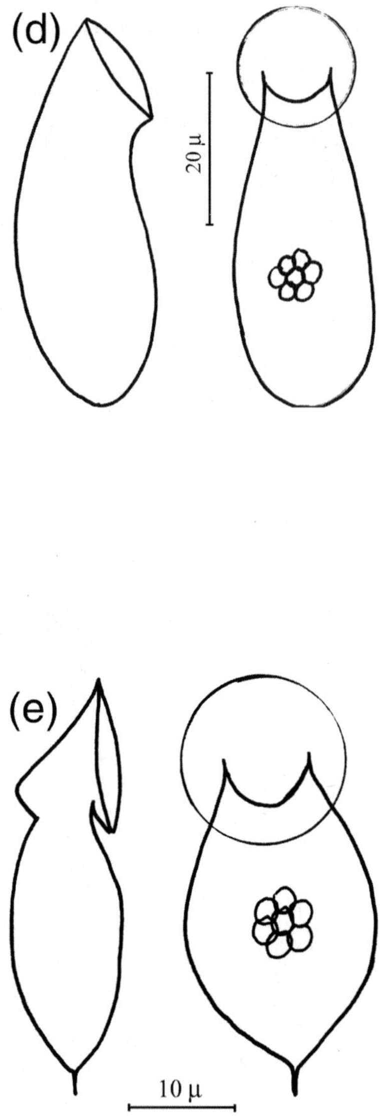

Fig. 4 continued

which test is longer than 100 μm, is very limited: *Cyphoderia compressa, Lagynis pontica, Micatuba flexilis*, and *Lagenidiopsis elegans*. The same trend was also observed in psammophilic testate amoebas. Chardez (1972) and Golemansky (1978) noted that the test dimensions of *Difflugia lucida, Cyclopyxis kahli* Deflandre, and *Hyalosphenia cuneata*, living in the marine supralittoral psammal, are smaller in comparison with the same species living in freshwater habitats such as lakes, peat bogs, etc. Undoubtedly, the reduced test dimensions facilitate life and movement of the organisms in the interstitial habitat.

- A laterally or dorsoventrally flattened test is another morphological adaptation to the limited space available in the interstitial habitat. Mainly the psammobiotic testate amoebas with a very small test size have a test, which is circular in a cross-section (*Micramphora pontica, Difflugiella psammophila, Ogdeniella lucida*, etc.).
- Presence of a fine organic test with a few xenosomes, disposed mainly at the region of the neck and the aperture. Some psammobiotic testate amoebas have a test without any apparent structure: *Chardezia caudata* (Fig. 4c), *Ogdeniella lucida, Psammonobiotus minutus*, etc. A light test is also an adaptation, which makes the movement between sand grains easier.

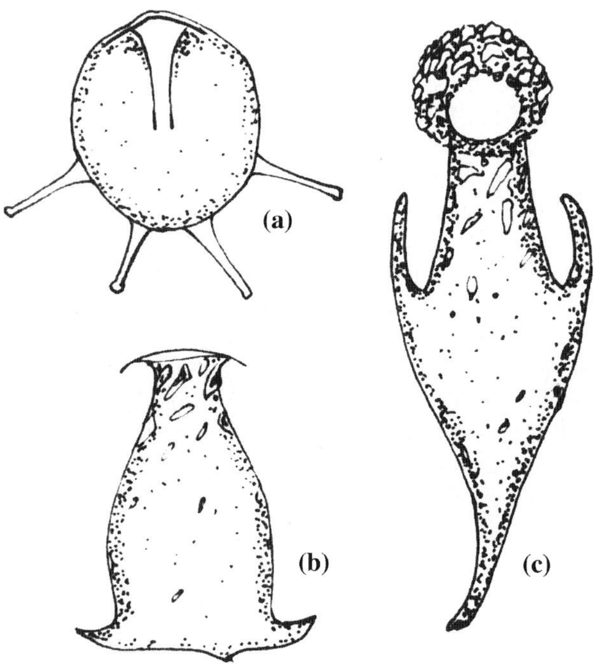

Fig. 5 Psammobiotic testate amoebas with shell spikes and horns (after Golemansky, 1978). (a) *Lagenidiopsis valkanovi*, (b) *Ogdeniella elegans*, (c) *Alepiella tricornuta*.

- Presence of certain test structures such as spikes and horns, which prevent drifting to the sea through the continental underground water or tidal waves. The psammobiont testate amoebas *Alepiella tricornuta* and *Ogdeniella elegans*, the foraminifera *Lagenidiopsis valkanovi*, and others have such lateral spikes (Fig. 5). Some other psammobionts such as *Pseudocorythion acutum* (Figs. 3a, 4a, 4b), *Chardezia caudata* (Fig. 4c), *Corythionella minima, Micropsammella retorta*, and others also have caudal horns.

5 Ecology of the Marine Interstitial Testate Amoebas and Foraminifera

The biodiversity and distribution of the marine interstitial amoebas in the supralittoral psammal depends on different ecological factors such as granulometry of sand, water salinity, dissolved oxygen, temperature, movement of underground water, presence of food (bacteria, algae), etc. The impact and role of these factors on the interstitial rhizopods have not been well studied. Golemansky (1980, 1994a) published the first detailed data, which gave information about the influence of three main abiotic factors: granulometry of sand, water salinity, and dissolved oxygen.

The shape and size of sand grains are very important for the ecological characteristic of the interstitial habitat. As shown by Frasser (1935), Ruttner-Kolisko (1962), Jansson (1966), etc., size of the sand grain determines the porosity of sand, which is of great importance for defining the quantity of water in a fixed volume of the sand and the rate of the dissolved oxygen. Our investigations, conducted on some sand beaches of the Black Sea coast, showed that the small-sized homogeneous sands with modality (Mo) of 0.10–0.80 mm were populated by a richer and more diverse testacean fauna than the heterogeneous sands with Mo values higher than 0.80–1.00 mm (Golemansky, 1994a). So far, a few psammobiotic testate amoebas have been found in coarse heterogeneous sands, e.g. *Ogdeniella maxima* and *Rhumbleriella filosa*, which live in sands with Mo $>$ 2.00 mm.

Delamare-Deboutteville (1960), Murray (1967), and others suggested that the dissolved oxygen in the underground waters rarely is a limiting ecological factor because the metabolism of organisms inhabiting these waters is very slow. Our investigations of the quantity of dissolved oxygen in the underground water of marine sand beaches demonstrated a trend towards a constant decrease of the oxygen content from the coastal water line to the continent (Golemansky, 1994a). This tendency is illustrated in Fig. 6 for two different sand beaches of the Bulgarian Black Sea coast. Variation of the species number is presented in the same graphs. In spite of the high concentration of oxygen in the neighboring coastal water line, the number of interstitial testate amoebas was relatively small. This is due to the limiting impact of the mechanical factors such as rapid filtration of water near the coastal line, movement of sand, temperature changes, etc. In the sampling points situated 3–4 to 10–15 m away from the coastal water line, number of psammobiotic

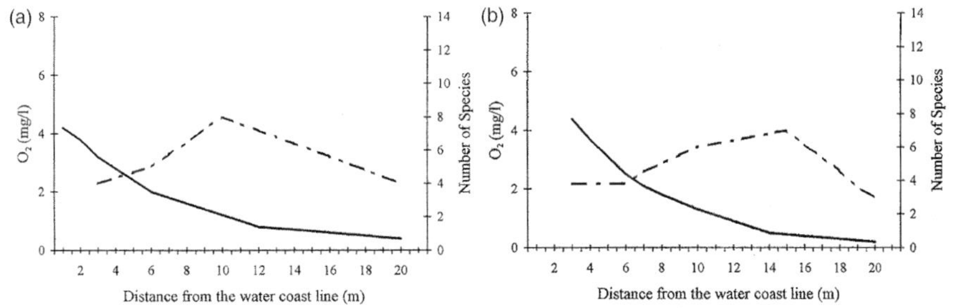

Fig. 6 Distribution of oxygen and number of testate amoebas on two sandy beaches of the Black Sea (after Golemansky, 1994a). (a) Slanchev Bryag, (b) Tsarevo. Solid line, oxygen; dashed line, number of species.

interstitial testate amoebas increased despite the decrease of the oxygen content. In the outlying sampling points the number of psammobiotic testate amoebas becomes smaller simultaneously with the oxygen decrease. Except oxygen, there are other limiting factors, such as filling of the interstitial space between sand grains with clay or organic particles, rapid refreshment of the underground water, lower temperature, etc.

The impact of salinity on biodiversity and distribution of the marine interstitial testate amoebas and foraminifera has not been well studied. The majority of the psammobiotic and psammophilic testate amoebas are considered to be euryhaline (Golemansky, 1980, 1994a). For example, psammophilic testate amoebas *Cyphoderia ampulla* and *Difflugia linearis* Penard were found in the North Sea at the salinity of 35‰ (Chardez, 1972); *Centropyxis constricta* and *Trinema enchelis* were found in the Black Sea at the salinity of 17–20‰ (Valkanov, 1970; Golemansky, 1980). The limits of salinity for the psammobiotic testate amoebas are broader. Some species survive at water salinity between 0.1–36‰. For example, *Psammonobiotus communis* and *Difflugiella psammophila* are widespread inhabitants of the marine supralittoral of many seas and oceans, but they have been also found in the relict freshwater lakes Ohrid and Prespa in Macedonia (Golemansky, 1994b). So far a few psammobiotic testate amoebas have been confirmed to be stenohaline. Such a stenohaline psammobiont is *Pomoriella valkanovi* (Fig. 3c), which has been observed always at the salinity of 13.5–18‰ in over 20 different points of the Black Sea coast.

References

Awerintzew, S., 1903. Beiträge zur Kenntnis der marinen Rhizopoden. Mitteilungen aus der Zoologischen Station zu Neapel 16(3):349–364.

Buchinski, P.H., 1895. [Protozoans from Hadzhibey and Kuyalnitski firths]. Zapiski Novorossiyskogo obshchestva estestvoispytatelei [Memoirs of the Novorossiysk Society of the Naturalists] 20(1):137–148 (in Russian).

Buchinski, P.H., 1897. [Fauna of Odessa firths]. Zapiski Novorossiyskogo obshchestva estestvoispytatelei [Memoirs of the Novorossiysk Society of the Naturalists] 21(2):135–216 (in Russian).

Cavalier-Smith, T., 1998. A revised six kingdom system of life. Biological Reviews 73:203–213.

Chardez, D., 1972. Etude sur les thécamoebiens des biotopes interstitiels, psammon littoraux et zones marginales souterraines des eaux douces. Bulletin des Recherches agronomiques de Gembloux, N.S., 6(3/4):257–268.

Chardez, D., 1977. Thécamoebiens du Mésopsammon des plages de la Mer du Nord. Revue Vervietoise d'Histoire Naturelle 34(4/6):18–34.

Delamare-Debouteville, C., 1960. Biologie des eaux souterraines littorals et continentals. Hermann, Paris.

Frasser, H. J., 1935. Experimental study of the porosity and permeability of clastic sediments. Journal of Geology 43:910–1010.

Golemansky, V., 1969. Sur une biocénose peu connue des eaux souterraines littorals des mers. In: Strelkov, A.A., K.M. Sukhanova & I.B. Raikov (eds), Progress in Protozoology. Proceedings of the 3rd International Congress on Protozoology, July 2–10, 1969, Leningrad. Nauka, Leningrad: 194.

Golemansky, V., 1970a. Rhizopodes nouveaux du psammon littoral de la Mer Noire. Protistologica (Paris) 6(4):365–371.

Golemansky, V., 1970b. Contribution à la connaissance de thécamoebiens (Rhizopoda: Testacea) des eaux souterraines littorales du golf de Gdansk (Pologne). Izvestiya na Zoologicheskiya institut s muzei (Bulletin de l'Institut de zoologie et musée) 3: 77–87.

Golemansky, V., 1970c. Thécamoebiens (Rhizopoda: Testacea) des eaux souterraines de quelques plages de Cuba. Izvestiya na Zoologicheskiya institut s muzei (Bulletin de l'Institut de zoologie et musée) 32:151–158.

Golemansky, V., 1970d. Thécamoebiens (Rhizopoda: Testacea) nouveaux des eaux souterraines littorales de la Mer Noire. Acta Protozoologica 8(2):41–46.

Golemansky, V., 1970e. *Chardezia caudata* gen. n. sp. n. et *Rhumbleriella filosa* gen. n. sp. n. – deux thécamoebiens nouveaux du psammon littoral de la Mer Noire (Rhizopoda: Testacea). Izvestiya na Zoologicheskiya institut s muzei (Bulletin de l'Institut de zoologie et musée) 32:121–125.

Golemansky, V., 1971. Taxonomische und zoogeographische Notizen über die thekamöbe Fauna (Rhizopoda, Testacea) der Küstengrundgewässer der sowjetischen Fernostküste (Japanisches Meer) und der Westküste Kanadas (Stiller Ozean). Archiv für Protistenkunde 113(3):235–249.

Golemansky, V., 1973. *Messemvriella filosa* n. gen. n. sp. – une nouvelle thécamoebienne psammobionte (Rhizopoda: Testacea) des eaux souterraines litoralles de la Mer Noire. Zoologisher Anzeiger 190(5/6):302–304.

Golemansky, V., 1974a. Psammonobiotidae fam. nova – une nouvelle famille de thécamoebiens (Rhizopoda: Testacea) du psammal supralittoral des mers. Acta Protozoologica 13(2): 137–141.

Golemansky, V., 1974b. Sur la composition et la distribution horizontale de l'association thécamoebienne (Rhizopoda, Testacea) des eaux souterraines littorales de la Mer Noire en Bulgarie. Izvestiya na Zoologicheskiya institut s muzei (Bulletin de l'Institut de zoologie et musée) 40:195–202.

Golemansky, V., 1974c. *Lagenidiopsis valkanovi* gen. n. sp. n. – un nouveau thécamoebien (Rhizopoda: Testacea) du psammal supralittoral des mers. Acta Protozoologica 13(1):1–4.

Golemansky, V., 1976a. Contribution à l'étude des Rhizopodes et des Heliozoaires du psammal supralittoral de la Méditerranée. Acta Protozoologica 15(1):35–45.

Golemansky, V., 1976b. Rhizopodes psammobiontes (Protozoa, Rhizopoda) du psammal supralittoral des côtes guinéenes de l'Atlantique. Acta zoologica bulgarica 4:23–29.

Golemansky, V., 1978. Adaptations morphologiques des thécamoebiens psammobiontes du psammal supralittoral des mers. Acta Protozoologica 17(1):141–152.

Golemansky, V., 1979. *Cyphoderia compressa* n. sp. (Rhizopoda: Arcellinida) – un nouveaux thécamoebien psammobionte du supralittoral des mers. Acta Protozoologica 18(3):429–434.

Golemansky, V., 1980. Interstitsialnata tekamebna fauna na pyasachniya morski supralitoral [La faune thécamoebienne interstitielle du psammal supralittoral marin]. Disertatsiya za doktor na naukite [DrSc Thesis]. Institut po zoologiya [Institute of Zoology], Sofia (in Bulgarian).

Golemansky, V., 1981. Description de trois thécamoebiens (Protozoa: Rhizopodea) nouveaux des eaux souterraines littoralles des mers. Acta Protozoologica 20(1:115–119.

Golemansky, V., 1986. Rhizopoda: Testacea. In Botosaneanu, L. (ed), Stygofauna Mundi. E. J. Brill & Dr. W. Backhuys, Leiden: 5–16.

Golemansky, V., 1994a. On some ecological preferences of marine interstitial testate amoebas. Archiv für Protistenkunde 144:424–432.

Golemansky, V., 1994b. Thécamoebiens (Rhizopoda: Testacea) d'hygropsammon littoral de trois lacs macédoniens d'origin téctonique: Ohrid, Prespa et Doiran. Archiv für Protistenkunde 144:309–313.

Golemansky, V., 1999a. Second observation of *Paramphitrema pontica* Valk., 1970 (Rhizopoda: Gromiida), and supplement to their morphometry. Acta zoologica bulgarica 51(2/3):3–8.

Golemansky, V., 1999b. *Lagynis pontica* n. sp., a monothalamous rhizopod (Granuloreticulosea: Lagynidae) from the Black Sea littoral. Acta zoologica bulgarica 51(1):3–13.

Gruber, A., 1984. Die Protozoen des Hafens von Genua. Nova Acta Leopoldina. Abhandlungen der Kaiserlich Leopoldinisch-Carolinisch Deutschen Akademie der Naturforscher 46(4):475–539.

Hausmann, K., N. Hülsmann & R. Radek, 2003. Protistology. E. Schweizerbart'sche Verlagsbuchhandlung, Berlin & Stuttgart.

Jansson, B.-D., 1966. Microdistribution of factors and fauna in marine sandy beaches. Veröffentlichungen des Instituts für Meeresforschung in Bremerhaven 2:77–86.

Lee, J.J., S.H. Hutner & E.C. Bovee (eds), 1985. An Illustrated Guide to the Protozoa. Society of Protozoologists, Lawrence, Kansas.

Lee, J.J., G.F. Leedale & P. Bradbury (eds), 2000. An Illustrated Guide to the Protozoa. 2nd ed. Society of Protozoologists, Lawrence, Kansas.

Levine, N.D., J.O. Corliss, F.E. Cox, G. Deroux, J. Grain, B.M. Honigberg, B.M., G.F. Leedale, A.R. Loeblich, III, J. Lom, D. Lynn, D., E.G. Merinfeld, F.C. Page, G. Polansky, V. Sprague, J. Vavra, J. & F.G. Wallace, 1980. A new revised classification of the Protozoa. Journal of Protozoology 27(1):37–58.

Margulis, L., J.O. Corliss, M. Melkonian & D.J. Chapman (eds), 1990. Handbook of Protista. Jones and Barlett, Boston.

Möbius, K., 1889. Bruchstücke einer Rhizopodefauna der Kieler Bucht. Abhandlungen der Preussischen Akademie der Wissenschaften Berlin vom Jahre 1888:1–31.

Murray, J.W., 1967. An ecological study of the Thecamoebina of Christchurch Harbour, England. Journal of Natural History 1(3):377–378.

Pennak, R.W., 1951. Comparative ecology of the interstitial fauna of freshwater and marine beaches. Annals of Biology 27:449–480.

Pérès, J.-M. & J. Picard, 1958. Nouveau manuel et bionomie béntique de la Méditerranée. Recueil de Travaux de la Station Marine d'Endoume 23:7–122.

Ruttner-Kolisko, A., 1962. Porenraum und kapillare Wasserströmung im Limnopsammal, ein Beispiel für die Bedeutung verlangsamter Strömung. Schweizer Zeitschrift für Hydrobiologie 24:444–458.

Schulze, F.E., 1874. Rhizopodenstudien. III. Archiv für mikroskopische Anatomie 2(1):94–139.

Sovinski, V.K., 1902. Vvedenie v izuchenie fauny Ponto-Kaspiysko-Aral'skogo morskogo basseyna [Introduction to the Study of the Fauna of Pontian–Caspian–Aral Sea Basin]. Kiev (in Russian).

Sudzuki, M., 1977. Protozoans in the marine beach interstices. II. Taxonomy and ecology of the Testacea from a sandy island recently constructed. Japanese Journal of Protozoology 8:23.

Sudzuki, M., 1979. Psammobiont Rhizopoda and Actinopoda from marine beaches of Japan. Acta Protozoologica 18(2):293–304.

Swedmark, B., 1964. The interstitial fauna of marine sand. Biological Reviews 39(1):1–42.

Valkanov, A., 1954. [Beitrag zur Kenntnis unserer Schwarzmeerfauna]. Trudove na Morskata biologichna stantsiya vav Varna (Arbeiten aus der Biologischen Meeresstation in Varna, Bulgarien) 18:49–53 (in Bulgarian).

Valkanov, A., 1970. Beitrag zur Kenntnis der Protozoen des Schwarzen Meeres. Zoologisher Anzeiger 184(3/4):241–290.

Valkanov, A. & T. Marinov, 1964. [Nachtrag zum Katalog der bulgarischen Schwarzmeerfauna]. Izvestiya na Zoologicheskiya institut s muzei (Bulletin de l'Institut de zoologie et musée) 17:51–59 (in Bulgarian).

Valkanova, Ch., 1964. Über eine neue Rhizopodengattung *Rhabdogromia*. Zoologisher Anzeiger 184(3/4):241–290.

Wailes, G.H., 1927. Rhizopoda and Heliozoa from British Columbia. Annals and Magazine of Natural History 20 (Ser. 9):153–156.

21 Florogeographical Elements and Evolutionary Trends in the Bulgarian Flora

Dimitar Peev[1] and Malina Delcheva[2]

[1] *Institute of Botany, Bulgarian Academy of Sciences, Acad. G. Bonchev St. 23, 1113 Sofia, Bulgaria, e-mail: dpeev@bio.bas.bg*

[2] *Institute of Botany, Bulgarian Academy of Sciences, Acad. G. Bonchev St. 23, 1113 Sofia, Bulgaria, e-mail: malinad@bio.bas.bg*

Abstract: The specific geographical location of Bulgaria, its geological history, and the varied relief underlie the contemporary richness of the Bulgarian flora. It includes 3700 species of vascular plants, among them 170 Bulgarian endemic species and 100 subspecies, and about 200 Balkan endemic species and subspecies. Analysis of the origin and the relationships between the various elements of Bulgarian flora demonstrates 92 types of florogeographical (floristic) elements. Nine Bulgarian mountain ranges and massifs may be ranked as follows according to the number of their endemic plant taxa: Stara Planina (90 species and subspecies), Rhodopes (over 80), Pirin (over 70), Rila (about 50), Vitosha (25), Strandja (19), Sredna Gora (12), Belasitsa (eight), and Osogovo (six). The complex nature of the Bulgarian flora, its high species diversity, and the relatively high percentage of endemism determine its high conservation value

1 Position of Bulgaria in the European Continent and the Balkan Peninsula

Bulgaria occupies 110,912 sq. km (ca. 1% of Europe and 20% of the territory of the Balkan Peninsula; Stoyanov, 1950). The specific features of this area are determined by the impact of three wide-ranging bioclimates in the region (Central European continental, Eurasian steppe, and Mediterranean), and the presence of transitional climatic conditions. The existing varied relief patterns (high mountains, spacious plains, narrow valleys, steep seashores, sand and dune complexes) and the geological history led to the formation of the rich and specific flora, which includes a high number of endemic and relict species. Considering the composition of the plant cover, the territory of Bulgaria belongs mainly to the southeastern province of the Central European Floristic Region (Stoyanov, 1950), or the European–West Siberian Region of the Palearctic Biophyletic Realm (Gruev, 1988), or to the Illyrian Province of the Circumboreal Region of the Boreal Subrealm of the Holarctic Realm, with the exception of the northeastern steppes and a small part belonging to the Euxinian Province (Takhtajan, 1978). The geological history of the Balkan Peninsula, which in the past used to be a peninsula of Asia Minor, has left traces in the plant geography. Evidence for that is the great number of Western Asian

V. Fet and A. Popov (Eds.), Biogeography and Ecology of Bulgaria, 571–579.

floristic elements. Pontian elements, characteristic of the northern Black Sea and the Caspian Sea steppes, dispersed to the territory of Bulgaria from the northeast. The floristic impact of the Central European Floristic Region is revealed by the analysis of the relationships between the Balkan Peninsula and the Alps. This connection can be traced to the east of the Alps through the Serbian mountain ranges, along the Stara Planina Mts. and to the southeast to the Rila and Rhodope Mts. The percentage of the Boreal floristic element is comparatively high. The adjacent position of the Mediterranean Region and the absence of substantial barriers to migration facilitate the dispersal of species from the Aegean Sea coast along the Maritsa and Struma valleys, as well as along the Black Sea coast (Stoyanov, 1950). These are only some of the factors determining the vegetation diversity in Bulgaria, a country small in size but comparatively rich in flora.

2 Taxonomic Richness and Structure of the Bulgarian Flora

Bulgarian vascular plants belong to 130 families, 872 genera, 3550–3750 species (depending on taxonomic interpretation), 847 subspecies, and 2000 varieties. Bulgaria is divided into 29 floristic regions and subregions (Fig. 1) (Assyov et al., 2002). Most of the species of Bulgarian flora are distributed in the lowlands, plains, and altitudinal belts reaching up to 700–900 m a.s.l. (xerothermic oak forests; mesophytic and xeromesophytic oak and hornbeam forests). The richest flora is that of the Stara Planina mountain range (about 2000 species), followed by the flora of Rhodope, Pirin, and Rila Mts. (1600–1800 species), the limestone mountains in western Bulgaria, Osogovo Mts., Strandja, and the lowland regions: the Struma valley, Thracian Lowland, the hilly plain of the Tundja River, the Black Sea coast and Dobrudja, and the Danubian Plain (from 800 to 2500 species). Biological diversity is related to the vertical distribution of species along the altitudinal belts and a variety of forest ecosystems: 2246 species are found below 500 m a.s.l.; 2137 species, from 500 to 1000 m; 780 species, from 1000 to 2500 m; and 114 species, above 2500 m. The richest flora is found on the limestone rocks. Most of the plant species found in Bulgaria belong to the following biological groups: perennial, 2264 species; biennial, 353; annual, 515 (Peev et al., 1998). There are new findings, which show that the total number of vascular plant species in the Bulgarian flora could be ca. 3900 (Petrova, 2002).

The flora of any given territory generally includes four groups of floristic elements: (a) edificators, the major builders of plant communities (which define the current status of the bioclimatic conditions); (b) dominants, the species, which are widely and evenly distributed in the major communities; (c) cosmopolites, the species with a great ecological amplitude, widespread over a large area; and (d) relict and endemic species, which reflect to the highest degree the history of the vegetation cover, even in the conditions of human interference. Endemic species are divided into different categories, which reflect the different stages of the speci-

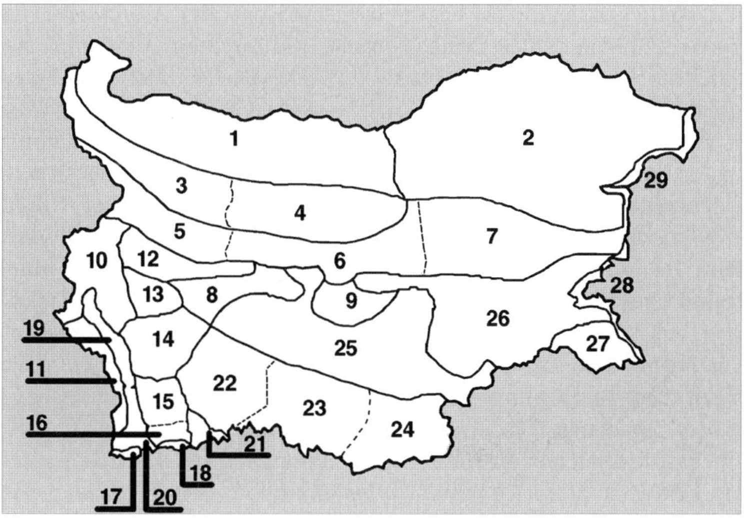

Fig. 1 Floristic regions and subregions in Bulgaria (according to Assyov et al., 2002). 1, Danubian Plain; 2, Northeastern Bulgaria; 3, Western Predbalkan; 4, Eastern Predbalkan; 5, Western Stara Planina; 6, Central Stara Planina; 7, Eastern Stara Planina; 8, Western Sredna Gora; 9, Eastern Sredna Gora; 10, Znepole Region; 11, Western Frontier Mts.; 12, Sofia Region; 13, Vitosha Region; 14, Rila; 15, Northern Pirin; 16, Southern Pirin; 17, Belasitsa; 18, Slavyanka; 19, North Struma Valley; 20, South Struma Valley; 21, Mesta Valley; 22, Western Rhodopes; 23, Central Rhodopes; 24, Eastern Rhodopes; 25, Thracian Lowland; 26, Tundja Hilly Plain; 27, Strandja; 28, Southern Black Sea coast; 29, Northern Black Sea coast

ation process, or different stages of selection and isolation. In the conditions of a temperate continental climate, which is characteristic for Bulgaria, the endemic species are most often found within the communities of widespread species, and rarely are edificators themselves (Kozhuharov, 1977).

Being the major component of floristic complexes, the endemic taxa in Bulgarian flora reflect its specific and genetic features, its vitality and stages of formation. The existence of strictly regional Bulgarian endemics proves that the territory of this country is rich in specific ecological niches and had a diverse phytogenetic and phytoclimatic history. The vascular plant flora of Bulgaria includes a considerable number of endemic species and subspecies. There are 170 Bulgarian endemic species and 100 subspecies, as well as 200 Balkan endemic species and subspecies. Bulgarian endemic species account for about 5% of the total flora (8% if subspecies are included). It is suggested that many Balkan endemic species probably have evolved within the modern territory of Bulgaria (mainly in the mountainous regions such as the Rhodopes, Pirin, or Slavyanka) and afterward have dispersed further over the Balkan Peninsula (Peev et al., 1998).

The endemic species in Bulgarian flora are divided in four groups as follows:

1. *Balkan endemics*: species that can be found only in the Balkan Peninsula but in more than one country. To this group belong *Aesculus hippocastanum* L. (found

in all countries of the Balkan Peninsula except Romania), *Acer heldreichii* Orph. ex Boiss. (Bulgaria, Macedonia, Albania, and Greece), *Alchemilla bulgarica* Rothm. (Serbia and Greece), *Crocus veluchensis* Herb. (widely distributed in the mountains of Bulgaria, Macedonia, and Northern Greece), *Fritillaria drenovskyi* Deg. et Stoj. (Bulgaria and Macedonia), *Gentianella bulgarica* (Vel.) Holub (Bulgaria, Albania, and Romania), *Geum bulgaricum* Panč. (Bulgaria, Bosnia, and Montenegro), *Haberlea rhodopensis* Friv. (Bulgaria and Northeastern Greece), *Lathraea rhodopea* Dingl. (Bulgaria and Greece), *Viola perinensis* W. Becker, etc. Most of the Balkan endemic species are distributed primarily in Bulgaria. They include: *Abies borisii-regis* Mattf., *Aquilegia aurea* Janka, *Astracantha thracica* (Griseb.) Podl., *Silene skorpilii* Vel., *Dianthus microlepis* Boiss., *Festuca valida* (Uechtr.) Penzes, *Pinus peuce* Griseb., *Satureja pilosa* Vel., etc. Some of these species form independent communities.

2. *Bulgarian endemics*: species that are found only in Bulgaria. In this group are included: *Anthemis orbelica* Panč., *Centranthus kellereri* (Stoj., Stef. et T. Georg.) Stoj. et Stef., *Chondrilla urumoffii* Deg., *Festuca pirinica* Horv. ex Markgr.-Dannb., *Geum rhodopaeum* Stoj. et Stef., *Jasione bulgarica* Stoj. et Stef., *Minuartia bulgarica* (Vel.) Griseb., *Ranunculus stojanovii* Delip., *Sedum kostovii* Stef., *Seseli rhodopaeum* Vel., *Silene velenovskyana* D. Jord. et P. Pan., *Tragopogon strybrnyi* Hayek, *Tulipa rhodopea* Vel., *T. urumoffii* Hayek, etc.
3. *Regional endemics*: species that are found only in a certain floristic region within Bulgaria. To this group belong *Anthemis strybrnyi* Velen. (Rhodopes), *Primula deorum* Vel. (Rila), *Rhinanthus javorkae* Soo (Pirin), etc.
4. *Local endemics*: species that are found in one, two or three adjacent areas of a single subregion. In this group are included: *Arenaria rhodopaea* Delip. (central Rhodope Mts.), *Brassica jordanofii* O.E. Schultz (in several adjacent dry valleys

Table 1 Families and genera of plants with the largest number of Bulgarian endemics

Family/Genus	Number of endemic species	Number of endemic subspecies
Asteraceae	38	12
Anthemis	10	16
Centaurea	8	–
Caryophyllaceae	18	10
Silene	6	1
Scrophulariaceae	11	14
Verbascum	8	4
Rosaceae	15	3
Alchemilla	6	–
Poaceae	12	4

of northern Pirin), *Silene velcevii* Jord. et Panov (Predbalkan), *Viola stojanowii* Becker (Belasitsa Mts., below Tumba Peak), etc.

The families and genera of plants richest in endemic Bulgarian taxa are listed in Table 1.

The Bulgarian endemic vascular plants are distributed mainly in the mountainous regions of the country. The richest in endemic taxa is the Stara Planina mountain range, which has over 90 Bulgarian endemic species, 20 of which are local. The following mountain ranges and massifs also exhibit Bulgarian endemics: the Rhodopes, over 80 species and subspecies (20% of which are found only in the Rhodopes); Pirin, over 70 species and subspecies (30 of which are local or strictly regional Pirin endemics); Rila, about 50 species and subspecies (ten of them are found only in Rila); Vitosha, 25 species and subspecies; Strandja, 19; Sredna Gora, 12; Belasitsa, eight; and Osogovo, six. The significantly lower number of endemic species and subspecies is found in the lowlands and low foothills (Peev et al., 1998).

3 Florogeographical Elements in Bulgaria

Flora and vegetation are the major components of the biota, and the analysis of their elements is substantially important for understanding the stability, evolution, and the required conditions for preservation of the ecosystems. So-called floristic elements combine species of similar formation within the limits of flora, or within certain periods of time. Florogeographical (floristic) elements include not only endemic species but also species whose biology and ecology best fit the conditions of the florogeographical region or province to which they are assigned. Therefore, a floristic element is a group of species originating from a certain region. Walter's classification (Assyov et al., 2002) allows suggesting the origin for ca. 3600 species of vascular plants in the Bulgarian flora. They are assigned to 92 florogeographical elements (Table 2). Some species cannot be clearly referred to a certain group.

The recent Bulgarian flora is a complex conglomerate; its endemic richness is determined by its complex evolutionary history and geographical conditions. In the analysis of that diverse picture, our attention will focus not only on those types of florogeographical elements which feature the highest percentage of distribution in our flora. The elements represented by individual species are also of great interest as they illustrate the broad range of evolutionary trends of the Bulgarian flora.

4 Centers of Speciation and Evolutionary Trends in the Bulgarian Flora

The most active speciation processes in Bulgarian flora take place in the regional and local centers of speciation. Such centers in Bulgaria are found at various altitudes above the sea level, and in nearly all botanical and geographical regions of the country. According to Kozhuharov (1977), there are three types of speciation centers:

1. The richest ones are found in the territories that are comparatively removed from the zones of probable Quaternary glaciers where vegetation was slow to change. Those speciation centers are referred to as refugia. To that type belong the Chervenata Stena Reserve near Bachkovo Monastery, and Zhdreloto Reserve near the village of Trigrad (Smolyan District). In those refugia, together with relict species exhibiting low variability that have remained from the Tertiary period such as *Lathraea rhodopea* Dingl., *Haberlea rhodopensis* Friv., *Trachelium rumelianum* Hampe, *Sideritis scardica* Griseb., *Morina persica* L., etc., one may come across comparatively young and not completely formed endemic species that have resulted from the action of natural selection on recent, postglacial settlement of plant species in those refugia.
2. The second type of speciation centers is found also on limestone rocks, but in the extreme climatic conditions of high mountains. In many respects these extreme conditions are mitigated by the limestone base; however, these conditions are considerably harsh, which results in much narrower ecological niches compared with the first type of speciation centers. They also feature a rich gene pool. To this group belong the speciation centers in the Rhodopes (Beglika), in the Vihren ridge of Pirin, in Slavyanka Mts., etc.
3. The third type of speciation centers is located on silicate rocks. Such centers can be found in the central part of Stara Planina (Botev, Ambaritsa, and Rusalka peaks), the Musala Peak in Rila Mts., etc. They do not feature a rich gene pool. Ecological conditions here are extreme: great temperature amplitudes; shallow, rocky, and cold soils; a short vegetation period; and few pollinators. These species are plastic, and most of them have clonal structure. In these conditions,

Table 2 Florogeographical elements in Bulgarian flora

Florogeographical elements	Number of species	% of the total species number
Submediterranean	479	12.5
European–Asian	355	9.3
Mediterranean	335	8.7
Balkan	294	7.7
European	257	6.7
European–Mediterranean	256	6.7
Boreal	215	5.6
Bulgarian	174	4.5
Pontian–Mediterranean	164	4.2
Eurosiberian	145	3.8
Cosmopolitan	120	3.1
Subboreal	95	2.5
Pontian	92	2.4
Carpathian–Balkan	49	1.28

continued

Table 2 continued

Florogeographical elements	Number of species	% of the total species number
Balkan–Anatolian	49	1.28
Mediterranean–Central Asian	48	1.25
Alpine–Carpathian–Balkan	39	1.02
Adventive	33	0.86
Pontian–Siberian	31	0.81
Alpine–Mediterranean	27	0.70
Balkan–Dacian	25	0.65
Arctoalpine	24	0.62
Pontian–Central Asian	23	0.60
European–Submediterranean	23	0.60
Apenninian–Balkan	21	0.55
North American	18	0.47
Mediterranean–Oriental–Turanian	17	0.44
South Pontian	16	0.41
European–Oriental–Turanian	15	0.39
Mediterranean–Asian	14	0.36
Alpine–Carpathian	13	0.34
Pontian–Submediterranean	11	0.28
Pannonian–Balkan	11	0.28
European–Central Asian	9	0.23
Pontian–Asian	8	0.20
Alpine–Balkan	8	0.20
European–North American	8	0.20
European–Pontian	8	0.20
European–South Mediterranean	7	0.18
Submediterranean–Asian	7	0.18
Alpine	6	0.15
South American	6	0.15
Pontian–Oriental–Turanian	5	0.13
South Euxinian	5	0.13
Pontian–Balkan	4	0.10
Central Asian	4	0.10
South Siberian	4	0.10
Asian	3	0.07
American	3	0.07
European–Mediterranean–Central Asian	3	0.07
South Mediterranean–Central Asian	3	0.07
Mediterranean–Siberian	3	0.07
Submediterranean–Central Asian	3	0.07
Paleoadventive	3	0.07
Hybridogenic	3	0.07
Pontian–Euxinian	2	0.05
Central American	2	0.05
Alpine–Apenninian–Balkan	2	0.05
Alpine–Apenninian–Carpathian–Balkan	2	0.05
European–American	2	0.05

continued

Table 2 continued

Florogeographical elements	Number of species	% of the total species number
European–West Asian	2	0.05
Mediterranean–South Asian	2	0.05
South Mediterranean–Siberian	2	0.05
Oriental–Turanian	2	0.05
Pannonian–Pontian	2	0.05
Japanese–Chinese	2	0.05
Alpine–Carpathian–Anatolian	1	0.02
Alpine–Caucasian	1	0.02
Apenninian–Alpine	1	0.02
Apenninian–Bulgarian	1	0.02
Balkan–Central Asian	1	0.02
Balkan–Dacian–Anatolian	1	0.02
Balkan–Caucasian	1	0.02
European–South Asian	1	0.02
European–Central African	1	0.02
European–Japanese–Chinese	1	0.02
Mediterranean–South American	1	0.02
Mediterranean–North American	1	0.02
South Mediterranean–Asian	1	0.02
Mediterranean–Chinese	1	0.02
Mediterranean–Balkan	1	0.02
Mediterranean–Atlantic	1	0.02
Submediterranean–Siberian	1	0.02
Submediterranean–Anatolian	1	0.02
Pontian–South Asian	1	0.02
Pontian–West Asian	1	0.02
Pannonian	1	0.02
Carpathian	1	0.02
Carpathian–Balkan–Caucasian	1	0.02
Carpathian–Balkan–Anatolian	1	0.02
Euxinian	1	0.02

the endemic apomictic species exhibit a number of advantages in the process of natural selection, which proves that they are competitive and possess the highest potential. The relict species here are much younger, of the glacial origin. Most often such centers include only neoendemic forms, intertwined in broad kinship circles, as for example the groups of *Poa alpina* L., *Myosotis sylvatica* Ehrh. ex Hoffm., many species of the genus *Hieracium* Mill., etc., whose closest relatives grow either in the immediate vicinity, or in the neighboring altitudes and florogeographical regions.

Two groups of speciation centers can be distinguished according to the extent of human impact and the contemporary floristic composition: (1) Primary speciation centers, which have been affected to a relatively small degree and have not been occupied by forest communities in the past; they are located in the middle and upper mountain belts (about 2000 m a.s.l.). Generally, the indigenous vegetation there has been preserved without alterations, or has undergone natural changes. (2) Secondary speciation centers, which have developed in places where the indigenous vegetation has been destroyed by man, namely, in the lowlands, the foothill belt and partially in the lower mountain zone (oak, black pine, and hornbeam forests).

The specific location of Bulgaria, the diversity of its relief, the rich geological history which contributes to the variety of the speciation centers, and the high percentage of endemism determine the diversity and the value of the Bulgarian vascular flora.

References

Assyov, B., D. Dimitrov, R. Vassilev & A. Petrova 2002. Conspectus of the Bulgarian Vascular Flora. Distribution Maps and Floristic Elements. Second Revised and Enlarged Edition. Bulgarian-Swiss Biodiversity Conservation Programme, Sofia.

Gruev, B., 1988. Obshta biogeografiya [General Biogeography]. Nauka i izkustvo [Science and Art], Sofia (in Bulgarian).

Kozhuharov, S. 1977. [Speciation centers and endemism in Bulgarian flora and conservation and enrichment of the gene pool]. Problemi na biologiyata [Issues of Biology] (Sofia) 10:34–61 (in Bulgarian).

Peev, D., S. Kozuharov, M. Anchev, A. Petrova, D. Ivanova & S. Tzoneva, 1998. Biodiversity of vascular plants in Bulgaria. In: Meine, C. (ed), Bulgaria's Biological Diversity: Conservation Status and Needs Assessment. Volumes I and II. Pensoft, Sofia & Moscow: 55–88.

Petrova, A., 2002. [Biosystematic and floristic studies in Bulgaria for the period 1993–2001]. In Temniskova, D. (ed), Trudove na Shestata Natsionalna konferentsiya po botanika, Sofia, 18–20 yuni 2001 [Proceedings of Sixth National Conference of Botany, Sofia, June 18–20, 2001]. Universitetsko izdatelstvo "Sv. Kliment Ohridski" [St. Kliment Ohridski University Publishing House], Sofia: 27–46 (in Bulgarian).

Stoyanov, N., 1950. Uchebnik po rastitelna geografiya [Textbook of Plant Geography]. Nauka i izkustvo [Science and Art], Sofia (in Bulgarian).

Takhtajan, A.L., 1978. Floristicheskie oblasti Zemli [Floristic Regions of the Earth]. Nauka [Science], Leningrad (in Russian).

22 Vegetation and Phytogeography: A Brief Characteristic

Tenyo Meshinev
Institute of Botany, Bulgarian Academy of Sciences, Akademik G. Bonchev Street 23,1113 Sofia, Bulgaria, e-mail: meshinev@bio.bas.bg

Abstract: The relatively small territory of Bulgaria ($111,000\,km^2$) is characterized by a high diversity of plant communities. Central European biological and climatic influence determines the main vegetation area type, deciduous forests. Biological and climatic influence of the steppe zone from the northeast, of the Mediterranean from the south, and of the South Euxinian Region from the southeast leads to the presence of certain specific vegetation elements. The diversity of plant communities is also determined by historical processes, by a variety of orographic and soil conditions, and by various forms of anthropogenic impact. Despite the anthropogenic changes, the natural environment within the territory of Bulgaria is well preserved

1 Vegetation and Phytogeography

Bulgaria is situated in the neighborhood of three European climatic and phytogeographic areas. To the northwest and west, mountain ranges connect the country's territory with the vegetation of Central Europe. To the north and northeast, Bulgaria borders the steppe zone covering southern Russia, Ukraine, and the Romanian part of Dobrudja. South of Bulgaria lies the Mediterranean landscape, typical of the Aegean Sea coast.

Deciduous (*Aestilignosa*) forests, the product of Central European bioclimatic influence, take up most of the Bulgarian territory. They are characterized, however, by the specific presence of a considerable number of South European and Submediterranean species. The close proximity to the Mediterranean area results in the limited presence of small patches populated by evergreen Mediterranean tree vegetation (*Durilignosa*). Mostly in the northeast of Bulgaria, as well as in its central parts, spread on rocky limestone terrain, there are stretches of steppe grassland (*Duriherbosa*).

The vegetation diversity is directly related to the orographic features of the country. The predominantly mountainous relief and the presence of high mountain ranges and massifs have served as a prerequisite for the spreading of the Central European phytoclimatic influence. At the same time, high mountains play the role of phytoclimatic barriers and create a wide range of diversity in natural habitats. The history of the evolution of the European vegetation and especially the history of glacial migrations shows that the mountain ranges of the Balkan Peninsula have

V. Fet and A. Popov (Eds.), Biogeography and Ecology of Bulgaria, 581–588.

provided migration routes to Boreal species, and such migration existed even prior to the glaciations (Stojanov, 1950; Palamarev, 1997).

The Alps, in their Dinaric part, and the Carpathian Mountains formed an important phytogeographic link between Central Europe and the southern part of the Balkan Peninsula. Today, we find a high percentage of common species in the high Bulgarian mountains and the Carpathians. Especially close in that respect are the South Carpathians, the Western Stara Planina Mts., Vitosha, and Rila, which probably constituted a major migration route for plants from Central Europe towards East Europe (Jordanov, 1924; Palamarev, 1997).

The Stara Planina Mts. play mainly the decisive role of a phytoclimatic barrier. Crossing Bulgaria latitudinally, it divides the country into the northern and southern parts. The northern slopes of Stara Planina are characterized with higher precipitation, more days with snow cover, and later spring. By limiting the influence of the northern and northwestern cold winds, Stara Planina provides conditions for a warmer and drier climate on its southern slopes and the adjacent Podbalkan valleys. The mountain climatic barrier determines the differences between North and South Bulgaria both at the floristic and at the phytocoenological levels. Despite the fact that the dominant vegetation type is solely that of deciduous forests, in North Bulgaria those forests are mainly dominated by *Quercus cerris* L. and *Q. frainetto* Ten., while in South Bulgaria, they are dominated by *Quercus pubescens* Willd. and *Carpinus orientalis* Mill. The presence of a large number of Mediterranean and Submediterranean species in the plant communities of South Bulgaria gives the vegetation there a more southern character as compared to that in North Bulgaria (Stojanov, 1941, 1950).

The Rila–Rhodopes Massif (Rila, Pirin, the Rhodopes, and Slavyanka) together with the massif of western Bulgaria, Serbia, and northern Macedonia, forms another considerable phytoclimatic barrier in the eastern part of the Balkan Peninsula. This barrier isolates the central parts of Bulgaria from the Mediterranean influence of the Aegean and Adriatic Sea coasts. The barrier is associated with the boundary between the evergreen sclerophytic tree vegetation and the deciduous forests. In their distribution to the north, a considerable number of plant species reach only as far as the southern slopes of the Rila–Rhodopes Massif (Lauraceae, Capparidaceae, Myrtaceae, etc.).

Consisting of the highest mountains on the Balkan Peninsula, this massif has remained closely connected with the Central European phytoclimatic influence. Approximately 85% of the high mountain species present in the Balkan Peninsula can be found also in Central or North Europe (Stefanov, 1943; Stojanov, 1950; Bondev, 1991; Velchev, 1997).

Along the valleys of Struma, Mesta, and Maritsa as well as at the Black Sea coast lie the migration corridors, where the climatic barriers have been partially crossed, and a number of Mediterranean elements have entered the inland areas of the Balkan Peninsula. Along the Black Sea coast, many southern species penetrate to Dobrudja and Bessarabia through Bulgaria. In this way, certain communities of evergreen shrubs have reached the southern slopes of Stara Planina (*Phillyrea latifolia* L.) or even North Bulgaria (*Pyracantha coccinea* Roem., to the town of Pliska).

Along the valleys of Maritsa and its tributary, Tundja, communities of *Phillyrea latifolia, Platanus orientalis* L., and *Juniperus excelsa* Bieb. reach the northern outskirts of the Rhodopes. The most important migration corridor, however, is associated with the Struma Valley. Through this valley, *Quercus coccifera* L. communities have reached as far as the village of Kamenitsa near Sandanski; *Phillyrea latifolia* and *Juniperus excelsa* communities have spread to the Kresna Gorge. A great number of Submediterranean species participate in the composition of various tree and shrub communities (*Colutea arborescens* L., *Jasminum fruticans* L., *Coronilla emeroides* Boiss., *Anthyllis aurea* Welld., *Juniperus oxycedrus* L., *Pistacia terebinthus* L., etc.) along the whole length of the Struma Valley. The Mesta Valley is of less significance for the dispersal of Mediterranean plant species.

Directly related to the mountainous character of the relief, six altitude vegetation belts are defined within the territory of Bulgaria (Velchev et al., 1982):

1. The belt of xerothermic oak forests, up to 500–600 m above the sea level (a.s.l.).
2. The belt of xeromesophytic oak and hornbeam forests, from 600 to 800 m a.s.l.
3. The beech belt, from 800 to 1500 (1600) m a.s.l.
4. The coniferous belt, from 1500 to 1800 (2000) m a.s.l.
5. The subalpine belt, from 1800 to 2500 m a.s.l.
6. The alpine belt, from 2500 to 2925 m a.s.l.

Deciduous forests are distributed throughout the country, and occupy the lowlands and mountain slopes up to 1500–1600 m a.s.l.

The area occupied by the xerothermic oak belt is mainly covered by forests, dominated by *Quercus cerris, Q. frainetto, Q. pubescens*, and *Q. virgiliana* (Ten.) Ten. In most cases these forests bear the marks of excessive human influence. Their undergrowth often contains *Crataegus monogyna* Jacq. and *Cornus mas* L. The herbaceous cover is characterized by the presence of spring ephemeres and ephemeroids such as *Scilla bifolia* L., *Isopyrum thalictroides* L., *Corydalis solida* (L.) Sw., *Mercurialis perennis* L., etc.

Mixed deciduous forests, where the oak plays a secondary role, occupy smaller areas, and their composition is dominated by *Fraxinus ornus* L., *Acer campestre* L., and *Tilia argentea* Desf. Shrub communities of *Carpinus orientalis, Paliurus spina-christi* Mill., *Cotynus coggygria* Scop., *Syringa vulgaris* L., *Rosa canina* L., and *Prunus spinosus* L. now occupy the place of former mixed deciduous and oak forests, which have been cut. Secondary grass communities differ in one major respect: depending on the amount of soil moisture and the economic use, they are divided into mesophytic meadows and xeromesophytic grasslands. The mesophytic meadows are used for hay harvesting and mostly occupy river terraces on rich, moist alluvial soils. These communities are dominated by *Festuca pratensis* Huds., *Poa pratensis* L., *Cynosurus cristatus* L., *Alopecurus utriculatus* Pers., *Bromus commutatus* Schrad., *Trifolium pratense* L., *T. repens* L., *Lotus corniculatus* L., *Leucanthemum vulgare* Lam., *Rumex acetosa* L., *Ranunculus acer* L., *Carex* spp., *Juncus* spp., etc. very much like in the similar communities in Central Europe.

Grass communities used for grazing occupy the hilly and drier parts of the relief. The plant formations of *Chrysopogon gryllus* (L.) Trin., *Dichantium ischaemum*

(L.) Roberty, *Festuca valesiaca* Schleich. ex Gaud., *Poa bulbosa* L., and *Haynaldia villosa* Schur. are widespread. In the northeastern part of the country as well as in the central territories where the terrain is limestone-based, one could find steppe elements in these communities.

The xerothermic oak belt includes plant communities with specific ecological features. The previously mentioned Mediterranean trees and other species such as *Juniperus excelsa, Phillyrea latifolia*, and *Quercus coccifera*, are fragmentarily distributed in the southern parts of the country, together with a large number of Submediterranean species. Narrow belts of *Salix alba* L., *Alnus glutinosa* (L.) Gartn., *Populus alba* L., and *Platanus orientalis* (in South Bulgaria) occupy the river banks and valleys. The most common communities near natural and artificial water basins are those of *Phragmites australis* (Cav.) Trin. ex Steud., *Typha latifolia* L., *T. angustifolia* L., and *Juncus* spp. The Black Sea coast sands host open communities of typical psammophytic species: *Leymus racemosus* (Lam.) Tzvel. ssp.*sabulosus* (Bieb.) Tzvel., *Ammophila arenaria* (L.) Link, *Eryngium maritimum* L., *Euphorbia paralias* L., *Silene thymifolia* S.S., *Jurinea albicaulis* Bunge ssp.*killaea* (Azn.) Koz., etc. Halophytic vegetation has a limited presence, mainly near swamps, lakes and along the Black Sea coast.

The longose forests around the mouths of the Kamchiya, Ropotamo, and Batova rivers are very specific elements of the Bulgarian vegetation. *Fraxinus oxycarpa* Willd., *Quercus pedunculiflora* C. Koch, *Ulmus foliacea* Gilib. are adapted to the regime of the long spring flooding together with the accompanying lianas *Smilax excelsa* L., *Periploca graeca* L., *Vitis sylvestris* Gmel., *Clematis vitalba* L., *Hedera helix* L., *Calystegia sepium* (L.) R. Br., etc. As a result of the overall drying of the climate and the additional amelioration activities, these forests are progressively shrinking.

The xeromesophytic oak belt is dominated by the forest communities of *Quercus dalechampii* Ten. and *Carpinus betulus* L. They mostly take up the northern slopes of the greater part of the Bulgarian mountains. Secondary shrub communities in this belt are most often formed by *Corylus avellana* L., while *Agrostis capillaris* L. dominates secondary grass communities.

Forest communities of the European chestnut, *Castanea sativa* Mill., have limited distribution and can be found only in single localities (one in Belasitsa Mts., above the town of Petrich, and another in Western Stara Planina, above the town of Berkovitsa). The same goes for the Balkan endemic species, horse chestnut (*Aesculus hippocastanum* L.), which can be found only at one site in the Preslavska Mts. (Eastern Predbalkan). The specific vegetation of Strandja Mts. is also attributed to the xeromesophilous oak and hornbeam belt. The combination of high soil moisture and air humidity with higher temperatures creates conditions for the development of typical representatives of the West Asian vegetation. The main tree species are *Fagus orientalis* Lipsky and *Quercus polycarpa* Schur. The undergrowth of these forests contains low shrubs of the laurel type such as *Rhododendron ponticum* L., *Prunus laurocerasus* L., *Ilex aquifolium* L., *Daphne pontica* L. The herbaceous cover is typically represented by *Scilla bithynica* Boiss., *Trachystemon orientale* (L.) D. Don., *Arum orientale* M.B., *Lathyrus aureus* (Stev.) Brandza, *Cicer monbretzii* Jaub. et Spach, *Primula acaulis* ssp.*rubra* (Sm.) Greut. Forests of *Fagus orientalis* and *Quercus polycarpa* are also distributed in the Eastern Stara Planina Mts.; however, they completely lack laurel-type shrubs, and West Asian herbaceous species are limited.

The beech belt is well defined on Stara Planina, Rila, Pirin and partially in other high mountains. *Fagus sylvatica* L. forests are remarkably similar to those of Central Europe. As compared to oak forests, they are better preserved, and in the Central Stara Planina Mts. one can find the largest territories of compact beech forests in Europe. Little, if any, grass species can be found under the thick covering of the beech. These are typical sciophytic species such as *Sanicula europaea* L., *Circaea lutetiana* L., *Oxalis acetosella* L., *Adoxa moschatellina* L., *Dryopteris filix-mas* (L.) Schott, *Phyllitis scolopendrium* (L.) Newm., *Luzula sylvatica* (Huds.) Gand., and *L. luzuloides* (Lam.) Dandy. Glades are most often dominated by *Agrostis capillaris* L. At a lower altitude (below 800 m), *Fagus sylvatica* is represented by the taxonomically poorly defined ssp. *moesiaca*, which takes up mainly the northern slopes and the southern parts of Stara Planina.

The coniferous belt is best expressed in Pirin, Rila, and the Rhodope Mts. Its presence is limited (fragmentary) in Stara Planina, Vitosha, and Slavyanka. *Pinus sylvestris* L. and *Picea abies* (L.) Karst. form the major plant communities in this belt, dividing among themselves the southern and northern slopes respectively. Quite often, however, these two species form mixed communities at higher altitudes. In terms of their composition and structure, silver pine and spruce forests are similar to those of Central and North Europe but in the Bulgarian mountains they grow faster, are more productive, and have less developed moss cover.

Abies alba Mill. communities delineate a thin fragmented strip between the spruce and oak wood borders. In the southern parts of the country, on limestone terrain, one could also find the limited distribution of *Abies borisii-regis* (Mattf.) Koz. et Andr. communities.

The Balkan endemic pines, *Pinus peuce* Grsb. and *P. heldreichii* Christ., add special feature to the coniferous belt. *Pinus peuce* grows either in pure stands or mixed with *Picea abies* in the highest parts of the coniferous belt of Rila, Pirin, and in one locality in Central Stara Planina. *Pinus heldreichii* grows only on Pirin and Slavyanka, mainly on limestone terrain.

The forests of *Pinus nigra* Arm. ssp.*pallasiana* (Lamb.) Holmb. (which some authors assign to lower mountain belts) have been reduced as a result of their intensive logging in the past. Today, only small fragments remain; almost all of them are distributed in South Bulgaria.

The secondary shrub and grass communities have emerged within the limits of the coniferous belt, where old forests have been logged. The Balkan endemic *Chamaecytisus absinthioides* (Janka) Kuzm. is widespread among the shrub species. Also the Subboreal species, *Potentilla fruticosa* L., is represented in the single locality in Western Rhodopes (its only record for the entire Balkan Peninsula).

Grass communities that are subjected to grazing are most often dominated by *Nardus stricta* L., *Festuca nigrescens* Lam., *Agrostis capillaris* L., *Bellardiochloa violacea* (Bell.) Chiov., and they are also widespread. Their composition includes numerous southern and endemic species such as *Viola rhodopaea* W. Becker, *Veronica rhodopaea* (Vel.) Deg., *Scabiosa rhodopaea* Stoj. et Stef., *Morina persica* L., *Linum capitatum* Kit. ex Schult., *Pastinaca hirsuta* Panè., *Geum rhodopaeum* Stoj. et Stef., and *Euphorbia myrsinites* L.

In moist habitats, one can find communities of *Molinia coerulea* (L.) Moench, *Deschampsia caespitosa* P.B., *Carex nigra* (L.) Rchb., and *Carex rostrata* Stockes, with *Eleocharis palustris* R. Br., *Galium palustre* L., and *Caltha laeta* Schott, Nym. et Kotschy as accompanying species.

The timberline in the Bulgarian mountains runs between 1800 and 2200 m above the sea level. Above this zone, up to 2500 m, there is the subalpine belt, consisting of shrub and grass vegetation.

In the high mountains, Rila and Pirin, there is a well-defined belt of *Pinus mugo* Turra communities. In many places the belt is fragmented as a result of the logging and fires. In Vitosha Mts. *Pinus mugo* has been completely destroyed for the same reason. *Juniperus sibirica* Burgsd. is exceptionally widespread in the subalpine belt. It populates enormous territories in Stara Planina, Vitosha, Rila, Pirin, and the higher parts of the Rhodopes. The Siberian juniper grows in rare and thick communities, massively intermingled with low shrubs such as *Vaccinium myrtillus* L., *V. vitis-idaea* L., and *Bruckenthalia spiculifolia* (Salisb.) Rchb. Grass communities in the subalpine belt are rather diverse and rich in species. Some of those species are endemic to the Balkan Peninsula and contribute to the special character of the subalpine belt. These communities are dominated by *Festuca balcanica* (Acht.) Markgr.-Dannb., *F. valida* (Uechtr.) Penz., *Sesleria comosa* Vel., *Festuca riloensis* (Hack ex Hay.) Markgr.-Dannb., and *Poa pirinica* Stoj. et Acht. Only in Rila, in more humid and limited localities, the local endemic *Primula deorum* Vel. forms communities.

Formations of *Calamagrostis arundinacea* (L.) Roth, *Brachypodium pinnatum* (L.) P.B., *Bellardiochloa violacea* (Bell.) Chiov., *Molinia coerulea* (L.) Moench, *Deschampsia caespitosa* P.B., and *Carex kitaibeliana* Deg. ex Bech. are widely distributed in the subalpine belt.

Above 2500 m of altitude, the alpine vegetation belt is represented by a limited number of plant formations, the most common among which are dominated by *Vaccinium uliginosum* L., *Juncus trifidus* L., *Agrostis rupestris* All., and *Carex curvula* All.

2 Human Influence on Vegetation in Bulgaria

The recent vegetation is determined by the specific phytogeographic features of Bulgaria; however, as everywhere around the world, here the anthropogenic influence has imposed its considerable impact. The changes caused by the human influence could be presented in the following main groups.

Deforestation. The process of deforestation began with the emergence of agriculture, probably back in the Neolithic period, and has progressed throughout history. There are old historical documents, testifying to the fact that in the early 18th century the territory of Bulgaria was covered by old mixed deciduous forests. Timber logging for construction and daily needs, the expansion of the tilled land, burning forest for the expansion of pastures, the random grazing of goats, and many more reasons have led to the decrease in the overall area occupied by forests.

Grazing. For centuries, the extensive form of cattle breeding in Bulgaria led to excessive utilization of the pastures, especially of the high mountain grasslands on the part of herds of sheep, cows, and horses. The number of such herds increased dramatically especially during the summer months, when the Karakachan sheep herds moved along their regular migration route from the Mediterranean to the high Bulgarian mountains and back. The excess grazing brought about changes in the species composition and decrease in the biodiversity. About 70% of the territory of the mountain grasslands was taken up by communities of *Nardus stricta*.

In the past several decades, various social and economic factors resulted in a considerable reduction of cattle breeding in Bulgaria. The number of herds in the high mountains sharply dropped, and over vast areas grazing practically ceased. As a result, grass communities are experiencing large-scale successional changes, which are leading to a decrease in the range of *Nardus stricta*, and an increase in the biological diversity. This process could be assessed as positive from the environment conservation point of view but it is relatively transient. Further successional changes lead to the quick advent of *Juniperus sibirica* in the high mountain zone and to the "invasion" of the forest in the lower parts of the mountains. The latter leads to loss of pastures and decrease in biodiversity. Recommendations have been made to the national parks management plans to control the successional changes in the vegetation (Meshinev et al., 2000).

Wetlands Destruction. In the conditions of the communist "planned economy" many of the wet areas along the Danube and Maritsa rivers, along the Black Sea coast and others were dried up in order to utilize this land for agriculture. That led to the destruction of considerable territories of hygrophytic communities and the disappearance of rare hygrophytic plant species. These unfavorable trends developed against the background of the overall drying and warming of the climate. The scientific institutions and non-governmental organizations have recently exerted a lot of effort to preserve the wetlands.

Pollution. Industrial, transport, and daily-life pollution of the environment, excess use of pesticides in farming and forestry, excess application of artificial fertilizers lead to unfavorable changes in the land and water ecosystems.

Notwithstanding the negative changes triggered by human activities, Bulgarian vegetation has preserved to a considerable extent its original, natural character. As compared to many countries of Central and Western Europe, Bulgaria has preserved a rich and diverse nature, whose conservation is one of the priorities of the government and the scientific institutions, of the non-governmental organizations and the entire society.

References

Bondev, I., 1991. Rastitelnostta na Balgariya. Karta v M 1: 600000 s obyasnitelen tekst [The Vegetation of Bulgaria. Map 1: 600000 with Explanatory Text]. Universitetsko izdatelstvo "Sv. Kliment Ohridski" [St. Kliment Ohridski University Publishing House], Sofia (in Bulgarian).

Jordanov, D., 1924 [On the phytogeography of Western Stara Planina]. Godishnik na Sofiiskiya Universitet. Fiziko-matematicheski fakultet (Annuaire de l'Université de Sofia. Faculté physico-mathématique) 20(1):1–102 (in Bulgarian).

Meshinev, T., I. Apostolova, E. Kachaunova, V. Velchev & I. Bondev, 2000. Flora and plant communities. In Popov, A. & T. Meshinev (eds), High Mountain Treeless Zone of the Central Balkan National Park. Biological Diversity and Problems of Its Conservation. Pensoft, Sofia.

Palamarev, E., 1997. [Basic stages of the development of flora and vegetation in the geologic history of Bulgarian lands]. I Yordanova, M. & D. Donchev (eds), Geografiya na Balgariya. Fizicheska geografiya. Sotsialno-ikonomicheska geografiya (Geography of Bulgaria. Physical Geography. Socio-Economic Geography). Akademichno izdatelstvo "Prof. Marin Drinov" (Prof. Marin Drinov Academic Publishing House), Sofia: 60–265 (in Bulgarian).

Stefanov, B., 1943. [Phytogeographische Elemente in Bulgarien]. Sbornik na Balgarskata akademiya na naukite. Klon prirodo-matematichen [Collected Works of the Bulgarian Academy of Sciences. Natural-Mathematical Series] 39(19):1–509 (in Bulgarian).

Stojanov, N., 1941. [Attempt on characterization of the main phytocoenoses in Bulgaria]. Godishnik na Sofiiskiya Universitet. Fiziko-matematicheski fakultet (Annuaire de l'Université de Sofia. Faculté physico-mathématique) 37(3):93–188 (in Bulgarian).

Stojanov, N., 1950. Rastitelna geografiya [Plant Geography]. Nauka i izkustvo [Science and Art], Sofia (in Bulgarian).

Velchev, V., 1997. [Main features and regularities of distribution of the modern vegetation]. In: Yordanova, M. & D. Donchev (eds), Geografiya na Balgariya. Fizicheska geografiya. Sotsialno-ikonomicheska geografiya (Geography of Bulgaria. Physical Geography. Socio-Economic Geography). Akademichno izdatelstvo "Prof. Marin Drinov" [Prof. Marin Drinov Academic Publishing House], Sofia: 265–269 (in Bulgarian).

Velchev, V., S. Ganchev & I. Bondev, 1982. [Vegetation belts]. In Galabov, Zh. (ed), Geografiya na Balgariya. Fizicheska geografiya. Prirodni usloviya i resursi (Geography of Bulgaria. Physical Geography. Natural Conditions and Resources). Izdatelstvo na Balgarskata akademiya na naukite (Publishing House of the Bulgarian Academy of Sciences), Sofia: 439–440 (in Bulgarian).

23 Landscapes and Climate of Bulgaria

Veliko Velikov[1] and Milena Stoyanova[2]

[1]*Episkop Konstantin Preslavski University of Shumen, Universitetska St. 115, 9712 Shumen, Bulgaria, e-mail: v.d.velikov@abv.bg*

[2]*Episkop Konstantin Preslavski University of Shumen, Universitetska St. 115, 9712 Shumen, Bulgaria, e-mail: stoyanovamilena@abv.bg*

Abstract: The landscapes of Bulgaria are briefly characterized. Landscape differentiation of Bulgaria is based on horizontal and vertical geographical zonality. Three major types of landscapes are described in Bulgaria: plains (below 300 m a.s.l.), mountains, and aquatic landscapes. Climatic features of the major landscape subtypes are described. Rivers and lakes of Bulgaria are also characterized

1 Introduction

Bulgaria is situated in the southeastern part of Europe. It covers 111,000 km^2 in the central-eastern part of the Balkan Peninsula, between 41°14′ to 44°13′ N, and 22°02′ to 28°36′ E. The territory of Bulgaria is delineated from the north by the Danube River, and from the east by the Black Sea. The southern border is mountainous and separates Bulgaria from Greece and Turkey. The western border is also mountainous, and separates Bulgaria from Macedonia and Serbia. The territory of Bulgaria is divided into two parts by the latitudinally running Stara Planina (Balkan) mountain range (highest summit, Mt. Botev, 2376 m a.s.l.). The territory is flat to the north from this mountain range, and mountainous to the south from it (Fig. 1).

Landscape differentiation of Bulgaria is determined by both horizontal (latitudinal and meridional) and vertical (altitudinal) geographical zonality. The major climax communities of plants and animals (biomes) change according to both horizontal and vertical zonality.

Latitudinal horizontal zonality is conditioned by the irregular distribution of the solar radiation according to the latitude. Almost every landscape component (climate, soils, vegetation, and animal world) changes with latitude. *Meridional horizontal zonality* is determined by the climatic changes according to the longitude. In Europe in general, and in Bulgaria in particular, the continental climate increases from the Atlantic Ocean eastward, affecting the landscapes. Both latitudinal and meridional zonality determine the major climatic differences and specific features of vegetation in the lowland parts of Europe.

Vertical zonality is expressed in the mountains above 500 m a.s.l., where the climatic elements change, especially as temperature decreases. The vertical zonality depends on the relief of the territory determined by the geology. The altitude, through its strong and direct influence on the local climate, affects soils, vegetation

V. Fet and A. Popov (Eds.), Biogeography and Ecology of Bulgaria, 589–605.

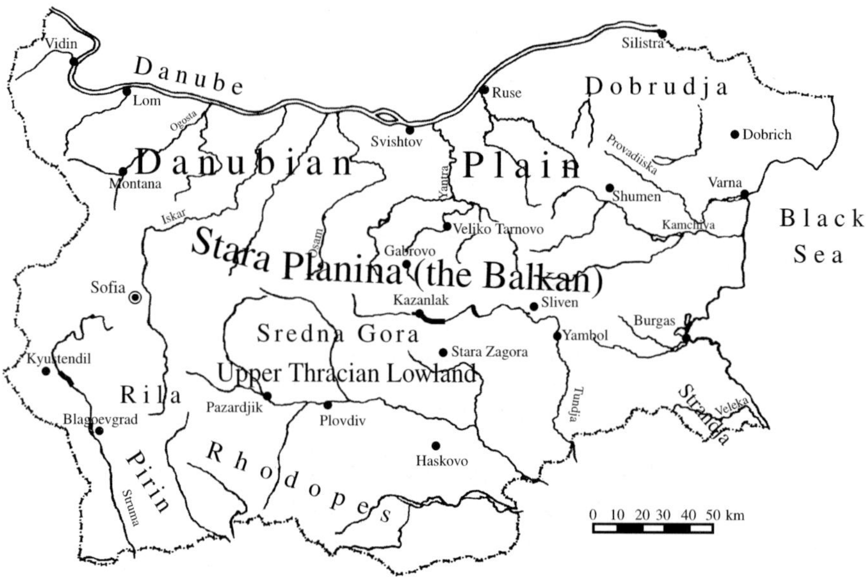

Fig. 1 Hydrogeographical map of Bulgaria.

and animals, and the entire landscape. In addition, the mountain slope exposition has a major effect on temperature and humidity. In the Bulgarian mountains, altitudinal belts of the northern macroslopes are shifted 100 to 300 m up compared to those of the southern slopes.

The degree of influence exerted by the slope exposition depends on the position of the mountain ranges, on their horizontal and meridional zonality, and on the predominant direction of air masses movement. Two kinds of exposition, solar and wind, produce similar kind of asymmetry in the altitudinal belts. The asymmetry of circulation is very pronounced when a mountain chain is perpendicular to the direction of the humid air masses movement. This kind of asymmetry is characteristic for the Stara Planinia, the Rhodopes, and the eastern and western slopes of Rila and Pirin (Mishev et al., 1989).

2 Classification of Landscapes

Out of many classification systems in landscape science (Petrov, 1990), some are typological (Gvozdetskiy, 1973; Harvey, 1974; Armand, 1975), while others are geochemical, based on the conditions of elements' migration (Glazovskaya, 1964).

Table 1 Classification ranks in typological systems of landscapes

Author	Gvozdetskiy (1973)	Harvey (1974)	Petrov (1980)
Taxonomic rank	Class Type Subtype Group Species	Class Subclass Group Subgroup Species	Class Type Subtype Group Species

A typological classification system is accepted here (Table 1) down to the subtype rank (corresponding to a group rank of Harvey, 1974).

Our typological system for the landscapes of Bulgaria is presented in Table 2. It was based on the classification according to Gvozdetskiy (1973) and Harvey (1974), with addition of diagnostic features.

Table 2 Typological classification system for the landscapes of Bulgaria

Taxonomic rank	Diagnostic features	Landscapes
Class	Physical condition of the Earth's surface: territory or aquatory; general geomorphological structure, determining zonality and correlation between temperature and humidity	Plain Mountain Aquatic (water)
Type (subtype of Harvey, 1974)	Characteristic zonal indexes for the hydrological and climatic conditions of a given landscape class; the type of vegetation and soils; the type of watershed	Mountain forest Mountain meadow Plain-steppe Plain-agrophytocoenoses Black Sea (aquatic) Rivers (aquatic) Natural and artificial water bodies
Subtype (group of Harvey, 1974)	Characteristic zonal indexes within the limits of a given landscape type	High-mountain alpine meadows High-mountain coniferous forests Low-mountain deciduous forests Danube River (aquatic) Black Sea (aquatic) Freshwater and artificial lakes (aquatic)

3 Characterization of Landscapes

3·1 Class 1. Plain landscapes

These landscapes include the lowland plains (below 300 m a.s.l.).

3·1·1 Type 1.1. Plain-steppe landscapes

Climate temperate-continental; only the Black Sea coast has the Mediterranean-continental climate. One pronounced precipitation minimum in January and one maximum in July. The lowest temperatures in January; average January air temperatures in North Bulgaria are from −1.5 to 3 °C. In the last 30 years, the average monthly air temperatures clearly increased, especially in North Bulgaria. In spring, increase of air temperature is due to the quick increase of radiation and thermal balance and frequent changes of western and northwestern warm and cold air masses. In April, the average air temperature is from 10.5 to 12.5 °C. Due to the Black Sea influence, the air temperature is lower near the coast (9.5–10.5 °C). The average air temperature in July in North Bulgaria is ca. 22 °C. In fall, the average temperature decreases but remains higher than in spring. The highest average October temperature is recorded at the Black Sea coast (Varna, 13.9 °C).

In the Danubian Plain, the annual precipitation increases from the Danube River toward the Predbalkan (Orehovitsa near Dolna Mitropoliya, 558 mm; Pleven, 578 mm; Lovech, 641 mm; Troyan, 767 mm). The lowest annual precipitation (150–200 mm) is recorded in northeastern Bulgaria. The temperate continental precipitation regime is connected with the Atlantic cyclones and has its maximum in June and May and minimum in February (the Danubian Plain). The snowfall comprises about 10–12% of annual precipitation in plains and lowlands of Bulgaria. Snow cover in the plains disappears during the first and second decade of March.

3·1·2 Type 1.2. Plain agrophytocoenoses

The agricultural terrains divide this landscape type into three subtypes.

3·1·2·1 Subtype 1.2.1. Plain agrophytocoenoses of the Moesian plain

These landscapes are found in Ludogorie and Dobrudja. Climate temperate-continental (the same as in the plain-steppe landscapes).

3·1·2·2 Subtype 1.2.2. Plain agrophytocoenoses of the Upper Thracian Lowland and the middle part of Tundja Valley

Here, the lowest temperatures are recorded in January. The average temperatures are between 0 and 1–2 °C in the Upper Thracian Lowland, and 1–3.2 °C near the Black

Sea coast and in the southernmost lowlands. The average temperatures in July in the lowlands of South Bulgaria are 23–24 °C. Higher average temperatures are recorded in October near the southern portion of the Bulgarian Black Sea coast (15 °C; Rezovo). November temperature differences are very pronounced and significantly decrease, by 1.4 °C, in South Bulgaria. The highest average maximum monthly temperatures are recorded in July and August in the Upper Thracian Lowland and in the southernmost border regions (31.2 °C in Harmanli, 32 °C in Svilengrad). In summer, the daily temperature rises to its extremes.

The Mediterranean precipitation regime is observed in the southernmost and easternmost areas of Bulgaria, which are influenced by the Mediterranean cyclones, especially during fall and winter. The precipitation maximum in these areas is in November and December, and the minimum is in August or September.

3·1·2·3 Subtype 1.2.3. Agrophytocoenoses of the depressions and broad valleys

Temperatures in the depressions are comparatively lower. The average winter temperatures there are lower by 1.5 °C than in the places with slope relief, due to frequent temperature inversions in the negative relief forms. The lowest average temperature in January in the depressions is −3.6 °C (Knezha), and the highest, 3.2 °C (Tsarevo). In April, temperatures are also lower in the depressions, by 1–1.5 °C. In Sandanski in the Struma Valley, the average July temperature reaches 24.9 °C. The lowest July temperature (18 °C) was recorded at Tran Station. The highest average temperatures in October are 11–12 °C in the rest of the lowlands; in Sandanski and Petrich they are 14.6 and 14.3 °C, respectively. The highest average maximum monthly temperatures are recorded in July and August in Petrich (31.8 °C) and Sandanski (31.4 °C).

In the depressions of Western Bulgaria, rain sums are the same as those in the southern part of the Danubian Plain. The precipitation regime here (e.g. in the Struma Valley between Zemen and Kresna gorges) is transitional compared to that of the other landscapes. It has two maximums and two minimums, due to the influence of both Atlantic and Mediterranean cyclones.

The snowfall comprises about 20% of precipitation in high depressions. The first snow cover there appears in December (end of December in the major part of Struma and Tundja river valleys). The average annual number of days with snow cover varies in different parts of Bulgaria. The Black Sea coast and the Struma Valley (to the south of Kocherinovo) have the lowest number of days with snow cover, less than five a year.

3·2 Class 2. Mountain landscapes

The mountain landscapes are characterized by the vertical zonality, expressed in the spectra of distinct vertical belts. The altitudinal spectrum of a separate mountain range or a separate macroslope is characterized by the number of vertical belts,

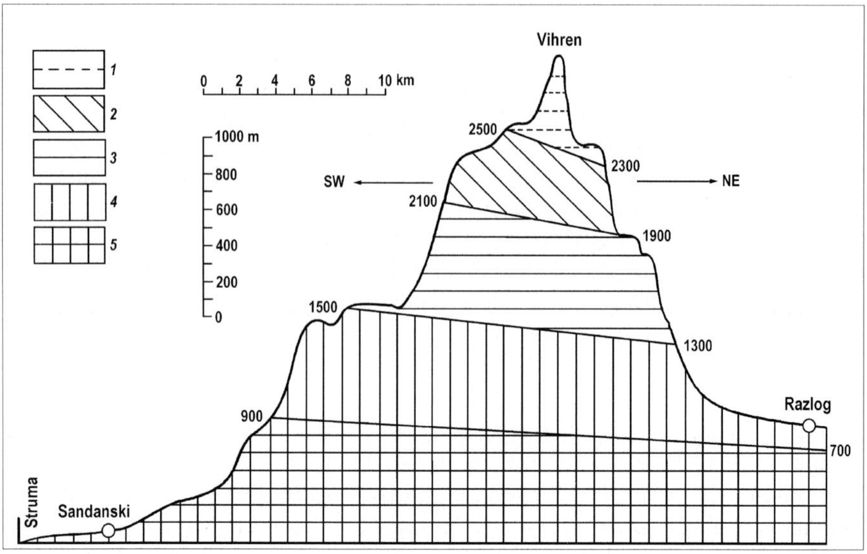

Fig. 2 Vertical landscape belts in Pirin Mts. (after Mishev et al., 1989). 1, Alpine belt; 2, High mountain belt; 3, Middle mountain belt; 4, Low mountain belt; 5, Foothill belt.

stages of their development, different belt altitude, asymmetry on the two opposite slopes, altitude limits, order of belts because of the reduction or inversion of certain belts, etc. (Mishev et al., 1989) (Figs. 2–6). According to the type (subclass) or subtype (group) of the mountain landscape, Mishev et al. (1989) distinguish several types of vertical belt spectra with their specific landscape belts (Table 3).

In the middle mountains (1200–1600 m a.s.l.), the average January temperature is −3 to −4 °C; in the high mountains (above 1600 m) it is under −5 °C; and on Musala peak (2925 m) it reaches −10.9 °C. In the highlands, April resembles a winter

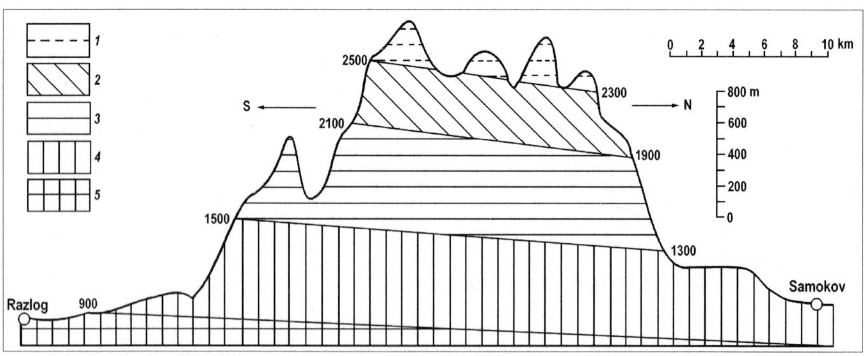

Fig. 3 Vertical landscape belts in Rila Mts. (after Mishev et al., 1989). 1, Alpine belt; 2, High mountain belt; 3, Middle mountain belt; 4, Low mountain belt; 5, Foothill belt.

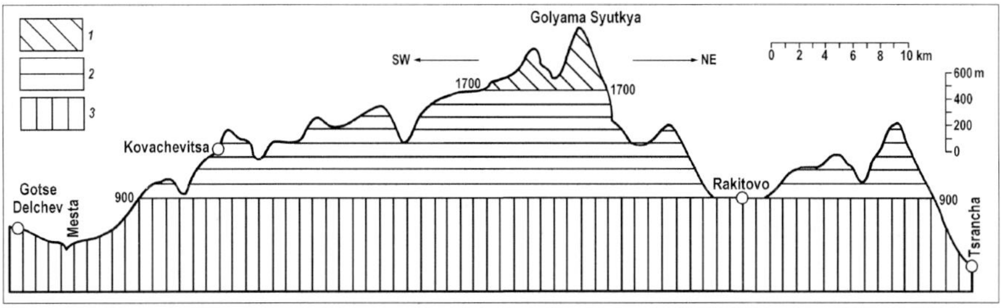

Fig. 4 Vertical landscape belts in Western Rhodopes (after Mishev et al., 1989). 1, High mountain belt; 2, Middle mountain belt; 3, Low mountain belt.

month. At 2000 m a.s.l., the average daily temperature in the end of May is 5 °C. In summer, temperature differences decrease in hilly and low mountain areas. July temperatures decrease with altitude to 5 °C (Musala). In the middle mountain areas October is warmer than April by 2–2.5 °C, and in high mountain areas, by 3–4 °C. November temperature differs by 0.8 °C in the areas above 1000 m a.s.l. Where the snow cover appears, average temperatures become negative (−5.3 °C on Musala peak). The average maximal monthly temperatures decrease with altitude to 9.1 °C (Musala).

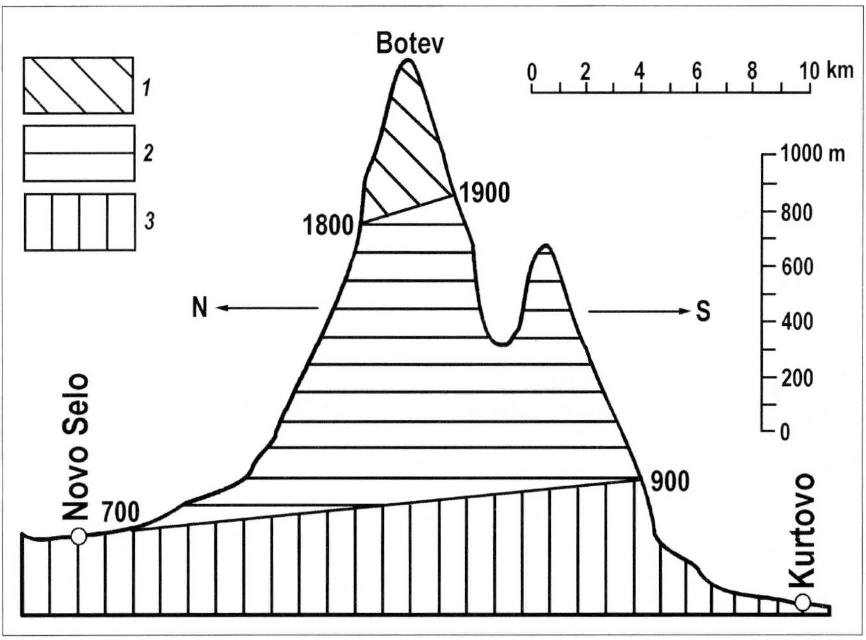

Fig. 5 Vertical landscape belts in Stara Planina (after Mishev et al., 1989). 1, High mountain belt; 2, Middle mountain belt; 3, Low mountain belt.

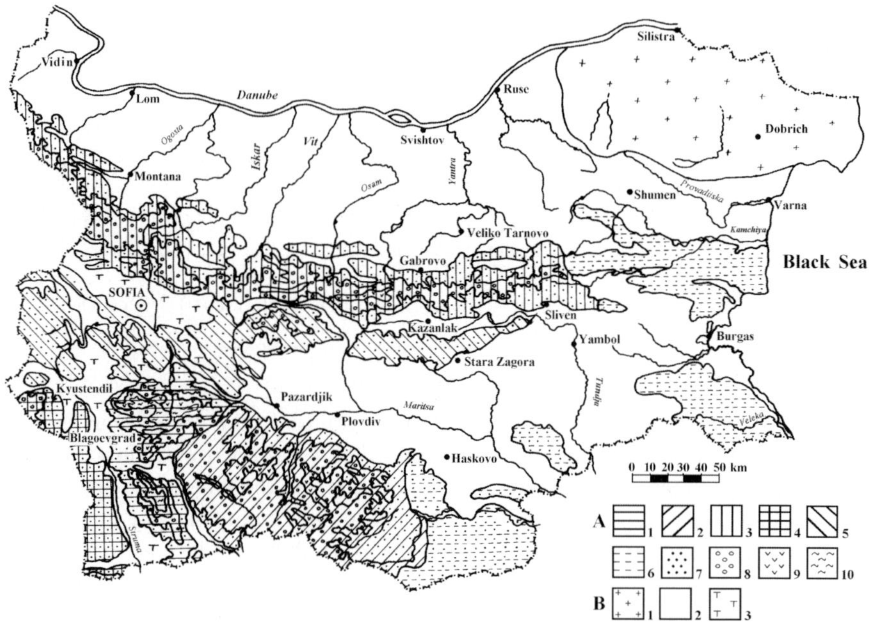

Fig. 6 Map of the vertical landscape belts and types of belt spectra (after Mishev et al., 1989, with additions). (A) Mountain landscapes: 1, Rila–Pirin; 2, Western Rhodopes; 3, Stara Planina; 4, Osogovo–Belasitsa; 5, Kraishte–Sredna Gora; 6, Eastern Bulgaria; 7, Low mountain landscapes; 8, Middle mountain landscapes; 9, High mountain landscapes; 10, Alpine landscapes. (B) Plain landscapes: 1, Plain-steppe landscapes; 2, Plain agrophytocoenoses; 3, Plain agrophytocoenoses of the depressions and broad valleys.

The lowest average minimal monthly temperatures are in January (−13.8 °C at Musala, and −12 °C at Beglika). Beglika has a record of average minimal monthly temperature in January −26.6 °C; it is considered the coldest place in Bulgaria.

The vertical gradient in precipitation is expressed up to a certain altitude, above which it decreases or totally disappears. This is true despite the fact that, in the mountain areas, precipitation includes snowfall, and strong winds winnow much of the snow so it cannot be measured. The vertical gradient in precipitation depends on the slope exposure; e.g. the difference in annual sum between the northern and southern foothills of Stara Planina is almost 300 mm. However, at 1500 m a.s.l. this difference is only 100 mm due to the altitude. The heaviest annual precipitation sums in Bulgaria (1700–2200 mm) are recorded in the high mountain stations facing the wind. Snow cover in Bulgaria forms annually, but it is stable (lasting over 30 days) only in the high mountains. The earliest snow cover in Bulgaria forms in the highest mountains in the second half of October, and in middle and low mountains, in November. The latest snow cover is observed from the end of March in the southeastern mountains (Strandja, Sakar, the Eastern Rhodopes) to the end of May (and even beginning of June) in high mountains above 1500–1700 m a.s.l.

Table 3 Vertical belts in the mountainous landscapes in Bulgaria (after Mishev et al., 1989)

Type of the vertical landscape	Altitude (m a.s.l.)	1	2	3	4	5	6	7	8	9	10	11	12	13
1. Rila–Pirin														
low mountain	700–900 to 1300–1500	44	11	24	800	43	7.3	100	20	563	65	*Fagus sylvatica* (50), *Abies alba* (20), *Pinus sylvestris* and *P. nigra* (30)	94	6
middle mountain	1300–1500 to 1900–2100	26	20	25.9	1000	63	4.3	150	50	426	97	*Picea abies* (60), *Pinus peuce* and *P. heldreichii* (30)	28	72
high mountain	1900–2100 to 2300–2500	18	31	26.9	1100	89	1.6	170	80	306	35.2	*Picea abies* (40), *Pinus mugo* (40)		
Alpine	above 2300–2500	12	38	23.2	1250	96	−1	180	85	196	34.5	*Sesleria comosa, Festuca*		
2. *Western Rhodopes*		100	20	100	960	66	5.1	135	46	437				
low mountain	300–900	34	9	21	700	41	10	50	8	723	94.7	*Quercus dalechampii* (40), *Q. frainetto* (30), *Fagus sylvatica* (10)	79	21
middle mountain	900–1700	58	16	63	850	59	6.8	100	30	502	96	*Fagus sylvatica* (10), *Abies alba* (10), *Picea abies* (50)	46	54
high mountain	above 1700	8.3	28	16	1100	80	3	150	70	339	38.4	*Picea abies* (40), *Pinus peuce* (50)	98	1
3. *Stara Planina*		100	15	100	820	56	7.6	87	25	564				
low mountain	500 to 700–900	52	11	32	750	46	9.7	60	10	600	55.7	*Quercus frainetto* (40), *Q. pedunculiflora* (10), *Carpinus orientalis* (30)	90	10

continued

Table 3 continued

Type of the vertical landscape	Altitude (m a.s.l.)	1	2	3	4	5	6	7	8	9	10	11	12	13
middle mountain	700–900 to 1800–1900	45	25	63	1100	72	6.5	100	30	482	62.5	*Quercus dalechampii* (50), *Carpinus betulus* (30), *Fagus sylvatica* (20)	98	0.2
high mountain	above 1800–1900	2.7	36	5	1250	91	1.8	160	75	248	79.8	*Fagus sylvatica* (20), *Picea abies* (50), *Pinus peuce* (20)	74	25.8
4. Osogovo–Belasitsa		100	18	100	920	60	8.5	80	21	537				
low mountain	700 to 1300–1400	81	9	66.6	800	36	9	65	15	749	65	*Fagus sylvatica* (40), *Quercus dalechampii* (40), *Pinus sylvestris* (20)	99	0.2
middle mountain	1300–1400 to 1900	18	18	29.1	1000	57	5.4	110	40	520	72.3	*Fagus sylvatica* (40), *Abies alba* (10), *Picea abies* (20)	58	42
high mountain	above 1900	1.8	25	4.3	1100	72	2.4	150	60	324	45.3	*Abies alba* (10), *Picea abies* (40), *Pinus sylvestris* and *P. nigra* (40)	46	54
5. Kraishte–Srednogorie		100	11	100	840	41	8.2	75	20	701				
low mountain	600–700 to 1200–1400	89	9	78.8	750	38	8	80	15	565	92	*Quercus dalechampii* (50), *Carpinus betulus* (20), *Pinus nigra* (30)	83	16.6
middle mountain	1200–1400 to 1700–1800	9.2	19	17.1	900	67	5.8	120	50	451	85.4	*Abies alba* (10), *Fagus sylvatica* (60), *Picea abies* (30)	48	52

high mountain	above 1700–1800	1.4	30	4.1	1100	86	2.2	150	85	304	32	*Picea abies* (60), *Pinus peuce* (10), *Pinus mugo* (30)	18	82.5
6. Eastern Bulgaria		100	10	100	770	42	7.7	85	19	551				
Eastern Stara Planina		36	6	26.3	650	29	11	35	8	747		*Quercus polycarpa* (20), *Q. frainetto* (40), *Fagus orientalis* (10)	83	16.7
Eastern Rhodopes		34	10	42.5	750	42	11	20	8	784		*Quercus dalechampii* (30), *Pinus nigra* (20), *Carpinus orientalis* (40)	65	35
Sakar Mts.		5.3	7.5	4.9	650	36	11	20	8	775		*Quercus frainetto* (50), *Tilia* sp. (10)	78	22
Strandja Mts.		25	8.5	26.3	800	34	11	28	6	700		*Quercus polycarpa* (40), *Q. hartwissiana* (20), *Fagus orientalis* (20)	76	24

Columns: 1, Share (%) of the belt in the mountain; 2, Average flow module (l/km^2); 3, Share (%) of the belt water resources in the mountain water resources; 4, Annual precipitation (mm); 5, Flow coefficient (%); 6, Average annual temperature (°C); 7, Duration of snow cover (days); 8, Thickness of snow cover (cm); 9, Evaporation (mm); 10, Share of the forest (%); 11, Edificator plant species, dominant (%); 12, Deciduous (%); 13, Coniferous (%).

The duration of snow cover varies with altitude from 2–3 to about 7 months. The stable snow cover is observed annually in the mountains above 1100–1400 m. The number of snow cover days increases with the altitude, and reaches its annual average of 190–200 days at the summits of Botev, Cherni Vrah, and Musala.

3·3 Class 3. Aquatic landscapes

This class includes all water bodies (rivers, lakes, artificial lakes) within Bulgaria as well as the Black Sea (see Golemansky, this volume, for the characteristics of the Bulgarian Black Sea).

3·3·1 Type 3.1. Rivers

Among the water resources of Bulgaria, the major part belongs to the rivers (total drainage length 19,761 km). The river network is formed by 1200 rivers of various size, with streams and small rivers prevailing; a large portion of those dries annually. Only 30 rivers in Bulgaria are longer than 100 km.

3·3·1·1 Subtype 3.1.1 The Danube River and its main tributaries

Twenty large rivers flow to the Danube River, most originating from Stara Planina and Predbalkan. All these tributaries of the Danube have similar conditions of formation and regime, except the streams of Ludogorie and Dobrudja regions, and the Iskar River.

Rivers in the western part of the Danubian Plain are short and shallow; their lowland portions are not fed constantly and dry up in summer. The length of the major rivers in this part of catchment area is highly variable. The most elaborate river system is that of the Ogosta River, with 40 tributary streams.

In the Central Danubian Plain, rivers are relatively large, with elongated watersheds and numerous meanders. Almost every river originates from Stara Planina, flows to the north and defiles Predbalkan as a typical canyon-shaped valley. Vit, Zlatna Panega, and other rivers are fed through karst formations. The longest Bulgarian river, Iskar (368 km), is an exception: it originates from Rila, flows along Western Sredna Gora, crosses Sofia depression, defiles Stara Planina and Predbalkan to the north and northeast, and enters the Danubian Plain.

The Danube tributary with the largest watershed (7862 km^2) is the Yantra River, 285 km long. It originates from Stara Planina, defiles the mountain and Predbalkan to the north, and forms a deep and narrow valley. In the Danubian Plain, Yantra increases its flow sizably after taking in the tributaries Lefedja and Rositsa.

The largest river in the Eastern Danubian Plain is Rusenski Lom. It has a well-developed watershed in the highest part of the plain with Popovo, Razgrad, and

Samuil hills, and the western parts of the Ludogorie Plateau, where the rivers have fan-shaped systems. The river cuts a deep canyon-shaped valley in the Cretaceous limestone with a constant karst flow. Other rivers are represented by dry valleys with inconstant flow. They originate from the numerous hills and plateaus in the northeastern Bulgaria.

3·3·1·2 Subtype 3.1.2. The Maritsa River and its main tributaries

The largest river of the Aegean Sea basin is the Maritsa River. Its length within the territory of Bulgaria is 321 km. Its watershed area amounts to $21,084\,km^2$ and is characterized by variable geomorphological conditions. The Maritsa watershed includes large portions of Rila, the Rhodopes, Sredna Gora, Stara Planina mountains, and Podbalkan basins, but its main portion occupies the Upper Thracian Lowland. Maritsa originates from the Marichini Ezera Lakes in Rila, and after Belovo it enters the Upper Thracian Lowland. There, it is joined by its left and right tributaries, which form a pinnate drainage pattern. The left watershed includes the tributaries of Sredna Gora, Stara Planina, Podbalkan basins, and the northern part of Upper Thracian Lowland. Downstream, the river takes in short and shallow streams from Sakar Mts. and the Eastern Rhodope foothills. Not far from Svilengrad, Maritsa leaves Bulgarian territory, entering Turkey, and near Edirne takes in its largest tributaries, Arda and Tundja.

The Arda River watershed includes a part of the high Western Rhodopes and a part of the hilly Eastern Rhodopes. Its higher portion is characterized by a well-expressed left drainage pattern and shorter right tributary streams.

The Tundja River is one of the largest Bulgarian rivers (watershed area $7884\,km^2$) and the longest (350 km within Bulgaria) tributary of the Maritsa River. Tundja originates from Stara Planina and flows to the southeast and south.

3·3·1·3 Subtype 3.1.3. The Kamchiya River

Kamchiya is the longest (254 km) and the largest (watershed area $5358\,km^2$) Bulgarian river among those running directly into the Black Sea. Its watershed embraces wide areas of the Stara Planina and Predbalkan ranges, and a small part of the Danubian Plain. It begins as the Golyama Kamchiya River; after passing the Preslav defile, it enters the Eastern Danubian Plain where it accepts the Luda Kamchiya tributary, and under the name of Kamchiya flows to the Black Sea. On its way, Kamchiya drains the wet lowland of Longoza.

3·3·1·4 Subtype 3.1.4. The Struma River

Struma takes the waters of the karst springs in the western foothills of Vitosha. It is the longest Bulgarian river, and the second after Maritsa in the size of its watershed area ($10,797\,km^2$). The long and deep left tributaries of Struma originate from Rila and Pirin; its short and shallow right tributaries originate from the Kraishte area.

The Struma tributaries system is asymmetric; it is characterized with pronouncedly meandering, deep, and picturesque gorges (Zemen, Kresna) along its longitudinal profile.

3·3·1·5 Subtype 3.1.5. The Mesta River

Mesta's watershed (2767 km^2) includes the southeastern slopes of Rila, northeastern slopes of Pirin, western slopes of the Rhodopes, and Razlog and Gotse Delchev depressions. It is the highest basin in Bulgaria (up to 1318 m), with an asymmetric shape. The right tributaries of Mesta are influenced by the mountain climate in Rila and Pirin with its high precipitation and prolonged snow cover. The left tributaries of Mesta originate from the foothills of the Rhodopes, and are shorter and comparatively shallow. The Mesta's main tributary, the Dospat River, joins it outside of the Bulgarian territory.

3·3·2 Type 3.2. Lakes

3·3·2·1 Subtype 3.2.1. Coastal lakes

The firths and lagoons are located in Bulgaria near the Black Sea coast, most of them around the two largest sea bays, Varna and Burgas. Some small lakes are also found along Dobrudja and Strandja coasts.

Durankulak, Ezerets, Shabla, Varnensko, Beloslavsko, Burgasko, and Mandra lakes are firths. They cover the firth mouths of Dobrudja's dry valleys, which are formed in the deep karst limestone. These firths are separated from the sea by narrow and low barrier beaches. They have low salinity (1–4‰) since they are fed by karst along their western and southern banks. The Beloslavsko and Varnensko lakes are typical firths formed in the mouths of the Provadiiska and Devnya rivers. The Beloslavsko Lake is 8 km long and 0.5–3.5 km wide, with low and marshy banks. Since 1975, a navigable channel connects it to the Varnensko Lake. Prior to that, the lake had fresh water; now its salinity has increased to 5‰ due to the influx of both sea water and chemical wastewater. The utilization of water from the Devnya karst springs resulted in the decreased flow to the Beloslavsko Lake and its volume.

The Varnensko Lake is largest in volume (170 billion m^3) and the deepest coastal lake in Bulgaria. It is limited by a barrier beach to the east, its northern banks are low and marshy, and the southern ones high and steep. It is fed by the river waters flowing from the Beloslavsko Lake, underground freshwater, and sea water. The Varnensko Lake is 15 km long and, in average, 1.3 km wide. It depth reaches 19 m; the regularly drainage works allow navigation. The water surface temperature in summer is 20–26 °C. The salinity is low (6.5–7‰) in the higher strata, and 8‰ below 10 m. At this depth, the oxygen disappears, being replaced by hydrogen sulfide.

The Burgasko Lake has the largest area, 27.6 km^2. It is located in the Burgas depression and covers the mouth of the Aitoska River. The lake is 9.5 km long; its depth reaches 1 m. Its salinity is higher than in other large firths and varies from 16 to 22‰. The rivers Aytoska, Sanderdere, and Chakarliika flow into its west part. The bottom of the Burgasko Lake is covered by a thick coat of clay ooze mixed with sand and shell material.

The Mandra Lake is a firth located in the eastern part of the Burgas depression. It has an elongated shape with length 11 km, maximum width 1.3 km, and volume 11 million km^3. The banks are hilly, low, and overgrown with reeds. The salinity is low because of the rivers Rusokastro, Sredets, and Fakiya flowing into the lake. The water is currently used for drinking.

The lagoon group includes Atanasovsko and Pomoriisko Lakes in the Burgas depression; marshy lakes Alepu, Arkutino, and Stomoplo along the Strandja coast; and Shablenska Tuzla, Nanevska Tuzla, and Balchishka Tuzla lakes along the Dobrudja coast. The last three are small, supersaline lagoons. The salinity decreases in spring, and increases in summer up to 150–200‰. Bottoms of the Dobrudja lakes are covered by black curative silt containing hydrogen sulfide. Life in these lakes is limited due to their high salinity.

The supersaline Pomoriisko Lake lies parallel to the sea and is separated from it by a 6 km long, narrow sand barrier. It has oval shape, and it is very shallow. This lagoon lake is fed by leaking seawater, and in summer its salinity increases to 100‰. Its volume is artificially regulated. Bottom of the Pomoriisko Lake is also covered by black curative silt containing hydrogen sulfide. Its water is used for salt production.

The Atanasovsko Lake is also a supersaline lagoon situated to the north of Burgas. The lake is very shallow (1 m) and receives sea water through an artificial channel. Its salinity reaches 200‰ during dry summers. The water is also used for salt production.

Lakes Alepu, Arkutino, and Stomoplo are small lagoons to the south of Sozopol, parallel to the sea coast. Their low salinity (1–4‰) allows for the existence of unique flora and fauna, e.g. the water lily (*Nymphaea alba*).

3·3·2·2 Subtype 3.2.2. Riverside lakes

These lakes are the abandoned meanders at the middle and low portions of the rivers. This type of lakes has been earlier widespread near the Danube, Iskar, Maritsa, Struma, Mesta, Tundja, and other rivers; now, most of these lakes dried out. The deepest riverside lake is Srebarna, which also has a very important location on the bird migratory way and possesses a unique ornithofauna. The lake is situated in the Aidemir depression, 1 km from the Danube River. It size decreased now to the area of 2.5 km^2 and depth of 2 m. The Srebarna Lake is fed by the Danube water through an artificial channel. Its bottom is covered with layers of clay and ooze deposits and bioorganic materials up to 7 m thick. The water is highly mineralized

(765 mg/dm^3), mostly with bicarbonate (322 mg/dm^3), chloride, and sulfate ions. The Srebarna Lake is a wetland zone of the international importance.

3·3·2·3 Subtype 3.2.3. Karst lakes

These are surface and underground lakes located in the karst regions. The surface karst lakes are small, and their area and depth frequently change. They are actively eutrophic, and fed by precipitation and water flow from the karst springs. The maximum water level is in spring, and the minimum, in summer and fall due to strong evaporation. As a result of their low depth, their temperature is high in summer and low in winter.

3·3·2·4 Subtype 3.2.4. Rockslide lakes

This type of lakes is widespread along the northern portion of the Bulgarian Black Sea coast and in the central part of the Rhodopes where the rockslide processes are present. The Smolyan Lakes in the Rhodopes are located at 1200 ma.s.l., about 10 km north of Smolyan. They are positioned in tiers and have small area and depth. The largest of these lakes is Milushevsko, 150 m long and 3 m deep. Another area with rockslide lakes is to the north of Cape Kaliakra, where the saline rockslide lakes Nanevska Tuzla and Shablenska Tuzla are located. The clear freshwater marshy lakes in Zlatni Pyasatsi and near Aladja Monastery are also positioned in tiers.

3·3·2·5 Subtype 3.2.5. Glacial lakes

These lakes were formed during the Quaternary Ice Age in the high parts of Rila and Pirin Mts. in the cirque bottoms and trough valleys. Glacial lakes are situated between 2000 and 2700 ma.s.l. due to the snowline position during the glacial period. Glacial lakes are found throughout Rila, and only in the north of Pirin. In Rila, glacial lakes are found between 2000 and 2600 ma.s.l., and are distributed by size as follows: 1–1.5 ha, 73.4%; 1.5–4.0 ha, 20%; over 4.0 ha, 6%. In Pirin, they are found between 2000 and 2700 m a.s.l.; they are from 0.1 to 2.0 ha in size and from 500,000 to 1,270,000 m^3 in water volume.

While glacial lakes usually are oval in shape, some have irregular shapes, e.g. Gorno Ribno Ezero and Babreka in Rila. The water transparency and color depends on lake's depth, content of dissolved materials, and aquatic vegetation. Shallow lakes with vegetation have a yellow-green or deep green color, and the deeper lakes have a blue-green or deep blue color. The deep lakes (which can reach 12 m) are more transparent. Surface temperatures vary from 9 to 18 °C in summer (up to 28 °C in shallow lakes). Temperatures become constant (4 to 7 °C) below 5–10 m. Glacial lake water is poorly mineralized and therefore is used for drinking. In addition, these lakes are attractive and important tourist objects.

3·3·3 Type 3.3. Marshes

Marshlands of various size, type, and regime are also found in Bulgaria. Most developed ones are located along the Black Sea coast and the Danube River; others are found in the mountains and in the low parts of depressions. The greater part of marshes is drained. The Choklyovo Marsh occupies the central part of Konyavska Mts. at 850 m a.s.l. It is drained partially and used as a peat source and for fishing; it has aquatic vegetation (reeds).

References

Armand, D., 1975. Nauka o landshafte [Landscape Science]. Mysl', Moscow (in Russian).

Glazovskaya, M.A., 1964. Geokhimicheskie osnovy tipologii i issledovaniy prirodnykh landshaftov [Geochemical Principles of Typology of the Natural Landscapes and Methodology of Their Study]. Izdatel'stvo MGU [Moscow State University], Moscow (in Russian).

Gvozdetskiy, N.A., 1973. Problemy fizicheskoi geografii [Problems of Physical Geography]. Izdatel'stvo MGU [Moscow State University], Moscow (in Russian).

Harvey, D., 1974. Nauchnoe obyasnenie v geografii [Natural Explanation in Geography]. Progress, Moscow (in Russian).

Mishev, K., M. Daneva, M. Yordanova, S. Velev, D. Gorunova & V. Velikov, 1989. [Vertical structure of landscape in Bulgarian mountains]. In Mishev, K. (ed), Prirodniyat i ikonomicheskiyat potentsial na planinite v Balgariya. Tom 1. Priroda i resursi [Natural and Economic Potential of Bulgarian Mountains. Volume 1. Nature and Resources]. Izdatelstvo na Balgarskata akademiya na naukite [Publishing House of the Bulgarian Academy of Sciences], Sofia: 412–456 (in Bulgarian).

Petrov, P., 1980. [Landscape regions in Bulgaria]. Godishnik na Sofiiskiya universitet "Kliment Ohridski". Geologo-geografski fakultet (Annuaire de l'Université de Sofia "Kliment Ohridski", Faculté de Géologie et Géographie) 71(2 – Geography):121–136 (in Bulgarian).

Petrov, P., 1990. Landshaftoznanie [Landscape Science]. Universitetsko izdatelstvo "Sv. Kliment Ohridski" [St. Kliment Ohridski University Publishing House], Sofia (in Bulgarian).

Systematic Index

Series Editor: **Prof. Henri J. Dumont**
State University of Ghent, Institute of Animal Ecology, Ghent, Belgium, henri.dumont@rug.ac.be

---

Published titles within the series:

Volume 85 **Monograph of the Urostyloidea (Ciliophora, Hypotricha)**
Berger, H.
2006, 1304 p., Hardcover, ISBN: 1-4020-5272-3

Volume 84 **Diapause in Aquatic Invertebrates: Theory and Human Use**
Alekseev, Victor R., De Stasio, Bart and Gilbert, John J. (Eds.)
Forthcoming 2007, approx. 3005 p., Hardcover,
ISBN: 1-4020-5679-6

Volume 83 **Bridging Divides - Maritime Canals as Invasion Corridors**
Gollasch, Stephan, Galil, Bella S., and Andrew N. Cohen
2006, 315 p., Hardcover, ISBN: 1-4020-5046-1

Volume 82 **Biogeography and Ecology of Bulgaria**
Fet, Victor and Popov, Alexi (Eds.)
2007, Hardcover, ISBN: 978-1-4020-4417-5

Volume 81 **Monograph of the Spathidiidae**
Foissner, W., and Xu, K.
2006, 486 p., Hardcover, ISBN: 1-4020-4210-8

Volume 80 **Nouragues**
Dynamics and Plant-Animal Interactions in a Neotropical Rainforest
Bongers, F.; Charles-Dominique, P.; Forget, P.-M.; Théry, M. (Eds.)
2002, 456 p., Hardcover, ISBN: 1-4020-0123-1

Volume 79 **Environmental Change and Response in East African Lakes**
Lehman, J.T. (Ed.),
1998, 250 p., Hardcover, ISBN: 0-7923-5118-5

Volume 78 **Monograph of the Oxytrichidae (Ciliophora, Hypotrichia)**
Berger, H.
1999, 1092 p., Hardcover, ISBN: 0-7923-5795-7

Published titles within the series:

Volume 77 **The Pantanal of Poconé**
Biota and Ecology in the Northern Section of the World's Largest Pristine Wetland
Heckman, C.W.
1998, 592 p., Hardcover, ISBN: 0-7923-4863-X

Volume 76 **An Introduction to Saline Lakes on the Qinghai-Tibet Plateau**
Zheng Mianping
1997, 328 p., Hardcover, ISBN: 0-7923-4098-1

Volume 75 **Rocas Alijos**
Schmieder, Robert W.
1996, 520 p., Hardcover, ISBN: 0-7923-4056-6

Volume 74 **Tropical Rainforest Research - Current Issues**
Proceedings of the conference held in Bandar Seri Begawan, April 1993
Edwards, D.S.; Booth, W.E.; Choy, S.C.
1996, 570 p., Hardcover, ISBN: 0-7923-4038-8

For more information on our ongoing series, please visit www.springer.com